The Victoria History of the Counties of England

EDITED BY R. B. PUGH, M.A., F.S.A.

A HISTORY OF CAMBRIDGE AND THE ISLE OF ELY

VOLUME IV

THE VICTORIA HISTORY OF THE COUNTIES OF ENGLAND

CAMBRIDGE AND THE ISLE OF ELY

PUBLISHED FOR
THE UNIVERSITY OF LONDON
INSTITUTE OF HISTORICAL RESEARCH
BY THE
OXFORD UNIVERSITY PRESS
AMEN HOUSE, LONDON, E.C.4

First published 1953
Reprinted 2002 with index to volumes I–IV
Reprinted 2015

A Victoria County History publication
in association with The Boydell Press
an imprint of Boydell & Brewer Ltd
PO Box 9, Woodbridge, Suffolk IP12 3DF, UK
and Boydell & Brewer Inc.
688 Mount Hope Avenue, Rochester, NY 14620–2731, USA
website www.boydell.co.uk
and with the
University of London Institute of Historical Research

ISBN 978 0 71290 244 1

ISSN 1477–0709

Printed and bound in the UK by 4edge Limited

INSCRIBED
TO THE MEMORY OF
HER LATE MAJESTY
QUEEN VICTORIA
WHO GRACIOUSLY GAVE
THE TITLE TO AND
ACCEPTED THE
DEDICATION OF
THIS HISTORY

THE VICTORIA HISTORY OF THE COUNTY OF CAMBRIDGE AND THE ISLE OF ELY

EDITED BY

R. B. PUGH, M.A., F.S.A.

VOLUME FOUR

PUBLISHED FOR
THE UNIVERSITY OF LONDON
INSTITUTE OF HISTORICAL RESEARCH
BY THE
OXFORD UNIVERSITY PRESS
AMEN HOUSE, LONDON, E.C.4
1953

CONTENTS OF VOLUME FOUR

CONTENTS OF INDEX TO VOLUMES I–IV

LIST OF ILLUSTRATIONS

LIST OF ILLUSTRATIONS

EDITORIAL NOTE

THE original plan of this volume was framed during the editorship of Mr. L. F. Salzman, F.S.A., and much of the material that it contains was commissioned by him. The topographical articles, however, incorporate the following features for which he is not responsible. Under each parish there is a history of each Nonconformist place of worship and each primary school. In addition the inclosure of the common fields, where effected by statute, is described. While it is hoped that these features, which in recent years have not formed part of the topographical articles in the *History*, will henceforth be permanent, their form is at present confessedly experimental. Owing to difficulties, which are unlikely to recur, it has not been possible in this volume to bring the history of charities down to the date of publication. A somewhat fuller account than usual has, however, been given of their origins.

Thanks are due to Mr. G. M. G. Woodgate, F.S.A., and the late E. J. Rudsdale for much assistance, particularly with the history of Wisbech. The Very Revd. S. J. A. Evans, F.S.A., Dean of Gloucester and formerly Archdeacon of Wisbech, kindly granted access to his own notes. The help given by the Clerk, Chief Education Officer, and Chief Planning Officer of the Isle of Ely County Council, the Ministry of Housing and Local Government, Post Office Headquarters, the Public Relations Officer of the Eastern Region, British Railways, and the incumbents of several parishes is also gratefully acknowledged.

The sources of the photographic illustrations appear in most cases on the plates themselves and thanks are returned to all those who gave permission for their reproduction. The blocks of the two plates facing page 90 and the lower plate facing page 91, which originally appeared in *The Mediaeval Fenland* and *The Draining of the Fens* by Professor H. C. Darby, were kindly lent by the Cambridge University Press, who also allowed the plan of Ely Cathedral and Priory on page 78 to be copied. Messrs. Faber & Faber supplied the block of Wisbech Market Place facing page 262. The reconstruction of the Ely pulpitum on page 67 appears by kind permission of the Society of Antiquaries of London. The 19th-century photographs of Wisbech, Elm, and Leverington, and the print of Wisbech facing page 243 are the property of Wisbech Museum, and are reproduced through the courtesy of the Curator, Mr. G. R. Stanton. Finally, the Editor is under many obligations to the Director and Staff of the National Buildings Record for allowing the use of photographic prints from their collection and for other assistance.

LIST OF CLASSES OF PUBLIC RECORDS

USED IN THIS VOLUME, WITH THEIR CLASS NUMBERS

Chancery
- C 1 Proceedings, Early
- C 2 „ Series I
- C 3 „ Series II
- C 5 „ Six Clerks Series, Bridges
- C 11 „ „ „ 1714–58
- C 47 Miscellanea
- C 54 Close Rolls
- C 60 Fine Rolls
- C 66 Patent Rolls
- C 78 Decree Rolls
- C 134 Inquisitions post mortem, Series I, Edward II
- C 136 „ „ „ „ Richard II
- C 137 „ „ „ „ Henry IV
- C 139 „ „ „ „ Henry VI
- C 140 „ „ „ „ Edward IV
- C 142 „ „ „ Series II
- C 143 Inquisitions, Ad Quod Damnum
- C 146 Ancient Deeds, Series C

Court of Common Pleas
- C.P. 25(1) Feet of Fines, Series I
- C.P. 25(2) „ „ Series II
- C.P. 43 Recovery Rolls

Exchequer, Treasury of the Receipt
- E 40 Ancient Deeds, Series A

Exchequer, Queen's Remembrancer
- E 134 Depositions taken by Commission
- E 150 Inquisitions post mortem, Series II
- E 178 Special Commissions of Enquiry
- E 179 Subsidy Rolls, &c.
- E 190 Port Books

Exchequer, Augmentation Office
- E 315 Miscellaneous Books

Exchequer, Lord Treasurer's Remembrancer
- E 357 Escheators' Enrolled Accounts

Home Office
- H.O. 52 Correspondence, Counties.
- H.O. 67 Acreage Returns
- H.O. 107 Various, Census Papers, Population Returns
- H.O. 129 Various, Census Papers, Ecclesiastical Returns

Justices Itinerant
- Just. Itin. 1 Assize Rolls, Eyre Rolls, &c.

Court of Queen's Bench
- K.B. 27 Coram Rege Rolls
- K.B. 28 Crown Rolls
- K.B. 122 Plea or Judgement Rolls

LIST OF CLASSES OF PUBLIC RECORDS

Exchequer, Land Revenue
L.R. 14 Ancient Deeds, Series E

Ministry of Agriculture and Fisheries
M.A.F. 1 Commons, &c., Inclosure Awards

Special Collections
S.C. 6 Ministers' Accounts
S.C. 11 Rentals and Surveys (Rolls)
S.C. 12 " " (Portfolios)
S.C. 13 Seals

State Paper Office
S.P. 12 State Papers, Domestic, Elizabeth I
S.P. 16 " " " Charles I

Court of Star Chamber
Sta. Cha. 8 Proceedings, James I

Court of Wards and Liveries
Wards 7 Inquisitions post mortem

NOTE

The Manuscript Collections of Mr. G. M. G. Woodgate, lodged in the Museum at Wisbech, are cited variously as 'Wisb. MSS.' and 'Woodgate MSS.'

The muniments of the Dean and Chapter of Ely are cited as 'D. & C. Mun.,' those of the Bishop as 'Bp.'s Mun.' The arrangement of the Episcopal manuscripts is described in A. Gibbons, *Ely Episcopal Records*. The capitular manuscripts are housed in one room but are not permanently arranged in shelves or cupboards: the numbering used here is that found on the documents when they were examined in 1949–50.

THE LIBERTY OF ELY

IN the Assembly Book of King's Lynn, for the year 1610, there is a note that 'Sir Robert Hitcham knight, the Queenes Majesties Attorney-generall is proposed to come to this towne from the Assizes at Norwich to take his jorny to Elie where he is judge of that County Palatyne'.[1] It seems to have been current usage at the time to describe in these terms the Bishop of Ely's liberty in the Isle; for, as Sir Edward Coke tells us, 'the royal franchise of Ely ... in divers statutes ... is named the county palatine of Ely'.[2] The description may not have been strictly correct; nevertheless, the liberty enjoyed by the bishops in the Isle of Ely was extensive enough in medieval and most of modern times to leave a decisive mark upon the administrative geography of Cambridgeshire. Even in the middle of the 19th century it could be said that 'Cambridgeshire virtually includes two shires or separate jurisdictions, the shire proper and the Isle of Ely'.[3] This peculiarity of modern Cambridgeshire is not a little due to the fact that the liberty acquired by the church of Ely in the Middle Ages was one of the very few franchises which resisted Edmund Dudley's rediscovery of *Quo Warranto* during Henry VII's reign and Thomas Cromwell's determination 'for the dissolution of all franchises and liberties throughout the realm'.[4] In fact the liberty of Ely lasted until the reign of William IV.

Equally important is the fact that the liberty of Ely is 'thought to have been very ancient'.[5] Coke traced back the bishop's peculiar powers in the Isle only to the time of Henry I;[6] but at the end of the 18th century the Rev. Mr. Warren could state with some certainty that the original of these powers lay in St. Etheldreda's concession in 673 to the monastery she founded at Ely of 'her temporal jurisdiction throughout the Isle'.[7] This privilege, lost for a time after the abbey was destroyed by the Danes in 870, was restored to the church by King Edgar a century later. Colour is given to this account by the 12th-century chronicle written in the cathedral priory at Ely,[8] and it has found support in a number of modern writings.[9] If it is correct, then the administrative geography of modern Cambridgeshire is grounded in arrangements made during the early days of Saxon England. This matter is of sufficient interest to constitute a first subject of inquiry.[10]

1 Hist. MS. Com., *11th Rep., App.* iii, 151, 177.
2 *The Fourth Part of the Institutes of the Laws of Engl.*, cap. 39 (ed. 1797, 219). The Isle is in fact described as, or implied to be, a county palatine in 33 Hen. VIII c. 10 and 5 Eliz. c. 23.
3 R. Gardner, *Dir. Cambs.* (1851), 28.
4 Helen M. Cam, *Liberties and Communities of Medieval Engl.* 216–19.
5 Naomi D. Hurnard, 'The Anglo-Norman Franchises', *E.H.R.* lxiv (1949), 317.
6 *4th Part of Insts. of Laws of Engl.* 219.
7 In J. Bentham, *Ely*, App., 21–22 (refs. are to the 2nd ed. of 1812).
8 See, for example, *Hist. Elien.* i, cap. 15, ed. D. J. Stewart (1848), 48.
9 e.g. my own remarks, *V.C.H. Cambs.* ii, 378.
10 In the earlier part of this chapter I have drawn upon material discussed in greater detail in my *Abbey and Bishopric of Ely*.

THE MAP

on the opposite page shows:

(*a*) The boundaries of hundreds as existing *c.* 1841.

(*b*) The boundaries of parishes as existing in 1932.

(*c*) The territorial disposition of the Isle after the foundation of the Bishopric of Ely in 1109.

(*d*) The extent of the 16th century alienations by both bishop, and dean and chapter.

NOTES

Benwick, March, Parson Drove, Thetford, Wimblington. These parishes, which were not manors in the 12th century, are not separately distinguished.

Redmere. This parish has been excluded from the map since it was not added to the Isle until 1895.

Chatteris, Leverington. The boundaries between the Ramsey and Chatteris, and episcopal and conventual manors in these respective parishes have been drawn conjecturally, on the basis of the documentary evidence.

Whittlesey. The boundary between the Ely and Thorney manors in this parish has been drawn arbitrarily, no documentary evidence surviving.

Ely. The extent of the episcopal and conventual interests here has been based on the parish boundaries as existing in 1932, which are probably manorial in origin. The common interests of bishop and convent in the cathedral precincts (the present parish of 'Ely College') have been omitted, the area (35 acres) being too small to be shown.

Outwell, Upwell. These parishes have been shown as belonging to the bishopric of Ely and abbey of Ramsey respectively, on the basis of documentary evidence, but it is possible that the Ely interests here were rather greater than shown.

Coveney, Manea. These are treated as a 16th-century alienation on the ground that, although the accustomed rent of 5*s.* was awarded in 1541 to the dean and chapter, the grant states that this manor 'belonged' to the Lords Scrope.

Mepal. This manor, which was held as a fee of the bishop in the 13th century and amortised to the prior and convent in the 14th, is shown cross hatched.

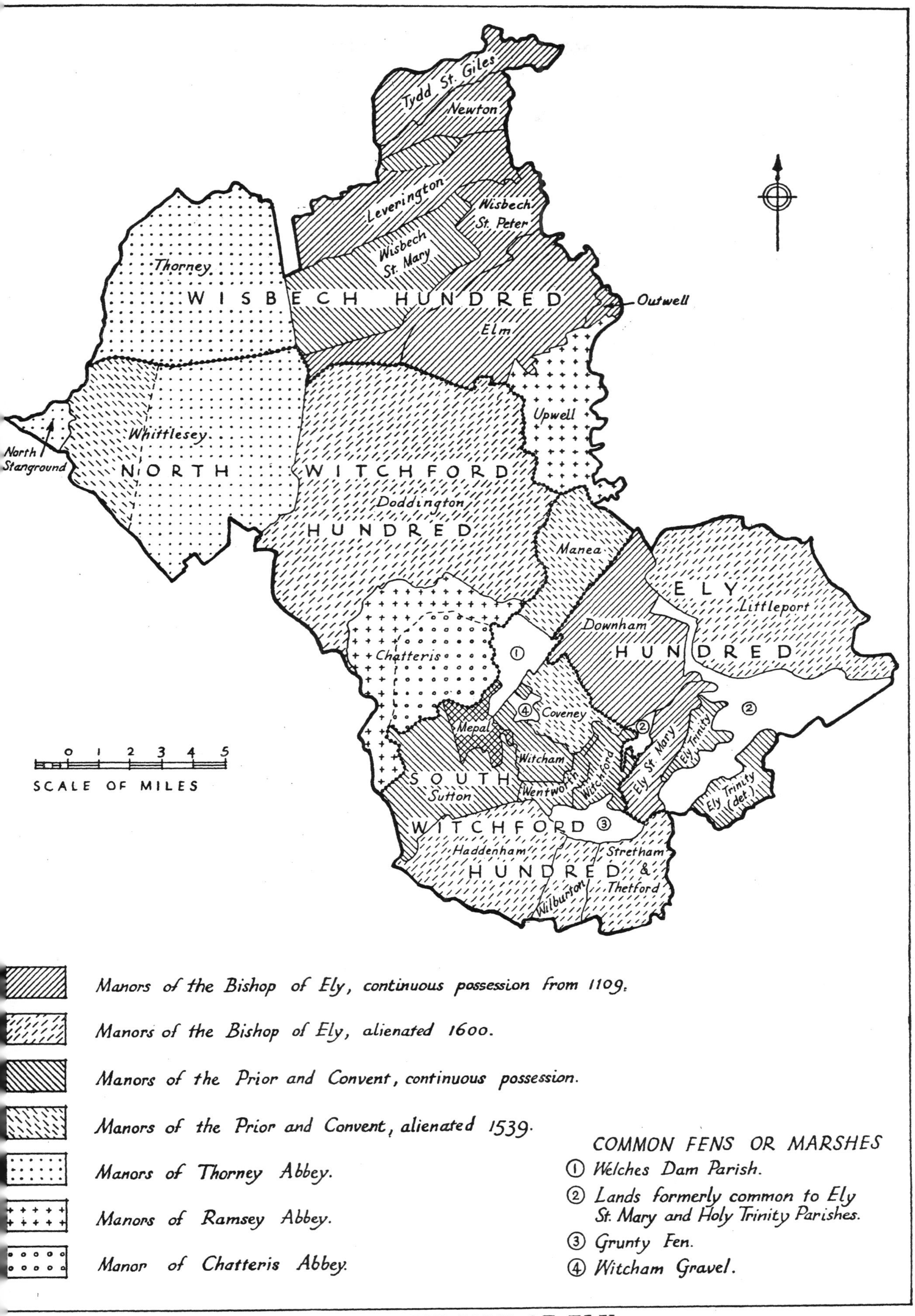

MAP OF THE ISLE OF ELY

Origins of the Liberty of Ely

The first presumably authentic and contemporary evidence for the liberty of Ely is contained in Edgar's charters of endowment for the refounded monastery in 970. In order to provide the monks with food and clothing, the king says, he has given them all the *soke* over the fenlands in the two hundreds, over the five hundreds meeting at Wicklaw in East Anglia, and over all lands now assigned to the monastery or that would be assigned to it in the future.[11] The two hundreds of the fenland are our immediate concern. Probably they had roughly the same bounds in 970 as had the two hundreds of the Isle surveyed by the Domesday commissioners in 1086 and as the Isle of Ely had in more modern times. The problems raised by Edgar's charter, therefore, are whether the king merely restored a state of affairs prevailing in earlier Saxon times and the meaning of the *soke* which he conferred upon the refounded abbey.

Clearly the 12th-century chroniclers of Ely did their best to convey the impression that Edgar did no more than restore rights enjoyed by the church from the time of St. Etheldreda onwards. It is, on the other hand, quite certain that no evidence has survived which permits a precise reconstruction of the administrative geography of eastern England before the 10th century, and that surviving accounts of St. Etheldreda's abbey are based mainly upon tradition rather than record. Further, some evidence does exist which tells heavily against the view that the Isle is an administrative area of extreme antiquity. It has been suggested that, in the 7th and 8th centuries, the district 'was normally partitioned into two spheres. The more southerly portions, including Ely, were part of East Anglia; while the district to the north of this, that is the district about Crowland, Thorney and Whittlesmere, were part of mid-Anglia or Mercia.'[12] The part of the Isle about Wisbech, in turn, retained for many generations a marked individuality of its own. Its social organization during the Middle Ages was closely akin to that of the Norfolk marshland; it was continuously referred to in terms which suggest that it had been originally an East Anglian 'ferthyng', an administrative area which in this district antedated the creation of the hundreds.[13] In short, the 'two hundreds' of the fenland of Edgar's charter bear evident marks of artificiality, of having been constructed out of more than one original piece. It may also be suspected that they were, in 970, of fairly recent creation—a product of reorganization, of reconstruction. This may have been the work of the Danish army based on Cambridge in the 9th century; or, perhaps more likely, of the West Saxon kings after they won back eastern England from the Danes.[14]

The liberty of Ely, therefore, though still ancient, may be less ancient than is sometimes supposed. It probably had a precise beginning in the year 970 in King Edgar's grant to the monks of *soke* over the two hundreds of the fen. This leaves the problem of defining the meaning of that *soke* which Edgar conceded to the church; and of tracing the steps whereby this *soke* came to

[11] Kemble, *Cod. Dipl.*, nos. 563, 564. It should however be noted that the Saxon version of Edgar's charter may, according to a recent suggestion, not have been composed until about 1006 (A. McIntosh, 'Wulfstan's Prose', *Proc. Brit. Academy*, xxxiv (1949)).

[12] H. C. Darby, *Antiquity*, viii (1934), 194.

[13] See the references to Wisbech as 'a quarter part of the hundreds of the Isle' and to a court at 'Modich' for the quarter part of those hundreds in *Hist. Elien.* ii, caps. 54, 75; and on East Anglian 'ferthyngs' F. M. Stenton, *E.H.R.* xxxvii (1922), 227, and D. C. Douglas, *Feud. Doc. from the Abbey of Bury St. Edmunds*, clxvi.

[14] Cam, *Liberties and Communities of Medieval Engl.* 90, and Stenton, *Anglo-Saxon Engl.* 295.

imply an immunity so privileged that it could be called a county palatine. Were the rights of the monks in the Isle such as to make the two hundreds an immunity from the very beginning? Did they even constitute a court-holding privilege at all? These are the questions which modern learning suggests we ought to ask about this Anglo-Saxon franchise.[15]

There is very little early evidence for interpreting Edgar's charters, but some of the records of late 10th-century land-pleas may be important in this connexion. Ealdorman Aethelwine of the East Angles, we are told, when disputes arose about the abbey's title to lands in Witchford, 'came to Ely, and within the cemetery at the north gate of the monastery held a plea with all the hundred present'.[16] Similarly Wulfstan of Dalham, sheriff of Cambridgeshire, presided over a plea to the north of the monastery, the two hundreds being gathered there.[17] Again, following a dispute about land at Stonea, 'Ealdorman Aethelwine came to Ely, and Begmund and others were summoned to a plea before the citizens and the hundredors once, and again, and many times'; because they failed to attend, the case was finally settled by the ealdorman in a 'great plea' at Cambridge.[18] It is not easy to know how to interpret these cases. They may be exceptional; the ealdorman and the sheriff may have been specially called in by the abbot to preside over his hundred courts. On the other hand, it is also possible that, in the beginning, the liberty of Ely was merely a fiscal privilege. The monks enjoyed the right to receive the *emendationes delictorum* from the courts of the Isle,[19] the fines and amercements in the terminology of a later time. They may not at first have had the right to hold the courts of the Isle themselves.

In the course of the 11th century, however, the liberty of Ely certainly became a court-holding franchise. In 1052 Edward the Confessor gave Ely Abbey to Wulfric 'in all things within borough and without, (with) toll and team and infangentheof, fihtwite and fyrdwite, hamsoke and grithbrice, let his man live where he lives and work where he works';[20] and it was in very similar terms that the Conqueror confirmed the liberty of Ely,[21] and later kings throughout the 12th century.[22] The use of the terms *team* and *infangentheof* in all these documents from the Confessor's writ onwards suggests that, wherever the church of Ely had lands, it had its own courts where defendants in cases of theft could vouch to warranty and where hand-having thieves could be dealt with.[23] That these private courts included the hundred courts of the Isle is implied, even for the time before the Norman Conquest, by two writs of Henry I. One expresses the king's will that the men of the Bishop of Ely should not plead elsewhere than they were accustomed to plead in the days of the Confessor, of the Conqueror and of William Rufus.[24] The other orders all barons holding in the bishop's hundreds to come to pleas held in the courts of those hundreds at the summons of the bishop's bailiffs, as they had been

[15] J. Goebel, *Felony and Misdemeanour*, i, 377 et seq., and N. D. Hurnard, *E.H.R.* lxiv (1949).

[16] *Hist. Elien.* ii, cap. 12.

[17] Ibid. cap. 18; H. M. Chadwick, *Anglo-Saxon Insts.* 231.

[18] *Hist. Elien.* ii, cap. 24.

[19] See the terminology of one of Edgar's charters: 'Elig cum omnibus sibi subjectis tam in emendatione delictorum quam in ceteris rebus' (Kemble, *Cod. Dipl.*, no. 564).

[20] Kemble, *Cod. Dipl.*, no. 885.

[21] E. Miller, 'The Ely Land Pleas in the reign of William I', *E.H.R.* lxii (1947), 455.

[22] Ibid. 445, 455–6; and see the confirmation of Henry II in Delisle-Berger, *Recueil des Actes de Henri II*, no. lx.

[23] *E.H.R.* lxiv, 292 n.

[24] Trinity Coll. Camb. O. 2. 1, f. 115*a*; printed in Bentham, *Ely*, App. xxiv.

accustomed to come in Henry's father's time and his brother's.[25] The significance of these writs is made quite clear by Bishop Hervey's concession in 1128 to Thorney Abbey in its manor of Whittlesey of 'all those customs and secular causes which the church of Ely has in its hundreds of Witchford'; and the elucidation of this concession by the Bishop of Rochester as implying exemption from suit to the bishop's court in Witchford hundred.[26]

The combined testimony of these documents makes it clear that the church of Ely possessed the right to hold the hundred courts of the Isle of Ely in Henry I's time, in the time of his brother and his father, and in the time of the Confessor. How much farther back this right went there is very little to tell us; nor is it easy to sum up the degree of immunity the church enjoyed in the Isle either before or after 1066. There are, however, two points worth making. First, none of the charters of the church of Ely, from King Edgar's down to that of Henry II, suggests that there was anything special or distinctive about the liberty of Ely in the Isle. The abbey's liberty extended over all the church's men and all its lands; the liberty in the Isle was merely a parcel of a liberty which the church of Ely had everywhere it had property. Presumably, therefore, at this early time the franchises of the abbots and bishops in the Isle were analogous to those which the church had later outside the Isle. These included the right to hold certain private hundred courts in Norfolk and Suffolk; to hold the view of frankpledge in all its manors and hundreds; to plead in its own courts all pleas the sheriff could plead in the public courts; to have the return of all writs; to arrest and imprison all criminals, bring them to trial and execute sentence upon them; to collect and retain for the use of the church all fines and amercements in whatever court they were imposed.[27] In short, the liberty of Ely everywhere consisted in a comprehensive immunity against the intervention of the sheriff and his subordinates within its lands and hundreds. It must have been a franchise which, in the Isle and elsewhere, excluded most outside interference so long as the control of local administration was concentrated mainly in the sheriff's hands.

During the 12th century, however, the sheriff's activities in local administration were being supplemented and in part superseded by those of other agencies, and in particular by those of the king's itinerant justices before whom, in 1166, no man was to fail to appear on account of any liberty. In face of these encroachments, the liberty of Ely outside the Isle became less of an immunity than in earlier times. The bishop and the prior made good a claim to receive the fines and amercements imposed on their men, lands, and hundreds by royal justices;[28] but this fiscal privilege conferred no exemption from the jurisdiction of the king's courts or from the performance of suit to them.[29]

[25] Bentham, *Ely*, App. xxiii.

[26] Camb. Univ. Lib. (hereafter C.U.L.) MS. Add. 3020. Red Bk. of Thorney, i, ff. 166–7. The latter charter includes the phrase 'Episcopus Elyensis super terram abbatis que in Wytles' erat huiuscemodi consuetudinem habuerat, quod ad hundredum suum apud Wycheford homines abbatis . . . venire consueverant'.

[27] This catalogue is compiled from the liberties claimed by the bishop before the king's justices in various counties in the later 13th century (Bp.'s Mun., G 3 ('Old Coucher', ff. 232–3, and Liber M, ff. 121–2); Just. Itin. 1/95 m. 52 d).

[28] e.g. Henry III's charter to Ely of 1233, confirming a charter of Richard I (1198) (*Cal. Chart. R.* 1226–57, 183–4).

[29] For performance of suit by the bishop's manors in 1251, see Caius Coll. Camb. MS. 485/489, ff. 324*a* (Hitcham), 333*b* (Barking), &c., and D. C. Douglas, *Social Structure of Medieval E. Anglia*, 151 et seq.; and for the appearance of the Ely hundreds before the justices of eyre, Just. Itin. 1/564 m. 3 and Just. Itin. 1/577 m. 75 (Mitford Hundred), and *E.H.R.* lxiv, 317 (the Suffolk hundreds).

Only the Isle of Ely came to enjoy such an immunity that the king's courts exercised no jurisdiction within its boundaries. At what precise date this franchise was established is less clear. Two charters of King Stephen, granting Becontree hundred (Essex) to the nuns of Barking, may have some relevance to this question.[30] The nuns were to hold the hundred 'as freely as the church of St. Edmund and the church of St. Etheldreda of Ely hold their hundreds'. But in both a qualification was attached: in one 'salvis aliis consuetudinibus meis regalibus'; in the other 'salvis tantum placitis corone mee que per justiciam meam debent placitari'. If these are genuine documents, and if the qualification also applied to the privilege of the church of Ely in the two hundreds of the Isle, the later exemption of that district from the jurisdiction of the king's justices may not apply in the first half of the 12th century. There is some support for this possibility in the tale of a miracle performed by St. Etheldreda in the reign of Henry I. She released from prison a certain Bricstan who, being accused of stealing the king's money, had been condemned 'Radulfo Basset sedente pro tribunali, congregatis eciam provincialibus universis apud Huntedoniam ut mos est in Anglia'. The important point is that Bricstan was a man of Chatteris in the Isle; yet he was taken before the justice sent out into Huntingdonshire from the king's *curia*.[31]

On the other hand, when the priory chronicle took final shape in the later 12th century, it was claimed that the church had the right to entertain all pleas of the crown in the Isle and to exclude all royal ministers from the Isle.[32] In 1189, finally, Richard I's charter for the first time drew a clear distinction between the liberty of Ely in the Isle and elsewhere.[33] The Isle was free and quit of all royal exaction; outside the Isle the church had its sake and its soke, toll and team and infangentheof and so on. By the early 13th century, the exemption of the Isle from the jurisdiction of the royal courts was firmly enough established to have given rise to institutions within it to administer higher justice. At least as early as 1220 there was a *curia Elyensis* capable of hearing a plea under a writ of mort d'ancestor.[34] By 1251 a number of tenures had been created in the vill of Ely specifically for the defence and exercise of this franchise. William the son of Elyas had to go with the lord's steward and knights to Cambridge on the coming of the king's justices there, to demand the church's liberties. Nicholas his brother and others held land for rent and services and for performing the office of coroner; while John de Marisco had 6 acres for which he paid nothing, but he too had to be a coroner and go to meet the justices.[35] By this date, the foundations of the medieval liberty of Ely in the Isle were fully laid.

Before discussing this medieval liberty, however, one further preliminary matter demands attention. One eventual result of the establishment of a bishopric at Ely in 1109 was that, in the Isle and elsewhere, the bishop and the cathedral priory became landowners more or less independent of one another.[36]

30 *Cal. Chart. R.* 1341–1417, 282, 284 (1383).

31 *Hist. Elien.* iii, cap. 33, in Trin. Coll. Camb. MS. O. 2. 1, ff. 122–3: 'Tempore Henrici regis . . . in provincia Eliensis ecclesie erat quidam homo in villa que vocatur Catariz', &c.

32 *Hist. Elien.* ii, cap. 54.

33 Cf. *E.H.R.* lxiv, 317. This charter survives as D. & C. Mun. Carta 13, and see *Cal. Chart. R.* 1226–57, 183 (1233).

34 D. & C. Mun. Carta 833, and cf. B.M. Egerton MS. 3047, f. 211*b*.

35 Caius Coll. MS. 485/489, f. 21*a*, and cf. B.M. Cott. MS. Tib. B. II, f. 87*a*, where Gilbert son of Azo, predecessor apparently of John de Marisco, in 1222 *debet ire contra justiciarios*.

36 For some of the problems arising out of the creation of the bishopric, see *V.C.H. Cambs.* ii, 203–4.

This had consequences also for the liberty. There is nothing in Bishop Hervey's charter of endowment for the monks to show that he gave them any share in the franchises of the Isle;[37] while Niel's well-known charter, purporting to come from about the year 1135 and granting to the priory judicial privileges similar to the bishop's, has been regarded as of dubious authenticity.[38] On the other hand, there is a later charter of Niel's which does concede to the monks the right to hold pleas arising in their own lands;[39] and this concession was confirmed by William Longchamp towards the end of the 12th century.[40] This agreement in principle, however, left a wide margin for disagreement about details. On at least three occasions during the 13th century the precise extent of the conventual liberty was submitted to the verdict of sworn inquests;[41] and there were still points at issue between the bishop and the monks which had to be put to arbitration in 1417.[42]

The final result of all this discussion, in general terms, can be summed up fairly concisely. The prior and convent had their courts and their leets in all their manors. They received the profits of justice done upon all their men, both in their own and in the bishop's courts. They enjoyed in their own manors administrative powers of arrest, attachment and execution of sentence. They had a prison at Ely separate from the bishop's prison. In short, they had all the lesser liberties the bishop had. The higher liberties, on the other hand—the right to hold hundred courts in the Isle and plead *placita vicecomitalia*; jurisdiction over pleas of the crown; the execution and return of royal writs—these were reserved to the bishop. Thus, in treating of the lesser franchises of the liberty, material can be drawn indifferently from conventual and episcopal records; but the conventual records contain little about those highest franchises which made the liberty of Ely very nearly a palatinate.

The Medieval Liberty of Ely

The franchises enjoyed in the Isle of Ely by the bishops and monks during the Middle Ages included three main groups of privileges. In the first place, the Isle, like the rest of the church's lands, was an administrative immunity. The bishop, the prior and their successors, said a charter of Edward III in 1343, shall make return of summonses of the Exchequer and all other royal writs in their hundreds and their fees, so that no sheriff or bailiff shall enter them save in case of default by the bishop, prior or their servants.[43] In other words, throughout the liberty all administrative duties normally carried out by sheriffs and their subordinates or by special royal commissioners were the responsibility of officers of the franchise. They arrested and imprisoned

37 The version of Hervey's charter printed in Bentham, *Ely*, App. xxvi is almost certainly a forgery (see *Sir Christopher Hatton's Bk. of Seals*, ed. L. C. Loyd and Doris M. Stenton, no. 420). I have published the genuine instrument (D. & C. Mun. Carta 51) in my *Abbey and Bishopric of Ely*, 282.

38 It is printed in *Sir C. Hatton's Bk. of Seals*, no. 415, with the comment that 'this charter bears on its face the suggestion of forgery'. On the other hand, it should be noted that Niel had presumably made some concession to the monks in view of the letter of Innocent II published by W. Holtzmann, *Papsturkunden in England*, ii, no. 21, and dated by him April 1139.

39 D. & C. Mun. Carta 55: 'sint omnes cause majores vel minores que in possessionibus vel ex possessionibus eorum orte fuerint ipsorum juri et potestati absque ulla excepcione semper subjecte.'

40 D. & C. Mun. Carta 57.

41 Bp.'s Mun., Liber M. ff. 138, 612; *Dugd. Mon.* i, 488. See also the charter of Bishop Robert of Orford to the monks in 1303 (D. & C. Mun. Carta 73 A).

42 Bentham, *Ely*, App. xxvii, pp. 27–34.

43 *Cal. Chart. R.* 1341–1419, 21–22 (1343, confirming charter of 1233).

criminals; executed sentence upon them[44] and collected amercements; distrained upon collectors of royal taxes to make payment at the Exchequer;[45] arrayed men-at-arms for the king's armies.[46] So the list might go on through all the infinite variety of activities of medieval local government. This right to exclude the king's ministers might not be absolute. The king issued commissions of sewers with a jurisdiction over the Isle;[47] and the surveyors of weights and measures appointed in 1350 were to exercise their office in the liberty of the Bishop of Ely as well as elsewhere in the county of Cambridge.[48] These are, however, small exceptions to a general rule.

Secondly, the Isle was a judicial immunity. No suit was owed from within it to the Cambridge shire court, nor did the sheriff hold any court whatsoever inside its boundaries. Further, in the words of letters patent of 1369, 'the abbots and afterwards the bishops . . . have always used that whenever any resident be impleaded within the Isle or without by denizens or foreigners of any cause or anything emergent within the Isle before secular justices, that plea shall be determined before the bishop's justices at Ely'.[49] This claim was asserted against the king's justices itinerant in the shires;[50] it was also interpreted as excluding the Isle from the jurisdiction of the central king's courts. Cases were successfully claimed by the bishop from the King's Bench[51] and Common Pleas;[52] the jurisdiction of the Exchequer of Pleas was denied[53] and the right of the steward and marshal to exercise any office in the Isle when the king was at Ely.[54]

There were, of course, exceptions once again to these broad rules. When the bishopric was in the king's hand the justices in eyre were no longer excluded;[55] and even at other times the sheriff might be empowered to enter the bishop's court by writ of *accedas ad curiam*.[56] In addition, the bishop's courts must commonly have taken action in accordance with royal writs;[57] and when the bishop claimed cognizance of cases begun in the royal courts, the plaintiff would be told to proceed in the bishop's court that justice might be done to him there, but otherwise he might return to the king's court.[58] In a like manner, the king retained the right to pardon offenders convicted by justices in the Isle.[59] Royal intervention might even be sought by the bishop or prior when faced

44 Execution might be a hanging matter. A number of tenants in Witchford hundred owed suit to the gallows at Aldreth, and had to stay until the man to be hanged was dead (Bodl. Libr. Laud. Misc. 647, ff. 169*b*–170*a*).

45 E 202/1/10.

46 *Cal. Close*, 1377–81, 302–3.

47 e.g. *Cal. Pat.* 1334–8, 283, 286, 291; 1340–3, 450; and see H. C. Darby, *The Medieval Fenland*, 155 et seq.

48 *Cal. Pat.* 1348–50, 533.

49 Ibid. 1367–70, 286.

50 The bishop's liberty was allowed by the justices in eyre (e.g. Just. Itin. 1/82, m. 3), the justices of the market (*Cal. Close*, 1272–9, 167), the justices of trailbaston (B.M. Add. MS. 9822, ff. 31*b*–32*b*), and the justices of oyer and terminer (C.U.L. Add. 2950, ff. 156–7).

51 e.g. B.M. Cott. MS. Claud. C. XI, ff. 321*b*, 358*b*; Bp.'s Mun. G 3 ('Old Coucher', ff. 200*b*, 218*b*); Caius Coll. MS. 485/489, f. 368*a*.

52 Bp.'s Mun. G 3 ('Old Coucher', f. 232*a*); B.M. Cott. MS. Claud. C. XI, f. 326*b*.

53 *Select Cases in the Exch. of Pleas*, ed. H. Jenkinson and B. Formoy (Selden Soc.), 65.

54 W. Petyt, *Jus Parliamentarium* (1739), 16.

55 e.g. the sessions of the justices of eyre in 1286 and 1299 in the Isle (Just. Itin. 1/90, Just. Itin. 1/95). In a like manner the king's court sat at Ely under King John during the Interdict (*Curia Regis R.* 1210–12, 52–53, 63, 72).

56 Just. Itin. 1/95, m. 30.

57 For an instance in the Wisbech hundred court, see Cam, *Liberties and Communities of Medieval Engl.* 200, and cf. *Close R.*, 1259–61, 162.

58 e.g. B.M. Cott. MS. Claud. C. XI, f. 321*b*.

59 e.g. *Cal. Pat.* 1327–30, 126; 1345–8, 144; 1350–4, 215.

with a situation that either found hard to manage. The prior asked for a special commission of oyer and terminer in 1292 when the Abbot of Ramsey and others had mowed his meadows and reed bed at Sutton, and felled his timber.[60] So did the bishop in 1328 when his bailiff was prevented from holding the view of frankpledge;[61] and when certain men at Ely erected a mill to the loss of the bishop's multure, and resisted the bishop's bailiff when he tried to take away the upper millstone.[62] There was much give and take between the king and the lords of the liberty of Ely. The king insisted upon the due performance of responsibilities, but he also was willing to defend the bishop and prior in the rights given by him and his ancestors to the church.

Finally, the liberty of Ely involved important fiscal privileges. From the 13th century these consisted in the main of the rights of bishop and prior to receive all judicial profits from their lands and men, both in the Isle and elsewhere. At an earlier date the liberty had also been exempt from national taxation, from gelds and Danegelds;[63] but this exemption never applied to the new taxation of the 13th and later centuries. In the early 13th century the bishop and prior might collect these royal taxes in the liberty,[64] or special royal collectors might be appointed by the king for the purpose;[65] but the liberty paid just the same. From 1270 onwards, moreover, the same commissioners seem to have acted both in the Isle and the shire proper.[66] The extension of the church's immunity to include lay subsidies may well have been something that no king felt he could afford. In 1327, the Isle contributed rather more than one-sixth of the total subscribed by the county of Cambridge to the twentieth levied in that year.[67]

In order to sustain the responsibilities of the liberty, there grew up in the Isle a series of courts, institutions and offices analogous to those which existed in the shire outside it. At the same time, though close, the analogy was not complete. Both the differences and the similarities will emerge soon enough from a description of the institutions of the liberty. Beginning with the courts and at the bottom of the franchisal ladder, the jurisdiction exercised in the shire in the view of frankpledge and the sheriff's tourn was, in the Isle, everywhere in private hands. But the arrangements for exercising this jurisdiction varied somewhat in the different hundreds. In Ely hundred the bishop held courts leet in his demesne manors,[68] and the prior may also have had a leet at Ely;[69] but there was no sheriff's tourn in the hundred, and presumably the business appropriate to it was transacted in the leets. Nor was there a sheriff's tourn in Wisbech hundred; and there, moreover, instead of a leet being held in each manor, a single leet was held for all the vills of the hundred.[70] It was held annually at Wisbech, and the juries of each vill came forward in turn to present the faults and failings of their own community. The business of the leets, how-

[60] *Cal. Pat.* 1281–92, 518.

[61] Ibid. 1327–30, 154.

[62] Ibid. 1385–9, 168.

[63] See the charter of Richard I (D. & C. Mun. Carta 13), and *Pipe R. 1130* (Rec. Com.), 46, 49, &c.

[64] In 1224 and 1225 (*Pat. R.* 1216–25, 494, 508, and *Rot. Litt. Claus.* (Rec. Com.), i, 593; ii, 40, 82).

[65] In 1232 and 1237 (*Close R.*, 1231–4, 160–1; 1234–7, 588).

[66] e.g. *Cal. Pat.* 1266–72, 418; E 179/81/6; *Cal. Close*, 1377–81, 321.

[67] E 179/81/6.

[68] For Ely leets in the reigns of Henry VI and Edward IV see Bp.'s Mun., C 6 (Court R.); and for Littleport, *The Court Baron*, ed. F. W. Maitland (Selden Soc.), 110, &c.

[69] There seems to have been a *curia prioratus* in Henry VIII's reign according to Bp.'s Mun., G 2 (Bp. Wren's Note Bk., f. 324).

[70] Bp.'s Mun., C 7, C 8 (Court R.).

ever, was very similar in both hundreds: the normal petty police matters, common nuisances, and offences against the assizes of bread and ale, together with a liberal admixture of domanial matters which might, more properly, have been dealt with in the halimotes.

The arrangements in Witchford hundred were rather different. The bishop held courts leet in his four demesne manors and in two or three other places in the hundred;[71] but elsewhere the leets were held by the prior or by other manorial lords.[72] In these circumstances (so unlike those of Ely and Wisbech hundreds, where most of the manors and courts were the bishop's) the bishop made arrangements more closely akin to those in the shire outside the liberty. Twice a year he sent his steward on a sheriff's tourn in the hundred; and each vill attended before him by its free jury and by the reeve and four men to make presentments about petty crime and breaches of the assizes.[73]

In the holding of the courts leet the bishop and the prior had equal rights in their own lands; but all the higher franchises in the Isle belonged to the bishop alone. First amongst these were the hundred courts, which, by the 13th century, were three in number. It is less easy to say when the practice of holding three separate courts was instituted. In Saxon times the 'two hundreds' seem to have been convened, at least on occasions, in a single assembly at the north gate of the monastery;[74] and in 1086 'the two hundreds which meet at Witchford' were certainly not two separate geographical areas.[75] By the later 12th century, however, some progress seems to have been made in the development of separate courts for the hundreds. 'All the men of the two hundreds, the chronicler says, ought to assemble fortnightly at Ely or at Witchford, or at *Modich* which is a quarter part of these hundreds.'[76] The court at *Modich* was almost certainly the original of the Wisbech hundred court; but it is hardly clear whether the courts which assembled at Ely and Witchford were already two separate courts, or merely alternate meetings of the same court. By the 13th century Ely and Witchford hundreds had become clearly separated from one another, and there remained only faint vestiges from the time when there had been only one court for the two hundreds. Yet still, in the 15th century, there was a court at Ely named the 'court of the palace' calling upon suitors from both the southern hundreds of the Isle. But the suitors did not come, and no business was done; and for all the mark it leaves upon the records, the court had done little or nothing for many generations.[77] Yet the list of those who should have come preserves some shadow of the court which assembled from the two hundreds to meet Ealdorman Aethelwine and Wulfstan of Dalham.

Of the three hundred courts, the surviving rolls of Wisbech hundred go back to the later years of Edward I.[78] At that period, it was still a very busy court. It met every three weeks and transacted all the normal business of a

[71] As well as the courts for his demesne manors, the bishop seems to have held leets in Coveney, Little Thetford and Chatteris (Bp.'s Mun., G 2 (Bp. Wren's Note Bk., ff. 278, 300), G 3 ('Old Coucher', f. 209).

[72] There is a splendid series of rolls for the prior's leet at Sutton from 20 Edw. I onwards amongst D. & C. Mun. In addition, the Abbot of Thorney held a leet for his manor in Whittlesey (Bp.'s Mun., G 3 (Liber M, ff. 387–9), Thorney Red Bk. i, ff. 176*b*, 177).

[73] Bp.'s Mun., C 7 (Court R. 18 Ric. II and 7 Hen. VI).

[74] Above, p. 5.

[75] Dd, i, pp. 191*b*, 192.

[76] *Hist. Elien.* ii, cap. 54.

[77] Bp.'s Mun., C 6 (Court R.). These date from the reign of Edward IV, but there is also a note in Wren's Note Bk. (f. 304) that the court was doing little business in the reign of Richard II.

[78] Bp.'s Mun., C 7 (Court R.).

hundred court; in addition, however, it exercised supervision over the leets and halimotes in the hundred and did a very great deal of business which was seignorial rather than franchisal. It exacted dues and obligations from the bishop's tenants, both bond and free; and spent a very great deal of time registering transfers of customary land. Indeed, in the reign of Edward II, a special court called the *curia bondorum* or *curia nativorum* seems to have been set up to deal with this business. It was a court for the whole hundred; and like the hundred court, at first at least,[79] it met every three weeks. Apart from seignorial business, moreover, it dealt with petty cases of trespass and the like very similar to those which are found on the rolls of the hundred court itself. These and other peculiarities of the Wisbech hundred are no doubt explicable by the fact that most of the men of the hundred were the bishop's tenants, and that convenience could dictate what business should be done in the different courts held in the hundred.

Witchford hundred court was very different, if only because its suitors were commonly men of other lords. There is, therefore, no domanial business recorded on its rolls. True, these rolls have not survived for any period before the later 14th century; and by that time their content is meagre, since the justices of the peace had taken away much of the business which earlier would have been done in the hundred court.[80] In fact, at this date, Witchford hundred court was concerned mainly with minor pleas of debt. In the case of Ely hundred, there is a preliminary difficulty of identifying the court in which the hundredal jurisdiction was exercised—particularly as there are no surviving records of any of the Ely courts of a date earlier than the 15th century. Of these records, those which bear most resemblance to the rolls of a hundred court derive from the Barton Court of Ely. They contain mainly pleas of debt and covenant—familiar residual business of the hundred courts in the later Middle Ages; and extracts from earlier rolls preserved by Wren make it clear that the Barton Court exercised jurisdiction over Ely and its hamlets, and over Littleport and Downham.[81] On the other hand, it was hardly an orthodox hundred court. It met weekly like a borough court; and (as at Wisbech) it transacted seignorial as well as hundredal business—for in it lands were let to farm, tallages and reliefs exacted, labour services enforced, and licences to marry given to the customary tenants.[82] The explanation of these peculiarities may be the same as that for the peculiarities of Wisbech hundred. In both cases, moreover, the hundredal business may have been tacked on to a pre-existing domanial court when the single administrative area of the 'two hundreds' broke up into its three constituent hundreds.

By contrast with the leets and hundreds, the courts in which the bishop exercised his highest franchises have left very few records, and no rolls at all, from medieval times; and their nature has therefore to be inferred from less direct evidence. The extent of their jurisdiction, however, was succinctly laid down on a number of occasions. The bishop had the right to determine before

[79] Later the court met only from 4 to 8 times a year (Bp.'s Mun., D 7 (Bailiffs' Accts., Wisbech Castle)).

[80] Bp.'s Mun., C 7 (Court R. 47–48 Edw. III and 13–14 Ric. II), and see Cam, *Liberties and Communities of Medieval Engl.* 198.

[81] Bp.'s Mun., E 1 (Gaol Delivery R. 16–20 Hen. VI), G 2 (Bp. Wren's Note Bk., ff. 256, 301, 317–20). It might also be added that, by the 15th century, the court was doing very little business.

[82] These are taken from Wren's notes, but even in the 15th-century rolls there are entries recording the transfer of customary land.

his own justices all pleas, including pleas of the crown, which arose in the Isle of Ely. Therefore, all commissions upon which the king's justices acted outside the liberty, and all writs touching cases in the Isle, should be handed over to the bishop's ministers. The bishop's justices would then execute the commissions and determine the cases.[83]

In the 13th century, this superior jurisdiction appears to have been exercised in a single court sitting at Ely, with powers as comprehensive as the courts held by the king's justices in eyre. This court in the early part of the century was called merely *curia Elyensis*;[84] later and more commonly it was termed the court of the bishop's itinerant justices.[85] It is clear, from extracts preserved in a conventual register, that these justices of the bishop did everything that the royal itinerant justices did in the shires. They heard civil pleas that were terminated by a final concord; executed criminal justice; heard presentments about escaped felons, claims to liberties and the verdicts of coroners' inquests.[86] If the rolls of this court had survived, they would differ little (if at all) from the rolls of the king's justices in eyre.

In the later 13th and 14th centuries, however, we begin to hear something of more specialized judicial commissions. As early as 1286 there is a record of the bishop's justices of jail delivery.[87] Though Edward I in that year claimed the right to appoint such commissions for the Isle,[88] in fact the bishop would appear to do so from time to time in the later Middle Ages,[89] just as he also seems to have felt empowered to appoint justices of oyer and terminer.[90] More important still, the bishop claimed the right to have his own justices of assize for the liberty. We hear of them holding a plea of mort d'ancestor in the reign of Edward II;[91] in the middle of the 14th century the Bailiff of Wisbech hundred was collecting amercements imposed by them;[92] and, at the end of the century, they were holding three sessions a year in the Isle just as the king's justices of assize were doing outside it.[93] These justices of assize, furthermore, seem gradually to have acquired a comprehensive jurisdiction in all common law matters; and to have stepped more or less into the shoes occupied by the bishop's itinerant justices of an earlier date. By the mid-15th century, they were called justices *ad omnia placita*, and equipped with a commission of assize in its comprehensive late medieval form and a commission to deliver the jails of the liberty.[94] Nor is it surprising that the active members of these panels of justices—the members designated for the quorum—should have included a number of professional lawyers. Such were John Prisott in 1448, who may have become Chief Justice of Common Pleas the following year; and Gilbert Haltoft (1448–56), already Chief Baron of the Exchequer; and Thomas Billings, first appointed a justice in the Isle in 1455 and continuing in the office

[83] e.g. Just. Itin. 1/82 m. 3 (1261); B.M. Add. MS. 9822, ff. 33*b*–34*b*; *Cal. Pat.* 1367–70, 286 (1369).
[84] B.M. Egerton MS. 3047, f. 211*b*; D. & C. Mun. Carta 140.
[85] e.g. Thorney Red Bk. i, f. 175: a final concord drawn up in 1246 'in curia domini Hugonis Elyensis Episcopi apud Ely . . . coram ipso Hugoni Elyensi Episcopo et Justiciariis suis itinerantibus et aliis fidelibus suis etc.'
[86] B.M. Cott. MS. Vesp. A. XIX, ff. 99*b* et seq.
[87] Just. Itin. 1/90 mm. 6, 9: 'Justiciarii episcopi ad gaolam de libertate Elyensi deliberandum assignati.'
[88] Ibid. m. 7*b*.
[89] C.U.L. Add. 2945, f. 140. Further, the commissions of jail delivery in the Isle in the 14th century seem to be headed by the bishop's steward (*Cal. Pat.* 1327–30, 126; 1345–8, 144; 1350–4, 215).
[90] B.M. Cott. MS. Claud. C. XI, f. 361*b*.
[91] Ibid. f. 318*a*.
[92] Bp.'s Mun., D 7 (Bailiffs' Accts.).
[93] C.U.L. Add. 2953, f. 3.
[94] Bp.'s Mun., Reg. Gray, ff. 5, 17, 42, 79*b*.

after he became a justice of the King's Bench in 1464 and Chief Justice in 1469.[95]

Finally, in the course of the 14th century, the Bishop of Ely also acquired the right to nominate his own commissions of the peace in the Isle. As early as 1338 Bishop Montacute was commanded to cause the statutes for maintaining the peace to be observed in the Isle, and to see that his bailiffs and ministers arrested persons transgressing against them.[96] In 1358, during the exile of Bishop Lisle, the king appointed special justices of labourers for the liberty.[97] By Richard II's reign the bishop's franchise in this particular was fully worked out,[98] and the justices in the shire were warned not to meddle with the keeping of the peace in the Isle, but to send copies of their commission to the bishop to execute as he had been wont to do.[99] At this period, the bishop's justices apparently held sessions quarterly at Ely, and others at Wisbech, Whittlesey, Sutton, and Haddenham.[1] By the end of the 15th century, however, the sessions had settled down at Ely and Wisbech, where the modern successors of the medieval justices continued to hold them down to the reign of William IV.[1]

This medieval liberty of Ely, naturally enough, called for a considerable administrative staff to shoulder the responsibilities involved in it. During most of the Middle Ages, the central figure of this staff was the bishop's steward. At least from the 13th century he was first and foremost a judicial official. He sometimes presided in the courts leet and the hundred courts;[2] and he was clearly the normal president in the higher franchise courts. It was to the steward that the itinerant justices at Cambridge in 1272 handed over their commission;[3] and in 1286 it was said that 'all the stewards of the bishop since the last eyre have held all pleas of the crown, in delivering jails, in pleas *de iuratis et assisis*, and in pleading all pleas which can be pleaded in the county court; and all this they have been able to do from time beyond memory'.[4] It is hardly surprising that, in Richard II's reign, the bishop's steward should have been called the bishop's justiciar.[5] Only in the 15th century was he beginning to be pushed aside by more narrowly professional men of the law, no doubt the ancestors of the later Chief Justices of the Isle of Ely.

Meanwhile, the steward had acquired a number of assistants. By the end of the 13th century, there was a deputy steward in the Isle;[6] and in effect the constable of Wisbech castle acted as a second deputy in the northern part of the liberty.[7] Each of the hundreds had its own hundred bailiff[8] and its own coroner;[9] and both bishop and prior had jailers looking after their prisons at

95 E. Foss, *Biographia Juridica* (1870), 94, 324, 540.

96 *Cal. Close*, 1338–40, 140.

97 *Cal. Pat.* 1358–61, 67.

98 Ibid. 1377–81, 599.

99 *Cal. Close*, 1385–9, 546.

1 C.U.L. Add. 2953, ff. 3, 9, and for the justices of the peace sitting twice during the year at Wisbech in 21–22 Edw. IV see Bp.'s Mun., D 7 (Bailiffs' Accts., Wisbech Castle).

2 e.g. Wisbech Court Leet, 28 Edw. I (Pentecost), and Wisbech Hundred, 33 Edw. I (Lent) and 35 Edw. I (Lent).

3 Bp.'s Mun., G 3 ('Old Coucher', f. 325).

4 Just. Itin. 1/90 m. 9.

5 C.U.L. Add. 2953, f. 9.

6 S.C. 6/1132/10. He was also called the bailiff of the liberty of the Isle (Bp.'s Mun., G 3 ('Old Coucher', f. 237), and G 2 (Liber B, nos. 73, 90, 91)), and is the direct ancestor of the High Bailiff of later times.

7 His judicial duties were clearly laid down in his commission (C.U.L. Mm. i. 49, ff. 177–9, and Bp.'s Mun., G 3 ('Old Coucher', ff. 237, 237*b*)). That he actually carried them out is clear from Just. Itin. 1/95 m. 71*b* and the rolls of the hundred court at Wisbech.

8 Just. Itin. 1/90 m. 6.

9 Just. Itin. 1/90 m. 9 and Just. Itin. 1/95 m. 65. At this date the coronership of Ely hundred was an

Ely[10] (the bishop's other jail at Wisbech was in the charge of the constable and his staff[11]). Finally, at the very bottom of the hierarchy were village and manorial officers: constables of the peace doing the work they did everywhere;[12] beadles and haywards, who made attachments and collected amercements;[13] and even the reeves, keeping the rolls of the tithings in the manor.[14] The total number of officials must have been quite large; but then they administered a very comprehensive and exclusive liberty.

This does not, however, necessarily mean that the liberty of Ely was a county palatine, even though 'divers statutes' might describe it so. In Coke's words:

> 'The power and authority of those that had counties palatine was king-like, for they might pardon treasons, murders, felonies and outlawries thereupon. They might also make justices of eire, justices of assise, of gaol-delivery and of the peace. And all original and judicial writs, and all manner of indictments of treason and felony, and the processes thereupon were made in the name of the persons having such county palatine.'[15]

Clearly the medieval bishops of Ely possessed a number of these powers. As indicated above, however, the right to pardon men convicted in the bishop's courts was retained by the crown; and the justices appointed by the bishop took action upon royal commissions and writs handed over to them by the king's justices sitting in the shire of Cambridge.[16] It may be significant in this respect that no chancery grew up in the medieval liberty of Ely comparable with the chanceries of Lancaster and Durham; for this liberty, however extensive, was never quite a palatinate.

The Liberty of Ely in Modern Times

After the end of the 15th century, some of the institutions of the medieval liberty of Ely can, for the most part, be left out of the reckoning. The business done by the hundred courts, though their rolls continue down to the 17th century, had already begun to decline quite rapidly in the 14th century;[17] while the hundreds themselves, as administrative areas, were commonly replaced by the new 'divisions' of the Isle established by the justices of the peace. Courts leet, similarly, though still held at Wisbech in the reign of James II[18] and at Ely in 1715 where they cost Bishop Fleetwood £6 10*s*. for the steward's dinner,[19] had likewise become of small importance. The institutions which

hereditary office in the Baret family. A 15th-century commission to the coroner of Witchford hundred has been preserved by Bentham (C.U.L. Add. 2951, f. 83*a*).

While at the end of the 13th century and apparently afterwards there was one coroner in each hundred, the earliest mention of these officers shows that in 1251 there were 3 coroners in Ely itself (Caius Coll. MS. 485/489, f. 21*b*). They may correspond to the 3 knights who were to keep the pleas of the crown in each county according to the Articles of the Eyre for 1194, but there is nothing to show whether their responsibility included the Isle as a whole or merely Ely hundred.

10 B.M. Cott. MS. Vesp. A. XIX, ff. 99*b* et seq. Bp.'s Mun., G 2 (Liber B, no. 70).

11 Bp.'s Mun., G 2 (Liber B, no. 62). The expenses for its maintenance are regularly entered in the castle accounts.

12 e.g. the appointment of a constable at Elm court leet, 13 Edw. II (Pentecost), and various presentments for hindering them in their duties in Turnus Vicecomitis, 50 Edw. III (Pentecost) (Bp.'s Mun., C 7 (Court R.)).

13 e.g. the rolls of Wisbech hundred, and *The Court Baron* (ed. Maitland), 120, 140–1.

14 D. & C. Mun., Court R. of Sutton halimote, 13 Edw. II.

15 *4th Part of Insts. of Laws of Engl.* 204.

16 Above, p. 13; and see also Gladys Scott Thomson, *Lords Lieutenant in the 16th Century*, 5 n. and S. and B. Webb, *The Parish and the County*, 313 et seq.

17 Above, p. 12.

18 Bp.'s Mun., C 10 (Court R.).

19 Bp.'s Mun., A 6 (Bp. Fleetwood's Commonplace Bk.). He reduced the cost of this dinner to a maximum of £5 just as, at the same time, he scaled down the cost of the Quarter Sessions dinners and the dinner to the judge at the Assizes.

mattered in Bishop Fleetwood's time and long before were the six-monthly assizes and the quarter sessions of the justices of the peace. They form the main subjects in the history of the liberty in modern times.

The Tudor period was a time of trial for the liberty of Ely, although it emerged with less abatement than most medieval franchises. The form which this trial took may best be discussed in connexion with particular institutions; here it is enough to say that there seems to have been an attempt by Henry VII and Henry VIII to assert closer royal control over the liberty, that the Privy Council down to 1640 at least made the royal will continuously felt in the Isle, and that the long vacancy from 1581 to 1600 tended to bring the administration of the liberty into closer connexion with that of the county of Cambridge. One consequence was that, throughout the modern period, the administrative history of the Isle of Ely is much like that of any of the English shires. It was governed by a very similar group of officials, exercising much the same powers in much the same way as any other local government area of the time. In what follows, therefore, illustrations of the administrative system of the liberty are testimony in the main to the similarities between it and the system which prevailed generally. This much at least was achieved by the Tudors and Stuarts and their Privy Councils.

Nevertheless, it was still precisely during this period that the liberty acquired the title of a County Palatine. If its institutions were more and more closely assimilated to those outside its bounds, the administrative distinctness of the Isle remained. Right down to 1836 it had its own assizes and chief justice; its own justices of the peace; its own jails and houses of correction; its own rates separate from the rates of the county; its own Chief Bailiff with powers equivalent to the powers of the sheriff in the shire.[20] In these respects the Isle was treated, nearly enough, as a separate county; and that is true in other respects too, even to the extent of giving to it in 1565 its own commission for the repressing of pirates.[21] On the other hand, although it had its own Chief Bailiff, the liberty never acquired a sheriff of its own; and (as in medieval times) it continued to pay taxes along with the shire, including the ship money of Charles I and the hearth tax of Charles II.[22] This half-way status of the liberty is perhaps best illustrated by its relations with the Lords Lieutenant of Cambridgeshire. In the reign of Edward VI Bishop Goodrich secured his own appointment as Lord Lieutenant for the Isle, an office treated as quite separate from the lord lieutenancy of Cambridgeshire.[23] The bishop combined the lieutenancy with power to nominate the custos rotulorum of the Isle;[24] and so might have prepared the ground for the union of the two offices, which became normal elsewhere,[25] in the bishop's hand. A period of obscurity follows. In the early years of Elizabeth there does not seem to have been a separate lieutenancy for the Isle; on the other hand the commission for the county does not specifically include the liberty,[26] and in 1565 Lord North was told that the mustering of

[20] On the general character of the Liberty in modern times see Webb, *The Parish and the County*, 313 et seq., Ethel M. Hampson, *Treatment of Poverty in Cambs.* 220, and W. Watson, *An Historical Account of . . . Wisbech* (1827), 104–5.
[21] *Acts of P.C.* 1558–70, 284, 289, 297.
[22] *Cal. S.P. Dom.* 1635, 375; 1635–6, 149; Bp.'s Mun., F 10 (Presentments of Constables, &c.).
[23] *Acts of P.C.* 1550–2, 259; 1552–4, 49, 276. Scott Thomson, *Lords Lieutenant in 16th Century*, 32–33.
[24] This was saved to the bishop by 3 & 4 Edw. VI c. 1.
[25] Webb, *The Parish and the County*, 286, 373–4.
[26] Despite the statement of *D.N.B.* xli, 170, Lord North's commission in 1569 does not seem to include the

great horses and geldings there should be referred (as had been accustomed) to the bishop.[27] During the long vacancy after 1581, however, the Isle almost inevitably was absorbed into the jurisdiction of the Lord Lieutenant of the shire,[28] and the latter's authority was effectively exercised within it.[29] In 1598, he obtained permission to appoint a special deputy for the Isle;[30] and even after the end of the vacancy in 1600 this pattern apparently persisted.[31]

If the medieval liberty of Ely was abridged in this way, there is nothing surprising in the fact that, under Cromwell's rule, there was a Major-General whose authority included the Isle of Ely and was exercised there by a special deputy.[32] Further, at least up to 1640, the Privy Council exercised a continuous administrative supervision over the liberty. It was continuously receiving reports on every conceivable matter from the justices of the Isle; it was ready to prescribe punishment for the author of 'slanderous reports touching some of the council'; it summoned before it a justice of the peace who had suffered the Bailiff of Wisbech hundred to utter with impunity 'seditious or very scandalous speeches touching her Majesty'.[33] In a like manner, the liberty did not exclude the jurisdiction of Star Chamber,[34] or the right of the crown to appoint special justices of assize to deal with the men who rioted in 1816 for higher wages, cheaper flour, and 2*d.* a pint off beer.[35] And, of course, the crown retained the right to pardon and revoke pardons,[36] and to commute a sentence of death imposed at the Ely assizes for the less serious one of transportation.[37]

In all these respects the crown was continuing to exercise an old authority, or refusing to allow the liberty to obtain an exemption from the authority of new agencies developed by the State during the 16th century. At least as important was the increased control exerted over the commissions of assize and the peace in the Isle. So far as the justices of assize were concerned, they were still sitting thrice yearly in Elizabeth's reign, but later sat only twice yearly at Ely and Wisbech alternately.[38] But large changes took place in the medieval practice. Then, the king's justices in the shire had handed over their commission to the bishop's representative, and this commission was then executed in the Isle by justices appointed by the bishop. This practice changed at any rate with the accession of Henry VII. In 1491, 1501 and 1515 commissions of jail delivery were issued by the crown for the liberty, in which the king

Isle: see S. Haynes, *A Collect. of State Papers . . . 1542–70* (1740), 560, and cf. Cooper, *Annals of Camb.* ii, 248.

27 *Acts of P.C.* 1558–70, 242, and cf. Scott Thomson, *Lords Lieutenant in 16th Century*, 52–53.

28 e.g. the commission to Lord North in 1596 (*Cal. S.P. Dom.* 1595–7, 296).

29 See Lord North's arrangements for the trained bands of the Isle (*Acts of P.C.* 1588, 134).

30 Ibid. 1597–8, 630, and Scott Thomson, *Lords Lieutenant in 16th Century*, 66.

31 The commissions for Cambs. in 1602 and 1603 extended to the Isle (*Cal. S.P. Dom.* 1580–1625, 419; 1601–3, 206), and for Sir Simon Steward's misdeeds as Deputy Lieutenant for the Isle, which brought him to the notice of the Privy Council and Star Chamber, see *Cal. S.P. Dom.* 1625–6, 331, 360–1, 529; *Acts of P.C.* 1626, 29.

It should, however, be noted that the bishop to a limited degree exercised some of the powers of a lord lieutenant, in nominating (for example) the panel of justices of the peace for the liberty (below, pp. 21–22).

32 *Cal. S.P. Dom.* 1655–6, 175, 235, &c.

33 *Acts of P.C.* 1571–5, 316; 1599–1600, 707–8, 731.

34 e.g. *Cal. S.P. Dom.* 1633–4, 69.

35 *A Full and Correct Report of the Trials for Rioting at Ely and Littleport*, by a Member of the Inner Temple (1816), esp. 1, 2, 37. The appointment of this special commission gave some offence to Edward Christian, Chief Justice of the Isle: see his *Charges delivered to the Grand Juries of the Isle of Ely* (1819), 386.

36 See *Cal. S.P. Dom.* 1679–80, 193; 1680–1, 441, 455, 464.

37 See below, p. 20.

38 Watson, *Hist. Wisb.* 104, 352 n.

nominated the justices who were to execute them. The bishop merely reissued the same commission to the same panel of men, although he still seems to have nominated his own justices of assize. Naturally enough, however, he commissioned a group of men who corresponded very closely with the justices of jail delivery nominated by the king.[39] This practice continued; for in 1574 Lord North set at the beginning of his roll for the Ely assizes the royal patent constituting himself and others justices to deliver Ely jail.[40] By the 18th century the assizes in the liberty were held by justices nominated by the crown in commissions of oyer and terminer and jail delivery of the sort delivered to the king's justices of assize outside it.[41]

This apparent assumption of control over the assizes of the liberty was not, in fact, quite so complete as it seems to be. In the early Tudor period (possibly even under the Yorkists[42]) the effective members of the lengthy list of justices who were nominated probably consisted of the bishop's steward of the Isle[43] and one or two important royal judges or officials.[44] The latter included, between 1487 and 1515, Sir James Hobart, the Attorney-General; and two Chief Justices of the King's Bench—Sir William Huse and the well-known Sir John Fyneux, 'steward of one hundred and twenty-nine manors and counsel to sixteen noblemen at once'.[45] If these royal judges were in fact the main figures at the bishop's assizes, then the king's control in the liberty had significantly increased during this period. Such a policy, moreover, is still suggested by the fact that in Mary's reign Sir Clement Heigham, one of her Catholic privy councillors, sat as the 'king's and queen's justice' at Wisbech when Robert Pygot was presented for heresy.[46]

The employment of such prominent royal judges as Flowerdew, Shute and Popham to hold the assizes in the Isle by Elizabeth,[47] on the other hand, may have a different explanation. They may have sat simply because the see was vacant, and there was no bishop to nominate his own justice; for as early as 1575 the Wisbech assizes were held 'before John Goldewelle, Chief Justice for pleas within the Isle of Ely'.[48] Except during the vacancy from 1581 to 1600, in fact, from that time forward the usual president at the assizes in the Isle was the Chief Justice of the Liberty. Some of them were men of distinction in the legal world—Hitcham in 1610 was queen's Attorney-General; Sir Edward Turner was Speaker of the Cavalier Parliament; and perhaps even Edward Christian ought to be mentioned here, for although Gunning said he died 'in the full vigour of his incapacity', he was an editor of Blackstone and Professor of the Laws of England in the University of Cambridge.[49] But these

[39] *Cal. Pat.* 1485–94, 360; 1494–1509, 247; Bp.'s Mun., Reg. West, ff. 1*a*, 1*b*.

[40] Bp.'s Mun., E 1 (Gaol Delivery R.).

[41] Ibid., F 5: commissions of 1692, 1761, and 1793. In 1714 the bishop paid for two commissions of oyer and terminer £5 to the clerk of the Crown and £1 1*s*. to the Chancellor for his fiat (Bp.'s Mun., A 6 (Bp. Fleetwood's Commonplace Bk.)). It does not appear that the bishop's justices received powers to entertain cases at nisi prius, perhaps because down to the 17th century at least the liberty was exempted from the jurisdiction of the central courts of common law.

[42] The fact that Thomas Billings remained a justice in the Isle even after he became a judge (above, pp. 13–14) may indicate that Tudor policy had been anticipated by the Yorkists.

[43] e.g. Robert Bradbery in 1491 and John Burgoyne in 1510 (*Ely Dioc. Remembrancer*, 1908, 180; Watson, *Hist. Wisb.* 110).

[44] *Cal. Pat.* 1485–94, 138.

[45] Foss, *Biographia Juridica*, 254, 359.

[46] Ibid. 337–8; *Ely Dioc. Remembrancer*, 1914, 279.

[47] Bp.'s Mun., E 1 and 2 (Gaol Delivery R.); Foss, *Biographia Juridica*, 257, 527, 612.

[48] Bp.'s Mun., E 1 (Gaol Delivery R., 17 Eliz.).

[49] Above, p. 1; Foss, *Biographia Juridica*, 680–1; Christian, *Charges*. For a very unfavourable portrait of

chief justices were not judges in the king's courts. Bramston, indeed, seems to have resigned the justiceship of the Isle when he became Chief Justice of the King's Bench in 1635, and Godbolt when he became a judge of the Common Pleas in 1647; so did Turner on his appointment as Solicitor-General in 1670, and the aged Salathiel Lovell even when appointed Recorder of London in 1692.[50] The office of Chief Justice, moreover, was very definitely conferred by the bishop. His duty was 'to hear and determine all pleas . . . real, personal or mixed, moved . . . by any person whatsoever within the Isle . . . which, according to the liberties of the bishop of Ely . . . ought to be held and impleaded . . . before the justice aforesaid in the Isle aforesaid, and to do justice therein to the parties according to the law and custom of the realm and my liberties aforesaid'.[51] He held office after 1689 'as long as he shall demean himself well therein', and previously, no doubt, at the bishop's pleasure. His patent stipulated for the payment of a salary of £10 a year by the bishop, but in fact at the end of the 18th century he was receiving the sum of 50 guineas for each attendance at the assizes.[52]

Thus, although the assizes at Ely and Wisbech were held under royal commission, from Elizabeth's reign down to the 18th century[53] the presiding judges in fact were the Chief Justice of the Isle and one or two of the justices of the peace. This was made possible by the fact that the royal commissions always included the Chief Justice and all or most of the justices of the peace in the liberty. A long list of judges might be nominated by the Crown, but effectively the list could be narrowed down to these few men who were in a direct manner the bishop's nominees. Thus, it would seem, the invasions upon the liberty by the early Tudors had not been sustained.

It was, therefore, in this manner that the Chief Justice of the Isle and his colleagues came to hold the assizes for the liberty. The occasions were not unmarked by ceremony. A coach would be sent to meet the judge, with a pair of leaders; at Wisbech there would be a nosegay for him, though not apparently at Ely; a preacher would be in attendance and trumpeters.[54] Before his arrival the Chief Bailiff would have summoned all justices, bailiffs, coroners, and constables to be there with their records and indictments; and commanded the presence of all prisoners in the jails of the liberty, and of such juries as were necessary for presentment and trial.[55] The beginning of proceedings would no doubt be delayed by the sermon; and further by the fact that Edward Christian at least 'twice during every year inflicted upon the Grand Jury of Ely a charge of unconscionable length' which 'rarely had any reference to the cases in the Calendar'.[56] He may have thought a more specific charge unnecessary: as he said, 'I will not detain you with any observations upon the indictments . . . because I see several magistrates upon the Grand Jury', and he was quite convinced that 'the enlightened magistracy of the Isle' knew well enough their routine business.[57]

Christian, including the final shaft quoted above, see H. Gunning, *Reminiscences of Camb.* i, 192–200 (1855 ed.). Gunning considers that Christian's appointment to the Chief Justiceship of the Isle was a matter of 'good fortune', though rather grudgingly he admits that the edition of Blackstone 'was very creditable to him'.

50 Foss, *Biographia Juridica*, 116–17, 166, 415, 618.

51 Patents of appointment in Bp.'s Mun., F 11 (1691, 1738).

52 Receipts in Bp.'s Mun., F 11. In 1836, apparently, the chief justice's salary was £155 a year (H.O. 52/28).

53 Bp.'s Mun., F 11 (Kalendars of Assizes, 1759–74).

54 Bp.'s Mun., F 6 (Chief Bailiff's Accts. 1798–9), and see A. Gibbons, *Ely Episc. Rec.* 128–9.

55 Bp.'s Mun., E 1 (Gaol Delivery R., Ely Assizes, 16 Eliz.).

56 Gunning, *Reminiscences*, i, 200.

57 Christian, *Charges*, 7–8, 218.

Much of the judicial business of the assizes was concerned with civil actions between neighbour and neighbour. In the early modern period the jurisdiction of the bishop's court in cases of this sort was probably as comprehensive as it had been in medieval times, and as late as Charles II's reign Bishop Wren could demand cognizance of such a case from the King's Bench.[58] In the 18th century, however, doubt was cast upon the legality of fines levied and common recoveries suffered in the bishop's court. Partly for this reason, and partly because the practice had grown up of allowing cases from the Isle to begin in the king's courts or to be removed there by writ of *certiorari*, it was noted that 'business on the pleas side' was on the decline in that century. None the less, even at this date, the liberty courts were still a great convenience for the inhabitants of the Isle since 'they have justice administered, as it were, at their very doors'.[59] It was no doubt similarly convenient that Courts of Requests, to deal more rapidly and cheaply with actions for debt under 40*s*., were set up at Ely in 1778 and Wisbech in 1811[60]—although this may have helped to reduce the amount of business transacted at the assizes.

The criminal jurisdiction of the bishop's assizes, on the other hand, remained little impaired right down to 1836. It seems to be true in the 18th century that serious crime was not very common. Time and time again (such episodes as the Littleport riots apart) Chief Justice Christian could congratulate the 'enlightened magistracy' of the Isle upon the few prisoners in the calendar; and, after sixteen years of office, he could say with satisfaction in 1816 that he had 'been called upon to pronounce judgement of death upon sixteen prisoners only'.[61] That is hardly a bad record in view of the severity of the criminal law of the time. The calendar of the Wisbech assizes in 1773 is not untypical. Four men were accused of stealing a wether; one was acquitted, and the others (together with a man who had stolen a red cow) were sentenced to death, but reprieved whilst application was made for commutation of this sentence to one of transportation. The only other prisoner, who had stolen two coats and some light coloured cloth, was sentenced to transportation for seven years. Most of the applications to commute the death sentence, furthermore, seem to have been allowed. Only four of the sixteen men Christian sentenced actually suffered the death penalty; ten he recommended to mercy himself, and two others 'by the recommendation and interference of others . . . obtained from the royal clemency that lenity which was refused them by myself'.[62] It may be added, however, that the business of the assizes in the 17th century was sometimes richer in picturesque incident. Even in a few published extracts,[63] we may read of Dorothy Ellis, to whom the Devil appeared in the likeness of a great cat; of John Neale who awaited the time when the Anabaptists would preach from the pulpits, called Charles Stuart a bastard and the seed of a bastard, and laid bets against the Restoration; and of William Gotobed and others who stole lead from the Cathedral roof.

Before leaving the assizes a word should be said about procedure. Much of the work had been done in advance: each indictment was accompanied by a file of depositions taken before the justices of the peace, the verdict of a coroner's inquest, and so on. It remained only for the grand jury of the Isle

[58] Bentham, *Ely*, App. 26–27.
[59] Ibid. 24–25.
[60] *V.C.H. Cambs.* ii, 108 n.
[61] Christian, *Charges*, 7–8, 218, 381, 388.
[62] Christian, *Charges*, 381; Bp.'s Mun., F 11 (Kalendar of Wisbech Assizes, 1773).
[63] Gibbons, *Ely Episc. Rec.* 112–14.

to certify the indictment as a 'true bill' for trial to proceed.[64] In addition, of course, both the grand jury and the constables of parishes and hundreds might present additional faults and failings in open court; but by the 18th century in fact they were presenting very little. At the Ely assizes in 1774, the constables of Wisbech hundred did present a causeway at Outwell which had become a public danger, but otherwise 'all was well'.[65] On at least one occasion, however, in 1764, the grand jury presented an address to the Chief Justice in return for his charge.

> 'The inhabitants of this Isle, they told him, having long endured much Sufferings from a corrupt Administration in this Island by the Infusion of Evil Council into the mind of the Prince of this Palatinate, rejoice not a little from knowing that a Man of your understanding and righteous spirit is joined to the Cabinet Council of our Great Good Lord Doctor Mawson Bishop over this Isle. . . . On first reflection we are a free born people. . . . The Magistracy . . . are made for the People. . . . We this day are more. We are the Grand Jury of this Island—and as such are—the Guardians of the Lives and Liberties of our Country. . . . From hence we are led to lay open—to remonstrate on Grievances too heavy to be borne patiently, silently. The Loss of our Liberty who are free born—False Imprisonment by the Suppression of one Assize of the two. . . . Barbarities under Imprisonment little inferior to the Inquisition abroad. . . . For that End We urge on your Lordship when in Council assembled to press on others our Prelates Council a speedy redress of those Grievances . . . before Christmas next . . . [otherwise] . . . We the Grand Jury are unanimously and stedfastly purposed in January next to fly for succour to the Guardians above all of the Commonwealth of England, The Honourable the House of Commons.'[66]

It is hardly a matter of surprise that the foreman of the jury on this occasion was that same James Collier who did so much to secure an improvement of conditions in the jail at Ely.[67]

The Chief Justice held the assizes only twice a year. By contrast and in Edward Christian's words: 'The justices of the peace constitute the broad basis of the pyramid of government of this country: they are acting perpetually in their towns, villages and neighbourhoods; by their constant co-operation and unceasing exertions, the good order and tranquillity of the country are preserved.'[68] This was as true of the liberty of Ely as it was elsewhere; and in this, as in other respects, the liberty had its own particularity. This was succinctly put in 1638, during the troubles which attended the draining of the fens. One Edward Powell had given *2d.* to the town crier of Ely to proclaim that all should meet next morning to go to the king to petition for their fens. When Mr. Goodrich, a justice, called Powell before him, the latter answered: 'If I deny it, the crier's evidence, being but one man's, is no evidence; and if I confess it, what harm? For what was he and the rest of the justices? They were but bishop's justices and not the king's.'[69] Yet Powell was not quite correct as to his facts. It is true that there was a separate commission of the peace for the liberty; but this commission was issued by the king, and the justices were 'the justices of the peace of our lord the king assigned to hold and keep the peace in the Isle of Ely'.[70] It is true, on the other hand, that in the 18th century the panel of justices to be appointed was submitted to the Lord Chancellor by

[64] Cf. Webb, *The Parish and the County*, 448, and for an indictment ignored, Bp.'s Mun., F 11 (Kalendar of Ely Assizes, 1761).

[65] Bp.'s Mun., F 1 (Plea and Gaol Delivery R.) and F 10 (Verdicts of Grand Juries, &c.).

[66] The whole document is printed by Gibbons in *Ely Episc. Rec.* 425–6.

[67] J. Howard, *State of the Prisons in Engl. and Wales* (1780), 248–9.

[68] *Charges*, pp. iii–iv.

[69] *Cal. S.P. Dom.* 1638–9, 301.

[70] Bp.'s Mun., F 5 (Commission of the Peace, 1793), F 11 (Q. Sess. Min. Bk., Ely Sess., 7 Oct. 1740).

the bishop,[71] and that to all seeming the Chancellor normally issued the commission as requested. But there might be difficulties. Bishop Fleetwood wrote, against eight of the names he suggested for the commission of the peace in 1714:

> 'These eight, and ten more, the Lord Chancellor had put into Commission, to serve some purposes, without acquainting the Bishop with it, contrary to immemorial Custom, and turned four of the Old ones out. I left out the ten to vindicate the Bishop's privilege, and to show some resentment for the ill usage of my Predecessor, and left the other eight in Commission, because the good of the Isle required it, as I was told.'[72]

Such troubles may have been infrequent, but once again a change had taken place since medieval times. In 1417 the Prior of Ely had complained about the treatment of his tenants by the justices whom 'the *bishop* has constituted to conserve the king's peace in the Isle'.[73] The change over to royal commissions of the peace is usually associated with the statute 27 Henry VIII c. 24;[74] but here again the second Tudor may have been following in his father's footsteps. In the commission which Bishop Alcock issued in 1487 there are the significant words: 'know that the king has assigned and constituted John, Archbishop of Canterbury, etc., to be keepers and justices of the peace'; and in 1490 the terms of reference of the commission were to be such as were 'contained in the letters of the lord king'.[75] It looks as though, as in the appointment of justices of assize, the bishop's liberty may have been abruptly curtailed as soon as the Tudors ascended the throne—even though Henry VIII allowed that the bishop and his steward should always be members of the commission of the peace for the liberty.[76]

It is true, of course, that the bishop recovered control of the commission by nominating its members, just as he recovered control over the commissions of assize by ensuring that they should be led by his own chief justice. The justices of the Isle remained the bishop's justices, even though they did have the king's patent. But under the early Tudors they may have been for a time in a more literal sense the king's justices; and as the king appointed them, so he might remove them, as Thomas Megges wished him to remove Alexander Balam in 1535.[77] Perhaps it was the control exercised by the Privy Council in the 16th and early 17th centuries which in the end led the Crown to acquiesce in *de facto* nomination by the bishop; for the bishop's justices fulfilled all the obligations of, and were as closely supervised as, the king's justices elsewhere. They certified Chancery about the behaviour of French denizens;[78] they investigated for the Council slanderous reports and treasonable rumours;[79] they regulated corn supplies, fixed prices, and dealt with 'the foul disorders committed by badgers and like engrossers of corn, grain and other dead victual';[80] they took measures to safeguard the recusants in custody at Wisbech castle;[81]

[71] Bp.'s Mun., F 5 for such a letter from the bishop to Henry, Earl Bathurst in 1776.

[72] Bp.'s Mun., A 6 (Bp. Fleetwood's Commonplace Bk.).

[73] Bentham, *Ely*, App. 31.

[74] e.g. Hampson, *Treatment of Poverty in Cambs.* 220 n.

[75] Bp.'s Mun., Reg. Alcock, ff. 12, 54.

[76] 27 Hen. VIII c. 18.

[77] *L. & P. Hen. VIII*, viii, no. 1088. Megges seems to have been an agent of Thomas Cromwell in the Isle, and to have hoped from his favour to have obtained a house and farm in the Isle which had belonged to the prior of Ely (ibid. ix, no. 754). No doubt he also hoped that his object would be brought nearer by the postscript to this letter: 'I send you a present of swans, cranes and herons. Your hundred pikes are ready when it is your pleasure to have them.'

[78] *Cal. Pat.* 1557–8, 13.

[79] *Acts of P.C.* 1571–5, 316; *Cal. S.P. Dom.* 1625–6, 42.

[80] *Cal. S.P. Dom.* 1581–90, 333; *Acts of P.C.* 1595–6, 132, 137.

[81] *Acts of P.C.* 1615–16, 383.

and, when they failed to take action about scandalous speech or any other matter, they could be brought before the Council to answer for their neglect.[82]

The number of men sustaining these burdens does not seem to have been large at an early date. In 1487 and 1489 there were a dozen names in the commission, in 1490 only eight: and since men like the Archbishop of Canterbury, the Chief Justice and the Attorney-General were included, the effective numbers must have been fewer.[83] In the late 16th and early 17th centuries, Council letters were addressed to groups of justices varying in number from three to seven;[84] while the Quakers reported, in 1659, that there were only eight justices in the Isle—a number they thought too low.[85] By the 18th century, however, they might have been satisfied. The commission of 1776 contained fifty-four names. Some of these (the half-dozen peers, the dean, and the archdeacon for example) may not have been active magistrates; but this would still leave a considerable number of effective members consisting of most of the twenty-five 'esquires' and the seven rectors or vicars named in the commission.[86] It is perhaps not inappropriate that the clergyman justice would seem to have played quite a prominent part amongst the enlightened magistrates of the Isle. There were at least two of them amongst the justices who tried to check the rioters at Littleport and Ely in 1816; and another, the Revd. J. Vachell, had perhaps been active enough to stimulate the rioters to plunder his house.[87] A little earlier, there were four clergymen amongst the magistrates who had committed the five prisoners in Ely jail on Boxing Day, 1805.[88]

In the 18th century, there is no sign whatsoever that these justices received any remuneration for their services, save for the Quarter Sessions dinner the cost of which Bishop Fleetwood was concerned to abridge.[89] Indeed, it is far from clear that they had ever received payment: even in 1586, during the long vacancy, there is no mention of wages paid to them, though much is said of the fees paid to other officials from the 'justiciarius ad assisas infra Insulam Eliensem' down to the Forester of Somersham and the keeper of Downham Park.[90] However, the important thing about the justices, as Christian said, was that, paid or unpaid, they were acting perpetually in their villages and neighbourhoods. In practice this meant there was a distinction between the general body of the justices, whose activities were more or less restricted to their own localities, and a kind of inner ring who undertook the greater and more specialized responsibility of holding the sessions. The work of the wider body did not lack variety. Justices sitting singly or in pairs in their own neighbourhoods compiled most of the depositions upon which the sessions and assizes acted;[91] and took security for appearance in court, or committed to jail, those accused of offences. Two justices convicted Benett Crabb 'for prophanely swearing of ten oaths' at Littleport; singly or in pairs they examined vagrants and issued passes for them to go even so far as Belfast; and generally supervised the day-to-day affairs of a society still literally parochial in its government.[92] It was in this real sense that the justices were 'the broad basis of the pyramid of government'.

[82] Ibid. 1599–1600, 707–8.

[83] Bp.'s Mun., Reg. Alcock, ff. 12, 54, 58.

[84] *Acts of P.C.* 1595–6, 132–3; 1615–16, 383; *Cal. S.P. Dom.* 1625–6, 42.

[85] *Cal. S.P. Dom.* 1658–9, 358.

[86] Bp.'s Mun., F 5 (Commissions of the Peace).

[87] *A Full and Correct Rep.* 37, 40, 49; *Camb. Chron.* 31 May 1816.

[88] Bp.'s Mun., F 11 (State of the Prisoners in Ely Gaol).

[89] Above, note 18.

[90] Bp.'s Mun., D 5 (5) (Sede Vacante Receiver's Acct., 29–30 Eliz.).

[91] For some 17th-century samples, see Gibbons, *Ely Episc. Rec.* 132–3.

[92] Bp.'s Mun., F 11 (Q. Sess. Min. Bk., Ely, 1740–1).

Benett Crabb's conviction, however, was submitted to the Quarter Sessions for confirmation; and the sessions spent much time, not only in dealing with offenders committed to their attention, but also in confirming or quashing vagrants' passes issued by local justices and otherwise supervising their acts. Formally, in the 18th century, these courts met quarterly—alternately at the White Hart in Ely and the Rose and Crown in Wisbech. In fact they met much more frequently (though hardly with absolute regularity) by means of the simple expedient of adjourning and reassembling periodically between the main quarter days. In 1740–1, for instance, there were no fewer than twenty meetings of the court; and the following time-table for the summer of 1741 seems to be quite typical. The main sessions met at Wisbech on 14 July; but further meetings were held at Wisbech on 25 July, at Ely on 1, 8, and 15 August, at Wisbech on 3 September, and again at Ely on 25 September.[93] In some cases it might be noted that there was 'nothing done at this adjournment',[94] but at least formal meetings of the justices were frequent enough to give adequate attention to business. These formal sessions might also be supplemented by less formal meetings: at Wisbech in 1827 these latter were held every Wednesday and Saturday in the new Town Hall from 12 o'clock to 3.[95]

The sessions and their adjournments were, as indicated above, very much the province of an inner ring of justices.[96] The work they did there had almost unlimited variety. They dealt, of course, with the usual range of offences, great and small. Francis Goodday had broken Benjamin Talbott's doors and windows at Sutton; Cyprian Day had feloniously stolen 5 lb. of candles, and Thomas Saunders two pieces of brass; John Harding was 'a loose, idle and disorderly fellow, being taken with wood on his shoulder'. John Woolaston, again, had been drunk but was discharged 'this time', even though Christian had advised the magistrates 'to pay particular attention to public houses' since he was convinced that the Littleport riots were 'the effect of a casual meeting of an idle rabble in an ale-house'.[97] Of course, too, from time to time, one like John Nunn would be ordered to appear 'concerning a bastard child begotten on the body of Catherine Cork, a single woman, with which the said Catherine Cork is now with child'. Some of these offenders might merely be bound over to keep the peace; a thief like Cyprian Day, on the other hand, was more likely to 'have his body stripped naked and be whipped around the market place in Ely on Saturday next . . . until his body be bloody and receive fifteen lashes and no more'. Women were no more exempt from this punishment than men.

Besides judicial work of this sort, the sessions had important administrative functions. They assessed highway rates and enforced statutory labour on the roads. They appointed drainage commissioners pursuant to a recent act of parliament. They assessed the poor rate, laid down rules for its collection, and audited the accounts of the treasurers who received and spent it. They scrutinized removal orders, and enforced upon churchwardens and overseers their responsibilities for the poor. They regulated apprenticeship; appointed constables of parishes, chief constables of hundreds, and treasurers of the poor rate; they dismissed the keeper of Wisbech House of Correction and appointed

93 Bp.'s Mun., F 11 (Ely and Wisbech 1740–1).

94 e.g. at Ely, 28 Feb. 1741.

95 Watson, *Hist. Wisb.* 352.

96 At Ely in 1740–1, for instance, two to four justices sat on the Bench, consisting of some or all of the following: Edward Partridge, Aungier Peacock, Henry Morley, and George Mayfield.

97 Christian, *Charges*, 8–9.

another in his place.[98] Further, though proceedings took place at the best inn available, there was a good deal of formality about them. The chairman might deliver a charge no less lengthy than that delivered by the Chief Justice at the assizes.[99] The Chief Bailiff summoned a grand jury and all the officers of the liberty as he did for the assizes;[1] and the grand jury and the constables made very similar presentments. True, in the later 18th century, they had little to say (save once, that Joseph Scorer was keeping a bawdy house at Wisbech); but a hundred years earlier it had been different. The grand jury had noticed that William Beesley had stopped a water-course, and that Jeremy Cutter had an inadequate fence between his house and the common hill next the river. The constables of Ely presented a lady who spoke evil words when asked to pay hearth tax; three men at Downham 'for walking about in the night and disordering themselves'; an Ely man who refused to lend his cart for removing vagrants, and three Downham men who refused to pay their rates. The constables of Witchford hundred had also noted a number of men who had absented themselves from their parish churches.[2] The administrative system which stretched down from the Quarter Sessions to these village constables was a very intimate affair.

There remains the administrative staff of the liberty which gave effect to, and supplemented, the occasional action of the Chief Justice and the continuous activities of the justices of the peace. The central administrative officer was the High or Chief Bailiff of the Isle, a life officer with powers equivalent to those of the sheriff in the shire. His duties included preparation for the assizes and sessions; the return and execution of writs; financial responsibility for assizes, quarter sessions, and Ely jail; and arrangements for the execution of condemned felons.[3] Under him he had a deputy bailiff in the 18th century who seems to have taken responsibility in the northern division of the Isle; and bailiffs in the three hundreds of the Isle, the bailiff of Ely hundred being also in charge of Ely jail. His custody of this jail seems hardly to have given satisfaction at this period,[4] despite a public subscription in 1724 to repair and improve it.[5] However, when Edward Christian visited the jail and the Bridewell in 1819, he was 'happy to find everything in each that could contribute to the health, comfort and reformation of persons who are placed there for punishment'.[6] The sting may lie in the last phrase of his verdict.

To some extent the Chief Bailiff and his subordinates still acted within the framework of the old hundreds of the Isle. This was less applicable in the case of officials serving under the justices of the peace. As early as the 15th century the sessions of the peace seem to have been settling down at Ely and Wisbech, and already to have exercised authority in what became the 'divisions' of the Isle.[7] Of these there were two only, a northern division consisting of Wisbech

98 On all this see Bp.'s Mun., F 6 (Treas. Accts.) and F 11 (Q. Sess. Min. Bk. 1740–7).

99 See J. Nasmith, *A Charge Delivered to the Grand Jury at the Gen. Q. Sess. of the Peace, . . . 2 Apr. 1799*—a very lengthy discourse on poor law administration, which bears some resemblance to Christian's charges.

1 Bp.'s Mun., F 4 (Notice of Q. Sess. at Wisbech July 1769).

2 Ibid. F 10 (Verdicts of Grand Juries, &c.).

3 Ibid. F 1 (Chief Bailiff's Acct. 1798–9) and F 11 (Pat. of Appointment for the High Bailiff, 1792). Gibbons, *Ely Episc. Rec.* 128–30.

4 *V.C.H. Cambs.* ii, 108–9.

5 Bp.'s Mun., F 6 (Receipts and Accts.).

6 *Charges*, 349.

7 See above, p. 14. In 36 Hen. VI a man from Whittlesey in Witchford hundred was presented before the justices at Wisbech by the constable of Witchford hundred, and there was a similar instance in 20 Edw. IV (Bp.'s Mun., G 2 (Bp. Wren's Note Bk. f. 350)—extracts from rolls which have since vanished).

hundred and the north part of Witchford hundred, and a southern division consisting of Ely hundred and the south part of Witchford hundred.[8] Each division in the 17th and 18th centuries acquired its own panel of active justices, a more or less self-contained staff, its own rates and its own Bridewell,[9] and its own coroner. In fact, by the 18th century, apart from the Chief Justice and the Chief Bailiff, only the Clerk of the Peace seems to have exercised his office over the whole of the liberty; and he enjoyed the additional titles of Clerk of the Crown, Clerk of the Assizes, Prothonotary, Chirographer, and Under-secretary in the Isle.[10]

At lower levels, each division had its own treasurer for administering the poor rate;[11] and lower still, of course, there was the great body of unpaid, part-time officials which ultimately shouldered most of the detailed work of local government. There was a chief constable for each hundred or part hundred in a division, responsible for the collection of poor rates as well as for police measures in his bailiwick (it may well have been appropriate, therefore, that the chief constable of Ely hundred in 1816 was also the local agent of Mortlock's bank).[12] At the real base of the pyramid were the parish officers: surveyors of highways; village constables who collected rates and taxes, maintained order, moved on 'strollers' and vagrants, and once in a while took a felon to London for transportation;[13] and, finally, the churchwardens and the overseers, responsible in particular for the administration of the poor law. In all these respects the liberty of Ely was very much like any other part of rural England in the 17th and the 18th centuries.

By the time of the Reform Bill of 1832, however, the external situation was changing. In the debates on that measure Macaulay said that it was time 'to pay a decent, a rational, a manly reverence to our ancestors, not by superstitiously adhering to what they, in other circumstances, did, but by doing what they, in our circumstances, would have done'.[14] From this rational point of view, the liberty of the 'prince of the palatinate' in the Isle of Ely, with his Chief Justice and his justices and his 'cabinet council', was doubtless anomalous and anachronistic. It could also, no doubt, be criticized in detail: there were no police in the Isle, and fifty special constables had to be enrolled to control an anti-Poor Law meeting near Ely in 1836; Ely jail, according to the Inspector of Prisons, was 'most inefficient' and its jailer 'old, respectable and now physically incompetent' after thirty-four years of service.[15] So the question of abolition was raised. Some opposition there was,[16] but none the less in 1836 an Act

[8] See Hampson, *Treatment of Poverty in Cambs.* 226 n. The north and south parts of Witchford hundred were already separately distinguished in 1640–2 (*Cambs. Subs. R., 1250–1695*, ed. W. M. Palmer, 17, 67). The south part had its own Chief Constable in 1741 (Bp's Mun., F 11 (Q. Sess., Ely, 31 Jan. 1741)), the north part apparently by 1769 (ibid. F 10, Verdicts of Grand Juries, &c.). The two parts are shown on the maps of Cambs. of R. G. Baker (1821) and J. & C. Walker (1837) as though they were separate hundreds but not *eo nomine*. In Pigot's *Commercial Dir.* (1830) and in the Census Returns of 1831 they are named as the separate hundreds of North and South Witchford.

[9] Wisbech Bridewell also served as a jail for the northern division after the castle ceased to be used for that purpose, presumably after it was leased in the later 17th century (Watson, *Hist. Wisb.* 131).

[10] Bp.'s Mun., F 11 (Pat. of Appointment for Edward Bellamy, 1802).

[11] Ibid. F 6 (Treas. Accts. 1741–9). There is no sign that there was ever a treasurer for the liberty as a whole, equivalent to 'county treasurers' elsewhere.

[12] *A Full and Correct Report of the Trials . . .*, 36.

[13] Bp.'s Mun., F 6 (Treas. Accts. 1741–9).

[14] T. B. Macaulay, *Writings and Speeches* (1871), 487 (Speech of 2 Mar. 1831).

[15] H.O. 52/28.

[16] A memorial praying for continuance of the separate jurisdiction of the Isle was presented by 350 'clergy,

was passed 'for the extinguishing of the secular jurisdiction of the archbishop of York and the bishop of Ely in certain liberties'.[17] The Chief Justiceship of the Isle was swept away; future holders of the offices of Custos Rotulorum and Chief Bailiff were to be appointed by the Crown. The Clerk of the Peace in the Isle for the future was to be nominated by the Custos Rotulorum, and coroners were to be elected by the freeholders of the Isle. Ely jail was abolished, the assizes were to be held in the Isle under the normal provisions laid down in 3 and 4 William IV c. 71, and the Isle was to be deemed a division of the county of Cambridge. It was a division, doubtless, which had some peculiarities. Its rates were separate from those of the county—assessed, levied, and paid as though the Isle were a separate shire. It still had its own commission of the peace. These anomalies were ironed out by making the Isle into an administrative county in 1888,[18] thus preserving into contemporary times the outline of the medieval liberty of Ely.

There remains only a footnote to the acts which abolished the bishop's liberty. In the course of debates on the Justices of the Peace Bill in 1949 in the House of Lords, Lord Merthyr moved an amendment which would have had the effect of ending the right of the Bishop of Ely and his steward to sit as justices for the Isle in virtue of the statute of 27 Henry VIII c. 18. But the amendment had been anticipated. Its mover 'earned a gentle rebuke from the Lord Chancellor . . . Lord Jowitt told him that by the Statute Law Revision Act passed last year the provisions of the old act passed in the reign of Henry VIII had been repealed'.[19]

gentlemen and yeomen' from all over the Isle. The corporation of Wisbech, on the other hand, seem to have accepted the extinguishing of the liberty but to have wished the Isle to be treated as a separate county. They stress the fact that its population was greater than some English and most Welsh counties; that rates would increase; that Wisbech is far from Cambridge and that more time would be taken at the assizes (H.O. 52/28).

[17] 6 & 7 Wm. IV c. 87, amended by 7 Wm. IV & 1 Vic. c. 53. By a further Act (1 & 2 Vic. c. 94) the records of the abolished jurisdiction were placed in the charge of the Master of the Rolls; but they had not been examined by any of his officers in 1853, and have not to this day been removed from the bishop's custody. See F. S. Thomas, *Handbk. of the Publ. Rec.* (1853), 356; *Rep. of the Royal Com. on Publ. Rec.* H.C. [Cd. 6361], App. ii, 15 (1912–13), xliv; H.C. [Cd. 7544], App. ii, 39, 142 (1912–13), xlvi; H.C. [Cmd. 367], 20 (1919), xxviii; H. Hall, *Repertory of Brit. Archives*, 79.

[18] Webb, *The Parish and the County*, 317.

[19] *Camb. Daily News*, 22 Oct. 1949, and see 11 & 12 Geo. VI c. 62.

CITY OF ELY[1]

'Elge', meaning eel-district, appears to be the correct origin of the name, though willow-place[2] and eel-island are explanations almost equally apt. An outlier of Lower Greensand, partially capped by boulder-clay and gravel, formed the island on the eastern edge of which the city of Ely developed, rising from the 'waste of waters opening wide' which, in early times, occupied the hollows of the underlying base of Kimmeridge clay. Though many species of fish abounded,[3] eels were much the most prolific product of the fenland waters and formed an important source of food and income from the beginnings of the monastic settlement.

The high ground on which the cathedral stands still permits the city to dominate the plain for many miles. Ely lies mainly on the western bank of the River Ouse, some 3 miles from its junction with the Lark and, until the advent of the railways, depended for its development largely on easy access by water, northwards to Wisbech and Lynn and southwards to Cambridge, numerous tributaries communicating with most of the Midland river systems.[4]

Ely is some 16 miles north by east from Cambridge and about 71 miles from London, on what was until 1948 the L.N.E. railway. It is an important rail centre. The main line from the south bifurcates here, one arm going northwards to King's Lynn and the other north-eastwards to Norwich. Lines also radiate north-westwards to Peterborough, south-eastwards to Newmarket, and south-westwards to St. Ives. A loop line (the Adelaide loop) leaves the line from Peterborough at Waterden Fen and joins the Lynn and Norwich lines south-west of their junction. This loop was constructed in 1890[5] presumably to enable traffic from Norwich and Lynn going northwards via Peterborough to by-pass Ely and avoid reversing engines. The railway station dates from 1845.[6] Its original 'Italian' design was much admired at the time of construction.[7] The building has subsequently been much altered.[8] British Railways is one of the more important employers in the city.

The city is traversed by the London to King's Lynn road (A 10) and is connected on the south-east with Soham and on the west with Chatteris by roads (A 142) of lesser importance. A road also leads northward to Downham and another north-eastwards to join the Littleport–Mildenhall (Suff.) road (A 1101). The second of these crosses the Lark at Prickwillow which is there spanned by a bridge—the successor of one which existed in 1677.[9]

Ely never acquired the status of an incorporated borough.[10] Since the Local Government Act of 1894 the city, with its environs, has been governed by an Urban District Council. The College Precincts formed an extra-parochial entity until 1933 and since then have been a separate civil parish though under the control of the Urban District Council. Only on one occasion, in 1295, did the city elect its own parliamentary representatives.[11]

The major significance of the city, despite some modern industrial development, is still its position as head of the episcopal see and centre of an agricultural area. The modern plan of the city follows remarkably closely the broad layout of medieval times.[12]

The built-up area of Ely itself comprises only about 1 square mile of the 24 within the boundaries of the Urban District. There are, however, several subsidiary settlements. Chettisham is a very small hamlet 2 miles north-west of the city on the road to Lynn. It comprises little but a church, a few farms and cottages, and a railway station. Stuntney stands on an abrupt island in the fen 1½ mile south-east on the Soham road. It is a compact little village of the ordinary East Midland type, with the church and the hall much in evidence. Two other hamlets lying north-east of the city are much more characteristic of the Fens. (Queen) Adelaide, about 2 miles out, is entirely modern, and named from a public house.[13] The Ely sugar-beet factory is here. Prickwillow, 4 miles east-north-east from the city near the junction of the Lark and Ouse, is mentioned in 1251,[14] but has only become a place of importance in the last 150 years. Situated in one of the lowest parts of the Fens, it has a curiously un-English look. Owing to the drainage operations the soil is shrinking at the rate of 2 inches a year, and the church, school, and other houses are built on piles. Where piles have not been used, subsidence and even collapse have often occurred. So high is the water table that the dead have to be buried in the city cemetery on the high ground 4 miles away. Farther out still are the scattered settlements of Burnt Fen and Shippea Hill, 8 miles from the city centre and on the Suffolk border. Here there is a railway station, opened in 1845 as Mildenhall Road. In 1885 its name was changed to Burnt Fen, and about 1904 to Shippea Hill.[15]

Nothing survives of Ely *CASTLE*, save the conspicuous mound (Cherry Hill) on the south-west side of the cathedral park. The remains belong to the motte-and-bailey type of fortress, and have been held

[1] For suggestions, corrections, references, loan of copies of documents and other source material, access to local archives, and for replies to innumerable questions the author of this account of Ely is indebted to many people, but in particular to the Ven. Archd. S. J. A. Evans, Lt.-Col. G. L. Archer, Major J. G. A. Beckett, Mr. Edward Miller, Mr. J. E. Watkins (Clerk to Ely U.D.C.), Revd. J. B. Rowsell, Vicar of Ely, Miss A. A. Taylor, Mr. H. B. Wells (Asst. Editor V.C.H.), and to the General Editor.

[2] Edmund Carter, *Hist. of County of Cambs.* 64.

[3] H. C. Darby, *Medieval Fenland*, 28; *Liber Elien.* 231–2.

[4] Darby, ibid., maps on pp. 95, 97. See also Gordon Fowler, 'Extinct Waterways of the Fens', *Geog. Journ.* lxxxiii (1934), 32.

[5] Infm. Brit. Rlwys., East. Region.

[6] *V.C.H. Cambs.* ii, 132–3.

[7] R. Gardner, *Dir. Cambs.* (1851), 482.

[8] Its original state is illustrated in C. Barman's *Introd. to Rly. Archit.*, 79.

[9] Gardner, *Dir. Cambs.* (1851), 483.

[10] The categorical assertions of John Ogilby in 1676, and of Celia Fiennes only twenty-two years later, appear to be erroneous. 'Ely . . . is govern'd by a Mayor &c.' (J. Ogilby, *Britannia Depicta*, ed. 1720, 86); 'They have lost their Charter and so are no Corporation but all things are directed by the Bishop' (C. Fiennes, *Journeys*, ed. 1745, 156). It is possible that the building known as the club house or guildhall (see below, p. 86) misled casual visitors.

[11] John le Palfreur and Nicholas Baret. A precept was also directed to the city, 27 Ed. III, but no return was made (Browne Willis, *Notitia Parliamentaria*, 2nd ed., 1730, i, 160; *Parls. of Eng., 1213–1702* (1878), i, 4). The two names erroneously given as M.P.s for Ely, in 1491–2, were really those of the M.P.s for Lynn. (*E. Angl. N. & Q.*, N.S., iii, 168, 257.) Presumably the prominent role played in the councils of the nation by the Bishops of Ely, together with the city's share in the election of knights of the shire, appeared to give ample representation (*V.C.H. Cambs.* ii, 398).

[12] See plan.

[13] *P.N. Cambs.* (E.P.N.S.), 223.

[14] B.M. Cott. MS. Claud. C. XI.

[15] Infm. Brit. Rlwys., East. Region.

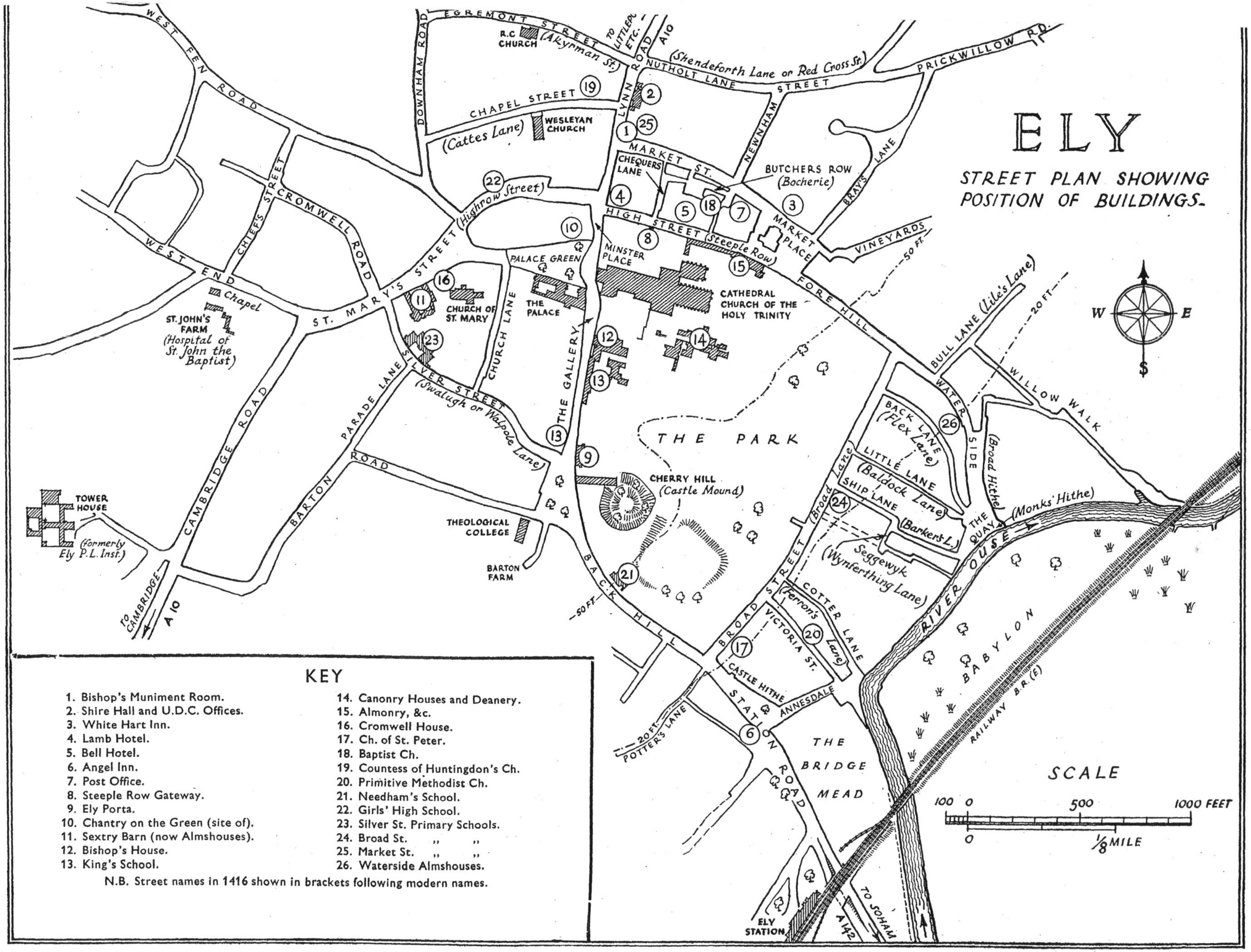

KEY

1. Bishop's Muniment Room.
2. Shire Hall and U.D.C. Offices.
3. White Hart Inn.
4. Lamb Hotel.
5. Bell Hotel.
6. Angel Inn.
7. Post Office.
8. Steeple Row Gateway.
9. Ely Porta.
10. Chantry on the Green (site of).
11. Sextry Barn (now Almshouses).
12. Bishop's House.
13. King's School.
14. Canonry Houses and Deanery.
15. Almonry, &c.
16. Cromwell House.
17. Ch. of St. Peter.
18. Baptist Ch.
19. Countess of Huntingdon's Ch.
20. Primitive Methodist Ch.
21. Needham's School.
22. Girls' High School.
23. Silver St. Primary Schools.
24. Broad St. " "
25. Market St. " "
26. Waterside Almshouses.

N.B. Street names in 1416 shown in brackets following modern names.

Based upon the Ordnance Survey Map with the sanction of the Controller of H.M. Stationery Office. Crown Copyright reserved

to be those of a royal castle built by William I to maintain the submission of the Isle.[16] There seems, however, to be no ancient testimony to a castle,[17] other than that at Aldreth, earlier than the time of Henry I. This king allowed the service of castle-guard, due from the Ely knights and hitherto performed at Norwich, to be given 'at the castle of the (recently created) bishopric of Ely'.[18] The site and nature of this castle are obscure. It has been suggested that it gave its name to the later well-known Castelbrigge and Castlehythe and was itself situated near the river and had no prolonged existence.[19] Certainly the distance of the precinctual mound from the river seems to rule out any connexion between this site and the bridge or wharf. During the Civil War of Stephen's reign, Bishop Niel is stated to have made 'a strong fort of lime and stone', within the precincts of Ely, but 'that fort being often destroyed, he built one of timber and walled in the round hill—called the keep'. The castle presently surrendered to Stephen. In 1144 Geoffrey de Mandeville was placed in charge of this castle and that at Aldreth, by authority of the representatives of Bishop Niel, who was absent at Rome.[20] A castle was said to have been taken and destroyed by Falk de Breauté, in 1216,[21] but during the baronial struggles half a century later, Ely with its fortifications was captured by Prince Edward, in 1268,[22] and it has generally been supposed that the castle built by Bishop Niel was demolished soon afterwards. According to Bentham, however, there was already a windmill on the mound in 1229.[23] Certainly no existing castle was recorded in the extent of 1251, but a messuage then stood 'where the old castle was'[24]—a statement which does seem to indicate a castle nearer to the river than the mound. In 1586–7, 6*s.* was spent on repairing 'the castle'—presumably the mound.[25] A windmill was depicted on Mount Hill on Speed's map of 1610. The hill was known as Mill Hill in 1649,[26] and by its modern name of Cherry Hill at least as early as 1821.[27] In 1779 it was crowned by a monumental pillar to the Revd. James Bentham, historian of Ely.[28]

The former *HOSPITAL OF ST. JOHN BAPTIST* is situated in West End.[29] The remains consist of three buildings of medieval origin now incorporated in a farmstead. In the 12th century there existed in close proximity the hospitals of St. John Baptist and St. Mary Magdalene, which were amalgamated by Bishop Northwold about 1240 under the title of the Hospital of St. John Baptist. By Northwold's ordinance the foundation was to consist of thirteen chaplains and brethren. It was suppressed in 1561, when the site was granted to Clare College, Cambridge, to which it belonged until sold in the present century to a Mr. Runciman. The last Master, Edward Leeds, was also Master of Clare College. The surviving remains date in origin from the early 13th century, before the amalgamation of the two foundations. Two of them are in fairly close proximity but the other is about 130 yards farther to the west. It seems probable, therefore, that portions of both the original hospitals are in existence, more especially as two of the surviving structures are obviously of ecclesiastical origin.

The most important of the remaining buildings, believed to have been the chapel of St. Mary, stands to the north of the farmhouse. It is now a simple rectangular structure measuring 41 by 25 ft., and was apparently the nave of the hospital chapel. There is now no clear evidence of a chancel. The lateral walls were probably originally somewhat higher than at present. The north wall alone retains definite pre-Reformation work *in situ*, the other walls being apparently rebuilt, but largely with old material. The windows, doorway, and stepped gables of Barnack stone and brick date from the 16th-century reconstruction, when the chapel was converted into a dwelling, and no trace remains of the original doorways and windows. The roof, which is tiled, seems to be an 18th-century reconstruction. A floor was inserted in the 16th century, with massive beams having stop chamfers. The north wall contains a blocked arcade of three bays, which proves the existence of an aisle now completely destroyed except for the foundations of the north wall, which show that it was 10 ft. 9 in. in width. After the destruction of the aisle massive stone buttresses were erected to support the nave wall; these buttresses are obviously composed of older material. The fact that the east arch of the arcade springs so close to the east wall that there is nothing to resist the thrust shows that the building was originally longer. The present east wall dates from the 16th century, but contains much old material including drums of Barnack stone, which must have formed part of columns. In the south wall over the doorway is a fluted piscina. The north arcade consists of two-centred arches with clustered columns and moulded caps and bases. All the windows belong to the 16th century; in the north wall there are three in the upper story of two uncusped lights, the heads of which are formed by the apex of the arches of the blocked arcade; below are three windows, two of which are of two lights and the other of four lights; in the east wall are two windows, the upper of three lights and the lower of two, both now blocked; in the upper part of the west wall is a three-light window. At the apex of the east gable is a double chimney-stack set diamond-wise on a square base. Within is a fireplace on each floor with chamfered stone jambs and wooden lintel. At the east end of the south wall is a plain 16th-century wooden doorway. The west gable is entirely of brick. The west wall has two brick buttresses, and in the thickness of the wall are two fireplaces; that on the ground floor is situated in the centre, while the upper is placed to the south of the window and has been partly destroyed by a later doorway; the chimney-stack has disappeared, and its place is taken by three steps of different brick to the rest of the gable.

To the south-east of the chapel is a structure of Barnack stone measuring 20 by 19 ft. Low down in the east wall are two blocked arches of clunch, and there are traces of a blocked arch in the north wall. The present roof, which is tiled, is probably of 17th-century date and

[16] *V.C.H. Cambs.* ii, 29–30.
[17] This conclusion is also reached by S. Inskipp Ladds, 'The Monastery of Ely', *Cambs. and Hunts. Arch. Soc. Trans.* v, 70–71.
[18] J. Bentham, *Ely*, app. xviii.
[19] D. J. Stewart, *Archit. Hist. of Ely Cath.* 193–4; *Pipe R. 1130* (Rec. Com.), 44; *V.C.H. Cambs.* ii, 202 & n. 49.
[20] J. H. Round, *Geoffrey de Mandeville*, 209; *Hist. Elien.* 623.
[21] *Ann. Prioratus de Dunstaplia* in *Ann. Mon.* (Rolls Ser.), iii, 48.
[22] *V.C.H. Cambs.* ii, 396, quoting *De Antiquis Legibus Liber*, 95.
[23] Bentham, *Ely*, ii, 53; *P.N., Cambs.* (E.P.N.S.), 216.
[24] B.M. Cott. MS. Claud. C. XI, f. 25*b*.
[25] S.C. 6/Eliz./282 (Bailiff's View of Acct., 28–29 Eliz.).
[26] *P.N. Cambs.* (E.P.N.S.), 216.
[27] Ibid.
[28] Gardner, *Dir. Cambs.* (1851), 477.
[29] For an account of the foundations see *V.C.H. Cambs.* ii, 308–10 and W. M. Palmer, 'The Hosps. . . . at Ely' in *Proc. Camb. Antiq. Soc.* xxxvi, 76. For an illustrated account of the buildings see L. Corbett, 'The Hosps. . . . at Ely' in *Proc. Camb. Antiq. Soc.* xxxvi, 58.

shows that the building has served at some time as a pigeon house; its original purpose is unknown.

The third structure of medieval origin, believed to have been the chapel of St. John, is situated about 130 yards to the west of the first. It is built mainly of rubble with Barnack dressings. The roof is tiled and hipped at the east and west; it is of modern construction. The building measures 53 by 24 ft., and it seems fairly certain that it was originally a chapel before the amalgamation of the two hospitals. It is of very early 13th-century date. In the north wall are four plain lancets, now blocked, with jambs of Barnack stone and heads of clunch; the internal arches are widely splayed. Between the third and fourth windows is a plain blocked doorway of one order with continuous chamfer and a hood terminating in masks. Over the doorway is a rude carving in relief depicting a man riding backwards on a cross-horned animal and blowing a horn. It has been claimed that this formed part of Etheldreda's monastery.[30] The west wall is quite featureless. In the south wall is a blocked doorway opposite to that on the north, and there are traces of similar lancets, all now blocked. A large wagon opening has been inserted in the south wall. In the east is a large shallow recess, which in all probability represents a former wagon entrance which was blocked when that on the south side was made. Traces of former lancets in clunch can be seen on either side of the recess. In a wall close to the building is a piece of Roman millstone.

The farmhouse seems to date from the second half of the 17th century, and probably replaced the desecrated chapel as a dwelling. It is largely constructed of old material.

The prior is said to have had a prison in 1229.[31] Certainly there are frequent references to it in the 14th century[32] and its existence is noted in 1417. It was situated in the porter's lodge of the Ely Porta.[33]

The bishop's prison, which was probably also his jail for the Liberty, is mentioned in 1276,[34] 1376,[35] 1394,[36] and early in the 15th century.[37] This prison, or a later one which stood at the corner of 'Jail Lane',[38] was partly rebuilt by Bishop Mawson in 1768[39] in consequence of a public inquiry into the severe measures taken to prevent escapes. It was visited by John Howard, who in 1777 catalogued its illegal deficiencies.[40] On James Neild's visit in 1810 marked improvement had been effected.[41] On the abolition of the Liberty in 1837 the building ceased to be used as a prison.[42] From *c.* 1842 until *c.* 1879 the Ely Mechanics' Institute appears to have been housed in it.[43] It is now used as the Diocesan Muniment Room, and the floor rings, by which prisoners were secured, may still be seen.

The city bridewell was built in 1651.[44] The conditions as observed by Howard were even worse than those prevailing in the bishop's prison[45] and at the time of Neild's visit little progress towards correcting them had been made.[46] The building presumably ceased to be used as a prison in 1820 when the new sessions house was built. Its subsequent fate is uncertain but it was perhaps converted into a barn prior to demolition in 1850–5.[47]

In 1820 a new sessions house or shire hall, and house of correction were built on the Lynn Road. Changes were made in these buildings in 1843 when a separate house of correction, containing 35 cells, was built at the back of the shire hall and the prison chapel in the south wing converted into a police station.[48] Between 1865 and 1869 the north wing was partly appropriated by the Volunteers for use as an armoury, and was taken over entirely for this purpose when, in 1878, the house of correction was closed under the Prisons Act of the previous year.[49] The Volunteers gave up their tenancy in 1908.[50] The shire hall, which is still used as a court, now also houses the U.D.C. Offices besides the police station.

The White Hart Inn on the north side of the market place is of 15th-century origin, but the original plan and details have been much obscured by rebuilding and later additions. It is at present L-shaped, but it is not improbable that originally there were two wings at the back. The roofs are covered with blue slates, except a gallery at the back, which is tiled. The front is now faced with brick and stucco, and a second story was added, probably at the end of the 18th century. The main portion of the north wing is entirely of brick and is of 19th-century date, but to the north of this portion is an early 17th-century structure containing the kitchens on the ground floor, beneath which is a cellar with a plain barrel vault of brick. All the fenestration is modern, but in the 17th-century section there are indications of the original windows, now blocked. The entrance passage from the street is the most interesting feature of the house. It retains late 15th-century timber framing, including one moulded angle post and principal with a bracket on the east side springing from a defaced stone corbel; there are wide joists and chamfered wall brackets. The passage has been curtailed on the west, and the brackets and angle posts on this side are no longer visible. At the back of the main block is a gallery on the first floor which, though modernized, is probably part of the original structure. The two rooms on the ground floor to the east of the entrance probably formed the hall of the medieval house, and there is a large cased-up beam in the ceiling. The 17th-century block retains an old chimney-stack of brick, and there is a stop-chamfered beam in the kitchen ceiling. The main staircase dates from the end of the 18th century and has a contemporary rail. In a room on the first floor of the main block is a good wooden fireplace of Adam character. The inn formed part of the property bequeathed by Catherine Needham in 1730, for the maintenance of her school (see below—Schools).

No. 5 Market Place, adjoining the 'White Hart', is a timber-framed house probably of 16th-century date, with a brick front added in the 18th century. There is a large wing at the back, also timber-framed except for the end wall which is of stone; there is a considerable

30 *Proc. Camb. Antiq. Soc.* xxxvi, 73–74.

31 Gardner, *Dir. Cambs.* (1851), 467, quoting 'an inquisition taken in 1279'.

32 B.M. MS. Cott. Vesp. A. XIX.

33 A chamber south of this gateway was being used as a prison in the 17th cent. (Atkinson, *Archit. Hist. Ely*, 152, 183).

34 *Rot. Hund.* (Rec. Com.), i, 49*b*.

35 *Ely Dioc. Remembrancer*, 1895, p. 55.

36 El. Ep. Reg. Fordham, f. 183.

37 El. Ep. Reg. Gray, f. 117.

38 Atkinson, op. cit., 166.

39 James Neild, *State of the Prisons* (1812), 203; *V.C.H. Cambs.* ii, 108, quoting *Gentleman's Mag.* (1804), 898.

40 Neild, *State of the Prisons*, 252.

41 Ibid. 203–6.

42 Atkinson (op. cit., 166) says it was used as a barn until demolished in 1850–66, but the reference is probably to the city bridewell (see below).

43 Gardner, *Dir. Cambs.* (1851), 480. It is last mentioned in Kelly, *Dir. Cambs.* 1879.

44 Neild, *State of the Prisons*, 205–6.

45 Ibid.

46 Ibid.

47 See footnote 42.

48 Gardner, *Dir. Cambs.* (1851), 479–80.

49 Kelly, *Dir. Cambs.* (1865, 1869); *2nd Rep. Commrs. Prisons*, [C. 2442], 28, H.C. (1878–9) xxxiv.

50 Kelly, *Dir. Cambs.* (1912).

overhang on the west side. Most of the windows are modern but there is a small two-light with leaded panes at the back of the main block. There is a large chimney-stack of brick, probably of 17th-century date. The building is now (1951) a fried fish bar.

No. 7 Market Place, adjoining the last, consists of a main block of early-18th-century date and of brick construction. There is an upper story surmounted by attics contrived in the high-pitched roof, which is tiled. One of the rooms on the ground floor has contemporary panelling smothered in brown paint. There is a long wing running at right angles at the back, which is timber-framed and of early-17th-century date, with a large contemporary chimney-stack of brick.

There is an interesting range of houses on the south side of the High Street to the east and west of Steeple Row gateway; they are partly timber-framed and partly of brick, and the roofs are tiled. The core of the structures is medieval, and it is probable that they originally formed part of the outbuildings of the monastery. After the Dissolution they were secularized and adapted for houses and shops. Most of the frontage appears to date from about the end of the 16th century, though greatly altered and faced up at a later date. Nos. 12 and 14 originally formed one house. Two rooms on the ground floor have chamfered beams and wide joists in the ceilings, and on the first floor there are some 17th-century doors. There is a large brick chimney-stack at the back, which probably dates from the end of the 16th century. No. 18, which extends over the gateway to the Close, has been much modernized with the exception of the portion over the gate, which is of two stories. The upper story has a considerable overhang and there are two carved late-15th-century bressummers on the north and one on the south; the arms of Ely occur on the lower bressummer on the north. There is a 15th-century wooden oriel window of three lights in the room on the first floor at the back, with moulded mullions and transom and a square head. No. 22 is built round a portion of 14th-century walling nearly 3 ft. thick, in the upper part of which are remains of two stone circular windows. There is a cellar beneath the back part of the house which is probably of late 16th-century date.

The Lamb Inn at the corner of High Street and Lynn Road was a coaching-inn in 1753.[51] The 'Bell' in the High Street itself was in existence in the 16th century.[52]

Nos. 3 and 5 Silver Street, originally one house, are of stone with continuous tiled roof. The structure probably dates from the beginning of the 16th century. There is a base course, and the east gable retains its original stone coping. In this wall is a three-light window in the upper story with stone mullions and a square head. At the back is a four-light window under a square head, partly cut through to form a doorway. There are two small rectangular openings, now blocked, in the east wall, and a similar one at the back which is in use. Inside, there is a stone newel staircase in the north-east angle, and in one of the rooms a beam with stop chamfers. A chimney, probably of the 17th century, has been inserted at the east end. The windows and doorways in the front are modern. A range of four cottages abuts on the west end. They are timber framed and probably date from the beginning of the 17th century. There are several large chimney-stacks of brick, the easternmost of which has three flues set diagonally. The windows are all modern except one, which has a wooden frame and leaded lights, and is probably of 18th-century date.

The White Lion Inn, at the corner of St. Mary's Street and Silver Street, is of brick and rubble and dates from the second half of the 17th century. The front is plastered and the roof, formerly thatched, is now slated. The brick gables are corbelled out.

Nos. 8 to 12 Church Lane are a row of timber-framed cottages of 17th-century date. The upper story has an overhang and is plastered. The lower story and the end walls are of brick. There is a large chimney-stack of brick in the centre.

No. 20 Church Lane is a detached timber-framed house, probably of 16th-century origin. It has been extensively restored within recent years, and the windows with wooden mullions and the chimney tops of brick are all modern. The roofs are tiled.

No. 22 Fore Hill is a mid-16th-century house with two stories above the ground floor; it is probably of timber-framed construction, but the exterior is now plastered and the end walls are of brick with corbelled gables; the roofs are covered with tiles. The structure has been considerably altered and extended at various dates and the windows on the front are all modern, with leaded panes. One old window of three lights with moulded wooden mullions and frame remains in the back room on the ground floor, and now opens on to the stairs. The front room on the ground floor has chamfered beams and plain joists, which are probably of 16th-century date.

The Angel Inn, at Castle Hithe, is a modern building on the site of an earlier inn.

At Barton Farm, about 500 yards south-west of the cathedral, there formerly stood a fine timber barn, probably of medieval date. The roof was thatched and the sides and west end were covered with modern weather-boarding. Two rows of upright timbers supported the roof, which had tie beams and collars and was carried down to within about 7 ft. of the ground. The timber framing of the sides and ends rested on a stone base. The barn was blown down in 1927 and the remains cleared away.

In Cromwell Road there is a stone-built barn probably of 16th-century date. The gables are of brick and the roof is tiled. There are narrow slits in all the walls and a large entrance towards the road with modern doors. The roof has old timbers.

There is a good early-17th-century house, known as the Manor, to the north of Stuntney church. The material is brick and the roofs are tiled. There is evidence on the south side that the structure was originally of greater extent. The front, which faces north, has a tall projecting porch, and the original plan was probably E-shaped. The windows have been modernized, as has the interior. On the north there are two large chimney-stacks, one with two tops set diamond-wise. A large brick barn is attached to the north-east angle, and the western elevation has two gables. There have been serious settlements and the building is not in very good condition (1950).

Ely Theological College,[53] on the Barton Road, was opened in 1881, for the training of students for the Anglican ministry. It is of red brick in 'Gothic' style with stone facings.

[51] Carter, *Hist. Cambs.* (1753), 64.

[52] *Cambs. and Hunts. Arch. Soc. Trans.* v, 373–84.

[53] *V.C.H. Cambs.* ii, 321–4, 336–7, 340, 342, 349, 351.

Aerofilms Ltd.

Ely from the Air, Looking South-east

Aerofilms Ltd.

Ely from the Air, Looking North-west

National Buildings Record

Ely Cathedral. West End from the South

National Buildings Record

Ely Cathedral. East End and Lady Chapel

The isolation hospital in St. John's road, the first hospital in Ely, was opened in 1917. It was erected at the sole cost of William and Mrs. Emily Ann Cutlack, and had 24 beds in 1937.[54] North of the town on Lynn Road is a large R.A.F. Hospital, established during the Second World War and retained since to serve the permanent air stations at Marham, Feltwell, Lakenheath, Mildenhall, and elsewhere.

There seems to have been some kind of post office in Ely as early as 1672. In this year unsatisfactory working from Cambridge is mentioned,[55] presumably owing to the dangerous state of the entries to the Isle at Stretham Ferry and Aldreth.[56] In 1851 the post office was in the market-place, and money order business was transacted.[57] A telegraph system was introduced in 1869,[58] on which the traffic had increased sufficiently by 1872 to make it necessary to appropriate the postmaster's sitting-room for the instruments.[59] A new post office was fitted up in 1884, after some public complaints about inadequate premises.[60] This post office was in Minster Place,[61] and an adjoining yard and stables were acquired in 1888.[62] These premises were held on lease only, and in 1896, two years before expiry, the existing (freehold) premises in Market Street were purchased. They were first occupied in 1897.[63] Ten years later a telephone service was provided by the post office.[64]

GENERAL HISTORY

THE MIDDLE AGES

There is archaeological evidence of prehistoric settlements in the neighbourhood of Ely,[65] and under Roman occupation an important road[66] passed through the later site of the city, but the history of Ely really begins with the religious settlement, traditionally about a mile from the present city, at a place called Cratendune, which gave its name to the later Cratendon field.[67] This church is said to have been founded by the Saxon King Ethelbert and to have disappeared before the onslaught of Penda, King of Mercia. The second attempt at religious settlement was that of Etheldreda, who eventually decided on a site nearer to the river and began her foundation about the year 673.[68] Almost from the outset a community of lay folk must gradually have gathered around the rising monastery, sharing its fortunes during the centuries which followed. Destroyed by the Danes in 870, it was refounded a century later and re-endowed by King Edgar. Threatened with Danish invasion again in 1010, Ely apparently held its own and a few years later welcomed King Canute in person. He revisited Ely on various occasions, treating the monks with peculiar favour; a period of marked prosperity followed.[69] The frequent visits of princes, noblemen, and statesmen to their abbot-kinsmen publicised the strategic and other advantages which Ely offered and helped to popularize the shrine of St. Etheldreda, thereby encouraging the growth of the city itself.

The part which Ely played as the last stronghold against the Norman invaders is discussed elsewhere, as is its important role during the tempestuous era of baronial and monarchical struggles in the 12th and 13th centuries.[70]

Whilst the geographical situation of Ely made it from the beginning a place of refuge, it was accessible enough to those who could command the approaches. Hence the care of the chief causeways was always of primary concern to the monastery—especially the causeway leading from the islet of Stuntney to the east of the main island on which the city stood, and the two causeways of Aldreth and Earith on the west.[71]

It would seem that the great political disturbances, of which Ely was repeatedly the vortex, were essentially baronial activities of outsiders and not initiated by the allegedly fierce fenmen themselves.[72] But prolonged experience of war and devastation must have toughened the fibre of Ely townsmen, though it also taught them to find protection beneath the shade of the monastery walls or behind the men at arms of warlike bishops. Certainly there was much to discourage initiative: the citizens were dependent on an abnormally powerful lordship, political as well as manorial, until well into the 19th century. Within the city and its environs organization was rigid compared with the northern part of the Isle,[73] though there were certain advantages enjoyed by the home manors: they suffered less continuously from absentee lordship than did those at a distance from the centre. Moreover, when marked distress existed at Ely it was within daily reach of the Almoner's aid—not slight aid in this monastery's heyday.[74] Extensive building activities, especially during the second quarter of the 14th century, spelt full, if arduous, employment for citizens, and more than one

[54] Kelly, *Dir. Cambs.* (1937).

[55] P.O. H.Q. Records. Peover Papers.

[56] See under Stretham and Haddenham in this vol. [57] Gardner, *Dir. Cambs.* 486.

[58] P.O. H.Q. file E 50861/869.

[59] Ibid. E 4248/1872.

[60] Ibid. E 7443/1884.

[61] Kelly, *Dir. Cambs.* (1888).

[62] P.O. H.Q. file E 1452/1888.

[63] Ibid. E 1748/1896.

[64] *P.O. Telephone Dir.*, 1907.

[65] *V.C.H. Cambs.* i, 265–7, 302–3.

[66] Ibid. ii, 2, map 1.

[67] *Anglia Sacra*, i, 41–42; Wm. Camden, *Britannia* (ed. 1806), 232; Bentham, *Ely*, 54 n.

[68] See *V.C.H. Cambs.* ii, 199 et seq. for history of abbey and cathedral priory; ibid. 377 et seq. for polit. hist. See also J. Bentham, *Hist. of Ch. of Ely* (1791), 45–59.

[69] *V.C.H. Cambs.* ii, 200–1.

[70] In 11th century as Hereward's centre; in 1139 and 1142; in 1216 and 1265. See *V.C.H. Cambs.* ii, 381–97; also H. C. Darby, *Medieval Fenland*, 143–6.

[71] The fen narrowed at the eastern and western ends of the main isle (Darby, op. cit., map 107). Tracks and fords possibly crossed here early. Stuntney causeway was constructed in the time of Bishop Hervey, 1109–31 (*Anglia Sacra*, i. 618). This passage played a vital part in the successive struggles centring on Ely. In 13th century part of the rectory of Wentworth was allocated to the Sacrist for the upkeep of Stuntney causeway and bridges (*Sacrist. R.* ed. F. R. Chapman, i. 61, 112; ii, 27, 82, 91; D. J. Stewart, *Archit. Hist. of Ely Cath.*, 155, 180–5). Aldreth causeway was traditionally built by Wm. I or by Bishop Hervey (Darby, op. cit. 110 & nn.). There was much activity about the fort of 'Alrehede' in Stephen's reign (*Anglia Sacra*, i, 629 et seq.). For the upkeep of this road Ramsey Abbey and various places in the Isle owed services, commuted by 12th century. On the home manor of the Bishop such services were still rendered in 1251 (see below, p. 37). The Earith road also suffered neglect in 13th and 14th centuries, despite the grant of manors to the Bishop for upkeep of both Aldreth and Earith causeways. These manors later passed to the Crown, whose grantees in 17th century proved neglectful of the roads (Darby, op. cit. 111–12). Episcopal initiative respecting drains, waterways, and roads was, however, often marked. Bishop Hervey (12th century) made a road from Ely to Exning (Camden, *Britannia*, ed. 1607, 216); Bishop Morton (15th century) gave his name to the famous 'leam'; Bishop Mawson (18th century) vigorously supported drainage schemes and the new turnpike road to Ely. (*Beauties of Eng. and Wales*, i and ii, ed. W. Brayley and J. Britton, 1801, 166.)

[72] *V.C.H. Cambs.* ii, 381–95. Pat. and Assize R. do not suggest lack of respect for law and order (Darby, *Medieval Fenland*, 143–6). [73] Infm. Mr. E. Miller.

[74] Almoners' R. of 1344–5, 1374–5, 1375–6, 1377–8, 1384–5, 1449–50, and 1473–4 record large distributions in kind (cloth, &c.). Accts. of Serjeant of Almonry Grange (1327–31) record very large quantities of mixed corn granted to the poor (*V.C.H. Cambs.* ii, 206 & n. 3).

wealthy bishop went down to posterity as a munificent donor to the poor.[75] Moreover, the issue of indulgences to those who contributed aid at times of disaster did afford very real succour to the town.[76] It was the ecclesiastical authorities of Ely who organized roads, bridges, and primitive drainage,[77] and who provided an educational ladder[78] for the gifted few: in later ages such activities were even more marked. Feudalism so paternal could neither easily be outgrown nor overthrown. Citizens, moreover, were early accustomed to the outward and visible signs which gradually inspired respect for law and order—though they might later inspire resentment: at Ely were held manorial courts for the township itself, gatherings of officials from the distant manors, courts for the hundred and courts of justice for the whole Isle. The courts, indeed, carried on activities elsewhere the function of borough courts.[79]

Nevertheless Ely was not isolated from the normal disruptive influences of a growing money economy: these influences appeared fairly early. There were markedly prosperous citizens, frequent buying and selling of old property and of new assarts from the waste, and beginnings of commutation of payments in kind by the 13th century.[80] The Black Death hit the monastery with severity:[81] it would have been miraculous had the citizens escaped. An official note of 1350 records the many uncultivated lands of the city—*propter ingentem mortalitatem hominum*.[82] It may well be that in the years which followed, bishop and prior had to exert more than a little pressure to secure a continuance of the heavy service of earlier times.[83]

The Revolt of 1381[84] saw Ely the centre of a far more violent uprising than in the rest of the shire. All the major motives existed here in concentrated form: the political power of great landowners, the oppressiveness of landlords, the growing economic control of administrators were personified in bishop and prior and their subordinates. The burden of the new Poll Tax set the tinder smouldering; the spark from risings elsewhere lit the flame. Robert Tavel of Lavenham provided a link with the wider movement, as did John Michel, an Ely chaplain who had been with Wrawe in Suffolk. The main leaders at Ely were local men—Richard de Leycester of 'Bocherisrowe', Robert Buk, a fishmonger, and Adam Clymme.[85] Clymme called upon the peasantry to refuse customary services and to behead lawyers, and made mysterious reference to the potency of the 'Great Society'. Leycester demanded the abolition of traitors to king and common folk—perchance here with some suggestion that episcopal courts usurped the powers of the Crown. On Saturday, 15 June, the revolt began: on Sunday Leycester, defying ecclesiastical authority, mounted the cathedral pulpit 'on behalf of the King'; on Monday the bishop's prison was the object of attack; the same day Leycester and Buk seized and executed Sir Edmund Walsyngham, a justice, placed his head over the town pillory and destroyed sundry rolls and documents.[86] From Ely bands of rebels moved on to other parts of the Isle. William Combe was appointed to hold the famous Stuntney Bridge, welcoming there Robert Tavel from Suffolk. On Tuesday the rebels marched to the abbey of Ramsey, there to meet their doom. Tavel was beheaded: the end of the rising was in sight. He was followed to scaffold or gallows by the principal leaders, but towards the rest a conciliatory policy prevailed.

There were sharp epidemics in the 15th century[87] and there were frequent floods, but no further serious unrest showed itself in medieval times. Ely citizens were aware of the prevailing crimes of lollardry and magic, but were duly warned: the victims of the law were outsiders. Their forced parade around the market-place, bareheaded and barefooted, carrying the mystic plates and books and wand,[88] or bearing faggots and candles,[89] lent colour to the drabness of native virtue: Ely was not too deeply concerned.

Ely is fortunate in the preservation of medieval surveys. Those of 1086, 1251, and 1416 depict development at three significant stages. In 1087 the settlement is purely rural; in 1251 it is still largely rural, though with marked urban beginnings; in 1416 the city is laid out much as in modern times, yet with the early possibilities of normal municipal development unfulfilled.

In 1086 Domesday Book[90] recorded Ely as assessed at 10 hides, with land for 20 plough-teams: 5 hides were in the demesne, with 5 plough-teams and capacity for a sixth. There were 40 villeins, each holding 15 acres; they shared the 14 plough-teams. There were 28 cottars and 20 serfs. The fisheries rendered 3,750 eels,

[75] Bishops Ridel (1174–89), de Fontibus (1220–5), Northwold (1229–54), Hotham (1316–36), and Arundel (1376–88); in later times Bishops West (1515–33) and Laney (1667–75); Bishop Andrewes (1609–19) was also revered for his general kindliness (Bentham, *Ely*, 145 et seq.).

[76] e.g. in 1394 on behalf of a blind man (*Ely Dioc. Remembrancer*, 1900, 54); in 1490 for a poor blind man (ibid., 1908, 165).

[77] Above, p. 33, n. 71.

[78] *V.C.H. Cambs.* ii, 201, 206–7. Among the free tenants in 1251 was a chaplain, the son of a local smith.

[79] Infm. Mr. E. Miller.

[80] *V.C.H. Cambs.* ii, 206; 321–3.

[81] There were 53 monks just before the plague: 28 just after (Stewart, *Archit. Hist. of Ely Cath.*, 207).

[82] A casual mem. (p. 89 of MS. 'Leases and Pattents' of Deanery of Ely in 1649), apparently copied from some then existing record of 1350 (Ely Cath. Libr.). 'The Manor Court Rolls of 23 Ed. III contain many incidental allusions to the ravages of the Black Death, as do the Rolls of the obedientiaries of the Priory; in one of these it is explained that no rent had come from a street in Ely, by reason of the houses being tenantless *propter pestilentiam* (Hist. MSS. Comm. *12th Rep.*, *App.*, viii and ix, 391). There was great mortality, if unequally distributed, in the diocese generally, necessitating the grant of papal indulgences (Bp.'s Mun., Reg. Lisle, f. 19*b*, 21). There were no fewer than 90 institutions in the diocese in 1349. At least 4 of the Chaplains on the Green died (*V.C.H. Cambs.* ii, 158–9). Bentham says 92 institutions (*Ely*, 161, n. 1).

[83] The services in 1251 were certainly heavy.

[84] *V.C.H. Cambs.* ii, 399–402.

[85] Leycester had a tenement with dovecot and two shops in 'Bocherisrowe', Ely, and goods valued at 40 marks. Buk had a messuage at Castlepath, four shops and other property in 'Walpolelane', worth £17 11*s.* 6*d.*, and £4 in money. Fish were included among his goods (E 357/8, m. 26*d.*). The name Buk appears in 14th century Subs. R.; a Walter Buck led the Brabanters besieging Ely in 1216 (*V.C.H. Cambs.* ii, 390). Adam Clymme was possessed of property worth £10 19*s.* 5*d.* (E 357/8 m. 26*d.*). Other Ely rebels mentioned were John Shethe, a glover, Thos. Litster, Robt. Plumer, and John, son of Nich. Gunmeld. Edmund Galon, an Ely lawyer, was killed; John Fedder and John Whyte of Ely apparently escaped. Walsyngham had a residence at Eversley, Cambs. (E. Powell, *Rising in East Anglia*, 1896, 46–48).

[86] Not all pre-1381 Ely rolls were destroyed; some were evidently in existence in the 17th century (Bishop Wren's Note Book, f. 318, Bp.'s Mun., G 2).

[87] e.g. in 1458–9 plague was widespread, affecting the monastery seriously, as shown by Infirmarer's R. (T. D. Atkinson, *Archit. Hist. Ely*, 106).

[88] e.g. in 1457 three men from Cambridge, Chesterton, and Reach, were condemned for Lollardry and penalized thus (*Ely Dioc. Remembrancer*, 1907, 43).

[89] In 1466 a Babraham man was punished at Ely for practising magic (ibid. 95).

[90] The whole of the manor was in demesne. According to *Inq. Elien.* 119, the number of cottars was 18, and there were 75 non-ploughing cattle, 268 sheep, 38 pigs, and 4 horses (*V.C.H. Cambs.* i, 369 and n. 2).

the tribute of fish amounting to 2*s.* 3*d.* There were adequate meadows and pasture for plough-teams and cattle. There were also 3 arpents of vineyard.[91] In all, the manor was worth £30; when received it had been worth only £20, though in the time of King Edward its value was £33.

Stuntney[92] appears in Domesday Book as a berewick of Ely, assessed at 1½ hide, with land for 3 plough-teams and the necessary meadow and pasture. It maintained 6 villeins, 5 cottars, and 3 serfs. Its great value to the abbot was its render of 24,000 eels and 18*s.* tribute in fish annually. Altogether the berewick was worth £10 14*s.*—over a third the value of Ely itself. At this date Little Thetford (Liteltedford)[93] was also a berewick of Ely, with 1 villein and 4 cottars there. Chettisham, the hamlet which, together with Stuntney, for long was and still is appendant to Ely, had not yet been assarted from the waste.

Including the abbot's servant, holding the little islet of Haneia,[94] and excluding the monastic inmates, there were 108 working members, or perhaps householders, recorded in the vill of Ely and its dependent settlements. There may have been unrecorded fishermen, but, on the analogy of the later medieval surveys of the city, other unspecified members of the community, such as swineherds and smiths, were almost certainly included in the cottar class.

The Survey of 1251[95] begins with the manorial demesne land: much arable land and pasture had been won from the waste since 1086. There were 240 acres 'in the field called Gruntifen', and 60 acres in Cratendon. Eleven other fen areas are enumerated, concluding with 35½ acres in 'the new assart of Chettisham'.[96] The total area of plough-land is stated to be 1,524¼ acres.[97] Here 10 plough-teams were occupied.[98] Of meadowland 11 blocks are specified, providing a total of 260 acres, of which 33¼ were newly assarted; 6 acres were covered with thorn-bush and 9 were enclosed for vini-culture.[99]

Wide areas of marshland stretched beyond and between arable land and meadow. The bishop's wastes of Ely merged into those of other townships, much intercommoning being practised. Cawdle Fen and 'Cloggesmere', 'Cowfen', Grunty Fen, 'Blythinghalefen', Middle Fen, Padnal Fen, and 'Northfen' are enumerated.[1] Over specified fens the bishop,[2] and occasionally the prior, enjoyed exclusive privileges; the rest were commonable. Interesting features are the ditch-inclosed holding of 'Brame' (Braham Farm),[3] in the Cawdle Fen area; the early settlement on the far side of the Ouse, between 'Cloggesmere' and the Great Bridge; and a part of 'Cowfen' held in severalty by the prior.

The seignorial stock at this date consisted of 500 sheep, 20 cows and 2 bulls, 100 pigs and 2 boars. The lord enjoyed fold of sheep of villein and even cottar tenants. There were 2 windmills, let at farm:[4] all rent-paying, customary tenants and cottars owed suit of mill. For fishing rights in 4 of the extensive stretches of mere and weir eel-renders had been commuted,[5] and partially so in the fifth. The total fishing rental amounted to 56*s.* 4*d.* in money and 14,500 eels per annum.

The tenants were classified as knights, free tenants, three grades of servile holders, and cottars.

There were 5 messuages occupied as knights' fees,[6] one of which, by gift of the bishop, was now held by the monks and was situated beyond their vineyard. These holders owed ward and suit of court and were required to notify the other knights within the bishop's liberty concerning the place and time where wardship was due. About 200 holdings were occupied by freemen or their sub-tenants: only 16 freemen held any appreciable amount of land.[7] Of the free holdings some 20 were mere stalls or booths, the beginnings of a shopping quarter,[8] mainly occupied by butchers; 12 other messuages and plots were held at the low rental of 4*d.* per annum; a further 13 very small holdings, rented at

[91] Approximately 3 acres; the French arpent, like the English acre, varied according to district.

[92] *V.C.H. Cambs.* i, 368. *Inq. Elien.*, 117, records 6 villeins, each having 10 acres, 9 non-ploughing cattle, 10 sheep and 1 horse (ibid., n. 3). Traces occur of occupation at Stuntney in the later Bronze Age (*V.C.H. Cambs.* i, 279, n. 1).

[93] It was assessed at 1 hide and had land for 1 plough-team on the demesne. The villein's holding was but of 6 acres. Meadow and pasture were adequate. 500 eels and 4½*d.* tribute of fish were rendered. When received the berewick was worth 20*s.*, T.R.E. 30*s.* (*V.C.H. Cambs.* i, 368). It was a hamlet of Stretham (q.v.) in 1327 (*P.N. Cambs.* (E.P.N.S.), 238).

[94] Consisting of ½ hide. It was part of Soham in 13th century (*V.C.H. Cambs.* i, 369 and n. 3).

[95] There are three copies of this survey, B.M. Cott. MS. Claud. C. XI, ff. 24–33, Caius Coll. Camb. MS. 485/489, ff. 19–29, and Bp.'s Mun. Ely, G 3 'Old Coucher'. The last two contain later insertions. The survey should be compared with that of 1222 (B.M. Cott. MS., Cott. Tib. B. II). Light is thrown on the subject by B. Dodwell, 'The Free Tenantry of the Hundred Rolls' (*Econ. Hist. Rev.*, 1944, 163–71), and by D. C. Douglas, *Medieval East Anglia*. The figures recorded for Ely, in the extent of 1251, are simpler than for some manors, but a number of free tenants have more than one holding or messuage, in addition perhaps to garden plots, shops, or tofts, some of which are commonly sublet, possibly to landless 'anilepimen', who escape enumeration. In other cases tenements and messuages are shared, though for Ely the joint-holders are usually named. Occasionally women are recorded as co-tenants with husbands, sons, or sisters. A few servile tenants appear in more than one servile group, but these are easily traceable; moreover, free tenants, even when avowedly also holding former customary land, are here unequivocally classified as free.

[96] Only part of this area was newly assarted (cf. below—Manors).

[97] Two-thirds of the whole arable lands were in demesne. Division into two or three arable open fields at this date is not clear.

[98] Six were composed of 6 steers, 4 of 6 oxen and 2 steers.

[99] Eight of these acres were said to be used as pasture.

[1] The spelling of the fen names is not uniform in the survey.

[2] In 'Northfen' and Padnal Fen the bishop had inclosures for peat-digging. In 'Blythinghalefen' and Middle Fen the bishop had sole right of mowing, 'not even shared with the Lord Prior of Ely', but after the hay was harvested the pasture was thrown open.

[3] It appears as 'Bramewere' in 1086; as 'Brame' about 1200 (cf. below—Manors). Prior Salomon (1169–73), probably a relative, granted a rent-charge of 5 marks on the estate to Salomon the goldsmith, his heirs and successors, confirmed by Hen. II (*Sacrist R.*, ed. Chapman, i, 151–66, App. C). The estate was one of those allocated to the sacrist (Bentham, *Ely*, 127; *V.C.H. Cambs.* ii, 204 & n. 2). In 1251 Salomon was said to hold the marshland of Brame in several, surrounded by ditches (B.M. Cott. MS. Claud. C. XI, f. 24*b*).

[4] One was let for 5 marks, the other for 6. In 1229 there was also a windmill in the priory grounds (Bentham, *Ely*, ii, 53).

[5] A century earlier, in the time of Bishop Ridel (1174–89), payment was still made in eels. Similar food-rents had been commuted before 1130 at Wells, on the Cambs.-Norf. border (H.C. Darby, *Medieval Fenland*, 31).

[6] Knights' fees had been created in place of the demesne knights quartered in the abbey buildings (*V.C.H. Cambs.* ii, 386, n. 87, quoting *Lib. Elien.* 274–5). The four knights, in 1251, were Philip de L'Isle, Ralph of Soham, Henry Pelerin, and Stephen of the Marsh.

[7] Most of these holdings with land had special services attached. One tenant held 36 'ware' acres; 6 held 18 such acres each (the Hosp. of St. Mary Magd. held this amount); 2 held 18 acres jointly; 1 held 12 acres; 6 held from 5 to 8 acres each. The 12-acre 'ware' unit is prominent. Half an acre of 'new feoffment', held with other property, exemplifies the assarting which provided for the growing town.

[8] There were four shops even a century earlier. J. I. Ladds, 'The Mon. of Ely' (*Cambs. and Hunts. Arch. Soc. Trans.* v, 75).

2*d.* or 4*d.*, lay 'across the water',[9] 30 town holdings were rented at 6*d.*, 21 at 8*d.*, 71 at 1*s.* per annum, and 19 at more irregular sums between 1*s.* and 2*s.* About a dozen messuages were shared by 2 householders; several were shared by more.[10] Some free tenants had, by 1251, built up considerable town property, composed of numerous messuages and plots. Agnes Fitzpayne and her 2 sisters thus held 10 messuages in different parts of the town, in addition to their own dwelling; they also held 18 'ware' acres of arable land, and portions of river meadow and newly assarted pasture.[11] Salomon, the famous goldsmith,[12] had 5 holdings in different parts of the town. The almoner, sacrist, and pittancer held between them 17 town tenements.[13] The educational ladder gave access to free tenure for Hugh the Chaplain, son of a city smith. Many of the lesser monastic and episcopal servants were city residents: the porter, gate-keeper, groom, the blood-letter, cook, and John of the refectory, also the bishop's baker and his surveyor of sedge-duties thus occupied freehold premises appurtenant to their respective offices.[14] Four freemen held messuages for service as coroners;[15] one for the hereditary duty *navigandi episcopum*. A larger holding was attached to the office of carrier of the bishop's equipage from Soham to Ely.[16] Other monastic employees, of higher status, had acquired independent property: Salomon the goldsmith was the most noteworthy. Both payment and receipt of money, for one purpose or another, were common among all tenant classes at Ely, as recorded in 1251. Strong seignorial pressure—easily increasable in periods of anarchy—could alone have obstructed widespread commutation here. Though all tenants paying money-rents were classified as freeholders in 1251,[17] the survival of very considerable seignorial claims upon them does seem to indicate an earlier status of villeinage.

By this date free holdings varied considerably not only in size, but in the liabilities attaching to the several small parcels of which some were composed. In certain instances it was categorically stated that a free tenant had acquired sundry formerly servile holdings which, though now rent-paying, still owed dues and services not universally demanded here. The prior held two messuages 'free from the service which had previously been owed'. Moreover, the considerable proportion of craftsmen who held at stereotyped rents of 6*d.*, 8*d.*, or 12*d.* per annum is perhaps indicative of earlier commutation.[18] About a quarter of the free holdings were liable for one day's digging in the bishop's vineyard;[19] a few gave 3 days' arable service annually and 3 days' carting; in one instance piece-work was required—tilling 3 acres of demesne annually; some free tenants paid marriage fines and heriot. *Omnes censuarii et consuetudinarii* owed multure; all free tenants owed suit of court.[20]

The diminutive size of many of the single holdings does not imply an impossible subsistence level. Apart from the fact that some tenants had more than one holding, most small freemen had a town messuage and plied a craft; others, as already stated, held some monastic or episcopal office.

Ely was better located for commerce than for industry but there were conspicuous rudimentary developments which might have led to the normal growth of guild organizations. In 1251 fishing and water-carriage employed many townsmen: the monastery was now purchasing a large part of its eel-supply. Moses, one of the hereditary fishermen, paid 3*s.* a year for his boat on the mere near Prickwillow and shared a tenement in the town; Henry, another fisherman, had a tenement across the river and a booth against the vineyard wall. William Mackerell, John the pilot, and John the steersman bore suggestive names. Still more interesting are glimpses of that suspect figure of medieval days, the 'mere merchant': *Symon mercator et participes sui* shared a messuage. Two other merchants had important tenements, whilst Reynold 'le seler' and William 'le achatur' were obviously traders. Other names show the variety of crafts already plied locally; in some cases there were several representatives of a craft—e.g. in the building, tanning, baking, and upholstery trades. The skilled cordwainer was even distinct from the humble cobbler. A whole row of 16 stalls appears under the heading 'the butchers' stalls'.[21] There was also a spacious market-place. Most of the glass purchased by obedientiaries still came from abroad even a century later, but there

[9] This area was known as Babylon at least from the 17th century onwards: it is so named in many of the documents relating to drainage schemes of that period.

[10] One was Bishop Northwold's chantry, occupied by four chaplains (see below, p. 82).

[11] From two of the properties a day's digging in the vineyard was required, 'from daybreak until nones', with no food supplied: it was not regarded as boon work.

[12] The names of Salomon and Alan were perpetuated down the centuries. The family had many wealthy branches in the city during the middle ages. One Salomon was prior in the 12th, another in the 13th century (*V.C.H. Cambs.* ii, 209). Prior Alan de Walsingham seems to have been the family's most distinguished member. (For an account of these hereditary goldsmiths, see *Sacrist R.*, ed. Chapman, App. C.) There were other wealthy families in the city in the 13th and 14th centuries. In 1313 Joan Bray lent money to the monastery (ibid.).

[13] Nine houses were stated to be held by the prior himself, but his holdings were only mentioned when vacant.

[14] The services were in the nature of petty serjeanty. The sense of dependence upon the lord must have been increased by the presence of such citizens.

[15] Combined with attendance upon the bishop's steward, when he travelled to Cambridge for the ceremonial granting of the bishop's jurisdictional liberties (see *V.C.H. Cambs.* ii, 64).

[16] A city house was leased to a surgeon appointed by the monastery in 1272 (Hist. MSS. Comm., *6th Rep.* (Leconfield MSS.), 290). In 1326 the bishop granted to his barber, Warin de Alnewyk, a place at the Steple (ibid., 295, quoted by Atkinson, *Archit. Hist. Ely*, 104 n.).

[17] There is no heading 'De Consuetudinibus Censuariorum' such as is common in the extent of 1222 (Douglas, *Medieval E. Anglia*, 73).

[18] Rather over half of the free holdings paid in addition to rent, 1*d.* per annum; a good many paid 2*d.*; some paid larger graded sums.

[19] Not commuted here, as was the case on the bishop's distant manors.

[20] It is possible, on the other hand, to argue that a degrading of former freemen or socmen had occurred at some period. This may perhaps be implied in the large proportion of acreages recorded as 'de ware' (cf. Douglas, *Medieval E. Anglia*, 96–116, 129). All the servile tenants of 1251, other than the cottar class, held standardized numbers of acres—18, 12, or 6—described as 'de ware', so did some of the free tenants. The graded sums paid annually, in addition to rent or services, would seem to represent commuted geld payments to the lord. It seems more likely, however, in the case of villein tenants for whose geld payments a lord was traditionally responsible, that, at Ely, the lord recouped himself by levying upon these tenants additional services or payments, originally assessed, as in the case of the free tenants, on an acreage basis. In such circumstances no degradation of status would be implicit in the term 'de ware'. Scandinavian influence is indicated here, as in East Anglia, in the 12-acre basis of the holdings, or some fraction or multiple of 12 (cf. Douglas, op. cit., 32–50). The joint geld-responsibility of the several villein tenants of such units appears clearly with respect to the 12-acre holders: whereas the 18-acre tenants pay 7*d.* per annum, in addition to giving service, all the 12-acre tenants pay two-thirds of this sum—4½*d.* and a third of 1*d.* 'in such manner that 3 such peasants give 14*d.*' (B.M. Cott. MS. Claud. C. XI, f. 31*b.*) The 6-acre holders pay 2¼*d.*

[21] The 'bocherie' and Butchers' Row of later days.

was a local glazier[22] in 1251. There was also a dealer in such foreign products as spices. Other trades mentioned are those of plumber, carpenter, quilter, tailor, sauce-maker, dyer, and webster. 'Master Alan of Swaffham' and 'Master Roger', two of the three masons, probably represent the highly skilled professionals brought in from a distance, but they were settled residents at this date. Nicholas and Everard Palmer recall the magnet which Ely proved to the pilgrim. Other townsmen bore place-names, indicative of the part played by immigration in the growth of this City of Refuge. It was ingress in the main from neighbouring regions.

The city survey mentions the *fratres hospitalis Magdalene*[23] and the Chantry-on-the-Green:[24] it also incidentally witnesses to the disappearance of the castle.[25]

As the record passes from free to villein and cottar tenantry, the atmosphere changes from mainly urban to rural. The two latter classes together formed less than half the working population of Ely.[26] There were 50 villeins, of whom 33 are graded as full holders of 18 acres; 13 as half-holders of 12 acres; 4 as holders of 6 acres each. There were also 95 tenants of cotlands, normally of 1 acre.[27] In this last group appear the plough-reeve, the two smiths, a swineherd, a webster, and that interesting figure the appleward—testifying to the early significance of fruit-growing here.[28] From the full holders 2 or 3 days' regular week-work was required, according to the season, but additional services, apart from boon-work, were numerous. There was much work in cutting, stacking, and carrying of sedge, digging and clearing of specified lengths of drain,[29] hurdle-making, sheep-washing and shearing, cutting, carrying, and malting of barley, and transporting by river and road.[30] Many of these labours formed no substitute for regular week-work.[31] When duties were nominally fixed by the day, the equivalent in piece-work was often specified. Ale and food—one loaf and two herrings per man per day—were usually supplied only for boon-work. Fines and fees, both in money and kind, were exacted: tallage, leyrwite, merchet, heriot, and commuted geld. Suit of mill was enforced and the sale of male breeding-stock forbidden. Three hens at Christmas and 30 eggs at Easter were demanded from each full tenant. The common use of money[32] is noteworthy: 6 days' harrowing per team was required, but this was wage-paid service—12*d.* or one sheep, and 8*d.* or 2 cheeses per team. Moreover, all villein tenants were liable for the monetary levy of 3*d.* per annum towards plough-repairs. Half-holders were subject to a similar body of regulations, save that their services and dues were proportional to their holdings. One-acre cottars gave only 1 day's week-work, but even they rendered additional services and payments in money and kind, less in amount, but largely of the same nature as those of villein tenants;[33] they were, however, excused carrying service, but provided a substantial part of the labour required in the lord's vineyard. The whole township of Ely, 'whether doing service to the Prior or the Lord Bishop', was required to 'make and maintain two furlongs of the causeway of Alderheye'.[34]

The labour regulations suggest meticulous organization, adapted to fenland circumstances and to the heavy demands of a community of consumers and a powerful lord. The services were not light and the courts were active to enforce them. Even on the very doorstep of this stronghold of the church no feast-days were to be observed as holidays, save one day at Christmas, and then only if it fell upon a normal working day. This picture of feudal exaction, however, has another side. To all tenants, in time of sickness, unpenalized leave of absence was granted, up to one month if before August, or fifteen days if subsequently.[35] Widows, moreover, were encouraged to retain their husbands' holdings and were exempted from payment of heriot for 30 days. Bishop Hugh of Northwold was not above turning the episcopal screw, demanding that the whole villeinage of Ely should ditch and fence his park at Downham, without abatement of normal duties. But Eleans had become familiarized with the organs of law and order to some purpose: 'the jurors say that they have never done this, nor are they lawfully bound so to do'.[36] Even the bishop's court had two facets.

Approximate population statistics derived from the survey, omitting residents in the claustral precincts, show some 345 householders or tenement-holders, living in Ely in 1251—a threefold increase since Domesday.

In the interim between this survey and that of 1416 the Poll Tax returns of 1377 perhaps afford the most satisfactory demographical source of information for Ely. Some light, however, on the city's development half a century earlier is thrown by the returns for the Lay Subsidy of 1327,[37] recording the name of each tax-payer and the sum paid by him. Only 96 people in the city, together with 14 in Stuntney, paid this tax,[38] at a date when the number of tenement holders must

22 'Le verer'.

23 They held one of the stalls in Butchers' Row and also 18 acres of land.

24 See below, p. 82.

25 'Toftum . . . quod iacet juxta placeam cast' (Extent 1251, f. 25*b*); cf. above—Castle.

26 145 out of about 345.

27 One 12-acre tenant also held an additional ½-acre, for which he paid 1*s.* yearly rent: he was not, however, classed as a free tenant. One cottar held 2 cotlands. Though the cotlands were very small, much casual agrarian, pastoral, and urban work, which received recompense, was obviously available.

28 Grounds near the abbey were laid out as gardens and orchards under the first abbot 970–81 (*Liber Elien.*, 168–9). The vineyard figured in Domesday Book: there was still nominally a vineyard in 1618 (D. and C. Leases, Ely Cath. Lib.), though decayed (Camden, *Britannia* (ed. 1607), 216). The built-up site of the bishop's vineyard is still known as the Vineyards.

29 One day's work on new drains, to be cut 5 ft. wide and 5 ft. deep; one day's work clearing old drains.

30 Especially by river to and from Lynn. Ale was an important Ely export to Lynn and Boston in 1319.

31 Though certain services—e.g. specified amounts of sedge-cutting and carrying —did excuse one day's week-work.

32 Or of estimation in terms of money: commutation here could obviously have occurred early but for the dominating lordship.

33 Cottars shared the normal sick-leave arrangements; they paid 1*d.* yearly; they were liable for leyrwite, merchet, and tallage; they owed suit of mill and fold of sheep, when they had sheep.

34 Aldreth. Works on the causeway due from Ramsey Abbey were already commuted in 12th century (*Cartul. Mon. Rames.* (Rolls Ser.) i, 437; ii, 187–9).

35 A less generous arrangement than, e.g., on the lay manor of Eversden, Cambs., where if a tenant were ill for a year and a day, all works were excused (*V.C.H. Cambs.* ii, 65).

36 B.M. Cott. MS. Claud. C. XI, f. 31*a*.

37 E 179/81/6 (reprinted from *E. Anglian*, vols. x, xi, xii, 1903–4. Transcript by J. J. Muskett, preface by C. H. Evelyn-White). A list of Subsidy R. relating to Cambs., compiled by Dr. W. M. Palmer, was published in *E. Anglian*, vii and viii. See also W. M. Palmer and H. W. Saunders, *Docs. relating to Cambs. Villages*, No. 6 (1926). The 1327 tax was 1/20th of personal income, imposed on movables (cattle, crops, stock-in-trade, &c.). Goods, other than farming stock, &c., must have been scanty at this date.

38 In Wisbech there were 126 taxpayers in the town and 66 in the hamlets.

have been more than three times this figure. Few cottars or humble craftsmen would have possessed the movable goods which even 7*d.*, the lowest sum levied, implies. Indeed, remarkably few of the taxpayers bore occupational names,[39] compared with the citizens enumerated only 50 years earlier. Five people paid over 5*s.*: Agnes Springenhait paid 12*s.* 8*d.*—the highest sum of all; the master of St. John's Hospital paid 10*s.* 1¼*d.*; the two holders of demesne farms, Simon de Keten and John de Brame, paid 10*s.* 11½*d.* and 10*s.* 8*d.* respectively; Roger Mariner, also obviously prosperous, paid 10*s.* 9*d.* Of the rest, 11 paid between 5*s.* and 10*s.*, 31 between 2*s.* 6*d.* and 5*s.*, 9 between 2*s.* and 2*s.* 6*d.*, 17 between 1*s.* 6*d.* and 2*s.*, 23 between 1*s.* and 1*s.* 6*d.*, and 15 paid under 1*s.* The total sum raised was £16 9*s.* 7½*d.* If these figures be compared with the corresponding groups at Wisbech, the higher percentage of Ely taxpayers in the wealthy and very prosperous groups and the much lower percentage in the very lowest group are noteworthy.[40] A comparison with a purely rural area—the hundred of Staploe[41]—reveals these distinctive features even more markedly. The total sum contributed by Ely amounted to over four-fifths of that paid by Cambridge and bore an even higher ratio to the contribution of urban Wisbech.[42]

The Poll Tax returns of 1377[43] extant for Ely show only the total number of taxpayers, 1,722, and the total sum collected, £28 14*s.*[44]

The Survey of 1416[45] depicts the medieval city, street by street, tenement by tenement, duly ascribing ownership to bishop or prior.[46] Upon a plan constructed on this data[47] could be superimposed one of the twentieth century with remarkable ease. Many of the medieval names are readily recognizable today.

The major gateways of the conventual enclosure and the walls of its buildings, above which towered the priory church, formed the most striking features of the narrow streets and tenements which huddled beside the precinct walls. Ely Porta, Steeple Gate, Sacrist's Yard Gate, Almonry or Monks' Gate, and a smaller gate in Broad Lane giving entrance to the monks' vineyard,[48] seemed integral parts of the city life. The little booths and tenements of this part of the town had increased in number since 1251, but the chief centre of activity had obviously moved to the wharves and quays and the streets leading thence to the higher ground of the older settlement. The focal point of this older part was St. Mary's church, built between 1198 and 1215.[49] To the north and north-east of the church lay the village green.[50] It was bounded by Highrow Street,[51] by the churchyard and the abbey precincts,[52] and by the western end of the abbey church. Upon this green the main roads converged. On it stood the 'chantry on the green',[53] as in 1251. A more striking landmark lay just to the west of St. Mary's church—the great Sextry Barn or grange,[54] in 1251 a fairly new structure. Here were stored the corn-tithes of the tenantry. A little farther along the road leading to the western outskirts of the town was the Hospital of St. John.[55] The whole conventual precincts formed a large enclosed area—the college and park of post-Reformation days. It was bounded on the east side by Fore Hill, which led up from the river to the market-place and continued thence under the name of Steeple Row[56] or High Street, reaching the village green at Kilby's corner.[57] The north-east boundary of the precincts, formed by Highrow Street, curved round to the Sextry Barn. This stood at the corner of 'Swalugh' Lane, or Walpole Lane,[58] which formed part of the western boundary of the enclosure, emerging into Back Hill. Broad Lane branched off at right angles to Back Hill, thus linking Back Hill and Fore Hill and completing the perimeter of the enclosure.

The newer part of the town consisted of a series of small streets running more or less at right angles to the river, between it and Broad Lane. Castle Hythe,[59] Monks' Hythe, Broad Hythe, and Stock Hythe were the most important of the medieval wharves.[60] Castle Hythe lay at the corner of Back Hill, some little distance from the river bank.[61] A continuation of Back Hill crossed the river by a bridge, becoming the famous causeway leading to the island hamlet of Stuntney and ultimately to Soham.[62]

In Broad Lane was a storehouse belonging to the monastery;[63] a better known storehouse, called 'Seg-

[39] Mariner, Pistor, Ferour, Braciator, Roper, and Hayward. Michel and Climme were taxpayers who bore names prominent in the Rising of 1381.

[40] In Ely 16 per cent. paid over 5*s.*; in Wisbech 10·6 per cent.; in Ely 13·5 per cent. paid under 1*s.*; in Wisbech 36 per cent.

[41] Mr. E. Miller's calculation.

[42] Ely with Stuntney (Chettisham is not recorded) paid £16 9*s.* 7½*d.* (the figure given in Muskett's transcript is slightly incorrect); Wisbech (excluding the numerous scattered hamlets) paid £12 16*s.* 6¾*d.* (£20 2*s.* 11¾*d.*, if hamlets be included); Cambridge paid £22 15*s.* 7¼*d.* A levy made apparently only a few years later (E 179/81/7) enumerates 157 taxpayers in Ely, including Stuntney and Chettisham. A larger percentage of small tradesmen was then assessed. The total sum raised was £33 1*s.* 6*d.*

[43] The enrolment has been published three times: J. Topham, *Archaeologia*, vii, 337 et seq.; C. Oman, *Great Revolt of 1381*, 163–5; R. T. Davies, *Docs. illustrating Hist. of Civilisation in Medieval Engl.*, 274–6.

[44] In Cambridge there were 1,902 taxpayers.

[45] B.M. Harl. MS. 329 (4 Hen. V). A transcript of this exists in D. and C., Mun., Ely: an indexed copy thereof, made by Archdeacon Seiriol Evans, is in his possession. Another version of the survey exists: B.M. Cott. MS. Vesp. A. xix, f. 76 et seq., varying considerably in detail. A somewhat abbreviated translation is published in *Cal. Pat.*, 1416–22, 183–95. The title of this runs: 'Inspeximus and confirmation to John, Bp. of Ely, and Wm. Prior, and the monks of the cathedral church of Ely, of an arbitration of certain disputes between them—by Henry Ware, Keeper of the Privy Seal, Wm. Hankeford, the King's Chief Justice, and Roger Horton, one of the Justices of the King's Bench, on the feast of St Nicholas the Bishop, 1417, confirmed by the said Bishop on the Feast of the Conception of St Mary, and by the said Prior and Convent on the Feast of St Lucy the Virgin.'

[46] The wording of the survey at times makes it obvious that various city properties, not specifically mentioned at the earlier date, were in fact allocated to the prior at the division of 1109.

[47] See plan, p. 29.

[48] See below, p. 81, and Atkinson, *Archit. Hist. Ely*, 16.

[49] See below, p. 83.

[50] The remains still exist.

[51] Now St. Mary's Street.

[52] Now the Bishop's Palace.

[53] See below, p. 82.

[54] See below, p. 81.

[55] See above, p. 30.

[56] Stepil Rowe, so called from the steeple connected with one of the entrance gates of the priory, on south side of the street.

[57] Richard Kilby occupied a tenement at south-west corner of 'Stepil Rowe': the house is now the corner house in High St., opposite the Lamb Hotel (Stewart, *Archit. Hist. Ely Cath.* 201). Other tenement holders whose names were to be associated with the town down the centuries, were Robt. Brame, John Duke, John Mepsale, and Thos. Parsons.

[58] Sometimes so called because of the gateway completed in the time of Prior Walpole (1397–1401), now Silver St.

[59] Mentioned in 1295 (*P.N. Cambs.* (E.P.N.S.), 216).

[60] Farther down the river were Quaney and Turbutsey wharves.

[61] With which it was connected by Auntresdale lane—now Annesdale.

[62] The Gallery, which now runs between the cathedral and the palace, was not mentioned in 1416: it was apparently not part of the original city plan (Atkinson, *Archit. Hist. Ely*, 35). It was known as Paley's Lane or Galely Lane in the early 15th century. It probably took its name from the Galilee, or porch, of the cathedral (*P.N. Cambs.* (E.P.N.S.), 215).

[63] Possibly survives as an existing medieval building.

wyk', gave its name to a street near the river.[64] Another landmark was the Red Cross,[65] opposite Braye's Corner,[66] which formed the angle between Akyrman Street[67] and the Littleport Road.

The monks had retained their old vineyard and gardens within the precincts, but an extensive enclosure on the east of the city, entered from the south-east corner of the market-place, was cultivated as the bishop's vineyard. It was bounded to the north by Brayes' Lane, which led to the suburb of Newnham, in the Prickwillow direction. The bishop's home manor of Barton lay mainly on the western side of the town.

One well, in Broad Lane's End, was recorded in 1416, and there were probably others.[68] There were at this date two mills mentioned as being the property of the bishop. The 'old Mill' dominated the road leading from Newnham towards Turbutsey; the other mill stood between Broad Lane's End and the south gate of the priory.[69] The 'Bocherie' was a little to the north-west of the market-place, on the north side of which was the almoner's granary. In addition to the stalls in the 'Bocherie' and along the wall of the priory vineyard, there were seven shops built against the almonry wall.[70] The market-place was now a prominent feature of town life, under episcopal control.

Particulars respecting the population and its housing, as given in 1416, are interesting but very ambiguous at times. There were apparently about 457 occupied premises in the city, outside the ecclesiastical precincts: 262 belonged to the bishop and 195 to the prior and monastic obedientiaries, of whom the almoner, sacrist, and infirmarian were the most considerable holders. These numbers seem to represent about 520 householders.[71] The size of certain tenements, gardens, and orchards suggests a considerable increase of very prosperous citizens since 1251.[72] On the other hand, overcrowding of the 'cotsetle' (*cotsetella*) class is obvious. Some 33 tenements[73] were shared by two or more families living 'under one roof' (*sub uno tecto*): in 10 of these cases 3 or more householders were so sheltered. A larger share of this poor property belonged to the prior and obedientiaries than to the bishop. There were 18 'little cottages' occupied by single families. The various stalls or booths do not appear to have been residential. There were a dozen or so unoccupied premises.

The nomenclature of the citizens in 1416 does not reveal very striking developments since 1251, though the variety, some 333 different names, is noteworthy. Many persons still bore neighbouring place-names. David Llanlidan and three men named Morgan[74] represent a Welsh element. Some 60 names are occupational, a few of which are purely rural in connotation—Hayward, Herdman, Coherd, Oxherd, Schereman, Shepherd, Thresshere. The name Taillour is borne by 7 tenants, Smythe by 6, Barber 6, Chaloner 6,[75] Wryght 4, Mason 3, Miller 3, and Baker 3. There are 2 representatives of each of a number of other crafts. The fisherman and boatman still figure in various guise. Skilled monastic crafts are suggested by Brevetour and Orfreyser.[76] Glovere and Bladesmythe have now appeared; the Goldsmyth family is still in evidence; there is only one Mercator. Not all who bear a craft-name at this date actually follow that craft: John Plomer, for example, is a baker; but John Cut, the butcher, is surely happily named!

In the Middle Ages there was always much activity, both by road and river, connected with the transport of sedge, corn, eels, and fish from the ecclesiastical manors. Even by the 13th century the two latter commodities were finding their way first to the open market.[77] Fruit-growing had begun early: viniculture may have produced poor quality grapes, but it was sufficiently successful to encourage extension: the 3 arpents of 1087 had grown to 6 monastic acres in 1272.[78] In addition to the bishop's city vineyard, 9 acres of waste had been inclosed for this purpose before 1251.[79] The monastic vineyard produce was not marketed, but certain obedientiary gardens provided surpluses for sale.[80] Town gardens and orchards had increased by the 15th century:[81] Brayes' orchard was by then a landmark. Of other commodities coming to Ely building-materials comprised the largest share. Luxury goods, too, were an early monastic demand and even to the townsman such products were familiar.

There were numerous sources of supply accessible to Ely. The important port of Lynn was but 30 miles by

64 Mentioned in Roscarius's R., 1392–4 (Atkinson, *Archit. Hist. Ely*, 13). Now leasehold property of D. and C. (Stewart, *Archit. Hist. Ely Cath.* 187).

65 Later giving name to Red Cross Street, i.e. Schendeforth Lane in 1416.

66 Braye's orchard was close to the site of the old cross. The property was a sub-manor with its own gateway. The family was wealthy enough to lend money to the monastery in 1313 (*Sacrist R.*, ed. Chapman, i, 155). Ketenes Place, on south side of Walpole Lane, was also a sizeable estate (see below—Manors).

67 Now Egremont St. Not the Roman Akeman St. (according to C. C. Babington, Ancient Cambs.' (*Cambs. Antiq. Soc.*, Octavo publ. iii, 16), though the Roman road did pass through Ely and Littleport, on the way from Cambridge to Brancaster (Norf.). Potter's Lane, running at right angles to Back Hill, opposite Broad Lane, contained a tenement, in 1416, formerly occupied by a potter.

68 The sites of several within the 'precincts' are known through ministers' accts. (Bentham, *Ely*, ii, 53). Defoe referred to the numerous wells early in the 18th century (below, p. 44.) The underlying boulder clay held the water, which came to the surface along the edges of the higher ground on which the city was built.

69 There was also a horse-mill, mentioned in this survey, between the south gate of the priory and Broad Lane, held by Edmund Cotermonger; there was probably still a windmill on the old castle mound, as there had been in 1229 (*P.N. Cambs.* (E.P.N.S.), 216), and was in 17th century (see above, p. 30).

70 Atkinson, *Archit. Hist. Ely*, 141.

71 The two versions of the survey occasionally supplement one another, and it is sometimes inferable whether tenants are freeholders, cottars, or other servile tenants, but no computation of the relative numbers is possible. Moreover, it is not always clear whether the sharing of messuages refers to the actual residents or to the legal holders, though in many cases the number of householders 'under one roof' is stated. Sometimes a widow shares a tenancy with her son, but there is no indication respecting the residence also of younger sons and daughters. (Many more tenements are specified in the Harl. MS. than in the printed *Cal. Pat.*) Some cottagers hold directly from bishop or prior; others are sub-tenants. Minor monastic and episcopal servants still figure among the citizens. The number of shops had increased since 1251.

72 The boundaries and measurements of properties are stated. There is a considerable increase also in moderately prosperous tradesmen, now holding several tenements: John Plomer, a 'bocher', holds 4; John Millere holds 2 cottages and 2 other tenements; a number of citizens have 2, 3, or 4 holdings. Henry, Richard, and John Smyth, 'bochers', have 7 tenements between them. Edmund Cotermonger has 8 tenements as well as a horse-mill.

73 Not including the hospitals or the Chantry-on-the-Green.

74 Philip Morgan was Bishop of Ely, 1426–35.

75 Châlons-sur-Marne gave its name to a closely woven woollen fabric, known in England as chalons or shaloons. A chaloner was a maker of such fabric.

76 Maker of gold-embroidered cloths.

77 See above, p. 36.

78 Quo Warranto Inquest, 1 Ed. I. (ref. given by Archd. S. Evans).

79 They were already so inclosed in 1251. (MS. 485/489, ff. 19–29, Caius Coll., Camb.).

80 Cellarer's Rolls (Atkinson, *Archit. Hist. Ely*, 157).

81 There were seven specifically mentioned, apart from numerous plots.

water. There were the fairs of Bury, Reach, St. Ives, and Stourbridge, as well as Ely's own fairs. From Bury and Newmarket came clothing-materials; from Reach, via the Ouse and its tributary, came clunch, flints, and 'skirt-land' timber.[82] From Barnwell Fair came wheels and other goods; from Boston large supplies of lead and wax; from St. Ives canvas; and from the great international fair of Stourbridge came a great variety of goods, including the products of the Far East.[83] Bricks, tiles, and pottery were early produced in and around Ely:[84] they were also brought from Wisbech, Wiggenhall, and Lynn.[85] Heavy stone came mainly by water: by river and coast to Lynn and thence by river to Ely.[86] Swaffham, Burwell, Reach, and the quarries of Peterborough supplied suitable clunch for internal sculpturing. Rubble was available locally.[87] Timber of many varieties reached Ely: oak came by water from Stourbridge and Reach and from Chicksands Priory, in Bedfordshire. Iron, steel, and lead came largely via Stourbridge, tin via Lynn, nails via Reach, Lynn, Newmarket, and even Derby.[88] Glass still came largely from abroad, through Yarmouth.[89] Rope-making was a local industry by the 14th century.[90]

SIXTEENTH CENTURY

The main developments of the 16th century were ecclesiastical and economic. Religious change at Ely[91] meant change in the flywheel around which city life revolved; yet there was less disturbance here than in many monastic centres. While two heretics were burnt, they were not townsmen.[92] There were considerable sales of monastic property[93] by the Crown, and within the actual college precincts change was, of course, pronounced. The demolition of much of the ancient monastery, the embodiment of parts of its remains in the new prebendal dwellings, and the extension of Bishop Alcock's palace by Bishop Goodrich[94] belong to this century.

In the economic sphere Ely's main concern was with the agrarian problems of the day. Inclosures here took on a local colouring. They did not provoke open riot, but danger was real. In 1548–9 William Saunders, the bishop's bailiff, accounted for expenditure on 'persons who diligently watched at the time of the commotion in Norfolk, lest there should be any rebellion or insurrection', and on 'arms bought by him for the Bishop'.[95]

In reply to the Commission on Inclosures, appointed in 1548,[96] over 70 complaints were lodged respecting Ely: they involved most of the demesne farms.[97] In all, in the period 1486–1548, about 329 acres were inclosed in Ely and Chettisham and 112 acres in Stuntney. In Bedwell Hay, Chettisham, Emery Barns, New Barns, Barton, and Stuntney, there were individual inclosures which embraced areas of as much as 40 to 180 acres, but most of the complaints related to small or very small inclosures. One of the commonest grounds of protest concerned lost rights of shack over inclosed lands. Many of the inclosures were for the purpose of converting arable land to pasture, but more commonly for dairy-farming than for sheep. There is no suggestion that the food supply was endangered. Moreover, dairy-pastures did not curtail employment so seriously as did sheep-runs. Nevertheless, both housing and employment problems did arise: there were farmers like the tenant of Northney, who, in Henry VII's time, had 'kept an honest house', but who had given place to one who 'used no tillage nor kept no house, but turned a dayry', and thus threw labourers out of work, it was said. In not a few cases two or more holdings were combined, former dwellings being thereby left empty, destroyed, or let without land. At least two of Ely's ancient 'mansion houses', as well as many cottages, suffered in this way: Ketons[98] manor house became a barn; the old Almonry grange was pulled down and a smaller tenement on the estate burnt. Some of the property mentioned in the complaints was copyhold; still more was held on lease; but nothing is said here about 'arrearing of rents or fines'. There is one glaring attempt to force copyholders to part with property.[99] Bishop, prior, and almoner emerge none too well from the inquisition;[1] the equity of the manorial courts is not above suspicion.[2] It is not without significance that Wolsey's Commission of 1517 awakened no sympathetic echo from the cloisters! There are many complaints of small intakes from the waste—sometimes merely a few square yards behind a town dwelling. For such there was the excuse of long tradition, and there was still much open fen available. It was urged, however, with abnormal sanitary susceptibility, that three small urban intakes were used as dunghills for the adjoining gardens, whereas, had they remained uninclosed, they would have been liable to prosecution. The most serious example of conversion of arable land to sheep-runs was at Stuntney; a number of landholders, however, were accused of surcharging the commons, with cattle if not with sheep—e.g. at Thorney, Northney, New Barns, and Stuntney. There are several instances of parts of common streams being inclosed, and 'hurdels' con-

82 *Sacrist R.* 1302–3, 1354–5. See also Atkinson, *Archit. Hist. Ely*, 37 et seq.

83 Atkinson, op. cit., 38 et seq.

84 Atkinson, *Cambs. Antiq. Soc. Proc.* xl, 46–47. They were to continue important industries for centuries.

85 In 14th century for building new bridge over Ouse (*Sacrist R.* 1339–40).

86 Fine quality Barnack stone probably travelled thus in early 14th century (*Sacrist R.* 1302–3, 1322–32).

87 Atkinson, *Archit. Hist. Ely*, 42.

88 *Sacrist R.* 1322–3; 1336–7.

89 Custos Capellae, 1356–9.

90 The name Roper figures in the Subsidy List of 1327 (E 179/81/6), and in later documents. Hemp was grown in the locality at various periods.

91 See *V.C.H. Cambs.* ii, 164–79.

92 William Wolsey, constable, and Robert Pygot, painter, both of Wisbech, were burnt at Ely as heretics. A detailed report of proceedings is given in Reg. Goodrich, 16 Oct. 1555. See also, Foxe, *Acts & Mons.* vii, 404.

93 *L. and P. Hen. VIII*, xiv (2), 542, and xvi, 1226 (11). See also Report of Commissioners of 1541, cited by Atkinson, *Archit. Hist. Ely*, App.

94 Ely Ep. Regs., Alcock and Goodrich.

95 B.M. Add Chart., 34274, m. 1. (bailiff's acct., 2 Ed. VI–3 Ed. VI).

96 Strype's record of the Inquiry is reprinted in Tawney and Power, *Tudor Econ. Doc.* (1924), i, 39. The Ely, Downham, and Littleport replies are apparently the only ones known save those of Warwickshire and Cambridge (W. M. Palmer, 'Enclosures at Ely, Downham and Littleport 1548', *Cambs. and Hunts. Arch. Soc. Trans.* v, 369–84). The original document is in Clare Coll. Treasury, Camb. Dr. Palmer's transcript has been fully utilized; also information drawn from documents in the care of Col. Archer, Solicitor, Ely. See also Maj. G. Fowler, 'Fenland Waterways Past and Present, Pt. II' (*Cambs. Antiq. Soc. Proc.* xxxiv, 25).

97 Bedwelhaye, West Fenne Farme, Ketons, Almory Barnes, Newbarnes, Beald Farm, Barton, Turbutsey, Brame, Thorney, Northney; the hamlets of Chettisham and Stuntney, and various small plots in the town itself or in the open fields. Stuntney and Thorney had been sold after the Dissolution.

98 Cf. below—Manors.

99 Symon Steward, purchaser of Stuntney manor, was guilty of this and of most other types of oppression exemplified in the Ely area.

1 Bishop West made a park of 180 acres at Chettisham Bushes, shortly before his appointment as an anti-inclosure commissioner! The ordnance map (1835) still shows this as woodland, despite ordinances against inclosure (*Cambs. & Hunts. Arch. Soc. Trans.* v, 372). Some of the lords' offences were not confined to demesne land.

2 Cf. below, p. 41.

structed for private fishing—e.g. at Turbutsey. Perhaps the commonest complaint is of the loss of public droveways, consequent upon inclosures. Some 20 rights of way had thus been obliterated or had led to illegal demands for payment.[3] The whole body of evidence conveys the impression of much petty irritation, occasioned as often by go-ahead small men as by large landholders;[4] of serious oppression in a few cases; and of dangerous loss of public trust in the paternal integrity of local rule.

The evidence throws light on the structure of the Ely demesne manors or granges, and also illustrates the transitional stages through which inclosure was passing in this region. By the end of the 15th century on most estates a good deal of the property was in more or less consolidated blocks, though there were often also scattered portions. Some of the 16th-century activity was directed towards assembling these isolated parcels. A certain amount of actual hedging or ditching is suggested, but holdings, for the most part, still lay physically open to the waste. Hence arose certain of the disputes. There was considerable confusion as to the rights of the general inhabitants over these semi-inclosed lands, and the continued rights of the inclosers over the commons.[5] The compromise agreement ultimately reached in the case of the New Barns estate, some decades later, admirably illustrates the transitional conceptions of severalty still obtaining.

As explained to the Commissioners, the New Barns farm was already in consolidated blocks in the late 15th century, let to Richard Baker, who maintained 5 plough-teams. He also held Chettisham Bushes. Pasture inclosures began under his successor, who converted about 110 acres in the western part of the farm. In 1537 Thomas Goderyk[6] held the lease and inclosed 70 more acres for pasture. Further closes—'Safron ground and Sprynge close'[7]—were presently included in the pasture. Part of the farm was sublet. Only two plough-teams were now employed. Trouble had also begun concerning the closing of five public droveways passing through Thomas Goderyk's lands, and a further droveway in the hands of John Goderyk, leading from Newnham through 'the pytts' to 'litell Tyrbesy and thence to Waterden'. For the use of this latter drift, in 1539, John demanded 14*s.* a year under threat of legal action. A certain Mr. Rudston—possibly the bishop's bailiff—posed as an intermediary and, by a piece of gross chicanery, persuaded the inhabitants to pay the rent once,[8] only to find the sum 'ever sythens required' of them! In the course of the next generation difficulties repeatedly occurred.[9] The estate was not physically inclosed, nor was it held in complete severalty. The sheep should, apparently, have been hurdled within the estate, but frequently strayed on to the surrounding fens, infringing the rights of the general citizens. Cattle legitimately grazing on the wastes wandered over the growing crops of the estate. The question of droveways was also still at issue. In 1566[10] it was agreed between the disputants to submit all matters to the arbitration of Bishop Cox. Under his decree the tenant was to inclose the manor by hedge and ditch, leaving specified free droveways.[11] Private breeding stock and swine of the estate were to be kept within the inclosure; other of the tenant's cattle, using the waste, were to be reasonably stinted, the tenant behaving 'as a neighbour among neighbours'. The citizens were to relinquish their former right of shack over the holding, in consideration of a thirty-year lease of 200 acres of its pasture land to representative citizens, at a fixed rent, this land to be separately inclosed at the cost of the users.[12]

Two other cases brought before the Commissioners merit notice: they relate to the Steward family.[13] The bishop's bailiff had evidently inclosed parts of the Barton manor (q.v.) in the late 15th century, using them for pasture. These included the 'Tylekyln close' and the 'Gravell Pytte close'. Nicholas Steward took over the lease in the 16th century, having 206 acres of arable, about 155 acres of pasture and an inclosure of 32 acres. He was interested in sheep, and added to his farm by attaching a part of the adjoining common of Cawdle Fen and by encroaching on the highway. For this last offence he had been 'divers times amercyed in the lord's courte, but it is thought his fine has been pardoned and soo no redress is had'. Symon Steward was a still more sinister figure. When he bought Stuntney manor (q.v.), part of its 130 acres had already been inclosed for pasture by his predecessor.[14] He himself

[3] This offence was not an official article of inquiry, but it was a common grievance in Cambs. (Cooper, *Annals of Cambridge*, ii, 38–39). Among the more important public ways affected at Ely were: a road off Chettisham highway, running through the western part of New Barns on to Waterden common; a road from Newnham to New Barnes and on to Redmore; a droveway from Newnham to the 'pytts' and thence to Little Turbutsey and Waterden Fen; a droveway, 'Brodebank by Turbesy hedge', to Roswell; a droveway leading from Potter's Lane to Cawdle Fen; a road between Stuntney and Bury Fen.

[4] Robt. Marche exemplifies the prosperous copyholder, as bent on taking advantage of the times as were his superiors. He began by inclosing an acre, with episcopal permission, in the late 15th century. He later secured 3 arable holdings, of 24 acres each, in Chettisham, turned out all the tenants and allowed one of the dwellings to decay. He also obtained three other previously inclosed plots in the hamlet, amounting to a further 8½ acres. Thos. Rydley and his heirs were even more successful, two of their numerous closes being the result of Prior Walpole's earlier inclosing activities. Thos. Chypley and Thos. Hadman hunted together, bringing '11 litell pyngells' (pightles) of half an acre each into the bag by 1519. The new 'Dean and Prebendaries' inclosed Beald Farm, containing 60 acres (*Cambs. & Hunts. Arch. Soc. Trans.* v, 369–84).

[5] e.g. Northney manor farm was turned into a dairy in 1511. It fell to the dean and chapter at the Dissolution and was presently leased to John Croplaye. He adopted the practice of taking in cattle to graze, at a price, and turning his own cattle out on the common! (*Cambs. and Hunts. Arch. Soc. Trans.* v, 379.)

[6] Probably a son of Bishop Goodrich (1534–54).

[7] Giving the name to the modern Spring Head (or 'Love') Lane. The saffron crops, hemp-yards and brick-and-tile making are points of 16th century economic interest mentioned incidentally in this inclosure inquiry.

[8] 'Mr. Rudston came to us wyllyng and movyng us to accomplish the said Mr Goodryk request in paying the said rent ones, and yf we so did he said he wolde paye y[t] of his own purse' (*Cambs. and Hunts. Arch. Soc. Trans.* v, 376.)

[9] The account prefixes the agreement drawn up by Bishop Cox in 1566. (Harold Archer, 'Ancient Ely Byeways', in *Cambs. Times*, 17 Mar. 1922).

[10] Ibid.

[11] One being that known as Spring Head Lane leading to Little Turbutsey (now Broad Pieces); the other leading from 'Ely Towns-end, by the Inclosures of New Barnes to Chittisham'. Under Ely Fields Inclosure Acts, the latter right of way was confirmed. It fell out of use after the construction of the modern highway from Ely to Chettisham.

[12] Presumably the 200 acres reverted to the bishop at the end of thirty years. They are still known as Little New Barnes (Archer, *Cambs. Times*, 17 Mar. 1922).

[13] Robert Steward was the last prior and first dean of Ely. Symon or Simeon Steward was one brother and Nicholas Steward another. Symon married the daughter and heiress of Edward Bestney of Soham. Nicholas Steward farmed Barton manor and was great-grandfather of Oliver Cromwell, whose mother was a Steward. (*Cambs. and Hunts. Arch. Soc. Trans.* v, 372.)

[14] Edward Bestney, of Soham, in 1530. Thus were 'plows laid down and the ferme fell in decay with great ruin of homes'. (ibid. 377–9.)

closed one highway, and his son Edward another one, which the tenants had used to gather their crops from Bury Fen. Moreover, the Commissioners found that whereas formerly 500 sheep, belonging to inhabitants and tenant, had found pasture, now 600 were fed, but that the tenants owned few or none of them. Steward permitted his sheep to trespass on the neighbouring common, but allowed no other sheep to use it—'to worry the tenants so that they shall be constrained to sell their copyholds that he may have the whole township in his own occupying'.

Bishop Cox's survey of Ely, in 1563, gives the total number of householders as 400: in Holy Trinity parish with Stuntney there were 246; in St. Mary's parish with Chettisham 154.[15]

SEVENTEENTH AND EIGHTEENTH CENTURIES

Whilst near-by Cambridge was the headquarters of the Eastern Association,[16] Ely was the home-town of Cromwell himself, who was in residence here[17] between 1636 and 1640: on Palace Green the New Model Army took shape.[18] Ely, however, was also the episcopal seat of 'The Laud of East Anglia', Bishop Wren, and was thus a centre of royalist sympathy.

When war broke out the bishop set about the raising of forces for the king. It was in January 1644 that Cromwell's famous interruption[19] of the cathedral services occurred. During the absence of the Eastern Association army the same year, the king planned an attack on Ely, and an attempt, in 1645, to open the city to the Royalists led Cromwell to return in person and to quarter a body of armed men there.[20] Both rich and poor bore a heavy financial burden which left its mark.

On the conclusion of the war, in 1649, a survey of the city and the demesne manors was made with intention to sell.[21] The bishop's palace was sold and some of the Precinct buildings were occupied by laymen; some loss also of city property was noted by the ecclesiastical authorities at the Restoration. The general neglect of the Cathedral during the Cromwellian régime[22] meant a serious loss of traditional employment for the citizens.

The forceful voice of Cromwell was familiar at Governors' meetings[23] in the city and as a champion of fenman rights[24] even before the war, and there are hints of a questing spirit on the part of other citizens.[25] But high politics were left to the Precincts in the later 17th century, as were the rivalries between the great Rutland and Hardwick parties in the eighteenth. It was not until the emergence of parliamentary reform, in 1780, that city freeholders showed any interest in their county vote. Public meetings and the popular activities of Archdeacon Watson[26] then proved novel attractions and might have resulted in a demand for local constitutional change had not the reactionary influence of the French Revolutionary Wars supervened.

The history of the several attempts to drain the South Level, in which district Ely lay, is told elsewhere.[27] This was a subject which touched the livelihood of the fenman, more alert to immediate loss of fishing-rights and commons than to long-period gains. In 1638 actual rioting occurred,[28] resulting in the arrest of an Ely man, Edward Powell, alias Anderson, and of five other active leaders. Powell's defiance of the magistrate, Goodrich, as being merely the 'Bishop's justice and not the King's', and an obstructor of the subject's right to petition the Crown, suggests a vague consciousness of constitutional issues. The subsequent report to the Privy Council makes much of the growing disrespect for local rulers.[29] The outbreak of the Civil War presently found other outlets for disgruntled fenmen, though the later drainage schemes, culminating in the Acts of 1659 and 1663, were not carried through without much local controversy. The 18th century, however, showed the material benefits of drainage, despite failures.[30]

Even before the Civil War Ely was facing unemployment[31] and more traditional types of poverty. Parochial authorities here were aided and to a considerable extent influenced by the dean and chapter.[32] Exceptionally large benefactions, moreover, eased the financial situation and brought into being a permanent administrative body which worked in frequent cooperation with the vestries, carrying on various public services. Valuable property in Stretham and Ely had been left to the inhabitants of Ely by Thomas Parsons, about 1445, primarily for the purpose of easing the burden of national taxation.[33] A reorganization of the charity was effected in 1633: moneys were henceforth to be devoted to more general relief of the city's needs, and control was vested in 12 feoffees, incorporated by royal charter, consisting of the Bishop, Dean, and Archdeacon of Ely for the time being, and nine other persons residing within the city, to be called 'Governors of the Lands and Possessions of the Poor of

[15] B.M. Harl. MS. 594, ff. 198–200.

[16] *V.C.H. Cambs.* ii, 407.

[17] M. Noble, *Memoirs of the Protectora House of Cromwell* (1787), i, 110.

[18] E. G. Punchard, *Cromwell and the Old House of Ely* (1906), 8.

[19] M. Noble, *Memoirs of . . . House of Cromwell*, i, 110.

[20] C. Cooper, *Annals of Cambridge*, iii, 387; A. Kingston, *East Anglia in the Civil War*, 149–50.

[21] Atkinson, *Archit. Hist. Ely*, App. ix; Stewart, *Archit. Hist. Ely Cath.*, 251 et seq.

[22] Some of the prebendal houses fell into disrepair, or into the hands of non-clerics. The ancient Knights' Lodging, given to the minor canons at the Dissolution, suffered the most heavily. It was finally demolished in 1694 (R. Gibbon, *Small Beer of History*, 102–3).

[23] He was one of the 'Governors of the Lands and Possessions of the Poor of Ely' (MS. Records of Parsons' Charity, Ely). See also M. Noble, *Memoirs of . . . House of Cromwell*, i, 107.

[24] In 1638 he strenuously opposed the drainage scheme, earning the popular title of 'Lord of the Fens' (Noble, op. cit. 108).

[25] Powell's defence (see below).

[26] He was also a prime mover on behalf of popular education at Ely. Bp.'s Mun., Ely, B 6, Schools; cf. also Watson's *Charge to the Clergy of the Archdeaconry of Ely* (1788); *V.C.H. Cambs.* ii, 414.

[27] *V.C.H. Cambs.* ii, 76–81.

[28] *Cal. S. P. Dom.* 1638–9, 301 et seq.; 503 et seq.

[29] The mob threatened to break open the bishop's jail. Some 200 Ely citizens seem to have taken part and to have enlisted sympathizers from Mildenhall, Lakenheath, Littleport, and Downham (*Cal. S. P. Dom.* 1638–9, 301 et seq.; 503 et seq.).

[30] Cf. Edm. Carter, *Hist. of Cambs.* (1819), 66.

[31] Always an agrarian rather than an industrial problem here. Floods were a familiar cause at all periods.

[32] In the Restoration period bishop, dean, and chapter provided much skilled employment on the neglected cathedral. Bishop Laney, Wren's successor, repaired the episcopal palace, which had previously been bought back after its temporary alienation. In 1663, a year of much distress, the dean and chapter made special contributions for bread for the poor. In the later 1670's, stress was laid on prevention of poverty: £20 p.a. was contributed by the dean and chapter to apprentice boys to good trades or to seamanship. Some boys were sent to Cambridge University. Bishop Laney's will left valuable funds for apprenticing children of Ely and Soham (R. Gibbons, *Small Beer of History*, 104 et seq.).

[33] A Thomas Parsons figured in the Survey of 1416. The name appears repeatedly in 15th-century documents.

Ely',[34] and to have perpetual succession. Vacancies were to be filled by co-optation.[35] The ecclesiastical representatives gave to the corporation high prestige, experience, and continuity, but they also maintained the influence of the powerful manorial and political lordship, effectively preventing any possible evolution of a municipal entity, such as occurred in similar circumstances elsewhere.[36] The activities approved in 1633 were the provision of work and dwellings, the furthering of apprenticeship, and the supply of corn, fuel, and other necessities, at controlled prices. Comprehensive functions, resembling those of a manor court, were gradually added, as extensive wastes and other property were acquired by gift, purchase, and lease. From the outset certain common abuses were obviated by careful restriction of the length of lease of trust property and by public inspection of accounts.[37]

The drainage of the South Level necessitated reallocation of lands. The Commissioners of the Bedford Level, set up under the Act of 1663,[38] upheld the claim of the Governors to some 600 of the 12,000 fen acres belonging to the city. The bulk of these 600 acres lay in West Fen and Padnal.[39]

From the 17th century onwards an exceptional proportion of the poorer citizens of Ely enjoyed the great blessing of a cow, as well as cheap turf and even coal. Other forms of public service developed in succeeding centuries, aid often being given to parochially sponsored schemes[40] or to supplement other charities.[41] In addition to an honorary treasurer, paid officials gradually evolved.

In the absence of any municipal authority, parochial organization[42] was no less important than were manorial courts and Governors of the City Lands.[43] The dissolution of the monastery had necessitated prompt appointment of parochial surveyors of the highways.[44] Though their unskilled efforts were singularly unsuccessful, these officials were prominent members of the parish vestries. Constables remained manorially appointed servants until the 19th century and do not figure here at ordinary parish meetings.[45] The activities of the churchwardens, as such, are not always distinguishable from those of the overseers in the 16th and early 17th centuries, church rates being often expended partly on poor relief.[46] Moreover, Ely churchwardens, usually of higher social status than the appointed overseers, pulled their full weight as ex-officio overseers.[47] From the later 17th century, though dominated by the cathedral clergy, the assembled vestry did include citizens[48]—sometimes as many as twenty—determined to record their mark beside the names of the usual coterie, composed of vicar, parish officers, and one or two magistrates. There were periods of negligence,[49] but the general impression is one of considerable vigilance and insight. 'There is a halfe crown in this money questionable', recorded St. Mary's vestry in 1654; in 1730 the meeting was lively: twenty-four signatures were appended to a protest respecting an overseer who had 'carry'd away the book so that noe rate could be made'—presumably the high rate necessitated to finance the recently erected workhouse.

In the 16th century the most interesting parish activities were attempts to help the small tradesman to find his feet, by granting loans from parish funds: from 1623 onwards the projects for employing the able-bodied were the outstanding features. Apprenticeship of children was also an important and long-continued policy: valuable benefactions facilitated the selection of trades which afforded genuine training and good prospects. A quantity of wool was purchased, in 1623, and substantial sums were paid for spinning and weaving. Some of this relief work was done 'in the house'. The enterprise almost paid its way at this date. Wool seems to have been procured locally, but 'oyle' was imported from Bury, Worlington, and London. The poor were also supplied with cheap turf and charcoal; the latter was brought by water from Worlington.[50] In 1675 the Ely Governors were the main sponsors of a new scheme. An elaborate four-year contract[51] was made with a Norwich man, who was to employ 'in spinning of jersey' any poor sent to him, paying them 'in mony and not in goods, the comon country rate'—important stipulations. The contractor was to receive £30 per annum, and also an interest-free loan of £50 in the trusteeship of the dean.[52] Ely was within reach of good textile markets, and spinning and combing, backed by the Feoffees,[53] were persistently carried on through the 18th century. The gap between receipts and expenditure grew unfavourably, but the provision of genuine work remained a determined, if unsuccessful, policy. A com-

[34] The infrequency of the levies of 10ths and 15ths made this desirable; changes in regulations had been permitted from time to time by Decrees and Letters Patent. *Rep. Com. Char.* (*Cambs.*), [103], 217 et seq., H.C. (1837–8), xxiv.

[35] The consent of at least 6 members was necessary, of whom 1 must be the bishop, dean, or archdeacon.

[36] S. and B. Webb, *Manor and Borough*, 132–3, 265; *V.C.H. Essex*, i, 369–74. In 1633 the property consisted of 93 acres of arable and 8 acres of marsh in Stretham, together with 7 tenements and small inclosures in Stretham and Ely.

[37] Modern public audit was inaugurated in 1835.

[38] 15 Chas. II, c. 17.

[39] A block of 270 acres lay in West Fen; 3 blocks amounting to 228 acres in Padnal; 42 acres in North Fen; the remaining 60 acres were distributed over Byal, Burnt, Cawdle, Milking Hill, and Great Middle Fens.

[40] e.g. in the establishment of workhouses.

[41] e.g. in the early 19th century towards the clothing of Nat. Sch. children. In the 1830's garden allotments on part of New Barns estate were leased by the Governors and let at low rentals to the poor. By 1835 the annual rental of the Stretham property was over £282, of the Ely property about £475. The latter included the Cross Keys and the White Swan Inns, 11 tenements and 2 warehouses, in addition to the fen commons. The total income of the trust reached nearly £1,150.

[42] No early vestry records are extant for Holy Trinity parish; those for St. Mary's begin in 1570 but are far from continuous: they form the main source of information on parish activities.

[43] This body, often referred to as 'The Feoffees', was closely associated with St. Mary's parish, with whose archives the Trust records were kept until modern times. There are many 18th-century 'Spinning-Books' extant among the rec. of St. Mary's parish.

[44] Under Acts of 2 and 3 Ph. & M., c. 8, and 5 Eliz., c. 13. The bishop still undertook the maintenance of the causeway to Stuntney and later bishops retained a special interest in communications.

[45] Their attendance was legally necessary for the annual election of surveyors.

[46] Thirty-eight of St. Mary's parishioners paid the church rate in 1584; 144 paid in 1604, 55 of whom were very small contributors. As the proportion of these latter remained high, there must have been strong temptation to allow some contributors to the Hearth Tax (e.g. in 1674) to escape.

[47] A duty often neglected elsewhere. Chettisham was clearly represented from the mid-17th century.

[48] Possibly thus seeking expression denied them elsewhere.

[49] A bequest to the parish, in the trusteeship of the Bedford Level Corporation, lay unclaimed for twenty-five years prior to 1738.

[50] A village on the south bank of the Lark, about a mile west of Mildenhall. Being between the fen and the breckland it was better wooded than Ely.

[51] Contractor might terminate the engagement at end of first year, if he desired.

[52] The materials were supplied to homeworkers, but there was certainly a workhouse in existence in 1687. The Dean and Chapter contributed £40 thereto in 1687. (R. Gibbons, *Small Beer of History*, 244.)

[53] Spinning Bks., St. Mary's parish rec.

prehensive programme for the relief of poverty was drawn up about 1729. It included a workhouse, but did not abolish domestic employment and other relief. In view of the harsh doctrines of the age, the programme was unusually sympathetic: even 'bacco' was regularly allowed to certain types of poor. The workhouse was managed directly by the overseers, under very shrewd vestry supervision.[54] The payment of rents on behalf of paupers was to be revised quarterly; workhouse children were not to be apprenticed before the age of 14 or 15, and in no case to blind-alley occupations or to mere drudgery. The aim of the whole programme was 'to provide for the poor in a more comfortable manner than has hitherto been done'. Public co-operation was frankly solicited: the weekly board-meetings and the quarterly audits were to be open to any two or more ratepayers.[55] A thatched building, in the gift of the dean, was leased to the parish as a new workhouse in 1747.[56] In 1776 there were 80 inmates in a workhouse probably shared by the two parishes. By 1785 the work programme was failing: in the next century the prospects of hand-produced textiles grew ever more dim. Publicly sponsored work was mainly in textiles and stone-gathering, but in the later 18th century employment was sometimes found in farming and brick-making—a policy strongly advocated, on various grounds, by the Revd. James Bentham.[57]

'The city of Ely', wrote Camden in 1586,[58] 'is not inconsiderable, nor yet to be boasted for beauty or populousness.' A century later Celia Fiennes[59] noted the willow-banked dykes and the commanding minster, but condemned both the civic rule by a bishop and the unenterprising citizens it engendered. Her opinion of drainage experts was no higher: a day begun by floundering through the flooded 'causey' and main streets of the city, and concluded in a slug-infested bed-chamber, doubtless darkened her view. The palace was left vacant[60] and the city was 'a perfect quagmire'. There was, she asserted, 'no tradeing in the town, their maine buissiness and dependance is on draining and fencing their grounds and breeding and grasseing cattle.' Defoe,[61] in 1724, was far less gloomy. Though also impressed by the overflowing wells—'almost every hundred yards' along Fore Hill—he was eloquent respecting the excellent market-gardens. Edmund Carter,[62] a generation later, was frankly adulatory. Drainage had made the city 'a most healthful place'. Needham's Charity School, erected 11 years earlier on Back Hill,[63] had begun to provide elementary education. The main interests were agricultural, but citizens were agog with energy. The vineyards had disappeared by the 17th century, but fruit-growing flourished, especially cherries and strawberries. A wide variety and 'prodigious quantities'[64] of vegetables found their way by water to many parts of Bedfordshire, Huntingdonshire, Lincolnshire, and Cambridgeshire, particularly to Stourbridge Fair. A mail coach left thrice weekly from the Lamb Inn for Cambridge, and a 'passage-boat' did the journey in six hours. The Saturday market was always well-stocked with victuals—meat, fish, poultry, dairy produce, and corn.[65] The two fairs flourished, hops being now a special feature at the October gathering. The city was plentifully supplied with spring water, but for washing and brewing soft water was carried on horseback from the river in large leather bags. Carter's picture is in strange contrast to that of Bentham only four years later. Bentham was more impressed by the inefficiency of drainage than by its success; by the trials of carriage-journeys to Cambridge than by their triumph. His one word of praise was for the recent paving of Ely High Street by aid of public subscription. He urged the adoption of the Turnpike system for roads between Ely and Cambridge and the major towns of Norfolk, Suffolk, and Huntingdonshire; regular public work on the roads, at fair wages, would go far to meet unemployment.[66] The depressing harvest of 1757 no doubt deepened the gloom of Bentham's picture, but his suggestions bore fruit.[67]

Some rise in the population of Ely appears to have occurred in the 17th century, and by 1638 a crowded area on the Lynn Road had been dubbed 'Little London'.[68] The plague of 1630–1[69] struck the city severely, and deaths were still above normal the following year. The Great Plague of 1665–6 visited Cambridge and towns near Ely,[70] but the toll of death was much heavier in the city itself in 1667–8 and 1680.[71] Smallpox probably explained the abnormal death-rates of 1720 and 1780. Infant mortality was high: on an

[54] A register was to be kept of all entrance to or discharge from the workhouse. Overseers were not to buy materials beyond the needs of their own term of office. Each overseer was to be individually responsible for any expense he sanctioned. Any officer purchasing for the workhouse goods above market price was to be compelled to keep and pay for these himself. No tradesman, whilst in office, was to supply any of his own commodities to recipients of parish relief. A public vestry was to be called before any changes in the regulations were made.

[55] Any officer hindering free public inspection was to be fined.

[56] Allegedly in conformity with the General Workhouse Act of 9 Geo. I, c. 7.

[57] J. Bentham, *Queries* (1757), 30 et seq.

[58] W. Camden, *Britannia* (1607), 215 et seq.

[59] *Through England on a Side Saddle*, 1695 (1888), 127–31.

[60] It had been restored at great cost, less than twenty years earlier, by Bishop Laney.

[61] *Tour through Whole Island of Gt. Britain* (1742), i. 77.

[62] Edm. Carter, *Hist. of Cambs.* 65 et seq.

[63] See below, p. 87.

[64] Asparagus, potatoes, onions, beans, peas, cauliflowers, cabbages, and artichokes.

[65] In 1768 Ely and Cambridge markets were setting the corn price for a wide area (E. M. Hampson, *Treatment of Poverty in Cambs.* 99).

[66] 'Whether setting them to work and regularly paying them just and reasonable wages (which Authority might appoint and limit) is not the best and most likely way to relieve their wants?' (Q. 18). He urged the desirability of offering genuinely needed work in hedging, ditching, embanking, gravel-digging, road-making, and brick-making; this would also benefit the health and morale of adults and children; the latter suffered badly from workhouse lack of sunshine and fresh air. He stressed also the need for the better education of parish children. There were, he said, several thousand acres 'very near us, and some of the best of our fens, that have been covered with water more than ten years past'. As for parish road-making: 'Are not common day works turned into so many days of relaxation from work?' (Q. 55). He made precise suggestions respecting the turnpiking of roads from Ely to Bury, Newmarket, Mildenhall, Cambridge, Huntingdon, St. Ives, Peterborough, Lynn, and Wisbech. He would begin with such a road from Earith or Chatteris, through Ely, to Soham: this would afford an admirable trunk for later branches to Norf., Suff., and Hunts. (Qs. 52–108).

[67] Bentham played a vigorous part in initiating the construction of the turnpike road between Ely and Cambridge, begun in 1769, and subscribed to largely by the ecclesiastical authorities and prominent townsmen of Ely and by the leading members of the town and university of Cambridge (B.M. Add. MS. 5813, f. 155).

[68] *Cal. S. P. Dom.*, 1638, p. 504.

[69] Cf. Cooper, *Annals of Cambridge*, iii, 243. The parish reg. of Ely—a small town, immediately under the episcopal eye and without any dissenting chapels in the mid-18th century—probably afford a fairly good basis for population statistics of the 17th and 18th centuries, though even here the marked increase in baptisms after Bishop Wren's reinstatement is suggestive. In the late 18th century episcopal surveys reveal the marked growth of dissent. (Schools, B. 6. & C. 2, Bp.'s Mun., Ely.)

[70] Cooper, *Annals of Cambridge*, iii, 520.

[71] Parish Reg., Ely.

average more than half the burials of the 17th century were specified as those of infants; from about 1730 this proportion dropped somewhat, save in bad years. In 1753 Carter computed the number of houses in the city as 609 and the population as 3,000.[72] By the end of the century Ely was experiencing the rapid rise in population general in England.

THE FRENCH WAR PERIOD 1793–1815

Two surveys made between 1793 and 1815 show the agrarian condition of the city.[73] Amid the fens which, in 1794, extended east of Ely for about a mile, the hamlets of Quaney, Thorney, Northney, and Stuntney formed a cultivated island, devoted equally to arable and pasture. To the north of the city lay about 1,200 acres of first-grade pasture, including 'the beautiful villa of New Barns'[74] and stretching to the hamlet of Chettisham. In this region also were 2 open fields of arable and many inclosed acres, large and small, of very fruitful orchards and nurseries. West of the city was 1 open arable field; to the south were 2 more arable open fields and some 700 acres of inclosed pasture. The arable fields thus lay in 5 distinct parcels or 'shifts', constituting about 2,100 acres, cultivated on the old strip system and using a 5-course rotation. About 500 acres of common were reserved for turf-digging and the very profitable sedge-growing. The rest of Ely's fenland was still often under water and 'miserable indeed'. Inclosure was strongly advocated by the surveyors.

By this date the ancient industries of coarse pottery and 'white Ely bricks' were important.

Two decades later Gooch deplored the conservatism of the arable farmers. Open-field cultivation still perpetuated a rotation 'foreign to every idea of modern farming'.[75] In contrast, the inclosed lands of Thorney attained perfection. In and around Ely a good deal of hemp was still grown, a profitable but exhausting crop. The New Barns estate had only recently ceased to cultivate woad[76] on a large scale, a dye-crop which had replaced the saffron-growing of earlier days. High tribute was paid to market-gardeners, both large and small. Turnpike roads were excellent, but other roads were still deplorably bad.

Despite the urgent pressure of the age, expense and clashing interests held up general inclosure of the open fields for yet a generation.

The abnormally high prices commanded by agricultural producers during the war gave good profits even without inclosure, for the wages bill lagged progressively behind. Population was growing rapidly and widespread distress of the labouring classes confronted both vestries and feoffees. The Ely authorities upheld the view expressed by the Revd. Dr. James Nasmith, Chairman of Quarter Sessions, when charging the grand jury in 1799,[77] that 'reception into the workhouse should not hastily be thought of': they had never favoured it as the best policy for the aged or temporarily sick. Nevertheless the workhouse changed its name to the 'Poor House' during these difficult years. Nasmith's novel dictum that parish surgeons should prescribe diet as well as drugs met with sincere approval: bread, dairy produce, and fuel were more generously provided than in most places, but distress was severe in the near-famine years. Nasmith's comments on the improving standard of living of the poor, during the decades immediately preceding the war, should be noted in conjunction with the rise in population. He did not share Bentham's fear of undue pampering.

In view of the manifest sensitiveness to suffering, it is strange to note the apathy respecting the state of affairs in both the bishop's jail and the Ely House of Correction. Some little had been done between Howard's visits in 1774–6 and Nield's in 1812, but conditions were still a terrible indictment of the private ownership of prisons.[78]

THE NINETEENTH AND TWENTIETH CENTURIES

The period 1815–37 was one of general depression. At the beginning of 1816 wheat was sold on the Ely market at 52*s.* 6*d.* a qr.; already in May it reached 76*s.*, and rose to 103*s.* in December.[79] Wages averaged but 8*s.* to 9*s.* a week and showed no response to soaring prices: many had no work at all. Such was the background of the Ely and Littleport riots of May 1816—the unorganized outbreak of a mob 'crazy with hunger',[80] designing, by raids upon food and drink stores,[81] to give point to their vague demand for 'the price of a stone of flour per day'. The outbreak began at a Littleport benefit club, meeting in a public house, and threatened to spread to Ely.[82] Special constables were sworn in at Ely and a small body of dragoons[83] was summoned from Bury to overawe the rioters. In fact little bodily injury was inflicted on either side. The Ely magistrates, led by one Metcalfe, minor canon and precentor, were conciliatory: enfeebled labourers would probably have capitulated forthwith, but for the unaccustomed access to free drinks. Some of the disorders were committed by mere boys, though there were also despairing men who 'might as well be hanged as starve'.[84] Certain of the instigators were men of better economic standing:[85] such was the unfortunate John Dennis, whose influence in checking the wilder elements marked him out as a ringleader. Much capital was made by Chief Justice Christian[86] out of the 'great wages' of the leaders and the expenditure of the moneys extorted upon liquor rather than bread. Between 70 and 80 rioters were arrested and a Special Commission was appointed for their trial at Ely. Ultimately only 5

[72] *Hist. of County of Cambs.* 64.

[73] Two reports made to the Board of Agric.: C. Vancouver, *Gen. View of the Agric. of the County of Cambs.* (1794); W. Gooch, same title (1811 and 1813).

[74] Vancouver, op. cit., 138–40.

[75] Gooch, *Gen. View of the Agric. of... Cambs.* 98 et seq.

[76] '100 hands have been at work at once; and 24 horses employed' (ibid., 172).

[77] *Duties of Overseers of the Poor, a Charge delivered to the Grand Jury, at the General Quarter Sessions for the Isle of Ely* (1799).

[78] See *V.C.H. Cambs.* ii, 108–10, and above, p. 31.

[79] *Camb. Chron.*, 1816, weekly price-reports.

[80] So described by an Ely citizen, who recalled the comments of his grandfather, Thos. Archer, respecting his youthful night-ride for assistance from Bury. (The manuscript letter and other correspondence from descendants of people concerned in the rising were lent to me by Col. Archer.) See also *Camb. Chron.*, 28 June 1816; *Trials of Rioters at Littleport and Ely* (anon.), 1816; Ed. Christian, *Charges* (ed. 1819), chap. xvii.

[81] Bakers and brewers were the chief victims, though landlords and others, including the Ely agent of Mortlock's Bank, also suffered.

[82] In a milder form also to Downham.

[83] A few members of the Volunteer Corps from Royston and militia men from St. Mary's Street Barracks, Ely, also took part in the rather light-hearted display of military force.

[84] Correspondence. (See n. 80, above.)

[85] These included several substantial farmers; possibly resentment of depredations allowed by the Game Laws played some part in their participation.

[86] Chief Justice of the Isle, who was included with Sir Charles Abbot and Sir James Burroughs in the Special Commission. Christian's lack of both humanity and perception is equally evident in his remarks elsewhere on the criminal law and on jail conditions at Ely (Edw. Christian, *Charges* (1819)).

suffered execution;[87] 5 were transported for life; 1 for 14 years; 3 for 7 years; and 11 were imprisoned in Ely jail.

Agrarian conditions were bad in 1829: Ely labourers 'no longer able to maintain themselves by the sweat of their brows', were driven 'to the scanty pittance derived from parish funds'.[88] The memory of 1816, however, deterred Ely citizens from any participation in the risings which occurred elsewhere in 1830.

During these difficult years parish authorities showed themselves experimentally progressive. Though only qualified success was attained, the way was paved for modern reforms. Paid officers increased in number and efficiency; the committee system grew; public meetings were more frequently summoned and at times demanded a poll; road questions and new public services evoked greater co-operation between the two parishes; lighting and watching improvements were discussed; under the new Poor Law the modern era in local government was inaugurated. Of these activities road problems and relief aroused most interest.

Road problems brought joint parochial action. The turnpike system, first adopted by the city in 1763,[89] for the road between Ely and Cambridge, proved unsatisfactory within the city itself and its immediate environs; hence, in 1817, the two parishes reassumed responsibility for St. Mary's Street. In 1832 and 1836 joint committees discussed the Old Mile Stone Road[90] and the road across Padnal Fen to Prickwillow. A new paid officer, the Collector of Surveyors' Rates, appeared in 1836. It was in connexion with St. Mary's Street that lighting by gas was first proposed. Much controversy ended in a final poll, postponing the imposition of a general gas-rate until 1838.

Serious increase in unemployment led St. Mary's parish, in 1822, to levy a labour rate.[91] Three years later the activities of churchwardens and overseers were submitted to the monthly guidance of a parish committee of twelve members. In 1834 weekly committee meetings were inaugurated.[92] In October 1835 the vestries were still awaiting instructions from the central Poor Law Commissioners. Though relief conditions hardened, inadequate wages forced St. Mary's vestry, aided by the Feoffees, to purchase bread and flour in large quantities, for distribution according to scale, to families with five children or more. The new Poor Law union was presently created,[93] and at first used the existing Ely workhouse.[94] In 1838 this building was sold, a much larger, white brick and stone institution having been erected on the Cambridge road.[95]

In the middle years of the 19th century Ely farmers had not that stimulus to rapid change which great industrial centres offered. General inclosure of the open fields—the 'Ely Fields', in St. Mary's parish—did not come here until the final award was made in August 1848.[96] The first steps were taken in 1844, when a public meeting was called[97] and three Commissioners were appointed, under the General Act of 1837.[98]

Grunty Fen, in which Ely had intercommoning rights, was inclosed between 1857 and 1861 (see Wilburton).

The practice of raising the clay from below the peat and spreading the clay on the surface had added markedly to the prosperity of Ely farms by 1860.

Population had grown more rapidly in the city of Ely than in the surrounding country-side: the heavy death-rate brought the city within the terms of the Public Health Act of 1848,[99] which compelled the appointment of a local Board of Health. An official investigation was ordered.

The Report[1] showed a rapid increase in mortality between 1841 and 1848,[2] from 24·1 per mille to 32·6 per mille. Of these deaths 61 per cent. were of children under five;[3] this fact was attributed, in part, to the wide use of narcotics. Of the total deaths 68 per cent. were preventable. Despite the improved agricultural situation, over half of the total 1,552 houses in the city were poor dwellings, rated below £5. The suburb of Little London, in Holy Trinity parish, had experienced a very rapid recent increase of small houses, partly because Ely had become the junction for three railways. Although situated on much lower ground, 'Potter's Lane and Common Muckhill' now 'yielded the palm' for insalubrity to Little London. Density of population, low diet, insufficient clothing and ventilation, contributed here to the ill effects of open ditches, into one of which drained 'all the piggeries and privies', through 'stinking gutters'. Even in many of the main streets of the city no drains existed. Water derived from the river was much polluted. The cathedral looked down upon an improving landscape, yet this agricultural community languished under a mortality to be expected only in 'the densely packed and ill-conditioned towns of the manufacturing districts'.

As a result of this official report, though in face of considerable opposition, a Board of Health for Ely, consisting of 15 members, was elected in 1850.[4] The Board assumed various functions hitherto performed by parish and manorial authorities. The abolition of the civil jurisprudence of the bishop, in 1836,[5] facilitated

[87] John Dennis and four others, who were buried in St. Mary's Churchyard, Ely.

[88] E. M. Hampson, *Treatment of Poverty in Cambs., 1597–1834*, 215.

[89] It ultimately formed part of the road between Cambridge and Lynn, passing through Milton, Stretham, Thetford, Ely, and Littleport. In the course of the next forty years the turnpike system was adopted on part of the Ely to Newmarket road and on the Ely to Wisbech road, passing through Witchford, Mepal, and Chatteris, where it joined the London to Wisbech road (Lysons, *Magna Brit.* ii (1), 41).

[90] 'Commencing at the corner of the Hospital Barn towards the Mills' (St. Mary's Vestry Ord. Bk., 1810–43).

[91] Ibid. At the same meeting the impounding of 'a cow belonging to a pauper of St. Mary's parish' was discussed: obviously parish officers at Ely did not follow the common practice of forcing the sale of a valuable possession before granting relief.

[92] Poor relief was evidently a heated question: expense was heavy. No fewer than 124 parishioners attended a meeting for the election of a new vestry clerk in April 1832.

[93] It included the city parishes and twelve others, covering 128 sq. miles.

[94] In St. Mary's parish.

[95] With accommodation for 340.

[96] A copy of the award, together with the details concerning it—commissioners appointed, letters of claim, applications regarding situation of allocated land, carriage-ways, &c.—is preserved in the office of Messrs. Archer & Beckett, solicitors, Ely. An article by Mr. W. E. Tate ('Cambs. Field Systems') in *Cambs. Antiq. Soc. Proc.* xl, 71 et seq., requires amendment on this point.

[97] *Camb. Indep. Press*, 30 Mar. 1844.

[98] 6 & 7 Wm. IV, c. 115, embodying 41 Geo. III, c. 109 and 1 & 2 Geo. IV, c. 23. Under the General Drainage Act of 1663 (15 Chas. II, c. 17) a certain amount of fen inclosure had taken place, checked by the Act of 1685 (1 Jas. II, c. 21). Small parcels of common field were inclosed from time to time (E. C. K. Gonner, *Common Land and Enclosure*, 219).

[99] 11 & 12 Vic., c. 63.

[1] The rep. is freely quoted, point by point, in *Statements and Observations in relation to the Report of Wm. Lee Esq., C.E., to the General Board of Health* (1850).

[2] Influenza and smallpox were very prevalent here in 1848. Population 6,754 in 1841; estimated as 7,610 by June 1849 (*Statements and Observs.*).

[3] Forty-four per cent. were children under 1 year (ibid.).

[4] Gardner, *Dir. Cambs.* (1851), 490; *Kelly's Dir.* (1950).

[5] 6 & 7 Wm. IV, c. 87, amended by 7 Wm. IV & 1 Vic., c. 53.

the gradual development of modern conceptions of local government, through a transitional phase of provision of certain public amenities by private enterprise. Gas-lighting had already thus begun in 1835. The Corn Exchange Company erected its building in the market-place in 1847, taking over on lease the age-long episcopal control of the market.[6] In the open, pillared front of the corn exchange a special butter-market was held. A weekly cattle-market also developed.[7] The city water-reservoir was completed in 1854 and was apparently under the aegis of the Board of Health.[8] A cemetery was laid out on a part of the New Barns estate under control of a Burial Board. A Mechanics Institute was established in 1842, under the patronage of bishop and dean. In 1851 it had a library of nearly 2,000 volumes. Party politics and controversy were 'rigidly excluded'. Elementary schools increased in number and offered improving facilities. The Feoffees, about 1850, adopted a permanent system of allotments on part of their fen property:[10] some 400 poor citizens were enabled to rent these cheaply. The Feoffees, about the same time, erected a double row of sixteen almshouses in St. Mary's Street; others were later built near the waterside.[11] Considerable modernization of houses and shops took place in this period.

Under the general Local Government Act of 1894 the present Urban District of Ely was created, with a council of thirteen persons in 1900.[12] The Urban District included the two city parishes, the College Precincts, and the hamlets of Stuntney, Chettisham, Prickwillow, and Adelaide Bridge.[13]

In the 20th century the principal interest of Ely remains agricultural, with stress on traditional specialities. Poultry-farming and pig-breeding have increased in significance. City industries are essentially ancillary to agriculture. Brewing[14] and brick-making have now extensive markets; boat-building and basket-making are of local import. More recent developments include the manufacture of jam and potato-crisps, portable buildings, farm gates and agricultural machinery, and, above all, beet-sugar. This last industry employs many hands, both in the culture of the beet and in its later processing. The beet-sugar factory,[15] opened in 1925, is one of the largest in England: during the summer months it deals also with the refining of imported sugar.

Changes in parochial structure and organization have occurred. The Isle of Ely Review Order, operative on 1 April 1933,[16] effected a net reduction of 2,083 acres in the area of Ely Urban District, but only diminished the population by 87. The area is, therefore, now 14,659 acres in 1951.[17] By the Ely Urban District (Union of Parishes) Order, operative on 1 June 1933,[18] the parishes of Ely Holy Trinity and Ely St. Mary, with the Intermixed Lands common to them, were combined to form the single parish of Ely Holy Trinity with St. Mary, having an area of 14,626 acres and a (1931) population of 8,199. Ely College (33 acres, 95 population) was constituted a separate civil parish, within the Ely Urban District. The Urban District Council numbered fifteen in 1950.[19]

In 1918, the Urban District Council bought from the then Ecclesiastical Commissioners the rights of the bailiff of the markets.[20] The rights of the bailiff of the fairs were similarly purchased in 1933.[21] The Council has acquired extensive free fishing facilities along the river. The far bank of the river, long known as 'Babylon', has been laid out as moorings for yachts and pleasure craft. The palace green is leased by the U.D.C. for use as a public open space. The Council controls the water-supply, obtained from the chalk 8½ miles away, and also owns an open-air swimming-pool.

During the first half of the 19th century Ely showed the marked rise in population characteristic of the whole country. Until about 1851 the population of Ely Trinity maintained the traditional proportion of about double that of Ely St. Mary; thenceforward the difference diminished, until in 1931 equality was almost reached. Changes in parish boundaries[22] affected population statistics, though not very markedly. Chettisham was included in St. Mary's civil parish, and Stuntney in that of Holy Trinity, as in early days. In 1951 the population of the city was 9,989—an increase of 20·4 per cent. over the population of 1931. The proportion of houses occupied by more than one family was not large at any date after 1801.

MANORS. In 1086 there was but the single manor of Ely and its berewicks.[23] When the see was created, in 1109, the bishop's charter[24] spoke of granting tithes of his 'manor of Barton' to the prior and convent, and of their right to hold courts in the manors allotted to them, but said nothing of a priorial manor in Ely.[25] Nevertheless, the berewick of Stuntney and the demesne manors of Beald, Turbutsey, and Shepey[26] were specifically allocated to the prior and convent by this charter. Moreover, there is no evidence that the pre-1109 grange of Brame ever passed into episcopal hands. Some central administrative body must have been necessary for these possessions and for the external priorial manors: it could readily have given rise to a local court, if none existed, as soon as town property began to be acquired by the monks. Many transfers of property, by grant, exchange, or purchase, had certainly occurred by the 13th cen-

[6] See below.

[7] The ancient market day was changed from Saturday to Thursday in 1801, on petition of the citizens to the bishop, who granted it after consultation with the Chief Justice (Bp.'s Mun., Ely, A. 5).

[8] Information given by Clerk to Ely U.D.C.

[9] Gardner, *Dir. Cambs.* (1851), 480.

[10] Each of ¼ or ½ acre. MS. Accts. of Parsons' Char.

[11] Ibid.

[12] *Kelly's Dir.* (1900).

[13] Embracing an area of 16,625 acres, exclusive of water areas.

[14] Brewing was an outstanding feature of Ely as early as the 14th century—'Cervyse de Ely' (*E.H.R.* xvi, 502).

[15] About 2 miles out of the city, at Queen Adelaide.

[16] M.H., No. 76940. *Gained*: from Witchford, 42 acres (pop. nil.). *Lost*: to Witchford, 204 acres (pop. 3); to Witcham, 399 acres (pop. 21); to Downham, 829 acres (pop. 11); to Littleport, 460 acres (pop. 23); to Little Thetford, 233 acres (pop. 29).

[17] Ibid.

[18] M.H., No. 77445.

[19] *Kelly's Dir.* (1950).

[20] The Corn Exchange Co. had only leased these rights, by arrangement with Chas. Newberry, bailiff of the markets by grant of Bishop Turton.

[21] On the death in 1933, of the last bailiff, Harold Archer. The fairs had a separate bailiff. (Infm. Lt.-Col. G. L. Archer, Ely.)

[22] The Census of 1851 showed a small part of West Fen, formerly in Littleport, as transferred to the lands common to both Ely parishes; areas formerly in Sutton and Witcham as included in Ely Trinity. Under the Divided Parishes Acts of 1876 and 1879 a small part of Prickwillow hamlet was transferred to Littleport, the remaining part being included in the joint lands of Holy Trinity and Ely St. Mary.

[23] See above, p. 34.

[24] Bp.'s Mun., Ely, Liber M., f. 143; see also n. 25.

[25] There is no mention of priorial lands in Ely in the charters of endowment cited in Dugdale, *Mon.* i, 462; Liber M., ff. 143–4; B.M. Cott. MS. Tib. A. VI, ff. 111–12; Titus A. I, ff. 35–35*b*.

[26] *P. N. Cambs.* (E.P.N.S.), 217, 219, 222. These granges are listed as in the jurisdiction of the Sacrist in the Award of 1417 (*Cal. Pat.* 1416–22, 183).

tury, and probably began earlier.[27] A manor called Brayes manor was in conventual hands not later than the 14th century: its gateway almost fronted the Porta Monachorum.[28] Possibly priorial courts were held in this gatehouse before the Ely Porta was built late in that century.[29] Whatever the date of the formal constitution of the monastic manor, it clearly existed by 1320.[30] Priory courts seem to have been held in the Ely Porta by 1423:[31] the name *ELY PORTA* manor appears in the 16th century.[32] The manorial courts of the dean and chapter were held here down to 1865.

The farm-buildings of the bishop's manor called *ELY BARTON* manor opened upon Back Hill, in 1416:[33] its fields extended westward. The bishop's vineyard lay on the eastern side of the city, approached from the market-place. In the built-up areas, by 1416,[34] the city properties of bishop and prior were much intermingled. Ely Barton was known in later centuries as the 'paramount manor',[35] and must have held pride of place from the outset. The extensive wastes appurtenant to the manor were claimed exclusively by the bishop, save where intercommoning rights of neighbouring townships had to be respected. Such sections of the waste as were enjoyed by the prior were granted, from time to time, by the bishop, except such rights as already pertained to Stuntney and the monastic demesne farms.[36] Barton farm was held on lease by the bishop's bailiff in the 15th century. At the Dissolution Nicholas Steward leased the manor.[37] It was sold to Richard Cromwell, the Protector's son, at the close of the Civil War, but later returned into episcopal hands and was henceforward granted on lease.[38]

Constant nibbling at the waste occurred from early times. Thus inclosed or semi-inclosed farms or submanors came successively into existence. Some were ancient monastic properties and remained so until the Dissolution: others were episcopal demesne farms.

BRAHAM or *BRAME*. The farm, or fishery, of 'Bramewere' appears as early as 1086.[39] Between 1166 and 1175 Salomon the Goldsmith acquired a rent-charge on the estate.[40] It figured in the extent of 1251, held by another Salomon.[41] Prior Fressingfeld eventually secured the property for the monastery early in the 14th century.[42] In 1417 Brame was held by the sacrist. Inclosure disputes arose here in the 16th century.[43] At the Dissolution Brame was granted to the dean and chapter, who thenceforward retained it.[44] Brame lay in St. Mary's parish.

The farm or hamlet of *TURBUTSEY* was a very early possession of the church[45] and was among those granted to the monks by the charter of 1109. In the arbitration award of 1417 it appears as one of the seven granges or manors, near to Ely, over which the sacrist had special jurisdiction: the others were Brame, Stuntney, Thorney, Northney, Quaveney, and Shepey. Turbutsey supplied a considerable part of the sacrist's food in the 14th century.[46] Inclosure disputes, especially over rights of way, arose in the 16th century.[47] The estate was granted to the dean and chapter at the Dissolution and repeatedly appears among the chapter leases subsequently.[48] It was called a dairy in the 16th and 17th centuries[49] and had fishing-rights, an ancient quay, a willow-holt, a kiln for brick- or tile-making, and a chapel. The remains of brick-pits and old brick buildings, near the present Thistle Corner, probably mark the site of what was known in the 16th century as 'Little Turbutsey'. Turbutsey manor house was pulled down to make room for the Ely beet factory in 1925. The remains of a tile-baking furnace, of Tudor or Jacobean date, were exposed at the same time.[50] The estate was in the parish of Holy Trinity.

SHIPPEA, allocated to the Convent in 1109, lay in Holy Trinity parish. Its name occurs frequently in the sacrist's rolls and, after the Dissolution, among the chapter leases.[51]

The farm of *THORNEY*, lying in a detached part of Holy Trinity parish, was in the possession of the monastery as early as 1206.[52] In 1527 it was inclosed for dairy-farming and passed to the Crown at the Dissolution, being then let on lease to a local man, William Clay.[53] It was sold in the later 16th century,[54] but was apparently recovered by the dean and chapter before 1649.[55] In the early 19th century it was the freehold property of Harry Spencer Waddington.[56]

NORNEA farm also was a detached part of Holy Trinity parish. It was mentioned in 1279 and subse-

27 B.M. Cott. MS. Claud. C. XI, esp. ff. 25*b*–29*b*, which give the details of present and past holders of 'free' tenements; cf. above, p. 35.

28 *Sacrist R.*, ed. F. R. Chapman, i, 145–6; Hist. MSS. Comm. *6th Rep.* 292 (Leconfield MSS., ff. 16*b* and 17) (1313). Stewart says that the Brayes manor was absorbed by the larger manor of Ely Porta. Stewart, *Archit. Hist. of Ely*, 192*c*.

29 It seems a traditional arrangement: Stuntney was granted to the abbey at a court held in one of the gateways in the time of King Edgar (*Liber Elien.* 133–7).

30 Perhaps by 1229: 'By inquisition taken in 1229, the Prior and Convent had a prison, a court and vineyard in the close of the priory' (Gardner, *Dir. Cambs.* (1851), 467); Ely Treas. R., 12–13, Edw. II, but no such reference in Treas. R. for 9–10 or 10–11 Edw. II. Mr. E. Miller suggests that this may mean that the conventual manor was formally constituted about 12 Edw. II.

31 Court 'ad Portas Prioratus' (Treas. R., Ely, 1423–4).

32 The name was used in 1526. In the 17th century both monthly manorial courts of D. & C. and Courts held by the surveyor of all the chapter manors were held in the Ely Porta. The Survey of 1649 calls the Ely Porta 'the Manor House' (Atkinson, *Archit. Hist. Ely* (1933), 153, n. 2).

33 Survey of 1416–17 (B.M. Harl. MS. 329).

34 Ibid.

35 D. and S. Lysons, *Magna Brit.* ii (1), 191.

36 'The soil of the wastes and marshes of Ely shall pertain to the Bishop as parcel of his manor of Ely, because nothing was said of them by the old separation' (*Cal. Pat.* 1416–22, 185). The extent of 1251 B.M. Cott. MS. Claud. C. XI, ff. 24*b*–25*a*) also assumes the episcopal claims to the wastes (above, p. 35).

37 W. M. Palmer, 'Enclosures at Ely . . . 1548', *Cambs. and Hunts. Arch. Soc. Trans.* v, 369–84.

38 Lysons, *Magna Brit.* ii (1), 192; Gardner, *Dir. Cambs.* (1851), 477.

39 *P. N. Cambs.* (E.P.N.S.), 217.

40 *Sacrist R.*, ed. F. R. Chapman, i, Appx. C. 151 et seq.

41 In 1280, in consideration of the bad roads, the bishop, with consent of prior and convent, permitted a chaplain, answerable to the sacrist, to conduct daily offices in the oratory at Brame. The property was then held by two knights, More and Waleham (Hist. MSS. Comm., *6th Rep.* (Leconfield MSS.), 291).

42 *Sacrist R.*, ed. F. R. Chapman, i, 155–6.

43 *Cambs. and Hunts. Arch. Soc. Trans.* v, 369–84.

44 Lysons, *Magna Brit.* ii (1), 192.

45 There are many refs. in *Liber Elien.* (*P. N. Cambs.* (E.P.N.S.), 319).

46 *Sacrist R.*, ed. F. R. Chapman, i, 120.

47 *Cambs. and Hunts. Arch. Soc. Trans.* v, 369–84.

48 *Leases and Pattents viewed at the Court of Survey, belonging to the Deanery of Ely* (1649), 33–34 (Ely Cath. Lib. MSS.).

49 *Cambs. and Hunts. Arch. Soc. Trans.* v, 369–84.

50 Ibid. 383–4; G. Fowler, 'Fenland Waterways Past and Present, Pt. II' (*Cambs. Antiq. Soc. Proc.*, vol. xxxiv, 25).

51 Lysons, *Magna Brit.* ii (1), 193; J. H. Clements, *Hist. of Cambs.* (1868), 60.

52 *P. N. Cambs.* (E.P.N.S.), 221.

53 *Cambs. and Hunts. Arch. Soc. Trans.* v, 369–84.

54 Ibid.

55 *Leases and Pattents* (1649), 33–34 (Ely Cath. Lib. MSS.).

56 Gardner, *Dir. Cambs.* (1841), 476; Lysons, *Magna Brit.* ii (1), 193.

quently, as a possession of the monks,[57] and was granted to the dean and chapter at the Dissolution.[58] It figured among the 16th-century dairy inclosures,[59] leased by John Croplaye of Ely. It was still a leasehold dairy property of the chapter in the 19th century.[60]

QUANEA was another outlying portion of Holy Trinity parish. It was monastic property in 1279,[61] and a valued perquisite of the sacrist in 1335.[62] It appeared in the award of 1417. Quaney hithe was the farthest downstream of the medieval city quays. At the Dissolution Quaney was granted to the dean and chapter. Early in the 19th century it was held on a chapter lease by the son of Bentham, historian of Ely.[63] A very ancient hall survived as a farmhouse until the mid-19th century. A portion of the original farm had by that date been sold; the remainder was let, partly on lease and partly as copyhold of Stuntney manor.[64]

BEALD FARM lay in the intermixed lands of the two parishes. It was allotted to the prior in 1109 and is frequently mentioned in 12th and 13th century documents.[65] It passed to the dean and chapter at the Dissolution. Sixty acres were directly inclosed by that body in the 16th century.[66] The farm remained a chapter estate.

BEDWELL HAY FARM, in St. Mary's parish, belonged to the monastery in 1302. It frequently appears in later documents and maps.[67] In 1548 it had become episcopal property, let on lease. The inclosure of 120 acres of the farm was a subject of complaint at this date.[68]

The almoner, in 1251, held scattered city tenements:[69] by 1327 *ALMONRY GRANGE*, a fair-sized property, was attached to the office.[70] In 1548 Almoner Cotes was accused of having pulled down the manor house and burnt a cottage, situated on the 100 acres which he had inclosed prior to the Dissolution,[71] at which date the estate passed to the dean and chapter. It remained in their hands.

KETONS[72] manor-farm lay in Holy Trinity parish. John of Keton was almoner before his selection as bishop in 1310.[73] Ten years later, among the fixed charges on the almonry, was a rent paid to Simon de Keten.[74] Probably ties of kinship had secured this lease, which continued to be held by Simon's heirs. Some time after 1362[75] the cellarer bought the manor of 'Ketens' with money received from Denny Abbey. The main building of the manor stood opposite the Ely Porta, in what was called Walpole Lane (Silver Street). The estate passed to the dean and chapter at the Dissolution and remained in their possession. In 1548[76] the farm was in the hands of the heirs of Thomas Rydley: Prior Walpole had previously converted to pasture two closes, called 'Orwell Pylk closes', and the 'mansion house' had become a barn. The modern Orwell Pit farm probably occupies the site of the old manor.

The demesne farm of *NEW BARNS*, in the parish of Holy Trinity, lay to the north-east of the city. It came into prominence in the Inclosure Inquiry of 1548 and was then held by Thomas Goodrich.[77] It figured as a valuable episcopal property of 700 acres in Bishop Cox's award of 1566, under which 200 acres were to be separately inclosed and let to representative citizens for thirty years. Presumably these acres reverted to the bishop at the end of that time and gave rise to the smaller estate known as Little New Barns.[78] The farm was sold at the close of the Civil War.[79] It was apparently bought back, for Great New Barns and Little New Barns were severally leased from the bishop, in 1799, by Edmund Tatersall, the horse-breeder. Both estates continued to be episcopal property.[80]

The hamlet of *STUNTNEY*, in the former Holy Trinity parish, was appendant to Ely long before 1087, when it was recorded as a berewick.[81] It was given to the secular clerks of Ely by Wolstan de Delham, in King Edgar's reign.[82] The eel-fishery was always highly valued, as was the strategic importance of Stuntney, a major entrance to the Isle. The upkeep of this passage was a constant care of the monastery.[83] On the creation of the see, Stuntney was granted to the priory and was appropriated to the cellarer's office.[84] Later it was transferred to the sacrist.[85] Stuntney was farmed, in 1527, by Edward Bestney of Soham, who did a considerable amount of inclosing.[86] In 1540 the manor was granted by the Crown to Sir Edward North;[87] thenceforward it held its own courts. Bestney's daughter and heiress married Simon Steward, brother of the first Dean of Ely; hence Stuntney passed to the Steward family. Edward, son of Simon, held the manor on a seven-year lease in 1548;[88] subsequently Sir Mark Steward settled there.[89] Oliver Cromwell inherited the estate in 1636. The manor passed out of the family's ownership in 1723 and repeatedly changed hands later.

The hamlet of *CHETTISHAM*, in the former parish of Ely St. Mary, first occurs in 1170,[90] and appears as a wooded assart in 1251.[91] In the late 15th century the land was held by the lessee of New Barns.[92] Bishop

57 *P. N. Cambs.* (E.P.N.S.), 219.
58 *Leases and Pattents* (1649), 27.
59 *Cambs. and Hunts. Arch. Soc. Trans.* v, 373–84.
60 Lysons, *Magna Brit.* ii (1), 193.
61 *P. N. Cambs.* (E.P.N.S.), 220.
62 D. and C. Mun., Sacrist R. 1335.
63 Lysons, *Magna Brit.* ii (1), 193; J. H. Clements, *Hist. of Ely and the neighbouring Villages*, 60.
64 Lysons, *Magna Brit.* ii (1), 193.
65 *P. N. Cambs.* (E.P.N.S.), 223.
66 *Cambs. and Hunts. Arch. Soc. Trans.* v, 369–84.
67 *P. N. Cambs.* (E.P.N.S.), 217.
68 *Cambs. and Hunts. Arch. Soc. Trans.* v, 369–84.
69 See above, p. 36.
70 Almoner's R., 1–2 Ed. III. The sacrist also had a city grange (*Sacrist R.*, ed. F. R. Chapman, i, 120).
71 *Cambs. and Hunts. Arch. Soc. Trans.* v, 369–84.
72 Mr. E. Miller supplied various refs.
73 B.M. Add. MS. 41612, f. 48.
74 B.M. Cott. MS. Vesp. A. VI, f. 120.
75 By 1374 the rent was paid to the cellarer for 4 ac., once belonging to Sir Henry Walpole and afterwards to the Ketenes (Almoner's R., 1374). A rent charge on Denny Abbey, devoted to hospitality for guests at feasts of St. Etheldreda and St. Luke, was sold to the abbey by the prior and convent of Ely (cf. Stewart, *Archit. Hist. of Ely*, 201, quoting Lambeth MSS. 448).
76 *Cambs. and Hunts. Arch. Soc. Trans.* v, 369–84.
77 Ibid.
78 See above, p. 41.
79 For £2,805—twice the sum paid for Barton Farm (Gardner, *Dir. Cambs.* (1841), 476–7).
80 See n. 79.
81 See above, p. 35.
82 *Liber Elien.* ii, c. 24.
83 A duty of the sacrist (Bp.'s Mun., Liber M., f. 184).
84 Wharton, *Angl. Sac.* i, 617; *Cal. Chart. R.*, 1226–57, 391 (1252); claim upheld, Award of 1417 (*Cal. Pat.* 1416–22, 185).
85 It frequently appears in Sacrist R. of 14th century, e.g. 33 Ed. 3.
86 *Cambs. and Hunts. Arch. Soc. Trans.* v, 369–84.
87 *Magna Brit.* ii (1), 198.
88 As for n. 86.
89 Probably Edward's brother, as he was also a nephew of the first Dean. Simon Steward was great-uncle of Oliver Cromwell (*Cambs. and Hunts. Arch. Soc. Trans.* v, 369–84). The Stuards, or Stewards, were descendants of a Scots family of Stewards of Dundevale. Dr. Steward was last prior and first dean of Ely. The family continued to live in the neighbourhood until mid-18th century (Bentham, *Ely*, 2nd ed., 190).
90 *P. N. Cambs.* (E.P.N.S.), 217.
91 Bp.'s Mun., G. 3 ('Old Coucher', f. 1), Caius Coll. MSS. 485/489, f. 19. Mr. Reaney's etymological problem seems thus answered; 'Checham bushes', in 1606 (E 134/4, Jas. 1/M, 14); and *Leases and Pattents* (1649) (Ely Cath. Lib. MSS.).
92 *Cambs. and Hunts. Arch. Soc. Trans.* v, 369–84.

West inclosed a park of 180 acres in 'Chettesham busshes', about 1515, and aroused deep resentment among former commoners. Other smaller inclosures in the hamlet were made by episcopal lessees at this period.[93] The wooded parkland was depicted in the Ordnance Map of 1835.

FAIRS[94] Henry I granted a fair to the abbot and convent for 7 days, beginning 3 days before the feast of St. Etheldreda;[95] in 1312 a fair was granted to the prior for 15 days at the festival at St. Lambert;[96] in 1318 a fair was granted to the bishop for 22 days, beginning on the Vigil of the Ascension.[97] Throughout the Middle Ages the fairs were marts of great activity, particularly that of St. Etheldreda. Booths were erected all round the precinct walls, at the gateways, in the streets, and on the wharves. By the 15th century agents of great commercial firms bargained here over the sale of iron and timber.[98] Townspeople took part in the lively scene. St. Audrey's ribbons, held in veneration even in the 16th century as having touched the shrine of St. Etheldreda, were in wide demand.[99] So profitable were the fairs to the grantee that a jealous eye was kept lest any other fair should queer the pitch.[1] In 1542 two annual fairs at Lynn were abolished, because the regrating of salt fish there was held to be detrimental to the fairs at Ely and elsewhere.[2] At the end of this century[3] the fairs were still an important source of revenue to the bishop, who appointed the special bailiff of the fairs. In 1665 they were prohibited owing to the Great Plague.[4] St. Lambert's fair had disappeared by the later 18th century, and the dates of the other two had been changed.[5] They were rapidly declining in importance by this period and were held for only a few days annually. The two fairs are still held in May and October, now under the aegis of the Urban District Council.[6]

CATHEDRAL *Preliminaries.* A religious house was founded on the present site by Etheldreda, a daughter of Anna King of East Anglia, in 673. The monastery was laid waste by the Danes in 870, refounded by Ethelwold, Bishop of Winchester, as a Benedictine abbey in 970 and dedicated to St. Peter and the Blessed Virgin by Dunstan in 974.[7] The building of the present church and monastic buildings was begun under Simeon the first Norman abbot. Bishop Northwold rededicated the church on the completion of a new presbytery about 1250 in honour of the Virgin, St. Peter, and St. Etheldreda.[8] At the time of the Dissolution the dedication was St. Peter and St. Etheldreda when it was changed to the Holy and Undivided Trinity.[9] An indenture between Henry VIII and the bishop has 'the convent of St. Audre's in Ely'.[10] The House was surrendered to the king on 18 November 1539. The New Foundation consisted of a dean, 8 prebendaries, 8 minor canons, deacon and sub-deacon, 24 scholars, 2 schoolmasters, an organist, and various singing men and boys; the prior Robert Wells or Steward became dean; the Almonry School was refounded as the 'Kings School' and still continues. This establishment continued, with the break from 1649 to 1661, with little change until the 19th century. By the Cathedrals Act (1840) the number of canonries was reduced to 6, and in the last few years has been reduced to 4. One canonry is annexed to the Ely Professorship of Divinity in the University of Cambridge, and one of the lapsed canonries was annexed to the Regius Professorship of Hebrew. There are now 24 honorary canons.

No fragment of the pre-Conquest church has been found but it is suggested on the evidence of early graves that it occupied the east part of the present nave: the bodies of the early abbesses were removed to make way for the new building in 1103, that is twenty-one years after its commencement. Definite architectural history opens with Simeon's rebuilding. Simeon had been prior of Winchester and was brother of Walkelin the bishop. The brothers were related to the Conqueror. Simeon's connexion with Winchester is important, for it influenced the lay-out of his own church, particularly in the two towers on the main axis and in the aisles of the main transept; on the other hand, the churches differ in the important matter of the treatment of the east end. It seems clear that the two are from the hand of the same master mason.

The monastic buildings lie on three sides of the church. The cloister and all the more important buildings are on the south, but one important guest house lay to the east while the almonry with its school and the workshops in the charge of the sacrist were on the north in Stepil Row now High Street: obviously a convenient situation being easy of access by mendicants, workmen, and merchants. The monks' cemetery was to the south of the presbytery and the parish cemetery to the north of the presbytery and nave. To the west of the claustral buildings and separated from them by a roadway spanned by a covered bridge was the Bishop's Palace. The great building periods for church and monastery were that immediately following the Conquest, the second quarter of the 13th century, and the second quarter of the 14th. Much less important was a short period on either side of the year 1500 which saw the rebuilding of the palace and half the cloister and within the church some very notable monuments. Until 1770 the medieval ritual arrangement remained almost intact.

THE ROMANESQUE CHURCH. Simeon was already 80 when in 1081 he became abbot and began the rebuilding. He continued the work until his death in 1093. The abbacy was vacant until the appointment of his successor Richard in 1100; Richard continued the work until his death in 1107. The progress made by either of these abbots is uncertain, but it was necessary to disturb the graves of the three successors of St. Etheldreda in 1102; the church was ready for her ceremonial Translation on 17 October 1106. The building itself is so uniform in character right up to the west end of the nave that it gives us no clue as to the rate of progress but we may suppose that that part was finished about 1170 and the west tower and transept about 1200. The plan was in several respects peculiar. It had

[93] As for n. 92. Bishop West sat on the Anti-Inclosure Commission in Cambridge, 1517.
[94] For the history of the market see above, p. 47.
[95] B.M. Add. MS. 5809, f. 97.
[96] *Cal. Chart. R.*, 1300–26, 185.
[97] Ibid., 403.
[98] D. and C. Mun., Sacrist R., 18 Ed. IV (1478–9); Stewart, *Archit. Hist. Ely Cath.* 191.
[99] A student of Clare Hall, Camb., seeking the favour of Thos. Cromwell, in 1533, sent him 'some hallowed Tawdreye laces' (*L. and P. Hen. VIII*, vi, 516). See also *Oxf. Engl. Dict.* s.v. tawdry lace.
[1] Matt. Paris, *Chron. Majora* (Rolls Ser.), v, 29.
[2] 33 Hen. VIII, c. 34.
[3] S.C. 6/Eliz./282 m. 1 (View of Acct.).
[4] Bentham, *Ely*, 131 n.
[5] Lysons, *Magna Brit.* (ed. 1810), ii (1), 189.
[6] Above, p. 47, n. 21.
[7] Stewart, *Archit. Hist. of Ely Cath.* 13, quoting *Liber Elien.* ii, c. 52.
[8] Stewart, op. cit., 68, quoting *Anglia Sacra*, i, 636.
[9] Bentham, *Ely*, 225.
[10] *L. and P. Hen. VIII*, xiii (2), p. 377.

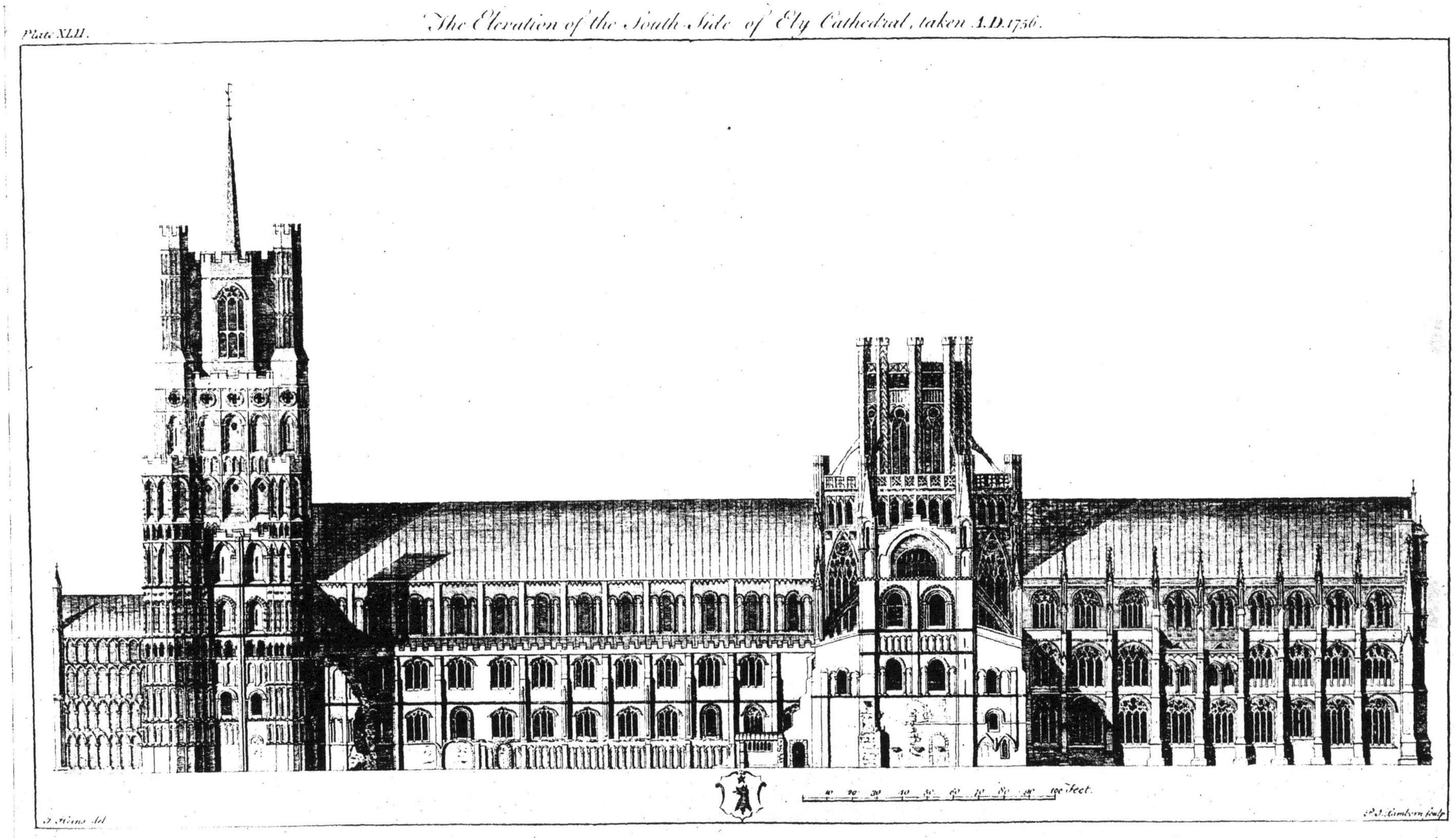

Ely Cathedral. South Side
(From Bentham's *Ely*)

Plate XLIII.

Ely Cathedral. Longitudinal Section, 1762

(From Bentham's *Ely*)

two steeples on the main axis, a central and a western, both flanked by transepts. The aisle was carried across the north and south ends of the main transept. In these respects the church followed the plan of Winchester of which house Simeon had been prior. It was to have been like Winchester in having an eastern apse but the idea was abandoned.

Bishopric of Ely. *Gules three crowns or.* (Seal of Bishop William of Louth (1290–8), S.C. 13/F 14.)

There seems to have been from the very first a detached bell tower. It was perhaps looked upon as a temporary expedient to provide for the period until either the central or western tower could be finished. As early as 1110[11] the tower over a 'porta' was struck by lightning.[12] The gateway and tower in question seem to have formed the entrance to the lay-folk's cemetery and thence to the church from the main street on the north.[13] Probably the whole was of timber.

The plan of the eastern termination as built is doubtful. Foundations uncovered in 1850 led Professor Willis, who examined and recorded them,[14] to think that the central division was intended to end with an apse the foundations of which were clear, and that the change to a square end was made at once, additional foundations being laid. The terminations of the aisles were still more doubtful. We may perhaps assume that the first plan was to have a central apse showing the round outside as well as within, and side apses showing square outside, a common English plan. Thus if the apse had been built it would have been very different from Winchester. The easternmost bay whether apsidal or square was deep, the length from east to west being nearly equal to the width. It was separated from the next bay by two half round columns which have survived all the alterations of later times. Between these columns and the central tower there were four bays. West of the tower there was a nave of thirteen bays with alternate compound and simple piers. The western tower is flanked by transepts with eastern apses, and a large decagonal turret at each angle. There is a large two-storied west porch called the Galilee. The aisles were continued on the east and west sides of the main transept and were intended to be returned across the north and south ends as at Winchester. But here again there was a change of intention or else a very early alteration. Instead of a continuation of the wide triforium gallery there are quite narrow bridges to connect the triforia on the two sides.[15]

The ritual arrangement of the Romanesque church was as follows. The high altar stood presumably on the chord of the apse-foundation. The bishop's throne (from 1109) may have occupied the primitive position behind the altar as it does at Norwich. But it must be remembered that at Ely the bishop has now no throne; he occupies the stall of the abbot on the south side of the quire door, the dean having the stall on the north side, formerly the prior's, and in other monastic cathedrals the sub-priors. This is an ancient, possibly the original, arrangement.[16]

The Lady Chapel was at the east end of the south aisle of the presbytery. The monks' quire was under the Tower and extended one bay beyond in both directions. It was enclosed by stone walls against which the stalls were placed. The northern of these walls contained in later times, and doubtless from the first, small sealed cavities in which were placed the bones of Saxon worthies. The eastern aisles of the transepts were divided by thick walls to form eight chapels. The west aisle of the north transept had a doorway in its north wall and a stone seat runs along the west wall; we may believe that this aisle formed an entry or cloakroom for pilgrims (cf. Winchester). The west aisle of the south transept was enclosed by walls rising to the arches and forming a sacristy (cf. Winchester and Peterborough). It had a doorway direct into the cloister, now blocked but visible, and one into the church, now destroyed. In the south wall of the transept there is a doorway, now blocked, into the monks' cemetery.

The bay west of the quire was enclosed on the west by a wall forming the pulpitum which separated the monks' quire from the nave forming the parish church. The pulpitum stood until near the end of the 18th century and some sketches of it have been preserved.[17] The screen wall was continued across the south aisle and presumably across the north aisle. The enclosed space was opposite the north-east corner bay of the cloister and into it was the doorway from the cloister. In the seventh bay west of the pulpitum was the second doorway for the return of the Sunday procession; it opened from the north-west corner of the Norman cloister. In the south wall of the west transept there is a doorway to a staircase leading to the upper parts of the transept and thence to the west tower. Another doorway at a higher level communicated with the raised 'gallery' or passage to the palace.

The elevational treatment of almost the whole of the Romanesque church can be recovered with fair certainty. There is much doubt, however, about the east wall. The flanks of the nave and east transept appear to have been the same in general arrangement and to have differed only in details such as mouldings. Externally

[11] Or 1111. S. Inskip Ladds, *Mon. of Ely*, p. 34, quoting *Ann. Mon. de Winton*, in *Ann. Mon.* (Rolls Ser.), ii, 44. 'Porta' is the word used for an outer gate.

[12] Stewart, *Archit. Hist. of Ely Cath.* 40, quoting *Liber Elien.* iii, c. 28, and *Anglia Sacra*, i, 617.

[13] See below, p. 81.

[14] Stewart, *Archit. Hist. of Ely Cath.* 24.

[15] Stewart (op. cit. 31) says that the evidences of the cross aisle having abutted on the side aisles were formerly clear, particularly in the south transept; it would seem that these evidences were removed within his own memory. We must accept this clear statement. But it is possible that what Stewart saw was the 'toothing' formed in the walls of the side aisles for bonding in the cross aisle at a later but not distant time. The architectural evidence points the other way; it is as follows. The alternate arrangement of the simple and compound piers in the nave is seen in the south transept; now the tower piers would of necessity be compound and so also, for preference, would be the responds against the south wall. The three intervening piers could therefore be alternate. If the cross aisle had been built its arches would have sprung awkwardly from round columns. We may therefore conclude that when this transept was built the idea of a south aisle had been abandoned. In the north transept the architectural evidence agrees with Stewart. The northernmost pier is compound like the respond; the two remaining piers are simple. The alternate arrangement had not been adopted; but the north pier was suitable for the arches of the cross aisle and these were probably built. This conclusion is supported by a slight kink in the shaft against the north-east pier rising from floor to roof. This shaft would be interrupted by the wall of the cross-aisle; when that was removed it was found that the upper part was not exactly over the lower part; thus the length of shaft to be built to connect the two had to be built slanting.

[16] For a fragment, possibly of the throne, see below, p. 59.

[17] See below, p. 66.

KEY TO THE PLAN OPPOSITE

1. Bishop Moore (1707–14).
2, 3. Brass indents.
4. Bishop Allen (1836–45) (monument).
5. Bishop Butts (1738–48).
6–8. Brass indents.
9. Sir Mark Steward (d. 1603).
10. Bishop Greene (1723–38).
11. Sir Robert Steward (d. 1570).
12. Dean Tyndall (1591–1614).
13. Brass indent.
14. Bishop Goodrich (1534–54).
15. Bishop Heton (1600–9).
16–20. Brass indents.
21. Robert Steward or Wells (Prior 1522–39, Dean 1541–57).
22. Bishop Gunning (1675–84) (grave slab).
23. Bishop Gunning (monument).
24. Bishop West (1515–33) (grave slab, removed from his chapel in 1885).
25. Canon Selwyn (d. 1879) (on site of entrance to Reliquary).
26. Altar of Bishop West's Chapel.
27. Canon Mill (d. 1855).
28. Bishop Allen (1836–45) (grave slab).
29. Bishop Niel (1133–69).
30. Pre-Reformation chest.
31. Indent of brass of Bishop Gray (1454–78).
32. Indent of brass of Sir William Thorpe.
33. Bishop Laney (1667–75).
34. Bishop Mawson (1754–71).
35. Bishop Patrick (1691–1707).
36. Bishop Chase (1905–24).
37. Brass of George Basevi (d. 1845); early medieval grave slab.
38. Canon Fleetwood (d. 1737).
39. Bishop Fleetwood (1714–23); grave slab of Thomas Stuart (d. 1744) on floor below.
40. Dean Caesar or Adelmare (1614–36).
41. Brass indent, probably Prior Walsingham (1341–64).
42. Bishop Hotham (1316–37) (restored brass).
43. Prior Crauden (1321–41) (restored brass).
44. Bishop Kilkenny (1254–7) (heart burial).
45. Bishop Barnet (1366–74).
46. ? Private Pew of Queen Philippa, enclosing fragments of Shrine of St. Etheldreda (formerly at 39).
47. John Tiptoft, Earl of Worcester (d. 1470) and two of his wives (cenotaph).
48. Bishop Northwold (1229–54).
49. Tomb of Bishop Hotham (1316–37) (removed from 42).

CATHEDRAL CHURCH OF THE HOLY AND UNDIVIDED TRINITY, ELY

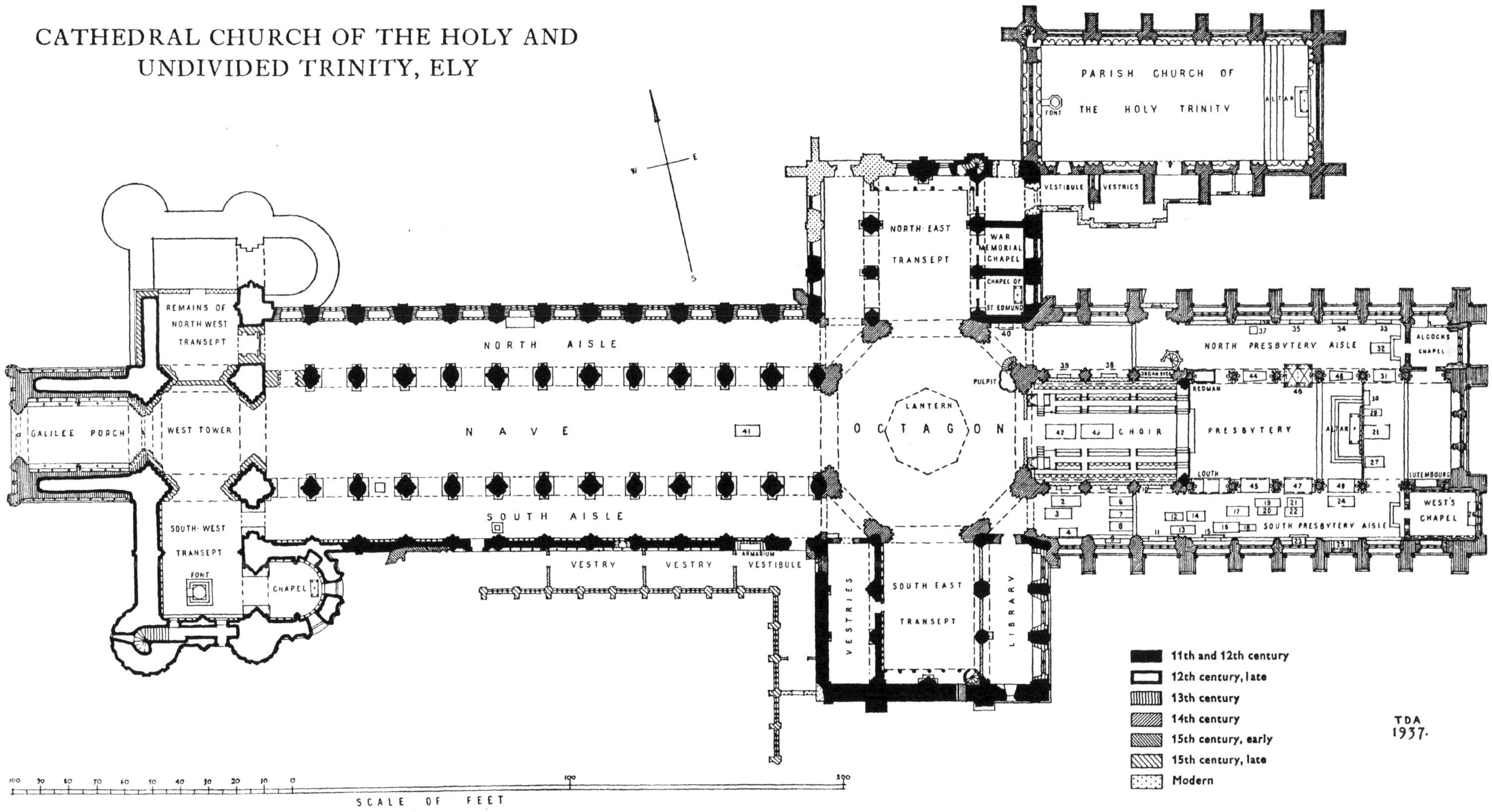

the design of the aisle windows is preserved in two examples in the south aisle which have shafted jambs; a window in the end of the south-east transept was doubtless the same before mutilation. The triforium bay had three equal arches on shafts, the central one being pierced as a large window; these are preserved in the east wall of the north-east transept. Immediately above the triforium windows is an arched corbel table carrying the parapet; it is stopped against the flat pilasters which divide the bays. The clerestory is preserved in the nave and transept. In each bay a wide window with shafted jambs is flanked by narrow blind arches also with shafts. A little above these is a corbel table and parapet similar to the aisle wall. There can be little doubt that the side walls of the presbytery were of the same general design and detail.

The transept ends were evidently alike except that in the south transept all the lowest windows apart from that of the east aisle were blocked by a building which abutted against the western part of the transept end: the one window of the east aisle was evidently the same as the windows of the aisles generally. The side windows and arches of the triforium were also repeated at the ends of the aisles. The same design is carried across the central part to a larger scale in two bays; above them is a row of small blind arches; the two large clerestory windows have two shafts in each jamb; immediately above these was the gable, but alterations at both ends have entirely effaced the original design. The central part of each transept end is separated from the aisles by square turrets, the eastern containing stairs. The upper parts of the four turrets vary; speaking generally the square becomes near to a round at the level of the triforium parapet and a little higher an octagon with an octagonal spire; these two top stages are arcaded. Of the central tower we know only that it contained bells, for they are mentioned in a letter at the time of the Dissolution. We may safely conclude that below the bell chamber and ringing chamber there was a stage with windows giving light to the quire.

The doorway to the vestry has been blocked and partly destroyed; the remains show a jamb-shaft and arch similar in section to the jamb, both with very rich scroll-work arranged between spiral bands; the keystone of the arch is carved with a human head; the capital is of cushion form carved with foliage. Above this arch is another which it superseded, carved with a slight and delicate enrichment. The eastern doorway from the cloister has a round head with cusps, the spandrels carved with kneeling figures holding crosiers. The arch is of four orders two of which are carved with scroll-work, one is a spiral with a head as keystone and one is moulded. The jamb has two nook shafts, the outer one plain and the inner a spiral. The flat faces of the jamb are carved with scroll-work. The western doorway is square-headed. Its lintel is on corbels from which human heads project. It is surmounted by a tympanum sculptured with a Majesty, under an arch of two orders. The inner arch is adorned with foliage arranged in spiral, and the outer with shallow carving standing on a grotesque, now decayed, and carved with spiral foliage like the arch above it. The innermost square part of the jamb is carved with a running scroll. The outer square has a succession of circles containing satirical subjects and finished at the top with a representation of a building; it carries a grotesque monster; the capital of the shaft is of cushion form carved with foliage, the abacus with a monster.

The west front is extremely rich. The transept is the same height as the nave; the west wall is of two bays and has two rows of windows corresponding with the triforium and clerestory windows. The wall below the lower windows, corresponding to the pier arches of the nave, is divided into three tiers, the lowest plain and the two upper with blind arcades. The lower windows are large and have a narrow blind arch on each side; the wall surfaces are diapered. Between these windows and the upper row there is another blind arcade. The top stage is the same height as the clerestory and corresponds closely with it in design, making allowance for later and richer detail. The arches are pointed and the jamb-shafts have three annulets. Each spandrel has a sunk quatrefoil. The parapet is carried on an arched corbel table; the parapet on the east side may be the original one, but that on the west side is 14th or 15th century.

The two lowest of the six tiers on the west wall are carried round the angle turrets and across the south end which is in two bays. There are slight variations in the upper parts of the south end; the lower windows are more lofty than the west windows and their sills are at a lower level; the windows of the turrets are necessarily narrower or are omitted. But the general character is preserved, most of the string-courses are continuous and the important quatrefoils are carried round. There is a third pair of windows in the transept end which lighted the space above the ceiling. The wall probably ended with a horizontal parapet level with the top of the roof. The top windows are repeated as blind arches in the turrets with rather greater height; those in the larger west turret have round arches; the east turret with shorter sides has pointed arches. The turrets are continued up for another two stages with slight variations between the two turrets. The lower stage is a small blind arcade, the upper has a large unglazed lancet with a pointed arch in each face. Above these there is an arched corbel table to carry the parapet. The present parapet is 14th and 15th century. The turrets have been already called decagonal.[18] Strictly speaking, however, their plan is twenty-sided. Each of the twenty sides forms an obtuse salient angle with the next on the one hand and a re-entering angle with the next on the other. In each re-entering angle there is a shaft standing on a pilaster in the lowest stage. These shafts are carried up passing in front of the large arches in the top stage but three. At the sill level of these arches the salient angles are pushed back making the angle more obtuse and obliterating the re-entering angle; the turret therefore becomes ten-sided. The shafts stop at this level in the east turret but are continued up another two stages in the west turret. The eastern apse of the transept is deep, like the great apse of the presbytery; the round is separated from the straight part by shafts. The apse fell and is shown in ruins in J. Heins's south elevation of 1764.[19] Enough was left standing to show the general design, and it has been rebuilt.

The fall of the north-west transept is not recorded but there is reason to suppose that it happened in the first half of the 15th century, for it seems that a beginning at rebuilding it was made at that time. An overhanging piece of the original wall is supported by masonry which, with its enriched plinth and its string-course, can hardly be mere utilitarian underpinning; moreover plans[20] and views of the 18th and 19th

[18] See above p. 51.

[19] Bentham, *Ely*, pl. xlii.

[20] Bentham, *Ely* and *Supplement*.

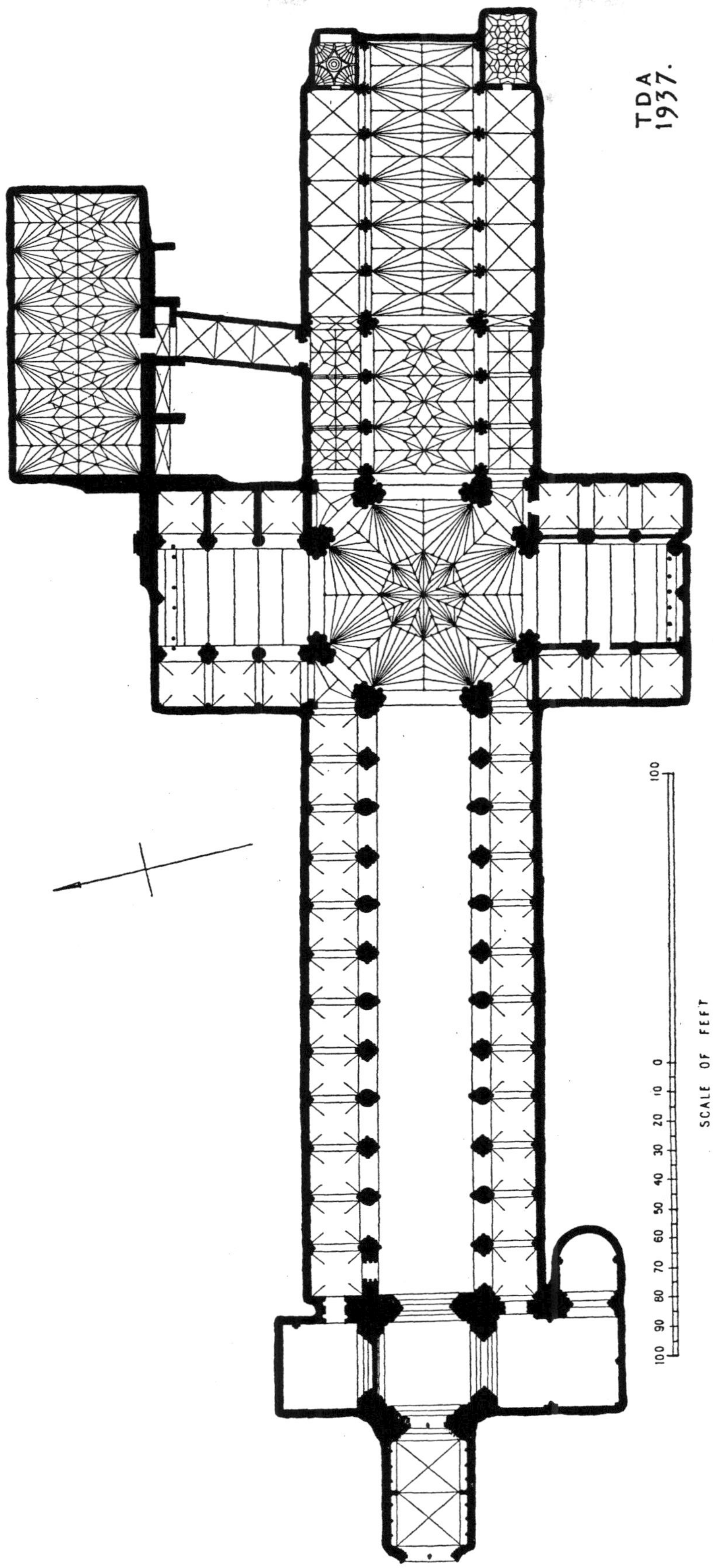

ELY CATHEDRAL: PLAN OF VAULTING

centuries show a complete transept with the walls standing several feet above ground. This would appear to be work actually carried out. If it had been a mere fancy of the draughtsman he would certainly have drawn it to correspond exactly with the south transept, whereas it is 15 ft. shorter from north to south. There are very small octagons treated in a 15th-century manner at the angles and there is no apse. The reason for the abandonment of the work is not known.

The west tower stands on four wide and lofty arches. To cover the west arch it was necessary that the porch should be as high as the present Galilee. Even if there had been no west arch, and instead a solid wall with a doorway, it would still have been necessary to build a porch or something equivalent in order that its walls might provide an abutment for the north and south arches. The west archway did not, however, begin at the floor level; the lower part was a solid wall, a part of which still remains and shows on its west face a small 12th-century blind arch just above the north haunch of the Galilee vault. The sill of the tower archway was above this small arch. It seems clear therefore that the original porch was of two stories like the present one. We must assume that the porch was covered, or designed to be covered, by a timber floor above the small blind arch and at about the level of the crown of the Galilee vault. In a word the original arrangement was exactly the same as that made in the 13th century,[21] whether the actual fabric is the original or a rebuilding.

The west tower is of three stages above the roof of the Galilee and two above those of the transept and nave; it has a large octagonal buttress or turret at each angle. Each stage is divided into two parts: a row of three lancets with pointed heads and a band containing in the two lower stages a low blind arcade of trefoil arches and in the top stage three cusped circles. In the two upper stages the wall treatment is repeated on the cardinal faces of the buttresses but in the lowest stage the buttress has three narrow arches instead of a wide one.

In the middle stage the jamb-shafts have two annulets. In the two upper stages the lancets are separated by a continuous shaft which is carried up to the arched corbel table and the parapet or eaves. The string-courses which separate and subdivide the stages are carried round the octagonal buttresses but the corbel-table is not. The intermediate sides of the octagons have rows of shafts running from bottom to top. The fine octagonal stage with its turrets forming the present top stage is later work.[22] Bishop Northwold (1229–54) erected a timber spire which was almost certainly covered with lead and was probably finished with an eaves.[23]

The tower has some features which are characteristic of a broad part of eastern England. Among them are the circles in the top stage seen in their most remarkable development at the top of the central tower of Norwich Cathedral. They may be seen as far south as Old and New Shoreham (Suss.), and may be compared with French examples such as St. Pierre-sur-Dives, Calvados. The closely set shafts on the octagonal buttresses occur at St. Margaret's Church, King's Lynn. The west front as a whole is very remarkable, the two lofty transept turrets being particularly striking; they are half as high again as the side wall of the transept and three-quarters the height of the tower.

The bay-design of the interior is much like that of Winchester. The heights of the pier arches, the triforium and the clerestory are in the proportion of 6:5:4. The bays are separated by a triple group of shafts. The pier arches are stilted and are of three moulded orders. The triforium arcade is also stilted; each main arch, of two moulded orders, embraces two stilted and moulded sub-arches and a plain tympanum. The clerestory arcade has three arches, the central one wide and stilted, the side arches half round, all of one order slightly moulded. The capitals throughout are of cushion form. The main transept is similar but the work is plainer; the bays are not separated by shafts and the arches of the main arcade are square edged.

The interior of the west transept contrasts as strongly with the earlier work as does the outside: a Byzantinesque architecture of colour has given place to a Western architecture of form. The west wall is covered with a succession of blind arcades of varying design. The ceiling was probably flat and of wood. The tower was doubtless open to a level above the middle row of lancets; that would still leave a height of 40 ft. for the ringing and bell chambers.

GALILEE AND PRESBYTERY. The Galilee was the gift of Bishop Eustace (1198–1215). The name was in use at Ely at least as early as 1340–2 and probably very much earlier; it occurs at Durham in 1186 which was about the time that the Ely porch was first planned.[24]

The present Galilee measures 44 ft. from east to west between the inner and outer doors; the vaulting covers a space 35 ft. by 20 ft.; the side walls are about 13 ft. thick. At the external angles there is a group of five attached shafts terminating in gablets and a spire. At a height of four-fifths of the whole height the shafts are diminished in diameter, the junction being masked by a row of gablets. Similar but smaller groups without diminution divide the west end into three. The central division contains the doorway and a group of three lancets, and finishes with a high, level parapet of three loopholed battlements. The side divisions are divided into four equal stages containing wall arcades of almost identical designs with pointed cinquefoil arches. They finish with a horizontal parapet at a lower level than the central division, each with one wide battlement of two embrasures. The four stages of arcading are carried with slight variations along the north and south sides of the porch. The sides are without windows and are finished with a plain parapet.

The doorway is divided by a central shaft supporting two cinquefoil arches; the tympanum contained a sexfoil panel, but this was destroyed early in the 19th century by Bernasconi, who filled the whole tympanum with the present tracery done in his artificial stone. Each spandrel of the outer arch is filled with three sunk panels of quatrefoil and multifoil, which are of rather earlier character than the rest of the work. The outer archways evidently had no doors but were probably provided with iron gates as they are at present. Harris's elevation of about 1700 shows the archways filled to the level of the capitals with a later wall containing two doorways; the drawing is not sufficiently clear to indicate the date. The engraving shows a clock face in the lower part of the central lancet.

All the capitals have carved foliage; it is of both spray and crocket sorts and is quite mature; dogtooth is used

[21] See below, p. 57. [22] See below, p. 65.
[23] Cf. Sutton St. Mary (Lincs.).
[24] The same is applied to the porch at the south-west corner of the transept at Lincoln and to the west porch at Byland (Yorks.).

between the shafts of the lancets and nail-head freely elsewhere. The curious form of the top of the west wall is probably due to the use of the 'kerb' roof, that is one with a truncated apex; it is of the 14th or 15th century.

ELY CATHEDRAL: GALILEE BEFORE 19TH-CENTURY ALTERATIONS

(Part of plate facing p. 57 in Stevenson's *Supplement* to Bentham's *History of Ely*.)

Internally the Galilee is of two bays covered with a quadripartite vault with cross ribs and diagonals but without ridge ribs. The bays are separated by a vaulting shaft rising from a continuous bench, but there are no shafts in the eastern angles of the porch corresponding with this. The side walls are divided into two stages of arcading. In the lower there are in each bay trefoil arches standing on shafts placed close against the wall along which their lower base mouldings are continued. Half-way up the columns the wall face is set back forming a continuous deep recess. This recess is covered by a vault springing from the four front columns and from a row of shafts against the back wall, placed opposite the centres of the front arches somewhat in the manner of the south quire-aisle of Lincoln. The spandrels of the front wall have sunk sexfoil panels. The upper tier has five cinquefoil arches, the recess behind being covered by one wide arch. The inner doorway is similar to the outer, but the whole is a restoration or renewal by Sir Gilbert Scott based on the remains of the original work and on an old drawing. The tympanum is pierced by a sexfoil; it had been altered by Bernasconi whose work is shown by Harris. The detail of the interior is similar to that of the outside: the mouldings are rich and refined; the dogtooth is used freely; all the capitals have carved foliage. A striking characteristic is the variety of stones: Purbeck, Barnack, and clunch are consistently used in positions which leave little doubt that it was done as decoration. This was a wise step, for in such an exposed position any applied colour would soon perish. At the same time it will be noticed that the softer and finer grained clunch is used for the arches which have the dogtooth enrichment.

It has been shown[25] that the room over the porch existed from the first, and was open to the church. For what purpose it was used or designed is unknown. It has been suggested that it was a chapel in honour of St. Michael the Archangel. Such chapels often occupied an elevated position. The objection to this theory is that the only access to the room is by a roundabout route and a very small doorway; there is a blocked archway in the north wall visible outside but this is merely a builders' access hole. Another suggestion is that the room was intended for a band who would sound a fanfare when a procession entered the church; the musicians could use the indirect approach or even a ladder.[26] The room was destroyed about 1800 when the roof was lowered to a level immediately above the vault and the tower-arch filled with a wall containing a window.

The first alteration of importance to the Romanesque church was the addition of a large presbytery or retroquire to the east of the high altar by Bishop Hugh of Northwold (1229–54), with help from the priory and from outside subscribers.[27] He took down the deep apse as far as but not including the two shafts which separated it from the work to the west and have survived later alterations. Northwold's work consists of nave and aisles of six bays in continuation of the four Romanesque bays east of the central tower. The work seems to have been begun in 1234 and finished in 1250. The whole was vaulted: the first vault that the brethren had seen except the small and low one over the Galilee. The only irregularity in the plan was a small Reliquary which was built between the third and fourth buttresses on the south side. The next bay but one is shown by Bentham as similarly treated: the enclosure was presumably medieval and formed a vestry for the Lady Chapel. On the north side the aisle wall shows outside in the second bay marks of a recess.[28] The third bay has a Purbeck threshold and the bases of two nook-shafts, but immediately above these the courses of the original ashlar run through without break or irregularity. This suggests that a door was planned but abandoned, although there are holes for roof timbers in the sides of the two buttresses. The next bay shows marks of a doorway cut since the wall was built and now blocked, and scars of roofs of several periods: indications of the feretrar's checker built about 1425[29]. Some work had been done in 1357–8 including fourth and fifth windows from the west, and also some carpentry and plumbing costing £46, but its character has not been determined.

Northwold's ritual arrangement was as follows. The high altar was placed on what would have been the chord of the Norman apse as the Norman altar had been. There was doubtless a screen at the back of Northwold's altar. The next three bays eastward were fenced round to form a feretory, in the centre of which was placed the Shrine[29] of St. Etheldreda with a small altar against its west face. Three of the bosses of the vaulting in this part are carved with figures. Northwold himself was buried at the feet of St. Etheldreda and farther east were the shrines of the three Saxon abbesses, Sexburga, Erminilda, and Werburga. St. Alban's shrine

25 See above, p. 56.

26 This gallery should be compared with that in the west bay of the north aisle at Winchester, for the two buildings have many resemblances. The Winchester gallery very likely takes the place of one in the destroyed west tower.

27 Stewart, *Archit. Hist. of Ely Cath.* 68.

28 Shown by Browne Willis, *Survey of Caths.* (1730), as a doorway with the window above it raised.

29 See below, p. 70.

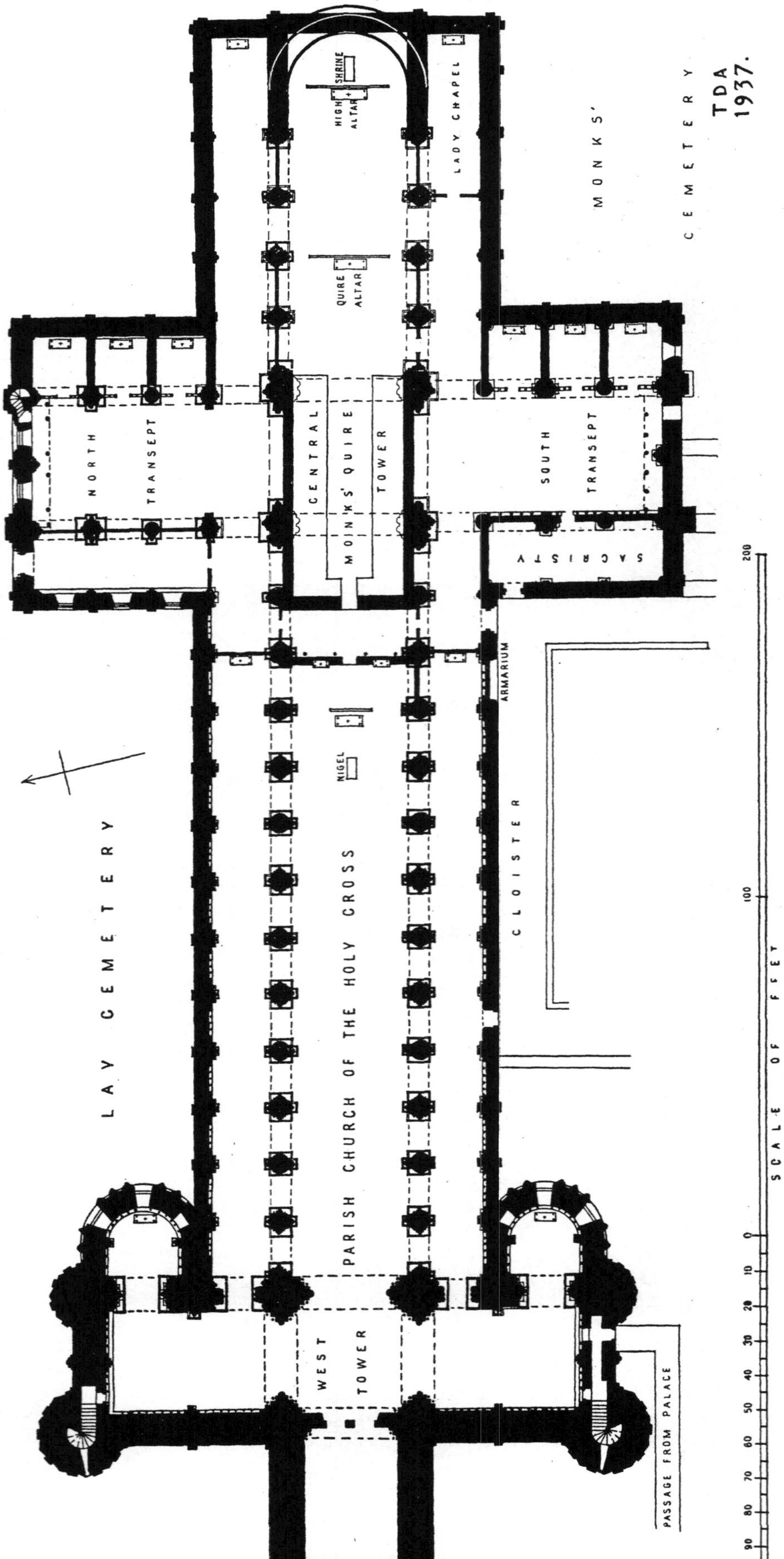

PLAN OF ELY CATHEDRAL, *circa* 1225

was in the north-east corner. The Lady Altar was placed in the south aisle opposite to the central pier. Here may be mentioned a fragment dating from Northwold's time. It is the massive standard and arm of an important stone seat; an animal clasps between its paws the head of a man, alluding probably to the legend of St. Edmund's head having been defended by a wolf; Northwold had formerly been abbot of Bury.[30]

Many alterations have been made in Northwold's work but enough has been preserved to show the original design in almost every detail. Four bays of the Romanesque eastern arm were left standing and the height of the several stories no doubt influenced the proportions of the new work. The clerestory is the same height as the clerestory of the nave but the triforium is rather lower and the main arches rather higher than those of the nave. The height of the stages had no doubt been preserved uniformly throughout the Romanesque church and there was therefore a step up in the level of the triforium sill where Northwold's work began; the rise was about 3 ft. The old and new work were, however, strongly separated by the half-round column which still exists, so that the break in continuity was unobjectionable. A capital and the *tas-de-charge* were inserted at the right level, and the 12th-century column became one of the vaulting shafts of the 13th-century work.

The main piers have eight detached shafts standing on a low octagonal plinth,[31] with annulets and with capitals carved with spray foliage. The bases have bold spurs, into some of which human heads are introduced. The whole is in Purbeck marble. The arch is well moulded and has one row of dogtooth spaced widely in the eastern bays and set close farther west. The spandrel has a sunk trefoil panel with a Purbeck back and two sprays of foliage. The triforium arcade has bluntly pointed arches containing two trefoil pointed sub-arches; the tympanum has a sunk quatrefoil panel with a Purbeck back; the jambs have five Purbeck shafts and the central columns show three. The clerestory has a group of three lancets; the central rear arch is much stilted, the flanking arches have one springing line much higher than the other and the difference is got over by making the high side trefoil; the jambs have two shafts and the columns show three in front. The east wall is of two stories; the lower is of equal height to the main arcade and triforium; it has three equal lancets with quatrefoils in the spandrels; the upper story, corresponding to the clerestory, has five lancets rising towards the centre to fit the line of the vault, the asymmetrical form of the arches being adjusted rather clumsily in the same way as those of the clerestory. The main vault is slightly plough-share in form; it has one intermediate rib between the diagonal and the cross rib and one between the diagonal and the wall rib; the longitudinal ridge rib is continuous, the cross ridge rib stops at the intermediate; the vault springs from a group of three shafts of the same height as the triforium; at sill level the shafts stand on a long conical corbel.

The detail is rich. All capitals are carved with the characteristic foliage of the period,[32] and the cusps of the trefoil arches and of the panels have small sprays. The bosses of the vaulting are carved with foliage which also appears on the three with figures, mentioned above. The admirable corbels carrying the vaulting shafts have a spiral of stalks breaking into small leaves at intervals until they reach the top, where they break into full foliage. The corbels opposite the shrine are richer but scarcely so happy; the spiral is abandoned and there are rings of heavy crocket foliage at frequent intervals from bottom to top.[33] There are three rows of crockets between the shafts of the triforium. The aisles are simpler but suitably finished; the vault has cross ribs and diagonals and wall ribs only; it springs from triple groups of Purbeck wall shafts which stand on a bench-table and have annulets and moulded capitals. No original aisle window remains.

The original triforium is preserved only in two bays on the south side. It has coupled lancets with shafted jambs, foliage capitals, and dogtooth enrichment. Above these externally there is a corbel table of trefoil arches and a low parapet.[34] The three windows of the clerestory are included under an arch; the capitals are carved. The buttresses are carried up as pinnacles with gablets and spires. The flying buttress consists of two arches giving not very steep slopes. The upper arch abuts on the very top of the aisle wall and was intended, Professor Willis says, to counteract the thrust of the nave rafters. This thrust would be met by the tie-beams which existed in 1760 when James Essex renewed the roof, but Essex considered that they were additions to the original work.[35] The lower arch abuts on the springing line of the nave vault and follows the line of the aisle roof. It carries a buttress of slight projection against the clerestory wall. It is probably for this reason that, instead of being a half-arch of the usual form, it springs from the clerestory wall and rises to an apex about 4 ft. away. The whole structure was not quite adequate; the arches pushed out the vertical part of the buttress. The upper part of every buttress, except the three where Northwold's triforium windows have been preserved, was rebuilt in the 14th century. The inner face of the vertical part was removed inwards and the pinnacle made higher; the lower arch was made much steeper and its point of abutment raised; the upper arch was also made steeper and designed to ride on the back of the lower arch.[36] The work was probably done at the same time as the rebuilding of the three western bays. The failure of the original flying buttress was probably accelerated by the ingenious way in which the upper flyer was made to serve as a channel to carry off the rain-water from the nave roof and through the buttress. Naturally the thing soon ceased to be water-tight and the wet soaked into the masonry. Moreover the whole work had undoubtedly been jarred by the fall of the central tower in 1321.[37]

The east end has not been very seriously altered. Externally the central part is separated from each aisle by a stair-turret and bold buttress, and similar turrets and buttresses flank the whole. The central part is divided into three stages: the two lower containing the windows described above, and the top stage corresponding to the part above the vault. The lowest stage has two shafts in the outer jambs and three in the piers between the lancets, with dogtooth between. The capitals are carved with foliage, there are two annulets, and the spandrels have quatrefoil panels. In the middle stage the central lancet occupies the whole height; over

30 See also below, p. 74.

31 The high plinth of the easternmost piers shows the original floor level.

32 All is of free spray form, there is none of the crocket sort.

33 Another variant is noted below, p. 72.

34 The subsequent treatment of these two bays is described below, p. 66.

35 B.M. Add. MS. 6772, f. 221.

36 This analysis is based on that of Willis (Stewart, *Archit. Hist. of Ely Cath.*, p. 78 and pl. iv).

37 The roof is described below, p. 66.

the next there are sexfoil panels and over the flanking windows niches with pedestals for statues; the jambs are similar to those of the lowest stage. The top stage has three equal lancets flanked by two statue-niches; above the windows there is a group of three sexfoil panels, the large central one having a pedestal for sculpture. The gable is crowned by a foliated cross. The buttresses have very little diminution as they rise, but in decorative treatment they are divided into five stages, in no case answering to those of the wall. The lowest is plain but all the others have a niche on the face and on each side, all with Purbeck shafts and dogtooth. The lowest niches are shallow as are the side niches of the upper tiers, but the front niches of the three upper tiers have coved backs and pedestals of varying size for statues;[38] the heads of the lowest and highest niches of the four are trefoil, the two intermediate cinquefoil. In the middle of the 18th century the east front was found to be leaning outwards nearly 2 ft. but was pushed back by the architect James Essex. The buttresses are shown in 1762 as being finished with a low gable and small cross, probably 17th or 18th century, as the northern one still does; this north stair turret is now flat topped. The south buttress has a high pitched gable and the turret has a crocketed spire on an arcaded octagonal shaft, the work of Sir Gilbert Scott.

The ends of both aisles have been altered but the original design can be recovered in part. They are in two stages corresponding with the two lowest stages of the main buttresses. The lower stage presumably contained a pair of lancets with their sill at the same level as the sills of the central group. The upper stage has a range of four arches, the two central pierced with windows, the lateral arches narrow and blank. The arcade was continued on the canted faces of the angle-turret and presumably round the buttresses; but these have been altered. The aisle wall no doubt finished with a slope to suit the original roof referred to above. The original finish of the top of the angles can only be conjectured; the buttresses now end with a gable and the staircases are continued up as octagonal turrets crowned with tall crocketed spires, characteristic of the district from the Eastern Midlands to the coast. The east end of the north aisle now contains a window of the 14th century and the south aisle one of the 15th century.

LADY CHAPEL. Until the present Lady Chapel was built the altar of the Blessed Virgin had stood in the south aisle of the presbytery. In the first quarter of the 14th century it was decided, in accordance with general ecclesiastical policy of the time, to build an enlarged chapel. The site chosen, almost detached, to the north of the presbytery is similar to that of Peterborough, built in 1274. The two were the same length, 97 ft., but Peterborough was rather narrow for its length, little over 30 ft., while Ely was exceptionally wide, being 42 ft. At Peterborough the ground did not admit of such a long chapel in the usual position to the east of the great church. At Ely there was land enough, but the important Outer Hostelry already occupied a part of it. It was necessary to remove a few buildings to clear the chosen site, but these were probably of no great importance, for the ground seems to have been originally a part of the parish cemetery. The first stone of the new chapel was laid on the feast of the Annunciation 1321. The moving spirit in the scheme was one of the brethren, John of Wisbech (d. 1349), who lived just long enough to see the work completed.

One angle of the chapel overlaps an angle of the north transept but there is here no communication with the church. The chapel was approached by a wide passage from the north aisle of the presbytery and was entered by a doorway in the centre bay. There was also a small doorway in the second bay on the south, clearly for the use of the chaplains and singing boys. This doorway was also reached by a staircase from the upper story over the passage and from a small raised gallery over the main entrance, to be described presently.[39] The chapel is divided into five bays separated by buttresses carried up as spired pinnacles and containing windows of four lights with tracery of geometrical form verging on flowing. Above these windows are two bulls'-eyes, formerly traceried, in each bay, to ventilate the space above the vault; the parapet is plain. The east end has a seven-light window, the gift of Bishop Barnet, and was made between 1371 and 1375. It has more advanced tracery in which the vertical line appears, including a transom with sub-arches, and the lights are continued down below the sill as panels. Above the window and filling the gable there are nine niches with pedestals for statues; the heads have four cusps and ogee hood-moulds; the central niche has two pedestals and presumably contained the Coronation of the Virgin. The low-pitched gable has battlements with sunk tracery. The two buttresses at each angle are carried up as pinnacles like those at the sides; they abut on a third and higher pinnacle. The contrast between this front and that of the great church well illustrates the change to greater concentration of interest which had taken place in the century which separates the two buildings. The west end has an eight-light window; each half of the window has tracery like the four-light side windows, the central space between filled with tracery in which the vertical line predominates, similar to but not exactly the same as the corresponding part of the east window. The wall is not dissimilar to the east wall but has many more niches for statues, for this end was very conspicuous from the parish cemetery and by people entering the church by the north transept door. Below the window sill there are six niches separated by pilasters rising from the ground; the niches have four cusps and ogee heads, and contain statue-pedestals; the backs are concave and retain clear traces of having been painted with a pattern. On each side of the windows there are three niches, one above the other, of similar character; the lowest is level with the niches under the window; the uppermost has a steep gablet, flanking which are shields bearing the arms of the see, of Montacute, and others. At top of all is a pair of small niches. Above the window is a row of seven niches with cusped heads and ogee hood-moulds like those at the east end. The centre niche has a small secondary spired niche above the canopy which may have contained a subject. The angles are treated similarly to the eastern, but the north-west angle contains a staircase and this demands a larger spire, and the spire-heads of the two buttresses are omitted. The face of each buttress has two niches, the lower one having a bold canopy carried on detached shafts standing on the weathering below. This end of the chapel thus showed, if the attached north buttress is included, twenty-nine statues.

[38] The arrangement of the pedestals for statues is as follows. Lowest tier, none. Second tier, on face and on outer side. Third tier, on face and on each side. Fourth tier, on face.

[39] See below, p. 62.

The vault[40] is a somewhat elaborate lierne. It belongs to the class which combines the cellular or ordinary vault with the barrel vault as seen in St. George's Chapel Windsor, Gloucester, Winchester, and elsewhere. This has one intermediate between the diagonal and the cross rib, and three between the diagonal and the wall rib. The vault is skilfully designed and carefully built and the construction is evidently light, for it has a span 7 ft. greater than the presbytery and the buttresses are of no great massiveness. The bosses[41] carved with subjects or recognizable heads are with one exception on the main ridge rib;[42] others have foliage or grotesques or unrecognizable heads. The wall-shafts which carry the vault stand on the stone bench of the stalls. They are partly masked by the stalls and by the two niches above, and must have been almost entirely hidden before the statues were removed. Nevertheless the scheme is carried out quite methodically and accurately from base to cap. But one mistake was made: the wall against which the shafts are placed is rather farther back than the plane of the window arch, and when the wall rib has risen about half its height it has to be brought forward to clear the window arch. This is done by a bend in the rib, rather awkwardly and not quite regularly.

The piers between the windows contain two tiers of niches above the stalls. Their hood-moulds are ogee in plan, bringing the pinnacles well forward in front of a base for another statue; each window jamb contains two tall narrow niches; thus there were in all ten statues to the pier. The feature for which the chapel is most famous is the series of seventy-four stone stalls. The normal bay contains three double bays under the window and there is a double bay against each pier. At the west end there are six stalls, wider than those at the sides, flat backed and not subdivided; they are flanked by narrow coupled stalls. The side stalls have rounded backs; the embracing arches are ogee, double-cusped and with crocketed hood-moulds, with finials; over them there are crocketed gables with finials. Each pair of stalls is separated from the next by a narrow pilaster with pinnacle top. In spandrels enclosed by the pinnacles and gables is the well-known series of subject sculptures. Above the cornice of the stalls there is a traceried parapet; a part of it has been removed and replaced by an arcaded corbel-table of the 13th century. The canopy work and cornice of the stalls are stepped up between the first and second bays; the seat is stepped to suit the altar steps; the easternmost niche on both sides of the chapel had a marble shelf, now broken away; the southern niches presumably contained piscina basins. The central part of the east wall is occupied by a stone reredos of later date. Next to the reredos on either side are two niches,[43] not coupled like the stalls; between them and the corners of the chapel there is one wide niche divided by a central moulding;[44] the niches were not intended for stalls, their sills being formed to a steep slope; this slope has been cut away, so it is possible that the recesses were used as seats at some time; the more important spandrels have no sculpture and the ruins of any pictures there were have been smoothed away. The work of the whole chapel is very rich in foliage and fanciful creatures exquisitely carved. The chronicler tells us that there were 147 statues, not including the reredos or the entrance [in the presbytery aisle], but including presumably the outside; a statement which the curious can verify or correct without difficulty. Walter Ymagour was paid £3 for making five images, Philip Wale £1 6*s*. 8*d*. for four, and John the painter £1 for seven.[45]

No fragment of the numerous large statues has been preserved and only a small proportion of the statues in the stall-work. The reliefs have been so much mutilated by the iconoclasts that out of 93 subjects 22 could not be identified even by M. R. James,[46] and many of the 71 which he was able to explain could be recognized only by the smallest fragments of evidence; hardly a head remains. The two easternmost bays and one pair of stalls in the third bay on the south side represent miracles wrought by the intervention of the Virgin after the death of her body. The remainder of the south side, the whole of the west end and the four westernmost niches of the north side are scenes from the lives of the Virgin and her parents; in the remainder of the north side the miracles are resumed. The first series of miracles is richly coloured; the rest of the sculpture shows only slight remains of colour.

James could not identify the literary source used by the artist of these pictures. He found one Bury MS.[47] containing a miracle which happened in the diocese of Ely and perhaps four miracles which are found in these sculptures. Yet the omissions outweigh the coincidences. He recognized the journey of the Virgin and St. Joseph to Bethlehem before the Nativity in the centre niche of the westernmost bay on the south, and thought it must be unique in Western art;[48] he had never seen any other attempt to figure the incident. One small figure on a corbel just outside the chapel was missed by James but not, unfortunately, by the iconoclast for it is now headless. The outside of the principal entrance, which is double, has over its central shaft a corbel for a statue, doubtless of the Virgin. On this corbel is carved the kneeling figure of a monk in supplication; this almost certainly represents John of Wisbech. At this statue of Our Lady or at that over the doorway in the aisle oblations were made in 1485–6.[49]

The reredos was made in the time of Bishop Fordham (1388–1425) who was a contributor to the cost. A roll of 1389–90 gives particulars of the work done under the direction of Master Robert Wodehirst in clunch from Burwell.[50] The reredos is 16 ft. wide and consists of a uniform row of thirteen niches with a narrow one at each end. The niches had rich vaulted canopies evidently of remarkable character, but the more projecting parts have been cut away, for the better accommodation of a wood reredos of the Corinthian Order[51] which was still standing in 1834. It is clear, too, that some important feature has been altogether removed from the top which is now a good deal lower than the stall-work. The central part of the parapet,

[40] The roof was renewed in 1762 (D. J. Stewart, *Archit. Hist. of Ely Cath.* 144).

[41] C. J. P. Cave, 'The Roof Bosses in Ely Cath.', in *Proc. Camb. Antiq. Soc.* xxxii (1932), 33.

[42] East bay: Head of Christ, Virgin and Child, Assumption (figure in a *vesica piscis* supported by St. Etheldreda and a monk). Bay 2: Coronation of the Virgin, Ascension, Annunciation, Visitation. Bay 3: Fall, Crucifixion. Bay 5: Translation of St. Etheldreda. Scrolls have words from the Magnificat and Ave Maria.

[43] The niches at the east end are flat-backed.

[44] See below.

[45] Acct. of 30–32 Edw. III (1356–9), J. H. Crosby's transcript.

[46] *Sculpt. Lady Chap. Ely.*

[47] Ibid., 7, referring to B.M. MS. Royal A VIII.

[48] Ibid., 31.

[49] Stewart, *Archit. Hist. of Ely Cath.* 144.

[50] About 10 miles to the south, and probably then, as now, accessible for barges by Reach Lode.

[51] G. Millers, *Description of Ely Cath.* 98.

which ran all around the chapel, has also been taken down.

The masonry generally outside is Barnack or Weldon or similar oolite stone, but the window tracery was worked in Burwell stone, which is unsuitable for outside work. The inside, including the vault and the reredos, seems to be entirely of clunch except the following parts: the pilasters between the niches are of Purbeck marble from the floor to the springing of the gables; the curved backs of the stalls are of oolite formerly plastered; the bench is of oolite and the shelves for sacramental uses are Purbeck. The pavement, also of Purbeck marble, cost the considerable sum of £59 16*s.* 8*d.*[52]

The glass has been almost entirely destroyed but what is left in the heads of eight lights and a dozen tracery lights is contemporary with the building and suggests a remarkable scheme.

A few charges occur in the accounts for 1356–9. William Pyrown was paid £22 for the glazing (*verrura*) of a window and Simon de Lenn £12 13*s.* 4*d.* for the Duke of Lancaster's window. Glass, costing with the carriage £39 11*s.* and probably from Flanders, was bought at Yarmouth.[53]

The Chronicler tells us that John of Wisbech died on 16 June 1349 and that he had continued the work for 28 years and 13 weeks from its commencement, which is a nearly correct statement of the period from 25 March 1321 when the foundation-stone was laid. At the beginning when his funds were low he had persuaded some of the brothers and some seculars to help with digging the foundations at night. He himself found a bronze urn full of coin, but he kept the fact secret and hid the money under his bed; by this means he made sure of the whole for his chapel. He sold private property of his own which the bishop had given him licence to retain. He seems to have been a shrewd man of business and left the building fund with a balance of over £100. He was buried at the door of the Lady Chapel.[54] He had finished, we are told, the stone shell, the east and west windows, the east gable and the images inside and out, including those at the entrance doorway, and he had made the timber roof and covered it with lead. Bishop Simon Montacute (1337–45), who had contributed liberally to the cost of the building, was buried at the altar. Bishop Fordham (1388–1425) was buried near the west end;[55] contributions were made at his tomb for a long period of years. Bishop Hotham bequeathed £100 to the chapel.[56] In 1478–9 the Guild of St. Mary gave 20*d.* in offerings (*oblacionibus*) at Bishop Fordham's tomb.[57]

The doorway from the church to the connecting passage is a wide and handsome arch with housings for statues in the casement moulding; on either side are very wide niches with two tiers of smaller niches above. In the spandrels of the arch are figures adoring or censing a large seated figure above, of which only the lower part remains; doubtless the Virgin and the Child. The passage-way was 48 ft. long by 12 ft. wide; this width would allow for presses to contain the gear of the chapel; it was paved with tiles, some of which are now laid in the south transept, making a panel 8 ft. wide. There was an upper story reached by a staircase at the north end; it probably formed a sacristy for the chapel and rooms for the custos; at the Suppression it was occupied by a monk, and in 1649 had a third story in the roof. The north end of the upper story was partitioned off to form a gallery looking into the chapel.[58] This gallery was reached from the church by a long and elaborately contrived approach; a stairway in the presbytery aisle led to a raised passage along the transept wall, and thence, turning along the Lady Chapel wall, passed through archways in the chapel buttresses on vaulting in the spaces between. The central window of the chapel has a transom, a little way below which are sub-arches; it was evidently against this transom that the roof of the upper chamber abutted.

The Lady Chapel had a pair of organs; there are charges for its repair before 1461. At the time of the Suppression, but after the removal of things required for 'the King's use', there were four laten candlesticks, a frontal, an altar-pillow, two cushions, two tappets, and a vestment. The 'Lady Chapel Chamber', the room over the passage, contained: four chasubles, four vestments, two frontals and some odds and ends; some chests, benches and chairs; an andiron, perhaps for baking the sacramental wafer, and a pair of tongs; 'a folding table' was no doubt a triptych; a psalter is the only book entered in the inventory. Nothing is entered for the passage itself.

In 1566 the use of the Lady Chapel was given to the parish of Holy Trinity (see below—Churches). The present roof dates from 1762. In 1938 the chapter resumed possession of the chapel and cleared it of its modern furniture.

OCTAGON AND HOTHAM'S BAYS. The central tower fell on the night of 12–13 February 1322. The quire had been abandoned for some time past, the services being held in the chapel of St. Catherine against the end of the south transept; so the catastrophe was not unexpected. The brethren had left the chapel after matins, had been in procession to the shrine of St. Erminilda and had returned; this had been done only, as it proved, at considerable risk. It is likely that St. Catherine's Chapel continued to be used as a temporary quire for the twenty years of the rebuilding. In the second year a cord was provided for the sanctus bell: *pro parva campana in choro.*

The credit for the idea of building an octagonal instead of a square tower has always been given to the sacrist, Alan of Walsingham, and very likely quite fairly. A general notion of this sort is, and presumably was in former times, often suggested by the employer. The sacrist was of an East Anglian family of wealth, capacity, and craftsmanship. But it was not a new idea. The polygonal form had come from the Continent and was already highly popular for several classes of building across central England; at Ely itself the large angle turrets of the west transept are sixteen-sided. Walsingham had to employ an architect, and secured the services of a certain 'Master John' and afterwards of a 'Master John atte Greene', masons. The timber lantern

52 J. H. Crosby's transcript.

53 Acct. of 30–32 Edw. III (1356–9), J. H. Crosby's transcript.

54 John Downham had leave of the prior and convent to be buried at the Lady Chapel door next to John of Wisbech's grave. The door was probably that from the church to the connecting passage.

55 When the flooring was renewed (before 1834) nothing was found marking the graves of Montacute or Fordham (Millers, *Description of Ely Cath.* 99) nor in 1806 (Stewart, *Archit. Hist. of Ely Cath.* 144).

56 Note by J. H. Crosby.

57 Stewart, *Archit. Hist. of Ely Cath.* 145, quoting Custos Capellae R. 18 Edw. IV.

58 The use of the gallery is discussed below, p. 68. The partition is perhaps referred to in the following: 'Solut' pro factura del parclos £1 16*s.*' (Acct. of 30–32 Edw. III, J. H. Crosby's transcript).

Plate XLIV.

ELY CATHEDRAL. CROSS SECTION SHOWING LANTERN CONSTRUCTION, 1766

(From Bentham's *Ely*)

Ely Cathedral. Interior of Octagon, 1763

(From Bentham's *Ely*)

was the work of 'Master William Hurle', or Hurley, a 'Master Thomas' being first engaged to put up a crane. That some departure was made from the customary procedure by the co-operation of an amateur seems possible and is suggested by the want of harmony between the parts, such as the plan of the octagon and of the stalls.

The eight piers are built against the first pair of old main piers left standing in each arm after the fall of the tower. This reduced each arm by one bay and gives an irregular octagon with the cardinal sides longer than the intermediate. This plan gives sufficient length for the quire which in the Romanesque church had extended east and west one bay beyond the tower. In each side there is a lofty arch all agreeing with the vault of the presbytery; in the intermediate sides there are large four-light windows overlooking the low aisle roofs; below the windows there are three niches of unconventional form with pedestals for statues.[59] The vaulting shafts of the octagon are small at the bottom but at about one-third of their height they are enlarged to adequate size, the change of plan being masked by a large niche: a mere architectural contrivance not capable of holding a statue; the corbels under them have minute sculptures illustrating the life and miracles of St. Etheldreda.

Externally the tower rises very little above the ridge of the main roof; it is then finished with a cornice, a parapet of pierced tracery and cresting. Bulls'-eyes light and ventilate the space between the vault and the floor immediately over it. Above the floor there is a ringing-room, 15 ft. high, covered by a flat lead roof, and lighted by square-headed windows of six and three lights with cusped ogee heads in the long and short sides; those to the east and west are partly blocked by the apexes of the roofs.

The lantern is a regular octagon on plan with its angles opposite to the centre of the sides of the stone tower. It is 30 ft. in diameter with angle-posts nearly 100 ft. above the floor. It has a large window on each side and a vaulted ceiling above which there is a room, formerly the bell chamber, 12 ft. high. This room has eight louvred windows, a flat lead roof, and a pierced and traceried parapet with a cresting. The angle-posts are carried up as pinnacles and finished with a level battlemented top; they are cased with boarding and lead; in one of them the space between the post and casing contains a vertical ladder. The octagonal lower room through which the lantern passes has shuttered windows which command striking views into the church below.

The designing of this great work forms almost a revolution in roof construction. The great strusses supporting the lantern are brackets and are in effect the germ of the hammer-beam system of fifty years later—a fact which places Hurley in the front rank of English architects or engineers. The posts are 63 ft. long, perhaps in two lengths, and of 3 ft. 4 in. by 2 ft. 8 in. scantling. Each is supported by two brackets from adjoining angles of the stone octagon; each of these brackets consists of two steeply raking struts springing from the same point low down in the angle of the stone tower, but the upper being the steeper abuts on the corner post of the lantern at a higher level. The inward thrust of these struts is met by horizontal beams from post to post forming a series of rings at several levels. All these timbers are below the lower chamber. The corner posts receive some further support and much aid in maintaining the vertical by a third series of raking shores at a higher level and almost wholly above the roof of the stone octagon. These are not in pairs like those just described but a single one to each post; nor from the corners of the tower, but from the middle of its sides. Springing from a point below the roof they break through it and strike the post half-way up the large windows. The comparatively small timbers of the roof and of the floor of the lower room are of use for other than their primary purposes: they hold the timber octagon with a firm grip. The roof rafters help to keep it upright; the floor joists, which radiate from the angles of the stonework and are secured directly or indirectly to the lantern, prevent a turning movement in the lantern. The upper part of the lantern is stiffened by the floor and roof of the bell-chamber. These were important matters. Although the dead weight is great it is constant and tends to tighten the joints. The wind pressure, on the other hand, is variable and is exerted in gusts, which tends to set up a rocking movement.

The Historian is emphatic on the care the sacrist took to secure a good bottom for his foundations.[60] The stone is from Barnack but some may be from Weldon or other Northamptonshire oolite quarry. It was finished as far as the upper *tabulatum* (either the cornice under the parapet or the sills of the square-headed windows) in 1328 and the timber lantern was, we are told, begun at once. Timber of ordinary scantlings was bought at Stourbridge or Reach Fairs; but for the great balks for the angle-posts much search had to be made. It is generally thought that they were those obtained from Chicksands Priory in Shefford (Beds.) where 20 oak trees were bought in 1322–3 for £9. The accountant was careful to give particulars of the transaction. Master Thomas a carpenter was fetched from Newport (Essex) and sent to fell the trees, for which he received £1, the incidental expenses amounting to 18*s.* 5*d.* Next year (1323–4) a carriage was made for 6*s.* 10*d.* and the wheels for 5*s.*, the sacrist providing the iron for the latter. The butts were then brought away. They went by land to the river at Barnwell, a distance of about 25 miles, and thence by water 15 miles to 'the cemetery' (of the parish, on the north side of the church), at a cost of £2 16*s.* for the whole journey. There are some difficulties about these particulars. The meaning of *tabulatum* is doubtful, the prices are low and the time is short. In regard to cost: although carriage and incidental expenses added about £4 14*s.* to the prime cost, the £9 which the Chicksand nuns got for their trees was a small sum. As to time: between February 1322 and Michaelmas 1323 the design of the octagon and lantern had to be made before suitable trees could be looked for. But they were found and felled, and brought away next year, and a small payment was made to Robert le Sawyer, carpenter, for roofing the stonework temporarily.

In 1334–5, six years after the stonework was ready, we find Master William of Hurley, carpenter, in receipt of the considerable yearly fee of £8 with board and lodging. In the same account there is an important memorandum that eight carpenters were engaged for nine weeks raising great posts, or *the* great posts, in the new quire. The roll for the next year is missing but that for 1336–7 shows Hurley still at Ely at the same fee and large charges for timber and for sawyers' and carpenters' wages; leadwork costs £12 10*s.*; the vaulting is painted for £12 19*s.* of which £10 is for the

[59] The present statues are modern.

[60] *Anglia Sacra*, i, 644.

painter Master William Shanks.[61] It is certain, therefore, that the fabric was then finished. After another gap in the rolls of two years Hurley's name does not appear in connexion with the new work and the cost of the carpenters' wages for 1339–40 is half that for 1336–7; the central boss of the lantern vault is carved; the cost of the leadwork is higher; there is the large charge of £6 for iron and nails; and an item of 45 yards of canvas at 3*d.* for the windows. By 1341–2 the whole was finished and the claim of the Historian[62] that the work was done in the twenty years that Walsingham, who now became prior, had held the office of sacrist is justified. He tells us that the cost was £2,408.[63] But there remained the numerous supplementary items which always take so long and cost so much. In 1345–6 four new bells were cast and hung in the west tower; windows were glazed and some more painting was done.

The foregoing particulars seem to suggest that the course of the work was somewhat as follows. An energetic start in the year of the fall and early purchase of timber including eight special posts; stone tower ready in 1328 for timber lantern which we are told was begun at once; yet in 1334–5, six years later, great posts are raised, much carpentry done and the leading carpenter of the country at work at high pay; decoration done in 1336. These two years 1334–6 would be time enough for the timber work which it would be unwise to dally over (for the timbers would warp from exposure and then the tenons would not fit); and time enough also for the decoration. We may therefore suppose a pause in the work of some years[64] between 1328 and about 1334 (notwithstanding the chronicler quoted above) due perhaps to lack of funds. It is to be noted in this connexion that twice (1334–5 and 1336–7) the sacrist had been obliged to borrow money apparently to meet the fees asked by his carpenter-architect.

Little is known of the history of the stalls. There is little doubt that they were designed by Hurley,[65] to whom in 1339–40 and perhaps in the two preceding years an honorarium of 6*s.* 8*d.* was paid. A large number of carpenters were at work in 1339–40 and large quantities of nails, chiefly of the smaller sizes, were bought. These items point clearly to the stalls being in hand. If there are similar charges in the lost rolls of the previous years the bench work would be finished and the stalls could be put up in 1340 or 1341. In 1339–40 the sacrist had contributed 13*s.* 4*d.* which had been allowed him by the convent for the annual autumn feast called 'Oet Olla'. In that year the plumbing cost £15. In 1341–2 there are a number of small charges for fittings in the quire and in connexion with Hotham's monument. We may therefore assume without much risk that the stalls were erected in 1340–1.

Before the erection of the woodwork it was necessary to build two stone walls across the octagon to which they could be fixed; perhaps some of the occasional references in the rolls to 'the parclose of the quire' refer to these. In the north wall seven chambers were formed, each 22 in. long, 7 in. broad, and 18 in. deep, to contain the bones of distinguished Saxons of East Anglia.[66] These cavities were no doubt in the same position that they had occupied in the Norman church and maybe were the identical chambers. At the fall of the tower the lower parts of the walls would at once have been buried in debris which would protect them from serious injury. The chambers were given new architectural fronts and paintings to represent the persons whose bones were within. These paintings would be the first things to meet the gaze of the pilgrim. When the wall was destroyed in the 18th century the relics were carefully preserved.[67]

The stalls are seventy in number; they have lofty arched canopies, over which is a continuous enriched cornice. Above this is a series of sculptured panels surmounted by projecting canopies with spires and pinnacles. Any sculpture or other decoration that there may have been in the panels in the Middle Ages has been destroyed and the present carvings were done by Belgian artists in the 19th century. They represent scenes from Scripture, those on the north being from the Old Testament, and those on the south from the New. The misericords have well-carved medieval brackets with scenes from the Old Testament (as the Fall, and the Return of the Dove to the Ark), from the New (as Herodias tumbling before Herod and the Decollation of St. John Baptist), from Legend (as St. Martin parting his cloak), grotesques and contemporary life (as men dicing). The sub-stalls and other seats and desks are of the 19th century, and were designed by Sir Gilbert Scott.

The octagon and lantern have been a good deal altered in detail since they were first built. A view[68] by Heins in 1756 shows the pinnacles of the stone tower without spires, and pinnacles at the angles only, without any in the centres of the longer sides, lantern windows of three lights and tracery of rather simple character and the lantern pinnacles finished square as at present. Most of these details are no doubt due to Essex, who found the lantern in a decayed condition in 1756 or 1759[69]

61 Perhaps a relative of Thomas Shank, working at the Palace of Westminster 1352–3.

62 *Anglia Sacra*, i, 644.

63 This does not correspond with any figure which can be calculated from the accounts with so many rolls lost.

64 If this explanation is accepted then Stewart (*Archit. Hist. of Ely Cath.* 99) had good grounds, which he did not particularize, for stating that the great posts raised in 1334–5 were indeed the angle posts of the lantern.

65 William Hurley (variously de Horle, Hurle, Hurleye in London and Ely), Horley (Sur.), Hurley (Berks.), and Hurley (Warw.) are possible places of origin. Working 'in responsible position' at St. Stephen's Chapel, Westminster, 1326—at £1 plus a robe per annum. In 1332 in charge at London Guildhall Chapel on the same terms. In 1334–5 and 1336–7 at Ely at £8 a year with board and lodging, probably for the lantern. Chief Carpenter to the King for all work south of Trent and Humber, 1 June 1336; again at St. Stephen's the same year. At Ely at 6*s.* 8*d.* in 1339–40 and at £1 in 1345–6 and 1349–50 perhaps on stalls and work in the monastery. In charge at Windsor 1344 and at St. Stephen's 1346. Making stalls for chapel of Windsor 1350 and for St. Stephen's 1351–2. Died in or before 1354–5 when the Sacrist of Ely paid 10*s.* to his executors. In 1337 his wife had joined him in disposing of tenements in the parish of St. Mary Aldermary.

66 Ulstan, Archbishop of York, d. 1023. Osmund, Bishop of Sweden, d. *c.* 1067. Alwin, Bishop of Elmham, d. 1029. Alfgar, Bishop of Elmham, d. 1021. Ednoth, Bishop of Dorchester, d. 1016. Athelstan, Bishop of Elmham, d. *c.* 996. Brithnoth, Duke of Northumbria, d. 991. See Bentham (*Addenda*, 34) who gives the principal bone-measurements; from these it is reckoned that Brithnoth was 6 ft. 6 in. or 6 ft. 7 in.

67 See below, p. 73.

68 Bentham, *Ely*, pl. xlii, an elevation. A perspective (ibid., frontispiece) shows spire with pinnacles instead of square turrets.

69 B.M. Add. MS. 5832, f. 93*a*. 'Going on Purpose to *Ely* to see the great work which the *Dean* and *Chapter* was then about, in pulling down the old *Lanthon* & rebuilding it, at a vast Expence, I took Notes of the following antient *Arms* in painted glass in some of the *windows* of the *Cathedral*'. The date is 19 Nov. 1756 (altered to 1759 or possibly *vice versa*). 19 Nov. fell on Friday in 1756, Monday in 1759. Add. MS. 5832 (Cole's vol. xxxi) is made up of miscellaneous gleanings and poems and is not arranged chronologically. 1759 is more likely, as Essex's report to

The Times

Ely Cathedral. Lantern from the South

(View taken from a helicopter, 1952)

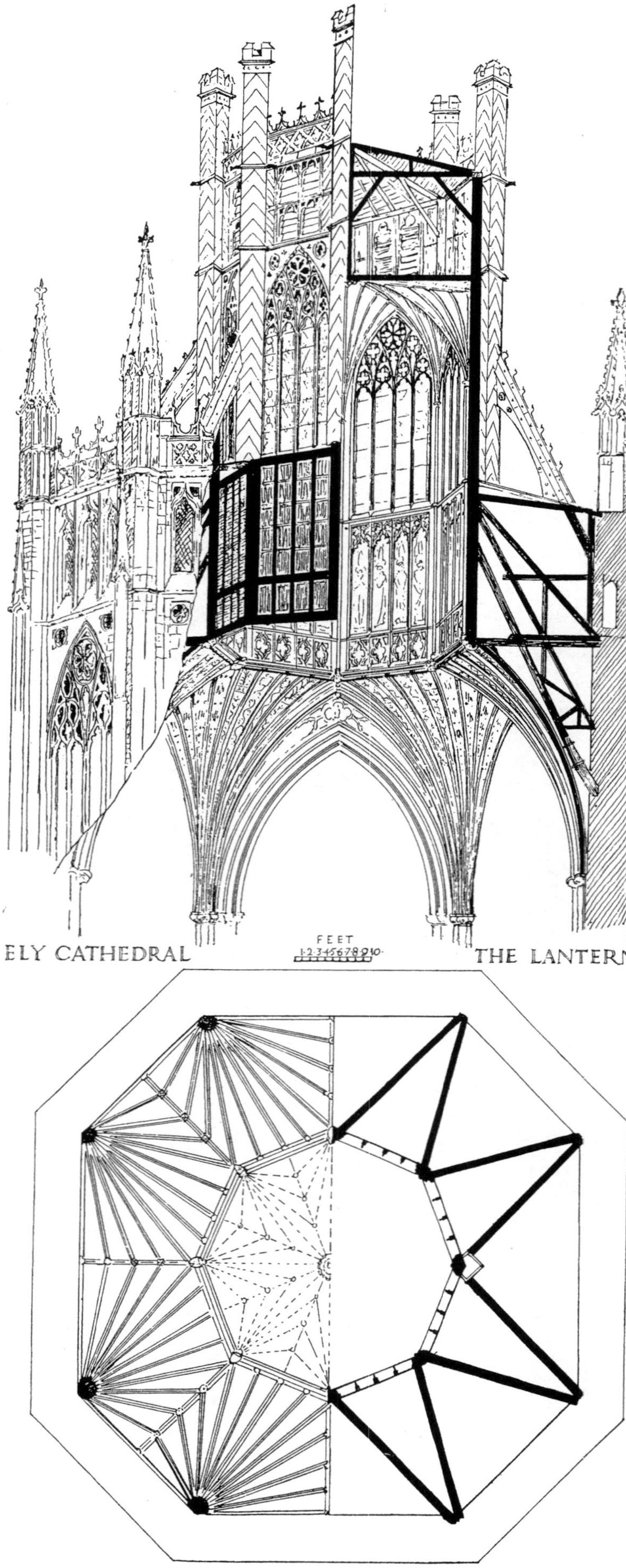

W. F. Haslop

Ely Cathedral. The Lantern. Constructional Drawing and Plan Prepared in Connexion with the Restoration of 1952

and carried out extensive repairs. Stewart[70] says that 'the original design or pattern of the tracery of the windows in the wooden octagon, has been irrevocably lost, but with this exception the strictly Decorative construction of this part of the church remains very much as it left the hand of the carpenters who put it together'. He states somewhat baldly[71] but unconvincingly that 'the roof of the bell-chamber is known to have been, from the first, protected with lead; but otherwise the timber was intentionally left freely exposed to the weather. The upper story was, in fact, intended to be (i.e. to show as) a wooden octagon, springing from a stone one.'

About 1860 a very thorough restoration was carried out by Sir Gilbert Scott as a memorial to Dean Peacock, to whose initiative the extensive works of repair of the church during the previous twenty years were largely due. Scott completed the shafts of the stone pinnacles and added spires; built additional pinnacles in the centres of the longer sides of the octagon, of which the old views give no sign, although there may have been some evidence in the fabric; added the picturesque cresting to the traceried parapet, perhaps from remains. He remodelled the casing of the angle-posts of the lantern and finished them without spires; reconstructed the parapet and added a cresting; and made new four-light windows with very elaborate tracery.

It is clear that the tower fell eastwards on the four remaining bays of the Romanesque presbytery between the tower and Northwold's work. One of these bays was absorbed by the Octagon; the remaining three were rebuilt at the expense of Bishop John Hotham (1316–37). There is a brief mention of the work at the very start when we find charges for sharpening the chisels and axes of the masons employed by the bishop and the priory in the same account. Thereafter we have no indication of the progress of the work, except that the west bay on the south side differs in some details from the others on either side and has been thought on these grounds to be the first to be built. The proportions are the same as those of Northwold's work. Neither Northwold's nor Hotham's rebuilding involved the destruction of the important pair of half-round columns of the Romanesque church which still stands. The piers of Hotham's work are coursed Purbeck marble with clustered attached shafts; the main arches have cusped spandrels; the triforium arch is subdivided by a shaft carrying very flat ogee arches elaborately cusped and with geometrical tracery; the clerestory has a wide four-light window with flowing tracery; Northwold's clerestory wall-passage is continued in Hotham's work. Above both the main arcade and the triforium there are enriched cornices with crestings. The vaulting shaft is carried on a high conical corbel with foliage in the spandrel of the main arcade. The detail of the whole is rich. In verbal description Hotham's work is thus very like Northwold's. The detail is, however, remarkably different. The century which separates the two is hardly an adequate explanation: there is as much difference between the art of Hotham's presbytery and the contemporary stone octagon as there is between Hotham's work and Northwold's. This shows that different architects were employed. The vault is a lierne-vault with one intermediate rising to the ridge between the diagonal and the wall and one between the diagonal and the cross rib stopped against the liernes before reaching the ridge. The bosses are carved with the foliage of the period. The vault of the south aisle is similar to Northwold's aisles with the addition of ridge ribs. The north aisle has a lierne-vault without intermediates; the liernes form an octagon round the centre boss and other liernes spring from the haunch of the wall rib and rise, parallel with the diagonal, to the ridge. The windows are of four lights except the westernmost which is of three lights on the south side, and on the north side of two lights (to leave room for a raised doorway); all are filled with flowing tracery. The buttresses have been already described along with the Northwold work.

WESTERN TOWER AND TRANSEPT, WINDOWS, ROOFS. West Octagon. The addition of the tall octagonal top stage gives much distinction to the dignified but rather monotonous square tower. It is one of the most important works at Ely and aesthetically is quite the most important single external feature. It must be considered along with two others: the strengthening of the Romanesque piers and arches below, and the work consequent on the fall of the north-west transept.

The new top stage belongs, like the central octagon, to the group of polygonal buildings characteristic of central England in the later Middle Ages. It follows closely the small steeple at Barnack. In both the octagonal stage is a late addition to an early tower and the designs of the two have a strong resemblance. At Ely the octagonal buttresses of the original work offer themselves as obvious bases for turrets, which follow the same plan and rise 75 ft., with panelled sides and high battlements. The turrets are connected with the alternate sides of the main octagon by two flying buttresses, the upper and heavier of which gives access from the stairs in the turret to the roof; they steady, not the tower, but the slender turret which could not have endured without them, and are, moreover, of great value aesthetically. The central octagon is set back about 10 ft. from the face of the tower and a battlemented parapet passes in front of it leaving a wide 'alure'; it is itself finished with a similar battlement 64 ft. above the corbel table of the tower. The construction is extraordinarily light. The intermediate sides are carried on squinch arches of the 12th century, suggesting that Northwold's timber spire was intended to have been of stone. The octagon was capped by a slender timber spire covered with lead. This elegant little structure, characteristic of the county, was removed in 1801, apparently only on the ground that it was not approved by the taste of the day.

The strengthening of the tower piers and arches must have been undertaken for one of four reasons: either as a general precaution against a repetition of the disaster of 1322, or because it had been decided to load the tower with a great additional weight, or because settlements had already occurred with visible results, or because of the scars raised by the fall of the north-west transept.[72]

Windows. Other medieval work consists chiefly of new windows. Late in the 13th century windows with simple geometrical tracery were made in the east chapels of the south transept: the only work of this character in the church. In the 14th century almost every aisle window in the eastern arm was made as wide as possible and the head filled with flowing tracery. The triforium windows were similarly treated; they were also made higher, the parapet was raised to allow of this, and the

the Dean and Chapter on the structural condition of the cathedral (copied by Cole in Add. MS. 5842, ff. 345–9) is dated July 1757.

70 *Archit. Hist. of Ely Cath.* 127.
71 Ibid. 125.
72 See above, p. 54.

roof made flatter accordingly. Two south bays alone were excepted, and were treated in a very peculiar manner: the original triforium windows were preserved, but the glass was removed; the arches of the triforium gallery were converted into windows, and the aisle roof was lowered upon the vault; the north side of these two bays was treated in the same way but the triforium wall and windows were not preserved; this work was done at the expense of Bishop Barnet, 1366–73, whose monument is in the easternmost of the two bays. Two of the original windows in the north aisle remained but these were renewed by Bishop Gray (1454–78) whose arms are seen on the jambs. All this work was no doubt done with the object of getting more light; the new windows were not only larger than the old but the new glass would be more translucent than that of Northwold's time.

The top window of the south transept, of six lights with tracery, dates from the 14th century; the north transept has two lofty three-light windows of the 15th century. In the nave the Romanesque windows of the aisle and clerestory were replaced by windows of three lights, without tracery, and with low four-centred heads in 1469–70.[73] Bentham's south views show two of the original windows remaining; the aisle windows have been destroyed again since his time and their place taken by imitation Norman.

Roofs. The nave roof is of trussed-rafter construction without tie-beams, probably the largest of the sort in England; it appears to have been continued in similar but not identical form[74] over the presbytery and possibly the whole was the work of Northwold. But the presbytery roof having become ruinous was renewed by Essex about 1760; it had tie-beams but it was the opinion of Essex that these were later additions to the original work; there would be some difficulty in getting suitable timbers long enough for the tie-beams (the clear bearing being 34 ft. 6 in.) and of sufficient scantling. A sketch[75] of the old presbytery roof has been preserved and shows diagonal cross-bracing cut by a collar half-way up the rafter and another collar above; the rafters were about 37 ft. long; the scantlings are not given. The names of the several members are given by Stewart.[76] The construction would produce a polygonal wagon ceiling of five cants.

In the 14th century a departure was made from the ordinary roof by the use of that with a truncated apex, sometimes called a 'kerb' roof, the predecessor of the French 'Mansard' roof. It was used on both sides of the main transept, and on the south wing of the west transept and on the Galilee.[77] The north side of the west transept was prepared for an ordinary roof, and such was actually put up for there are signs on the stonework of an alteration in it. The small hall of the Prior's House also has a kerb roof. This type is peculiar to the eastern parts of England and to this period. In the 15th century the transept roofs were replaced by the present hammer-beam roof. They are of the simplest possible construction of Norfolk type: that is, without collars or other strengthening features except the hammer-beams.

RITUAL ARRANGEMENT AND MONUMENTS AT THE CLOSE OF THE MIDDLE AGES. The ritual arrangement of the whole church has been so much altered from time to time that it is difficult to appreciate the significance of what remains. It is therefore necessary to describe in some detail its leading features and monuments as they were just before the Dissolution.

The pulpitum of the Romanesque church survived until 1770 when Essex destroyed it. Essex, however, left some rough sketches[78] of its west face and a short description, while the general arrangement is shown by Browne Willis in his plan of the church. Essex's description is as follows. 'The front of it was a solid wall pierced with three doors, and decorated with small pillars and feint arches, behind which was a low arcade which supported the roof loft, the walls or battlements of which are composed of open-work of little pillars and circles. The way up to this gallery was by a stone staircase, on the north side, still remaining.' The pulpitum was therefore an elaborate architectural composition 14 ft. 6 in. high of three arches with narrow stilted arches between and a rich parapet above. The central arch had a clear span of 2 ft. 2½ in. and was open to the floor; all the columns stood on a plinth about 1 ft. 6 in. high, which was carried across the side arches; these arches were 2 ft. 10 in. wide; they had a low segmental sub-arch with joggled joints and a tympanum of masonry jointed diagonally, lattice-wise. The four or six small columns shown in the plan against the east face of the wall are explained as a low arcade. Another wall crossed the church 8 ft. farther east. Essex mentions only one stairway to the loft, on the north side, whereas the plan shows a second and smaller one on the south. In the 18th century the loft contained the organ, and seats for the use of the bishop, dean, and canons when sermons were preached in the nave.[79]

The first pair of piers west of the octagon are in part of clean masonry showing where the pulpitum abutted before the original whitewashing was done. Between the south pier and the next to the west there was a wall 1 ft. 2 in. thick up to the soffit of the arch, and there is some evidence of a wall under the corresponding arch on the north side. On the second pair of piers there are marks of a screen: holes have been cut in the bases and at heights of 5 and 11 ft., clearly for attaching woodwork; these have been filled with modern stone. The northern colonnade has remains of painted decoration, consisting of a 7-in. band of scroll work from 7 ft. 10 in. to 8 ft. 5 in. high with masonry lines above; below the band the stonework is bare.

Taking into account all the foregoing facts it would appear that the west wall of the pulpitum had at the close of the Middle Ages a central doorway and an archway on either side with a sill 18 in. from the floor. These side arches may have been windows to give a borrowed light to the entry of the quire under the low floor of the pulpitum. One bay west of the pulpitum there was a light screen with the nave altar in the centre and a doorway on each side.[80] The principal nave altar,

[73] Atkinson, *Archit. Hist. Ely*, p. 11, quoting roll of steward of lands 9 Edw. IV.

[74] Perhaps additional bracing has been introduced into the nave roof giving the ceiling its present profile.

[75] B.M. Add. MS. 6772, f. 221, a careful tinted drawing. There is also a rough sketch with figured lengths in a letter from William Wiles at Ely to James Essex at Cambridge, 29 Feb. 1760.

[76] *Archit. Hist. of Ely Cath.*, p. 75, quoting B.M. MS. Tib. A VI, f. 248.

[77] Bacon thought that the west tower showed evidence of the nave roof having been the same.

[78] Sir W. St. John Hope, 'Quire Screens in Engl. Churches with special reference to . . . Ely' (*Archaeologia*, lxviii, 43). The sketches are in B.M. Add. MS. 6768, ff. 122–4.

[79] Ibid.

[80] These conclusions are at variance with those of St. John Hope. He sees in the central doorway a recess for the altar and in the side arches doorways from the nave to the quire; thus his reconstruction involves a good deal of alteration in Essex's description and sketches.

A. Vault boss with coronation of Virgin.
B. Vault boss with figure of St. Etheldreda.
C. Vault boss with figure (?) Bishop Northwold.
D. Door in Screen of Feretory.
E. Piscina of Second Lady Chapel.
F. Remains of Consecration Cross.
G. 'The Boies'.
H. Doorway (blocked).
J. Cupboard (blocked).
K. Position of Pulpit in 18th century.

1. Grave of Bishop Ketene (1310–16).
2. Grave of Bishop Balsham (1257–86).
3. Grave of Bishop Fountains (1220–5).
4. Grave of Bishop Walpole (1299–1302).
5. Grave of Bishop Orford (1302–10).
6. Monument of Bishop Kilkenny (1255–6).
7. Monument of Bishop Barnet (1366–73).
8. Monument of John Tiptot, Earl of Worcester (?1427–1470) and wives.
9. Position of monument of Bishop William Gray (1454–78).
10. Positions of Shrine of St. Erminilda.
11, 12. Graves of St. Withburga and St. Sexburga (exact positions unknown).

13. Monument of Cardinal Luxemburg (Bishop 1438–43).
14, 15. Grave of Bishop Eustace (1197–1215), alternative sites.
16–20. Grave slabs of priors, ? 15th century.
21. Grave of Bishop Kirkby (1286–90).
22. Grave of Bishop de Burgh (1225–8) and heartburial of Bishop Longchamp (1189–97).
23, 24. Graves of John Downham and John of Wisbech.
25. Grave of Bishop Montacute.
26. Grave of Bishop Fordham.
27. Grave, believed to be Prior Alan of Walsingham (1341–64). Slab 12ft. by 4ft. 6in. with sinking for brass of mitred ecclesiastic, mid. 14th cent.
28. Grave of Bishop Niel (1133–69).

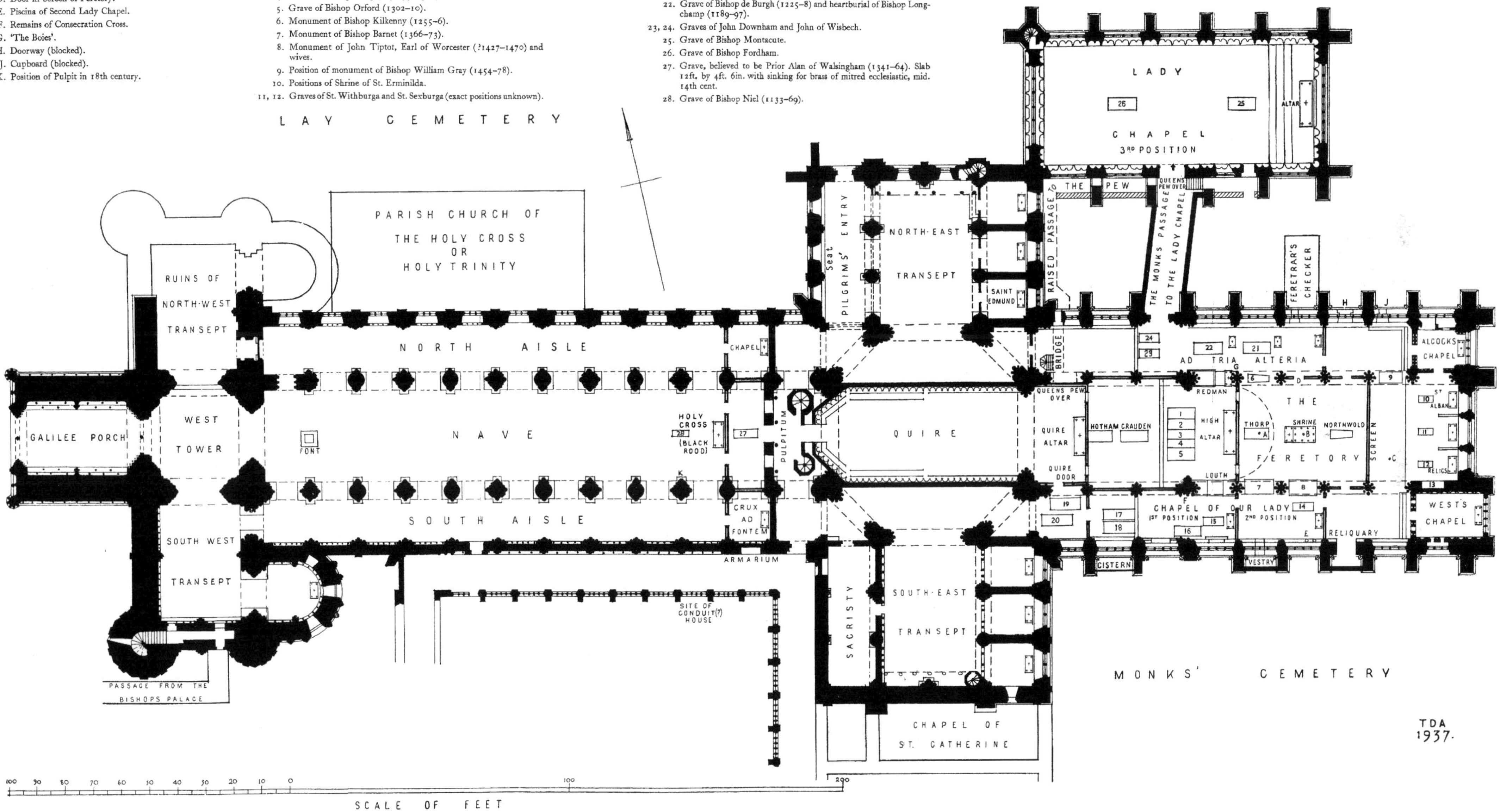

ELY CATHEDRAL. CONJECTURAL PLAN (*c.* 1530) SHOWING RITUAL ARRANGEMENT

dedicated to the Holy Cross, was also known as the Black Rood; perhaps the figure was of silver which had tarnished.[81] With the building of a separate church for the use of the parish in the 15th century the principal use of the altar was gone but it remained and was called *altare ad crucem*. There is a charge in the Sacrist's Roll for 1359–60 which would seem to be for the Lenten Veil hung in front of the Cross. In front of the central doorway of the pulpitum there is a grave slab with a sinking for a brass, believed to mark the grave of Alan of Walsingham who was buried *ante chorum*. The figure wore a mitre, and the right to wear a mitre was granted to Walsingham's successor William Powcher; the brass may have been made subsequent to 1413, the date of the grant. A little farther west is a stone which is believed to mark the grave of Bishop Niel (1133–69).

In the same bay as the pulpitum is the monks' entrance to the church. In the aisles this bay is cut off from the next to the west by the extension of the west wall of the pulpitum. In the south aisle the next bay to the west was the chapel called 'Crux ad Fontem'. This is one of the three altars that received offerings in 1291–2. In 1354–5 the sacrist paid a goldsmith 3*s*. 4*d*. for making the arms (*manus*) of this cross. The east wall of the chapel was 2 ft. 6 in. thick. There are remains of the painting on the vault; on the north-east pier there is a corbel and canopy for a statue. The floor of the aisles was until the mid-19th century several inches lower than the nave.[82] The corresponding bay in the north aisle, perhaps the chapel of St. Peter, was similarly enclosed and there is a niche for a statue.

The west aisle of the north transept has an external doorway; there is a stone seat against the west wall; it seems likely that this aisle was enclosed to form an entry for pilgrims.[83] The east aisle is divided by stone walls into three chapels. The southernmost is presumably that of St. Edmund as it contains a painting of his martyrdom; the west arch is enclosed by a wooden screen of about 1350 said to have been brought from the south transept. The remains of paintings on this chapel will be described below. The north, originally the centre chapel, contains the memorial to those who fell in the First World War; one of the piers has scars of the old screen. The transept was enclosed on the south by a stone wall against which the quire stalls were built. This wall contained small sealed cavities in which had been placed the bones of several Saxon bishops and of Duke Brithnoth.[84]

In the south wall of the south transept there was a doorway of Simeon's time into the monks' cemetery; it was blocked early in the 13th century when the building against the transept-end was enlarged eastwards. In the 14th century a small stair turret was built within the church under the narrow south gallery, to which and to the triforium it gave access.

The central part of the transept must have been used chiefly for such purposes as marshalling processions. It is believed that the clock, which was smashed in the fall of the tower, was in this transept; it is mentioned in 1291; the brethren were enjoined to be punctual for service *per horologium*.

The east aisle of the south transept was divided by

ELY CATHEDRAL. RECONSTRUCTION OF PULPITUM

[81] The name is found elsewhere, probably under like conditions.

[82] H. Goodwin, *Ely Gossip*, 47. Sir Gilbert Scott thought the arrangement original.

[83] The corresponding aisle at Winchester Cathedral has an external doorway and until the 19th century was enclosed by thick walls; shafts had been cut away to give more space and there was a 'sink', probably a stone washing trough; all indications of a use like Ely.

[84] For further details of the wall and cavities see above, p. 64.

cross walls but these were removed in 1814[85] to form the present library, the fittings of which cover the walls and hide any ancient features; the south bay has a window which has been cut about and may at some time have formed a doorway into a room over St. Catherine's Chapel. The west aisle was enclosed to form a vestry 50 by 18 ft.; it was entered from the cloister until the doorway was blocked and partly covered by a buttress of the octagon; the present entrance has its medieval door of latticed framework. The vestry appears to have contained 'seven great chests' in which were kept the vestments and jewels. An inventory of the former made in 1539 of what remained exclusive of what had been taken for the king's use may be summarized as follows: copes 61, albs 82, suits 16, 'vestments' 4, tunicles 10, altar fronts 5, besides various small things; probably Henry had taken few vestments.[86] Of the precious metals the king had taken 344 ounces of gold and 5,040 ounces of silver, making with what was left a total before the Suppression of 344 ounces of gold and 6,228 ounces of silver.[87] A few more vestments but no more jewels are described as being kept in other places.[88] Some of the vestments received proper names, as a chasuble called *Pascha Floridum*, a *capa* called *Gloria Mundi*, a chasuble called *Summa Confessorum*, and 'Theofile' with his story embroidered on it given by Bishop Kilkenny.[89] One valuable robe was given to Prior Crauden by Queen Philippa; a robe of purple-red velvet powdered with golden squirrels, which she had worn at her Churching after the birth of the Black Prince, was not included in the inventory of 1539.

The east triforium of the south transept was occupied as living-rooms which were reached by the small 14th-century stair turret on the inside of the south wall. There is a blocked fireplace, the character of which is hidden; the windows are near the floor; on the wall above them there is the mark of a ceiling; there is a sink or washing-basin in a recess lighted by a small quatrefoil window. The rooms, which were perhaps two or three in number, were probably occupied by watchers; the Ordinances of 1314 lay it down that two servants of honest conversation must guard the church day and night.

Entering the quire we find on either side a range of thirty-five canopied stalls.[90] Those near the west end are placed at a canted angle to avoid the piers of the octagon; a plan not found elsewhere. The actual return stall on the decani or south side was occupied by the bishop as abbot; there is no throne at Ely. The prior's stall, now the dean's, is therefore on the cantoris side. The entrance used by the monks as they took their places in the quire was on the south side.[91] Near to it was a small set of organs and a lectern; the great organs seem to have been in the pulpitum[92] as was the large organ of the 18th century. The quire was decorated at Whitsuntide with ivy which the homagers of Wentworth were bound to bring;[93] and the precentor paid the children of the almonry 2*s.* a year for strewing flowers in the quire.

The quire altar of St. Peter[94] stood against a screen[95] between the first pair of piers east of the octagon. Browne Willis shows the screen with two doorways and an altar 11 ft. long[96] standing on a dais of three steps which are very likely medieval. A table covered with a white cloth is shown without the steps and in incorrect perspective in the picture in the Palace of the funeral of Bishop Cox in 1581. The one bay between the stalls and the screen is enclosed on either side, by return screens, that on the south containing the doorway referred to above. The furniture about the quire altar at the time of the Dissolution consisted of two frontals, four large laten candlesticks before the altar and two others of iron, six iron stools with leather seats: probably high folding stools for the ease of assistants of the celebrant. It may be that it was over the quire altar that the 'folding table' or triptych was placed that was confiscated by Henry VIII and described by his Treasurer of the Jewels as: 'A table of silver and gilt, with two leaves, the inside plated with gold and garnished with saphires, balases [like rubies], small sparks of emeralds, and small coarse pearls lacking many of the stones with the collets [bezels, sockets] and part of the pearls in the back of the said table, plated with gilt plates of silver, weighing with the wood and stones 104 oz.';[97] no hint of the subjects of the pictures.

The north arch contained a raised gallery.[98] Although there may be no direct evidence of this, such as actual mention in a document, it may be inferred from a complete if slight chain of circumstantial evidence. The gallery in the Lady Chapel was reached by a long, raised passage from a doorway in the wall of the north aisle. Below the threshold of this doorway there is a skewback indicating the springing of a segmental arch 5 ft. 9 in. wide crossing the aisle very much after the manner of that at Norwich Cathedral. If it had been necessary to gain access to the Lady Chapel gallery only, a staircase close to the wall and leading directly up to the doorway, or one in an external turret reached by a doorway at floor level, or even the stairs against the Lady Chapel wall, would have sufficed: a bridge gives a strong presumption of some gallery[99] looking into the quire and of the importance of direct communication between it and the Lady Chapel gallery. There is room for difference of opinion on the use[1] of the galleries, but all the explanations have their difficulties; to almost all there is that presented by the elaboration of the connecting passage. For this passage vaulting was made between the buttresses[2] of the Lady Chapel when they were only some 10 ft. above ground; and the chapel was begun

[85] Stewart, *Archit. Hist. of Ely Cath.*, 33.

[86] Using the word in its ordinary sense.

[87] See also below, p. 76.

[88] See above, p. 62, and below, pp. 69, 70, 73.

[89] Cf. the contemporary inventory made by Cromwell's visitors at Peterborough (*Northants. Rec. Soc.* xii, 21, 22).

[90] See above, p. 64.

[91] The *solarium juxta chorum* referred to in the Precentor's R. of 1329–30 was probably in the temporary quire, in use during the building of the Octagon.

[92] Precentor's R. 1374–5, J. H. Crosby's transcript. 'Waynscot pro parvis organis et pro lecterno ad hostium chori 2*s.* 4*d.* . . . pro quodam lecterno in pulpite faciendo et pingendo 4*s.* . . . pro j cathedra in choro emendanda.'

[93] A like custom was observed at Winchester. The church was cleaned once a year.

[94] 'Altare Sancti Petri in choro. Magnum altare ex opposito [? tumuli].'

[95] *Sacrist R.*, ed. F. R. Chapman, ii, 117—bars and hooks 'pro le parclos juxta altare in choro'.

[96] Accuracy in a matter of this sort is not to be expected in the plan and he naturally calls it the high altar.

[97] Condensed from two passages in the acct. quoted in Atkinson, *Archit. Hist. Ely*, 177, from *Acct. of the Mon. Treasures confiscated* . . . (Abbotsford Club, 1836).

[98] See above, p. 62.

[99] Other galleries in churches are: 1. Cambridge: several college chapels. 2. Durham: in quire north of high altar. 3. Durham: in nave south of Jesus altar. 4. Ely: over Galilee. 5. London: St. Bartholomew's, Smithfield: in quire north side. 6. Westminster Abbey: in nave, south side, near west end. 7. Malmesbury Abbey: nave. 8. Norwich Cath.: in quire north of high altar. 9. Winchester Cath.: in nave, north side west end.

[1] Suggestions put forward for the use of the Ely gallery include: place for an anchorite; pew for prior, or for his guests; pew for bishop; place for exhibition of relics; loft for organs; for singing boys; pew for Queen Philippa.

[2] See above, p. 60.

Plate XVIII.

Ely Cathedral. Bishop Hotham's Monument, 1767
Before the Separation of Tomb and Canopy

(From Bentham's *Ely*)

twelve months before the fall of the central tower. The bridge at Norwich is said to have been for the exhibition of relics and perhaps for their preservation. The position of our quire gallery is like that near the high altar at Durham, which was first an anchorage and later the prior's pew; the gallery in Durham nave was also a pew for the prior; the Cambridge examples are analogous to the Durham galleries, being for the use of the master. In regard to its possible use for the quire organ which has been proposed there is some evidence that this was on the south side. A suggestion has been put forward that both the quire and the Lady Chapel galleries were pews for Queen Philippa.[3] As such, the elaboration and cost of the connecting passage would not be considered. Philippa arrived in England for her marriage in 1328; the provision made in the fabric of the Lady Chapel would have to be after that date which was seven years after the beginning of the work and twenty-one years before its completion. There was a personal friendship between the young Queen and the prior. She used to visit him, and it was almost certainly for her that he built the hall to the west of and connected with his own lodging. She gave him the valuable robe described above.[4]

The high altar was almost certainly 16 ft. long, for a large Purbeck marble slab now lying in the paving of the south aisle as a post-Reformation gravestone is quite certainly a part of an altar of that length; a *mensa* of such a length must have been the high altar or the Lady altar, which latter is shown by the reredos to have been of that length. The ornaments of the high altar at the time of the Suppression, exclusive of what had been taken away for the King's use, were as follows: 'A fronte of bawdkyn imbrodred with swannes. Thirteen altar clothes good and bad. Two rede tappetts [small pieces of carpet] to lye afore the altar, with roses and flowres, and other of blew. A pall of silke for an altar. A standyng lecturn of laten with an egle. Two great candlesticks of laten and 2 little candlesticks of laten.' In front of the altar were buried five bishops of the 13th and 14th centuries and a dean of the 16th.

In the middle of the three bays which he had built was the monument of Bishop Hotham, of interest for its architecture but more remarkable for its curious arrangement. The two parts of which it consisted, the tomb and the canopy, have both survived though they are now separated. The tomb is of clunch, with figures of weepers, of which a part of one remains, with a Purbeck marble top 9 ft. 8 in. long. In the slab there are six iron staples,[5] the spacing of which suggests that at the head of the figure there was the common form of a recumbent niche-head or that that part of the slab was covered in some other way. It is recorded that the effigy of the bishop was of alabaster. The canopy was a box-like thing, open at the top, on a vault supported by eight Purbeck columns. The upper part, of Barnack stone or the like, is ornamented with arched shallow panels formerly painted with coats of arms.[6] The detail is refined and delicate and has both natural and conventional foliage—oak, maple, and rose appear; two label stops are monks' heads with their black hoods pulled forward; one stop is a mitred head; some of the mouldings have and others had metal paterae fixed with ornamental studs; a good deal of colour and gilding remains on the north side.

The lower, arched, part of this canopy is 7 ft. 7½ in. long between the columns and 9 ft. 2½ in. outside. This was to cover a tomb some 3 ft. too long. The canopy is clearly shown, in a view in Bentham's *History*,[7] covering the eastern part of the tomb, the head of which disappears into the stone screen against which the canopy is built. It is curious that none of the historians and other writers ancient[8] or modern[9] who have described or referred to the monument has made the slightest reference to this unusual arrangement. Several possible explanations may be put forward: (1) that it was the bishop's intention that his tomb should be actually built into the choir altar which stood on the other side of the screen; (2) that a mistake in the measurements had been made, perhaps through tomb and canopy being made in different and widely separated places, and that Hotham or his executors accepted the work and hit upon this solution; or (3) that at some later time alterations in the quire included moving the screen eastwards, thereby cutting across the monument, destroying one-third part of the canopy and burying the head of the tomb in the altar, much in the same way as was at a later time done with Stephen Langton's tomb at Canterbury Cathedral, and for much the same reason. On the other hand, doubt has been expressed on the canopy being a part of Hotham's monument; it has been held by some since *c.* 1850 to be the shrine of St. Etheldreda.[10] That is impossible but it has been claimed that it was perhaps the private pew for Queen Philippa, and that the Elizabethan or Jacobean inscription,[11] or indeed any such labelling,

[3] See below. For admission of women to the quire, see Statutes of Bishop Walpole 1300, ed. S. J. A. Evans, *Camden Miscell.* xvii, p. 11, referring to Stat. Gregory IX (Matt. Paris, *Chron. Maj.* (Rolls Ser.), vi, 235).

[4] See above, p. 68.

[5] The staples are of stout iron but have very small eyes, and are hardly suitable for securing a grating like that over the Beauchamp effigy at Warwick. They may be holding-down irons. A charge in the Sacrist Rolls may or may not refer to these: '1341–2, In ydemptitate x barr' supra tumbam domini Johannis de Otham Episcopi 7*d*.' (*Sacrist R.*, ed. Chapman, ii, 117).

[6] Common-place book (ii, 894) of Thomas Gray the poet preserved in Pembroke College, Cambridge: 'Altar tomb under a low square canopy which has two scalloped arches open below and two close above on three of its sides with niches and weepers that have lost their heads. It has been painted of various colours; in one place the arms are visible ermine a fess duguy. The statue is torn from the tomb, nothing but the recumbent tabernacle remains upon it. Near it stands a Bishops figure of grey marble without a head which does not seem to have belonged to Hotham's monument.' Gray died 31 July 1771 and had probably completed the second of his three volumes long before 1770, when the monument was removed. There is a similar description in Charles Parkin's *Collections* in Pembroke. He suggests Clifford for the coat of arms. Hotham was barrully on a canton a chough.

[7] See plate, opposite.

[8] 'Ipse autem sepultus est in ecclesia sua cathedrali apud Ely et honorifice collocatus ad partem orientalem altaris in choro versus magnum altare sub quadam pulchra structura lapidea cum imagine Episcopi de alabastro super tumulum ipsius erecta cum vii candelabris ex uno stipite decentissime procedentibus et circa siquidem imagines de creatione hominis et ejectione ejusdem de Paradiso quatuor etiam imagines regum armatorum et iv dracones ad iv partes ejusdem structurae', *Anglia Sacra*, i, 648.

[9] Bentham, *Ely*; Gray and Parkins as above (n. 6); G. Millers, *Description of Ely Cath.* 89, 100. Stewart, *Arch. Hist. of Ely Cath.*, has a very slight reference to it without comment.

[10] It is said that this view was supported by Sir Gilbert Scott, but it would be hardly fair to hold him to that perhaps casual remark, even if he is correctly reported.

[11] Bentham's description is as follows: '... his Effigies in Alabaster; it was adorned with a sumptuous Branch, for seven Tapers, on the top of it; and in the several compartments on the sides and east-end, adorned with sculpture, representing the History of the Creation and Fall of Man; and many other decorations; the Tomb still remains, but much defaced, and divested of the above-mentioned ornaments: instead whereof is placed on the top of it a kind of wooden ornament, and a modern inscription (wrote about the time of Q.

almost proclaims the thing as a reconstructed antiquarian exhibit by Bishop Wren (1638–41 and 1660–7) or one of his school.

Under the arch immediately to the south of the altar stands the monument of Bishop Louth (1290–8) consisting of a table tomb[12] under a canopy of three arches with gables over them and tall pinnacles between. The monument is remarkably like the Westminster monuments of Edmond Crouchback, Earl of Lancaster, and Aymer de Valence (d. 1296) assigned[13] to Master Michael of Canterbury. Like those it has brackets for statuettes from the central gable; the panels of the tomb under the side arches are carved with the symbols of the Four Evangelists: the destroyed central part may have had a row of 'weepers'. The slab has an indent for a brass under a niche-head. There seems to have been a close grating standing on its south edge.[14]

In the northern archway is the tomb and chantry chapel of Bishop Redman (1501–5). It consists of a high tomb on which is an effigy of the bishop under a hanging canopy (one having a series of arches without supporting columns); the risers of the tomb have traceried panels with shields bearing the Instruments of the Passion and the arms of the see, and of Redman and of the see of St. Asaph of which he had formerly been bishop. The monument is placed under the western part of the canopy leaving a clear space to the east sufficient for a small altar of St. Andrew and for a celebrant. There is a reredos formed of three niches with a space for a picture below, the masonry now restored; at the west end there are corbels for three statues. The standards supporting the ends of the canopy have on their north and south sides four niches for statues (eight in all). The canopy arches are double cusped and have ogee hood-moulds with lofty and elegant finials; the cornice is stepped up over the middle arch. At the Suppression there was a vestment, an old diaper altar cloth, a frontal, and a little chest.

Prior Crauden was buried 'at the foot of the tomb of Bishop Hotham towards the high altar'. One day when the bishop 'had celebrated mass at the high altar and was returning to the vestry to take off his robes it happened that his pastoral staff broke on the very place where he was afterwards buried and turning to the prior who was walking with him he said: "Prior, this will be the place of my burial and you will afterwards be buried at my feet."' A grave slab, believed to be Crauden's, with the sinking for a brass of a kneeling figure at the foot of a cross, has been filled by a good modern brass and laid very nearly in the right position.

The screen behind the high altar doubtless had a doorway at each end leading to the feretory which appears to have included three bays and to have been enclosed on the north, south, and east sides by an iron paling. The existence of this fence is proved by the sacrist's account for 1349–50: 'Item in j pare garnet pro le Wyket versus feretrum'[15] and by the feretrar's account[16] for building his checker which is described as 'opposito hostii feretri S. Etheldr' ex parte boreali'.

Bishop Hugh of Northwold, the builder of the retrochoir, was buried in 1254 at the feet of the Saint and in the middle of his building. The monument was moved in 1770 to the arch north of its former position and north of the present altar. It consists of a low sarcophagus (modern) with a very elaborate effigy in Purbeck marble. The figure lies in a niche with a cinquefoil arch and canopy and richly carved columns. On either side of the niche is a series of three small niches containing figures in relief; on one side St. Etheldreda, an abbess or queen and a nun; on the other a king, a bishop, and a monk. Above the canopy is a sculpture representing the soul being lifted up to Heaven in a sheet by two angels; a motif going back to Hellenic times; an early Ely example will be noticed below. The bishop had formerly been Abbot of Bury, and at his feet is cut the scene of St. Edmund being shot by the Danes; thus the effigy was made to represent Northwold and was not merely a stock design.

The monument of Bishop Kilkenny (d. 1256), Northwold's immediate successor, is under the next arch; perhaps its original position.[17] Practically contemporary with that of Northwold's it has a strong likeness to it; in its greater severity it is perhaps even better than Northwold's; it has the niche-treatment, the arch of the niche being simple. Under the corresponding arch on the south side of the shrine is the monument of Bishop Barnet (1366–73). It is a simple high tomb without canopy but of great size for a monument of this type, filling as it does the whole space between the piers. The design is remarkable for its simplicity, the vertical sides and ends being covered with sunk quatrefoils; on the slab is the indent for a brass: a demi-effigy of the bishop under an elaborate canopy. In the next bay eastwards is the fine monument of the scholarly 'Italianized Englishman' of sinister record, John Tiptoft, Earl of Worcester (d. 1470), and two of his three wives; the three effigies lie on a high slab.[18] The hanging canopy, the second of the Luxemburgh type, has a level cornice and cambered lower transom. To the west of the shrine was the grave of Sir William Thorpe covered by a large slab with an elaborate brass.[19] Sir William was brother and heir of Sir Robert Thorpe, Chancellor of England; he died in or before 1398 and stipulated in his will that he should be buried near the tomb of St. Etheldreda.[20] The feretrar had, in addition to his primary duty, the keeping of the relics, and he took the offerings, averaging £15 in 1420–9, made at the relics altar.[21] Presumably in this part of the church and within the feretory enclosure there was the statue of St. Ursula for which the feretrar bought a crown from the goldsmith in 1423–4.

The remains of St. Etheldreda had been translated 'into view' from the old church on 17 October 1106, by Richard the last abbot (1100–1107). Neither the

Elizabeth or K. *James* I) . . .' (*Ely*, 158). This description is evidently based partly on observation and partly on the passage in *Anglia Sacra* quoted above (n. 8) in which, however, he failed to distinguish the parts applying respectively to the monument and the candelabrum.

12 The centre part has been cut away and the slab placed on the floor.

13 W. R. Lethaby, *Westminster Abbey and the King's Craftsmen*, 182, 184.

14 Indicated by the groove and by the inscription immediately to the north of it being cut to be read from the north. Cf. the well-known grating of the monument of Queen Eleanor who died in the same year as Louth's accession. The inscription is now illegible. The indent shows the niche widening upwards. The colouring on the north side has been renewed.

15 Garnet is a hinge; the word is still in use. No money is given.

16 The year is not known; it is believed to be between 1420 and 1429.

17 Thought to be so by Millers.

18 The earl was buried in the tower after decapitation; his wives are buried here.

19 The slab is now opposite to the entrance to Alcock's Chapel; the brass has been removed.

20 *Cal. of Wills . . . Court of Husting, London*, ed. R. R. Sharpe, ii, 326. Thorpe completed the Divinity School at Cambridge. (R. Willis and J. W. Clark, *Archit. Hist. Univ. Camb.* iii, 10, and *D.N.B.* s.v. Thorpe, Robert.)

21 J. H. Crosby, MS. notes.

Suggested Reconstruction, showing Cover in Outline

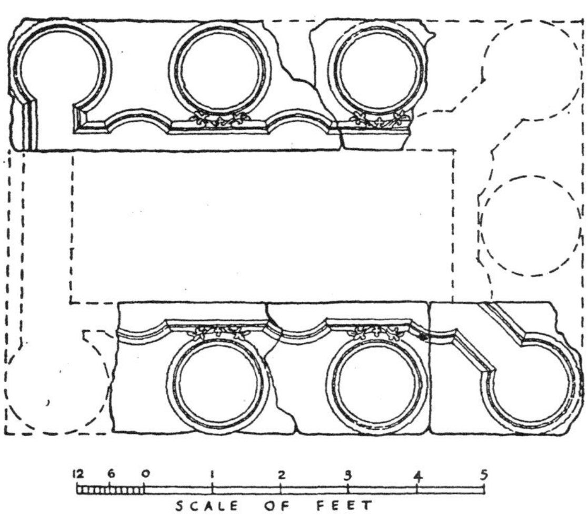

Plan as at Present

Section through Base of Column (reconstructed)

ELY CATHEDRAL, SHRINE OF ST. ETHELDREDA

old position nor the new is known. The new position must have been to the west of that which is here assumed (see plan) for the next renewal (by Northwold), for otherwise it would have been outside the Romanesque church. Bishop Niel (1133–69) robbed the shrine of much precious metal but it was repaired by his successors Ridel and de Burgh who covered the 'cumulus' with silver. Of Northwold's rebuilding there are some slight but important remains, and an account for work done on it at a later time, and statements of the offerings made. The remains consist of: fragments of a slab of Purbeck marble on which are worked a series of shaft-bases; some loose upper bases which stood on the slab-bases and are prepared for 3-in. columns; and (hardly more doubtfully) an arch cut in one stone with another, and projecting, trefoil arch and gablet below it. The lower bases are in two rows; the back row are arcs of circles joined by a plinth with the same mouldings; they are placed opposite to the intervals between the front bases; the whole plan is the same as that of the internal arcades of the Galilee. The arcades clearly covered three sides of a rectangular mass of masonry, one of the short sides being plain. This arrangement exactly suits the pedestal of a shrine, the plain end being to the west and forming a reredos for the shrine altar. The restoration here attempted is based on these fragments; the height of the columns, the capitals, the spandrels, and the cornice are conjectural. The degree of richness indicated is justified by the treatment of the bases; the upper bases have sprays of foliage growing downwards from the necking and lapping round the roll; the lower bases are connected with the plinth behind by delicate sprigs of foliage. The feretory itself, the coffin containing the relics of the saint, which stood on this high podium has, of course, been entirely destroyed; it is shown in the sketch as hidden by its protecting cover (*cooperatorium*). The account for making and decorating the cover in 1455 includes the following payments: for the expenses of John Soham, B.C.L., going by land and water to 'Welles' and Wisbech for a carpenter and 'graver', 2*s.* 3*d.*; for making the cover, 6 marks; for the labour of Robert Pygot 'peyntour' of Bury and of one Henry, 12*d.*; to them for a picture 12 marks with their keep; for 11 ells of linen cloth to hang within the cover 6*d.*; for 300 gilt nails, 7*d.*[22] There is also a payment of 12*s.* to T. Glaswryche. Perhaps he was a glasswright who glazed peepholes in the cover. In the latter half of the 14th century the annual offerings at the shrine sometimes amounted to £40 as against about £25 from the five other similar sources, including the high altar, acknowledged by the sacrist. The shrine was violated by robbers in 1386.[23] Opposite to it was the feretrar's checker, described below.

The architectural detail of the church was somewhat enriched in the part immediately surrounding the shrine. The boss in the vaulting over it is carved with a seated figure of the foundress wearing wimple, hood, and crown and holding in her right hand a crozier and in her left a book; the boss to the west has the Coronation of the Virgin. In the second bay from the east there is one other boss in the same style; it has the seated figure of a monk, hooded, holding in his right hand two keys and in his left, which is in his sleeve, the model of a cruciform church with central tower and spire; this might be Bishop Northwold who was a monk, or Prior Ralph (1229 until after 1235). The corbels of the vaulting shafts are also enriched and one capital of the main arcade on the south side has leaves growing up round the necking.

The bay east of the feretory contained, under the north arch, the monument now destroyed of Bishop William Gray (1454–78). Bentham shows it as a slab with the indent of a brass on the floor with two short wing-walls about table-height at head and foot; the canopy is a very flat arch under a level cornice; on the standard at the head there are indents for brasses. The monument was of 'grey marble', presumably Purbeck; the slab remains on the floor. Gray had contributed to the strengthening of the west tower and gave two windows in the north aisle, near his monument on the jamb of which are shields of his arms (Gray[24] impaling the see) and of Gray impaling a saltire.[25] On the pier adjacent to the monument is a fragment of a paper notice, with the red lion of the Gray family at its head.

There were three altars against the east wall. On the north the altar and shrine of St. Alban. The saint's relics had been sent to Ely by the St. Albans monks for safety during the Danish incursions and when peace was made they were returned. The monks of St. Albans protested that they were not those of the saint, but it did not matter, for the bones they had sent to Ely were not St. Alban's but those of some nameless monk. The dispute was ultimately settled in favour of St. Albans and the Ely shrine does not seem to have acquired a reputation, for there is no record of oblations being made at it. The dedication of the central altar is unknown; the southernmost is that of the relics,[26] so called, it is thought, from its being near the reliquary. The lower part of the east wall has traces of thirteen painted panels with ogee-arched cusped heads which evidently formed a continuous series and contained figures of saints.[27] In this part of the church were buried saints Sexburga, Erminilda, and Werburga, abbesses; the positions of their graves are not known either actually or relatively to one another. The monument of the Cardinal Bishop Lewis of Luxemburgh (1438–43) fills the easternmost arch on the south side. It is a high tomb with an effigy and has a hanging canopy of three arches; the panelling above has a cambered sill and lower transom. The effigy lost its head in the 18th century.

The easternmost bays of the aisles are the chantry chapels which Bishops John Alcock (1486–1500) and Nicholas West (1515–33) made during their own lifetimes. Alcock's Chapel is formed in the north aisle by enclosing the west and south sides with screen walls; no outside work was required except the blocking of the north window; the tracery of the east window is of the 14th century. A foundation or consecration stone was found in the 19th century near by and is now preserved in the chapel; it states that Alcock caused the chapel to be made in 1488. The chapel is characteristic of its late period in including a vestry 7 ft. by 3 ft. 3 in. The monument is in a recess on the north side and consists of a high tomb with a cadaver below; over it is a hanging canopy of two arches embraced under an ogee; the grave is in the centre of the chapel. The niches for

22 Stewart, *Archit. Hist. of Ely Cath.* 130. 'Welles' is either Wells (Norf.) or, more probably, Outwell or Upwell, which are near Wisbech.

23 *Ely Dioc. Remembrancer*, no. 131, p. 59.

24 Gules a lion rampant and a border engrailed argent.

25 Possibly for Neville.

26 Cardinal Luxemburgh is described as being buried near the Altar of the Holy Relics, which must have been this one as that at the end of the south aisle would be at its back.

27 Cf. the row of statuettes facing the shrine at Winchester.

statues[28] form a single tier; they start from the floor and contain high pedestals; the canopies are triangular in plan and are solid but covered with ordinary architectural forms; they are surmounted by high openwork spires. The vault is of fan form with a very bold openwork pendant boss; the wall ribs are cusped. The chapel has been censured as heavy and dull; it seems to be the design of a west-country mason, perhaps a Gloucester man. The small windows in the screen-walls have the original painted glass displaying the rebus of the bishop: a cock perched on a globe, frequently repeated in Alcock's work. The west entrance has its original iron gates; there is now no gate in the south doorway. At the Dissolution the chapel contained 'a gilted table upon the Altar' presumably either a portable altar or a picture, two altar cloths, four 'vestments', and a chasuble.

West built his chapel in the south aisle, 1525 to 1533. He enlarged the area by taking down the east wall up to the level of the vaulting and rebuilding it as far east as the projection of the buttresses would allow. His monument is on the south side; he rebuilt the vault in two bays with a rich lierne vault without diagonals and with small deep cells decorated with Italian carving; the central bosses are particularly remarkable, each consisting of a descending angel bearing a shield of the bishop's arms; the chapel is enclosed by a beautiful stone screen, unpierced except by the doorway. The walls of the chapel are covered with two tiers of tabernacle work for statues with panels above filled with sculpture in relief.[29] The lowest niches start from the floor but contain high pedestals for the statues, the capitals having carved foliage in the highest relief and very delicate, all now woefully damaged. The niches are separated by groups of pinnacles or buttresses among which are small decorative tabernacles containing figures in relief 6 in. and 3 in. high; even these small niches have gables on which minute men and animals gambol. The bishop's motto: GRACIA · DEI · SUM · ID · QUOD · SUM · appears over the doorway outside and is repeated many times. The Italian detail will well repay close study; of heraldry there is very little, the bishop's arms[30] appearing but few times, and then in association with others, as France and England, and the three crowns of Ely diocese. The grave is presumably under the centre of the floor;[31] the monument is in a low recess on the south side, and formerly had thirteen figures sculptured or painted in niches on the front and seven at the back of the recess; the panelling above the monument is pierced so that the window behind it is seen through with admirable effect. Many fragments of bright colour remain showing that the whole was brilliantly painted. The iron entrance gate has a fine flourish of briar rose naturally treated; it is of Flemish character and accomplished technique, and is probably the work of a Fleming settled in England.[32]

The destruction of these statues, the work of an Englishman, Edmund Moore,[33] and of all the small figures and reliefs, is probably the greatest single loss that Ely has suffered. The chapel contained at the Dissolution four 'vestments', an altar frontal, a hearse-cloth of black damask with a white cross, possibly that used at the bishop's funeral six years before, and a few other things. When the quire was removed from the octagon and the stone walls against which they were placed taken down, the relics which had been sealed up in cavities in the north wall were brought to West's Chapel and placed in holes prepared for them in the south wall, in the same order as they had held in the transept.[34]

In the third bay of the south aisle the wall contains a wide archway contemporary with the building, and outside there is evidence to show that a small building, 12 by 8 ft., filled the space between the buttresses. There can be little doubt that this was the reliquary. A roll in the arch-moulding contains a series of small iron pins let into holes in the stone; they are very slight and can have held nothing heavier than metal enrichments. The following are charged for by the feretrar, showing that he had custody of the reliquary: a leaden laver (*lavacro*) standing at the mouth of the reliquary with the octroi (*dacione*) thereof, 2*s.* 2*d.*; a large round bowl (*pelve*), 3*s.*; a large table for the reception of relics on Ascension Day, 2*s.*[35] The two next bays formed Northwold's Lady Chapel. The easternmost has two piscina basins (plan E). The western bay had between its buttresses a vestry similar in plan to the reliquary; it is shown by Browne Willis and referred to in the Ordinances of 1330.[36] Bishop Eustace, who built the Galilee, was buried 'prope altare sancte Marie in capella veteri'. Projecting mouldings of the Worcester monument have been cut away, evidently to allow some piece of furniture to be pushed back. Bays 6 and 7, west of Northwold's Lady Chapel, were the site of Simeon's 11th-century Lady Chapel. The pier between these two bays marks the junction of Hotham's three western bays with Northwold's presbytery. At the height of 4 ft. from the floor there are some remains, in red paint, of one of Hotham's consecration crosses: the only trace of a consecration cross in the church.

West of these bays the aisle is crossed by two Purbeck steps. Below these there are three Purbeck grave slabs 11 ft. 6 in. long[37] in a row. The two southernmost are said by Browne Willis to mark the graves of priors and this is not unlikely; they are of the 15th century. The third stone has the indent of a late monumental brass and is post-Reformation; but it also shows clearly two of the small incised consecration crosses of an altar: one of the angle crosses and the central one: these give us the size of the altar, namely 16 ft. by 3 ft. 6 in.; one end and one side have a hollow chamber underneath. The slab must have belonged to the high altar or to the 14th-century Lady altar which is shown by the reredos to have been of that length. Between the buttresses of this bay there appears to have been a cistern in former days, but we have no knowledge of the period of its construction.

Some part of the north aisle, nowhere precisely defined, was commonly called *ad tria altaria*. Here were the altars of Saints Martin, Benedict, and John Baptist,

28 The inside contains niches for 13 large statues, 4 medium, and 80 small, and in addition 12 small at the back of the tomb recess; the outside 8 large, 80 medium, and 66 small; total 263. There are occasional medieval fancies, such as snails in combat.

29 The number of sculptures was approximately as follows. Inside: large statues 38, medium 2, small 70. Figures carved in relief: 3 in. high 65, the same 6 in. high 38. Sculptured panels 19. Outside: large statues 9, medium 10, small 9. Total 260.

30 Argent between three slipped roses, a cheveron sable. The spread-eagle appears once.

31 'A marbell stone with carriage £4 7*s.* 10*d.* To Sheryff for paving the chapel £3 16*s.* 6*d.*'

32 To the smith of Ipswich for the ironwork, £7 13*s.* 4*d.* (D. and C. Mun., Paper R.).

33 To Edmund Moore upon a reckoning for the images, £9 (ibid.).

34 See above, p. 64.

35 R. of feretrar, 5 Hen. V (1426–7), J. H. Crosby's transcript.

36 The passage is given in Atkinson, *Archit. Hist. Ely*, 33.

37 The greatest length obtainable.

but their relative positions and how they were placed is unknown.[38] Here also, equally vaguely, were the graves of Bishops Longchamp, Burgh, and Kirkby. The wall of the bay west of the Lady Chapel door has an incomplete inscription asking for prayers for deceased benefactors; it is evidence of paintings of benefactors on the wall below, without which it would make small appeal to pilgrims who would pass it at the beginning of their progress. It was partly destroyed by a mason in the 19th century in the process of 'refreshing' the surface of the wall. The legible part runs '. . . pro animis benefactorum ecclesie Eliensis et omnium fidelium defunctorum'. On the column to the north of the high altar hung 'the Boies' (from medieval Latin *boia*): the fetters (or copies) which St. Etheldreda struck off the criminal Britstan. They were hung here by Bishop Niel (1133–69), and offerings are recorded yearly from 1302 till 1420; a cross was put up over them and renewed in 1352–3.[39] The Boies were the origin of the well-known toy St. Audrey's (tawdry) chains, sold at Ely Fair.

In the fourth bay from the east, opposite to the shrine, is a blocked doorway which formed the entrance to the feretrar's checker. This building was of two stories, and the quantity of the lead, 2 foders, charged for the roof suggests a length of about 20 or 25 ft. with a width of 12 ft. (the space between the buttresses). Bricks called 'waltyl' were brought from Wiggenhall (Norf.) about 20 miles down the river; as only 4,700 were used, it is clear that they were for an inner lining only, as in the great barn at the south end of the College. The windows were glazed and barred, and there were two privies. The timber bought includes 'overwayes' [floor joists], 'bemys', 'spars' [rafters], and 'stodys' [upright studs]; poplar boards for beds, a tripod, a pair of bellows, a pair of tongs, and some napery indicate that the upper story formed the feretrar's checker and private chamber; it seems also to have been arranged to form a watching loft. The shrine had been robbed in 1386.[40] The same account includes some alterations to the aisle windows. At the Suppression the rooms were allotted to a 'discreet' monk who remained; they seem to have been destroyed before 1649. In the next bay, the third from the east, there is seen on the outside the bases of two nook shafts of a contemporary doorway.[41] In the second bay from the east there is a rough doorway of unknown date, now blocked, which may indicate the position of the ankerhold of the hermit John Growe who is known to have been living in 1434–5; a little hut between the buttresses would have sufficed for him. In this part of the church several fragments are preserved. (1) The massive arms of the stone seat shaped as an animal grasping a human head. It is of 13th-century character and may be Northwold's work. If so, it probably represents the wolf which guarded the head of St. Edmund after his martyrdom. (2) The headless Purbeck marble grave slab of a bishop with a finely sculptured effigy of the first half of the 14th century; the raised hand stands out clear, the staff is across the body, the scarf wound round it, the feet rest on a bird; the bishop wears fringed dalmatic, chasuble, and stole. Around the edge of the slab there is a running scroll of ball-flower. This is perhaps the monument of Bishop Ketene. (3) An early sculptured grave slab brought from St. Mary's church, where it was found buried in 1829. The Archangel Michael, under an elaborate canopy, supports in a figured cloth the soul of a bishop, represented by a small naked man with a crosier. Prior[42] thinks the slab is from Tournai in Belgium and dates it at about 1150.

POST-DISSOLUTION HISTORY. The Main Transept. The north-west corner of the north transept fell without warning on 29 March 1699 owing to an earthquake felt some years before. The chapter accounts for 1698–9 show an expenditure 'upon the occasion of the Breach' for £97, presumably for shoring and so forth. The chapter seems to have begun by consulting Mr. Grumbold[43] and some masons of the name of Boston. In August (1699)[44] Dean Lambe was in London: he was going to get the opinion of the 'ablest builders' and to consult the archbishop. The latter commanded him to consult Sir Christopher Wren. Wren and Mr. Fulkes could not exactly understand Grumbold's plan and thought his charges excessive (and Wren would write to him) but thought the Bostons' account and demands plain and reasonable. On 14 June 1700 the chapter resolved to 'agree with Mr. Grumbal of Cambridge freemason to reform and do all the mason's work . . . which shall be built to the other part from which it fell, exactly in the same manner, and on the same foundation it stood before and that Mr. Grumbal shall receive . . . what it shall be adjudged to be worth (by the foot) by Sir Christopher Wren, Mr. Banks his Majesty's master carpenter and Mr. Fulks overseer of the mason's work on the Cathedral Church of St. Paul's in London. . . .' Notwithstanding the deliberation in arriving at this decision it was not adhered to and the work was done in Wren's manner and of new Ketton stone. The sum of £443 was spent in this first year and the work was finished in 1702 at a cost of £1,667.[45] There is nothing to show who was responsible for the design: there is no record of a fee. It is strongly held by some that the doorway is Wren's and in his early manner, say 1665;[46] a favourite design used at Bow Church, finished 1680. But Wren's style had become familiar to many by 1700, and if built in 1665 the masonry would certainly bear traces of the earthquake.

There was a balance of £385 on the north transept account. This may be said to have been earmarked for general repairs, for the chapter had so ordered in June 1700 before they knew that there would be any overplus. It was probably spent on the south face of the

[38] The precentor charges for 3 missals *ad tria altaria* in 1360–1, J. H. Crosby's transcript.

[39] *Sacrist R.*, ed. Chapman, ii, 156.

[40] See n. 23.

[41] See above, p. 57.

[42] E. S. Prior and A. Gardner, *Med. Figure-Sculpture in Engl.* 84.

[43] Robert Grumbold, of Cambridge, a good builder-architect, b. 1639; master mason Trinity College Library 1676; designed and built hall of Clare Hall 1687.

[44] One of Dean Lambe's letters is dated 20 Aug. without year and his second is without date. (See article by Revd. R. H. Gibbon in *Church Quarterly Rev.*, Jan. 1935). As to the cause of the fall, Wren thought that 'the decay of the peer was on the inside of the isle because it fell inward'. Maybe the dean (1693–1708) had not informed him of the earthquake which 'was observed particularly to affect that part of the church' (Memorandum by Mr. John Dowsing, one of the minor canons and Deputy Receiver). All the visible mischief due to the earthquake had been made good.

[45] Wren had advised the dean about the raising of money and about having the window frames of new stone while the rest might be of old stone. He seems to have thought that the work would take two years but that there need be no fear as to the decay of the scaffolding.

[46] In writing to Grumbold about his charges Wren, who hoped to effect a reduction, may have enclosed a sketch design. The separation of the functions of architect and builder was not then complete. The employment of Wren to advise, perhaps to design wholly or in part or not at all, and to value the finished work would be in accordance with the practices of the Middle Ages.

south transept. The chapter had ordered that £100 be set aside 'towards the refraising and beautifying the south side of the Church' in 1696, and in 1697–8 Grumbold was paid 5*s*. 'to view the South Wall'. We may assume that both entries refer to the transept, part of which has been refaced with Ketton stone. The north part of the east pane of the cloister was enclosed to form a vestibule to the church. Both doorways are provided with medieval iron gates taken evidently from the east part of the church. The part of the north aisle wall against which the parish church had abutted was refaced in 1662.

The west aisle of the south transept, built as a sacristy, was originally or at some later time divided by cross walls, and is so shown in Browne Willis's plan. The south bay was the 'Archive Room and Chapter House'; it was entered from the transept; in the middle of the 19th century Statutable Chapters were opened here and immediately adjourned to the deanery dining-room[47] where they are still held. The wall between the vestry and transept was destroyed in 1840, its place being taken by the present arcaded stone screen;[48] it was probably at that time that the handsome oak door with Bishop Alcock's rebus was hung; it was brought from Landbeach, having been sent thither from Jesus College, Cambridge. The eastern aisle, originally three chapels separated by walls, are so shown by Browne Willis, who calls the southern bay 'The Library' and the two northern bays 'Vestreys'. The dividing walls were doubtless stone and 12th century, like those of the north transept, and the enclosing screens wood and of later date. All have been destroyed to form a larger library, which now consists of three bays enclosed by walls.

In 1849 the chapter library consisted of 4,300 volumes, mainly theological but with an admixture of works on English history and the classics. The larger part had been bequeathed by Bishop Patrick (1691–1707) and the Revd. Dr. Ralph Perkins (d. 1751), Canon of Ely 1715–51.[49]

The Quire. In 1768 it was decided to move the quire to the east end of the church. It had been first suggested by Bishop Gunning (1675–84), but his scheme not taking effect he bequeathed money for a new pavement. It has been thought that the idea of the removal was prompted by the extreme cold of the building in winter; that it was felt that the east part with its stone vault and low aisles would be at least less uncomfortable in a building without any artificial heating than the old position under the lofty lantern and with the open-timber roofs of the nave and transepts on three sides. But there is no suggestion of this in Bentham. He says that it was agreed that the removal would be 'one of the most useful and ornamental Improvements that could be effected; the Design was worthy of that Taste and spirit of Improvement, which so eminently distinguish this Age and Country'. He points out that the east part had been built for 'the reception of the High Altar, but more particularly to make room for the Magnificent Shrine of St. Etheldreda, and for such-like gainful and superstitious purposes', and consequently became useless at the Reformation; which is very true. It had also been found better for reading and hearing. Moreover, the octagon 'being already laid open is now seen as it deserves', like St. Paul's in London.

The work was carried out in 1770. Of course, it was necessary first to remove any monuments that remained in the central space including Hotham's and the reredos of the old quire altar. There are two plans of the new quire in the second edition of Bentham: one, dated 1 November 1770, is described as 'the Choir as proposed... which Work is now carrying on'; the other, dated 1811, is said to be 'as designed by Mr. Essex'. The first plan, which no doubt was also by Essex, shows a sacrarium of two bays, stalls occupying four bays and an open space with an organ-loft above in the easternmost of Hotham's three bays. The front of the loft is supported on two columns with responds against Hotham's piers, and these columns probably carried arches; at the back there is a narrow enclosed space with entrance to the quire in the centre, and a staircase on either side. The work, of course, involved repaving, the total effacement of any traces of graves, and the removal of several monuments, notably Bishop Hotham's and Dean Caesar's. The latter stood against the seventh pier from the east on the north side; it was moved into the north aisle and put against the blocked doorway to the Lady Chapel whence, about 1850, it was moved about 50 ft. to the west. The space occupied by the stalls is about 65 ft., against 85 ft. shown by Browne Willis as the length of the medieval stalls, but the number, thirty-five each side, is the same, a discrepancy not explained. The pair of arches beyond the stalls are filled with screens with doors. The easternmost bay is panelled at the sides and end, the corners coved. The new work was evidently done in the Classical[50] style of the period. Some remains—good work of the Corinthian Order—preserved in the triforium are believed to be part of the reredos. When the stalls and the stone walls across the octagon and the pulpitum had been cleared away the octagon was paved.

Restorations. The thorough restoration of the church was begun in 1845, under the direction of Dean Peacock (1839–58) to whose initiative the work was in great measure due. A floor above the main arches in the west tower was removed in 1845, bringing into view the fine 12th-century masonry above. It was then that George Basevi, who was paying a casual visit to the cathedral, was killed by falling through a hole in the scaffolding. Basevi was not there professionally, for neither he nor apparently any other architect was consulted in the earlier stages of the restoration.[51] The present ceiling of the tower was put up and painted by Henry le Strange,[52] and finished in 1855; the floor was laid in 1870, and the ruined apse rebuilt in 1848.[53]

The most important single item was the second removal of the stalls. The question had long been mooted, and the work was carried out in 1852. At one time it was proposed to replace the stalls in the octagon. The ultimate decision was, of course, a compromise: they were put in their present position in Hotham's three bays. They were enclosed by a screen designed by Sir Gilbert Scott, and elaborate stalls for the bishop and dean, and sub-stalls were added. The reredos was the gift of John Dunn Gardiner in 1858; it was designed by Sir Gilbert Scott, and executed by Rattee and Kett of Cambridge; it is a very rich and costly piece of tabernacle work, containing sculptures of scenes from

47 Goodwin, *Ely Gossip*, 13.

48 This was no doubt done to admit more light and air to the two north bays.

49 B. Botfield, *Notes on the Cath. Libr. of Engl.* 124–31; *Rep. of Select Cmee. on Public Libr.*, p. 229, H.C. 548 (1849), xvii.

50 A plan and elevation by Essex dated 1811 (Bentham, *Ely*, 2nd ed.) show panelling in the 18th-century 'gothic' style.

51 Goodwin, *Ely Gossip*, 31.

52 Henry l'Estrange Styleman le Strange of Hunstanton Hall (Norf.), 1815–62.

53 The chapel was dedicated to St. Catherine in the mistaken belief that it was the medieval chapel with that dedication.

scripture; on either side of the reredos is a traceried screen made at the same time.

The principal altar ornaments when they were catalogued in 1901 were the following. A cross of silver-gilt with gold and enamel ornament and set with jewels, strengthened at the back with copper, on a copper-gilt base and shaft, at the back the arms of Woodford and of the see and an inscription; the whole 37 in. high; made in memory of Bishop Woodford, 1886. Two alms-dishes, silver-gilt, 1795. Two flagons, silver-gilt, without hall-marks, probably 18th century; given in 1844. Two handsome candlesticks, silver-gilt, marked 1661; weight (inscribed) 129 oz. 12 dwt. and 132 oz.; height 22½ in. With the exception of the two flagons all the communion vessels are 19th century; the whole, with the flagons, may amount to 200 oz.

A new floor of several kinds of stone and marble, designed by Sir Gilbert Scott, was laid in the nave and aisles in 1868–70. Before that time the aisle floors were several inches lower than the nave, and though the actual paving was of common brick, Scott considered this to be the original arrangement;[54] it had been relaid in 1676. The nave floor was also higher towards the west end, and this part was lowered to show the bases of the piers.[55] The paving of the north transept was finished in 1876. The octagon pulpit is by Scott; the lectern by Thicknesse, 1897, in commemoration of the Diamond Jubilee of Queen Victoria; the reredos of St. Edmund's Chapel is by J. A. Reeve; the memorial of those who fell in the First World War by Guy Dawber.

The trussed rafters of the nave roof had until now been exposed to view. The present boarded ceiling was begun in 1858. The painting was begun by Henry le Strange, without payment, and after his death in 1862 on the completion of the six western bays, was carried on by Gambier Parry[56] on the same terms; it was finished in 1865. The scheme of the subjects is to illustrate 'the sacred history of man from his creation by the Word of God, to the final consummation in the glorified humanity of the Son of Man reigning in Majesty'.[57] Le Strange maintained that the painting of the ceiling could only be a first step to colouring the walls and arches of the nave.[58] The restoration of the lantern was done by Sir G. Scott in 1874, as a memorial to Dean Peacock. The monument to Bishop Woodford (1873–86) in the north aisle hides the blocked doorway which led into Trinity Church; it was designed by G. F. Bodley. The font was given by Canon Selwyn in 1866, and was placed in its present position in 1895; it replaces one which had been bought with a bequest of Dean Spencer in 1693, and is now in Prickwillow church. In 1870 the west tower was strengthened by Scott with iron tie-rods and the foundations of the south-east transept and the south side of the quire were repaired. In 1897 vestries were formed in a part of the ruined north pane of the cloister. Advantage was taken of a double doorway shown in Browne Willis's plan as that through the back wall of the cloister armarium. In the present century the cloister vestry was extended westwards, and curtailed eastwards so that the bay containing the armarium was thrown into the open part of the cloister, and a new doorway made to the church from the vestry; the lower part of the armarium was then filled in and the doorways blocked to restore it to its original form. An oak screen has now been placed across the vestibule which had been formed in the 18th century by roofing and glazing the remains of the east cloister.

There is no medieval glass except some fragments unfortunately collected into one window from several parts of the church and the slight though important remains in the Lady Chapel. Most of the painted glass is of the middle of the 19th century. The east window of Alcock's chapel was given by Bishop Lord Alwyne Compton as a thank-offering for the escape of everyone in what might have been a serious accident in the Palace. The open arch between the nave and the room over the Galilee was converted into a window when the Galilee roof was lowered in 1807. The window was filled with Flemish glass of a late date, stonework and glass being the gift of Bishop Yorke. The window in the south-west apse had some crude glass by Wilmshurst, 1855, which was at least better than any other modern work, except the Compton window; it has recently been replaced. The great east window was executed in 1857 by Wailes, a well-known Victorian stained-glass artist. The easternmost window of the south nave aisle, the middle window in the west wall of the south transept, and the east window in the middle tier of the north wall of the north transept are by the Revd. A. Moore, Rector of Walpole St. Peter (Norf.).[59]

Among the restorations of this period may be recorded Ovin's cross, standing in the south aisle near the western door from the cloister. It consists of the base and part of the shaft of a stone cross which was brought in the 19th century from Haddenham, where it had long been used as a horse-block. The base bears the inscription: *lucem tuam ovino da deus et requie(m) amen*. Ovin was steward to Etheldreda. The monument is therefore of high importance as the only object that has survived from the foundress's time.

Paintings. There are remains of 12th-century paintings[60] in the north transept and nave. In the transept the south-east chapel has on its north wall a band of scroll-work at the level of the springing of the vault and above this the Martyrdom of St. Edmund (now almost disappeared); below the band there is a square, 9 ft. 6 in. wide by 8 ft. high, filled with vertical stripes 3 in. wide and 3 in. apart with a sort of cusping on a white ground, with a 9-in. border; below this there are faint traces of hanging drapery. On the south wall there is a square of the same size filled with circles 20 in. in diameter, alternately red with a white centre and rim, and white with a red centre and rim; the spaces between the circles are filled with rings alternately red and white; in every case what appears as white may have been yellow or some other colour; the red is deep Indian. The vault has bold scroll-work; the arch of the entrance has disks 24 in. across. The next chapel has drapery in bold red lines on the round column and on the walls (the walls have also 20-in. circles) from a band about 3 ft. below the capital; the band consists of a sort of honeysuckle enrichment and above it there is a diagonal diaper; the above described patterns on this column have been painted over an earlier painting of broad

54 Goodwin, *Ely Gossip*, 47.

55 *Handbook to Cath. Ch. of Ely*, ed. C. W. Stubbs (20th ed.), 58.

56 Thomas Gambier Parry, of Higham Court (Glouc.), inventor of the 'spirit fresco' process (*D.N.B.*).

57 *Handbook to Ely Cath.*, ed. Stubbs (20th ed.), 59. Some of the figures are portraits; Dean Peacock is said to have been the model for Isaiah (Goodwin, *Ely Gossip*, 39).

58 Goodwin, *Ely Gossip*, 45.

59 See E. Milner White, *Rep. on Windows of Ely Cath.*, in 1st Rep. of Friends of Ely Cath.

60 Many of these are included among the invaluable copies of English wall-paintings made by Prof. E. W. Tristram and now in the Victoria and Albert Museum.

spirals; the groins of the vault are painted. The south transept has remains of painting on the scroll-work of the capitals and the arches have alternate voussoirs coloured, but all was repainted in the 19th century.

The east part of the nave and south aisle have remains of contemporary painting. Parts of the masonry are clean and fresh, showing where the walls of the pulpitum abutted; the rest was whitewashed and decorated with patterns; this was the part occupied by the principal nave altar; the vault of the south chapel was elaborately painted. The wall between the quire and north transept containing the bones of the Saxon bishops was decorated in the 14th century with paintings representing the persons whose remains lie in the cavities; over each is canopy-work. There are few traces of colour in Northwold's work; the back of the piscina in the south aisle has painted masonry-lines. Of the 14th-century colouring a good deal remains; the vaulting of the octagon preserved much of its original colour but it has all been repainted. The colouring of the Queen's pew ('Hotham's monument') has not been retouched. At the east end the wall below the windows was painted with thirteen panels with cusped ogee heads, containing figures and with an inscription above. There was probably a similar painting on the wall of the north presbytery aisle.

The Great Guest Hall (now the Deanery) is said to have contained a painting of which the 'Tabula Eliensis' in the Bishop's Palace is a copy. In Prior Crauden's Chapel there is a large blank wall-space opposite to the entrance, showing signs of having been a picture, which was recognizable as a Crucifixion when Professor Tristram made his notes. There are said to have been slight remains of painting in the Prior's Hall visible when alterations were being made in the 19th century. The north-east wing which Alan of Walsingham added to the infirmary was sometimes called the 'Painted Chamber'. Barclay in his *Eclogues*, v, 515–42, describes a wall-painting of the Nativity in the church but he does not say in what part. The painted panel[61] now in the possession of the Society of Antiquaries has four pictures: The Marriage of St. Etheldreda; her parting from her husband; building her abbey at Ely; her burial. The painting is of the 15th century and of high excellence; it probably formed the 'Table' above an altar; the panel is believed to have been found in a cottage in Ely during the 19th century.[62]

The Organ. The earliest reference to the organ is in the gift made by Bishop Niel (1133–69). In the 14th century the great organ stood on the pulpitum but there were smaller sets in the quire and in the Lady Chapel. The quire organ (on the south side) was renewed in 1396.[63] The great organ was rebuilt in 1662; the case is described in 1843 as being 'of Renaissance design, and of excellent workmanship in oak, much enriched with carving and surmounted by reclining figures of angels, blowing long gilded trumpets'. The organ was rebuilt by Elliot and Hill, and fitted into the old case in 1849, and ingeniously arranged in its present position in the north triforium. The existing case was designed by Sir Gilbert Scott, being partly imitated from that in Strasbourg Cathedral. The instrument has been enlarged several times since, notably by Harrison about 1918.

Bells. In the early part of the 14th century there were bells hanging in both the central and western steeples. In writing to the king in 1322 announcing the fall of the central steeple Prior Crauden particularly mentions the breaking of the bells. In the same year new clappers were bought for the tenor bell called Baunce and for another called Peter; whence it would appear they hung in the west tower; perhaps the repairs were consequent on the destruction of the bells in the central tower. On the completion of the rebuilding in 1341 the famous Master John of Gloucester was engaged to cast four bells; this was presumably a recasting of old bells as no material but one blome of tin and some charcoal fuel was brought;[64] no doubt the metal from the old bells had been salved from the ruin; the cost was £7 1*s.* 7*d.* Baunce was again repaired.[65] In 1345–6 John of Gloucester was again employed, this time to cast four new bells which were named IHS, John, Maria, and Walsingham; they were hung in the west tower at a total cost of £63 8*s.* 2*d.*[66] There were six bells hanging in the Lantern about 1539 and six in the west tower.[67]

The bells were removed from the Lantern in 1669 and the frame 100 years later.[68] Dean Harvey Goodwin[69] maintained that they had been rung from the floor of the church in monastic times, for he found that some of the timbers were scored by the ropes, and that these scorings pointed to the base of the eastern column of the arch of the south transept; he assumed that this was done before the Suppression. In 1723 'a small peel of five the treble of which is now used for a clock-bell' was cast by Henry Penn of Peterborough.[70] There are two bells in the tower standing hard on the High Street used by Holy Trinity parish:[71] (1) the alarm bell cast by a Bury St. Edmunds foundry in the latter half of the 15th century, and (2) 'a good bell' by Thomas Norris of Stamford in 1648.[72]

MONASTIC BUILDINGS AND PALACE

The remains of the monastic buildings are very extensive, probably the most extensive in the country if we consider not only the extent of the ground plan that is recoverable but also the large number of buildings that stand practically intact with vaulted undercroft and medieval roof.

The *Cloister* is on the south side of the nave. The gross measurements of the Norman cloister were 130 ft. east to west by 145 ft. north to south. It was enlarged by the rebuilding of the west pane in the 15th century about 19 ft. farther west. The window walls of the north and east panes are practically entire and date apparently from about 1510. These two panes were covered by flat roofs, as they still are in part. The north-west bay was vaulted, as was probably the whole of the

[61] Figured in Atkinson, *Archit. Hist. Ely*, frontispiece.

[62] There is a small figure of St. Etheldreda in glass, about 20 in. high, in a window of Sandringham ch. (Norf.), and the saint was represented on a panel in the screen (now destroyed) of the neighbouring ch. of Babingley (*Norf. and Norw. Arch. Soc.* xix, 131 and fig. 13). She also appears on the Ranworth screen and in nine other places in Norf. named by M. R. James in *Suff. and Norf.* He also names the places where SS. Sexburga and Northburga are pictured in Norf.

[63] *Sacrist R.*, ed. Chapman, ii, 28, 29.

[64] Ibid. 122. A 'blome' is a bloom which appears to be the bar hammered from one lump taken from the pudding furnace.

[65] Ibid. 118.

[66] Ibid. 138.

[67] Inventory in Atkinson, *Archit. Hist. Ely*, 179.

[68] Stewart, *Archit. Hist. of Ely Cath.* 119.

[69] H. Goodwin 'Recollections of a Dean' in *Essays on Cathedrals*, ed. J. S. Howson (1872), p. 21 n.

[70] J. J. Raven, *Church Bells of Cambs.* 53. The tenor only is inscribed with the founder's name.

[71] Inscribed ✠ AVE GRACIA PLENA (ibid. 14).

[72] Ibid. 32.

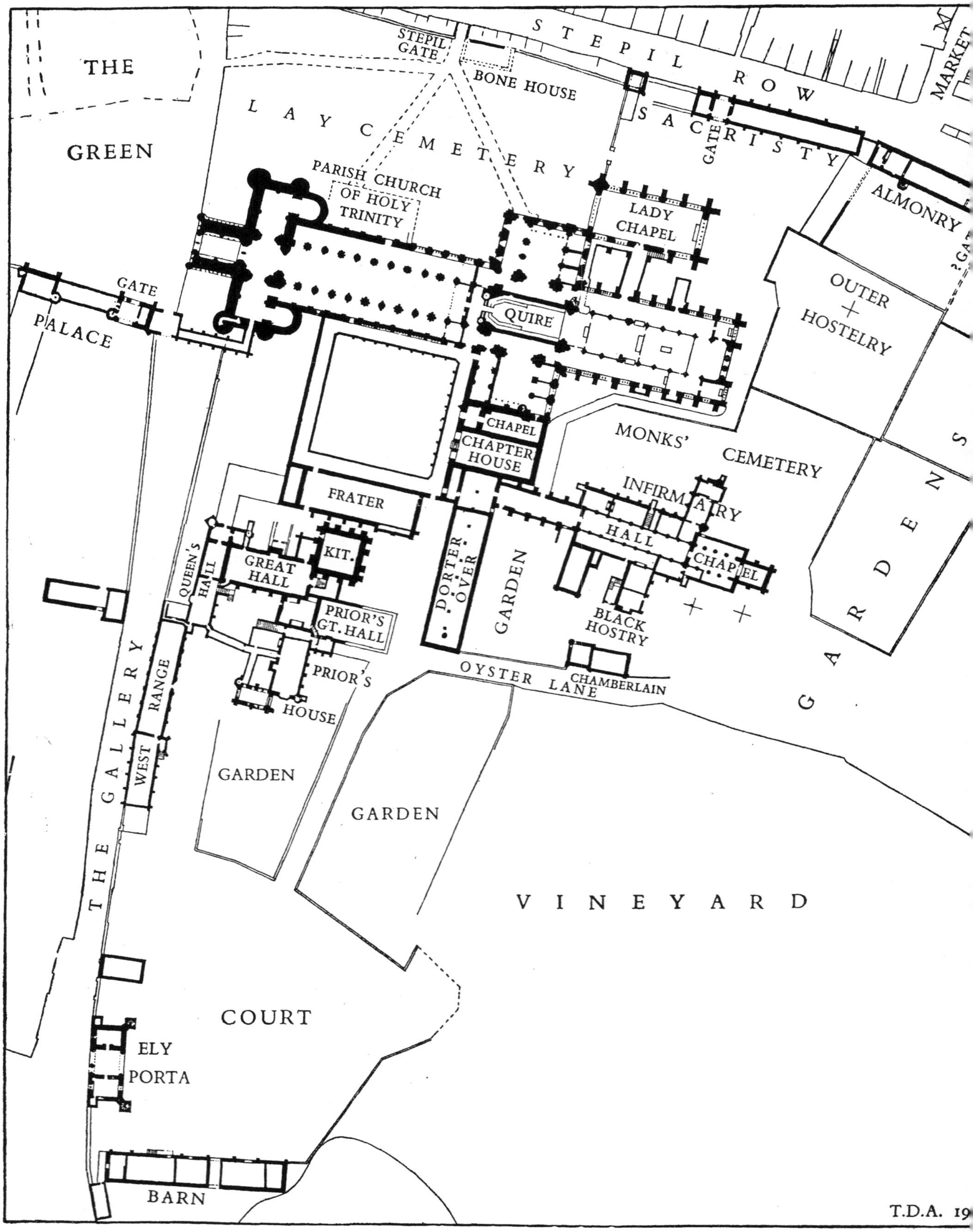

PLAN OF ELY CATHEDRAL AND PRIORY

west pane.[73] In the north pane, in the bay next to the north-east angle, there is an armarium recess of the 13th century, formerly rich work but now much mutilated. It takes the place of one contemporary with the wall, of which there is a fragment in place. The cloister garth now forms the deanery garden. Some of the floor joists in the Deanery are probably made up from roof timbers of the cloister.[74]

From the east pane there was originally a passage to the monks' cemetery, with a doorway into the transept. The passage was converted into a chapel of St. Catherine, and was vaulted early in the 13th century. Next to this passage was the chapter house, of which only parts of the foundations remain. Excavations in 1892 showed that it measured about 76 ft. by 34 ft. It had been demolished by 1649.[75] All the dorter range has disappeared except for one of the late-12th-century columns with the springing of the vault of the undercroft, near the north end. Of the frater only the lower part of the south wall remains. This was built in the 13th century. It contains evidence, at the east end, of the staircase up to the lectern, and there are the shafts, with foliaged capitals, of a wall arcade. The frater measured 110 by 35 ft. To the south of it there are massive remains of the Norman kitchen, consisting of about two-thirds of its south and west walls. These survived when the rest was destroyed during the Commonwealth because they had been used as end walls for the guest hall and its buttery, later part of the Deanery. The south wall contains a very large round-headed window, now blocked, and two clustered responds with cushion caps. These must have carried arches to support the roof, which may have included a stone lantern in the centre, like the one remaining at Glastonbury.

Behind the kitchen was a large guest hall,[76] 80 by 30 ft., consisting of five bays and raised on a vaulted undercroft. The undercroft is divided by a row of columns down its length and by a wall separating the western pair of bays from the others. The vaulting of the western bays, without intermediate or ridge ribs, is earlier than that of the rest, which has these features and is particularly fine. The guest hall itself has a 14th-century open timber roof with arched principals springing from corbels with fine crouching figures. The building dates from the 13th century, and was reconstructed during the 14th. It has a porch of complicated plan, forming entry to (*a*) the hall itself, (*b*) the undercroft, (*c*) the prior's guest hall. The building has been cut up into several stories to form what was once the *Deanery* and is now the Bishop's House. On the ground floor the doorway leading to the former buttery (now the kitchen) retains its original 14th-century woodwork and ironwork. The study at the west end of the ground floor has a stone party wall, on the line of the one in the undercroft; all the other party walls are of timber. The rooms over the entrance porch were probably in monastic times the offices of certain obedientiaries; after the Dissolution they were used as an audit chamber and later by the registrar. Dean Pearce (1797–1820) replaced them by a Georgian drawing-room, whose sash windows were removed to make room for mullioned ones from Landwade Hall inserted by Dean Goodwin (1858–69).[77] The east wall of the Deanery, which was of modern brick, was rebuilt in stone between 1858 and 1862 by Styleman le Strange, who designed the north window of the drawing-room.[78]

To the south of the guest hall was the prior's house, now the *Eighth Canonry*. It has an early-12th-century vaulted undercroft, but the hall itself was rebuilt by Prior Crauden in the first quarter of the 14th century and the windows are later still. It has a good roof of the 14th century, over the hall, which is of 'curb' or mansard type, like those of the transepts of the cathedral. There is a fine fireplace, the only remaining fragment of the prior's study, an interesting timber building demolished in 1883 before its antiquity was realized. Portions of its screenwork are preserved in the north pane of the cloister. The most remarkable feature of the house is the chapel built by Crauden between 1321 and 1341. Raised on an undercroft to the level of the hall, it was reached by a bridge from the main building. It has been much restored, having become very much mutilated and almost ruinous since the Dissolution. The flowing tracery of the main windows is particularly fine, and there are low-side windows; also a curious contemporary tile pavement with a picture of the Fall, and faint remains of a painting of the Crucifixion. The internal arrangements of the house have been much altered, and the entrance door has been moved from the west to the east side. Some restoration seems also to have taken place in the last years of the monastery, many of the existing windows dating from the early 16th century.

The space between the prior's house and the guest hall was filled by the prior's kitchen. There are remains of two of its three large fireplaces; the third is intact but blocked. From the east side of the kitchen there ran a large hall, about 75 ft. by 37 ft., on the ground story, of the side walls of which there are remains. This seems to have been at the same time the prior's great hall and the misericord, there being entrances on both sides, on the north for the monks and on the south for the laity. The porches of these entrances are the chief architectural remains of this hall; the one on the south has several square-headed windows with cusped sub-arches, probably late 13th century, and an early-16th-century fireplace in cut brickwork on its upper floor. The hall may almost certainly be identified with the 'Bougre', a name peculiar to Ely. Attached to it was the 'knights' lodging', the sleeping chamber of the prior's squires or knights, who dined in the great hall; it probably stood over the gateway leading to the cloister.[79]

West of the prior's house and formerly connected with it by a bridge stands the 'Queen's hall', built by Crauden for the reception of Queen Philippa about 1330.[80] It still stands almost perfect, on a vaulted undercroft, but has been cut up internally to make a house for the headmaster of the King's School. The undercroft is in two parts. The northern, of three bays, has very thick walls, but the upper stories have been destroyed. The southern part, of four bays, supports the hall itself,

[73] A fragment of the vault is preserved in the triforium.

[74] Atkinson, *Archit. Hist. Ely*, 50. The cloister was one of the buildings destroyed during the Commonwealth, about the time the old guest hall was divided by floors to form the Deanery.

[75] Ibid. 55–6.

[76] A post-medieval name. See Atkinson, op. cit. 80.

[77] Ibid. 86.

[78] Goodwin, *Ely Gossip*, 3.

[79] When Atkinson published his *Archit. Hist. Ely* the use of one hall as misericord and prior's great hall was only suspected (op. cit. 135) for it was not known that the Bougre was a dining-hall. That such was the case has since been shown by the researches of Archdeacon S. J. A. Evans; thereafter its identification with the misericord became fairly certain.

[80] The name dates from 1258 at least. It is probably eulogistic—a hall fit for a queen—though most English sovereigns from Canute onwards visited Ely, many no doubt with their wives. See Atkinson, op. cit. 87.

which was 47 ft. long by 20 ft. wide, and, with a height of about 22 ft. to the tie-beam, unusually lofty. Its fireplace, on the west side, is blocked, but the corbelling of the chimney-stack may still be seen on the former outside wall. The great south window, and two on the east side, survive, with attractive tracery. The corbels of the low-pitched roof are similar to those of the guest hall (Deanery) roof; subsequent to the Reformation the original leaded roof was replaced by a tiled one of high pitch, providing an extra story. Attached to the south end of the Queen's hall is a long range of 12th-century granaries and storehouses, now used by the school. This was converted in the 14th century into a two-storied building by the introduction of vaulting, the upper floor forming chambers for guests, and about the same time it was prolonged southwards. It has been mutilated in detail but not radically altered. Traces of the original round-headed lancets, cut across by the 14th-century vaulting, exist in the east wall of the northern half, and the 14th-century windows to both floors remain on the west (street) side. The 14th-century extension to the south has also been much altered by modern windows on the east side, but retains its original covered outside staircase to the upper floor. The internal arrangements are probably entirely modern.[81]

To the east of the cloister are the extensive remains of the large infirmary, now forming, with some additions, the *Second*, *Third*, *Fifth*, and *Sixth Canonries*, and a *Minor Canon's House*. The infirmary was approached through the undercroft of the dorter, and thence through a long vaulted passage of the 13th century called the 'Dark Cloister', of which the south wall still stands. This exhibits good mid-13th-century work, but the vaulting and upper story have entirely disappeared. The Norman infirmary consisted of a large nave with aisles, forming the hall, and an aisled chapel with a vaulted chancel, the whole being about 175 ft. long. The nave is now roofless and forms a blind lane between houses, in the walls of which can be seen the blocked arcades of the 12th-century hall. These are elaborate work, with fine enrichments on the arches and interesting scalloped capitals; they date from about 1175. The walls which block the arcades are medieval, for during the 14th century the aisles were cut off from the nave and divided into two stories and into separate rooms.

At the south-west corner of the hall there stands a projecting wing, now the *Sixth Canonry House*, consisting of a 12th-century building of two stories. The lower one, with a barrel vault, was probably the infirmary kitchen, the whole building being known as the 'cellarer's lodging'. Large modern additions have been made to the house on the south and west, but there remains a 16th-century fireplace in the east wall of the first floor, and the roof is probably medieval. As in so many cases at Ely it has been ceiled underneath and attics introduced.

Another wing to the east of the above and parallel with it was added early in the 13th century for the accommodation of visiting monks from other Benedictine houses. Hence it was called the 'Black Hostry'; it is now the *Third Canonry House*. It also is of two stories, the lower one subdivided and vaulted; the upper story, forming a hall, has its original open-timber roof. The vaulted lower floor retains two original windows on the east side, of two lights with plate tracery. In the southern of the two dividing walls is an ingenious three-way serving-hatch of medieval date. The upper story is timber-framed on the east and south sides. The present oriel window near the north-east corner is modern, but probably reproduces a medieval feature. There is a very fine 15th-century brick chimney-stack, with three octagonal chimney shafts, at the south end. The chimney-stacks at the north end of the main block of this house, and on the east wing, are probably 17th century, with blind arches in moulded brickwork. This latter wing forms part of the south aisle of the infirmary hall and chapel. It was converted in the 15th century into two sets of rooms, the lower being occupied at the Dissolution by the cellarer.[82] The windows are modern but a 15th-century fireplace remains on the first floor.

The *Fifth Canonry House* is largely modern, but incorporates the chancel of the infirmary chapel, which has a vault and other excellent late 12th-century details, and a house of uncertain medieval date which stood south of the chancel and was in 1541 occupied by the sub-prior.

The north aisle of the infirmary was taken down, probably about 1320, and replaced by a wider building of two stories, each containing several rooms. The use of these rooms is not known, but many persons, including the convent's physician,[83] were lodged in and about the infirmary. In the 19th century this wing was called the 'blood-letting house',[84] but that seems to have been south of the infirmary chapel and detached. All the windows of the north aisle have been altered, but fragments of the original are preserved in each case, so that the whole design can be recovered. A timber-framed top story was added, probably early in the 16th century, making a three-story building—an unusual feature at Ely. It retains its original roof, and some exposed beams on the north side. This building is now the house of one of the minor canons.

About 1335 Alan of Walsingham, then sacrist, built a wing extending northwards from the north-east corner of the infirmary hall; this is now the *Second Canonry House*. It is of two stories, the lower one vaulted. Walsingham stipulated that he was to have the use of it for life, but the house, or at least the ground floor, was also to be a place where the brethren might see their women relatives. It was called the 'painted chamber'. The building still stands almost complete, but has been a good deal restored, not always correctly. The almost flat roof, if original, is somewhat unusual for its date. The buttresses, which splay outwards to meet the wall at an oblique angle, are of an exceptional form.

A little distance to the south of the infirmary was the chamberlain's checker, of which building only two fragments remain, embedded in the garden walls of the Third Canonry House. It contained store-rooms for clothing and the laundry; in close connexion with it was the bath house.

To the east of the church stood the outer hostelry, the foundations of which may be traced on the turf in a dry summer. It was a self-contained establishment with its own chapel, and was perhaps used chiefly by the commercial classes, standing as it did fairly near the river, which in the Middle Ages carried a good deal of traffic between Cambridge and Lynn. A wall running from the south-east angle of the church, which existed

[81] Atkinson, *Archit. Hist. Ely*, 96.

[82] The upper room may have been appropriated to the infirmarer (ibid. 114).

[83] Ibid. 120, 136.

[84] Atkinson accepted this view but withdrew it at the last moment (op. cit., p. xxxi) in favour of that here put forward.

until the present pathway was made round the east end of the cathedral about 1854, was probably connected with the *ostium versus cimeterium monachorum*, a gate near the outer hostelry dividing the more public part of the monastery from the monks' own quarters.[85]

Along the High Street, to the north of the monastery, was another range of buildings. The easternmost was the almonry, still intact though much altered internally and with modern windows on the north towards the street. The lower story has vaulting dating from the end of the 12th century. The blocked lancet windows in the east wall suggest that the almonry chapel, which was dedicated to St. Martin, was at this end.[86] The almonry included a school for the boys who took part in the Lady Chapel services. To the east or the west of it was a gateway, now destroyed, referred to as *porta monachorum*.

Farther west was the sacrist's office (the building department of the monastery) greatly enlarged by Alan of Walsingham in *c.* 1325.[87] It also had a gateway from the street, but both range and gateway have been completely remodelled. The room over the gateway was converted into the dean and chapter muniment room under Dean Goodwin (1858–69); the muniments had previously been housed in the canons' vestry. To make the room fire-proof a stone floor and vaulted roof were inserted, heat being provided by running the flue of the porter's lodge fireplace behind the wall; 'it [i.e. this heating system] has . . . turned out to be a failure'.[88]

At the west end of the sacristy range is a square building. Here the goldsmith retained by the monastery had his workshop on the ground floor and the sacrist his counting-board on the upper. It is said that the third story was added during the reign of Elizabeth,[89] when the building was converted into a bell tower for the use of the parishioners of Holy Trinity worshipping in the Lady Chapel. Beyond the goldsmith's workshop was the lay cemetery, fringed along the street with tenements belonging to the bishop. Half-way along the row was a tower gateway into the cemetery, which with two tenements on each side was the property of the prior and convent. The present gateway is of timber, and dates from the close of the Middle Ages. It has long been known as 'Stepil Gate', and from it High Street was called 'Stepil Row'. Under the gateway and the house to the east there are vaulted cellars of *c.* 1330. Adjoining the gateway on the south-east was a bone-house, in which were deposited any bones disturbed in digging graves, with a chapel of St. Peter over it. The north wall of this building still stands to a considerable height and seems to date from the middle of the 14th century. If so, it replaces an earlier building, for the statutes of 1300 order that the pyx containing the Body of Christ and the oil for the dying shall be kept 'in capella nova juxta vetus campanile situata'—clearly a provision for the parishioners, for whom it would be conveniently situated. It was apparently this steeple which was taken down in 1354–5 at a cost of 7*s.* 8*d.*[90] It may be that this tower served the parish when the cathedral nave was used as the parish church, and would not be required if the new church on the north of the nave had a bell-cote. Judging by the small expense of its demolition, it was of timber.[91]

The last group of monastic buildings to be noted includes the great gateway known as Ely Porta and the two-story range running east and west along the south side of the courtyard entered through the Porta. This gateway was begun in 1397 and finished about 1405; its designer was John Mepsale or Meppushall, the then architect of the convent.[92] It is of three stories. The outside has hardly been altered, except for the substitution of a high-pitched tiled roof for a lead flat and, almost certainly, of the present low plain parapet for high battlements.[93] The chief architectural features of the simple but effective façades are the three-light traceried windows in the centre of the top floor, flanked on the west by four empty niches, and the boldly projecting square turrets at each corner. The entrance, with a large and a small arch, is flanked by rooms, formerly a porter's lodge on the south[94] and a prison on the north. The carriage way reaches up through two floors, there being two rooms, over the porter's lodge and the prison, on the first floor, and three on the top floor, the centre one over the archway. The three top floor rooms have been thrown into one as a classroom for the King's School. The whole of the upper floors were in monastic times given over to the offices for manorial business.

The range south of the courtyard contained storerooms at the west end and granaries at the east. It dates from about 1375 and preserves most of its original features, including traces of an outside staircase on the north side. The passage through the centre, judging by the arches at its ends, dates from about 1475. The cottage south-west of this range is medieval, but all architectural features have been altered except one roof truss. It was probably occupied by the storeman; in middle of the 19th century it was called 'The Hermitage'.[95]

The monastic grange, commonly called the *SEXTRY BARN*, lay to the west of St. Mary's churchyard.[96] The barn was a stone building 219 ft. 6 in. long by 39 ft. 5 in. wide internally. It was divided into nave and aisles by two rows of oak posts on stone bases, ten on each side, and was covered by a continuous roof, the nave being thatched and the aisles tiled. There were two large doorways on the north side, the eastern of which had a large threshing-porch with an upper story. The building, which appears to have dated from the middle of the 13th century, was destroyed in 1842.

In front of the grange was a yard surrounded by a wall built in 1302–3 at a cost of £7 18*s.* 1*d.* In the north-east corner of the yard stood, and still stands, the house of the steward, now *ST. MARY'S VICARAGE* but commonly called *CROMWELL HOUSE*.[97] The office of tithe-collector or steward to the priory and later to the chapter became hereditary in a family which assumed the name of Steward or Styward. In their time the house was probably enlarged. The last of these Stewards was Sir Thomas (d. 1636), the maternal uncle of Oliver Cromwell. Cromwell succeeded to the tithe-office in 1636 and lived in the house for the next ten

85 Ibid. 137.
86 Ibid. 140–1.
87 *Sacrist R.*, ed. Chapman, ii, 57.
88 Goodwin, *Ely Gossip*, 12–16.
89 Unpublished MSS. of Mr. Bacon, Cathedral Clerk of Works *c.* 1870.
90 *Sacrist R.*, ed. Chapman, ii, 165.
91 See p. 51.
92 J. H. Harvey, *Gothic England*, pl. 48. Mepsale's name is preserved in Mepsale's Corner, at the junction of St. Mary's Street and Downham Road.
93 Shown in J. Bentham, *Ely*, pl. 1.
94 Used for the brewing of chapter and audit ale by the verger, William Southby, up to Dean Goodwin's time (1858–69) (Goodwin, *Ely Gossip*, 5).
95 Atkinson, *Archit. Hist. Ely*, 155.
96 R. Willis, 'A Description of the Sextry Barn at Ely' in *Proc. Camb. Antiq. Soc.* i, 5. Sketches were made by Prof. Willis just before demolition.
97 See generally E. G. Punchard, *Cromwell and the Old House at Ely* (Ely, 1906).

years.[98] Farmers of the capitular tithe continued to occupy the house until 1840 when the office was abolished under the Tithe Commutation Act of 1836. The house was sold by the dean and chapter in 1843. It changed hands several times, at one period being used as an inn—the Cromwell Arms. It was bought by the Revd. E. G. Punchard, D.D., Vicar of St. Mary's, in 1905 and bequeathed to the chapter for use as a vicarage.

The house has a lower story of stone and brick and a timber-framed upper story with a reed and clay filling. The roofs are tiled and the chimney-stacks are of brick. It is probably of late-15th-century origin, and was enlarged and altered about a century later. The original plan was apparently L-shaped with the main block facing north, as now, and a wing at the south-east angle. Early in the 17th century another wing was added on the south-west and a connecting gallery added at the back of the main block. Sash windows and various internal alterations were effected, probably in the 18th century. On the acquisition of the house for a vicarage a thorough restoration was effected and the sash windows replaced by leaded lights with wooden mullions; at the same time the west wing was extended and many of the minor timbers were renewed. The east wing has massive 16th-century buttresses of brick and ashlar. The tops of the chimney-stacks have been much renewed, except that of the kitchen in the east wing, which is of 16th-century date. Only one old fireplace is visible; it is in the first floor room of the west wing and has plain chamfered jambs of stone and a chamfered wooden lintel. The drawing-room at the east end of the main block has early 17th-century oak panelling, which has been partly renewed in pine. The ground floor of the west wing has similar panelling, but less complete. The staircase retains its wooden newel post, but the treads have been renewed. At the back of the upper part of the main block is an original three-light window with wooden mullions and iron stanchions, now blocked. In the study at the west end of the main block is a plain two-light stone window, possibly medieval, which now looks into the west wing. Several of the rooms retain their oak floor boarding. At the north-west angle of the main block is the jamb of a large gateway of Barnack stone.

The *BISHOP'S PALACE*[99] stands quite close to the west end of the cathedral, from which it is separated by the street called the Gallery,[1] and on the south side of what was the village green of the original settlement at Ely. It consists of a three-sided court facing north, with a long gallery running west from the north-west corner of this. The building is entirely of brick. The most striking features are the two towers of four stories, terminating the wings of the courtyard. The eastern of these, that nearest the cathedral, was built by Bishop Alcock (1486–1500) as a gateway tower. In its north face was the entrance, probably a large and a small archway like Ely Porta; the large arch would have been under the group of three niches which still remain. The gateway passage, now enclosed to form a room, has a handsome vault; on the west was the porter's lodge, and in the back wall a single arch still remains.

The gateway tower was flanked on each side by a narrow wing. That on the east was short and was pierced by an archway through which the road ran. From the end of this wing a narrow raised gallery gave the bishop a private covered way to the church through a 12th-century doorway still to be seen in the south wall of the south-west transept. The gateway tower, the east wing, the covered way, and the west end of the church are shown in a painting preserved in the palace of the funeral of Bishop Cox (1581); in this picture the positions of the buildings are reversed, and it is more easily understood if looked at in a mirror. The three-storied west wing seems to have included the present western tower, and at that point to have turned south to form a wide main building containing the hall and other apartments.

An important addition was made to Alcock's palace by Bishop Goodrich (1534–54), who built a long west gallery in 1549–50. This is on the upper floor, its windows looking northwards upon the Green. It was originally carried on a loggia with open arches towards the south, but these have been blocked and the loggia converted into kitchen offices. It appears that when this gallery was built the western tower was remodelled to match Alcock's gateway tower.

The present house is principally the work of Bishop Laney (1667–75). He took down almost the whole of Alcock's work, leaving probably nothing but the two towers, and closed the gateway in the east tower. He planned his building to form a three-sided court, using the towers to terminate the wings. Along the three sides of the court was a one-story corridor, which in the 19th century received sham Tudor windows. The house itself is a fine brick building, a good example of the architecture of its time, with a high hipped roof. The interior was remodelled by Bishop Keene in 1771, when the upper story was probably added to Laney's corridor. The stable range is probably of Bishop Goodrich's time, and has reused fragments of Norman windows.

The palace is now used as a school for crippled girls.

Bishop Northwold founded in 1250 a chantry of four priests to sing masses for the souls of Henry III, Queen Eleanor and their children, and of the bishops of Ely, the monks, and their benefactors. He provided for them a yearly stipend of 20 marks and a hostel in which to live, on the green at the west end of the cathedral, whence it was called the Chantry-on-the-Green. It is mentioned in the extent of 1251 and frequently in subsequent documents.[2] At the Dissolution in 1548 the value of the chantry was £21 15*s.* yearly.[3] The building clearly stood actually on the green and was in fact an encroachment on the common. The exact spot seems to have been the east side of the group of houses to the south of St. Mary's Street. Bishop Wren, in whose time the buildings were still in existence,[4] describes it as 'right against the gate of the demolished palace', Bentham as being on the site of 'the house of the late John Waddington Esq', and Stewart as 'on the site of a house now in occupation of J. Muriel Esq.' An 18th-century house now occupies the site.

CHURCHES Presumably there was originally but a single parish, whose spiritual needs were adequately met by use of a part of the abbey church.[5] The chapel of the Holy Cross, in the mother church, was so used before the division of the monastic property between bishop and prior in 1109. Even at this date, however, there existed the second parish of St. Mary: a specific grant of its church

[98] Two of his children, Mary and Frances, were born at Ely and apparently baptized at St. Mary's (M. Noble, *Memorials of Cromwell Family* (ed. 1789), i, 359; Punchard, op. cit. 10).

[99] A fuller description, with illustrations, is in *Country Life*, lxiii (1928), 850.

[1] See n. 62 on p. 38.

[2] e.g. *Cal. Pat.* 1385–9, 310.

[3] Ibid. 1548–9, 32–33.

[4] *Fenland N. & Q.* ii, 95.

[5] Atkinson, *Archit. Hist. Ely*, 8.

was then made to the convent.[6] The chapel of the Holy Cross was merely included in the general grant to the mother church of all the emoluments and oblations of her altars.[7] The church of St. Mary was apparently rebuilt and endowed by Bishop Eustace (1198–1215).[8] For a short period the monastery granted this new church to a priest named Baldwin, but recovered it in 1232 and reallotted it to the sacrist's office. Thenceforward both parishes continued until the Dissolution to be so appropriated, each being served by a chaplain appointed by the sacrist.[9]

The name of the original parish underwent changes, as the following account will show. The parishioners used the nave of the monastic church with the altar of the Holy Cross immediately west of the pulpitum as their principal altar. Early in the 14th century complaints were made of the inconvenience of services conducted simultaneously in nave and monks' choir.[10] A commission of inquiry was appointed by the archbishop in 1315. The return spoke of the two parish churches as 'the church of St. Peter or the greater parish' and 'the church of St. Mary or the lesser church'.[11] A new church for the greater parish was decreed and work was begun in 1341–2.[12] Prior to its consecration, between 1362 and 1367,[13] it was referred to either as the church of 'the greater parish' or as that of Holy Cross.[14] The dedication may have been changed upon consecration: at least by 1373 the parish was known as that of the Holy Trinity.[15] This church was not finished until 1459–60. A lean-to against the north wall of the north aisle without sun or cross ventilation, the building was, it would appear, unsatisfactory from the first.[16] It served the parish, however, until 1566,[17] when the Lady Chapel of the cathedral (q.v.) was devoted to this use, by day but not by night, and so continued until 1938. Entrance was by a new doorway from the transept; the passing-bell for a departed parishioner was rung by the sub-sacrist. The old church appears to have been taken down about 1566.[18] The ruins were removed and the cathedral aisle refaced in 1662.

At the Dissolution the parsonage and emoluments of both churches were granted to the dean and chapter,[19] whose custom it was to appoint prebendaries or minor canons to the two curacies. In 1929 the benefices were united under one incumbent. The rectory and tithes of the two parishes were frequently leased. The most famous lessees were the Stewards, who acquired the rectory of Ely, in addition to Stuntney manor (q.v.), in the 16th century. Oliver Cromwell inherited the property, on the death of his uncle, Sir Thomas Steward, in 1636.[20]

In 1605 Robert Walden left rent charges of £2 per annum to the minister of St. Mary's church for the preaching of two sermons a year.[21] In 1724 William Cole left a rent charge of £3 per annum to the churchwardens of Holy Trinity, £2 of which was to provide sermons.[22] In 1772 John Howard left £100 to the dean and chapter to provide an afternoon sermon in the same church.[23]

The ancient chapel at Stuntney was served in the Middle Ages by 'archpriests' and later by the curates of the parent parish of Holy Trinity.[24] The patronage remained with the dean and chapter after the alienation of the secular property.

Chettisham had a chapel at least as early as the 15th century, under the same patronage as the parent church of St. Mary.[25] Chettisham became a separate ecclesiastical parish in 1876.[26]

The church of *ST. MARY* consists of chancel, clerestoried nave, aisles, south chapel, north porch, west tower and spire, and vestry. The material is mainly rubble with stone dressings, and the spire is of Barnack stone; the tower and south clerestory are plastered externally. The roofs are covered with slates except those of the south chapel and vestry, which are tiled. The chancel, nave, and aisles were erected at the beginning of the 13th century; a few years later the south chapel was added. The tower and spire date from about 1300 and the porch from the first half of the 14th century. The clerestory belongs to the first half of the 15th century, and towards the end of the same century the aisles were provided with new windows. A drastic 'restoration' was begun in 1877, when the whole church was reroofed. Later, a vestry was erected against the south side of the tower.

The chancel has an east window of three lights with rectilinear tracery, which dates from the end of the 14th century. There are the original clamped angle buttresses round which the string-course extends. In the north and south walls are two plain lancets. There is a doorway in the north wall with a two-centred arch of two orders with a continuous chamfer. The north-west window is a 15th-century insertion with three cinquefoiled lights under a square label. The south-west window, dating from the end of the 14th century, has early rectilinear tracery, which has been renewed on the exterior. The original internal string-course has been cut away at the west end on both sides to accommodate the inserted windows. The lofty chancel arch is two-centred and of two orders with keel-shaped responds having modern foliaged caps and bases concealed by the modern floor. There is a much renewed double piscina of the 13th century with a central shaft and turned responds and deeply moulded arches. The sedilia without divisions has an acutely pointed trefoiled arch.

The nave has arcades of seven bays with two-centred arches of two orders having keel-shaped mouldings and a hood-mould; the columns are round with square caps and bases. Two of the caps (those of the eastern

[6] Bp.'s Mun., Ely, Liber M, f. 143.

[7] Ibid., f. 143.

[8] Ibid., f. 165.

[9] *Sacrist R.*, ed. F. R. Chapman, i, 141, ii, 121, and many other entries; *Cal. Pat.* 1416–22, 183–5.

[10] *Sacrist R.*, ed. Chapman, 101–2.

[11] Injunction of Archbishop Reynolds, B.M. Add. MS. 41612, f. 34*b*; cf. Baker MSS., vol. 38, f. 123. The original dedication of the monastic church was to St. Peter and B.V.M. (*Liber Elien.* ii, c. 52).

[12] *Sacrist R.*, ed. Chapman, ii, 121. Stone for the foundations came from 'Overtone'—perhaps Overton (Orton) Longueville or Waterville on the south bank of the Nene just above Peterborough.

[13] *Sacrist R.*, ed. Chapman, i, 104.

[14] Ibid. ii, 148, 171, &c.

[15] Visitation of Holy Trinity by Thos. de Wormenhale, Cant. Reg. Whittlesey, f. 152*b*. The churchyard of Holy Trinity is mentioned in Bishop Fordham's Survey of 1416–17 (B.M. Harl. MS. 329), but this is not, as Chapman says, the first use of the name Holy Trinity for the greater parish. (*Sacrist R.*, ed. Chapman, i, 104). The name occurs also in 1383 (Reg. Arundel, f. 47) (Archd. S. J. A. Evans has furnished these refs.). See also *Sacrist R.*, ed. Chapman, ii, 148, 171, and Bentham, *Ely*, 162.

[16] The doorway from the nave was in the seventh bay from the east and its remains are hidden by the monument of Bishop Woodford.

[17] *Sacrist R.*, ed. Chapman, i, 103.

[18] R. Gardner, *Dir. Cambs.* (1851), 471.

[19] MS., C.C.C., Camb., Misc. XX, 391 et seq.

[20] Gardner, *Dir. Cambs.* (1851), 485.

[21] *Rep. Com. Char.* (*Cambs.*), [103], p. 230 H.C. (1837–8), xxiv.

[22] Ibid., p. 231.

[23] Ibid., p. 229.

[24] Reg. Gray, f. 224.

[25] Ibid. 'Chesham Chapel' is mentioned also in S.C. 6/Eliz./282, m. 1.

[26] *Clergy List* (1915).

responds) have been entirely renewed and one on the south has been partially renewed. The columns rest on rough stone basements. The clerestory windows have two cinquefoil-headed main lights with a quatrefoil above. Beneath the clerestory is a plain string-course. The parapets are plain and apparently much renewed. The nave opens to the tower by a two-centred arch of two orders with moulded caps and bases to the semi-octagonal responds; it is of early-14th-century date.

and much renewed except the second one from the east. The plain parapet has been renewed. The north doorway is particularly notable. It has a two-centred arch of three orders with elaborate dogtooth and other mouldings and banded jamb shafts with stiff foliaged caps and moulded bases; some of the mouldings have been repaired in plaster. Inside the doorway is a plain stoup recess. The porch has a two-centred outer arch with a continuous chamfer. In the east and west walls

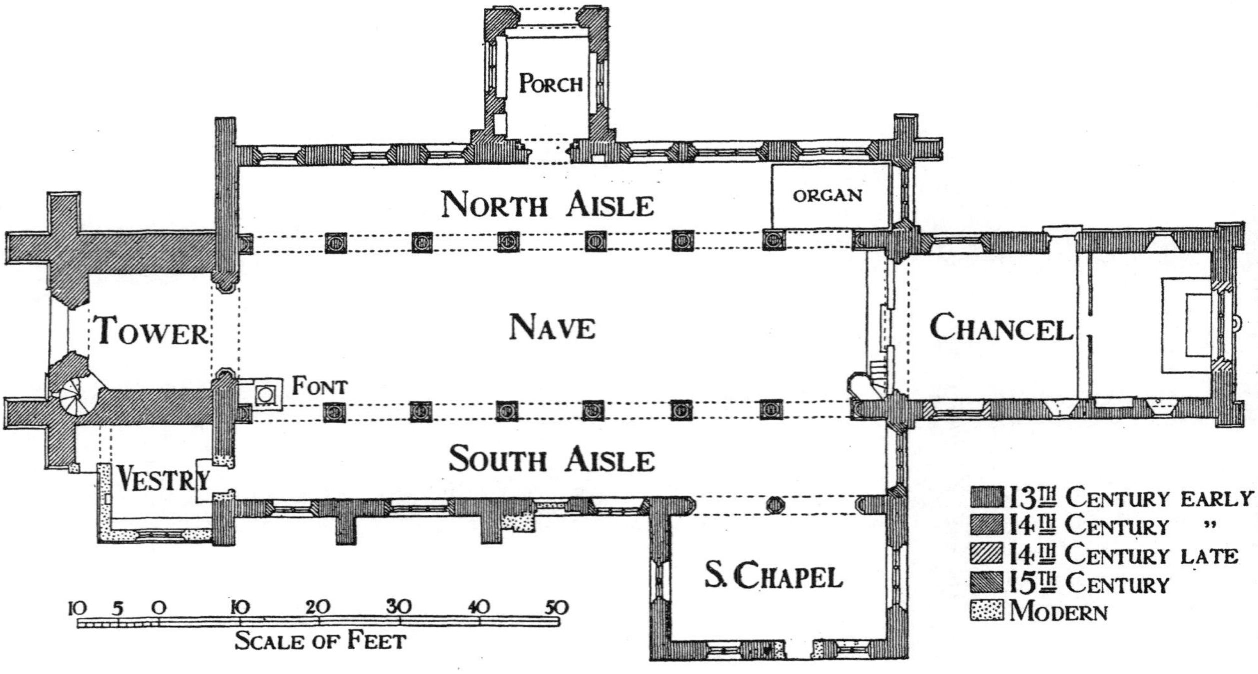

Plan of St. Mary's Church

The south aisle has three lateral buttresses with one set-off, the upper parts of which have been repaired with brick, probably in the 16th century. There is a blocked doorway in the south wall with a two-centred arch having zigzag mouldings and caps with stiff foliage; the jamb shafts are missing. In the 15th century a doorway with continuous mouldings, now also blocked, was inserted within the earlier entrance. There are two windows of three lights and one of two, all with cinquefoil heads under a square label. The plain parapet has modern coping. The aisle opens to the chapel by an arcade of two bays with two-centred arches of two orders having moulded caps and bases to the octagonal column and semi-octagonal responds. The chapel has an east window consisting of three plain lancets under a containing arch. In the south wall are two windows of two cinquefoiled lights under a square head. The west window consists of two lights with uncusped heads and a quatrefoil above and is of 13th-century date. There is a plain modern doorway in the south wall. The gables have old coping and broken crosses on the apex. There is a good double piscina similar to that in the chancel but in a more genuine state. There is a plain image bracket in the north-east angle.

The north aisle has an east window of three cinquefoiled lights under a depressed arch. There are two angle-buttresses with one set-off at the north-east and one at the north-west corner. The fenestration of the north wall consists of two windows of three lights and four of two lights, all cinquefoiled under a square label is a two-light window with cinquefoil-headed main lights and rather clumsy tracery of 14th-century date.

The tower has angle buttresses with four set-offs extending to the parapet. The plain west doorway is of three orders with continuous chamfer and a hood-mould. The west window consists of three lights with plain intersecting tracery. The belfry windows are of two trefoil-headed lights with a quatrefoil above and a plain hood-mould. There is a plain parapet with pinnacles at the angles, which are embattled and finished with conical caps. The newel stair is contrived in the south-west angle and is entered by a plain doorway having a continuous chamfer. The octagonal spire has two-light gabled openings at the base on alternate faces and single lights similarly arranged towards the top. On the south-west buttress is a tablet commemorating the burial of five men executed for complicity in the Littleport riots of 1816.

All the roofs are modern.

The fittings are modern and mediocre. In the churchyard is the bowl of an ancient font of Barnack stone, of irregular shape.

The plate consists of a communion cup of silver, 1684, a chalice of silver, 1870, a paten of silver, 1684, a paten of silver, 1873, a flagon of silver, 1878, a cruet of glass with silver mountings, 1878, and an alms-dish of pewter.

The tower contains eight bells, all by Edward Arnold of St. Neots, 1781, except the third, which is dated 1766. The sanctus bell was cast by Joseph Eayre of

St. Neots, 1778. The oak bell frame was reconstructed in 1882, but retains some timbers of 17th-century date.

The registers begin in 1670 and are complete except for the entries of baptisms and burials between 1702 and 1773 and of marriages between 1702 and 1763.[26a] The registers of Holy Trinity (see above) date from 1559.

The church of *ST. PETER,* Broad Street, was erected in 1890. It is a stone structure in the Early 'Decorated' style and consists of chancel, nave, south porch, and south-west bell turret with one bell. There is a particularly fine wooden screen and rood loft decorated in colours which was designed by Mr. J. N. Comper. It originally served as a chapel of ease to Holy Trinity.

The church of *ST. ETHELDREDA,* Queen Adelaide, erected in 1883, is a brick structure consisting of nave, south porch, and west turret containing one bell. It originally served as a chapel of ease to Holy Trinity, Ely.

The church of *ST. PETER,* Prickwillow, was erected in 1868. It is a brick structure on a foundation of wooden piles in the 'Early English' style, and consists of chancel, nave, transepts, south porch, and a central bell-turret with a short spire. There is one bell, dated 1691, and a handsome font of Italian marble, dated 1693, both formerly in the cathedral. The font has an elaborately carved wooden cover, which is suspended from the roof by the figure of an angel. The ecclesiastical parish was formed in 1878 from Holy Trinity and St. Mary Ely, Littleport, Lakenheath, and Mildenhall (Suffolk). The register of marriages dates from 1864, of baptisms from 1874. There is no burial register as the dead are buried in Ely.

The church of *ST. MICHAEL AND ALL ANGELS,* Chettisham, consists of chancel, nave, south porch, north vestry, and west bell-cote. The nave and chancel are continuous without any structural division.

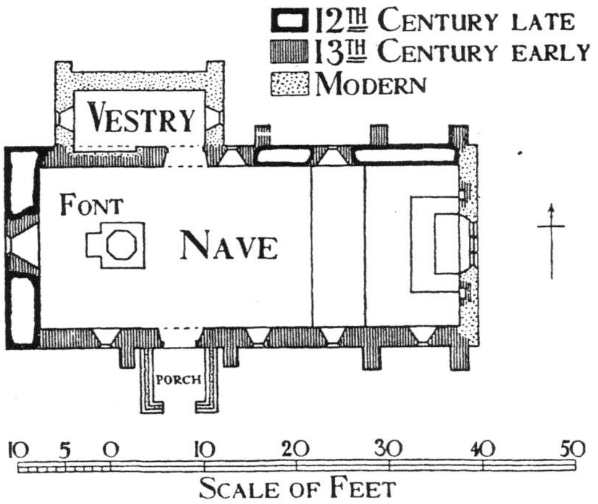

PLAN OF ST. MICHAEL AND ALL ANGELS' CHURCH

The fabric is of late 12th-century date, but was drastically restored in the 19th century and the porch, vestry, and bell-cote are modern. The east wall contains three modern lancets with a trefoiled opening above, also modern. There are four plain lancets in the south wall, two in the north, and one in the west, several of which have been much renewed. In the west wall above the lancet is a modern sexfoiled opening. There are four buttresses on the south and three on the north wall, all with two set-offs, which have been considerably renewed. The north and south doorways are very plain, with continuous chamfers, and of the first half of the 13th century. The roofs are modern, that of the chancel being of the braced-rafter type, and that of the nave couple close with collars and tie-beams; the collars and tie-beams are apparently old, but of uncertain date. The bell-cote is tiled and shingled. The porch is of wood on a stone base. There are two plain aumbries with modern doors in the east wall. The font is of 15th-century date and cup-shaped, with an octagonal bowl having quatrefoils enclosing blank shields on each side. In the vestry are portions of three stone capitals and a corbel of late 13th-century date.

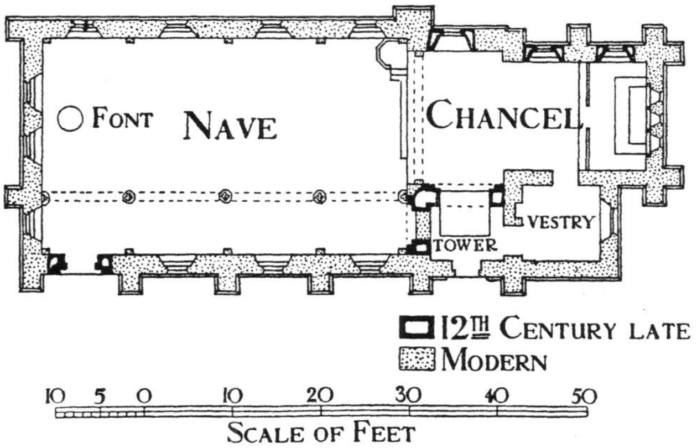

PLAN OF THE CHURCH OF THE HOLY CROSS

The plate consists of a communion cup of silver, 1569, with a paten cover of the same date, and two pewter alms-dishes. There is one bell without inscription. The registers for baptisms begin in 1701, those for marriages in 1754, and burials in 1842, but the first interment in the churchyard was in 1854.

The church of the *HOLY CROSS,* Stuntney, consists of chancel, south vestry and tower, nave, and quasi-south aisle. The material is rubble with ashlar dressings and the roofs are tiled. The fabric is of 12th-century origin, but a disastrous restoration in 1876, which virtually amounted to rebuilding, has robbed it of most of its interest. Before these alterations occurred the church consisted of chancel and nave only, with a double opening in the west gable for the bells. Owing to insecure foundations and the wide span of the nave roofs it was necessary to undertake further restoration in 1903, when the modern chancel arch was removed and a wooden arcade inserted to form a quasi-south aisle.

The chancel has three round-headed windows in the east wall and angle buttresses with one set-off, all modern. On the north are two round-headed windows, which appear to be partially ancient but which have been reset. The rest of the fenestration is of similar design but is entirely modern. There is now a wooden chancel arch inserted in 1903. The western portion of the chancel is wider than the eastern and seems to occupy part of the site of the original nave. The fine 12th-century chancel arch is now on the south side of the chancel and opens to the base of the tower, which serves as an organ chamber; it is of two orders and semicircular

26a Infm. Major J. G. A. Beckett. Diocesan transcripts of all types of register exist for 1606–40, 1642, 1648, and 1661–9.

in form with angle shafts having stiff foliage; there are slight traces of red on the arch. The ugly modern tower has round-headed belfry lights and a gabled top. There is a large 12th-century corbel carved with a human head now lying on the ground outside the tower. The original north doorway is now inserted on the west side of the base of the tower at the east end of the south aisle; it has a round arch with zigzag mouldings and renewed angle shafts. The south doorway, which still serves as the principal entrance to the church, is similar in design but has billet as well as zigzag mouldings. All the roofs are modern.

The 12th-century font has a cup-shaped bowl with large scallops; the oak cover is of 17th-century date. All the other fittings are modern with the exception of a small 17th-century alms-box. The plate includes a chalice of 1700 inscribed 'Stuntney in Trinity parish in Ely' and 2 silver patens.

The tower contains one bell by W. Dobson of Downham, 1807.

The registers begin in 1545, but those for burials are missing from 1813 to 1849.

GUILDS Ely was a great centre for guilds, of which eleven existed at the time of the 1389 inquiry, and another (St. Anne) by 1458.[27] Two others, of St. Withburga and St. Sexburga, are mentioned in 1516–17.[28]

The earliest foundation was that of St. Etheldreda, which was said to have been founded 'a hundred years and more' before 1389. It was thus the oldest guild in the county except perhaps that of St. Mary at Cambridge.[29] It also seems to have always been the most important of the Ely guilds. In 1389 it possessed £5 worth of goods,[30] and at the Dissolution was the only guild in the city known to have had any real estate. This consisted of an inn called the Crown, with houses, shops, cellars, solars, stables, and garden. It was in the tenure of Richard Mey, and was granted in 1549 to John Harford of Bosbury (d. 1559)[31] and Richard Willison of Ledbury (Herefs.). The value of this property is not known, but from the Crown Inn 12*d.* rent was payable to the bishop. From the remainder of the premises 6*d.* was payable to the bishop, 5*s.* 8*d.* to the dean and chapter, and 8*s.* to Sir Edmund Knyvett.[32] A messuage called the 'Guyldhall in Ely', perhaps the former hall of this guild, was held by Robert Warden in 1609.[33]

The guild of St. Peter had been founded eighty or more years before 1389, and that of All Saints in St. Mary's church in 1331; otherwise the dates of foundation, where stated in the certificates of 1389, are late in the reign of Edward III or in that of Richard II. The guild of Corpus Christi is interesting as having had a monk, Robert Aylesham, as its co-founder.

Most of the guilds served the usual objects of such institutions—saying masses for deceased members, maintaining lights before certain altars, and, in a few cases, distributing small sums to the poor. Three of them (All Saints in St. Peter's church, Corpus Christi, and St. Peter) made allowances to indigent members according to their means. Most of the Ely guilds admitted women to their membership.

The guilds already existing in 1389 were: in St. Mary's Church, those of the Assumption (in the 'new chapel' of the Blessed Virgin), All Saints, St. John Baptist; in Holy Trinity, those of Holy Trinity, Holy Cross, St. Peter, St. Etheldreda, St. Katherine; in St. Etheldreda or St. Peter, those of All Saints, Corpus Christi, St. John Baptist. It is noteworthy that the oldest guilds, those of St. Peter and St. Etheldreda, were in Trinity, whereas others, including the Guild of Corpus Christi, were allowed to remain in the nave of the cathedral where the parishioners of Holy Trinity had formerly worshipped.[34] It is not known to which church the later guilds of St. Anne, St. Withburga, and St. Sexburga were attached.[35]

NONCONFORMITY The Roman Catholic Church of St. Etheldreda in Egremont Street dates from 1891.[3]

It was not to be expected that Protestant Dissent would make an early appearance in a city so literally overshadowed by its cathedral as was Ely. No licences were issued in Ely under the Declaration of Indulgence (1672), and four years later only 33 Protestant Nonconformists and 1 Papist were reported.[37] About 1700 there was a General Baptist congregation in the city, with a Mr. Clack as pastor, but this seems to have been short-lived.[38] In 1753 Edmund Carter asserted that there were no dissenting chapels in Ely[39] but a rapid increase occurred before the end of the century.[40]

About 1797 a section of the Countess of Huntingdon's congregation in Ely (see below) adopted Strict Baptist doctrines, split away and began to hold services in two licensed rooms.[41] The breach was not irrevocably widened, and in 1840 a chapel (Salem) was built in Chequer Lane and used by the Independents and Baptists in mixed communion. In 1851 Salem chapel had no resident minister or Sunday School.[42] The absence of a resident minister seems to have led to weakness, and in 1853 the Baptists finally seceded and built a chapel of their own (Zion) at the corner of Butcher Row and High Street Passage. This chapel still exists.[43] The congregation is not, however, in affiliation with the Baptist Union. Salem chapel continued to be used by the Independents until *c.* 1875.[44]

From about 1780 itinerant preachers of the Countess of Huntingdon's Connexion visited Ely.[45] A regular congregation was formed in 1785[46] and in 1793 the existing church, in Chapel Street, was built by a certain Morgan James and opened by the Revd. Thomas Wills of London, the Countess's chaplain. In 1797 it was closed as a result of the Strict Baptist schism (see above). It was reopened in 1802 under John Sheppard, a minister from Cheshunt College, and became a vigorous institution. In 1819 galleries had to be erected to provide extra accommodation. The congregation was then estimated at 500 to 600.[47] It remained at this level for at least a generation, and in 1851, when there were over

27 Mentioned in a will of that date (*Fenland N. & Q.* v, 112).

28 *Sacrist R.*, 8 Hen. VIII (infm. Archd. S. J. A. Evans, F.S.A.).

29 *Camb. Gild Rec.*, viii.

30 C 47/38/23. In 1389 St. John Baptist Guild in St. Mary's Church had £5 'in hand' (not necessarily goods to this value). None of the other guilds had more than £3 0*s.* 6*d.* of goods.

31 His monument, one of the remarkable series by John Guldo of Hereford, still exists in Bosbury Church.

32 *Cal. Pat.* 1549–51, 27, 29, 30.

33 C 142/323/51.

34 See above—Churches.

35 C 47/38/17 to 23; Palmer, *Village Gilds of Cambs.*, esp. 374–6.

36 Kelly, *Dir. Cambs.* (1900).

37 Bp.'s Mun., Ely, A 6.

38 *Cong. Mag.* ii, 502 (1819).

39 *Hist. Cambs.* (1753), 66.

40 Bp.'s Mun., Ely, B 6 and C 2.

41 *Cong. Mag.* ii, 502.

42 H.O. 129/7/190.

43 Kelly, *Dir. Cambs.* (1937); *City of Ely Red Bk.* (1950).

44 It is last mentioned in Kelly's Directory of that year.

45 *Cong. Mag.* ii, 502.

46 *Cong. Yr. Bk.* 1949.

47 *Cong. Mag.* ii, 502.

400 adult worshippers and 200 Sunday scholars, was the largest congregation in the city.[48] This church is still in existence.

John Wesley visited Ely in 1774 and 1786. On the former occasion he preached in a house 'well filled with plain, loving people' and admired the cathedral; on the latter he was only passing through on his way from Lynn (Norf.) to Hoddesdon (Herts.).[49] It was not, however, until 1818 that the Wesleyans built a chapel in Ely. In 1851 there was a Sunday School.[50] The chapel building, in Chapel Street, near the Countess of Huntingdon's, was restored in 1891[51] and is still in use.

The Primitive Methodists built a chapel in Victoria Street in 1847. There was a Sunday School in 1851.[52] The chapel is still in use.

The Salvation Army began work in Ely, also in Victoria Street, between 1892 and 1896. Their work ceased between 1908 and 1912.[53] The Railway Mission Hall in Silver Street was opened in 1901 and is still in existence.[54]

Conditions in Prickwillow were much more favourable to Nonconformity, and the Baptists (1816), Wesleyan Methodists (1826), and Primitive Methodists (1846) all had organized congregations there before the established church. In 1851 the last named had as yet only a licensed schoolroom (1849) with a service held alternately in the morning and the afternoon. The congregations were larger than at any of the chapels individually, but each of the latter had a flourishing Sunday School.[55] The Baptist chapel was rebuilt in 1875 and the Primitive Methodist in 1894. They still exist. The Wesleyan congregation had more or less died out by 1894, and the chapel, under the title 'The People's House' (now St. Peter's Hall), became a centre for religious and social meetings.[56]

There are no Nonconformist congregations in Adelaide, Stuntney, and Chettisham.

SCHOOLS Catherine Needham, by her will dated 1730, left to trustees lands and tenements in Witchford, Wentworth, and Downham; the White Hart Inn, Ely, together with a cherry orchard; 12 acres in Further Middle Fen, Ely; 6 acres in Burnt Fen; a tenement in Newnham, Ely; and a close of pasture, for the provision of a school and school-house for Ely boys. In 1741 funds derived from her estates amounted to nearly £900 and with them school buildings, which still stand, were erected in that year on Back Hill.[57] The scheme provided for the teaching of English, writing, and accounts, for 8 hours daily, to 24 Ely boys, not under 8 years of age, 'born of poor parents'. The boys were not to remain at school above 4 or 5 years. Surplus funds were to go to the maintenance of the boys and to apprenticing them.

When the Charity Commissioners reported (1837) the master was receiving a salary of £20 and his house, and the boys still numbered 24, though in 1819 there had been 36, some of them no doubt fee-payers.[58] By 1846–7 they had increased to 50, of whom 30 attended both weekdays and Sundays.[59] In accordance with the will of the foundress, the boys received a suit of clothes every year,[60] and were, when leaving at 14 years, apprenticed with a premium of £20; nothing, however, was taught except reading, writing, and arithmetic.[61] In 1872 the (uncertificated) master received a salary of £55 a year. The school was enlarged to take nearly 150 boys, and by a revised scheme of the Charity Commissioners of 1909 the managers were authorized to charge a fee of 6*d.* a week, reserving 25 per cent. of free places, and to provide a modicum of higher education in a 2-year course for boys of 12 years of age and upwards. At the same time the buildings were enlarged to take, on the revised scale, 102 boys, and a science laboratory was provided.[62]

Efforts were made from time to time to convert the school to secondary status, but it was considered that secondary education was sufficiently provided for by the King's School, the Girls' High School (1905), and Soham Grammar School, to which boys from the south of the Isle were admitted in return for similar facilities for girls at Ely High School. The somewhat anomalous position of Needham's School, intermediate between primary and secondary, was rectified in 1933, when it was enlarged to take 160 boys and all those of 11 years and upwards in Ely were transferred to it. Shortly before 1939 there were 115 boys on the roll, 45 from outside the city. The school became a Secondary Modern school under the 1944 Act, but is handicapped by its obsolescent, if picturesque, buildings. The trustees still endow apprenticeships.

In 1813 the dean and chapter granted a building in the College, formerly used by the King's School,[63] as a school for 200 boys to be conducted in union with the National Society. One of the lay clerks of the cathedral was engaged as master. The corresponding girls' school, which in 1813 held 20 children and was run in connexion with a Sunday school of 40 girls, was in 1814 moved to a room over the Shire Hall capable of accommodating 100. There was already a satisfactory mistress[64] who undertook to make herself familiar with the 'Madras system'. In 1816 there were 50 boys and 80 girls attending these schools, and £130 was subscribed locally for their upkeep.[65]

In 1820 the Shire Hall was replaced by the present building, and the girls were removed to a new school in Market Street, towards the cost of which the National Society granted £100. The boys remained in their building in the College until the Silver Street schools (see below) were erected in 1859.[66]

Progress was not very rapid. In 1820, when the new girls' school was built, 500 Ely children were going untaught, and a generation later (1846) the number was no smaller. At the latter date about 650 to 700 children (in a total population of over 7,000) were getting some kind of education, but half of the children were at dame schools, of which there were said to be some 20 in the city.[67]

At this period some of the National School children were, like those at Needham's School, provided with

48 H.O. 129/7/190.
49 *J. Wesley's Jnl.*, ed. Curnock, vi, 51; vii, 215.
50 H.O. 129/7/190.
51 Kelly, *Dir. Cambs.* (1900).
52 H.O. 129/7/190.
53 Kelly, *Dir. Cambs.*
54 Inscr. on bldg.
55 H.O. 129/7/190.
56 Kelly, *Dir. Cambs.* (1900 and 1933).
57 See *V.C.H. Cambs.* ii, 342, 349.
58 *Rep. Com. Char.* (*Cambs.*), [103], pp. 217 et seq., H.C. (1837–8), xxiv.
59 Archives of Nat. Soc.
60 Boys on the foundation continued to be clothed until the late 19th century (*V.C.H. Cambs.* ii, 342). A photograph dated 1871 of a boy clothed in uniform still (1951) hangs in the schools.
61 Gardner, *Dir. Cambs.* (1851), 481.
62 M. of Ed. file No. 12/22.
63 At this time the National School was drawing boys away from the grammar school on account of its more practical curriculum (*V.C.H. Cambs.* ii, 322).
64 'A person much approved' (2nd Ann. Rep. of Nat. Soc., p. 112).
65 Archives of Nat. Soc.
66 The Silver Street school was 'about to be erected' as early as 1850 (Gardner, *Dir. Cambs.* (1851), 481).
67 Archives of Nat. Soc.

free clothing. Sums of £30 and £10 were appropriated from Parsons's Charity for the boys' and girls' schools respectively.[68] In 1851 70 of the 150 boys and 80 of the 100 girls at the National Schools had free suits or 'cloaks, frocks, aprons, tippets and bonnets'.[69]

The next step forward was taken in 1859, when new National Schools for boys and girls were built in Silver Street at a cost of £2,577, about half of which was subscribed locally.[70] This school has remained the principal primary school in Ely to the present time. The buildings were enlarged by additional classrooms (1888) and technical instruction rooms (1892 and 1901) and accommodated 678 children in 1904, reduced to 538 (268 boys, 270 girls) after 1909. In 1933 it was reorganized for boys from 7 to 10 and girls from 9 to 14 (now 7 to 11 and 11 to 15 respectively). There were (1949) 184 boys and 134 girls on the books.[71]

The Broad Street school was built in 1859 for infants, and enlarged to 200 places by a new classroom in 1900. The accommodation was scaled down to 177 in 1910. Under the 1933 reorganization this school was allotted to junior girls. There were 101 on the books in 1936, increased in 1950 to 155, as the older girls were not able to proceed to the Silver Street school owing to congestion. A temporary hutted classroom was built (1947) at Broad Street to accommodate them.[72]

The Market Street school, on the site of the old girls' National School, was rebuilt as a mixed school for 135 at a cost of £430 in 1868.[73] It was enlarged by the conversion of an adjoining house in 1886, and in 1910 was an infants' school with 246 places. It has an awkward site surrounded by higher buildings and has long been due for closure, though it was re-equipped with new desks and chairs in 1938 for £128. The average attendance in this year was 152.[74]

No provision was made in the city hamlets until the middle of the 19th century. At Prickwillow, the largest of them, a temporary mission church and school was put up about 1855, but soon became unsafe owing to the sinking peat soil. A permanent school with 150 places was built in 1862–3 at a cost of £1,018 including a teacher's house, to which £691 was subscribed locally and £40 received from the National Society. The building was enlarged in 1871 and at the turn of the century accommodated 220 children, reduced in 1910 to 179 (130 mixed, 49 infants). It was reorganized as a junior mixed and infants' school in 1945 and granted controlled status in 1948.[75]

Prickwillow St. James (formerly Mildenhall Road) School stands in a very isolated position, near Shippea Hill Station but 8 miles from Ely city and 4 miles from the nearest schools at Prickwillow village and Kenny Hill over the Suffolk border. It was built in 1870 for £926 (128 places, reduced in 1910 to 102). In 1932 a parish room known as the Burnt Fen Institute was approved as temporary accommodation for an extra 40 children; the school was at that time overcrowded.[76] In 1943 it became a junior mixed and infants' school with 49 pupils, 12 boys and 11 girls aged 11 and more being moved to Littleport. The teacher's house was badly damaged in the great gale of 16 March 1947 and the school accepted controlled status in that year.[77]

Stuntney School dates from 1864, when the National Society granted £12 towards the total cost of £321. The land was given by the dean and chapter. Originally it held 60 children; in 1886 it was enlarged and in 1900 a teacher's house was built; the school now has 75 places (51 mixed, 24 infants). It became a junior mixed and infants' school in 1946, and a controlled school in 1950. There were 32 children on the books in 1948, all present when the school was inspected; this school has a long-standing reputation for high average attendance.[78]

A school was built at Chettisham in 1880 for 30 children at a cost of £479 including the teacher's house. The National Society granted £42. It was enlarged in 1897 to take 48 children and closed in 1934,[79] when only 21 children were on the books.

Adelaide Bridge School was built in 1872 for 64 infants and enlarged in 1885 (new classroom) to provide 82 places. In 1893 a teacher's house was built. It became a junior mixed and infants' school in 1935, and in 1944 had 46 on the roll. Since 1950 it has been a controlled school.[80]

CHARITIES[81] Parsons's (*c.* 1445), Robert Walden's (1605), and Needham's (1730) charities and Howard's preaching charity (1772) are discussed above.[82] Margaret Walden (1607) left, for the benefit of the poor, two inclosures—one, worth £13 10*s.* per annum, to Holy Trinity parish; one, worth £6 per annum, to St. Mary's parish, to be administered by the respective churchwardens. John Usher (1626) left a rent charge of 10*s.* per annum to the churchwardens of Holy Trinity parish, for the use of the parish poor. Thomas Lamb (1654) left land worth £7 3*s.* per annum to the churchwardens of Holy Trinity, for the use of the parish poor. Anne Bell (1670) left two tenements for two poor widows of Holy Trinity parish, to be administered by the Governors of Parsons's charity. In the same year David Offley gave £2 per annum for the poor of St. Mary's parish. This charity was lost after 1812. Lady Norton (1713) left £50 to the churchwardens of St. Mary's parish, for the use of the poor of that parish. Bishop Laney, in 1674, left valuable lands at Soham, in the hands of feoffees, for the benefit of the poor of Soham and for the apprenticing of Ely children.[83] Trilton and Cockain (1714) gave a plot and several tenements to be let to poor widows of Holy Trinity parish, at low rentals, the benefaction to be administered by the Governors of Parsons's charity. In the same year Dr. Turner gave interest on £100 to apprentice poor boys of Holy Trinity parish. The charity was lost after 1836. Wm. Cole (1724) left a rent charge of £3 per annum to the churchwardens of Holy Trinity parish: £1 was to go to ten poor widows of the parish; £2 to provide sermons. Samuel Cole (1728) left a rent charge of £5 per annum to the poor of Holy Trinity parish, to be administered by the Governors of Parsons's charity.

Mary Austin (1746) left to the poor of St. Mary's

[68] *Rep. Com. Char.* (*Cambs.*), [103], pp. 217 et seq., H.C. (1837–8), xxiv.
[69] Gardner, *Dir. Cambs.* (1851), 481.
[70] Archives of Nat. Soc.
[71] M. of Ed. file No. 12/16.
[72] M. of Ed. file No. 12/17.
[73] Kelly, *Dir. Cambs.* (1900).
[74] M. of Ed. file No. 12/18.
[75] M. of Ed. file No. 12/23.
[76] It had been proposed to move the older children to a new senior school at Littleport, which was not built owing to the economy campaign of 1931.
[77] M. of Ed. file No. 12/24.
[78] M. of Ed. file No. 12/21.
[79] M. of Ed. file No. 12/20. The extension of the footpath along the Lynn Road to Chettisham was made for the benefit of the children, as the County Council were not at first willing to provide conveyance. In the end, after strong local protests, they made arrangements for taking the younger children by bus to Ely.
[80] M. of Ed. file No. 12/19.
[81] *Rep. Com. Char.* (*Cambs.*), p. 217.
[82] See above, pp. 42–43 and 46–47, 83, 87, 83 respectively.
[83] Feoffees' Accts., Vestry MSS., Soham.

parish lands near Chettisham, worth £6 per annum, to be administered by the Governors of Parsons's charity. Gotobed East (1752) left to the churchwardens of Holy Trinity parish 5 cottages in Newnham, to be occupied by 5 aged widows of the said parish. John Howard (1772) left to the Governors of Parsons's charity, together with the ministers of St. Mary's and Holy Trinity, £400, to supply fuel, clothing, or food to poor inhabitants not receiving public relief. The Revd. Thomas Watkins[84] left £53 10*s.* 2*d.* to the churchwardens of Holy Trinity for aid of the parish poor. Thomas Edge (1814) left to the churchwardens and overseers of Holy Trinity two freehold cottages for the poor to dwell in; also two freehold plots in Cawdle Fen, valued at 21 guineas annually, for aid of the industrious poor of the parish. Elizabeth Morley (1817) left £500 to the churchwardens of the two parishes of Ely, for distribution in sums of 10*s.* each, to 30 poor persons of Holy Trinity and 20 of St. Mary's parish. Ann Bull (1843) left £350 for small annual distributions to Ely poor.[85]

Except for Offley's and Turner's charities, all these charities survive, though for the most part now grouped, for administrative purposes, under three bodies of trustees—Parsons's Governors, Thomas's charity, and the Trinity Dole charity. The Dispensary charity, founded in 1867, came to an end under the recent National Health legislation. Two groups of 19th-century almshouses, replacing earlier ones, are maintained from joint charitable funds: one group is situated at Waterside and the other in St. Mary's Street, on the site of the old Sextry Barn.[86]

84 Date of will unknown. Sum not received 1812–24; arrears then paid.

85 R. Gardner, *Dir. Cambs.* (1851), 438.

86 I am indebted to Revd. J. B. Rowsell, Vicar of Ely, for information about the present position of Ely charities. He includes Diana Brown's Charity, administered by Parsons's Trustees: it is not recorded by the Charity Commissioners.

ELY HUNDRED

DOWNHAM

Downham, sometimes since the 14th century called Downham in the Isle or Little Downham to distinguish it from Downham Market in Norfolk, is situated at the northern end of the Isle proper. The scarp of the high ground is comparatively steep at this point, the road from Downham village to the hamlet of Pymore descending 50 ft. in about ½ mile. Cole remarks upon the fine views obtainable from the point, half-way along the ridge, where the remains of the bishop's palace stand. Wisbech, 15 miles away, is clearly discernible, and the view extends across the fen into Huntingdonshire and Norfolk.[1]

A secondary road from Ely to Wisbech (B 1411) forms the main street of the village and takes a zigzag course via Pymore to the Hundred Foot Bank, which it follows to Welney Bridge; another road leaves this one at the east end of the village to connect Downham with Littleport, and there are minor roads across the fen. The Ely–Peterborough line of the Eastern Region, British Railways, constructed in 1846,[2] crosses the parish from south-east to north-west and has a station at Black Bank (formerly named Downham), about a mile north-east of the village. The Old Bedford River forms the north-west boundary of the parish, with the New Bedford or Hundred Foot River running parallel about ¼ mile to the south-east.

About 3,000 acres of the parish are upland and about 7,000 fen.[3] The differences in the pattern of settlement of the two types of land are well shown by contrasting Downham village with Fodder Fen in the north of the parish. The former is fairly compact, with a closely built-up main street, and back ways to which the gardens of the houses formerly extended; in Fodder Fen the houses are scattered at considerable intervals along four parallel droves, each drove from ½ to 1 mile long and about ¼ mile from its neighbour. There are other fen hamlets at Pymore, on a small 'island' rising a few feet above the general level of the fen, and at Oxlode and Hundred Foot Engine by the New Bedford River bank. The latter hamlet derives its name from an important drainage station, which was built in 1830 and has the following inscription:

> These *Fens* have oft times been by *Water* drown'd.
> Science a remedy in *Water* found.
> The powers of *Steam*, she said, shall be employ'd
> And the *Destroyer* by *Itself* destroy'd.

The soil in both fen and upland is too valuable for any land to be left as woodland, but there is a small remnant of uninclosed common—almost the only one left in the Isle—on the road to Pymore. There are several orchards in and around Downham village, the northernmost extension of the mid-Cambridgeshire fruit-growing area. There was fruit-growing at Downham in 1286, when a rent of 9*s.* from apples and vines was payable.[4]

In the 18th century the value of fen farms fluctuated violently. A farm of 293 acres in Downham parish by the Hundred Foot River illustrates this fact. Its value increased at successive sales from £800 to £1,200, and then, owing to the floods of the 1760's, declined to £200. In 1805, however, the owner stated that he would not take £1,500 for it.[5]

The inclosure of the common fields took place under the General Inclosure Act of 1801, by resolution of a meeting of landowners in 1844. The award was made in 1850.[6] The fields were West Moor and Mill Piece north, Clayway south of the Pymore road, and Hawkley Field south of the road to Littleport. South of the village lay Ely Lane Field, and south of it and west of the Ely road Hurst Field. Hither and Further Beald Fields, with Pit House Field to the south of them, lay east of the Ely road. Memories of these fields are preserved in Mill Hill, the section of the Pymore road where it falls to the fen, Hurst Lane, Bield Drove, and Orwellpit Farms—the last named over the border in Ely. The nine fields comprised 582 acres, and were divided between thirty-seven proprietors. The Ecclesiastical (now Church) Commissioners, representing the bishop as lord of Downham, were much the most important, their two lessees William Martin (135 acres) and the Revd. J. H. Sparke (69½ acres) each obtaining more than any other person or corporate body. Martin and Sparke were also awarded 31 and 34 acres respectively in their own right. William Tingey (61 acres), Charles Finch (25 acres), and J. T. Waddington and Clare College (22 acres each) also received fairly large allotments.

The 18th-century school building, with the rectory and church, and Bury House on the opposite side of the street, make a prominent and attractive feature at the east end of the village. Bury House, also an 18th-century building, has 'Gothic' casements to the windows. Other houses in Downham of some architectural merit include the premises occupied by F. Gibson (17th century and later), and the Club Inn, the Chestnuts, and the Laurels, all 18th century. At Pymore there is a fine tower windmill with sails still in position.

Among the rectors have been Thomas Peacock (? 1516–? 1582), President of Queens' College, a staunch Catholic who lost all his preferments at the accession of Elizabeth, and Isaac Barrow (1614–80), Bishop of Sodor and Man and uncle of the mathematician.[7] Cole mentions an 18th-century rector, Thomas Jones, who married a rich heiress from Chelsea and ruined her in three years by 'adventurous dealing in fen farms'. In 1777 Jones was imprisoned in the Fleet, owing over £38,000.[8]

[1] B.M. Add. MS. 5821, ff. 13*b*, 14*a*.

[2] *V.C.H. Cambs.* ii, 132–3.

[3] Under the Isle of Ely Review Order, 1933 (M. of H. no. 76940) 829 acres, with a 1931 population of 11, were transferred from Ely to Downham. The present area of the latter parish is 10,952 acres.

[4] S.C. 6/1132/9.

[5] A. Young, *Annals of Agric.* xliii, 556. Young considered the 1805 estimate an inflated one, owing to the improvements forecast as a result of the proposed Eau Brink Cut near Lynn.

[6] Incl. Awards, County Hall, March.

[7] *D.N.B.*

[8] B.M. Add. MS. 5821, f. 14*a*.

A. J. Ball

Morton's Leam from Dog and Doublet Sluice

Lilian Ream

Sea Bank at Newton

Rex Wailes

Windmill at Stretham
(Now Sailless)

H. C. Darby

Forty Foot Drain from Welches
Dam Sluice

P. L. L. Keiller

Pumping Station, Hundred Foot River, Downham

MANOR Six hides in Downham were purchased by Aethelwold and Abbot Brithnoth *c.* 970 for the monastery of Ely, and 2 hides were given by Siverth.[9] In 1086 the monastic property here was rated at 4 hides or 8 ploughlands, 2½ hides (4 ploughs) being in demesne. There were 15 villeins, 8 cottars, and 8 serfs, sufficient meadow for all the plough teams and pasture for the cattle of the vill, woodland for 100 pigs, and a fishery producing 300 eels and 2*s.* rent. When received it had been worth £5, £12 T.R.E., and £10 in 1086. This estate had always formed part of the demesne of Ely.[10] After the foundation of the see Downham was allotted to the bishop, and the manor house became one of the main episcopal palaces for more than 500 years. Judging by the number of ordinations, charters, and letters dated at Downham[11] it was the most popular residence. Centrally situated in the diocese, but outside the cathedral city, it must have had special attractions for secular bishops who wished to remain aloof from monastery politics. Five medieval bishops (de Fontibus, Northwold, Orford, Fordham, and Gray) died here.[12] The bishops held free warren in this manor from time immemorial, a right which was confirmed in 1399.[13] The park is first mentioned in 1250–1, when it contained 250 acres;[14] in 1251 it was, with two marshes held in severalty by the bishop, a league long and nearly half a league broad.[15] Its custodian from 1445 was Henry Botorcawbe, who resigned his office on Bishop Gray's appointment (1455).[16] At the time of the Northwold survey (1251) the demesnes amounted to 444½ acres,[17] rated at 3 carucates, stocked with 60*s.* worth of livestock, besides more than 100 sheep belonging to the tenants. There was a newly constructed windmill, worth £3,[18] at which all the tenants had to grind their corn. The fisheries were increasing in size and value. In 1222 there had been one, rendering 7*s.* yearly and 35 sticks of eels;[19] in 1251 there were two, Downhamhythe and Manea, rendering 12*s.* Their boundaries were defined as from Coveney Bridge to 'Godrige beche' (near Downhamhythe) and thence to 'Cokeslode' (in Manea).[20]

The prosperity which was general in the Isle during the 13th century is illustrated in the case of Downham by the increases between 1222 and 1251 of free tenants from 4 to 17, of customaries from 17 to 20, and of cottars and toftmen from 25 to 35. In the later survey there were also 9 *novi feoffati*; the 13 new free tenants were holding 69 acres of reclaimed land in Apesholt, on the northern edge of the parish bordering on Littleport, at 1*d.* an acre. The total rents in 1251 were £2 15*s.*, besides the fisheries and a rush ground worth 10*s.* 6*d.* Among the services due from customaries were work on Aldreth causeway, where each tenant was required to repair 15 ft. or pay 2*d.* annually, and the production of 40 bundles of thatch from the marsh.

In 1286, 737 summer and autumn works were commuted for £2 8*s.* 2*d.*, and 994 for £2 1*s.* 5*d.* in 1302, but only 214 (8*s.* 11*d.*) in 1316. The gross values of the manor at these three dates were £11 5*s.* 10*d.* (1286), £10 18*s.* 11*d.* (1302), and £21 9*s.* 1*d.* (1316); the great increase at the last date was due mainly to a very large hay crop, sold for £11 17*s.* 6*d.* at 1*s.* an acre.[21] In the early 14th century the total value of the manor remained fairly constant—for example, £15 3*s.* in 1337 and £13 6*s.* 4*d.* in 1345, but that of particular items fluctuated considerably. The windmill was rated at £1 5*s.* in the former year but at only 13*s.* 6*d.* in the latter, whereas the fisheries increased from 8*s.* 11*d.* to 14*s.* 7*d.*[22]

In 1356 another survey[23] shows a less prosperous state of affairs. The manor house was ruinous, and its accompanying dovehouse and orchard of no value. The demesne had shrunk to 160 acres, worth £2 13*s.* 4*d.*, and rents from tenants brought in only £2 a year. The two fisheries produced 11*s.* 8*d.* The windmill and the deer park were still in existence. The movable property of the bishop, with the stock and produce of the demesne, was worth £21 4*s.*, and perquisites of court produced 13*s.* 4*d.* yearly.

The value of the manor in 1540–1 was £27 17*s.* 9½*d.*,[24] and in 1548–9 £37 7*s.* gross, excluding arrears.[25] At this time it was held, apparently on a lease for lives, by the Megges family,[26] who were also tenants of the episcopal manor of Wisbech Barton. In 1548 Thomas Megges was reported to have inclosed 63 acres, in portions, for conversion from arable to sheep-farming, leading to the decay of a cottage. Another cottage was turned into a barn and common ways were stopped up. Thomas St. Aubyre had let to Robert Megges the lands on which the villagers of Downham had been accustomed to pasture sheep and from which they were now excluded. Megges had let them to others.[27]

During the long Elizabethan vacancy in the see John Graunge received a twenty-one-year lease of the site of the manor and the demesnes.[28] In the reign of James I and up to the Civil War the bishops resumed occupation of this palace. Bishop Andrewes (1609–19) spent considerable sums on its repair,[29] and Bishop Wren (1638–67) was arrested there in 1642 by order of Parliament.[30] In 1632, when William Marsh was tenant of the manor house at £3 a year, there were quit rents of £10 5*s.* 5½*d.* payable twice yearly, and fines certain on copyholds which amounted to £5 8*s.* During this year £3 'recognition money' was paid to the newly appointed Bishop White (1631–8) on his first arrival in Downham manor, which was one of those where there was a regularly used episcopal residence.[31]

The Parliamentary Commissioners for the Sale of Church Lands quickly turned their attention to Downham manor, which was disposed of in 1648. The purchaser was Samuel Warner, citizen and grocer of London, who had already advanced half the price (£3,208 8*s.* 9*d.*) and was required to produce the

9 Jas. Bentham, *Ely*, 75.
10 *V.C.H. Cambs.* i, 369.
11 e.g. *Cal. Pat.* 1361–4, 337 (confirmation of a charter of 1308); 1381–5, 97, 231; 1391–6, 53; *L. & P. Hen. VIII*, iii, 565, 1030, 2615, 3476; xi, 1375.
12 Bentham, *Ely*. Of these, Fordham and Gray were from the secular clergy.
13 *Cal. Chart. R.* 1341–1417, 381.
14 B.M. Add. MS. 5847, f. 94, quoting Bp.'s Mun., Ely, G 3 ('Old Coucher').
15 B.M. Cott. MS. Claud. C. XI, ff. 34, 35*b*. Wood from this park was taken to Ely by the customaries in two-horse carts for the kitchen of the guest house. The wages of the keeper of the park were 3*d.* a day in 1358 (*Cal. Pat.* 1358–61, 130).
16 *Ely Dioc. Remembrancer*, no. 206, p. 141; no. 229, p. 131.
17 B.M. Cott. MS. Claud. C. XI, f. 34*a*.
18 Bp.'s Mun., Ely, G 3 ('Old Coucher').
19 B.M. Cott. MS. Tib. B. II, f. 94.
20 B.M. Cott. MS. Claud. C. XI, f. 35.
21 Bp.'s Mun., Ely, D 5 (1, 2).
22 S.C. 6/1132/9, /11, /13.
23 B.M. Add. MS. 6165, f. 276.
24 A. Gibbons, *Ely Episc. Rec.* 13.
25 B.M. Add. Chart. 34274, 4*b*.
26 *L. & P. Hen. VIII*, xii (1), 129, 245, 361, and xvii, 256; C 3/124/1 and /128/9; B.M. Lansd. MS. 21, no. 44 (Lease by Bp. West to Thos. Megges, 1521–2).
27 *Cambs. and Hunts. Arch. Soc. Trans.* v, 379.
28 C 66/1379, m. 34.
29 Bentham, *Ely*, 198.
30 Wharton, *Anglia Sacra*, i, 675.
31 Bp.'s Mun., Ely, D 6.

remainder within six months. The appurtenances of the manor included the palace with a private chapel and demesne of 291 acres, fishing rights, court leet, and view of frankpledge.[32] In 1650 a subsidiary portion of the manor, in which the episcopal tenant had been Thomas Marsh,[33] was sold to Richard Turner of London, tailor, for £686 7*s*.[34]

At this time Downham Palace, like many of its kind, fell into disrepair, and it was not used by the later 17th-century bishops. Simon Patrick, who was translated to Ely in 1691, obtained statutory powers for himself and his successors, to lease the manor house and demesnes.[35] As a result a varied succession of persons exercised manorial rights in Downham: Dr. William Cave in 1710,[36] Mrs. Arabella Thomson of York in 1733,[37] George Grant of Piccadilly, London, in 1808,[38] and William Smith Simpson in 1851[39]—only the last named seems to have actually resided at Downham. In 1778–9 the principal tenants were William Martin of Downham, who held the Park for £100 a year, Sir John Lade, bt., who held Caterclose at £7, and Philip Cawthorne, who held Downham Plain (22 acres) at 16*s*. 6*d*. The quit rents now amounted to £20 17*s*., and 6 capons, or 6*s*. in lieu, were produced annually for the episcopal table.[40] In the middle of the 19th century the bishops still retained some property in Downham in their own right, but the ownership of land had become, and has since remained, much divided. The Church Commissioners, representing the Bishop, are the present lords of the manor.[41]

The prior and convent of Ely were tenants of the Bishop in Downham; in 1291 their property, which was appropriated to the almoner, was worth 15*s*. 4*d*.[42] In addition to this several small estates in Downham were bequeathed to the convent by various persons. Thus in 1309 and 1310 16½ acres were received in seven lots, of which the largest (5½ acres) was allotted to the sacrist.[43] In 1314 there was a grant from William de Fresyngfeld, a relative of the then prior,[44] and in 1370 a messuage was appropriated to the sacrist.[45] In 1539 the convent's temporalities in Downham amounted to £1 12*s*. 4*d*.[46] From 1454 the convent had the right to 4 deer annually, 2 in summer and 2 in winter, from the bishop's parks at Downham and Somersham.[47] The property was formally transferred to the dean and chapter in 1541.[48]

A sub-tenant of the prior in 1222 was Fulk son of Ivo, who held a carucate;[49] in 1170–1 the lands of Ivo of Downham, presumably his ancestor, had been rated at 10*s*.[50] The property seems to have remained in the family, as in 1251 the tenant was Clement son of John Folke.[51] Another 1222 tenant was Roger le Grand, who held 3½ acres by the service of working on Aldreth causeway and in the vineyard, and of entertaining the bishop's messengers.[52] By 1251 the hospital of St. John Baptist, Ely, had succeeded to this property, rendering the same service.[53] In the next century this hospital acquired more land in Downham,[54] which lay in scattered portions and was worth 12*s*. 8*d*. in the 16th century.[55] The dissolution of the hospital was delayed until 1562, when all its property was given to Clare Hall, Cambridge, to endow scholarships.[56] The college was among the principal landowners in Downham in 1851.[57]

The remains of the former Bishop's Palace are incorporated in farm buildings about ½ mile west of the parish church. The palace was rebuilt by Bishop Alcock (1486–1500), and the surviving remains are all of this period. The building fell into decay while it was alienated from the see during the Commonwealth, and was never readapted as an episcopal palace after 1660. The material is brick, with stone dressings in certain portions of the fabric; the roofs are (1949) of green-painted corrugated iron. There is some doubt as to the original purpose of the surviving fragments, but a detached structure on the east is believed to have been the chapel. It has a fine south doorway with four-centred arch and continuous mouldings, with an ogee hood terminating in a large finial flanked by shields, one of which is charged with the arms of Ely and the other blank; the rebus of Bishop Alcock, a cock standing on a globe, is shown below the shields. The palace was originally of two stories, the lower of which was vaulted, but only the springers remain. The north wall has been partly rebuilt, but there is a portion of a fine doorway opposite the one on the south side. In the south wall on the ground floor are a three-light window, with uncusped heads under a square label, and two single lights, and in the upper story two single lights and a large window of five lights under a square label. All the windows are of stone, as are the doorways. There is a mutilated sundial near the south doorway. To the north are remains of a structure of which little more than the base of the walls has survived. To the west of the present farm-house is a large building now used as a barn. It has good stepped gables and an original chimney at the west end. Built into the south wall is an enormous oven, and above a drying-chamber. It is probable that this section of the palace contained the hall and kitchen. There is a walled garden, and at some distance to the south of the supposed chapel a large rectangular pigeon-house with thatched roof, now in a dilapidated state. The present farm-house seems to date from about the beginning of the 19th century, but it is largely built of older materials and it may incorporate portions of the walls of the palace.

CHURCH. The advowson has always descended with the manor, and remains with the Bishop of Ely at the present day.[58] The church was never appropriated, but as early as 1291 a portion was paid to the prior, which in 1539 amounted to £2.[59] The total value of the rectory was £4 10*s*. in 1217, £13 6*s*. 8*d*. in 1254, and £20 (including the

[32] C 54/3406, no. 28.

[33] Probably a descendant of the William Marsh who was lessee of the manor house under Bishop White.

[34] C 54/3515, no. 8.

[35] 9 & 10 Wm. III, c. 2. See also B.M. Add. MS. 5821, f. 12*b*.

[36] C.P.25(2)/911/Mich. 2 Anne. He was a sub-lessee of William Needham and others for ninety-nine years.

[37] Gibbons, *Ely Episc. Rec.* 21.

[38] D. and S. Lysons, *Magna Brit.* ii (1), 178.

[39] R. Gardner, *Dir. Cambs.* 491.

[40] Bp.'s Mun., Ely, D 6.

[41] Gardner, *Dir. Cambs.* (1851), 491; Kelly, *Dir. Cambs.* (1900, 1933).

[42] *Tax. Eccl.* (Rec. Com.), 271.

[43] D. & C. Mun., Ely, Cartae 157, 948, 963–6.

[44] *Cal. Inq. Misc.* iii, no. 784 (p. 295).

[45] *Cal. Pat.* 1313–17, 121.

[46] Dugd., *Mon.* i, 492.

[47] *Ely Dioc. Remembrancer*, no. 219, p. 179.

[48] *L. & P. Hen. VIII*, xvi, g. 1226 (12).

[49] B.M. Cott. MS. Tib. B. II, f. 94.

[50] *Pipe R. 1170* (Pipe R. Soc. xv), 96.

[51] B.M. Cott. MS. Claud. C. XI, f. 35.

[52] B.M. Cott. MS. Tib. B. II, f. 94.

[53] Bp.'s Mun., Ely, G 3 ('Old Coucher').

[54] *Cal. Pat.* 1358–61, 15; 1391–6, 144.

[55] *Cambs. Arch. Soc. Proc.* xxxvi, 97, 102.

[56] *V.C.H. Cambs.* ii, 310.

[57] Gardner, *Dir. Cambs.* (1851), 491.

[58] P.R.O. Inst. Bks.; *Crockford's Clerical Dir.*

[59] *Tax. Eccl.* (Rec. Com.), 265; Dugd., *Mon.* i, 492.

prior's portion) in 1291.[60] In 1535 it was £17 2*s.*, probably without the portion.[61] In 1718 a lawsuit between Thomas Jones, the rector, and the parishioners regarding tithes was decided in favour of the former. The rectory thus benefited by £100.[62]

In 1405 the missal of Downham church was worn out, and it was agreed to purchase a new one by a general levy on the parishioners. Some refused to pay their share and were given a month's grace before being cited before the bishop.[63]

A guild of St. Mary of the Porch and St. Leonard was in existence before 1464,[64] and was refounded on a rather elaborate scale in 1475. Its officers were to consist of an alderman and two deans, elected annually, and it was licensed to acquire property to the annual value of 10 marks.[65] In 1547 £4 14*s.* 6*d.* worth of land belonged to this guild, including 30 acres in March let to John Handfortes.[66] Two years later the main portion of the guild's property was granted to Robert Carr. At this date (1549) 2*s.* 1*d.* was derived from house rents in the village, for which with other Downham property the guild paid 11*s.* 8*d.* yearly to the bishop as lord of the manor. The guild owned property as far away as Littleport and Witcham.[67] The guildhall itself went to Thomas Jennings in 1571.[68] It stood on the same site as the present village school, and it is possible that parts of the building were in the 18th century used as a workhouse which in its turn was converted into a school in 1779.

A hermitage at Downham Hythe is mentioned in 1398, when John Skillynge was its 'warden'.[69] He and his fellows probably made their living by ferrying travellers from this promontory of the Isle to Doddington and other Fenland villages.

The church of *ST. LEONARD* consists of chancel, north vestry and organ chamber, clerestoried nave, aisles, south porch, and west tower. The fabric dates from the end of the 12th century with later additions and alterations. The chancel was provided with new windows about the middle of the 14th century, and a hundred years later the porch was added. In the second half of the 19th century the church underwent a drastic restoration, when the porch was rebuilt, partly with old materials, and new roofs were provided; at the same time the vestry and organ chamber were added and the top stage of the tower rebuilt. Much of the lateral wall of the north aisle had been previously rebuilt in yellow brick. The material is mostly flint with stone dressings, except the porch and top stage of the tower which are of ashlar. The chancel, nave, and north aisle are tiled and the porch and south aisle are leaded. All the internal roofs are modern and poor. There is an absence of buttresses, except three modern ones with two set-offs to the vestry and organ chamber, and two to the porch.

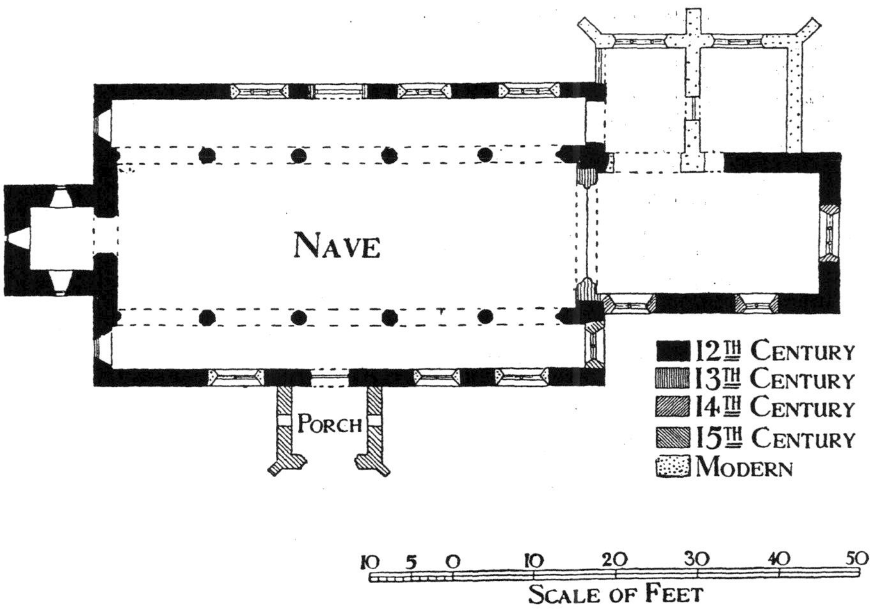

PLAN OF ST. LEONARD'S CHURCH

The chancel has an east window of three lights with modern tracery of mid-14th-century character, evidently a renewal of old work. In the south wall are two mid-14th-century windows of two cinquefoiled lights under a square head, with a hood-mould terminating in carved heads; they have both been considerably renewed. There is a double piscina with turned central shaft and responds with moulded caps, the western section of which has been cut away for the insertion of the adjacent window. The chancel arch is two-centred and of two orders with plain moulded caps partly renewed; it dates from the 13th century.

The nave arcades are of five bays with piers alternately round and octagonal and moulded caps. The clerestory consists of small lancets set over the piers and having wide internal splays. The tower arch is small and pointed and is considerably recessed; it is probably

[60] W. E. Lunt, *Valuation of Norw.* 216, 539.
[61] *Valor Eccl.* (Rec. Com.), iii, 499. At the Dissolution the £2 pension was appropriated to the sacrist (D. & C. Mun., Ely).
[62] B.M. Add. MS. 5821, ff. 13*b*, 14*a*.
[63] *Ely Dioc. Remembrancer*, no. 188, p. 20.
[64] W. M. Palmer, *Village Gilds of Cambs.* 387, quoting will of Richard Cok (P. C. C. Godyn, 84).
[65] *Cal. Pat.* 1467–77, 518–19.
[66] Palmer, loc. cit.; *Cal. Pat.* 1547–8, 323.
[67] *Cal. Pat.* 1548–9, 348–50.
[68] C 66/1077, m. 38.
[69] *Cal. Pat.* 1396–9, 320.

contemporary with the arcades. There are two small lancets high up in the west wall of the nave to the north and south of the tower respectively. The original fenestration of the aisles has been destroyed with the exception of the west windows, which are plain lancets; there is also a blocked lancet towards the west end of the lateral wall of the north aisle. The east window of the south aisle is of the 15th century and of two cinquefoiled lights under a depressed head. The lateral windows of the south aisle are all modern and similar to the 14th-century windows on the south side of the chancel. There is a plain rectangular piscina in the south wall. The north aisle has three lateral windows of three cinquefoiled lights under a square label, all modern. The windows of the organ chamber and vestry are also modern and of three lights with intersecting tracery. The south doorway is of transitional Norman character, but probably contemporary with the arcades; the arch is pointed with zigzag and beakhead ornament; the shafts are cheveroned and cabled and the caps carved; the original door survives though partly renewed, and it retains its hinges and straps terminating in fleurs-de-lis, and also the closing ring. There is a plain 13th-century doorway in the north aisle with continuous chamfer. The organ chamber opens to the chancel and north aisle by modern arches.

The rebuilt porch retains a 15th-century outer doorway with continuous mouldings and a hood with renewed finial. It has side windows consisting of four quatrefoils in a square frame. The buttresses are diagonal.

The tower consists of four stages without buttresses. The base has three 12th-century round-headed windows, the second stage three plain lancets, and the third three trefoil-headed lancets; the modern belfry stage has two-light windows under a square head, and there is an embattled parapet with pinnacles at the angles.

The font is of the 15th century with octagonal bowl and shaft, all richly panelled with quatrefoils and shallow trefoiled niches. There is a late-16th-century oak chest with shallow carving on the front. On the west wall of the nave over the tower arch is a large painting of the royal arms of George III on canvas.

The plate consists of a chalice of silver, 1882, a paten of silver parcel-gilt, 1883, and two glass cruets with metal mountings.

The tower contains four bells: 1st by Thomas Newman of Norwich, 1702; 2nd by Joseph Eayre of St. Neots, 1784; 3rd and 4th by Robert Gurney of Thetford, 1659.

The registers begin in 1558, but the volume containing burials from 1678 to 1729 is missing.

The rectory is a late-18th-century house of local yellow brick with roof of variegated tiles. It is of two stories and attics, and has sash windows and a pedimented doorway with fanlight. The date 1790 appears on the west chimney-stack with the initials T. W. for Thomas Waddington, rector 1787–1805.

There is an Anglican chapel between Oxlode and Pymore (Holy Trinity) and another at Third Drove (St. Owen). They serve both as mission churches and as schoolrooms. They were built at the cost of rectors of Downham, the Revd. Frederick Fisher and Canon F. F. M. S. Thornton, in 1865 and 1895 respectively.[70]

NONCONFORMITY. In 1676 there were 15 Dissenters in Downham.[71] Licences for meeting-houses of Protestant Nonconformists in Downham were granted to Thomas Burges (1737, renewed to Ann Burges, 1741), Alice Carlile (1748), Philip Atkin (1749), John Cole (1754), and Francis Morrey (1786). None of these gives the denomination of the congregation.[72] The Particular Baptists were established in Downham at an early date, with a chapel built in 1788 and rebuilt in 1858.[73] In 1819 they had 12 regular adherents, and it was stated that 'though the church is not numerous, public worship is well attended'.[74] The average congregation in 1851 was said to be 100;[75] a separate Sunday-school building was erected in 1930.[76] A small congregation of Christian Baptists claimed to have been established before 1800; it was probably an offshoot, and averaged 25 in 1851.[77] It does not seem ever to have had a chapel building of its own. Wesleyan Methodist chapels were established in the village in 1811 and 1816, and one at Oxlode, in the out-parish, in 1845. A Primitive Methodist chapel was established on the Hundred Foot Bank in 1836 and rebuilt in 1878. These four chapels had congregations totalling nearly 300 in 1851.[78] All of them are still in existence except the later of the two Wesleyan chapels in the village. The existing Wesleyan church in the village dates from 1907, the former building now being used as a schoolroom.[79]

SCHOOLS. Downham was one of the few villages in the Isle to have a schoolmaster at the end of the 16th century. George Archer was licensed to act in this capacity in 1582,[80] and a schoolmaster is mentioned in the Visitation Return of 1596.[81]

In 1779 the feoffees of the extensive and valuable Town Lands of Downham opened a school in the old village workhouse, which stood on the north side of the main street—traditionally the site of the guild hall. In 1837 the schoolmaster received £6 a year from the feoffees for teaching 15 boys.[82] A scheme made by the Charity Commissioners in 1872 divided the proceeds of the Downham charities into moieties, and laid it down that not less than £150 should be devoted annually to education. In 1877 the feoffees opened the school to government inspection and increased the master's salary to £80 a year.[83] They also reported that the 70 infants, who had had a separate schoolroom since 1855,[84] were being taught on the kindergarten system.[85]

Until 1910 the accommodation was assessed at 272 mixed and 81 infants, from 1910 to 1927 at 219 and 70, and from 1927 at 170 and 70. The school was for a long time handicapped by its antiquated though picturesque building. Thus in 1924 three classes were being held in one room. In 1926, however, alterations were carried out at a cost of £850. These consisted in the provision of a new classroom and the enlargement of the windows. The numbers attending have never reached the very generous provision made; they were 146 in 1901–2, 135 in 1919, and 144 in 1938. They have now been reduced to about 100 by the transfer

70 Kelly, *Dir. Cambs.* (1900).
71 Bp.'s Mun., Ely, A 6.
72 Ibid.
73 H.O. 129/7/190; inscription on building.
74 *Cong. Mag.* ii, 501 (1819).
75 H.O. 129/7/190.
76 Kelly, *Dir. Cambs.* (1933).
77 H.O. 129/7/190.
78 Ibid.
79 Inscription on building.
80 Gibbons, *Ely Episc. Rec.* 183.
81 *V.C.H. Cambs.* ii, 338 n.
82 *Rep. Com. Char.* (*Cambs.*), [103], 232, H.C. (1837–8), xxiv.
83 M. of Ed. File no. 12/9.
84 Kelly, *Dir. Cambs.* (1900).
85 M. of Ed. File no. 12/9.

(1948–9) of the senior children to the Cromwell School, Chatteris.[86]

Since 1896 the educational portion of the Downham Feoffee Charity has amounted to £160 a year, distributed as follows: Feoffees' School in Downham village, £90 (mixed school £67 10*s*., infants £22 10*s*.); Oxlode and Pymoor, £42; St. Owen's, £28.[87]

The Oxlode and Pymoor school was established at the sole cost of the Revd. Frederick Fisher, Rector, the trust deed being dated 1864 and the school opened the following year. The building was designed to be used as a school on weekdays and a mission church on Sundays; this practice, more common perhaps in the Fens than elsewhere, has not contributed to efficiency in either aspect. In this particular case the difficulties have been increased by the peculiar situation in the fields between the two hamlets, approached only by a footpath. The original accommodation was for 80 children, enlarged before 1910 to 144 and then scaled down to 118 (83 mixed, 35 infants). An additional classroom (1929) brought it up to 158, but the average attendance was only 64 in 1938 and has been further reduced by the transfer in 1948–9 of the senior children to Chatteris.[88]

St. Owen's School, Third Drove, was built in 1895 by Canon Thornton, Rector; it also serves both as school and mission church. There are places for 90 (60 mixed, 30 infants), but the average attendance was only 57 (1901–2), 70 (1919), and 35 (1938). Here also the senior children have been transferred to Chatteris.[89]

CHARITIES In 1837 the Downham Town Lands consisted of 135½ acres of copyhold land, held at a quit-rent of 19*s*. 2*d*. an acre, and 44½ acres freehold, taken as 'overplus land' when the fen was reclaimed and subdivided. The average net rent was £154, with £4 for a cottage. Out of this sum about £50 a year was expended in turves for the poor at the rate of 3,000 a family, £11 in coals (1½ bushel a family), and an unspecified amount in drainage and other taxes and repairs to the eighteen town tenements. Four of these tenements had been built in 1813 for £500, of which £195 was still outstanding.[90] A salary was also paid to the schoolmaster. (See Schools.)

Schemes of the Charity Commissioners in 1872 and 1896 increased the proportion to be made available for educational purposes. The total proceeds of the Downham Feoffee charity were about £300 and £370 a year at these dates.[91]

LITTLEPORT

Littleport is a large parish on the eastern border of the Isle, 5 miles north of Ely. The village stands on a patch of high ground, rising to 65 ft., separated from the Isle of Ely proper by the narrow Wood Fen. Inclosure, as in most Fenland parishes, has been gradual, as reclamation proceeded, but though some 1,500 acres had already been so treated in the early 17th century and nearly 1,000 more were ready to be inclosed, the area surrounding the village is still known as Littleport Fields. In or before 1548, 28 acres of arable were unlawfully inclosed and converted to pasture by eleven persons, including Mr. Hopkyns, then lessee of the manor, and Christopher Bryggs, vicar.[1] The final inclosure of the common fields did not take place until 1840.[2] There were five common fields, all south of the village—East Field on the river side of the Ely road, and Knowle, Mill, Down, and Gray Fields west of it. An area of 546 acres was divided between thirty-seven proprietors, by far the largest allotment (134 acres) going to Henry Martin. Clare College and Thomas Gillett (40 acres each), the Earl of Hardwicke, lord of the manor (39 acres), and John Cutlack (35 acres) were the other principal allottees. The first two and the earl were among the 'principal landowners' in Littleport in 1851.[3] The small Hemp Field, already inclosed, just south of the road to Sandhill, is a reminder that Littleport was one of the few parishes in the county where, at the beginning of the 19th century, this crop was grown, along with the typical fen crops of potatoes and mustard. In 1801 the total cultivated area amounted to just over 5,000 acres (rather less than one-third of the total of the parish). Nearly 4,000 acres were under oats.[4]

In the middle of the 18th century Carter, the historian of Cambridgeshire, remarked that it was as rare to see a coach at Littleport as a ship at Newmarket,[5] but the village is now well served with communications. It is on the main road (A 10) from Ely to King's Lynn, which is here crossed by an important fen cross-country road, that from Wisbech to Mildenhall and Bury St. Edmunds (A 1101), on the canalized River Cam,[6] which is crossed just below the village by a bridge carrying the above roads; and on the Ely–Lynn branch of the Eastern Region, British Railways, which was opened in 1847 and has a station here.[7] The Littleport–Wisbech section of A 1101 was turnpiked under an Act of 1824 (5 Geo. IV, c. lx) through the efforts of the Revd. W. G. Townley of Upwell, at whose cost the suspension bridge at Welney (now replaced by a concrete structure) was erected in 1826.[8] The village is a very large one, with a population greater than that of several neighbouring market towns such as Downham Market and Mildenhall.[9] It is one of the few in the county to have a non-agricultural industry, the shirt factory of Messrs. Hope Bros., which employs some 300 persons.[10] A tradesman's token of 1668 has been recorded.[11] Two serious fires occurred in 1707 and 1727. Collections for relief were made by brief. The

86 Ibid.
87 Ibid.
88 M. of Ed. File no. 12/10.
89 M. of Ed. File no. 12/11.
90 *Rep. Com. Char. (Cambs.)*, [103], 232, H.C. (1837–8), xxiv.
91 M. of Ed. File no. 12/9.
1 *Cambs. and Hunts. Arch. Soc. Trans.* v, 379.
2 K.B. 122/1341.
3 R. Gardner, *Dir. Cambs.* (1851), p. 493.
4 H.O. 67/9.
5 Quoted by E. Conybeare, *Highways and Byways in Cambs. and Ely*, 389.
6 Until the cutting of the Bedford rivers the Great Ouse, and still sometimes so miscalled.
7 Illustrated in C. Barman, *Introd. to Rly. Archit.* 61.
8 Diary of Revd. J. Jackson, 4 June 1824 (Wisb. Mus.); White, *Dir. Norf.* (1864), p. 850.
9 In 1931, 4,709 as compared with Downham Market (2,465) and Mildenhall (3,235).
10 *V.C.H. Cambs.* ii, 375. Four hundred were employed at this factory at the opening of the 20th century (Kelly, *Dir. Cambs.* (1900)).
11 S. H. Miller and S. B. J. Skertchly, *The Fenland Past and Present*, 478–80.

damage caused by the former fire was nearly £4,000.[12] In the middle of the 19th century the place was making considerable progress. The Removal of Nuisances Acts (9 & 10 Vic., c. 96, amended by 11 & 12 Vic., c.123) were vigorously enforced, and 'the town, which once appeared to be immersed in mud and mire, now wears the face of cleanliness and comfort, possessing good roads, well-drained premises, paved footpaths and other evidences of civilization . . .'.[13] Such urban features as a Gas Company (1867), Savings Bank (1868), and a working-men's club have appeared. The third of these is the Alexandra Institute, erected by Hope Bros. for their employees. In 1879 a town hall was erected by the Town Lands Trustees with some of the proceeds of the charity estates.[14] In 1900 there was a town crier.[15] The village boasts a weekly newspaper, the *Littleport Gazette*, established 1879, and a cinema in Hempfield Road. At the Constitutional Hall (1890) films are also shown. The British Legion premises in Silt Road[16] are a very handsome and substantial brick building. The Grange, Ely Road, a house built about 1855 by Canon Sparke of Ely, and later occupied by the Hope family, was used in 1914–18 as a home for Belgian refugees and later for the internment of German prisoners of war. In 1920 it was purchased by the Transport and General Workers' Union for a convalescent home. It has so continued except during 1939–45 when it was used as a R.A.F. hospital. The various rooms were fitted up at the cost of Areas of the Union.[17]

The rise of Littleport to the status of an urban village is a comparatively modern feature. The recorded population of 31 in 1086 approximates to the average of the Isle vills. The 1251 Ely cartulary shows a large number of *novi feoffati*, but, judging by the number of 'works' commuted for money payments round about 1300, Littleport was at that time the smallest of the six episcopal vills in the southern part of the Isle.[18] A return of 1563[19] shows 80 householders, the same number as in Downham but distinctly below many neighbouring villages such as Haddenham, Stretham, and Sutton. The ship money rating of 1639–40 was £42 10*s.* for Littleport, again the same as Downham, while Wisbech, Whittlesey, Ely, Haddenham, Elm, and Chatteris were assessed more highly.[20] By 1676, when there were 556 persons of communicant age in Littleport,[21] it had become the largest village except Haddenham, and the 1801 census returns show that it had outpaced Haddenham by a considerable margin and Thorney by a very slight one.[22] Thereafter development was rapid, a 150 per cent. increase being recorded in the next half-century; this rate was considerably above the high average for the Isle as a whole.

At the 1851 census, 2,622 of the 3,832 inhabitants of Littleport were natives of the parish. The proportion of natives (68 per cent.) is higher than might have been expected, in view of the exceptional increase of the previous half-century; it may be compared with 71 per cent. in Potterne (Wilts.), 68 per cent. in Deddington (Oxon.), and 59 per cent. in Bottesford (Leics.) at the same date.[23] Slightly earlier figures for Castle Acre (Norfolk), a village more akin to Littleport in its social structure and rate of growth, showed in 1843 that at least 103 of the 249 families had come to the village from elsewhere.[24] The distribution of the Littleport immigrants is interesting. Four hundred and sixty-four, or 38 per cent. of them, came from the immediately surrounding parishes. Ely (126) and Downham (74) come first, but it is remarkable that 71 and 44 came from as far away as Mildenhall and Welney respectively, and that the numbers from Lakenheath (38), which to this day is very difficult of access from Littleport, should almost equal those from Southery (42), the first village across the fen along the Lynn road. One hundred and thirty-eight came from other villages in the Isle, all of which were represented except Thorney, Newton, and Tydd, and 83 from elsewhere in Cambridgeshire. Three hundred and eighty-five came from other parts of the eastern counties (Norfolk, Suffolk, Essex, Huntingdonshire, the Soke, and parts of Holland) as against 63 from the rest of England. This bears out Professor Redford's conclusion that migration in the early 19th century was mainly short-distance—from one county to or from its immediate neighbours.[25] It is interesting to note the presence of 12 Irish, 4 of whom were described as 'vagabonds'.[26]

There are very few secular buildings of any antiquity, and none of much architectural distinction. The following, all dating from the late 18th or early 19th centuries, may be mentioned. The Old Turk's Head Inn, Granby Street; Nos. 7 and 8 Victoria Street; No. 14 Wellington Street; and the house in Main Street opposite the town hall.

The greater part of the parish is fen land, and north and east from the village stretches an expanse that is one of the loneliest pieces of country within a hundred miles of London. It includes Burnt Fen (said to be so called from having been burnt by Hereward the Wake)[27] on the Cambridgeshire, and Feltwell Fens on the Norfolk side of the Little Ouse. There are no made roads between the one from Littleport to Shippea Hill and Mildenhall and the one from Southery to Feltwell nearly 6 miles away north-east, except the by-road to the remote village of Little Ouse on the county boundary. Much of this potentially fertile land lay derelict from lack of drainage and sheer inaccessibility until the Second World War. 'A desolate country except in harvest time under blazing sun, and quite without interest except to the farmer who gets his living there.[28]

[12] *E. Anglian*, N.S. vii, 146; *Wilts. Arch. Mag.* xlvi, 522.

[13] *Camb. Chron.* 6 Jan. 1849.

[14] Kelly, *Dir. Cambs.* (1900).

[15] Ibid.

[16] A reminder of the long tongue of silt land that intrudes into the peat fen, following the course of the river.

[17] A souvenir booklet on the Convalescent Home and its companion at Buxton has been issued by the T. & G. W. Union.

[18] Doddington, Downham, Haddenham, Littleport, Stretham, Wilburton. See B.M. Cott. MS. Claud. C. XI, ff. 38–43; S.C. 6/1132/11, /13.

[19] B.M. Harl. MS. 594, ff. 198–200.

[20] *E. Anglian*, N.S. vi, 40, quoting S.P. 16/445/75. Doddington, Wimblington and March, and Stretham and Thetford also came higher on their combined assessments.

[21] Bp.'s Mun., Ely, A 6.

[22] Strict comparison is difficult as Thorney, a donative of the dukes of Bedford, was not recorded in the returns of 1563 and 1676, which were for ecclesiastical purposes.

[23] These are large agricultural villages; higher proportions were recorded in industrial villages in Leicestershire, e.g. Barwell (77 per cent.), Earl Shilton (73 per cent.), Wigston Magna (71 per cent.).

[24] Hasbach, *Hist. Engl. Agric. Labourer*, 197 n. The figures are quoted from the Report of the 1843 Commission on the Employment of Women and Children in Agriculture. The 103 'foreign' families were all agricultural labourers, and it is not quite clear how many of the ninety-seven non-agricultural families in Castle Acre also came from elsewhere.

[25] *Labour Migration in Engl.* 1800–50, esp. pp. 157–64.

[26] H.O. 107/1766.

[27] *P.N. Cambs.* (E.P.N.S.), 226.

[28] Doreen Wallace, *East Anglia*, 67.

In 1638 the local justices had information of a gathering of 600 men who were to meet in Whelpmore (now Wheltmore)[29] Fen at 'a foot ball play or camp, which camp should be called Anderson's camp, who should bring an hundred strong with him'. Rain prevented this assembly, but 200 men, including some from Lakenheath across the Suffolk border, met the following day and threw down the Undertakers' drainage dikes. They did not hurt any man's person or goods. Six men in all were committed to jail as a result of these disturbances.[30] Whelpmore was at that time common to Littleport, Ely, and Downham, and as late as 1782 there was considerable doubt as to where the boundary ran. One of the witnesses before an Exchequer commission, who was then 64 years of age, stated that as a child he had been rowed to a house called Lees Hill in a gunning boat.[31] Farther east, the former extra-parochial place of Redmere,[32] originally on the Norfolk but now on the Cambridgeshire side of the Little Ouse, was transferred in 1895 from the former to the latter county. It had at various times in its history been associated with Methwold and Hockwold in Norfolk and Lakenheath in Suffolk. In 1933 Redmere was abolished as a civil parish and absorbed into Littleport.[33] Another change involving county boundaries took place in 1885, when a detached part of Hilgay (Norf.) was added to Littleport or to Southery according as it lay on the left or right bank of the Little Ouse.[34]

Among the vicars or curates of Littleport have been Edward Leeds (d. 1590), Master of Clare Hall and a noted civil lawyer, George Townsend (1788–1857), miscellaneous author, and G. B. Jermyn (1789–1857), whose collections for a history of Suffolk are in the British Museum.[35] Henry Moore, clerk to Sir Edmund Berry Godfrey, was living at Littleport in 1684; Cole says that he 'was made use of by the Presbyterian faction to stifle and conceal all evidence that could be brought to prove [Godfrey's] self-murther'.[36]

In 1381 Thomas Ixning and Thomas Lyncoln of Littleport led the rioters who burnt the Bishop of Ely's court rolls at Balsham.[37]

Owing to discord amongst the inhabitants Thomas Tawney, overseer, found it impossible to make an assessment to the poor rate during the greater part of 1698–9. In consequence he spent £135 5*s.* in relief but recovered in rates only £72 1*s.* 3*d.* An appeal was made to the Queen's Bench, whence in 1703 a *mandamus* issued enjoining the churchwardens and overseers to make an assessment of the lesser or the greater sum upon the assessable inhabitants and to pay Tawney what was due to him, or else to show cause why not.[38] Further proceedings have not been traced.

Littleport was one of the principal scenes of the rioting that took place in the Isle in 1816.[39] The disturbances broke out at a benefit club meeting at the Globe Inn on Wednesday, 22 May, when several hundred people assembled, besides the 50 or 60 club members. Various houses were broken into, including those of the vicar (the Revd. J. Vachell, a magistrate),[40] Messrs. Clark, Sindall, Speechley, and Mrs. Waddelow, and a post-chaise passing through the village was robbed. Two days later, when the rioters withdrew from Ely, they barricaded themselves in the George and Dragon Inn at Littleport, and in the fight to dislodge them one rioter, Thomas Sindall, was killed, and another and one of the militia severely wounded. The pursuit and arrest of the rioters occupied several days, some of them hiding in the fens. One of them took refuge at Apes Hall, whence the then occupier, Mr. Crystal, drove him 70 miles in his trap, thus enabling him to escape completely.[41]

MANORS

LITTLEPORT, acquired by the church of Ely at an unknown date, was in 1086 rated at 2½ hides (6 ploughs), 1 hide with 2 ploughs being then in demesne. There were 15 villeins, 13 of whom held 9 acres each and two, 12 acres. There were 8 cottars and 8 serfs, and sufficient meadow for the plough-teams and pasture for the cattle. The value had been £7 when received, £6 in 1066, and £10 in 1086. These high values, for which the hidage alone could not have accounted, were made up by very important fisheries, rendering 17,000 eels and 12*s.* 9*d.* in tribute. The manor belonged, as it had always done, to the demesne of Ely.[42]

Littleport was allotted to the bishop on the formation of the see of Ely in 1109, and remained in the bishop's hand until the Elizabethan alienations. The survey of Bishop de Fontibus (1221)[43] shows a demesne of 277½ acres arable and 29½ meadow, 29 customary tenants with holdings ranging from 18 to 5 acres, 15 cottars without land, and 12 fisheries rendering 26,500 eels. Thirty years later[44] the demesne was of approximately the same area, rated at 2 carucates and stocked with 10 cows, a bull, 20 pigs, a boar, and 300 sheep. There were 8 customary tenants *cum plenis terris* and 23 *cum dimidiis terris*, and 17 cottars. While these figures show little increase upon those of 1221, there were over 60 *novi feoffati*,[45] mainly at 'Apesholt', holding nearly 500 acres of reclaimed land, usually at 1*d.* an acre.[46] With these last, the rents due in 1251 amounted to £3 4*s.* 7*d.*, besides 40*s.* from the *roseria* or rush ground of Rack Fen. The fisheries were now called upon to produce 36,000 eels, besides 2,000 'bedrepeeles' from 'Mudyke' and 'Burewere'. These fisheries had now become much subdivided, as was usual by 1251 on the episcopal manors. There was also a windmill,[47] valued at £3 6*s.* 8*d.*, which dated from the episcopate of Bishop Northwold (1229–54). The tenants owed suit of multure.[48] In return for 40*s.* a

29 Cf. Weltmore Farm on Mildenhall Road, about half-way between Littleport and Shippea Hill.

30 *Cal. S.P. Dom.* 1637–8, 503–4; *Fenland N. & Q.* ii, 355–9.

31 E 134/22 Geo. III/E. 6.

32 It seems to have become a separate parish between 1864 and 1883 (White, *Dir. Norf.*), probably in 1868, when the Poor Law Amendment Act (31 & 32 Vic. c. 122 S. 27) directed that extra-parochial places should be annexed to their bordering parishes unless an overseer had been appointed.

33 By the Isle of Ely Review Order (M. of H. no. 76940), which also transferred 460 acres from Ely to Littleport.

34 Kelly, *Dir. Cambs.* (1900) (under the Divided Parishes Act, 1882 (45 & 46 Vic. c. 58)).

35 *D.N.B.*

36 B.M. Add. MS. 5821, f. 24*b*.

37 E. Powell, *Rising in E. Anglia 1381*, p. 46.

38 K.B. 28/6.

39 See *V.C.H. Cambs.* ii, 116, and C. Johnson, *An Account of the Ely and Littleport Riots in 1816*, from which the following particulars are taken.

40 Vachell successfully claimed £708 9*s.* damages against the Hundred of Ely at the next Cambridge Assizes (Johnson, op. cit. 77).

41 Johnson, op. cit. 14–15.

42 *V.C.H. Cambs.* i, 368.

43 B.M. Cott. MS. Tib. B. II, ff. 95*b*–97*a*.

44 B.M. Cott. MS. Claud. C. XI, ff. 38–43.

45 Sixteen more are entered in the 1251 (Northwold) cartulary in a later (probably 14th-century) hand.

46 The rents and areas are not given in every case.

47 Let for 5 marks (inserted in a later hand).

48 Bp.'s Mun., Ely, G 3 ('Old Coucher').

year Bishop Northwold had granted his tenants the right to cut sedge and rushes in the marsh of Rackfen, provided they were not sold outside the manor.[49] The cottars had the duty of thatching the demesne corn stacks with these reeds and rushes, and were responsible for damage if it was badly done.[50] Customary services included work on Aldreth causeway, and the inclosing of the bishop's park at Downham with a ditch and his garden at Ely with a wall. In 1286, 116 summer and 490 autumn works were commuted for 4*s.* 10*d.* and £2 0*s.* 10*d.* respectively; in 1302 19*s.* was received for 477 works, in 1316 10*s.* 3*d.* for 246 works.[51]

In 1298–9 a total of £36 8*s.* 6¼*d.* was obtained from Littleport manor.[52] Fisheries (£10 1*s.* 6¾*d.*) continued to be the most profitable source of revenue, but rents had gone up to £4 1*s.* 9*d.* and perquisites of court amounted to £4 12*s.* 5*d.* The windmill brought in £3 18*s.* 8*d.*[53] In 1302 the total receipts were £19 6*s.* 1*d.* and in 1316 £23 16*s.*[54] The usual mid-14th-century decrease in value is shown in the survey of 1356.[55] Fisheries then accounted for only £4, and the rents of customary tenants for £3 13*s.* 7*d.* The mill was let for £1, and the demesne (let to farm for £2 3*s.* 4*d.*) had shrunk to 125 acres of arable and a 10-acre meadow, of which the stock and produce were valued at £23. The manor house, not being an habitual residence of the bishop, was small (*parva aula*) but was in better condition than many, e.g. Doddington and Wisbech Castle, that were recognized palaces. All its defects were stated to be repairable. Its farm buildings included a windmill.[56] The right of free warren in Littleport was confirmed to the bishop in 1399.[57] A warren is mentioned in 1468, when Elias Cliderow and his son Clement were put in charge of it.[58]

In 1541 the value of the manor was said to be £36 7*s.* 5*d.*,[59] and in 1548–9 it produced £55 15*s.* 9*d.* gross, excluding arrears.[60]

During the Elizabethan vacancy (1581–1600) a survey was taken,[61] showing the 'Berrystead'[62] with 280 acres of demesne let for £5 6*s.* 8*d.*, rents of assize of £28 4*s.* 10*d.*, fisheries worth £6 a year, and ferry tolls of £1 13*s.* 4*d.* The total capital value of the manor, at forty years' purchase, was put down at £1,640, but there were considerable differences of opinion as to certain items. Thus the value of the fisheries was variously estimated at £260 and £340, of the ferry at £71 13*s.* 4*d.* and £76 13*s.* 4*d.*, and of the demesne at £273 6*s.* 8*d.* and £308 6*s.* 8*d.* There were large commons, but 2,500 acres in the parish were worth only 3*d.* an acre, presumably owing to poor drainage. The occasion for this survey was probably the lease (1592) of the manor to Richard Drake.[63] Previous lessees had been John Cox and Richard Arkenstall, son and nephew respectively of the late bishop.[64]

Littleport was one of the manors granted away from the see by Bishop Heton on his appointment;[65] in 1602 it was purchased with all its appurtenances, except the advowson, by Sir John Peyton, Lieutenant of the Tower.[66]

During the reign of James I another survey was made, with less ambiguous valuations.[67] The manor house, with two barns, a stable, 160 acres of arable land in the common fields, and 70 acres of good meadow (worth nearly twice as much as the arable) was £158 in capital value. Fen ground, 1,490 acres in extent, had been embanked and planted with willows and divided into twelve plots of meadow. It was worth £391 7*s.*, the value of an acre varying from 6*s.* 8*d.* to 4*s.* There were also 294½ acres inclosed but not embanked, mostly in hand, worth 5*d.* an acre, and 200 acres (described as 'decreed to the lord but not yet several') ready for inclosure. There were also 400 acres capable of inclosure about which agreement had not yet been reached. All these inclosed or inclosable lands were valued at 4*s.* an acre. The vaccary of the medieval bishops was represented by a dairy farm of 10 acres with 100 cattle. There were 14 miles of fisheries (£20), sheepwalk in the common fields for 500 sheep (£10), 2 windmills (£8 together), and manorial perquisites of £40. The total value was computed at £740 0*s.* 4*d.*, the fee farm rent to the king being £41. There were 'very many cottages' on the waste.

In 1618 Sir John Peyton, with his son John and Alice John's wife, sold Littleport manor, with appurtenances including 5 mills and 2,000 acres of (arable) land and 10 of wood, to Sir Thomas Josselyn for £1,400.[68] Josselyn was lord of the manor four years later,[69] but became financially embarrassed and had to hand it over to Sir Miles Sandys, bt., of Wilburton, and his son, Josselyn's creditors. The Sandys in turn had difficulty in collecting the rents owing to the detention of deeds by Thomas Wadelowe and other tenants, and failed to pay the king's fee farm rent.[70] It was probably on the extinction of the Sandys baronetcy in 1654[71] that the manor was purchased by Sir Edward Partheriche or Partridge, who in that year refused to pay an annuity of £50 (dating from the Josselyn tenure) from Littleport manor to Eleanor, daughter of Sir Robert Brooke.[72] The Partheriche family held the manor until 1734,[73] when Edward, High Bailiff of the Isle of Ely from 1726 to 1749, sold it to Lord Chancellor Hardwicke,[74] whose family retained the lordship until 1851 or later.[75] In the 1820's the net profit averaged about £20 a year. Quit-rents produced between £25 and £30 annually and fines up to about £2. The expenses of the bailiff and steward were £8 7*s.* In the early 1840's an annual court dinner was held in March or April, for which £2 was set aside.[76]

49 B.M. Add. MS. 5846, f. 148. A commission was appointed by Bishop Hotham in 1334 to inquire into the destruction of sedge and other products of the marsh at Littleport, Ely, and Downham.
50 B.M. Cott. MS. Claud. C. XI, f. 41*b*.
51 S.C. 6/1132/11 and /13.
52 S.C. 6/1132/10.
53 The court was usually held six times a year, one meeting (in Dec.) being a leet. See Selden Soc. iv, 107–47.
54 S.C. 6/1132/11 and /13.
55 B.M. Add. MS. 6165, ff. 281, 292–3.
56 So described; the one mentioned as let for £1 was a *molendinum* only, possibly a water mill on the Ouse.
57 *Cal. Chart. R.* 1341–1417, 381.
58 *Ely Dioc. Remembrancer*, no. 244, p. 195.
59 A. Gibbons, *Ely Episc. Rec.* 13.
60 B.M. Add. Chart. 34274, ff. 3, 25.
61 S.C. 12/33.
62 A local term for the manor house, found also in the contemporary surveys of Haddenham and Stretham, and still surviving at Sutton in the Isle and at Wellingborough (Northants.).
63 C 66/1394, m. 33.
64 E 134/25 Eliz./H. 7 and /26 Eliz./E. 2.
65 B.M. Add. MS. 5846, f. 149.
66 C 66/1591, m. 2. He paid £4,507 15*s.* 11½*d.* for Littleport and Doddington together.
67 S.P. 14/190/14.
68 C.P. 25(2)/278/East. 6 Jas. I.
69 B.M. Add. MS. 5821, f. 25.
70 C 3/381/1.
71 G. E. C., *Compl. Baronetage*, i, 90.
72 C 5/32/12.
73 C.P. 25(2)/634/East. 33 Chas. II; C.P. 43/488, ro. 76; C.P. 25(2)/1000/ Trin. 2 & Mich. 3 Geo. I.
74 W. Watson, *Hist. Wisb.* 110; D. and S. Lysons, *Magna Brit.* ii (1), 231.
75 C.P. 43/889, ro. 176; C.P. 43/920, ro. 223; R. Gardner, *Dir. Cambs.* (1851), 493. The Earl of Hardwicke gave a site for the gas works in 1867 (Kelly, *Dir. Cambs.* (1900)).
76 B.M. Add. MS. 36231.

In 1900 Mr. A. H. Irvine of Meldreth held the manor, and in 1933 Mrs. J. Spearing and Mrs. W. L. Raynes. By the opening of the 20th century no fewer than 13 'principal landowners' were enumerated, 6 being non-resident.[77]

The convent of Ely, as in other of the bishop's manors, retained certain interests. In 1251 parcels of land in Littleport, valued at 6*s.* 7*d.* and 5*s.*, were appropriated to the infirmarian and to the almoner respectively.[78] In 1538–9 the convent's property, specified as a mill called Monks Croft and an acre of arable land, was worth 6*s.* 8*d.*[79] It was formally transferred to the dean and chapter in 1541.[80]

A tenement called the Priesthouse has a long recorded history. It is first mentioned by implication in 1301 when an agreement was made between St. John's Hospital, Ely, as rector, and the vicar, concerning tithes of hay in Presthous fen.[81] In 1426 Henry Parys held of the Bishop of Ely a messuage in Littleport called the 'Presteshous' and died so seised.[82] Fines were levied upon the same property in 1590 and 1625, when the appurtenances included a free fishery.[83] The Priesthouse was perhaps the old rectory house which was rendered superfluous upon the appropriation of the rectory.

REDMERE is mentioned in the 1251 cartulary of Ely.[84] It lay between the bishop's manors of Littleport and Feltwell and Methwold (Norf.). In 1347 it was described as a marsh belonging to Methwold and was held by John de Warenne, Earl of Surrey of the king in socage.[85] It followed the descent of Methwold for about a century[86] as parcel of the Duchy of Lancaster. At an uncertain date it was granted to Fotheringhay College (Northants.), founded in 1411.[87] It was in the possession of the college in 1535, and was then worth £3 6*s.* 8*d.* and described as a lordship.[88] The post-Dissolution grantee was Sir Richard Lee,[89] who in 1553 had licence to alienate in fee a moiety of this 'manor' with its 874 acres of moor, marsh, and fishery, to Nicholas Bacon.[90] In the same year Lee conveyed the other half to Sir Ambrose Jermyn of Rushbrooke (Suff.).[91] The Bacon half passed in 1590 to Sir Robert Wingfield, who died seised six years later;[92] the Jermyn half was conveyed by Sir Robert, who succeeded his father in 1577, to Henry Warner six years later.[93] By 1646 the Warner family were in possession of the whole,[94] after which its history is unrecorded until its constitution as a separate parish about 1868[95] and transfer from Norfolk to Cambridgeshire in 1895.

Bishop Gray in 1465 granted 20 acres and 1 rood in his marsh of Littleport to the hospital of St. John the Evangelist, Cambridge, in frank almoin at a yearly rent of 1*s.* This parcel of land, called 'le Fenpole', lay in Westmore, near plots belonging to St. John Baptist Hospital, Ely.[96]

'Apesholt', now Apes Hall in the north of the parish, the area newly reclaimed in the 13th century, was in the 16th and 17th centuries in possession of the Guibon family. In 1560 Anthony Guibon of Cockfield (Suff.) leased it to Paul Gresham of Little Walsingham (Norf.) together with tenements called 'Thameshouse' and 'Cowpers'.[97] His nephew William Guibon died seised of these in 1612, being succeeded by his second son Thomas, a minor.[98]

CHURCHES During the episcopal vacancy of 1169–73 Geoffrey, Archdeacon of Canterbury, accounted for tithes from Littleport to the value of 11*s.* 5*d.* yearly.[99] The church was appropriated to the hospital of St. Mary Magdalene, Ely, by Bishop de Burgh in 1225,[1] but a vicarage was not ordained until the latter part of the 13th century (see below). This hospital was united shortly afterwards with the hospital of St. John Baptist, Ely.[2] After the suppression of the hospital the rectory must in the first instance have gone to Clare Hall,[3] but in 1720 a seventeen-year lease of it was made by Richard Bigg and others to Thomas Marsh.[4] Cole records the final verdict (1772) in a lawsuit between the vicar and lay rector, the latter having claimed corn tithes from the severals and fens, basing his claim on his rights in common law as impropriator. The vicar won the case, which was regarded as a leading one in the neighbouring parishes.[5]

In 1217 and 1254 the church was valued at £10.[6] Another valuation, probably of 1276, raised it to £20.[7] It was probably shortly after this date that the vicarage was instituted, at the customary third (£6 13*s.* 4*d.*) of the total value.[8] By 1291 the vicarage had been reduced to £5; at this date and again in 1535 the appropriated rectory was returned with the spiritualities of the Ely hospital.[9] In 1414 a further reduction, made by order of the Archbishop of Canterbury, reduced the value of the vicarage to £4,[10] but in 1535 it had risen again to £8.[11] In 1301 it was agreed between St. John's Hospital, Ely, and the vicar that the tithes of hay in 'Presthous' fen should be paid in kind and equally divided between the two parties. All other hay tithes were to be paid in money, of which the hospital was to take 3*s.* 4*d.* and the vicar the remainder. Another composition in respect of tithes was made between the master and brethren of the hospital and the vicar in 1384.[12]

77 Kelly, *Dir. Cambs.* (1900 and 1933).
78 *Tax. Eccl.* (Rec. Com.), 271.
79 Dugd., *Mon.* i, 492.
80 *L. & P. Hen. VIII*, xvi, g. 1226(12).
81 *Ely Dioc. Remembrancer*, no. 322, p. 56 (document copied into Reg. Goodrich).
82 C 139/29.
83 C.P. 25(2)/93/854 and /400/East. 1 Chas. I.
84 Bp.'s Mun., Ely, G 3 ('Old Coucher'); B.M. Cott. MS. Claud. C. XI.
85 *Cal. Inq. p.m.* ix, 54 (p. 47).
86 *Cal. Pat.* 1413–16, 357; *Cal. Close*, 1413–19, 386; Blomefield, *Hist. Norf.* ii, 202.
87 It is not mentioned in the charter of Edward IV (1462) (*Cal. Chart. R.* 1427–1516, 167–71).
88 *Valor Eccl.* (Rec. Com.), iv, 288.
89 *Cal. Pat.* 1547–8, 158–9.
90 Ibid. 1553, 262. The area of the former civil parish of Redmere was 632 acres.
91 C.P. 25(2)/68/558.
92 C 142/244/116.
93 C 142/179/87; C.P. 25(2)/93/846; C.P. 43/1.
94 C 142/778/120, /146. Certain fisheries 'lately leased to Sir Edward Warner' were in 1616–17 in possession of the Earls of Kent (C 142/349/172; C.P. 25(2)/386/Trin. 15 Jas. I).
95 See note 32.
96 B.M. Add. MS. 5827, f. 90; Add. MS. 5846, f. 148*b*. The Ely hospital was drawing £9 from Littleport at its dissolution (*Cambs. Arch. Soc. Procs.* xxxvi, 102). This was probably mostly accounted for by the impropriate rectory.
97 E 40/12157.
98 C 142/336/51 and /60.
99 *Pipe R. 1170* (Pipe R. Soc. xv), 96.
1 Jas. Bentham, *Ely*, 140.
2 *V.C.H. Cambs.* ii, 308.
3 It accounted for most of the £9 ex-hospital property of Clare in 1576 (*Cambs. Arch. Soc. Procs.* xxxvi, 102).
4 C.P. 25(2)/1000/Trin. 6 and Mich. 7 Geo. I.
5 B.M. Add. MS. 5821, f. 24*b*.
6 W. E. Lunt, *Valuation of Norw.* 216, 539.
7 Ibid. 553, quoting B.M. Cott. MS. Claud. C. XI, ff. 22–23.
8 *V.C.H. Cambs.* ii, 308.
9 *Tax. Eccl.* (Rec. Com.), 265.
10 *Ely Dioc. Remembrancer*, 1906, p. 199. This reduction brought the vicarage within the category of those exempt from the payment of tenths.
11 *Valor Eccl.* (Rec. Com.), iii, 499.
12 *Ely Dioc. Remembrancer*, no. 128, p. 14.

The advowson of the parish church has always been with the bishops of Ely, as has that of the daughter churches of St. John Evangelist, Little Ouse (1869), and St. Matthew, Littleport (1878).[13]

The patronal festival (St. George's Day) falls in spring. The date proving an inconvenient one, the village feast was altered in 1768 to the Tuesday after 5 November.[14]

Littleport had four guilds—St. George and Corpus Christi (1377), St. John Baptist (1380), and St. Mary (1382). The first two were unusual in making allowances of 3*d.* a week to indigent members whilst funds lasted.[15] St. Mary's guild hall was in 1571 granted to Richard Hill and Robert Don.[16] This was probably the messuage of that name granted in 1747 by Aungier Peacock for charitable purposes.[17] Certain property of the Downham guild, specified as 50 acres of land and 6 'holtes' (18 acres) of marsh, with a tenement called Bartrams, 4 cottages, and a croft called Chamber Croft, with 2*s.* 6*d.* rent, lay in Littleport.[18]

The church of *ST. GEORGE*, Littleport, consists of chancel, north organ chamber and vestry, south chapel, clerestoried nave, two north aisles, north porch, south aisle, south porch, and west tower. The fabric was almost entirely rebuilt late in the 15th century with the exception of the west bay of the north arcade, which is of 14th-century date. In 1857 the church was drastically restored, when the north aisle was rebuilt and enlarged and a second north aisle added. The vestry and organ chamber, which are eastward extensions of these aisles, are also modern. The whole church was re-roofed at this time: the roofs are slated. The chancel has a modern east window of five lights, and there are diagonal buttresses with three set-offs. There is a window on the north and south of the chancel of three lights with cinquefoiled heads. The chancel opens to the south chapel by a four-centred arch with moulded caps and bases. There is a plain piscina in the chancel with a four-centred head, chamfered, and a stone credence shelf. The chancel opens to the organ chamber by a modern two-centred arch. The chancel arch is four-centred and of two orders; the outer order has continuous mouldings, and the inner shafts with moulded caps and bases. The fenestration of the south chapel is similar to the lateral windows in the chancel. In the south wall there is a plain doorway with a four-centred and chamfered head. The chapel opens to the south aisle by a four-centred arch of two orders dying into the wall.

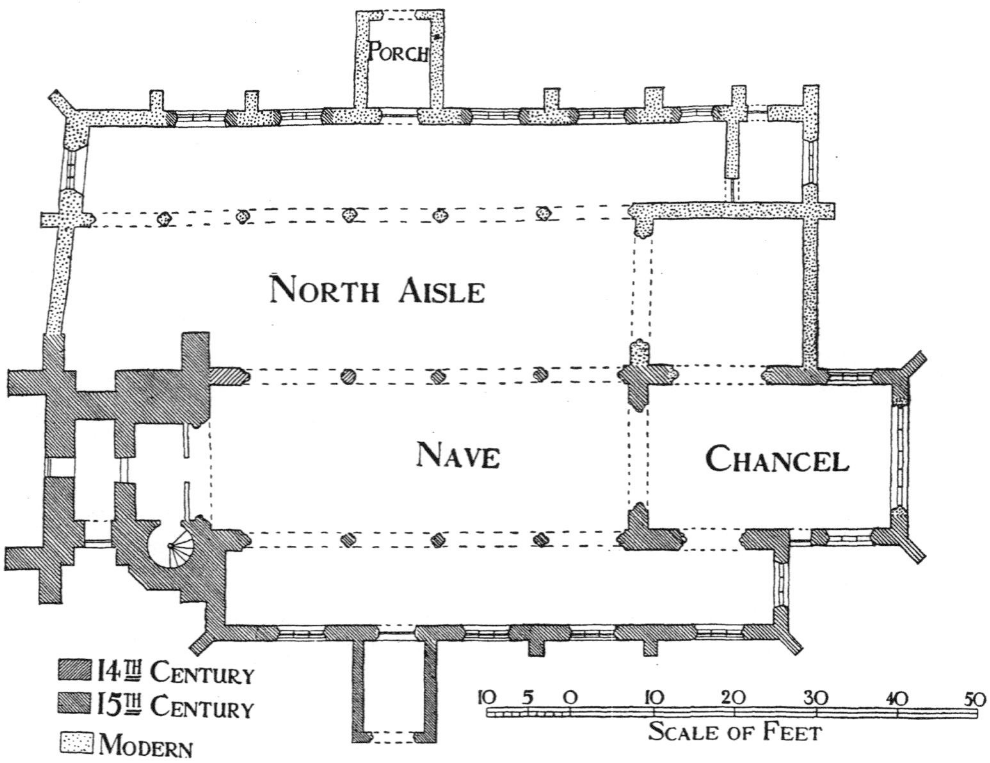

Plan of St. George's Church

The nave has arcades of four bays, the arches of which are similar to the chancel arch except the west bay on the north, which has an octagonal column and semi-octagonal respond with moulded caps and bases supporting an arch of two orders, and is of 14th-century date. The clerestory has windows of two lights with cinquefoiled heads; the tracery on the north has been removed to the inner north aisle, which is also clerestoried. The two-centred tower arch is of two orders with moulded caps and bases to the inner order of the responds. The two north aisles terminate in a line with the west face of the tower. The arcade between the two aisles is modern and similar to those of the nave. The windows of the south aisle are of three lights with cinquefoiled heads, without tracery, and date from the

[13] P.R.O. Inst. Bks.; *Crockford's Clerical Dir.*

[14] B.M. Add. MS. 5821, f. 24*b*, quoting *Camb. Chron.* 9 July 1768.

[15] W. M. Palmer, *Village Gilds of Cambs.* 344, 394.

[16] C 66/1074, m. 5. They also received lands in Littleport and Welney that had provided endowments for lights in Littleport Church.

[17] See—Charities.

[18] *Cal. Pat.* 1548–9, 348.

15th century. The windows of the old north aisle, which are similar, are now in the outer north aisle, to which they were removed when the church was enlarged. The aisle buttresses have two set-offs, those on the north being modern.

The roofs are modern, but that of the nave has eight medieval wooden angels, which have lost their wings, attached to the bases of the intermediates. The south porch is of 15th-century date with a four-centred arch of two orders, the inner with moulded caps and bases. The south doorway has continuous mouldings. The north porch is entirely modern.

The tower, a fine example of 15th-century design, is of four stages with an embattled parapet and angle buttresses with four set-offs. There is a stair turret at the south-east angle reaching to the parapet, the upper part of which is semi-octagonal. There is a west window of three cinquefoiled lights in the second stage; the third stage has three two-light windows with cinquefoiled heads on the north, south, and west sides. The windows of the belfry are of three lights with cinquefoiled heads, and the parapet is of brick. There is no west doorway. There was formerly a passage running through the base of the tower from north to south, separated from the church by a thin partition wall with a ringers' gallery above; the north entrance is now completely blocked but the south still exists, though it has had a modern doorway and door inserted. The entrance to the tower stairway is by a door in the south-east angle, which has a moulded arch and chamfered jambs. The ringers' gallery has a 17th-century rail serving as a parapet on the east.

The font is of 15th-century date with an octagonal bowl having quatrefoils in the panels. There is a massive iron-bound chest with rounded cover dated 1672. There are two massive benches with moulded poupée heads having a shield charged with a lion rampant, and another bench with poupée heads of less massive construction, all being of 15th-century date. At the west end of the inner north aisle is preserved the old north door, a fine piece of medieval carpentry, with a band of leaf carving round the frame and plain old hinges.

The plate includes a communion cup and paten cover of silver, 1570; a paten of silver, 1830; an alms-dish of silver, 1829; a flagon of glass; a silver-plated memorial chalice of 1916; and a paten of 1928. The 1830 paten and the alms-dish were the gift of Bishop Sparke of Ely (1812–36).

The tower contains eight bells by Warner, 1891. There were formerly four, three by John Draper of Thetford (1st, 1640; 2nd, 1622; 3rd, 1624).

The registers begin in 1754 (marriages), 1756 (burials), and 1783 (baptisms). The earlier registers were destroyed during the Littleport Riots in 1816.

The church of *ST. JOHN EVANGELIST*, Little Ouse, stands in a remote position on the Norfolk boundary, 2 miles from the Littleport–Mildenhall road. It was erected in 1869 at the cost of Canon E. B. Sparke, Rector of Feltwell and son of Bishop Sparke. It is a building of flint with stone dressings and slate roof, consisting of chancel with north vestry and organ chamber, nave, and north-west tower with pyramidal slated roof. The ground stage of the tower forms a porch. The weather vane is in the form of a dragon.

The tower contains three bells. The registers begin in 1867. The parish was formed the previous year from Littleport and the following Norfolk parishes—Feltwell St. Mary, Feltwell St. Nicholas, Hilgay, and Southery.

The church of *ST. MATTHEW*, on the road to Welney and Wisbech, stands about 2 miles from Littleport town. It was erected in 1878 and is a brick structure consisting of chancel, transepts, nave, south porch, and west turret containing one bell. The registers begin in 1878, in which year the parish was formed from Littleport, Ely Holy Trinity, Ely St. Mary, and Downham.

NONCONFORMITY There was a Baptist congregation at Littleport during the Interregnum. About 1655 its elders, Samuel and Ezekiel Cater, seceded to the Friends with 10 of the members, leaving 22 in the congregation.[19] In 1676 there were 67 Dissenters amongst Littleport's 556 persons of communicant age, but no Papists.[20] Licences for various Nonconformist meeting-houses were issued—to Thomas Young in 1747, to Rebeccah Godson in 1750, to James Clarke in 1754.[21] The denominations of the licensees, however, are not recorded, and it is not possible to connect them either with the Baptists mentioned above, or with the numerous Nonconformist chapels which were later established in Littleport.

A Particular Baptist church was founded in 1834 and a Calvinistic Baptist church in 1841. Neither of these was in association with the Baptist Union, and in 1851 the congregations were quite small. Both are now closed.[22]

The Methodists were established somewhat earlier. Their church in High Street dates originally from 1806, and was rebuilt in 1890 when the old building became a Sunday school. In 1851 its congregation of 380 (including 100 Sunday scholars) was equal to those of all the other Nonconformist chapels put together, and was about two-thirds of that of the parish church.[23] This was a Wesleyan church, and other chapels of this denomination were established at Dairy Houses (1840) and Black Horse Drove (1843). Both of these are in use today (1951), though that at Black Horse Drove was rebuilt towards the end of the 19th century. Another chapel was opened about 1916 on Mildenhall Road near the Little Ouse turning; this was closed during 1939–45, and the iron building has since been sold and demolished.[24] The Primitive Methodists opened a church in Victoria Street in 1845, having been established as a congregation ten years earlier. This church was restored in 1871. The Primitive Methodist chapel at Little Ouse (Morley Theobald Memorial) dates from 1910. Both of these are still in use.

The Independent church (Salem) in Globe Street dates from 1850. The Salvation Army barracks in City Road were established between 1904 and 1908.[25]

SCHOOLS In 1789 Littleport was one of the sixty-eight Cambridgeshire parishes without a school, but nine years later three small unendowed day schools had sprung up, whose unlicensed masters taught from the Bible and Prayer Book.[26] One of these schools, or a successor, adopted the monitorial system of the National Society as early as 1819.[27] Nevertheless the parish lagged behind many

[19] Adam Taylor, *Hist. Engl. Gen. Baptists* (1818), i, 141–3.
[20] Bp.'s Mun., Ely, A 6.
[21] Ibid.
[22] H.O. 129/7/190.
[23] Ibid.
[24] *Meth. Ch. Bldgs. Stat. Return as at 1 July 1940* (privately printed 1947).
[25] Kelly, *Dir. Cambs.*
[26] Bp.'s Mun., Ely, B 6.
[27] *V.C.H. Cambs.* ii, 350*a*.

in its educational equipment, and when the society made its inquiry in 1846–7 only 130 children, in a population of over 3,500, were being taught—about 50 of them on Sundays only.[28] In 1847, however, a National School, 'a handsome building',[29] was erected in Wisbech Road, near the church, and a master appointed at a salary of £36 a year, without house. This school, which cost £1,756, provided 172 places in the first instance, and was in 1871 supplemented by an infants' school for 100 at a cost of £413.[30] Attendance in 1867, by which date the school seems to have been enlarged, averaged 151 boys and 117 girls during the winter. In summer, owing to the gang system, the numbers were only 85 and 67 respectively.[31] Provision for the outlying districts was made by other National Schools at Fen and Dairy Houses, at the end of Hale Drove (1869), and Little Ouse (1870); these buildings were combined schools and mission churches. The total accommodation provided (about 400 places) was, however, still inadequate, and in 1874 a School Board was formed,[32] which provided two more outlying schools at Black Horse Drove (1874) and at Westlands in Littleport St. Matthew (1900). Between 1874 and 1885 the Board took over the four National Schools.

The Town Schools[33] were enlarged in 1885 to take 537 children. Further enlargements were made in 1909–10 and 1922 to keep pace with the population and to counteract the reduced capacity that would have resulted from the adoption of improved standards of accommodation. After the second enlargement there were places for 552 (207 boys, 212 girls, and 133 infants). In the early part of the Second World War the presence of evacuees made it necessary to hire additional temporary accommodation in the Salvation Army Hall (boys), Alexandra Hall (girls), and Primitive Methodist Hall (infants). To replace the halls upon their surrender, two additional classrooms and a staff room were provided in the shape of a hut at the infants' department in 1950.

In 1894 an infants' school for 80 was erected at the Littleport end of Ten Mile Bank. Before 1910 it was enlarged to provide 108 places, reduced to 94 in that year. In 1923 there were only 25 children in attendance and the school was closed. The buildings were sold nine years later for £300.[34]

Black Horse Drove and Little Ouse Schools are both near the Norfolk border, and take children from that county. This has led to overcrowding. At Black Horse Drove in 1932 there were 114 children on the roll and recognized accommodation for only 75. The overflow was taught in the Wesleyan Sunday school. The school had been on the black list for some time owing to its dangerous structural condition. A new school was therefore built, providing 80 places at a cost of over £3,000, and renamed at its opening (1937) the Coronation County Primary Junior Mixed and Infants' School. There were 38 children on the books in 1949.[35]

The Little Ouse School was housed in very poor premises—an iron hut used also as a mission room, which still exists. Shortly before the First World War a new site was bought and a teacher's house erected, but the school itself was not furnished with new buildings until 1927. They are an attractive example of semi-permanent timber construction.[36]

Also in Little Ouse village, on the Norfolk side of the river, is Feltwell Fen School, known until 1929 as Feltwell Anchor School. This was built by Feltwell School Board in 1889 for 60 children at a cost of £650, and was enlarged by a classroom for 27 infants in 1914, after a merger with the Little Ouse (Isle) School had been rejected. In the great floods of 1915, 1916, and 1917 the buildings were used as a refuge for homeless families.[37]

The Westlands School also started in converted mission-room premises (1900), but unlike Little Ouse was a provided school from the outset. The original accommodation was for 48. In 1921 the County Council purchased a new site slightly nearer Littleport, on which an army hut to hold 80 was erected as temporary accommodation. No permanent buildings, however, were ever erected, and in January 1932 the seniors were transferred to Littleport Town; the remaining 28 juniors followed in September of the same year.[38]

Fen and Dairy Houses School originally (1869) provided 100 places, reduced in 1910 to 83. This school was leased by the church authorities to the School Board in 1874. It stood in a very remote position in the fen, more than 6 miles north of the village, and the average attendance had dropped to 31 in 1938 and 25 in 1943. The school was therefore closed the following year, and the children transferred to the Town Schools.[39]

CHARITIES In 1747 Aungier Peacock gave a messuage called the guildhall[40] and 27½ acres of land for the maintenance of a causeway and the relief of the poor. Under the Bedford Level Act of 1677, 158 acres were allotted as Town Lands. By 1837 these Lands with Peacock's bequests were being let for £430 a year. The Commissioners of 1837 also recorded the existence of Fleet's charity, of unknown date, valued at £100, and two small charities producing £5 10*s.* a year. In 1837 all these charities were given indiscriminately to the poor in clothing, coals, and money. There were also 21 tenements inhabited rent free by the poor, one of which was said to have been given to the parish under the will of Robert Allison (1593).[41] In 1875 the Town Lands were regulated by a Scheme of the Charity Commissioners, and in 1879 the Trustees built a town hall with some of the proceeds of the charitable estates.[42]

[28] Archives of Nat. Soc.
[29] Gardner, *Dir. Cambs.* (1851), 493.
[30] Archives of Nat. Soc.
[31] *Rep. Com. Empl. Children in Agric.*, [4068–I], 341 (1867–8), xvii.
[32] *Lond. Gaz.* 1874, 1770.
[33] M. of Ed. File no. 12/34.
[34] M. of Ed. File no. 12/33.
[35] M. of Ed. File no. 12/30.
[36] M. of Ed. File no. 12/32.
[37] M. of Ed. File no. 27/139.
[38] M. of Ed. File no. 12/31.
[39] M. of Ed. File no. 12/35.
[40] See p. 100.
[41] *Rep. Com. Char.* (*Cambs.*), [103], 232, H.C. (1837–8), xxiv.
[42] Kelly, *Dir. Cambs.* (1900).

NORTH WITCHFORD HUNDRED

CHATTERIS

Chatteris is a small town on the western border of the Isle, about 9 miles south of March and 12 miles north-west of Ely, and on the main road between them. The Ely road is known as Ireton's Way, having been constructed by the Parliamentary general in the Civil War to convey troops from Chatteris to Ely.[1] It was a private road, with toll gates, until taken over by the County Council in 1902.[2] Other roads connect the town with St. Ives, Huntingdon (via Warboys) and Wisbech (via Upwell). The Wisbech road runs east from Chatteris for about a mile. It then runs north-east along the Sixteen Foot Drain, which is the only important watercourse in Chatteris except the Forty Foot or Vermuyden's Drain crossing the north of the parish. There is a station on the March and St. Ives branch, Eastern Region British Railways, opened in 1848.[3] A Local Board, now replaced by an Urban District Council of nine members, was formed in 1873. The market, apparently held by prescription, had been discontinued before 1808 owing to the bad state of the roads and the frequent floods in the district. In 1834 it was revived, on the petition of nearly 100 inhabitants, and was held on Fridays.[4] By 1851 it was well attended and the centre of a fair trade in corn.[5] It became obsolete early in the present century, but fairs are still held on the last Friday in April and the Friday before Michaelmas.[6] From 1778 until 1846 Chatteris alternated with March as one of the meeting-places of the Court of Requests for the Isle of Ely for the recovery of small debts in the Isle.[7]

In 1563[8] there were 206 householders in Chatteris, a figure comparable with the undivided parish of Doddington (201) and Haddenham (188), and much below that for Whittlesey (355) and Ely with its hamlets (400). Chatteris must then have contained about 1,000 inhabitants. The assessment for ship money in 1639–40 shows little change, Chatteris being rated at £50 2*s.* as compared with £75 15*s.* for Haddenham and £69 9*s.* 11*d.* for Doddington with its hamlets. Chatteris, however, ranks with the villages rather than the towns; it is much below Wisbech, Whittlesey, and Ely, and about in the middle of a group, assessed at between £40 and £60, which contains also Elm, Stretham with Thetford, Littleport, Downham, and Sutton.[9] Bishop Compton's 'census' of 1676[10] is almost certainly inaccurate as regards Chatteris, where it shows only 271 persons of communicant age (including 43 Dissenters), perhaps representing a total of about 400; the comparable figures for Doddington and Haddenham are 1,762 (949 in March) and 700 respectively. At the opening of the 19th century Chatteris, with 2,393 population, had far outpaced Haddenham (1,090), but was by a slight margin the smallest town in the Isle, March having 2,514 inhabitants.[11] The subsequent trend has been the usual one in Fenland towns—a rapid increase up to 1851, as the surrounding countryside was converted from pasture to tillage, followed by a decrease of some 10 per cent. in the second half of the 19th century and subsequent return to about the 1851 level. At this date (1851) the town was said to be 'considerable', to be 'lighted with gas', and to contain 'some handsome houses and good shops'.[12] There were at that time 15 inns and 43 beerhouses to serve a population of about 5,000.[13] Chatteris suffered badly in the 1832 epidemic of cholera; the General Baptist congregation lost 3 out of about 50 members.[14] A serious fire which broke out on Mr. Billips's farm on 14 September 1864 caused the death of a 7-year-old girl and the destruction of upwards of 100 houses. The total damage was over £10,000, and a relief fund of £300 was subscribed for the sufferers. The fire was said to be the work of an incendiary.[15] An earlier fire (1706) caused £1,787 worth of damage.[16]

A halfpenny tradesman's token of 1670 is known.[17] The first record of a post office in Chatteris is in 1782.[18] In 1851 Mrs. Mahala Rowley, confectioner, of High Street, was postmistress.[19] The post office was opened in its present premises at No. 16 Park Street in 1924;[20] the building is of the late 18th century, with a pedimented doorway with fanlight and bow window above, and a Mansard roof. Telephone service was first provided by the National Telephone Company in 1910. The present exchange dates from 1924.[21]

There are some small engineering works in Chatteris, but agriculture is the chief industry and the town is of importance mainly as a local shopping centre. It has not the industrial interests of other Fenland towns like Wisbech, Whittlesey, and March, and though the population (5,153 in 1931, 5,528 in 1951) is large compared with some neighbouring centres (e.g. Huntingdon and St. Ives) it is spread over some 22 square miles at a density not much greater than that of some of the rural parishes in the north of the county. In 1921 just over half (50·6 per cent.) of the male workers were employed in agriculture, a proportion above that for the Isle as a whole. At that date, however, all the towns, including Wisbech,[22] returned more than 25 per

1 *P. N. Cambs.* (E.P.N.S.), 27.
2 Kelly, *Dir. Cambs.* (1933).
3 *V.C.H. Cambs.* ii, 133.
4 R. Gardner, *Dir. Cambs.* (1851), 517.
5 Ibid.
6 Kelly, *Dir. Cambs.* (1933).
7 18 Geo. III c. 36; 9 & 10 Vic. c. 95.
8 B.M. Harl. MS. 594, ff. 198–200.
9 *E. Anglian*, N.S., vi, 40, quoting S.P. 16/445, No. 75.
10 Bp.'s Mun., Ely, A 6.
11 *V.C.H. Cambs.* ii, 139.
12 Gardner, *Dir. Cambs.* (1851), 518. The gas works were established in 1838.
13 Ibid.
14 W. H. Clarke, *Hist. Sketch W. Park St. Bapt. Ch. 1785–1935*, p. 8.
15 *The Times*, 16, 19, 27 Sept. 1864. Between April and August of this year the rainfall fell below the average by 39 per cent., and serious fires also occurred at Billinghay (Lincs.) and Thaxted (Essex). Chatteris is one of the few places in England which have experienced fewer than 100 rain days in a year (93 in 1921) (A. B. Tinn, *This Weather of Ours* (1946)).
16 E. Conybeare, *Hist. Cambs.* (1897), 263.
17 S. H. Miller and S. B. J. Skertchly, *The Fenland Past and Present*, 478–80.
18 P.O. H.Q. Rec. Min. 2426/1924.
19 Gardner, *Dir. Cambs.* (1851), 519.
20 P.O. H.Q. Rec. Min. 2426/1924.
21 Ibid. E 17104/1908.
22 It should be remembered that at this date Wisbech Borough included a considerable portion of fenland west and south of the town, and that New Walsoken was then returned with Norfolk.

cent. of their male workers as engaged in agriculture. In 1931 the Chatteris proportion had risen to 57·7 per cent., a figure unsurpassed among English urban areas except by Holbeach (Lincs.) and Ramsey (Hunts.).[23]

Inclosure was effected by Acts of Parliament in 1793 and 1809.[24] The employment of women and children in agricultural gangs was said to have come into existence in Chatteris soon after the latter Act.[25] The award under this Act was not made until 1819. The delay was no doubt due to the large amount of land inclosed (5,669 acres), and to the fact that with six separate manors in the parish, the 139 holders of common rights could hold them in seven different capacities, each of which might require a separate set of allotments.[26] William Dunn Gardiner was much the greatest beneficiary. He received 582 acres, 469 in his own right or that of his wife, 104 as lord of the Ramsey manor and 9 as lord of Bartles. Anne Waddington and Mary Chafy, who received 259 acres jointly, were also important proprietors. Charles Cholmondeley (the impropriator of the great tithes) and the vicar each received about 300 acres.

In 1851 there were two recently established libraries: the Literary Institution with 70 members and 1,100 volumes and the Mechanics' and Working Men's Book Society. Their subscriptions were 12*s.* 6*d.*, and 1*s.* or 1*s.* 3*d.* respectively.[27]

Some pieces of domestic architecture are worth noting. Chatteris House (No. 17 High Street) is a fine early-19th-century building, with portico porch sheltering a good doorway and door, an iron balcony above, and a central pediment. Inside, the staircase is of stone with an iron balustrade, and there are some moulded plaster ceilings. No. 84 High Street has a pedimented doorway with a fanlight, and a bow window above. The Red Lion Inn (No. 2 London Road) is plastered and thatched; it has a carved wooden eaves cornice and dates probably from the late 17th century. No. 1 Market Hill, 18th century, has a hooded doorway with fanlight. Westwood (No. 9 Park Street) is a late-18th-century building of some merit. No. 8 St. Martin's Road, also 18th century but with a chimney brick dated 1605, has stepped and Dutch gables, a roof partly stone-slated, and another good doorway. Ivy House (No. 19 Station Street) has similar features, except for the gables. The manor house also has a pedimented doorway; its stables and outbuildings are very extensive and have a picturesque cupola, but are very dilapidated; the date is late 18th century. There are several other houses of less distinction but pleasing design, especially in High Street. In Angood's, Skeels' and Cole's Yards at the north end of High Street there are several one-storied labourers' cottages of a type once extremely common in the Fenland but now passing away.

The nunnery[28] and the 17th-century mansion built on its site by the Gascoyne family have vanished except for a few walls. Some of the last remains were pulled down to make way for the Empress Cinema in 1935.[29]

MANORS Part of Chatteris was given to Ramsey Abbey by Athelstan Mannessone, and confirmed by Edgar (974), Edward the Confessor and other kings.[30] In 1086 this portion was assessed at 3 hides less 1 virgate, with land for 4 ploughs, 1½ hide (1 plough-land) being in demesne. There was meadow for the 4 plough teams and wood for 100 pigs, also 3,000 eels from the fisheries and 27*d.* fish tribute. The recorded population was 10 villeins, 5 bordars, and 2 serfs. It had been worth 20*s.* when received, £4 T.R.E. and £6 in 1086. The Abbot of Ely, who also had an estate in Chatteris (see below), had the soke.[31]

This manor remained with Ramsey Abbey to the Dissolution. In 1254 it was valued at £20 and in 1291 at £68 14*s.* 2*d.*[32] In 1396–7 it had fallen to £34 18*s.* 4*d.*[33] In 1535 the value had risen to £46 5*s.* 2*d.* and was £48 4*s.* 5*d.* five years later.[34]

During the abbacy of Robert of Reading (1202–7) Chatteris produced £12 a year in rents;[35] there is a full extent of the manor, undated but *c.* 1240.[36] The cultivable land amounted to 374 acres, of which 48 were in demesne. There were about 23 acres of high meadow in several scattered portions, and an unspecified amount of meadow in the fen. Some of the localities and bounds of these meadows survive in present-day place-names, as Wenny, Wyliethe (Willey Fen), Hollode (Holwoods Farm), and Hunneye (Honey Hill).[37] The fishery of 'Wlweyewere', half the fisheries of 'Redestecheswere' and 'Hachenwere', produced 30*s.* a year. There was a small wood of 2 acres, and an alder grove at Honey Hill. The marsh was held in common. No free tenants are recorded, but there was a customary tenant with as much as 50 acres, another with 16 and 32 tenants with 8 acres each. There were also 31 croftmen without land who in general performed half the services of the customaries. This manor seems to have approached a money economy more nearly than many monastic estates at this period. Customary tenants, who were actually performing their services, rendered 6*s.* 3*d.*, but if they were not working in person as much as £7 3*s.* 5¼*d.* By 1396–7 the commutation process seems to have been complete.[38] In 1409, 2,306 works were commuted at 1*d.* and 64 at 3*d.*—a total of £10 18*s.* 2*d.* The total revenue of the manor (£58 7*s.* 2*d.*) also included £21 9*s.* 10*d.* from rents, £13 13*s.* 4*d.* from Hollode and £5 1*s.* from Crowlode, and £7 1*s.* 2*d.* from perquisites of court. There were also heavy arrears amounting to over £90.[39] In 1471, 36 tenants paid £5 1*s.* 11*d.* for 98 holdings.[40]

The general history of this manor whilst in monastic hands is uneventful. Up to the time of Bishop Geoffrey (1177–89) the bishop or the prior of Ely had the right to the services of Ramsey tenants in Chatteris for work on Aldreth causeway;[41] by 1240 this had been commuted for 2*s.* yearly.[42] The suit of court owed at Ely

[23] Since 1931 Holbeach has lost its urban status.

[24] 33 Geo. III c. xii and 49 Geo. III c. cvii.

[25] *Children's Employment Com. 6th Rep.* [3796], Evidence, p. 52–5 H.C. (1867), xvi. If this statement is correct, the gang system was in force at Chatteris some time before its adoption at Castle Acre (Norf.), which is generally considered its birthplace (J. H. Clapham, *Econ. Hist. Mod. Britain*, i, 469). Chatteris was one of its most prominent centres in Cambridgeshire.

[26] Incl. Award at County Hall, March (in a copy in the county 'enrolment book', occupying 272 pages).

[27] Gardner, *Dir. Cambs.* (1851), 518.

[28] For the history of the nunnery see *V.C.H. Cambs.* ii, 220–3.

[29] Eliz. M. Harland, *Norf. and Isle of Ely* (Penguin Guides), 228.

[30] *Cartul. Monast. Rames.* (Rolls Ser.), ii, 56, 72, 93.

[31] *V.C.H. Cambs.* i, 371.

[32] W. E. Lunt, *Valuation of Norw.* 117; *Tax. Eccl.* (Rec. Com.), 269.

[33] S.C. 6/765/21.

[34] *Valor Eccl.* (Rec. Com.), iv, 271; Dugd., *Mon.* ii, 591.

[35] *Cartul. Rames.* (Rolls Ser.), iii, 328.

[36] Ibid. i, 429–37.

[37] *P. N. Cambs.* (E.P.N.S.), 248–50.

[38] S.C. 6/765/21.

[39] S.C. 6/765/23.

[40] S.C. 11/Roll 91. This roll is defaced in parts and apparently incomplete. The areas of the various tenants' holdings are not always given.

[41] *Cartul. Rames.* (Rolls Ser.), ii, 187–9.

[42] Ibid. i, 437.

and at Witchford hundred was in 1294 fixed as three-weekly, to be performed by two tenants; in 1339 it was reduced to an annual suit, with homage of 2*s.* payable at the hundred court at Witchford.[43] In 1391, to settle long-standing disputes, the Abbot of Ramsey 'granted' his manor of Chatteris with all its appurtenances to the Bishop of Ely. Presumably this means that he recognized the bishop as his overlord in this, the only Ramsey manor wholly within the Isle.[44] The grant was inspected and confirmed six years later.[45] In 1229 an assize of novel disseisin was brought against the abbot and others by Katherine de Chatteris regarding a tenement in the vill.[46]

The manor remained with the Crown for some time after the Dissolution. When William Whiteakers was appointed its bailiff and collector in 1546–7 it brought in about £50 a year gross.[47] In 1558 it was sold, for £300, to Thomas Wendy the queen's physician, to be held in chief at $\frac{1}{40}$ fee. The appurtenances included the marsh or pasture of 'Holliwood farme' held by the heirs of Sir Richard Cromwell, and 10*s.* rent from 'Honye pasture' which had been granted to William and John Sewster.[48] Thomas Wendy died in 1560, his heir being his brother John, aged 50; in his will he desired his nephew Thomas to marry one of his own daughters.[49] John survived until 1589. The manor itself was then valued at £18 12*s.* 6*d.* Non-manorial property included Holwood and rent from Honey Hill, bringing the Wendy estates in Chatteris up to £34.[50] The last of the male line in this family was Thomas Wendy (d. 1673), who with his wife Lettice (Willoughby) and other members of the family was dealing with the manor in 1655.[51] By this time Honey Hill was reckoned independently, the manor of *HONNEY FEN* or *HONNEY PASTURES* being recorded in the Sewster family, its original grantees, in the early 17th century.[52] The property of this family in Huntingdonshire passed by marriage to the Peytons of Doddington in the later 17th century,[53] and it is probable that this small 'manor', which was in the north-east of Chatteris parish towards Doddington, followed the descent of and became merged in the latter. In Cole's time the lessee (at £300 a year) of Honey Fen was Porter Thompson of Trumpington.[54]

The main Ramsey manor is next found in the Caryl family.[55] In 1682 Malbone Caryl settled it on his mother-in-law Sarah Hart (d. 1707), with remainder to other members of the family.[56] In the early 18th century the manor became divided between two coheiresses, Malbon Worsley being vouchee as to 1 moiety in 1730.[57] This half was still in the Worsley family in 1808.[58] The other half was being dealt with by John Lees and Deborah his wife in 1759.[59] Later it was further divided, a third part of a moiety being passed by John and Mary Grayson to Thomas Moore in 1802.[60] Six years later the joint lords were stated to be Thomas Caryl Worsley and William Dunn Gardiner, the latter having purchased his interest from the Terry family, descendants by marriage of the Caryls.[61] Subsequently the manor became reunited in the Gardiner family, John Dunn Gardiner being lord in 1851 and 1900.[62] Major-General C. E. Lawrie, C.B., D.S.O., was lord in 1933.[63]

In 1086 the church of Ely held 2 hides and half a virgate in Chatteris, with land for 3 ploughs. Half a virgate was in demesne. There were 6 villeins, 2 bordars, and 2 cottars, meadow for 3 plough teams, woodland for 20 pigs, and fisheries providing 1,500 eels. The estate had been worth 30*s.* when received, 50*s.* T.R.E. and 40*s.* in 1086.[64] Neither the bishop nor the convent of Ely are afterwards recorded as immediate owners of a manor in Chatteris, and this is probably the origin of the manor of *CHATTERIS NUNNS*, held by the Benedictine nunnery in the town which was given to Ely by Henry I.[65] The 15th-century cartulary of this nunnery[66] shows a very large number of small grants, usually amounting to not more than one or two messuages or up to an acre of land. The lands granted lay mostly in Chatteris, where the nuns' estates were worth £16 4*s.* 5*d.* in 1291, about a quarter of the value of those of Ramsey.[67] Licences to acquire in mortmain were granted at intervals in the 14th century,[68] but most of the property so obtained was of very little value; for example, a grant in 1387 by John Dreng of 15 messuages, a toft, 21a. 1½r. of land and 2*s.* 2*d.* rent in Chatteris, Doddington, and Wimblington was in all worth only 10*s.* yearly. At the Dissolution, however, the value of the nuns' manor had risen to £28 11*s.* 8½*d.* Of this sum £20 11*s.* 4½*d.* was derived from the demesne, which included a windmill, and 4*s.* 10*d.* from 'autumn works'.[69]

The manor and rectory were granted in 1551 to Edward Lord Clinton and Saye at yearly rents of 29*s.* and 14*s.* 5*d.* respectively.[70] The site and demesnes of the nunnery, and the rectory, were then held by Walter Cromwell, and the windmill by John Townesend. All these were almost immediately regranted by Lord Clinton to Thomas Rowe, merchant tailor of London.[71] Rowe in his turn soon had licence for another alienation, which was made in 1554 to William Bettys.[72] His family, who gave the manor an alternative title of *CHATTERIS BETTS*, held it for about a century.[73] During the Commonwealth Scipio Cokayne, John Tripp, and Edward and John Hobart, with their wives, probably sisters of Thomas Betts the last in the male line,[74] were dealing with the manor.[75] George Gas-

43 Ibid. i, 216–17; iii, 59.
44 The Ramsey portion of Upwell was mainly in Norfolk.
45 *Cal. Pat.* 1391–6, 73; 1396–9, 142.
46 Ibid. 1225–32, 307.
47 *L. & P. Hen. VIII*, xxi (2), 435. See also S.C. 6/765/21.
48 Ibid. xix (1), g. 1035 (106); *Cal. Pat.* 1557–8, 450–1.
49 C 142/125/6. Thomas actually married Elizabeth, daughter of Richard Atkins of Gloucestershire (*Visit. Cambs.* (Harl. Soc. xli), 40).
50 C 142/221/90.
51 C.P. 25(2)/539 (Hil. 1655).
52 C 142/291/140, 293/61.
53 *V.C.H. Hunts.* ii, 200.
54 B.M. Add. MS. 5809, 107.
55 The M.I. to John Caryl or Carill, dating from 1670, existed in the ch. at the time of Cole's visit (1745).
56 C.P. 25(2)/634 (Trin. 34 Chas. II).
57 C.P. 43/590, ro. 376.
58 Lysons, *Magna Brit.* ii (2), 161.
59 C.P. 25(2)/1104 (East. 32 Geo. II); C.P. 43/704, ro. 183.
60 C.P. 25(2)/1283 (Mich. 42 Geo. III).
61 Lysons, *Magna Brit.* ii (1), 161. The manor house, which stood on a moated site known as Caryl's Grove, had recently been pulled down.
62 Gardner, *Dir. Cambs.* (1851), 516; Kelly, *Dir. Cambs.* (1900).
63 Kelly, *Dir. Cambs.* (1933).
64 *V.C.H. Cambs.* i, 368.
65 Ibid. ii, 220.
66 B.M. Cott. MS. Jul. A. I, esp. ff. 76–95 for Chatteris.
67 *Tax. Eccl.* (Rec. Com.), 268. The Norwich valuation of 1254 had been £5 only (W. E. Lunt, *Valuation of Norw.*, 117).
68 e.g. *Cal. Pat.* 1330–4, 242; 1385–8, 298.
69 Dugd., *Mon.* ii, 620.
70 *Cal. Pat.* 1550–3, 191–2.
71 Ibid. 59.
72 Ibid. 432; 1553–4, 338.
73 C 142 111/14, 184/13; C 3/233/4; C 5/385/12; C.P. 25 (2)/400 (East. 4 Chas. I).
74 *Visit. Cambs.* (Harl. Soc. xli), 79.
75 C.P. 25(2)/401 and 538 (Mich. 24 Chas. I and East. 1653); C.P. 43/288, ro. 83.

coyne, one of the parties to the above transactions, converted what remained of the nunnery buildings into a country house, seen by Dugdale on his visit in 1657.[76] From Gascoyne the manor descended to Michael Holman, who with his wife Eleanor (Gascoyne) executed a settlement in 1661.[77] After this the manor changed hands frequently.[78] In the early 18th century it was in the Hake family.[79] From about 1757 to 1772 it belonged to Robert Fawcett of London. Fawcett bequeathed it to John Seymour,[80] from whom it descended to the Revd. T. C. W. Seymour, his grandson, who received 51 acres as lord of Nunns manor under the 1819 inclosure award. Some time before 1851 it was sold to John Ord Hall.[81] Thomas Jennings of Newmarket was lord of this manor in 1900; it has since remained in his family.[82]

In 1386 certain *iniquitatis filii* withheld 2*s.* 1*d.* rent, due to the church of Ely, from three tenements. Payment was ordered within 15 days on pain of excommunication.[83]

As late as 1809–10 the Bishop of Ely claimed to be lord paramount of the manors of Chatteris Ramsey, Nunns, and Benet College, all of which did suit and service at his palace and hundred courts.[84] Until the middle of the 19th century the bishop held courts leet on St. Matthew's Day at a house called the Guildhall.[85]

The manor of *CHATTERIS LISLES* is first mentioned in 1498, when it was worth £3 6*s.* 8*d.* and was held by John, Lord Scrope of Bolton, of the Bishop of Ely by fealty.[86] John's great-grandson, Ralph Scrope of the Vine (Hants.), sold it in 1564 to Symeon Stywarde and his son Robert. At Symeon's death in 1568 it was held of the queen in free socage.[87] The descent for the next three centuries is identical with that of Coveney (q.v.). Manorial rights had lapsed by 1900, possibly on the extinction of the Rokeby barony in 1883.

A manor of *TYNDALLS* is mentioned in 1498, when Sir William Tyndale died seised thereof. It was then worth 30*s.* and was held of the Bishop of Ely by unknown service.[88] By 1571 it had come to the Styward family[89] and subsequently descended with the manors of Coveney and of Chatteris Lisles (q.v.).

The manor of *BARTLES* or *GARDINERS* was devised by Thomas Gardiner (d. 1566) to his son Humphrey in tail male, with contingent remainders to his brothers and sisters and their heirs male.[90] Humphrey left a life interest to his wife Elizabeth and appointed her brother Whitehill Audley as a trustee in the event of her death, his son William having predeceased him and his grandson Humphrey being a minor. Humphrey II had livery of this manor in 1639.[91] The subsequent descent is obscure; William Needham and his wife Katherine were vouchees in 1705.[92] A century later this manor was inherited by John Gardiner (d. 1804) through his marriage with the heiress of a Marriott of Chatteris.[93] Its descent during the 19th century was identical with that of the Ramsey manor (q.v.). Mr. Alexander H. M. Wedderburn was lord in 1933.[94]

Another manor of *ANDREWS* was held in the early 14th century by the St. George family. In 1369 it was settled by Baldwin St. George on trustees, from one of whom, John Andrew of Eltisley, the manor acquired its name. It seems that the estate was intended by the original donor to be applied to charitable purposes. Further trustees were appointed in 1377. In 1395 the manor was made over to John Necton, master, and Thomas Kelsale and John Titleshale, fellows of Benet (now Corpus Christi) College,[95] whose successors have since held it.[96]

CHURCH The church of Chatteris was granted to the nunnery by Bishop Niel (1133–69) and confirmed by Prior Roger (*c.* 1210–29).[97] Its value was £4 in 1217, £13 6*s.* 8*d.* in 1254, and £16 13*s.* 4*d.* in 1291.[98] The church was styled *ecclesia* in 1291, so that a vicarage would not seem to have been ordained at that time. In 1535 the vicarage was valued at £10. The appropriated rectory was not entered separately, but included among the other spiritualities of the nunnery.[99]

The advowson descended with the manor of Chatteris Nunns or Betts to the end of the 18th century. Some time before 1803 John Seymour, then lord of this manor, sold it to Robert Chatfield of Croydon (Surrey), who presented in that year.[1] In 1851 Revd. M. A. Gathercole was patron and incumbent.[2] In 1909 the advowson was obtained by Caius College, in exchange with Revd. C. H. Brocklebank for that of Ashdon (Essex), and it has since remained with the college.[3]

The rectory also descended with Nunns manor until 1615, when Bestney Betts and his brother John sold the greater part, including Wylley Close, to Simon Bennett of Swineshead (Beds.) for £1,900,[4] and in 1624 the remainder, consisting solely of the great tithes, a barn and an acre of land, was conveyed by Bestney with his wife and other members of the family to Sir Thomas Hetley, serjeant-at-law, and his son Francis.[5] Shortly after this (*c.* 1632) it was stated that the Nunns manor and impropriate rectory had been held together for so long that the glebe was not distinguished from the demesne. The Hetleys had difficulty in collecting their tithes, obstacles being placed in their way by Betts, George Carter the vicar, and the Earl of Suffolk who held 400 acres, valued at £200 a year, in Honey Fen.[6] Francis died in 1638 and was succeeded by his brother William, then a minor,[7] who was dealing with the rectory in 1651.[8] Carrina Hetley, widow, was in possession in 1676–7,[9] and it continued with the Hetley

76 B.M. Lansd. MS. 722, ff. 29–38, quoted in H. C. Darby, *Draining of the Fens*, 277.
77 C.P. 25(2)/632 (Trin. 13 Chas. II).
78 The last Holman presentation to the living was in 1701; the Hake family presented in 1737 and 1749 (P.R.O. Inst. Bks.).
79 Mr. Hake, the owner of the priory site, who died in 1746, was 'looked to be somewhat disordered in his intellects' (B.M. Add. MS. 5809, 107).
80 W. Watson, *Hist. Wisbech*, 573.
81 Gardner, *Dir. Cambs.* 516.
82 Kelly, *Dir. Cambs.* (1900 and 1933).
83 *Ely Dioc. Remembrancer*, no. 131, p. 59.
84 A. Gibbons, *Ely Episc. Rec.*, 78.
85 Gardner, *Dir. Cambs.* (1851), 516.
86 *Cal. Inq. p.m. Hen. VII*, ii, no. 219 (p. 140).
87 C 142/150/188.
88 *Cal. Inq. p.m. Hen. VII*, ii, no. 11 (pp. 17–18).
89 C 142/156/11.
90 C 142/383/84.
91 C 60/531.
92 C.P. 43/487, ro. 51.
93 Lysons, *Magna Brit.* ii (i), 162.
94 Kelly, *Dir. Cambs.* (1933).
95 B.M. Add. MS. 5809, 105*b*, 106*b*.
96 Kelly, *Dir. Cambs.* (1900).
97 B.M. Cott. MS. Jul. A. 1, 77*b*, 78*a*.
98 W. E. Lunt, *Valuation of Norw.*, 218, 539.
99 *Valor Eccl.* (Rec. Com.), iii, 500.
1 Watson, *Hist. Wisbech*, 573. P.R.O. Inst. Bks. Various transactions in the advowson were made by both Seymour and Chatfield. C.P. 25(2)/1282 (East. 32 Geo. III); C.P. 25(2)/1283 (Trin. 40 Geo. III); C.P. 25(2)/1490 (Trin. 1 and Mich. 3 Geo. IV).
2 Gardner, *Dir. Cambs.* (1851), 517–18.
3 E. H. Gross, *Biog. Hist. Gonv. and Caius Coll.* iv (2), 33.
4 *E. Anglian*, xiii, 268.
5 C.P. 25(2)/278 (Mich. 22 Jas. I). See also C.P. 25(2)/400 (Mich. 5 Chas. I); and C 142/483/84; C 2/Jas. I/H. 20/57.
6 C 78/1199, pt. 6.
7 C 142/487/126.
8 C.P. 43/274, ro. 80.
9 C.P. 43/376, ro. 130; C 5/512/81.

family to the end of the century.[10] Charles Cholmondeley was returned as impropriator in 1819.[11]

In 1562 the Bishop of Ely was awarded a pension of 20*s.* from the vicarage.[12] The rectorial tithes were commuted in 1818 for £600, and there were 297 acres of glebe. In 1851 the glebe was the property of J. D. Gardiner, lord of two of the Chatteris manors. The vicarial tithes were commuted for £900, and 344½ acres of glebe, including 24½ in lieu of common rights.[13]

There were in 1389 guilds in Chatteris of the Holy Trinity, Blessed Virgin and All Saints (1334), and of Corpus Christi.[14] In 1549 a cottage called the Guildhall, let at 2*s.* 6*d.* yearly, with a yard, and 2 acres and 1 rood of land in tenure of the churchwardens, was granted to William Warde.[15] The Guildhall was regranted in 1571 to Richard Hill and Robert Don.[16]

The church of *ST. PETER AND ST. PAUL* consists of chancel, north and south chapels, north vestry, clerestoried nave, aisles, two south porches, and west tower. The material is rubble with stone dressings. The roofs of the chancel, north chapel, and nave are tiled and the other roofs leaded. The church had originally a chancel, nave of five bays, aisles, and west tower. In 1910 it was in need of repair and a drastic reconstruction then began which involved the rebuilding of the whole fabric with the exception of the nave arcades, western south porch and tower. The nave was extended one bay to the east and the old chancel arch re-erected; the aisles were widened and this involved the destruction of part of the porch. Fortunately many of the old windows were reset in the new walls.[17] The sum of £5,000 for the restoration was bequeathed by Robert Wright, son of a former sexton. He emigrated to America, found work in the Presidency gardens at Washington, and later became proprietor of a laundry. He died at Walthamstow in 1903.[18] The old work is all of mid-14th-century date with the exception of the porch, which is of the 15th century, and the modern additions are in keeping.

The chancel has an east window of five lights with flowing tracery, and is entirely modern with the exception of part of the jambs. There are diagonal buttresses with one set-off, the northern of which is composed of old material. There are north and south arcades of two bays with two-centred arches, all modern and of 14th-century character. The lofty chancel arch is two-centred and of two orders with hood-mould and rounded responds with embattled caps and moulded bases. On the north is the rood stair with upper and lower doorways, the former square-headed and quite plain and the lower with shouldered lintel, both much restored. The south chapel has an east window of three lights, entirely modern. There is a diagonal buttress with one set-off and another similar one on the south, both modern. There are two 14th-century two-light windows in the south wall, cinquefoil headed with quatrefoils above and hood-mould. The north chapel has two reset 14th-century windows of two trefoiled main lights with quatrefoil above, and hood, in the east wall, and two similar windows in the north wall above the vestry with trefoils in the head, the eastern of which has renewed tracery. The vestry is entirely modern with the exception of the outer doorway, which is of the 14th century, reset, with continuous mouldings and hood.

The nave has arcades of six bays with two-centred

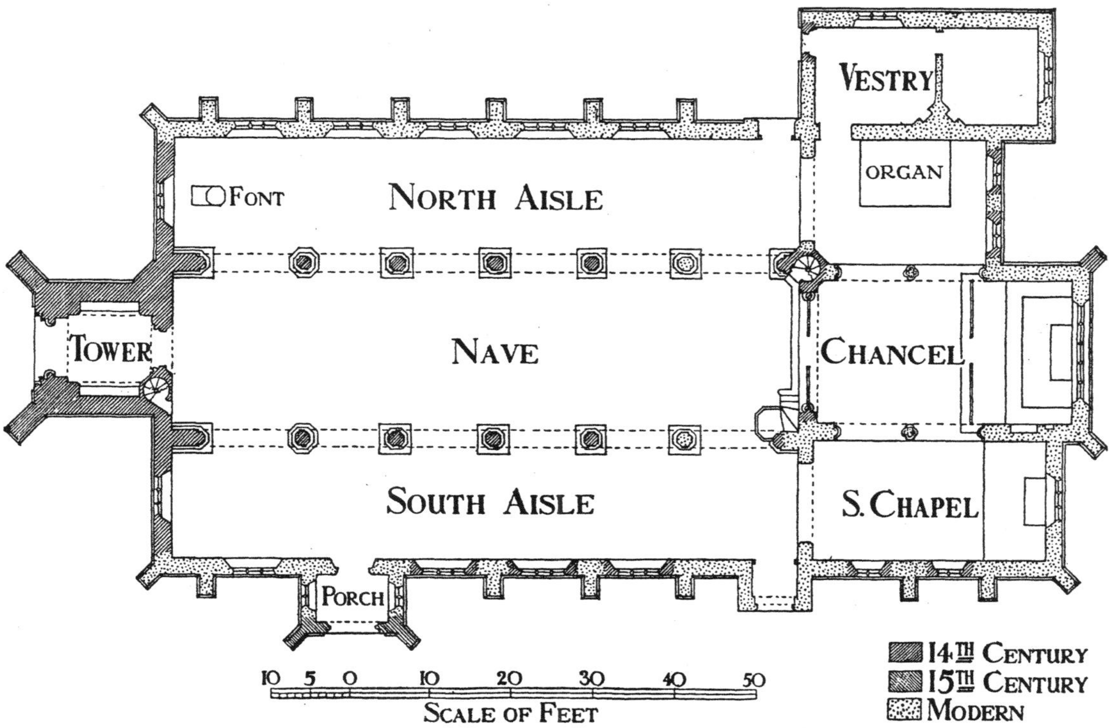

Plan of the Church of St. Peter and St. Paul

[10] C.P. 43/448, ro. 38.
[11] Incl. Award, County Hall, March.
[12] *Cal. Pat.* 1560–3, 226.
[13] Gardner, *Dir. Cambs.* (1851), 517.
[14] W. M. Palmer, *Village Gilds of Cambs.* 386.
[15] *Cal. Pat.* 1548–9, 326.
[16] C 66/1074, m. 5.
[17] A complete set of photographs of the old church taken in 1910 is kept in the vestry.
[18] Infm. Revd. J. C. Hawthorn, Vicar of Chatteris.

arches of two orders, slightly stilted, with hood-moulds terminating in heads; the columns are octagonal with moulded caps and bases. The easternmost arch and column are modern, but the eastern responds, which are semi-octagonal, have been re-used. The western responds are similar. There are six clerestory windows on each side, of two cinquefoiled heads with external hood-moulds. The internal head consists of a shouldered lintel. All the tracery has been renewed and the easternmost pair of windows is entirely modern.

The modern south aisle has four lateral windows, all of similar design, of which the three easternmost are of the 14th century; they are of three cinquefoiled lights with hood-mould to the arch. The west window is of similar design; part of the west wall is original. The eastern porch is entirely modern, but the western is of the 15th century, though curtailed by the widening of the aisle. The outer arch has jamb shafts with moulded caps and bases. There are diagonal buttresses with one set-off, and a modern two-light window with square head in the east and west walls. The parapet with moulded coping and gabled end is modern. The north aisle is entirely modern except for part of the west wall and the west window, which is similar to those in the south aisle. All the buttresses of both aisles have one set-off and are of 14th-century character. The aisles open to the chapels by plain modern two-centred arches of two orders.

The tower is of three stages with diagonal buttresses with four set-offs. The newel stair is contained in the south-east buttress, which is enlarged for the purpose. In the interior of the base of the tower there is a large plain arch in each wall. Each of these arches is blind except that on the west which is fitted with doors and serves as the entrance; on the exterior the entrance has continuous mouldings and a hood terminating in heads. The tower communicates with the nave by a doorway set in the east arch with deep mouldings and jamb shafts with moulded caps and bases and a hood terminating in a male and female head. Above the doorway there is a niche. There is a large stoup recess to the south of the inner doorway. The second stage has a lancet on the west face. The belfry windows are of two trefoiled lights with a hood-mould terminating in heads. The tower is finished with stepped battlements much renewed. There is a small leaded spire.

All the roofs are modern.

The font has a plain octagonal bowl and shaft and is of 14th-century date. The other fittings are modern. In the vestry is a panel painting of the royal arms as borne between 1814 and 1837.

There are some good 18th-century headstones in the churchyard, of local type.

The plate includes a paten marked 'Ex Dono Sarah Carill 1708', a chalice and a flagon inscribed 'Ex dono Sarah Dilke to the Parish of Chatris 1728', a paten marked 'Chatteriss parish 1744', a chalice given in memory of H. K. Bagshaw by M. and C. Burnett, 1919, and a paten given by Hilda Francis in memory of Eliza Richardson, 1922.

The tower contains six bells: 1st to 4th by Joseph Eayre of St. Neots, 1735; 5th by Mears and Stainbank, 1886; 6th by Taylor of Loughborough, 1911. The peal was re-tuned and the 5th re-cast in 1911. The 1st is inscribed 'Utile dulcis vivitur ingenio, grata sit arguta resonans campanula voce', the 2nd has the last six words of the above inscription, and the 3rd 'labor ipse voluptas'.[19]

The registers begin in 1614 and are complete except for the years 1665 to 1671 inclusive which are missing.

There are Anglican mission-rooms at Horseway and Slade End, the latter established 1908.

NONCONFORMITY Though there were some Baptists living in Chatteris in 1654 there was apparently no permanent Baptist church in the town until the late 18th century.[20] Chatteris people were baptized as members of the Fen Stanton Baptist congregation at various dates in the latter half of the 17th century.[21] In 1785 Joseph Scott and about 30 others were baptized at Wisbech and began meetings in a hired barn in Lyon's Yard, West Park Street.[22] In 1818, when this meeting had 21 members, it joined the New Connexion of General Baptists. Membership increased, reaching 58 in 1845 and 140 in 1851.[23] In 1835 a permanent building was erected in West Park Street at a cost of £800, on land bought from Theophilus Seymour for £110.[24] This church was enlarged in 1869 and renovated in 1885; schoolrooms were added in 1898.[25] There is a Particular Baptist church (Salem) at Hive End, Huntingdon Road,[26] founded in 1800 and endowed by the Clarett family with land. The endowment brought in £100 per annum in 1851.[27] Another such church (Zion) in Park Street was founded in 1819. In 1851 the congregation numbered 485.[23] These two churches are still in existence. A daughter chapel of the West Park Street church on the Forty Foot Bank was established in 1860, rebuilt in 1889[28] and closed during the Second World War.[29] The Congregational church in East Park Street dates from 1838, and is a good example of early Nonconformist 'Gothic'. At the 1851 census it had just been taken over by the Wesleyan Methodist Reformers,[30], who had a congregation averaging 315.[23] The Wesleyan Methodist church, in New Road, was originally built in 1815 and was enlarged in 1855.[31] The Wesleyans, who first came to Chatteris in 1806, used the General Baptist meeting-room during the first years.[32] The congregation numbered 360 in 1851.[23] The Friends' Meeting, established in 1703, had in 1851 20 members:[33] it was dissolved between 1929 and 1933. The graveyard still exists behind the meeting-house in High Street.[34] A Salvation Army hall was built between 1900 and 1904.

SCHOOLS A new school building seems to have been erected about 1738, according to a reference in the Churchwardens' Accounts of that year. This building is now a social club, and the former school, used as a storehouse during the re-roofing of the church in 1747, stood on the site of the present parish room, just south of the church.[35]

[19] J. J. Raven, *Church Bells of Cambs.* (1869), App.

[20] Adam Taylor, *Hist. Engl. Genl. Baptists* (1818), i, 146.

[21] *Fenstanton Records* (ed. Underhill, 1854), 253, 284.

[22] W. H. Clarke, *Hist. Sketch W. Park St. Bapt. Ch. 1785–1935*, p. 5.

[23] H.O. 129/7/191. The 1851 figure is that of the average congregation, not all of whom may have been baptized members.

[24] Clarke, *Hist. Sketch W. Park St. Bapt. Ch. 1785–1935*, p. 6.

[25] Ibid. 11, 14.

[26] *Baptist Handbk.*, 1950.

[27] Gardner, *Dir. Cambs.* (1851), 518. The value had decreased to £50 by the end of the century (Kelly, *Dir. Cambs.* (1900)).

[28] Clarke, *Hist. Sketch W. Park St. Bapt. Ch.*, p. 11.

[29] *Bapt. Handbk.*, 1950.

[30] Now again a Congregational church.

[31] Kelly, *Dir. Cambs.* (1900).

[32] Clarke, *Hist. Sketch W. Park St. Bapt. Ch.*, p. 5.

[33] H.O. 129/7/191; ex-inf. Librarian, Friends' House.

[34] Infm. Revd. J. C. Hawthorn, Vicar of Chatteris.

[35] Ibid.

At the end of the 18th century there were four private schools in Chatteris, kept severally by Thomas Camps the parish clerk, Christopher Walters, a former exciseman, Joseph Scott the Baptist minister and Thomas Kid, a Baptist layman. None of these teachers was licensed.[36]

No further provision for primary education was made in Chatteris until 1819 when application was made to the National Society for a grant for a two-roomed school for 200 boys and 150 girls to be held weekdays and Sundays. It was stated that there were about 200 boys and 200 girls between 7 and 13 who required education, which was already supplied to 80 boys and 50 girls in Dissenters' schools. An evening school had recently been set up which functioned during the winter months. This was attended by 80 to 100 young people and was supported by voluntary subscriptions. The site chosen for the new school was by the churchyard, and was held copyhold of Corpus Christi College, Cambridge. The National Society granted £200 towards the school, for which £835 was subscribed locally, including £175 and a yearly endowment of 10 guineas from Dr. Robert Chatfield, vicar. The master's salary was fixed at £70 and house, and 'if the master were acquainted with land measuring, it is very probable that during the month of harvest, he might be able to increase his profits by that manner'. In 1821 112 boys and 128 girls were attending the school. The numbers increased in 1845 to 123 boys and 168 girls. In the latter year the National Society gave £40 towards the estimated cost of £250 of a school for 120 infants.[37]

The boys' National School was rebuilt in 1856 to accommodate 200 boys and remained open until 1927: it had long been more than half empty. The average attendance was 68 in 1907, and 37 in 1919. The boys were eventually transferred to the King Edward School (see below).[38]

Another infants' school was built by trustees at Slade Lode, near the old workhouse on the March Road, in 1858. The cost was £589 and 98 places were provided. On the death of the last trustee in 1887 this school was taken over by the School Board, which had been established in 1873. The Board enlarged it (1895) to take 146 infants—a figure reduced in 1910 to 129. This school is still (1950) in use.[39]

A British School was founded in Chatteris about 1820.[40] In 1830 it had 118 boys and 80 girls on the books, with average attendances of 70 and 50 respectively.[41] It still existed in 1851,[42] but failed to attract sufficient voluntary support and was taken over by the School Board (1873). The Board built in 1873 boys' and girls' schools in New Road and an infants' school at Hive End. The last named was rebuilt in 1888 at a cost of £325, for 138 children. It was enlarged in 1904 to accommodate 207 and scaled down at the 1910 reorganization to 192, and to 130 infants in 1939. From 1928 it accommodated the infants who had previously been taught at New Road. This school is still (1950) in use.[43]

The New Road Boys' Board School, which was built for 200, existed until 1902. It was then replaced by the King Edward School for 305 boys which was enlarged in 1912 by the addition of a woodworking centre and in 1928 by the provision of an extra classroom. Central heating was installed at a cost of £495 in 1934. The educational system in Chatteris has been re-cast in each generation since the establishment of the School Board. The culminating developments occurred in 1938–9, when the Cromwell Secondary Modern School (see below) was built, to give general education with a rural bias to 500 pupils in the 11–15 age-groups from Chatteris and the surrounding villages. The King Edward School was at the same time converted into a junior mixed school for 330 children, the older boys who had gone on to the Cromwell School being replaced by the 62 remaining girls from New Road school (see below) and 89 and 52 pupils of each sex from the older children at Hive End and Slade Lode schools respectively.[44]

Girls and infants were accommodated in a Board School in New Road, which provided 341 places and was enlarged early in the 20th century to take 287 girls and 198 infants. It was never more than about two-thirds full (240 in 1899, 323 in 1919) and by the 1920's was considered to be dilapidated in appearance and inconvenient in arrangement. It was closed in two stages, the infants and some girls being transferred to Hive End in 1928 and the remaining girls to King Edward School in 1939.[45]

Proposals for a senior elementary school in Chatteris were put forward as far back as 1930, but were not taken up until 1934 owing to the economic crisis and the prior claims of March and other places. The site finally selected was of 6 acres, later increased to nearly 9 acres, at the corner of Wenny Road and Matting Close, with an additional frontage on Wood Street. The school, which has places for 320 boys and 320 girls, was opened in January 1939. The total cost was nearly £40,000.[46]

CHARITIES In 1837 the charities consisted of (i) the 'Poor's Land', of 34¼ acres in Slade End and Nightlayer's Fen and 5¾ acres at Purrant's Pound, let for £82 per annum, (ii) the Acre Close, let for £3 3*s.*, (iii) the 'Poor's Pieces', 7½ acres in Great Acre Fen, let for £12 5*s.*, and (iv) a house and land left by John Dring in 1682 which produced £25 6*s.* yearly. There was also (v) the Town Meadow of 9 acres, and (vi) the Town Yard, on which 10 houses, worth £1,000, had recently been erected. There were no documents, however, to prove that the former belonged to the parish. The total available for charity after expenses had been met was £97 8*s.* which was given in sums of from 2*s.* 6*d.* to 12*s.* to about 300 persons, preferably non-paupers. The same persons could benefit two years in succession, and the trustees published a list of the intended beneficiaries.

Chatteris shared with Doddington and March an area of 235 acres which was used by the poor of these parishes indiscriminately for turf cutting. The Commissioners stated that this land had been so much dug up as to be almost useless for this purpose, but if properly cultivated could be very productive, and that its capital value as agricultural land was £8,000.[47]

36 Bp.'s Mun., Ely, B 6.
37 Archives of Nat. Soc.
38 M. of Ed. file no. 12/2.
39 Ibid. file no. 12/6.
40 The 1821 *Rep. of the Brit. Soc.* is the first in which this school appears.
41 *25th Ann. Rep. of Brit. Soc.* (1830), App. 33–34.
42 Gardner, *Dir. Cambs.* (1851), 578.
43 M. of Ed. file no. 12/3.
44 Ibid. file no. 12/4.
45 Ibid. file no. 12/5.
46 Ibid. file no. 12/78.
47 *Rep. Com. Char.* (*Cambs.*), [103] 299–301, H.C. (1837–8), xxiv.

DODDINGTON

The original parish of Doddington, with its hamlets of March (q.v.), Benwick, and Wimblington, covered 37,801 acres. It was the largest parish in Cambridgeshire, and one of the largest in England. The nucleus of this area was an island of clay and loam rising out of the fen to about 25 ft. above sea level, running more or less north and south for about 4 miles, on which the villages of Doddington and Wimblington, and the town of March, are situated. Benwick, about 3 miles west of Doddington, is unique among ancient Fenland villages in not being on an actual 'island'; here, however, the gravel comes to within a few inches of the surface, providing a fairly secure foundation for buildings.[1] North-east from Doddington and Wimblington, and about 4 miles from the former, is Stonea. This hamlet lies on another very small 'island', and is the site of one of the bishop's vaccaries in the Middle Ages, and of an outlying grange. At the farm in Stonea known as 'The Stitches' is an ancient earthwork, probably a camp of the Early Iron Age period.[2] Benwick and March are on the old course of the River Nene, and the south-eastern part of Doddington and Wimblington is intersected by the Sixteen Foot Drain. Otherwise the waterways are not so important as those in most Isle parishes. The Ely–Chatteris–Wisbech road (A 142 and 141), making a circuit around the South Level, enters Doddington at Primrose Hill on the south edge of the parish and takes a winding course north-eastwards. It forms the main street of Doddington village and continues through the outskirts of Wimblington towards March. Second-class roads connect Doddington with Benwick, whence there are connexions to Whittlesey (B 1093) and Ramsey (B 1096), and with Manea by a road (B 1093) running eastward across the fen; another road from Wisbech to Chatteris (B 1098) follows the Sixteen Foot Drain on its east side. There is a station for Stonea, formerly named Sixteen Foot Station, where the direct Wisbech–Chatteris road crosses the railway from Ely to Peterborough (1846). Another station is at the south end of Wimblington village on the line from St. Ives to March (1848). The goods line from Three Horseshoes to Burnt House, opened in 1897, was extended to Benwick in 1898.[3] The district is mainly one of large farms growing the typical fen products of wheat, potatoes, and sugar-beet, but there are a few orchards, and small patches of woodland north of Doddington.

An Inclosure Act (31 Geo. III c. lxxxi) was passed for Wimblington in 1791. The award was not made until 1805,[4] when 676 acres of land were divided amongst fifteen proprietors. Of these John Waddington received 136 acres and Sir Henry Peyton 133, 45 as lord of the manor. In 1830 another award was made under a separate Act (6 Geo. IV c. lxv).[5] By this 81¼ acres were allotted to 11 persons in Doddington, and 166½ to 23 persons in Wimblington. Another small inclosure took place in Turf Fen, Benwick, under the general Act of 1845.[6] About 68 acres were then dealt with; 6 were allotted to the poor and half an acre for a school building; 7 acres were sold to defray the costs; the remainder was divided between 14 proprietors.

Doddington tradesmen's tokens of 1669 have been recorded.[7]

A troop of yeomanry cavalry was raised here in 1798, to whom a memorial was erected in the church in 1828.[8]

In the Middle Ages Doddington was one of the most important places in the Isle, but very few memorials of the past now exist. Apart from the church, the only important pieces of architecture are the village cross (14th century, with a modern head) and the fine early-19th-century tower windmill.[8a] Wimblington has a few houses of some distinction. These include the 18th-century 'Manor House' and Addison House. The latter is of the early-19th-century date and is of three stories, with pedimented doorway and fanlight. Stonea Farm, 100 yards due east of Stonea Grange, dates from the 18th century and is roofed with old tiles of local origin. Doddington Rectory, an admirable example of the large Victorian villa, replaces a 15th-century building which in the middle of the 19th century contained some interesting relics. These included a gilt wooden font designed for a private chapel or oratory, and an oak carving with figures of the Virgin and Child, St. Luke and St. Mark, Justice, and a boar hunt. On the east side of the house was 'one of the most delightful gardens in the kingdom'.[9]

Doddington has been an enterprising village. The Diamond Jubilee of Queen Victoria was commemorated by a clock tower with four dials, and as early as 1924 the 'George' and 'Three Tuns' Inns had installed wireless sets for the entertainment of their customers.[10]

MANORS Brithnoth the first Abbot of Ely obtained 60 acres in Doddington and Wimblington by exchange with Wine, son of Osmund. This land he gave to the monastery with a weir, producing 1,000 eels yearly. This weir he had purchased from Gunulph for 100s. From Turketil, Abbot of Ramsey, Bishop Aethelwold bought a hide of land at Doddington for the monastery, with a half of 'Weremere' and the marshes belonging to it. In its early days the monastery was also granted Stonea and its marsh. The donor was Wulstan of Dalham who had obtained it from the widow Ascuven. This marsh was let to farm for 2,000 eels. After King Edgar's death Begmund and other relatives of Ascuven seized it from the abbot, who took the case to the hundred court. The jury decided in his favour, and Begmund had to pay six years' fish rent for the duration of his illegal seizure.[11]

In 1086 the Ely property in Doddington was assessed at 5 hides, with land for 8 ploughs. Two and a half hides (3 ploughs) were in demesne. There was meadow for the 8 plough-teams, pasture for the cattle of the vill, and wood for 250 pigs. The population included 8 sokemen 'of 1 hide', 14 villeins, 8 cottars, and a serf. There were fisheries yielding no fewer than 27,150 eels, and 24*s.* was also paid in fish tribute. When received, the property had been worth £10. It

[1] *The Cambridge Region*, ed. H. C. Darby, 111.
[2] *V.C.H. Cambs.* ii, 46–47.
[3] Ibid. 132–3.
[4] Award at County Hall, March.
[5] Ibid.
[6] Ibid.; M.A.F. 1/365.
[7] *Fenland N. & Q.* iv, 329.
[8] Ibid. ii, 71–72.
[8a] See plate facing p. 206.
[9] R. Gardner, *Dir. Cambs.* (1851), 522.
[10] *Ely Standard*, 21 Oct. 1949 ('25 years ago' column).
[11] T. Gale, *Historiae . . . Scriptores* (1691), xv, 477–8.

was worth £12 T.R.E. and £16 at the time of the Survey.[12]

Doddington was one of the manors allotted to the see on its foundation, and became one of the bishops' principal residences. Bishop Balsham died there in 1286. The survey made for Bishop de Fontibus (1221) shows a demesne of 262½ acres, with half the meadow land of the vill in the bishop's hands. There were three fisheries in severalty, which with others, including 'Weremere', produced £4 10*s.* 10*d.* There were also three vaccaries[13] with 100 cows and 5 bulls.[14] Against the name of Robert de Marisco, a free tenant holding 5 acres for 2*s.* yearly, is written 'John de Aulton', who was perhaps his son-in-law; in 1229 de Aulton was granted a carucate of land belonging to the manor of Doddington for 30*s.* yearly by Bishop Northwold.[15] In 1251 the vaccaries were let to farm at £6 a year, and the value of the fisheries had risen to £9 5*s.* 5*d.*; except for the three held in severalty they were all much divided, in some cases among as many as ten persons.[16] In 1251 there was also a rush ground (*roseria*) at Benwick, worth £4, and though the amount of demesne land within the 1221 bounds was slightly smaller, a good many assarts, specially at Eastwood,[17] had brought the total area of the demesne up to 331 a. 2 r., rated at 3 carucates. There were 64 *novi feoffati* among the tenants, including 8 at Wimblington and 39 at 'Cuniwode' (Coneywood Fen).[18] Doddington was too far inland to be affected by the great flood of 1236 which caused so much damage at Wisbech and Elm, but like most of the episcopal manors it passed through a period of depression in the 14th century. The decline in prosperity seems to have set in about 1300. In the two-year vacancy between Bishops de Luda and Walpole (1298–9) the manor brought in £101 0*s.* 7*d.* In 1302 the receipts were £48 2*s.* 8*d.* and in 1316 only £35 8*s.* 5*d.* The population was smaller than might have been expected, seeing that March, Benwick, and Wimblington were all included with Doddington; the number of commutations (803 in 1286, 847 in 1302, and 155 in 1316) was on the whole lower than in the much smaller manor of Downham.[19] A survey of 1356[20] shows some recovery, but not to the level of the 13th century. There were only 140 acres worth £2 6*s.* 8*d.* in demesne. The value of the fisheries was £8 14*s.* 11*d.* and of the movable property (crops, livestock &c.) £15 17*s.* 10*d.* The tenants' rents, in Doddington and the three hamlets, stood at £35 4*s.* 5*d.* The manor house itself consisted of a hall, principal chamber, cloister (*claustura*), pantry, buttery, and other chambers. Enclosed within the same encircling wall were a kitchen, brewhouse, chapel, lodgings for knights and esquires, dovehouse, granary, and stables. In the wall was a gatehouse. All these buildings were in good repair. The rest of the manorial buildings, however, which consisted of a grange, ox-house, two windmills,[21] a newly erected grindery, the park wall, and the vaccary at Stonea, were all more or less ruinous.

As in other manors where the bishop had an actual residence, the customary tenants of Doddington had the duty of ferrying him to his next manor—in Doddington's case as far as Somersham, Willingham, Downham, or Ely.[22]

The bishops had free warren in Doddington from time immemorial. Towards the end of the 14th century this right was called in question, but was confirmed in 1399.[23] In 1300 the area of the great park was 80 acres and of the little park 60 acres;[24] there was also a wood, called Northwood, of 30 acres. In 1445 Philip Fysshewyke was appointed keeper of the parks and manor of Doddington,[25] and in 1490 Robert Rowden or Royden was made keeper of the parks and warrens at 2*d.* a day with perquisites.[26]

In 1389 certain miscreants living in Bardeneymoor (cf. the present Bradney House and Farm in the fens north-west of Doddington village) drove away the bishop's cattle. Bishop Fordham ordered them to be publicly denounced with the full medieval curse 'cum cruce erecta, pulsatis campanis, candelis accensis et postmodum extinctis ac in eorum vituperacionem in terram projectis et pedibus conculcatis'.[27]

In the later Middle Ages the bishops ceased to use the manor house as an official residence. In 1493 Bishop Alcock leased it to Robert Rowden, the keeper of the parks, at £3, and in the reign of Henry VIII Alexander Balam was tenant at the same rent.[28] The manor as a whole was valued at £58 14*s.* 8*d.* in 1540–1,[29] and was bringing in from £55 to £125 later in the same decade.[30] In the second half of the 16th century the Crown made a series of attempts to force the Bishops of Ely to hand over some of their manors, and Roger, 2nd Lord North, representing Queen Elizabeth, drew up a set of accusations of corrupt and illegal practices in Doddington and elsewhere against Bishop Cox.[31] The bishop was allowed to keep Doddington palace, which after his death (1581) was leased to his son Sir Richard for three lives,[32] but the manor was finally alienated from the see by Bishop Heton, Cox's successor, after a long interval (1581–1600) during which the Crown was *de facto* lord of the episcopal manors.[33] Heton's alienations included Conny Wood Land, Wadland, Bradney Moor, Stonea Grange, the Great and Little Parks, the Burystead (manor house) and Woodley Field, both in the tenure of Sir Richard Cox, and purpresture granted to William Cottney by copy of court roll.[34]

A survey made during the vacancy in the see shows rents of £63 (£20 freehold, £43 copyhold), a block of land which included the two parks, Woodley Field and a place called the Athys, comprising 300 acres, let at £7 a year and stated to be worth £150 in capital

12 *V.C.H. Cambs.* i, 367.

13 At Stonea, Westrea, and Dereford; the name of the last is preserved in Dartford Road, March (*P.N. Cambs.* (E.P.N.S.), 253).

14 B.M. Cott. MS. Tib. B. II, ff. 97*b*–102.

15 Ibid. f. 98*a*; Add. MS. 5847, f. 88.

16 Bp.'s Mun., Ely, G 3 ('Old Coucher').

17 Now Eastwood End just north of Wimblington village. See also Eastwood manor below.

18 B.M. Cott. MS. Claud. C. XI, ff. 61–68; Bp.'s Mun., Ely, G 3 ('Old Coucher').

19 S.C. 6/1132/9, /10, /11, and /13.

20 B.M. Add. MS. 6165, ff. 275–9.

21 One dated from 1300 (B.M. Add. MS. 5847, f. 89).

22 B.M. Add. MS. 5837, ff. 90–91.

23 *Cal. Chart. R.* 1341–1417, 381.

24 B.M. Add. MS. 5837, ff. 90–91. The park was broken into during an episcopal vacancy in 1302 (*Cal. Pat.* 1301–7, 42).

25 *Ely Dioc. Remembrancer*, no. 206, p. 141.

26 Ibid. no. 283, p. 179.

27 Ibid. no. 144, p. 97.

28 B.M. Lansd. MS. 21, no. 45. See also A. Gibbons, *Ely Episc. Rec.*, 415. Balam also held a close called Estreford and other properties in Doddington (B.M. Add. MS. 5837, f. 205).

29 Gibbons, *Ely Episc. Rec.*, 13.

30 B.M. Add. Chart. 34274. Two accounts survive, on fols. 3 and 25.

31 J. Strype, *Annals*, ii, 572–95.

32 B.M. Add. MS. 5847, f. 89.

33 Jas. Bentham, *Ely*, 196. Doddington, at £74 16*s.* 3*d.*, ranked second in value (after Haddenham) among the manors in the Isle that were alienated (B.M. Add. MS. 35171, f. 173).

34 B.M. Add. MS. 5847, f. 89.

value. 'Stony Farm' was let for 40 years at £6 1*s.* 8*d.*; 100 acres of land, worth £30, had recently been inclosed here. There were 2,800 acres of common belonging to the manor.[35] The gross receipts from the manor in 1591 and 1598 were £101 4*s.* 2*d.* and £79 1*s.* 8*d.* respectively.[36]

When Heton was appointed bishop the manorial rights were already on lease to Sir John Peyton,[37] and they were granted to him outright in 1602.[38] He was Governor of Jersey, and made Doddington his principal residence in England after the death of his wife in 1605.[39] In 1613 he changed the manorial customs from fines arbitrary to certain.[40] At his death in 1630 his son, another John, was already 51.[41] John II and his wife Alice, daughter of Sir John Peyton of Isleham (his uncle), settled the manor in 1633[42] for the benefit of their eldest son Robert, who succeeded two years later.[43] From 1641 Robert was insane. He died in 1658 without issue, and was succeeded by his brother Algernon, rector of the parish.[44] The latter outlived his two sons John and Algernon. Sewster, son of Algernon II, succeeded to his father's baronetcy and grandfather's manor. Until he came of age in 1693 his interests were looked after by John Skelton, who was dealing with the manor in 1677[45] and represented him in a suit brought by Richard Wakefield, a tenant, over escheated premises.[46] Later on Sir Sewster[47] was Master of the Buckhounds to Queen Anne. His two sons Sir Thomas and Henry both died without heirs, and the manor came to the son of his third child Margaret, Henry Dashwood, who took the surname Peyton by Act of Parliament on his succession in 1771.[48] The manor continued in this family until the end of the 19th century.[49] Mr. C. W. King was lord in 1933.[50] The manor house had ceased to be used as such before 1808[51] and was converted into a farm, but courts leet and baron were being held in 1827.[52]

The manor of *EASTWOOD* originated in an $\frac{1}{8}$ fee held of the bishop for 5*s.* In 1302–3 John de Estwode was tenant and in 1346 Sir William de Thorpe.[53] The manor was in the Lexham family in the following century. Margaret, relict of William Lexham, and Margaret (Rosse) her daughter, were involved in lawsuits over the detention of manorial deeds and over a forged will of William Lexham, which was produced by John Farewell and Richard Vowell, priors of Walsingham (Norf.), and purported to bequeath Eastwood to that house.[54] From this period (between 1501 and 1532) this manor is alternatively known as *LEXHAMS*. The descendants of William Lexham were successful in their suit, and the manor passed to Anthony Hansard (d. 1533),[55] a nephew of Lexham, whose relict Alice passed it in 1541 to Sir John Hynde, serjeant-at-law.[56] His son Francis was a leading freeholder of the manor of Doddington in 1553–5.[57] Elizabeth Hynde, relict of William, Francis's son, conveyed it in 1606 to Sir John Peyton,[58] since when it has descended with the main manor.

A manor of *SOUTHWOOD* existed in the 13th century. It was granted *c.* 1275 by Ela, Countess of Warwick to Reading Abbey,[59] with 20 acres of marsh in Northwood. By an agreement of 1284 between the Countess and the monastery, 20 marks from the manor were to be devoted to providing spices (*ad species faciendas*) for the monastery.[60] The value of this manor was £6 13*s.* 4*d.* in 1291.[61] In 1534–5 it was rated at 3 carucates and held in socage, and was still recorded as having formerly belonged to Reading.[62] It does not seem to be known at what date the abbey relinquished its interest; in 1340 Sir Nicholas Franceys was licensed to hear mass in the chapel of his manor of Southwood,[63] but he may have been only a tenant of the monastery. The next references (1440 and 1443) are to an inclosure called Roulowe or Rowloo in Southwood, in possession of William Lexham,[64] so that the manor must have become merged in that of Eastwood.

The subsidiary hamlets of Doddington did not, with the exception of March (q.v.), rise to manorial status.

BENWICK had fifteen tenants in 1221 and 32 in 1251.[65] In 1275–6 the Bishop of Ely was charged with appropriating a portion of the king's forest in Hurstingstone hundred (Hunts.), including, as it was said, the part of Benwick belonging to Huntingdonshire.[66] Bishop Balsham retaliated seven years later. He complained that John Pycard the king's forester forcibly entered his free chase in Benwick marsh to inquire whether it was within the bounds of the forest.[67] Letters patent were dated at Benwick in 1314, presumably when Bishop Ketene was entertaining Edward II at Doddington Palace.[68]

Thomas Burgess (d. 1634) devised his house, known as Betts, to his relict Ann for life, and also a half share in his 'great boate called a liter', of which the other half went to his son Thomas.[69] This seems to be a reference to navigation on the old Nene, as at March in 1566.

WIMBLINGTON, as the nearest hamlet to the parent village, is not separately distinguished in the 13th-century surveys, though it is mentioned by name. It had three guilds—of the Holy Trinity, the Purification, and St. Peter—which were sufficiently important to own between them two guildhalls and a chapel.[70]

35 S.C. 12/30/33, f. 11. The manor is called 'Dunnington'.
36 S.P. 12/268/110.
37 W. Watson, *Hist. Wisb.* 522. According to C. E. Walker, *Recs. of a Fen Parish* (1888, reprinted 1930), p. 39, the Peytons had been lessees under the bishops throughout the 16th century.
38 C 66/1591, m. 1.
39 *D.N.B.*
40 Watson, op. cit. 583.
41 C 142/741/133.
42 C.P. 25(2)/527 Trin. 9 Chas. I.
43 C 142/573/62.
44 C 142/622/17.
45 C.P. 25(2)/633 Mich. 29 Chas. II.
46 C 5/91/17.
47 He derived his unusual Christian name from his mother Frances, daughter and heir of Sir Robert Sewster of Raveley (Hunts.).
48 G. E. C., *Compl. Baronetage*, v, 190.
49 C.P. 43/866, ro. 442; C.P. 43/973, ro. 359; R. Gardner, *Dir. Cambs.* (1851), 521; Kelly, *Dir. Cambs.* (1900).
50 Kelly, *Dir. Cambs.* (1933).
51 D. and S. Lysons, *Magna Brit.* ii (1), 176.
52 W. Watson, *Hist. Wisb.* 583.
53 *Feud. Aids*, i, 151, 173.
54 C 1/249/64; C 1/669/32.
55 C 142/55/43.
56 C.P. 25(2)/4/21; C.P. 40/1123 m. 713.
57 C 1/1347/58.
58 C.P. 25(2)/277 Hil. 4 Jas. I. See also C 142/249/52.
59 J. B. Hurry, *Reading Abbey*, 179; B.M. Harl. Chart. 54 D 15.
60 Hurry, op. cit. 186; B.M. Add. Chart. 19633.
61 *Tax. Eccl.* (Rec. Com.), 271.
62 B.M. Add. MS. 5837, f. 206.
63 *Ely Dioc. Remembrancer*, no. 59, p. 404.
64 C 146/1391; C 146/2990.
65 B.M. Cott. MSS. Tib. B. II, f. 97*b* and Claud. C. XI, f. 68.
66 *Rot. Hund.* (Rec. Com.), i, 197. No part of Benwick has been considered to be in Huntingdonshire since 1801, but part of Ramsey, estimated at 100 acres in extent, was considered to be in Cambridgeshire up to about 1861.
67 *Cal. Pat.* 1281–92, 41.
68 Ibid. 1313–17, 195, 198. The letters are dated 5 Nov., the next dates before and after being 4 Nov. (at Ely) and 9 Nov. (at (Fen) Ditton).
69 *E. Anglian*, N.S. xiii, 268.
70 W. M. Palmer, *Village Gilds of Cambs.* 388.

One of the guildhalls had come into the possession of the lord of the manor (Sir John Peyton II) by 1635.[71]

By 1669 the commons of Wimblington, which included Stonea and Horsemoor Fens, had become overstocked. Sir Algernon Peyton, as lord of Doddington manor, therefore made an agreement with his 42[72] Wimblington tenants which defined the system of common pasturing at great length.[73] Each tenant was to be stinted to 6 bullocks or cows, or 3 horses or mares (with foals), or 30 sheep or ewes. 'Poor persons' were allowed to keep one or two cows or heifers each. The 800 acres of Stonea Fen were to be left 'fresh', without stock of any kind, from 14 February to 23 April each year, on penalty of 6*d.* for a horse, 2 cows, or 8 sheep, in the case of a first offence, and 1*s.* for subsequent offences. Half the fines were to go to the parish, half to the impounder of the stock or hayward. Ten trustees were appointed to administer the scheme. Closes varying in size from 13 to 7 acres were to be inclosed for the use of each homestead, to a total of 432 acres. The sites and bounds of the closes are set out in great detail.

The cartulary of Chatteris nunnery records a fishery and several small properties in Doddington. They were probably given during the 14th century, as there is no mention of them in the *Taxatio*.[74] They were worth only 4*s.* 4*d.* in 1539, the fishery not being mentioned separately.[75]

CHURCHES The advowson of the undivided rectory of Doddington descended with the main manor, but the right of presentation to this very valuable living was sometimes granted out *pro proxima vice*. For example, John Snell presented in 1719 and Edward Beresford in 1751.[76] In 1750 the right to one presentation was sold to Dr. Baptist Proby for £1,000.[77]

The history of the division of Doddington rectory is as follows. The first Act (1847)[78] provided for separate rectories of Doddington, Benwick, and March. This Act was amended by another of 1856,[79] which provided for three additional rectories in March (q.v.) and another at Wimblington. Under this latter Act the stipend of the rector of Doddington was to be £1,700 a year. The Acts did not begin to come into force until the death in 1868 of the Revd. Algernon Peyton, the last rector of the ancient parish of Doddington, who had held his lucrative office for fifty-seven years. Benwick rectory was established in that year, and Wimblington in 1874. The three advowsons were at first reserved to the Peyton family, who retained Doddington itself until *c.* 1922, when this living was transferred to trustees. The patronage of Benwick was early transferred to the Revd. W. A. Wood, rector 1868–*c.* 1915, and has since 1920 been with the Bishop of Ely. That of Wimblington was held by the Peytons until between 1900 and 1905. It was then transferred to the Hill family, and in 1927 to the Ratford family. The executors of the Revd. H. Ratford held it in 1940.[80]

The value of Doddington rectory showed a continuous rise during the 13th century, from £24 in 1217 to £33 6*s.* 8*d.* in 1254 and £36 13*s.* 4*d.* in 1291.[81] In 1535, however, the value, including the chapelry of March, was only £22 4*s.* 11*d.*[82] Even so the rectory was more valuable than any living in the Isle except Wisbech vicarage and Leverington rectory. In 1341, after a tithe dispute between John de Herle, rector, and the monastery of Ely, it was laid down that the tithes from 'Wyshammouth', 'Plantecroft', and 'Wattes atte Barre' belonged to Doddington rectory, those of 'Henlondes' to the monastery.[83] In 1562 a pension of £2 4*s.* 6*d.* was awarded to Bishop Cox out of the rectory.[84]

With the draining of the fens, the value of the tithes from the 60 odd square miles of the undivided parish rose greatly, and in the late 17th and early 18th centuries caused a series of lawsuits between the rectors and their parishioners.[85] In 1808 the tithes were said to be let for 5,000 guineas a year.[86] The rectors were usually non-resident, partly owing to the dampness and unhealthiness of the climate. Their duties were performed by curates at a stipend of £150 a year with house.[87] By the middle of the 19th century the value of the rectory had risen to £7,306 yearly, considerably more than that of many bishoprics.[88]

There have naturally been many distinguished holders of this wealthy living, among them Hugh Cressingham (d. 1297), treasurer of Scotland; Robert de Bardelby (d. 1323), a Chancery clerk; Nicholas Hawkins (d. 1534), resident ambassador at the court of the Holy Roman Emperor and designated for the see of Ely just before his death; Christopher Tye (?1497–1572), musician; Hugh Bellot (1524–96), Bishop of Bangor and of Chester; Thomas Neville (d. 1615), Dean of Canterbury; Edward Martin (d. 1662), President of Queens' College and later Dean of Ely; John Nalson (1638–86), royalist pamphleteer and historian; and Baptist Proby (d. 1807), Dean of Lichfield.[89] The work in the East End of London of the Revd. Richard Ridge, rector 1922–49, is commemorated by an early William Morris window in the church, which was unveiled by General Townshend of Kut in 1923.[90]

In 1561 Sir Edward Warner and Ralph Shelton were granted lands in Doddington which had been used to provide lights before the image of St. Mary (the patron saint) in the church.[91] William Stanbrighe of Doddington in 1489 left 2 lb. of wax to the guild of the Holy Trinity here.[92]

The Church of *ST. MARY* consists of chancel, clerestoried nave, aisles, south porch, west tower, and spire. The material is rubble with Barnack stone dressings. The chancel roof is covered with slates and the others with lead. The chancel and part of the north

[71] B.M. Add. MS. 5838, f. 200.

[72] Forty-two male tenants are named, with the wives of three. In these cases the tenant presumably occupied his land in the right of his wife.

[73] C 78/1292, no. 13. The agreement occupies nearly twelve membranes.

[74] B.M. Cott. MS. Jul. A. 1, ff. 137*a*, 167*b*–171*b*.

[75] Dugd., *Mon.* ii, 620.

[76] P.R.O. Inst. Bks.

[77] W. Watson, *Hist. Wisb.* (1827), 584.

[78] 10 & 11 Vic. c. 3.

[79] 19 & 20 Vic. c. 1.

[80] *Crockford's Clerical Dir.* (1926, 1935, and 1940); *Clergy List* (1905, 1915); Kelly, *Dir. Cambs.* (1900, 1933).

[81] W. E. Lunt, *Valuation of Norw.*, 217, 539.

[82] *Valor Eccl.* (Rec. Com.), iii, 500.

[83] *Ely Dioc. Remembrancer*, no. 65. In 1320 considerable difficulty had been experienced in collecting tithes from the parishioners.

[84] *Cal. Pat.* 1560–3, 226.

[85] E 134/3 Wm. & M./M. 44; E 134/3 and 4 Wm. & M./H. 7; E 134/12 Wm. III/E. 5; E 134/12 Geo. I/ E. 7.

[86] D. and S. Lysons, *Magna Brit.* ii (1), 177.

[87] A. Gibbons, *Ely Episc. Rec.*, 3; *Fenland N. & Q.* i, 158.

[88] Gardner, *Dir. Cambs.* (1851), 521; *Clergy List* (1859).

[89] *D.N.B.*

[90] Kelly, *Dir. Cambs.* (1933).

[91] *Cal. Pat.* 1560–3, 106.

[92] Gibbons, *Ely Episc. Rec.*, 203.

aisle belong to the middle of the 13th century. There was a great reconstruction about a century later, which included the nave arcades, the tower and spire, the south aisle and the western portion of the north aisle. Early in the 15th century the south porch was added, and later in the same century the lateral windows of the chancel were altered. In the second half of the 19th century there was an extensive restoration, which included the rebuilding of the east gables of the chancel and nave. The roofs of the nave and aisles were reconstructed in the present century, some of the old material being retained. The chancel has an east window of four lights with modern tracery of 14th-century character. later. Both aisles terminate in a line with the west face of the tower. The east window of the north aisle is modern and of three lights, set in a 13th-century opening. There is a clamped 13th-century buttress at the north-east angle, and an internal string-course. The lateral windows are of various periods; the first from the east is of two lights with plate tracery, and has a hood-mould; it is of mid-13th-century date. Next there is a 14th-century window with three trefoiled main lights and three quatrefoils above, then a 15th-century one of three lights with rectilinear tracery; the north-west window is of the 14th century with two trefoil-headed lights under a square label terminating in heads. The

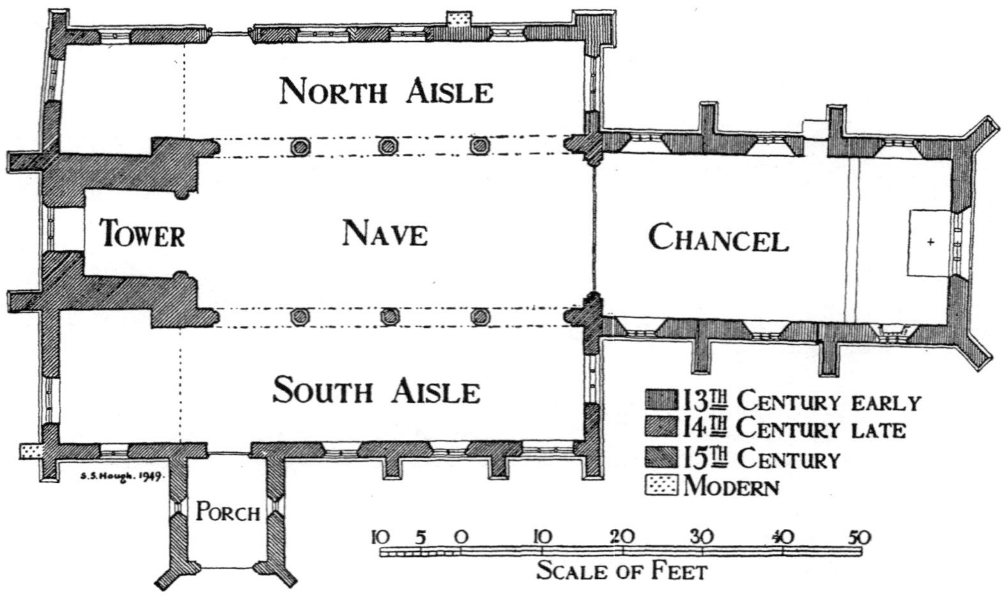

Plan of St. Mary's Church

There are diagonal buttresses with one set-off and two similar on the north and south. In the north wall is a doorway with continuous mouldings and protected by a gabled pent surmounted by a finial and supported on large brackets in the form of grotesque figures. The lateral walls have three windows of three lights, the openings being of 13th-century date with 15th-century tracery inserted; the original internal hood-moulds, terminating in heads, are retained. The eastern window on either side has cinquefoiled lights, and the others cinquefoiled lights with cusped tracery in the centre light; all have embattled transoms. The lofty chancel arch is two-centred and of two orders, the outer resting on shafts with moulded caps and bases, and the inner springing from semi-octagonal responds with moulded caps and bases. There are squinches in the western angles of the chancel, and the stone corbels for the rood beam remain on the west side of the chancel arch. There is a double piscina in a modern recess. The nave has arcades of four bays dating from the third quarter of the 14th century, with two-centred arches of two orders and octagonal columns with moulded caps and bases. The clerestory windows are of two cinquefoiled lights with a quatrefoil above and external hood-moulds terminating in heads. The east gable has been rebuilt. The nave opens to the tower by a 14th-century two-centred arch of two orders with continuous chamfers.

The north aisle is of two periods. The eastern section is mid-13th century and the western a century west window has three lights with rectilinear tracery and a hood-mould terminating in male and female heads. The north doorway, which is of 13th-century work, has a deeply moulded two-centred arch and jamb shafts with moulded caps and bases. The lateral buttresses are modern, with one set-off, except the westernmost which is of the 14th century. There is an image bracket to the north of the east window. The south aisle, which belongs entirely to the second half of the 14th century, has an east window of three lights with rectilinear tracery. The south-east window is of similar design. These two windows must be somewhat later in date than the rest of the aisle. The three remaining lateral windows are all of two cinquefoiled lights with a quatrefoil in the head and a hood-mould terminating in heads. There are four buttresses with one set-off. There is a trefoil-headed piscina recess in the south wall. The early-15th-century porch has an outer doorway of two orders, the outer with continuous hollow mouldings and the inner springing from semi-octagonal jambs with moulded caps and bases; there is a hood-mould terminating in male and female heads. There are diagonal panelled buttresses, and in the east and west walls a window of two trefoiled lights with a quatrefoil above, and hood-moulds terminating in male and female heads. There is a modern parapet with pinnacles. The inner doorway has continuous mouldings and a hood terminating in crowned heads.

The tower, which is engaged by the aisles, has angle

buttresses with two set-offs. It is of three stages. The west window is of three lights with rather clumsy flowing tracery of the third quarter of the 14th century. The second stage has a cinquefoiled lancet on the north, south, and west sides. The belfry windows are of two trefoiled lights with a quatrefoil above and a hood-mould terminating in heads. There is a plain coped parapet. The low octagonal spire of Barnack stone has a gabled trefoil-headed light on alternate faces and terminates in a moulded finial, which has been renewed.

The nave roof is of the queen-post type and of 15th-century origin, but some years ago it was largely renewed; there are full-length angels with outspread wings at the intersection of the purlins and intermediates; the wall posts rest on stone corbels carved with heads. The aisles have plain lean-to roofs, the north modern and the south much renewed with the exception of some of the principals. The chancel and porch have modern roofs. The 13th-century font has a plain octagonal bowl resting on a central and four angle shafts, with moulded caps and bases. The chancel screen is of 15th-century origin but has been so extensively restored that little old work now remains; it is of eight bays, the two in the centre being occupied by the doorway, and has rather elaborate rectilinear tracery; there is an ugly modern cornice.

The plate includes a chalice, paten, and alms-dish, none of any great age or interest.

The tower contains eight bells: 1st, 2nd, 3rd, and 8th by Mears and Stainbank, 1936; 4th by Thomas Gardiner of Sudbury, 1737; 5th, 6th, and 7th by the same, 1736.

The registers begin in 1681 and are complete from that date. In *c.* 1938 there was a broken series of registers of earlier date going back to 1599. These now appear to be missing.

There was an unconsecrated chapel at *BENWICK* in 1637–8.[93] The church of *ST. MARY* was erected in 1850–1 on its site. It is a structure of carr stone consisting of chancel, nave, aisles, south porch, and north-west tower and spire. It is an early work of S. S. Teulon, but not very characteristic of that extremely mannered Victorian architect. Many walls are out of plumb owing to uneven subsidence of the fen soil. The church was restored and the spire rebuilt in 1902. There are two bells. The plate includes a silver chalice and paten given in 1942 in memory of a former churchwarden and his wife, an electroplated chalice and paten and a brass almsdish. The registers begin in 1851 and are complete.

The church of *ST. PETER*, Wimblington, erected in 1874, is a stone structure consisting of chancel, nave, south porch, and central tower with short square spire. The tower contains one bell. The registers begin in 1874 and are complete.

NONCONFORMITY The number of Dissenters in Doddington was fairly large in 1676, and the proportion (43 out of 813 inhabitants) higher than in such larger places as Whittlesey and Wisbech.[94] In 1851 there were three chapels in Doddington. One of these, with an average congregation of 30, was for Latter Day Saints; the date of foundation is unknown, and it has long ceased to exist. The two Methodist chapels were of very recent foundation in 1851 (Wesleyan, 1848; Primitive, 1850). The former is still in existence; the latter was closed between 1869 and 1875.[95] A Baptist chapel was erected at Doddington in 1882, and closed in 1937.[96]

The Wimblington chapels were Wesleyan (1809) and Primitive (1834) Methodist. The latter was at the hamlet of Hook, a mile north-east of the village. Another Wesleyan chapel was built at Stonea in 1829, on the Sixteen Foot Bank. The Benwick chapels were Calvinistic Baptist (1818) and Wesleyan Methodist (1833);[97] the former was endowed by Gideon Gascoyne with 2 acres of land, let at £8 in 1851.[98] All these chapels are still in existence.

SCHOOLS Lionel Walden, by his will dated 1719, directed that his property should be sold and the interest on the proceeds allowed to accumulate until it reached £500. This sum was to be used to build a school at Doddington. Nothing was done until 1787, when the estate amounted to £2,804 8*s.* 6*d.* In that year Dr. Baptist Proby, rector, offered a house and outbuilding for the school for £300. The house was bought in 1790 and converted into a school at a cost of £80; nine trustees, including the Bishop of Ely *ex officio*, were appointed. There were to be 30 free scholars, who were to be taught reading, writing, and arithmetic; they were to be allowed to remain up to the age of 14 in the case of boys and of 12 in the case of girls. The schoolmaster was given a salary of £40 and was forbidden in the first instance to take private pupils. Owing to the failure of the trustees to execute the deed of declaration of trust a lawsuit ensued, as a result of which the schoolmaster obtained the right to take paying pupils at the rate of 6*d.* to 8*d.* a week.[99] When Bishop Yorke made his inquiry (1789–98) into the schools of Ely diocese there were 15 such pupils.[1] In 1837 there were about 20 fee-paying pupils on the books, as well as the 30 free place holders, but owing to the rise of the agricultural gang system in the neighbourhood not more than 6 paying and 12 non-paying pupils attended in summer. At this date the endowment, invested in 3 per cent. Consols, amounted to £1,817 17*s.* 8*d.* This brought in £54 10*s.* 8*d.* a year.[2] By 1867 the free place system had been abolished, and a fee of 1*d.* a week was now universal. At this date there were 56 boys and 63 girls attending. There was still only one master, who was unqualified. As in most of the endowed schools of the Isle, no children were learning anything beyond reading, writing, and arithmetic. None had left to take up apprenticed employment during the past year. The school was not subject to Government inspection.[3] In 1876 a Doddington School Board was set up,[4] which took over Walden's School and its endowments; two years later new buildings were provided for 220 children, at a cost of £1,851, on the site of the old school. In 1899 the average attendance was 140.[5] After the 1910 reorganization the recognized accommodation was for 146 mixed and 97 infants; the two departments were amalgamated in 1924. In 1939 the school was reclassified as a junior mixed and infants' school with 135 places; there were 101 children in 1948.[6]

There was no provision for elementary education

93 A. Gibbons, *Ely Episc. Rec.*, 87.
94 Bp.'s Mun., Ely, A 6.
95 Kelly, *Dir. Cambs.*
96 *Bapt. Handbk.*, 1949.
97 H.O. 129/7/191.
98 Gardner, *Dir. Cambs.* (1851), 524.
99 *Rep. Com. Char.* (*Cambs.*), [103], 301, H.C. (1837–8), xxiv.
1 Bp.'s Mun., Ely, B 6.
2 *Rep. Com. Char.* (*Cambs.*), 301.
3 *Schs. Inq. Rep.*, [3966–XII], 514, H.C. (1867–8), xxviii (10).
4 *Lond. Gaz.* 1876, 5067.
5 Kelly, *Dir. Cambs.* (1900).
6 M. of Ed. File no. 12/8.

in Benwick until 1871,[7] when a School Board was established. Her Majesty's Inspector considered that a school for 90 boys and girls and 60 infants would be adequate, but the Board claimed that 200 places were required.[8] The school actually built contained accommodation for 216 (154 mixed and 62 infants) and cost £1,590 including the teacher's house. In 1939 the school was reorganized as a junior mixed school for 128 children, 81 attending in 1948.[9]

In 1715 Thomas Eaton bequeathed two houses and 8½ acres of land for providing a school in Wimblington. There was even greater delay than at Doddington in carrying the bequest into effect. A new trust for Eaton's charity was set up in 1734, but the last trustee died in 1771 without anything having been accomplished. A new trust was established in 1802, to start a school for not more than 40 children. The inclosure award of 1805 allotted 42 acres in respect of the Eaton estate. At length in 1815–17 a school was built at a cost of £800. When the Charity Commissioners reported in 1837 the endowment consisted of 71 acres of freehold and 4 of copyhold land, producing £144 8*s*. a year.[10] The master's salary was £60, so that a considerable surplus was available. Attention was again called to this surplus in 1867, when £83 of the total of £149 derived from the charity was devoted to the school. At the latter date there were 78 boys and 65 girls on the books, the children of 'agriculturalists'.[11] This school seems to have been no more efficient than Walden's at Doddington, and it was reorganized under a scheme of the Charity Commissioners made in 1877. The buildings were altered and enlarged to take 150 children. By another scheme (1885) the transfer of the school to the Wimblington School Board (established 1881) was sanctioned, but the transfer did not actually take place until 1896. Soon after this date the Board enlarged the school to take 196 children, and in 1910 the recognized accommodation was reassessed at 160 (100 mixed, 60 infants). From 1909 the buildings, which appear to have had insufficient foundations, were on the 'black list' of the Board of Education. They obtained a reprieve owing to the First World War, but by 1922 had become so dangerous that one classroom had to be closed. A new school was built in 1924 at a cost of about £3,000, providing 202 places. This school is now used for junior mixed and infants and had in 1949 114 children on the roll, including 40 from Stonea (see below). A kitchen-dining room has recently been provided.[12]

The school at Stonea was built by Wimblington School Board in 1882, and enlarged in 1898 to provide for 94 children. It was restricted to juniors and infants in 1939 and closed altogether in 1947.[13]

CHARITIES[14] The Doddington Town Lands consisted in 1837 of Beezling Drove and about 61 acres, let for £136 10*s*. 6*d*., and a mill, let for £9 15*s*. and subject to the obligation upon the miller to grind corn for parishioners at a reduced fee. Of this rent £5 were spent annually upon a Sunday School, 12 guineas upon the clerk's salary, £1 13*s*. upon the sexton's salary, and about £25 on an average upon church repairs. No church rate was levied.

Nicholas Spencer, by will dated 1597, left house property in the City of London from which £2 12*s*. was given yearly through the Merchant Taylors' Company[15] to the clergy and churchwardens to provide bread for the poor of Doddington. The conditions of this charity were strictly observed in 1837 and all the proceeds distributed in kind.

The Benwick Town Lands in 1837 consisted of 6¾ acres let by auction. They produced £9 yearly when in pasture and £18 to £20 when arable. The proceeds were appropriated to the poor rates, and also used for the repair of the chapel. There were also 6 old cottages in poor repair, occupied rent free by paupers, and one larger house called the 'Great House' which accommodated 8 pauper families. These families were nominated by and removable at the will of the overseer.

Doddington and its hamlets, including March, shared with Chatteris (q.v.) the privilege of cutting peat in Turf Fen.

MARCH

March, the county town of the Isle of Ely, once formed a chapelry in the northern part of the parish of Doddington. A Local Board of Health, set up in 1851,[1] was converted in 1894 into an Urban District Council of twelve members, by which the town is now governed. Small adjustments of boundary were made in 1933 under the Isle of Ely Review Order.[2] An outlying portion of Wisbech Borough was brought into the district, and a mutual exchange made with the parish of Elm.

The town probably owes its origin to the ford on the old course of the Nene,[3] where the road between Ely and Wisbech, the two chief towns of the Isle, crossed the river. It early outstripped the parent settlement of Doddington and in the 15th century possessed several guilds;[4] the church also is as large as and finer than that of Doddington, and its wonderful hammer-beam roof testifies to the affluence of the place at the end of the Middle Ages.[5]

A market, with two annual fairs, was first granted in 1670,[6] in spite of the opposition of Wisbech Corporation.[7] In 1785 the tolls were assessed at £6.[8] Soon after this the market appears to have lapsed, and an attempt to revive it in 1821 was not very successful,

[7] *Lond. Gaz.* 1871, 5145.

[8] The 1871 population of Benwick was 857.

[9] M. of Ed. File no. 12/1.

[10] *Rep. Com. Char.* (*Cambs.*), [103], 313, H.C. (1837–8), xxiv.

[11] *Schs. Inq. Rep.* xiii, pt. x, 3966–XI, 518 (1867–8), xxviii (10).

[12] M. of Ed. File no. 12/59.

[13] M. of Ed. File no. 12/60.

[14] *Rep. Com. Char.* (*Cambs*), [103], 303–5, H.C. (1837–8), xxiv. For Walden's Charity (Doddington) and Eaton's Charity (Wimblington) see above—Schools.

[15] Spencer was Master of the Company in 1588 (C. M. Clode, *Early Hist. . . . Merchant Taylors*, i, 322, 404).

[1] 14 & 15 Vic. c. 103.

[2] M. of H. no. 76940.

[3] It was usually known as March(e)ford in the Middle Ages, e.g. in the inquisition *post mortem* on Sir John Colvylle in 1360 (*Cal. Inq. p.m.* v, 655 (p. 551)).

[4] W. M. Palmer, *Village Gilds of Cambs.* 388.

[5] The 13th-century north arcade shows that the church had at least one aisle at that date. Local tradition holds that the site first chosen for the church was that of the existing wayside cross in the Avenue, and that the stones were moved farther away from the town by the devil (*Fenland N. & Q.* i, 16).

[6] C 66/3120, no. 8.

[7] Mann Hutcheson's MS. Hist. Wisbech, 107 (Wisb. Mus.).

[8] *Rep. Royal Com. Mkts. and Fairs*, [C 5550–I], 221, H.C. (1888), liii.

though the fairs continued at this period to be prosperous. The development of the market was impeded by the absence of a covered hall and by the fact that market day in March and several neighbouring towns fell on the same day (Friday). After the opening of the railway in 1847 another attempt was made to increase the market. The want of a market house was remedied, in a makeshift fashion, by Sir Henry Peyton, 2nd baronet of the 1776 creation, who was lord of the manor and holder of market rights. His building, however, was only 40 ft. long by 17 ft. broad,[9] and provided 14 stalls under cover as compared with the 54 available at Wisbech and 74 at Peterborough. The difficulty of the clashing with other market days was solved in December 1856 by changing the day to Wednesday 'by private arrangement and without any formalities'.[10] In 1851 the market had been stated to be 'making some progress'; £150 was subscribed to give a treat to the poor at its reopening.[11] The tolls were, however, collected in an arbitrary and haphazard way; they were assessed for poor rate purposes at £10, but the toll-keeper in 1888, though he had no fixed scale of charges and kept no record of receipts, was said to be taking about £50 a year. An attempt by the Local Board to purchase the market rights to mark the Jubilee of 1887 was a failure,[12] but they were secured by the Urban District Council in 1898, and the market is now well attended. The fairs are held on the Monday before Whitsun and the third Tuesday in October.[13]

The first record of a post office is in 1793. In 1832 March was made a post town, and in 1851 the post office was situated in High Street.[14] It was rebuilt in 1887, and was moved to Broad Street in 1901;[15] the present building dates from 1936.[16] A telegraph service was first provided in 1870, and telephones in 1908.[17]

The 1563 list of householders[18] does not distinguish March from Doddington, but the evidence of the registers at this period suggests that March may have had a population of 1,000.[19] The ship-money assessment of 1639–40 rated March at £35 5*s*. The rate is a little greater than that set upon Doddington and Wimblington combined, but lower than that upon several of the larger villages, for example Downham, Elm, Haddenham, Littleport, Stretham, and Sutton.[20] In 1669 there were 165 holders of common rights in March as compared with 42 in Wimblington,[21] and 7 years later Bishop Compton's 'census' showed 949 persons of communicant age in March as compared with 813 in the rest of Doddington parish.[22] In 1808 March was described as a 'considerable town'.[23] It was in fact slightly larger than Chatteris, but about a third smaller than Ely and Whittlesey and only half the size of Wisbech. It had long been consolidating its position as the main centre in Doddington parish, and the 1801 Census showed about twice as many people in March as in the three villages. The growth of Doddington and its hamlets in the first half of the 19th century was exceptional even for the Fens, and the proportion of about two to one between March and the rest of the parish still obtained in 1851. The usual heavy decrease occurred in 1851–61, amounting in the case of March to over 12 per cent. Since 1861, however, March has forged ahead, not only in comparison with the rest of Doddington parish, but with the other towns of the Isle. By 1891 March had outstripped Whittlesey, and by 1911 Ely also, and the 1931 figure for March (11,266 persons) was more than twice that of 1861. The 1951 population was 12,993.

In the reign of Queen Elizabeth March was a minor port. In 1566 eight boats, capable of carrying one, one and a half, or two cartloads, were used in the coal and grain trades.[24] A certain amount of traffic in coal and other commodities, carried in barges, was observed by Dugdale in 1657.[25] Tradesmen's tokens of 1669, and a silver shilling token of 1811, have been noted.[26] A 'town hall' existed in 1669.[27] From 1778 to 1846 the 'Guildhall' was used by the Court of Requests, which had been established in the Isle for the recovery of small debts.[28] The Guildhall was rebuilt in 1827 and still stands in High Street, but is no longer used for any civic purpose.[29]

The main cause of the rapid growth of March in the past hundred years is its emergence as the chief railway centre of the Isle. The first line was that from Ely to Peterborough, opened in 1846, from which branches were laid to Wisbech and to St. Ives respectively in the two following years. Another important line (to Spalding) was opened in 1867 and provided through communication between the north and east of England.[30] The situation of the railway station at the north end of the town has quickened the consistently northward trend of the town plan. From its original nucleus round St. Wendreda's Church the town has spread past the cross and along High Street to the bridge. Thence it has fanned out northwards between the two roads to Wisbech. Land is not so valuable round March as farther north in the county, and the modern development is rather scattered and untidy as compared with that of Wisbech.

The sidings at Whitemoor on the Spalding line, constructed in the 1930's, are the largest in England and among the largest in Europe, and in its industrial structure the town is more akin to Peterborough or Wellingborough than the other towns of the Isle, whose industrial interests are closely tied to the land. In 1921 a remarkably high percentage of the population (22·3 per cent. of occupied males) was engaged on the railway. Proportionately, this was three times as many as in the railway town of Swindon. The startling fact that the Isle of Ely had then a greater proportion of railway workers than any other county except Cumberland and the Soke is mainly explained by this notable concentration in March. In 1931 the Isle had surpassed Cumberland also. The percentages then were: the Soke, 9·2; the Isle, 4·8; Cumberland, 3·6. The percentage for March was 24·5. But the Urban District contains some 30 square miles of farm land as well as the town itself, and even in March agricultural

9 Ibid.
10 Ibid.
11 R. Gardner, *Dir. Cambs.* (1851), 527.
12 *Rep. Royal Com. Mkts. and Fairs*, [C 5550–I], 222, H.C. (1888), liii.
13 Infm. Clerk to March U.D.C.
14 Gardner, *Dir. Cambs.* (1851), 531.
15 Kelly, *Dir. Cambs.* (1900, 1933).
16 G.P.O. H.Q. Min. 7712/1932.
17 Ibid. E 3710/1873; N.T.C. Dir.
18 B.M. Harl. MS. 594, ff. 198–200.
19 C. E. Walker, *Recs. of a Fen Parish* (1888, reprinted 1930), 17.
20 *E. Anglian*, N.S. vi, 40, quoting S.P. 16/445, no. 75.
21 C 78/1415, no. 3.
22 Bp.'s Mun., Ely, A 6.
23 D. and S. Lysons, *Magna Brit.* i (1), 177.
24 *Fenland N. & Q.* vii, 95.
25 H. C. Darby, *Draining of Fens* (1940), 277, quoting B.M. Lansd. MS. 722, ff. 29–38.
26 *Fenland N. & Q.* iv, 330; vii, 49–50.
27 C 78/1292, no. 13; C 78/1415, no. 3.
28 18 Geo. III c. 36; 9 & 10 Vic. c. 95.
29 Nos. 80 and 82, now an ironmonger's shop.
30 *V.C.H. Cambs.* ii, 133.

workers approach railwaymen in numbers and proportions.[31]

It was mainly due to the excellent railway communications that March became the county town of the Isle. The first meeting of the County Council took place, 24 January 1889, in the County Court House at March, 'this being considered neutral ground' as opposed to the older centres of Ely and Wisbech where alternately the Quarter Sessions were held.[32] At the next meeting, also held at March, a motion was put forward for meetings at Ely, March, and Wisbech in rotation, but an amendment to make March the permanent meeting-place was carried. It was pointed out that, although March was the most convenient centre for only 18 of the 52 members who expressed a preference, it was more convenient than Ely or Wisbech for most of those who preferred the other of these towns. Experience at Quarter Sessions had, moreover, shown that meetings at Ely were poorly attended from the north of the Isle and meetings at Wisbech attracted few from the south.[33] In its early years the County Council met in the Temperance Hall, erected in 1885 by the Peckover family of Wisbech;[34] the County Hall dates from 1908 and as a late comer among the public buildings of the town has been crowded out from the centre. The original cost was £4,764; it has been enlarged in 1927 and 1937 for considerably larger sums.[35]

ISLE OF ELY COUNTY COUNCIL. *Argent three bars wavy azure and a pile gules charged with three open crowns or.* (Granted 1931.)

An Inclosure Act for March was passed in 1792.[36] The award, dealing with about 2,760 acres, was not made until 1805.[37] There were 95 beneficiaries. Of these Sir Henry Peyton was the chief. He received 300 acres, 138 of them as lord of the manor.

In the early 19th century a 'pretty little theatre' was built in Bridge Street by a Mr. Smedley, but did not flourish, and in 1844 the building was converted into a British school and a Mechanics' Institute.[38] Another feature of Early Victorian March was the Exotic Gardens, laid out in 1836 by Mr. Fuller with 'choice flowers, evergreens, American plants etc. . . . giving them a very pleasing aspect'.[39]

In a field immediately south of Eastwood Avenue, adjoining the golf course, is a small star-shaped sconce, almost obliterated but showing up in air photographs, formerly known as the 'Battery Hills'. It stands on land which was once part of Cavalry Barn Farm, so named according to local tradition because 'Oliver Cromwell kept his horses there'.[40]

About 600 yards north of St. Wendreda's Church, on the west side of the road into the town, is the base of a wayside cross on three steps. The base is square with a sunk panel on each face, two with blank shields and two with roses.

There are no medieval secular buildings remaining in March, but several of later date have some architectural distinction. Some of the best are: (*a*) The Ship Inn, Nene Parade, dating from the 17th century, timber-framed with an overhanging upper story and thatched roof; (*b*) Elwyn House, in the market place, a Regency building with a good doorway; (*c*) No. 3 West End, 18th century, also with a good doorway and a roof partly stone-slated; (*d*) No. 38 West End, dated 1626, with elaborately carved beams in the parlour ceiling and an Adam-style fireplace in the dining-room. West End contains other old houses of less architectural importance, and there are some also in High Street and at Knight's End. The banks of the Nene, near the bridge, recall the Brinks at Wisbech on a small scale, and the Urban Council Offices, with a conspicuous and picturesque tower, are good for their date (1900). The clock in the tower is a Diamond Jubilee memorial.[41] During the 19th century a fashion for ornamental structures of cast iron developed in March and the neighbourhood. An early example is the Regency-style porch of No. 36 High Street, which may be compared with that at Wade's Hotel, Station Road (*c.* 1870). Another building in this local style is the 1911 Coronation Fountain in Broad Street.[42]

MANORS March was given to the monastery of Ely *c.* 1000 by Oswy and Leofleda when their son Aelfwin was admitted as a monk.[43] In 1086 the berewick of March contained 12 villeins, each with 12 acres. It was valued as part of Doddington (q.v.). The abbot of Bury St. Edmunds also held 16 acres, land for half a plough, 3 bordars and woodland for 4 pigs, stated to be worth 3*s.*, which had always belonged to the demesne of that abbey. The abbot of Ely had the soke.[44] The Bury estate is recorded in 1291 as worth 16*s.*,[45] but the greater part of March was throughout the Middle Ages included in the Bishop of Ely's manor of Doddington. Its importance *vis-à-vis* the present settlement soon began to increase. Already in 1221 one of the bishop's tenants was Alan the merchant (*mercator*), who held 'Stretreche', probably in March, for 500 eels and 3*s.* 3*d.* yearly. There were at the same time 11 customary tenants specifically described as holding in March.[46] By 1251 March had become a fair-sized village with 77 messuages;[47] there were 35 free and about 50 servile tenants, including 9 *novi feoffati* and 9 others enfeoffed since the time of Bishop Geoffrey.[48] The bishop's vaccary of Dereford or Dernford lay in March, forming a minor but continuous item of manorial economy in the early 14th century.[49] The name is preserved in the present Dartford Road.[50]

A manor of March is mentioned for the first time in 1328, when Geoffrey de Colevill sued Roger Huse for unlawful entry into the manor of *HATCHWOOD*. The case was removed from the king's to the bishop's court and must have resulted in[51] favour of Geoffrey,

[31] In 1921 28·6 per cent. of the total population were agricultural workers, in 1931 22·9 per cent.

[32] *Camb. Chron.* 25 Jan. 1889.

[33] Ibid. 22 Feb. 1889.

[34] Kelly, *Dir. Cambs.* (1900). The rent was £21 a year.

[35] Infm. Clerk, Isle of Ely C. C.

[36] 32 Geo. III c. cviii, amended by 50 Geo. III c. lxxviii.

[37] Award at county hall, March.

[38] Gardner, *Dir. Cambs.* (1851), 527.

[39] Ibid. 529.

[40] Infm. the late E. J. Rudsdale, Curator, Wisb. Mus.

[41] *March Official Guide*, 10.

[42] Other examples of this style are the Leverington Road Cemetery gates at Wisbech, and the churchyard gates (now removed) in the same town.

[43] Jas. Bentham, *Ely*, 86.

[44] *V.C.H. Cambs.* i, 367, 370.

[45] *Tax. Eccl.* (Rec. Com.), 270. There were three tenants of the fee of St. Edmund in 1251 (B.M. Cott. MS. Claud. C. XI, 68–74).

[46] B.M. Cott. MS. Tib. B. II, 98.

[47] In 1251 sixty-seven messuages were enumerated in 'Mercheford' (Bp.'s Mun., Ely, G 3).

[48] B.M. Cott. MS. Claud. C. XI, 68–72.

[49] S.C. 6/1132/9, /11 and /13.

[50] *P.N. Cambs.* (E.P.N.S.), 253.

[51] C.P. 40/275, m. 165d.

for in 1361 Sir John de Colvile, before his departure overseas, settled property in 'Marcheford' on trustees for the benefit of his son John, or of himself should he return. It consisted of a messuage, 10 acres of inclosed ground, 2 acres of meadow and 2*s.* 6*d.* from the rent of a cottage.[52] In 1407 Geoffrey Colvile and his wife had licence of private oratory in their house at Hachewood near March.[53] This manor of Hatchwood[54] was conveyed by John Colvile to Humphrey Gardner in 1586;[55] it then contained 40 messuages. Gardner and his wife Elizabeth passed it to William Hynd (1559), whose relict conveyed it to Sir John Peyton in 1606,[56] since when it has followed the descent of Doddington (q.v.). Robert Masters was tenant under the Peytons in the middle of the 17th century.[57]

The growth of population in the 17th century resulted, as at Wimblington, in the overstocking of the commons, and in 1669 Sir Algernon Peyton entered into an agreement with his 165 March tenants somewhat similar to that made at Wimblington (q.v.). About 4,440 acres were set aside as common and cow pasture in two portions: (*a*) Burrowmoor and Great and Little Binnimoor, and (*b*) Stow Fen, Great and Little Hurst, part of Witch Fen, Joan Sades Hole, Poutsherne, Low Fen, and Town End, Knight's End, Tyburn, Northwood, and Crawford Greens. From May Day to Michaelmas each tenant was stinted to 2 horses or 4 cows or 16 sheep, double that number being allowed between Michaelmas and Lady Day. The fines for overstinting or pasturing during the close season were similar to those levied at Wimblington, but in view of the greater number of tenants a steward or bailiff was appointed to collect the fines. He was allowed a salary not exceeding £10 a year, and was chosen by the fifteen trustees who regulated the scheme, every Easter Monday at the 'town hall' of March. A 'register' or town clerk was also appointed to set down the trustees' decisions in writing, and to copy them into the Doddington court rolls. He was to have £5 a year. The commoners were responsible for keeping the gates, bridges, and dikes in repair. The lord of the manor was entitled to nominate twenty poor persons to the privilege of grazing one or two cows or heifers, and the trustees empowered to grant temporary licences for one year to other poor persons. An area of 200 acres on Burrowmoor was set aside for the poor to dig turves. The commonable houses were defined as those existing in 1613, when Sir John Peyton altered the customary fines from arbitrary to certain; to discourage further building, future houses were to pay 10*s.* monthly as a fine. An unstated amount of inclosure had already been carried out, and proving 'very beneficiall and advantageous' for the provision of winter feed, a further 1,567 acres, in 175 parcels, were decreed for inclosure. An inclosure by Roger Jennyns, one of the chief tenants, of 600 acres in West Fen, was recognized. The Couledge, Sheppard, Southwell, and Walsham families are prominent among the names of the consenting parties.[58]

Manorial courts were held as recently as 1900, annually at the Griffin Hotel.[59]

The priory of West Dereham (Norf.) had property in March worth £1 6*s.* 8*d.* in 1291.[60] By 1535 the value, including a fishery,[61] had gone up to £5, when this estate ranked as a fee and involved suit at the bishop's court at Doddington twice a year.[62] The post-Dissolution grantee was John Huddilston.[63]

In 1364 Adam Orchard and William atte Lofte were licensed to alienate to Walden Abbey (Essex) 2 messuages, a toft, and 190 acres of land in March, Doddington, and Elm.[64] This property was worth only 14*s.* in 1535,[65] and with the other possessions of this monastery was granted three years later in fee to Sir Thomas Audeley.[66]

In 1249 Bishop Northwold granted 12 acres of marsh in March to the priory of St. Neots (Hunts.).[67] This small piece of land is not noticed in the *Taxatio*, but in 1535 its value, with a fishery, was £1 6*s.* 8*d.*[68] In 1560 the reversion in fee simple was granted to John Norden of London and Clement Robertes of Little Braxted (Essex),[69] Thomas Barrett and John Wrighte *alias* Goldwell being tenants on a twenty-one-year lease dating from 1548.

CHURCHES The Doddington Rectory Division Act (1856)[70] provided for the separation of the chapelry of March from Doddington and for its elevation to the same rectorial status as its parent. The sum of £1,300 was reserved as the stipend of the rector of March (St. Wendreda's), and provision was made for the formation in due course of three other rectories in the town. The Act did not come into force until the death (1868) of the Revd. Algernon Peyton, last rector of the undivided parish. The following year March St. Wendreda's became a rectory, and the rectories of St. John, St. Mary, and St. Peter were created in 1872, 1873, and 1881 respectively. The four advowsons were at first reserved to the Peyton family, with whom those of St. John and St. Mary have remained. That of St. Wendreda was before 1900 transferred to the trustees of the Revd. G. E. Walker (a former Rector of Doddington), and between 1915 and 1926 to the Martyrs' Memorial Trust, who now hold it. That of St. Peter was transferred between 1905 and 1915 to the Revd. D. A. Stewart; in 1940 Mrs. Stewart was patron.[71]

The Act of 1856 also provided that a chapel should be built in the parish of St. Mary, which was to comprise the expanse of fen west of the town. Such a chapel was built in 1891 and dedicated to St. Mary Magdalen.[72]

One of the March guilds, that of St. Wendreda,[73] dated from the 14th century.[74] Its hall was sold in

52 *Cal. Inq. p.m.* x, 655 (p. 551).
53 *Ely Dioc. Remembrancer*, no. 191, p. 67.
54 First mentioned in 1251, when Stephen of the Marsh (*de marisco*) held 80 acres in the marsh below Hachwood for 10*s.* (Bp.'s Mun. G 3).
55 C.P. 25(2)/93/850.
56 C.P. 25(2)/93/863; C.P. 25(2)/277.
57 C 78/1415, no. 3.
58 Ibid.
59 Kelly, *Dir. Cambs.* (1900).
60 *Tax. Eccl.* (Rec. Com.), 269.
61 This fishery, called Marcheferry, was granted in 1388 by John Charite for an annual payment of 4*d.* (*Ely Dioc. Remembrancer*, no. 135, p. 122).
62 B.M. Add. MS. 5837, f. 206.
63 *Cal. Pat.* 1554–5, 27.
64 Ibid. 1364–7, 31.
65 *Valor Eccl.* (Rec. Com.), vi, App. xii. There was also a fishery, not included in the 14*s.*
66 C.P. 25(2)/4/20; *L. & P. Hen. VIII*, xiii (1), g. 1115 (23).
67 B.M. Add. MS. 5847, f. 89.
68 *Valor Eccl.* (Rec. Com.), iv, 261.
69 *Cal. Pat.* 1558–60, 396–7.
70 19 & 20 Vic. c. 1, which amended an Act of 1847 (10 & 11 Vic. c. 3).
71 *Crockford's Clerical Dir.*, 1926, 1935, and 1940; *Clergy List*, 1905, 1915; Kelly, *Dir. Cambs.* (1900, 1933).
72 Kelly, *Dir. Cambs.* (1900). It is not now used.
73 Patroness of the church. She was a canonized Saxon lady, of whom little is known except that her body was enshrined at Ely and later at Canterbury (Bentham, *Ely*, 81, 89).
74 The 1389 certificate of this guild is printed in *Fenland N. & Q.* v, 206.

1571 to Richard Hill and Robert Don, the former receiving other properties of the guild including an inn called the 'Swan' in tenure of Henry Cornewallis.[75] There were also five smaller guilds of Holy Trinity, St. Anne, St. John Baptist, St. Mary, and St. Peter.[76]

The church of *ST. WENDREDA* consists of chancel, clerestoried nave, aisles, south porch, and west tower and spire. The material is flint and rubble with Barnack stone dressings. The chancel roof is tiled. The roofs of the nave, aisles, and porch are covered with lead. The earliest surviving work is the north arcade, which belongs to the middle of the 13th century. A rood loft; it has plain upper and lower doorways. The clerestory consists of nine windows on each side, of two cinquefoiled main lights with trefoil-headed openings above. The external face of the clerestory is panelled with flushwork in the East Anglian manner. On the east gable of the nave is a sanctus bell-cote with crocketed pinnacles.

Both the aisles were completely rebuilt in the first quarter of the 16th century and are of uniform character. The east and west windows and those in the north and south walls are of three lights cinquefoiled with trefoil-headed tracery lights. There is blind

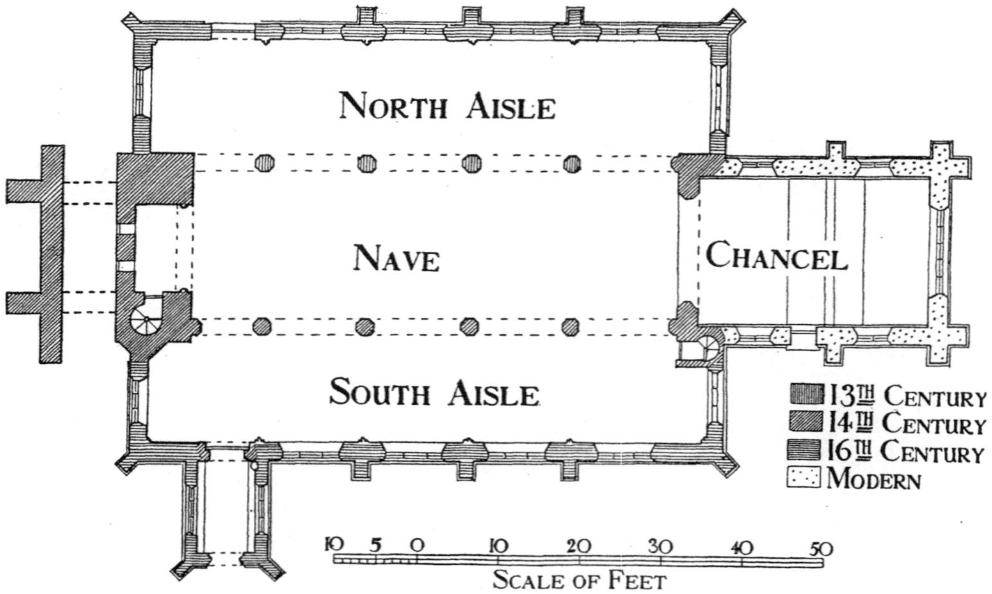

Plan of St. Wendreda's Church

general reconstruction took place, doubtless in pursuance of the indulgence granted by Clement VI in 1343 for the rebuilding of the chapel.[77] To this period belong the south arcade and the tower and spire. Early in the 16th century the church was completely transformed by the addition of a clerestory and the rebuilding of the aisles and south porch. New roofs were also provided at this time. In 1874 the chancel, which had been modernized in the 18th century with rounded windows and plaster ceiling, was completely rebuilt in 14th-century style.

The chancel, which is entirely modern, has an east window of five lights with flowing tracery. In the lateral walls are three windows of two lights of 14th-century character. The chancel arch is two-centred and of two orders with semi-octagonal responds having moulded caps and bases; it dates from the middle of the 14th century. The nave arcades are of five bays, the westernmost on each side being truncated. The north arcade is the sole survival of the 13th-century fabric and has plain two-centred arches of two orders with octagonal columns having bell-shaped moulded caps and moulded bases. The south arcade has two-centred arches of two orders with octagonal columns having moulded caps and bases, and dates from the middle of the 14th century. In the south-east angle of the nave is the newel stair which formerly led to the arcading on the internal face of the lateral walls, within which the windows are set, with demi-figures at the apex of the arches, a feature characteristic of East Anglian work of the period. Beneath the windows on the interior is a string-course. The buttresses are uniform, with one set-off, and there is a panelled base course with sunk quatrefoils which is continued round the porch. Both the nave and aisles have embattled parapets, which in the case of the nave and north aisle are furnished with crocketed pinnacles. The north aisle parapet has a band of sunk quatrefoils beneath the battlements. There is a good series of gargoyles. The north doorway has continuous mouldings, and the south doorway, which belongs to the 14th century, has continuous mouldings and a hood-mould terminating in heads. The porch has an outer doorway with four-centred arch of two orders beneath an ogee hood-mould; the outer order has continuous mouldings and the inner has engaged jamb shafts with moulded caps and bases. In the east and west walls is a window of three cinquefoiled lights under a depressed head. The parapet is embattled and there is an original gable cross and good gargoyles. There is a stoup recess in the north-east angle.

The tower and spire are of mid-14th-century date. The tower is of three stages. There is a passage through the ground stage from north to south, which has

[75] C 66/1074, m. 5; C 66/1083, m. 16.
[76] W. M. Palmer, *Village Gilds of Cambs.* 388; *Fenland N. & Q.* iv, 158–61. Lands of St. John's guild, comprising 3½ acres and a messuage, were granted in 1571 to William James and John Grey (C 66/1088, m. 31).
[77] B.M. Add. MS. 5804, f. 1.

ribbed vaulting with carved bosses. This passage occupies only the western half of the ground stage and is separated from the other half by a wall pierced by two narrow slits. The space above the passage serves as a ringing-chamber open on the east to the church. The west window is of three lights with flowing tracery and an external hood-mould terminating in heads. The lofty tower arch is two-centred and of two orders, the outer dying into the wall and the inner springing from shafts with moulded caps and bases. In the second stage is a window of two trefoil-headed lights with a quatrefoil above and a hood terminating in heads, on the north and south faces. The belfry windows are of two trefoil-headed lights with a sexfoil above. There are angle buttresses with four set-offs and an embattled parapet. There is a newel stair turret at the south-east angle, which is entered from the church by a plain doorway with continuous chamfer. The ribbed spire is octagonal with two tiers of openings, the lower of two lights and the upper of one.

The nave has a fine timber roof[77a] of double hammer-beam construction, which was erected early in the 16th century. There are demi-angels with outspread wings on the hammerbeams, collars, and wall-plate, and on the corbels which terminate the wall-posts. The wall-posts have niches with a series of figures holding musical instruments. All the principals, rafters, and purlins are moulded. The aisles have lean-to roofs with moulded principals and purlins and wall-posts resting on stone shafts, which spring from the floor and have moulded caps and bases. The porch has a plain cambered beam roof. The chancel roof is modern.

The 14th-century font has an octagonal bowl with crosses having splayed ends in low relief on the alternate sides; there are slight traces of colour. A portion of the 18th-century altar rails with turned balusters is now at the west end of the south aisle. There are two brasses: (1) William Dredeman and Joan his wife, 1501, with effigies of a civilian and lady, much worn; (2) Anthony Hansart and Catherine his wife, 1517, with effigies of man in armour with tabard and lady and one child, mutilated; above is a representation of the Annunciation and beneath a shield of arms.

The plate includes a silver chalice of 1752, the gift of James Collier, and a silver paten, 1703, the gift of Lora, wife of John Walsham.[78]

The tower contains six bells by Thomas Osborn of Downham, 1802.[79]

The registers begin in 1558 and are complete. Some loose sheets going back another eleven years are preserved.

The church of *ST. MARY THE VIRGIN*, erected in 1874, is a stone structure in 14th-century style consisting of apsidal chancel, nave, south porch, vestry, and west turret containing one bell. The registers begin in 1874 and are complete.

The church of *ST. JOHN*, erected in 1872, is a stone structure consisting of chancel, nave, aisles, south porch, and west turret containing one bell. The registers begin in 1872 and are complete.

The church of *ST. PETER*, erected in 1881, is a stone structure consisting of chancel, nave, aisles, north-west tower, and spire; the base of the tower serves as a porch. The tower contains one bell. The registers begin in 1881 and are complete.

NONCONFORMITY March was an early centre of Nonconformity, with 139 Dissenters and 1 Papist in 1676.[80] The proportion of over 13 per cent. was the highest in the Isle,[81] and surpassed only by Barrington, Litlington, Oakington, and Orwell in the county; the number was surpassed by Cambridge alone. The Baptists, as usual, were early in the field, and established in the town by about 1700.[82] The house of Stephen Coward, occupied by Josiah Black, was licensed for their worship in 1703.[83] Their pastor, Thomas Speechley, had 160 adherents in 1715.[84] A Baptist Sunday school was started in 1808.[85] In the middle of the 19th century there were three Baptist churches in the town.[86] The oldest dated from 1799, and was rebuilt in 1870 as the Centenary Baptist Church, in High Street. Providence Strict Baptist Church, Burrowmoor Road, was established in 1821 and rebuilt in 1835, and there was a Particular Baptist Church in the town dating from 1849. The two former churches are still in existence, and also the chapel at Chain Bridge (1859), which was an offshoot of the now closed West Fen chapel (1845). The Independent or Congregational Church in Station Road dates from 1836; this sect had previously worshipped in a granary near the present church.[87] The Wesleyan Methodists opened a church in High Street in 1829, which was rebuilt in 1889, and the Primitive Methodists one in Station Road (1848).[88] There is a third Methodist church at Floods Ferry, west of the town. The Salvation Army began work in March between 1900 and 1904,[89] and the Roman Catholic Church of Our Lady and St. Peter, in St. John's Road, was built in 1912.[90]

SCHOOLS A schoolmaster is reported in March in the Archdeacon's Visitation Return of 1596.[91] Besides the Grammar School,[92] there were in 1798 six Anglican and one Nonconformist schools at March, teaching about 230 children. None of the teachers was licensed, but the condition of the schools was said to be good.[93] In 1827, after certain of the March charities had been consolidated for educational purposes, the Guildhall in High Street was rebuilt and enlarged. The new building contained rooms for boys' and girls' schools conducted on National Society principles;[94] provision was made for 200 boys and 150 girls. In 1837 the master and mistress received £80 and £50 respectively, and there was an average attendance of 100 boys and 80 girls; this was in the winter months—in summer the numbers were lower.[95] No return was

77a See plate facing p. 186.

78 Infm. Mr. H. E. Morton, Churchwarden.

79 The inscription on the treble bell is as follows:

'In wedlock's bands all ye who join,
With hands your hearts unite.
So shall our tuneful tongues combine
To laud the nuptial rite.'

This inscription is also found on the 9th bell at Wisbech St. Peter, which was cast in the same foundry about twenty years later (J. J. Raven, *Church Bells of Cambs.* (1869), App.).

80 Bp.'s Mun., Ely, A 6.

81 The proportion in Chatteris is about 15 per cent. according to the figures given, but the recorded total number (271) is almost certainly too low.

82 *250th Anniversary of the Centenary Bapt. Ch., March, 1700–1950* (not paged); *Fenland N. & Q.* i, 158.

83 *250th Anniversary of . . . Bapt. Ch., March.*

84 Evans MSS., f. 9*b*, Dr. Williams's Libr.

85 *250th Anniversary of . . . Bapt. Ch., March.*

86 Gardner, *Dir. Cambs.* (1851), 527–8; H.O. 129/7/191.

87 *Cong. Year Bk.* 1949; Gardner, op. cit. 529; H.O. 129/7/191.

88 Gardner, op. cit. 528; H.O. 129/7/191.

89 Kelly, *Dir. Cambs.*

90 *Catholic Year Bk.* 1950.

91 *V.C.H. Cambs.* ii, 338 n.

92 See *V.C.H. Cambs.* ii, 333–4.

93 Bp.'s Mun., Ely, B 6.

94 Gardner, *Dir. Cambs.* (1851), 528.

95 *Rep. Com. Char.* (*Cambs.*), [103], 305–13, H.C. (1837–8), xxiv.

received from Doddington and March to the National Society's inquiry of 1846–7, but it was estimated that 400 children were being taught, weekdays or Sundays or both, in the undivided parish of Doddington, which at this date had about 9,000 inhabitants.[96] From 1844 the Society's efforts had been supplemented by the British and Foreign Schools Society, which opened a school in the old theatre in Bridge Street in that year.[97] In 1851 the Charity Commissioners introduced a revised scheme for the town charities, which allotted five-ninths of the total revenue to education.[98] A good deal of the money was swallowed up in building and maintaining a new Grammar School, and in 1872 'there being a great insufficiency of elementary school accommodation' the Trustees agreed to lease their primary schools to the School Board which had been set up the previous year[99]—the first in the Isle.[1]

The Board built a new girls' department, which was enlarged in 1902 to provide 190 places. In 1907, when the original buildings of this school (the South District Council School) were eighty years old, a new boys' and girls' school was built in Burrowmoor Road, with 300 places for each sex. The old girls' school was left to the infants, who still occupy it,[2] and the boys' and infants' departments converted into a cookery centre. Handicraft rooms were provided at Burrowmoor Road in 1913, and a new master's house the following year. Special practical instruction rooms were built in 1925, but in 1933, owing to great pressure on the available accommodation, they were used as ordinary classrooms. The position was eased the following year, when the Hereward School was opened. From 1922 to 1934 Burrowmoor Road had been the town's senior school, exchanging its younger children with the seniors from Dartford Road. In 1949 there were 211 boys and 218 girls at Burrowmoor Road.[3]

The North District Schools in Dartford Road were built by the March School Board in 1873–4 to take 510 children. They were enlarged in 1895 to provide 700 places, scaled down to 666 in 1910. After 1934, when all the senior children were transferred to the Hereward School, all the juniors, including those who had always attended Dartford Road, were transferred to Burrowmoor Road school which became the town's junior school, and Dartford Road was reserved for the infants from the northern part of the town. There were 388 on the roll in 1949.[4]

Another school was built by the Board in 1889,[5] in West Fen near St. Mary Magdalen's Church. It had accommodation for 100 children. The isolated position, some 3 miles out of the town, led to difficulty in finding staff, and in 1926 this school was closed. At this date there were only 26 children on the books. The buildings were valued at £550, but an auction sale failed to achieve the reserve price, and they were finally sold in 1930 for £100 only.[5]

The first step towards the reorganization of the March schools was taken in 1922, when the three principal schools at Burrowmoor Road, Dartford Road and by the old Guildhall were allocated to seniors, juniors, and infants respectively. This was soon found to be inadequate. Between 1925 and 1932 the number of school children in March increased 25 per cent., and at the latter date there were 1,838 on the books, but only 1,374 places for them. A site for a new senior school was purchased in 1931, in Robin Goodfellow's Lane, at the end of County Road. The plans were prepared during the depression, and, for economy's sake, Mr. R. D. Robson, the County Architect, introduced some interesting and unusual forms of construction, flat roofs among them. The school was completed in 1934, and provided 640 places for less than £20,000 including the equipment. At its opening it was named the Hereward School.[6]

CHARITIES[7] The oldest of the March charities was that of John Fringe, who in 1579 left 12 acres in Leverington and Wisbech St. Mary for the benefit of the poor of March. The deed of feoffment seen by the Charity Commissioners was dated 1689. This charity, with those of William Neale (1696) and Henry Wade (1713),[8] was consolidated in 1826. The trustees then had at their disposal 268½ acres, producing £471 12*s.* a year. This included 12 acres allotted in respect of the Guildhall under the March Inclosure Award of 1805. Most of the proceeds were devoted to education in accordance with Neale's bequest, but in 1837 about £50 a year was being spent in apprenticing, £10 each was given to two decayed non-pauper householders, and an unspecified amount devoted to buying heifers or young cows to be given to householders on Easter Monday at the discretion of the trustees, in accordance with Wade's will.

A number of independent charities also existed in 1837.

Thomas Coward bequeathed 6 acres in Oldfield, Elm, in 1674, to provide coats for 10 poor men and 3 chaldrons of coal. The 1837 revenue of £10 was insufficient for these purposes.

In 1672 Roger Jennings built 4 almshouses on High Dyke, near the centre of the town. These were unendowed.

Thomas Rumsey in 1798 left £1,000 3 per cent. Consolidated Bank Annuities to provide bread for poor persons not on the rates. There were seventy-five recipients of this charity in 1837.

Dr. Abraham Jobson, sometime curate of March, in 1822 gave stock bringing in £5 yearly, to buy bibles, prayer books, and other religious books for the poor.

Martin Pierson, at an unknown date, gave £100, which was used to buy 6 acres in Norwood, which brought in £10 yearly at the time of the Commissioners' report. This was given in twopenny loaves to the poor.

John Walsham gave a rent charge of £4 10*s.* on 17 acres in Cow Fen, Norwood, which was given in crown pieces to thirty-six persons in alternate years, the amount being too small to buy heifers as was the donor's intention.

Beacham Walsham in 1675 gave 3 acres for the benefit of the poor, which brought in £2 in 1837.

Thomas Walsham gave a rent charge of £2 on 30 acres of fen in Norwoodside, of which £1 was devoted to a sermon, and £1 to bread for the poor, on St. Thomas's Day.

96 Archives of Nat. Soc.
97 Gardner, *Dir. Cambs.* (1851), 527.
98 *V.C.H. Cambs.* ii, 333–4.
99 Archives of Nat. Soc.; M. of Ed. File no. 12/38; *Lond. Gaz.* 1871, p. 2222.
1 Kelly, *Dir. Cambs.* (1900). The date of establishment was 1 June 1871.
2 There were 174 children on the roll in 1949.
3 M. of Ed. File no. 12/38.
4 M. of Ed. File no. 12/37.
5 M. of Ed. File no. 12/39.
6 M. of Ed. File no. 12/79.
7 Except where otherwise stated this section is based upon *Rep. Com. Char.* (*Cambs.*), [103], 305–13, H.C. (1837–8), xxiv.
8 For these charities see *V.C.H. Cambs.* ii, 333.

James Shepherd in 1684 gave 3 acres in Norwood, which in 1837 brought in £5 5*s*. and was applied to the Sunday School.

The Revd. Philip Williams gave £50 to buy land, the rent of which was to be applied to buying books for the poor. In addition, Michael Wyldbore in 1737 gave £10 for bread for the poor, and the following at various dates between 1635 and 1761 gave sums totalling £50 for the general benefit of the poor: John Neale, Reynold and Sarah Walsham, Thomas Shepherd, Stephen and Elizabeth Shepheard, Leonard Cremor. These nine charities together brought in £5 a year in 1837.

There was also a 'town fishery', let at 10*s*. a year, and 200 acres at Pillard's Corner, Burrowmoor, to provide 300,000 turves annually for the poor at the discretion of the minister and churchwardens at the rate of not more than 3,000 per person. The value of these lands in 1837 was £60.

All these charities were consolidated in 1851, when almshouses were built near St. Wendreda's Church for four married couples and four single persons, and five-ninths of the total revenue devoted to educational purposes. The scheme was revised in 1888 and 1898, when the amount reserved for general charity was fixed at £220 a year.[9]

NORTH STANGROUND

This wedge-shaped tract of land represents the northern portion of the former parish of Stanground, which until 1905 was partly in the Isle and partly in Huntingdonshire. In that year it was divided into North Stanground (Isle) and South Stanground (Hunts.).[1] North Stanground has no separate history,[2] but although it contains less than 2 square miles and a population of only 32[3] it presents some interesting administrative features. Historically, it forms part of North Witchford hundred, but it is now included in Thorney Rural District—one of the small local government areas constituted in 1894 to regularize the position when a Poor Law Union extended into more than one county.[4] Ecclesiastically it still forms part of Stanground and of the archdeaconry of Huntingdon, and its connexions with the Soke of Peterborough have been close, for the Peterborough Corporation established a Smallpox Hospital[5] and Sewage Farm within its bounds. Its 1,117 acres are divided amongst no fewer than six 'principal landowners'. There is no made road through the parish, but it is crossed by the Peterborough–Ely section of British Railways, Eastern Region,[6] and by two important waterways—Moreton's Leam and its successor the modern River Nene.

There are no churches, chapels, schools, or charities in this parish.

WHITTLESEY

Whittlesey is a town in the north-west of the Isle, 6½ miles east of Peterborough and 11 miles west of March. Its 'island' rises to 26 ft. above sea level, and is about 3 miles long and ½ mile broad. It is therefore larger and slightly higher than that of Thorney. The 'island' is composed of a thin capping of gravel on a subsoil of Oxford Clay, which here reaches its farthest extension to the north-east. This fact has led to the development of important brickworks, whose forty chimney-stacks render Whittlesey conspicuous from a great distance.[1]

The town is situated at the intersection of the road from Peterborough to March with that from Ramsey to Thorney and Crowland.[2] The latter (B 1040) is planted with trees on the Thorney side of the town, a pleasant but unusual feature in this part of the country; from it fine views of Peterborough Cathedral may be obtained. The former is a first-class road (A 605) on the Peterborough side, but east of Whittlesey it is only a second-class road (B 1097) and becomes very crooked, crossing the railway three times in less than 2 miles. Branch roads diverge from it at Eastrea, a mile east of Whittlesey, to Benwick and Doddington (B 1093), and beyond Coates (2½ miles) to Ring's End and Wisbech (B 1072). The Ely–Peterborough section of the Eastern Region, British Railways, opened in 1846,[3] provides railway communication. From Three Horseshoes, 3 miles east of the town, a goods line (1897–8) runs south to Benwick.

The street plan of the town itself is complicated and irregular. An interesting feature is the 'back way' formed by Wallcroft Road on the west, Stonald and Bassenhally Roads on the north, and Cemetery Road and Inham's Lane on the east. This arrangement, representing perhaps the line of an earthen rampart and stockade behind the gardens and yards of houses fronting inwards to the main streets,[4] suggests that Whittlesey may have been settled very early. On the other hand, the names of certain streets, e.g. Scaldgate and Briggate, point to Danish influence and therefore to a late date for settlement. Some of the waterways traversing the parish are of considerable antiquity, such as Whittlesey Dyke, a continuation of Cnut's Dyke and King's Dyke, and Moreton's Leam. The modern course of the Nene forms the northern boundary, and the tides come up as far as Dog in a Doublet Bridge on the Thorney road. Here a 'floating bridge' was erected in 1787.[5]

Several travellers have recorded their impressions of Whittlesey. These have not always been compli-

[9] Kelly, *Dir. Cambs.* (1900).

[1] By L.G.B. Order No. 47,601.

[2] The manorial, &c., history of Stanground will be found in *V.C.H. Hunts.* iii, 212–17.

[3] Seventeen males and 15 females (including a staff of 2 at the Smallpox Hospital) in 4 farm-houses and 3 cottages.

[4] Until 1933 this Rural District was divided into two detached portions by an intrusive area of Whittlesey (q.v.). It may also be noted that North Stanground, although a 'rural' area, is closer to Peterborough City—whence its letters are delivered—than South Stanground which forms part of Old Fletton Urban District.

[5] This institution had 38 beds in 1936 (Kelly, *Dir. Northants*). There were no patients at the date of the 1931 Census.

[6] Opened 1846 (*V.C.H. Cambs.* ii, 132–3).

[1] For further treatment of the brick-making industry see below.

[2] The first meeting of the turnpike trust for the Ramsey section of this road was held at the Falcon Inn, Whittlesey, in 1793 (*Fenland N. & Q.* iv, 184).

[3] *V.C.H. Cambs.* ii, 132–3.

[4] This arrangement is more commonly found in the north of England, e.g. Heighington (co. Durham) and villages in the Eden valley under Cross Fell.

[5] *Fenland N. & Q.* vi, 226.

mentary. Cole, in 1745, thought St. Mary's spire the most beautiful he had ever seen, but stated that the townsmen were 'reckoned but a boorish and rough kind of people'.[6] At the end of the 18th century a local farmer, when noting in his diary that the Volunteers paraded for the first time on 12 May 1797, remarked that they were 'very few'. He added that 'they like drinking better than fighting at Whittlesea'.[7] About this time the town had the reputation of being a 'wild and dirty' place, though Jeremiah Jackson, on his visit in 1816, noted a 'rapid improvement from the substitution of gravel for mud in the streets'.[8] A generation later rioting occurred regularly in the town on Guy Fawkes' Night, and in 1834 140 special constables were enrolled for the period 4 to 7 November to put a stop to it.[9] By 1851, however, Whittlesey was stated to have 'a cleanly appearance'. It had recently 'been much improved by the erection of some good shops, the slabbing of the pavements, and the introduction of gas'.[10] As at Chatteris, a literary movement was apparent in the first half of the 19th century, with societies known as the Old Book Society and the Whittlesey Institution. The former, it was said, 'is more select, but the principles of the latter are more in keeping with the spirit of the times—that is, it partakes more of the nature of the Mechanics' Institutes'.[11] In 1900 there was a Science and Art School in Whittlesey.[12]

At the present day, with its two fine churches and numerous old stone-built houses, Whittlesey is by no means an unattractive town. The great brick-pits also, whether active with their smooth steep sides of blue clay, or worked out and forming miniature descendants of the great Mere, form an agreeable contrast to the trim fenlands.

Judging by the value of the manors, Whittlesey was a prosperous place in the Middle Ages, and it is rather surprising that it did not attain corporate status. The division of the town into two lordships, however, must have been a handicap to independence. The numerous guilds, moreover, were all very small, and, when they were dissolved, none was in a position to act as a Corporation, as was the case at Wisbech. The economic importance of the town continued in the 17th century. Several tradesmen's tokens of that period are recorded,[13] and in 1639–40 the town was assessed at £115 9*s.* 7*d.* —more highly in fact than any place in the Isle except Wisbech.[14]

In 1563 there were 355 householders in Whittlesey, 266 in St. Mary's parish, and 89 in St. Andrew's.[15] About a century later (1676) the town contained 2,021 persons of communicant age.[16] At both dates Whittlesey ranked second amongst the towns of the Isle, Ely coming first. The actual population in 1563 and 1676 may have been rather more than 1,500 and 3,000 persons. This would represent an increase of about 100 per cent. and was similar to that which apparently occurred in both Ely and Wisbech. The increase, however, was not maintained during the 18th century. In 1801 the actual population of Whittlesey was 3,841, showing an increase of only about 25 per cent. in 125 years as contrasted with one of about 130 per cent. in Wisbech.[17] The doubling of the Whittlesey population in the first half of the 19th century and its slight decline in the second half show a trend similar to that of the Isle as a whole. The subsequent recovery, however, began earlier and, owing to the brickmaking industry, has been more marked in Whittlesey than elsewhere in the Isle. The preliminary figure of the 1951 census was 8,609, showing an increase of 8 per cent. since 1931.[18]

The right to hold a market, and three annual fairs on 25–27 January, 11–13 June, and 25–27 October, was granted in 1715 to George Downes, steward to Richard Price and Nathaniel Webb, the lessees of the manors.[19] The market had been discontinued for about twenty years in 1808, but the June fair, limited to one day only, was still kept up. Horses were the chief item of trade at this fair.[20] Shortly before 1851 the market was revived 'and bids fair to become an excellent corn market',[21] but in 1868 it was said that only 'the tradition of a market lingers about the place on Friday afternoons'.[22] Friday is still market day in Whittlesey, but the town is too close to Peterborough for the market to be of much importance.[23] The horse fair on 13 June still exists.[24]

The National Provincial and Gurney's Banks had established branches in Whittlesey by 1851;[25] the latter is now absorbed by Barclays Bank.

A 'swanner' and deputy swanner of the manor are mentioned in 1595.[26] At the beginning of the 19th century there were four parcels of land, totalling 42 acres, known as Constable's, Churchwardens', Bellman's, and Herdman's Grasses, enjoyed by the respective parish officers.[27] The Constable's Grass existed *eo nomine* in 1625, when a lawsuit occurred over the illegal seizure of its hay crop.[28] This piece was also known as Bull's Grass, as two bulls for the use of the commoners were kept on it.[29]

There was some hostility in Whittlesey to the inclosure and drainage movement of the early 18th century. In 1703 George Goulding of Whittlesey and seventeen others were accused of unlawful assembly *axibus et securis*, breaking into the close of Francis Keate and overthrowing a windmill or 'dreyning engine'. The defendants pleaded not guilty. The Attorney-General cited the Liberty of the Bishop of Ely and prayed a *venire facias* for a jury of inhabitants

[6] B.M. Add. MS. 5810, f. 18*b*.

[7] *Fenland N. & Q.* vi, 226. The diary is anonymous.

[8] Diary, 16 July 1816 (Wisb. Mus.).

[9] Letter dated 4 Nov. 1834 from two J.P.s for the Isle to the Home Sec. (H.O. 52/24).

[10] R. Gardner, *Dir. Cambs.* (1851), 540. A Gas Company was established in 1834.

[11] Ibid. 539–40. The fact that the Whittlesey Institution had 70 members in 1851 'argues well for the literary taste of the town's-people, but little can be said for their fellow parishioners and neighbours in the fens, from whom these excellent institutions receive little support; truly the strides of the *march of intellect* are not very *rapid* here—the schoolmaster is still abroad *from* many of the fen districts' (italics as in original).

[12] Kelly, *Dir. Cambs.* (1900).

[13] *E. Anglian*, 1st ser. iii, 40; *Fenland N. & Q.* iv, 330.

[14] *E. Anglian*, new ser. vi, 40, quoting S.P. 16/445, no. 75.

[15] B.M. Harl. MS. 594, ff. 198–200.

[16] Bp.'s Mun., Ely, A 6.

[17] It seems probable that Whittlesey was overtaken by Wisbech as regards population about the middle of the 18th century; Cole, in 1745, mentions Whittlesey as being 'exceedingly populous', with 140 parliamentary voters (B.M. Add. MS. 5810, f. 18*b*), while the number at Wisbech in 1780 was 188 (Poll Bks. in Wisb. Mus.).

[18] The latest population figures are not strictly comparable with pre-1931 figures owing to the boundary changes of 1933, which reduced the 1931 population of the Urban District from 8,301 to 7,971.

[19] *Fenland N. & Q.* i, 290.

[20] Lysons, *Magna Brit.* ii (1), 278.

[21] Gardner, *Dir. Cambs.* (1851), 538.

[22] W. D. Sweeting, *Parish Churches around Peterb.*, 98.

[23] *Whittlesey Official Guide*, 27. Personal observation of present writer.

[24] *Whittlesey Official Guide*, 21.

[25] Gardner, *Dir. Cambs.* (1851), 538.

[26] E. 134/38–39 Eliz./M. 16, 17.

[27] *Rep. Com. Char.* (*Cambs.*), [103], 316 et seq., H.C. (1837–8), xxiv.

[28] Sta. Cha. 8/111/23.

[29] Gardner, *Dir. Cambs.* (1851), 541.

of Soham, as the nearest town to Whittlesey outside the liberty.[30] Goulding and his party were found guilty at the Cambridgeshire assizes.[31] Shortly after this (1707) it was stated that 750 acres in the Whittlesey parishes had already been inclosed, and there were 18,000 acres which might be so treated.[32] At the end of the 18th century Vancouver estimated that there were 1,550 acres of open arable field, 2,400 acres of pasture, and 20,000 acres of fen.[33] The final inclosure by Act of Parliament did not take place until 1840–1.[34] The award (1844) divided 4,051 acres amongst 332 proprietors.[35] The distribution of common rights was more even than in the other peat-fen parishes of the Isle, approaching nearly to that of marshland parishes like those near Wisbech, where minute subdivision of the land had taken place as far back as the later Middle Ages. Only six Whittlesey proprietors received more than 100 acres. These were Alice Hemmant (190 acres), the trustees of William Blades Fawn (180), the trustees of the Ladies Horatia Elizabeth and Ida Anna Waldegrave, in respect of their rights as ladies of the manors (130), James Read 'the farmer' (123), Alice Haynes (112½), and James and Elizabeth Ainger (102 acres). Thirteen other proprietors received between 50 and 100 acres each. Bassinghally, Church, Coates, Eastrea, Lattersey, and Stonald Fields are mentioned in the award, all of which are marked at the present day on the 1/25,000 Ordnance Survey map. It was stated that of the 4,051 acres partitioned, 892 were in respect of freehold and 3,159 of copyhold property.

As in most Fenland towns, a very large portion of the area within the urban boundaries is farmland, and the proportion of agricultural workers in Whittlesey (41·9 per cent. in 1921, 39 per cent. in 1931) is roughly the same of that of agriculturists in English rural districts as a whole (41·5 per cent. in 1931).[36] But for more than half a century[37] Whittlesey has been one of the chief centres of the brickmaking industry. In 1921 361, and in 1931 183 persons in Whittlesey were employed in the brick- and tile-making industry, forming 12·3 per cent. and 6·1 per cent. of the male working population. The decrease, caused mainly by increased mechanization, is not confined to Whittlesey and the Isle, though a contributory factor in this neighbourhood is that the local 'seams' are strictly limited in extent, and some of the best have been worked out.

Whittlesey was constituted a separate Poor Law Union under the 1834 Act.[38] By the Whittlesey Improvement Act, 1849,[39] an area of about half a square mile (the present East Central and West Central Wards of the Urban District) was placed under Commissioners and the provisions of the Towns Improvement Clauses Act, 1847,[40] applied to it. The Commissioners were replaced in 1894 by an Urban District Council of 18 members; at this date the remainder of the combined parishes of St. Andrew and St. Mary became a Rural District and civil parish under the name of Whittlesey Rural. In 1926, by the Isle of Ely (Whittlesey Urban District) Confirmation Order,[41] the boundary of the Urban District was extended to include the whole of Whittlesey Rural, making it in area one of the largest urban districts in England. In 1933 an area of nearly 4,000 acres along the northern border was transferred to Thorney.[42] This alteration, which left the North Side Ward of the Urban District as an area of nearly 2,500 acres with no population, linked up the Thorney Rural District, which had previously been in two detached pieces. Petty Sessions have been held at Whittlesey since 1797;[43] the Division also comprises the parishes of Thorney and North Stanground. From 1778 to 1846 Whittlesey was also one of the meeting-places of the Court of Requests for recovering small debts within the Isle.[44]

A postal service was first provided in Whittlesey in 1807.[45] By 1851 the post office was in the market place,[46] but the present office, one of the attractive 18th-century buildings which are a feature of the square, was not brought into use until 1913; it was extensively renovated in 1933.[47] Telegraph service was first provided in 1872,[48] and telephones in 1906.[49]

A cemetery was opened in 1850, and has since been enlarged to contain 8 acres. The older burial grounds attached to the churches and chapels were closed by Order in Council, 1873.[50]

The town was probably the birthplace of William Whittlesey, Archbishop of Canterbury 1368–74.[51] Another famous native was General Sir Harry Smith (1786–1860), who had a successful military career in the Napoleonic and Sikh wars, and was Governor of the Cape of Good Hope. After his victory at Aliwal he was given a great reception in his native town,[52] and his still existing birthplace in London Street is known as Aliwal House.

Whittlesey is situated just off the limestone belt of England, and has an extensive display of good domestic architecture. The most important buildings are the manor house and the Butter Cross.

The manor house, immediately to the south of St. Mary's Church, is of medieval origin, but was considerably enlarged at two periods in the 17th century,

[30] Soham is one of the few parishes in the county proper with much fen land. Juries with knowledge of fen conditions could have been found much nearer Whittlesey, e.g. from Farcet or Stanground, but these villages are in Huntingdonshire and not Cambridgeshire.

[31] K.B. 28/6 m. 20.

[32] B.M. Lansd. MS. 1039, f. 177.

[33] C. Vancouver, *Gen. View Agric. Cambs.* (1794). The acreage of Whittlesey before the alterations of 1933 was 26,199 acres.

[34] 3 & 4 Vic. c. vi, amended by 4 & 5 Vic. c. xiii.

[35] Award at County Hall, March. It consists of 483 parchment pages rather larger than foolscap. It will be noted that the amount inclosed agrees closely with Vancouver's estimate of 1794 as regards open arable and pasture.

[36] T. Sharp, *Anatomy of the Village* (1946), 39. This figure is the average of forty counties, derived by choosing the smallest rural district in each county with more than 30 per cent. engaged in agriculture.

[37] Kelly, *Dir. Cambs.* (1900), states: 'The brick-making industry here has considerably increased during the past few years, and the four yards now employ a large percentage of the labouring population of the town. The bricks are made by the semi-dry process; there are also some excellent beds of gravel.'

[38] Gardner, *Dir. Cambs.* (1851), 540. A rare example of a one-parish union, outside the large towns. It is probable that the peculiar geographical position of Whittlesey dictated this step; the town lies in North Witchford hundred, but was too isolated from the other places in that hundred to be included in the Union of that name, whereas to have included it in Peterborough Union, like Thorney, the other parish in the north-west corner of the Isle, would have made that Union too large, besides offending the inhabitants of a struggling market town.

[39] 12 & 13 Vic. c. xxxii.

[40] 10 & 11 Vic. c. 34.

[41] M.H. no. 70490.

[42] M.H. no. 76940.

[43] *Fenland N. & Q.* iv, 245, quoting note-books (1780–98) of John Lamb.

[44] 18 Geo. III c. 36; 9 & 10 Vic. c. 95.

[45] P.O. Minute Index 1807.

[46] Gardner, *Dir. Cambs.* (1851), 543.

[47] P.O. H.Q. Rec. E 33288/1913.

[48] Ibid. E 6022/1872.

[49] Nat. Telephone Co. Dir.

[50] *Whittlesey Official Guide*, 29.

[51] *D.N.B.*

[52] Ibid.; Gardner, *Dir. Cambs.* (1851), 540–1.

and at a subsequent date the interior was altered and refashioned. It is now L-shaped in plan and may have been so always, but the north wing is now of 17th-century date. The west wing is of 15th-century origin and retains several features of this period. Early in the 17th century this wing was extended on the east and new windows inserted in the original portion. Later in the same century, probably about 1680, the north wing was added or reconstructed. In the first half of the 19th century some new windows were inserted, and then or previously the interior was considerably altered. Though there is no visible evidence to show the full plan of the medieval house, it is almost certain that it was of greater extent than the existing west wing. The house throughout consists of two floors.

On the south front of the west wing there is a 15th-century two-light window with trefoiled heads under a square label in the upper story. There are contemporary angle buttresses with one set-off and a similar buttress towards the west end of the south front. On the north side of the west wing is a large medieval chimney-stack with a later brick top. In the north wall in the upper story is a late-15th-century window with square head and formerly of two lights, but now deprived of its mullion. The original coping remains on the east and west gables. Early in the 17th century, when this wing was extended towards the east, several windows were inserted in the old portion, mostly of four lights with stone mullions and square heads. In the west wall there is a two-storied oriel, the lower window of which has been cut through to form a doorway. The eastern extension of this wing has been much modernized, but it retains one four-light window at the east end in the upper story. At the junction of the north and west wings is a 17th-century stone chimney-stack. The north wing, which appears to date from the latter part of the 17th century, has some contemporary windows at the back with wooden mullions and transoms. There is also on this side a large external chimney-stack of brick. All the roofs are covered with stone slates. The interior has been much modernized and the original arrangements have been obscured. There are two staircases; that in the west wing is of medieval origin and runs up straight between two walls from north to south; it is lighted by the two-light 15th-century window at the top; the treads are modern. The other is in the north wing and is of late 17th-century date with turned balusters and moulded rail. There is an old cellar under part of the north wing.

'Portland House', in Station Road about 50 yards south-east of St. Mary's Church, was a fine stone building of two stories with attics and mullioned windows. The original structure dated from about 1600 and was H-shaped on plan; an addition on the south between the arms of the H dated from the later 17th century. Inside were several panelled rooms and one or two good fireplaces, contemporary with the house. The building was damaged by fire in 1949[53] and subsequently demolished, but the massive garden walls remain, with an outer entrance on the north consisting of a large pair of gate piers surmounted by stone vases, and an inner entrance facing east with gate piers surmounted by eagles. The site is now (1951) being developed as a residential estate of detached houses.

The Butter Cross is a good late 17th-century example of an open market house; it is square in plan, and the roof, which is covered with stone slates, is supported by one square and two round stone columns at each angle and a single round column in the middle of each side.

The Black Bull Inn in Market Street and No. 8 Market Place are good examples of stone-built houses of mid-17th-century date, with stone slates and chimney-stacks and mullioned windows of Northamptonshire character.

Besides the foregoing, the following buildings in Whittlesey can be ascribed to a date earlier than 1714. No. 3 Delph Lane, a one-story stone-built cottage with thatched roof; 'The Wheatsheaf', Eastgate; No. 6 Gracious Street, also stone-built; Nos. 59 and 61 Gracious Street, of a similar type to the Delph Lane cottage; the Falcon Inn, London Street; 'The Wilderness', London Street, a three-story house of red brick with stone dressings, with a good doorway with shell canopy; Nos. 6, 12, 13, and 15, a set of one-story thatched cottages, timber-framed with cob walling; the post office and the house and shop at the south-east corner of the market place; 'Tudor House', Market Street, a 17th-century stone building now converted into a shop; No. 23 Market Street, one story, dated 1710; the National Provincial Bank premises in Market Street, with a modillioned cornice and classical doorway; Grove House, Station Road, late 17th century with mullioned windows and a doorway with a shell canopy; Nos. 1 and 2 Turner's Lane; a stone-built barn at No. 15 West End; the Pack Horse Inn, Whitmore Street. Other later houses of some merit are No. 4 High Causeway, with a well-designed Georgian front, and No. 7 Horsegate, which has a Doric portico in stone and modillioned cornice. St. Andrew's Vicarage is a pleasant building for its date (1861). The Town Hall (1857) in Market Street is of unusual and not unattractive design. The Market Place as a whole forms a group of buildings of some architectural importance.

MANORS There were two important manors here, held in the Middle Ages by the Abbot of Thorney and the Prior of Ely severally and distinguished by the dedications of the two churches.

WHITTLESEY ST. MARY, the Thorney manor, was purchased by Bishop Aethelwold for £90 in pure silver, and given to the abbey. The vendors were three Saxons, Leofsig who had held one half, Leofwine who held a third, and Ufa who held a sixth, with part of Whittlesey Mere.[54] King Edgar confirmed the gift in 973.[55] This manor was in 1086 assessed at 4 hides with land for 6 ploughs, 2 hides with 2 ploughs being in demesne. There were 16 villeins each with 8 acres, 6 cottars, and 1 serf. The value, which had been 20*s.* when received, £7 in 1066, was £6 at the Survey. The 1086 value included 4*s.* from a weir and 20*s.* from fish. Though the manor was part of the Thorney Abbey demesne, the Abbot of Ely had the soke.[56]

This right was transferred to the Abbot of Thorney by Hervey, first Bishop of Ely, when he consecrated the abbey in 1128,[57] but disputes between the two monasteries as to their exact rights were frequent for a long time after. In 1251 the abbot was allowed view of frankpledge of his tenants in Whittlesey at Easter and Michaelmas, and assize of bread and ale, but in cases involving judgement of life and limb the bishop

53 Infm. Isle of Ely Fire Service.

54 For the history of this mere, which has always been in Huntingdonshire, see *V.C.H. Hunts.* iii, 185–7.

55 *Cart. Sax.*, ed. Birch, iii, no. 1297.

56 *V.C.H. Cambs.* i, 371.

57 *Cal. Chart. R.* 1300–26, 244.

was to preside through his bailiff. The weights and measures used in Whittlesey were to be those used in the Liberty of Ely.[58] Another serious dispute occurred in 1341, when an assize of novel disseisin was brought against the Abbot of Thorney and men of Whittlesey, who had forcibly seized 3,800 acres of marsh which was held to belong to Ramsey.[59] In the later Middle Ages the manor was frequently leased by the abbots, apparently in the first place to Sir John Holt, a justice of the common pleas, who was impeached in Parliament in 1383 and forfeited his lands, rents, fisheries, and other services in Whittlesey and elsewhere. The forfeited lands were given to his son John in 1391,[60] but restored to him in person by Henry IV ten years later.[61] Later lessees were Ralph Grey (for sixty years from 1527) and Thomas Lyname (for eighty years from 1538).[62]

This manor was very valuable to the abbey, bringing in £67 14*s.* 6*d.* in 1291[63] and £83 8*s.* 7½*d.* in 1539–40.[64] The latter sum included £4 2*s.* from perquisites of court and £2 6*s.* 8*d.* from mills.[65] Whittlesey and Yaxley (Hunts.) were by far the most valuable manors belonging to Thorney.

Lyname does not seem to have been disturbed in his tenancy after the Dissolution, and no outright grant of the St. Mary's manor is recorded until 1603 (see below). In 1589 a presentation was made to the church on behalf of Henry Coney, a minor.[66] In 1611 Coney with his wife Martha passed the manor of Whittlesey St. Mary, with property in both parishes, to Sir Humphrey Orme, who in turn with his wife Frances conveyed it to Thomas Glapthorn two years later.[67] Glapthorn acted as bailiff to Lady Elizabeth Hatton in the St. Andrew's manor, and got into trouble with Bishop Andrewes of Ely (1609–19), who as king's almoner claimed as deodand against the lord of the manor a cart loaded with 'hassocks' that had caused a fatal accident.[68] Glapthorn's son George and grandson Robert were engaged in lawsuits over the property as late as 1671,[69] by which date the manors had become merged. George Glapthorn was at this time the occupier of Stonald Close.[70] The Lysons brothers, writing in 1808, state that the boundaries of the two parishes could not then be ascertained, and that the manors had long been held together.[71]

The Ely manor of *WHITTLESEY ST. ANDREW* was given in *c.* 1000 to the monastery of Leofwine son of Adulf, in atonement for the murder of his mother.[72] In 1086 it was assessed at 2 hides, with land for 4½ ploughs, 1 hide with 1½ ploughland in demesne. There were 8 villeins, 4 cottars, and 3 serfs. The value was £3 when given to the monastery, £5 in 1066 and £4 in 1086. The last sum included 2*s.* from a weir.[73] The manor was confirmed to the prior and convent by Bishop Hervey.[74] The exact status of the prior as regards manorial perquisites, however, was not defined until 1417, when after a long and costly lawsuit, arbitrators awarded the convent all fines, amercements, forfeitures, goods and chattels of felons and fugitives, and the wastes and marshes belonging to certain manors, including Whittlesey.[75] Free warren had been granted to the prior and convent in 1252.[76]

This manor, which lay on the west side of Whittlesey, intruded between the Thorney home property in that vill and Whittlesey St. Mary and the large compact block in Yaxley, Farcet, and Stanground (Hunts.). Boundary disputes were therefore common. One between the men of Whittlesey and Stanground was settled in 1241. The Stanground men were allowed 70 perches for hay in Whittlesey marsh between Easter and Michaelmas, after which this piece was to be in common.[77] A dyke was to be raised at the cost of the tenants of Whittlesey to mark the boundary of their exclusive portion of the marsh, which in turn was to be in common between the two manors of Whittlesey. Later in the century the rights and boundaries of these two manors were more clearly defined.[78]

St. Andrew's like St. Mary's was a valuable manor, assessed at £51 13*s.* 3½*d.* in 1291.[79] In 1324–5 it was farmed at £57 (including the rectory); perquisites of court amounted to £2 17*s.*[80] In 1539–40 it produced £51 2*s.* 5*d.* gross, including 26*s.* 8*d.* perquisites of court, from which £2 16*s.* 8*d.* had to be deducted as bailiff's fees to Roger Wylson, then lessee of the manor and rectory.[81] Wylson was replaced in 1546 by Thomas Walles,[82] to whose family the manor continued to be leased by the Crown up to the end of the century.[83] In 1597, towards the end of a twenty-one-year tenancy of the manor, Garnet Walleys and Bridget his wife sold 240 acres in the marsh to Edmund English for £80.[84] The reversion was granted in 1588–9 to Richard Beamond, John Pasfield, and Miles Barker.[85] They did not, however, long retain their interest, for in 1598 Sir Thomas Cecil, 1st Earl of Exeter, was holding both manors in trust for his daughter Lady Elizabeth Hatton.[86]

From this time the manors are usually treated together, although separate dealings in the St. Mary's manor continued for about seventy years. The manors were formally entrusted to Exeter in 1603,[87] and formed part of Lady Elizabeth's dowry on her marriage to Sir Edward Coke. In 1625 the earl and his co-trustees Sir Edmund Wittipole and Thomas Bellot sued William Lincklove, William Caveney, and other of their tenants for the illegal seizure of hay from the town meadows.[88] Lady Elizabeth brought an action in 1639 against her stepson Sir Robert and others for breach of an agreement made before her marriage 'damnifying' her for £30,000 derived from Whittlesey and other manors.[89] By this time, however, she had parted with her interest, having with her husband conveyed the manors to Richard Lord Weston of Neyland,

58 *Cal. Pat.* 1348–50, 30; B.M. Add. MS. 5810, f. 12*b*.
59 *Cal. Pat.* 1340–3, 170.
60 Ibid. 1388–92, 285–6.
61 *Cal. Close*, 1399–1402, 352.
62 L.R. 14/892/7 and /24.
63 *Tax. Eccl.* (Rec. Com.), 268.
64 Dugd., *Mon.* ii, 612.
65 Two mills were leased in 1522 for ten years to Robert Brown (L.R. 14/892/25).
66 A. Gibbons, *Ely Episc. Rec.*, 446.
67 C.P. 25(2)/278 Mich. 9 & Hil. 10 Jas. I.
68 Sta. Cha. 8/2/24.
69 C 5/418/177, C 5/591/11, C 5/608/6, C 5/634/89.
70 C 78/1090/1.
71 *Magna Brit.* ii (1), 277–8.
72 Jas. Bentham, *Ely*, 82. His son Edelmer became a monk of Ely.
73 *V.C.H. Cambs.* i, 367.
74 B.M. Harl. Chart. 43 H 4.
75 Bentham, op. cit. 167; App. 30, 33.
76 *Cal. Chart. R.* 1226–57, 391.
77 This is probably the origin of the portion of Stanground in the Isle of Ely, made into a separate parish in 1905 under the title 'North Stanground'.
78 *Cal. Pat.* 1348–50, 28–30.
79 *Tax. Eccl.* (Rec. Com.), 270.
80 D. & C. Mun., Ely.
81 Dugd., *Mon.* i, 493.
82 *L. & P. Hen. VIII*, xxi (2), p. 433.
83 C 66/1025, m. 22, and 1177, m. 36; E 134/29 Eliz./E 3.
84 C.P. 25(2)/93/861.
85 C 66/1332, m. 9.
86 *Cal. S.P. Dom.* 1598–1601, 77.
87 C 66/1603, m. 1–5.
88 Sta. Cha. 8/111/23.
89 *Cal. S.P. Dom.* 1638–9, 226.

later Earl of Portland (1632).[90] He died seised of them in 1635.[91] His son Jerome was a co-adventurer with the Earl of Bedford, receiving 2,000 out of 5,500 acres of newly drained fen in Whittlesey, which was to be held of the king in chief by $\frac{1}{40}$th knight's fee.[92] Jerome conveyed part of this land, in North and South Fens, to his brother Nicholas to secure him an annuity of £300 a year. During the Civil War Nicholas was disturbed in possession.[93] The manors remained in the family for another two generations,[94] and were passed by fine and recovery in 1671 by Thomas, 4th Earl, to Jeremiah or Jerome White,[95] whose successor George White conveyed them in 1685 to Philip Waldegrave of Borley (Essex).[96] The latter devised his estates to his kinsman James, 1st Earl Waldegrave.[97] In 1707 the value of the three manors of Whittlesey (St. Andrew's, St. Mary's, and the rectory manor) was estimated at £1,900 a year. This sum, it was stated, might be increased by 'improvements' to £2,520, since only 750 acres had yet been inclosed and there were 18,000 acres of fen whose value might be doubled by being so treated. The manorial perquisites were worth £80.[98] The occasion of this survey seems to have been an action brought by Francis Walmsley, nephew of Thomas, 4th and last Earl of Portland, against Philip Waldegrave and various members of the Downes family[99] regarding the payment of an annuity of £400 out of the manors, which had been granted by the Earl of Portland to his niece in 1671. Walmesley was successful in his suit.[1] In 1712 Berrystead House, probably the manor house, with 17 acres of high land, was valued at £9 for taxation purposes. The average value of land in Whittlesey at this time was about 5*s.* an acre, but it varied from 10*s.* to as little as 6*d.* 'Acre lands' with cottages were valued at £1.[2]

The united manors were held by the Waldegraves for about a century and a half[3] but did not follow the direct male line. By a series of family settlements[4] the lordship became vested in the representatives of Elizabeth, relict of the 4th earl (d. 1816), and her sisters Charlotte Duchess of Grafton and Lady Hugh Seymour. In 1843 it was held in trust for the Ladies Horatia Elizabeth and Ida Anna Waldegrave, sisters of the 7th earl, from whom it was purchased shortly before 1851[5] by John Walbanke Childers of Cantley (Yorks.), sometime M.P. for Cambridgeshire. At this date courts leet and baron were held twice a year, at which fines certain were levied.[6] The six-monthly manorial courts were still being held in the present century.[7] Childers died in 1886, but his trustees continued to be principal landowners for nearly fifty years longer.[8]

The £400 annuity from the manors which was claimed by Francis Walmesley at the beginning of the 18th century was later diverted by the Earls Waldegrave to the Petre family, who were related to the Waldegraves by marriage.[9] This annuity continued to be paid up to the beginning of the 19th century at least.[10]

The *RECTORY* or *COQUINARY* manor is dealt with under the Churches.

Lands called Prior's Fen and 'les Willougrowse' were leased by Ely priory shortly before the Dissolution to Henry Parker and later to Giles Smartefote, who was confirmed in his tenancy (at £4 a year for twenty-one years) in 1547. The reversion was granted in 1554 to John Pratt of Great Weldon (Northants.).[11]

The abbey of Waltham Holy Cross (Essex) acquired, apparently after 1291, land and pasture in Whittlesey called West Fen. The last tenants were successively Robert Everard and Thomas Wren. The latter obtained the reversion in 1553.[12] Fifty years later this small estate was held in socage by Robert Bevill[13] of Chesterton (Hunts.), whose family is commemorated in Bevill's Leam, one of the outlets of the former Whittlesey Mere.

CHURCHES

St. Mary's, the church of the Thorney manor, was from an early date appropriated to the monastic kitchen.[14] Its value was £13 6*s.* 8*d.* in 1217 and 1254,[15] and £22 13*s.* 4*d.* in 1291.[16] By 1535 it had risen to £46 7*s.* 1*d.*[17] Of this sum, £26 13*s.* 4*d.* was accounted for by the farm of the rectory.[18] The vicarage seems to have been ordained between 1254 and 1291. At the latter date it was worth £6 13*s.* 4*d.*;[19] in 1535 £19 13*s.* 9*d.*[20]

The advowson of this church, which had been held by Thorney Abbey up to the Dissolution, continued to follow the descent of the abbey manor,[21] although in 1622 a presentation was made by Robert Cage, who was apparently only a sub-lessee of the manorial rights.[22]

The rectory of St. Mary's was granted in socage in 1550 to Thomas Mildmay, whose son Sir Walter received a sixty-year lease of the tithes of hay in Northey in 1562.[23] Probably at Sir Walter's death (1589) the grant of the rectory was revoked; from the 17th century the rectory has followed the same descent as the united manors.[24] It is occasionally referred to as a separate manor, as in a survey of 1707,[25] sometimes under the name *COQUINARY* on account of the original appro-

90 C.P. 25(2)/527 Trin. 8 Chas. I.
91 C 142/542/141.
92 *Cal. S.P. Dom.* 1639, 354.
93 Hist. MSS. Com., *6th Rep. App.* 18, 112.
94 C.P. 25(2)/401 Trin. 18 Chas. I; C.P. 25(2)/633 Hil. 19–20 Chas. II.
95 C.P. 25(2)/633 Trin. 23 Chas. II; C.P. 43/354, ro. 151.
96 C.P. 25(2)/806 Mich. 1 Jas. II. Waldegrave was a recusant (see Gibbons, *Ely Episc. Rec.*, 49). His family was related by marriage to that of the Earl of Portland.
97 P. Morant, *Essex*, ii, 318.
98 B.M. Lansd. MS. 1039, f. 177.
99 George Downes was steward of the manors.
1 C 78/1090/1.
2 *Fenland N. & Q.* iv, 185–7.
3 C.P. 43/923, ro. 239; C.P. 43/934, ro. 157.
4 C.P. 25(2)/1282 East. 33 & 34 Geo. III; Lysons, *Magna Brit.* ii (1), 278.
5 Under Act of Parliament, 6 & 7 Vic. c. 18.
6 Gardner, *Dir. Cambs.* (1851), 536.
7 *Whittlesey Official Guide*, 21.
8 They are still mentioned as such in Kelly, *Dir. Cambs.* 1933 ed.
9 *Compl. Peerage*, x, 506.
10 C.P. 43/876, ro. 284. See also C.P. 25(2)/1088 Trin. 5 Geo. I; A. Gibbons, *Ely Episc. Rec.* 50.
11 *Cal. Pat.* 1553–4, 401.
12 Ibid. 1553, 218.
13 C 142/281/61.
14 *V.C.H. Cambs.* ii, 215 n.
15 W. E. Lunt, *Valuation of Norw.* 217, 539.
16 *Tax. Eccl.* (Rec. Com.), 265.
17 *Valor Eccl.* (Rec. Com.), iii, 500.
18 Dugd., *Mon.* ii, 612.
19 *Tax. Eccl.* (Rec. Com.), 265. In the Norwich Valuation of 1254 the only vicarage separately distinguished is that of Wisbech, but it seems that some others in the Isle (e.g. Sutton, Witcham) were ordained by Bishop Northwold before he died in that year.
20 *Valor Eccl.* (Rec. Com.), iii, 500.
21 Ecton, *Thesaurus*, 103; Bacon, *Liber Regis*, 238. At the bishop's visitation in 1638–9 it was presented that Lady Hatton, lately lady of the manor, had made a door into the churchyard and blocked up the church with pews for herself and her servants (A. Gibbons, *Ely Episc. Rec.* 88).
22 P.R.O. Inst. Bks.; C.P. 25(2)/278 Hil. 12 Jas. I.
23 *Cal. Pat.* 1549–51, 242; 1560–3, 261.
24 C.P. 25(2)/400 Trin. 7 Chas. I; C 66/2600, no. 9; C.P. 25(2)/634 Mich. 30 Chas. II.
25 B.M. Lansd. MS. 1039, f. 177.

priation of the church.[26] It was never of much consequence; in 1844 only 60 acres were awarded in respect of copyholds of Coquinary manor, as against 1,970 for the copyholds of St. Mary's and 1,129 for those of St. Andrew's manor.[27]

In 1404 the dedication feast of St. Mary's was changed from the morrow of St. Bartholomew to 21 September, because on the former day the vicar and parishioners were busy 'circa defalcacionem que illo die secundum consuetudinem observatam solet fieri'.[28]

St. Andrew's church was given to Ely priory by Bishop Niel (1133–69) to provide for the precentor and the care of the library.[29] This church was worth £10 in 1217 and 1254,[30] and £13 6*s.* 8*d.* in 1291.[31] Of this last sum, £10 was accounted for by the rectory. A vicarage was ordained some time between 1254 and 1291,[32] which in 1535 was valued at £4 13*s.* 4*d.*[33] At this last date the rectory, let to farm, was included with the other spiritualities of the cathedral priory.

In the Middle Ages the advowson of St. Andrew's was held by the prior and convent of Ely. The conventual property in Whittlesey, probably because it lay separated from the liberty proper by the Thorney estate of Whittlesey St. Mary, was excepted from the formal transfer[34] to the Dean and Chapter of Ely in 1541, and the advowson of St. Andrew's, the Ely church in Whittlesey, has been retained by the Crown ever since.[35]

A messuage and lands called a Full lande in Whittlesey St. Andrew which had been given by a member of the Folyatt family for an obit in the church were in 1571 granted to William James of London and John Grey of Nettlestead (Suff.).[36]

The two churches stand less than a quarter of a mile apart, and from 1570 to 1815 were served by the same incumbent.[37] St. Mary's was reckoned the senior church, and it was customary for the Lord Chancellor, who exercised the patronage of St. Andrew's, to accept the nominee of the lord of the united manors, the patron of St. Mary's.[38] In 1712 the 'field tithes' and the parsonage house were worth £150, the 'fen tithes' and small tithes £100. The hamlet of Eastrea paid £9 in tithes in respect of St. Mary's and £4 10*s.* in respect of St. Andrew's.[39] The fact that the lordship of the manor and the impropriation was identical in the case of St. Mary's, but not in the case of St. Andrew's, led to certain anomalies. It was stated at the beginning of the 19th century that 'it is worthy of remark that the vicar of St. Andrew's, being entitled to the small tithes of that parish, can only obtain them by uniting with the lords of the manors, who have all the tithes of the other parish, in a joint claim; the vicar is then allowed a third part of the small tithes of the whole, as it has always been supposed that St. Andrew's parish is not more than half so extensive as the other'.[40]

The anomalies were removed by an Act of 1850, which confirmed the union of the two parishes for civil purposes but divided them, and defined their boundaries, for ecclesiastical purposes.[41] The great and small tithes of St. Mary's continued to be united to the manor; those of St. Andrew's were commuted for a rent charge of £456 10*s.* 5*d.*[42]

In the Middle Ages there were chapels in the hamlets of Eastrea and Eldernell, east of the town. Licence to found the former was granted in 1403.[43] In 1561 its site and lands, in the tenure of the churchwardens, were granted to Ralph Shelton of Deopham (Norf.) and Sir Edward Warner, Lieutenant of the Tower of London, in fee simple.[44] Its site is unknown, but in the middle of the 19th century 'traces . . . of its existence' might be seen 'in many a sculptured stone now built into the walls of barns and cottages'.[45] Eldernell chapel was licensed in 1525, and consecrated by Abbot Blythe of Thorney in his capacity of Bishop of Down and Connor. It contained an image of the Blessed Virgin which is said to have wrought a miracle; Robert Whyt of Whittlesey St. Mary, a bedridden man, was cured by invoking its aid.[46] A tenement called the 'Chappell house in Eldernayle' was granted in 1571 to William James of London and John Grey of Nettlestead (Suff.).[47] Anthony Cave subsequently leased it, with a croft, poleyard, fisheries, and holts, for 40*s.* yearly.[48] The ruins of the chapel survived into the 18th century, and the site is still known.[49]

The ecclesiastical district of Coates was formed out of Whittlesey St. Mary in 1850.[50] The living was endowed with £600 and declared a rectory in 1867.[51] The patronage was originally exercised alternately by the Lord Chancellor and J. W. Childers as patrons of St. Andrew's and St. Mary's respectively; the rights of the latter are now exercised by the Bishop of Ely.[52]

The church of *ST. MARY* consists of clerestoried chancel, north and south chapels, clerestoried nave, aisles, south porch, and west tower with spire. The material is mainly rubble and Barnack ashlar, and the roofs are covered with lead. The chancel arch and the three eastern bays of the north arcade of the nave date from the middle of the 13th century. About fifty years later the north aisle and chancel and possibly also the chancel arch were rebuilt. In the 14th century the nave was extended by one bay and the wide south chapel and aisle and the porch were added. In the 15th century there was much building activity, when the chancel was extended and the tower, spire, and clerestory added. Restoration was carried out in 1862 by Sir Gilbert Scott at a cost of £3,000. It involved the reconstruction of the east end of the chapels and the refacing of the porch. Further restoration took place in 1926–7 when the roofs of the chancel and north

26 *V.C.H. Cambs.* ii, 215 n.

27 Whittlesey Inclosure Award (1844), County Hall, March.

28 Gibbons, *Ely Episc. Rec.* 402 (Reg. Fordham, f. 198).

29 Jas. Bentham, *Ely*, 128. Hence the name 'Canter's Doles' given to fields south-west of the town.

30 Lunt, *Valuation of Norw.* 217, 539.

31 *Tax. Eccl.* (Rec. Com.), 265.

32 See note 19 above.

33 *Valor Eccl.* (Rec. Com.), iii, 500.

34 *L. & P. Hen. VIII*, xvi, g. 1226 (12).

35 P.R.O. Inst. Bks.; *Crockford's Clerical Dir.*

36 W. M. Palmer, *Village Gilds of Cambs.* 371; C 66/1088, m. 27.

37 W. D. Sweeting, *Parish Churches around Peterb.*, 102.

38 Ibid. 99.

39 *Fenland N. & Q.* iv, 185–7.

40 D. and S. Lysons, *Magna Brit.* ii (1), 277 n. The population figures of 1563 (see Topographical section) suggest that it was then much smaller than St. Mary's, but in 1844 the copyholds of St. Andrew's manor were a little over half those of St. Mary's (see above, note 27).

41 12 & 13 Vic. c. xxxii.

42 R. Gardner, *Dir. Cambs.* (1851), 539.

43 B.M. Add. MS. 5810, f. 11*b*.

44 *Cal. Pat.* 1560–3, 106.

45 Gardner, *Dir. Cambs.* (1851), 542.

46 B.M. Add. MS. 5810, ff. 9*b*, 10*b*; *Fenland N. & Q.* ii, 228. The date of the miracle is not stated in the printed broadsheet seen by Cole. Its author was Roger Dodynton, a Thorney monk. He was not a pensioner at the Dissolution, so must have been dead by 1539.

47 C 66/1088, m. 27.

48 B.M. Add. MS. 5810, f. 11*b*.

49 An undated print exists in Wisb. Mus. The architecture of the ruins as shown in this print suggests an earlier date than 1525 when the chapel was consecrated.

50 12 & 13 Vic. c. xxxii, S. 9.

51 *Lond. Gaz.* 1867, 3975, 7012.

52 Kelly, *Dir. Cambs.* (1900, 1933).

aisle were repaired and a pier in the north arcade reconstructed.[53]

The chancel has a 15th-century east window of five cinquefoiled lights under a depressed and panelled head with an internal hood-mould. There is a canopied niche above and on either side of the window of the exterior. There are panelled diagonal buttresses with three set-offs having a canopied niche in the second stage, and a moulded base course. The north and south windows of the sanctuary are of two cinquefoiled lights with quatrefoil above and a hood-mould. There are three clerestory windows on either side of three cinquefoiled lights under a depressed head, and an external hood-mould. The walls are finished with a plain coped parapet. The chancel opens to the south chapel by a 14th-century two-centred arch of two orders with semicircular responds having moulded caps and bases and a hood terminating in heads. The arch opening to the north chapel is also of 14th-century date with semicircular responds having moulded caps and bases and a hood on both sides, that on the south terminating in heads the eastern of which is modern. To the east of this arch is a doorway of the 15th century with four-centred arch, continuous mouldings, and a hood. The 13th-century chancel arch is two-centred and of two orders with moulded caps and bases to the responds and a hood on both sides. There is a 15th-century piscina with a cinquefoiled head, and farther to the west a 14th-century piscina and a double sedilia with cinquefoiled heads and ogee hood-moulds with finials and heads as stops. In the north wall is a plain oblong aumbry, which may have served for the Easter sepulchre.

The north chapel, which is structurally part of the aisle, has a three-light east window with plain intersecting tracery and angle buttresses with one set-off; most of the east wall and the buttresses have been rebuilt. On the south wall above the arch is a corbel table consisting of masks. The north window is of three uncusped lights under an almost rounded head. There is an aumbry in the north wall and a niche with a trefoiled head in the east wall.

The south chapel has two east windows of three trefoiled lights with reticulated tracery. There is an angle buttress with one set-off, which like the east wall has been rebuilt. The south wall has two windows of three trefoiled lights with a pair of quatrefoils above and a hood-mould; beneath the western of these windows is a doorway with depressed ogee head formerly crocketed, and with continuous mouldings: the plain door is probably contemporary. Between the windows is a buttress with one set-off. The chapel communicates with the aisle by a two-centred arch of two orders with semi-octagonal responds. There is a piscina with trefoiled head in the south wall. The eastern part of the floor is considerably raised to accommodate a contemporary crypt of two bays with quadripartite vaulting. The crypt is lighted by a trefoil-headed opening in the east and south walls.

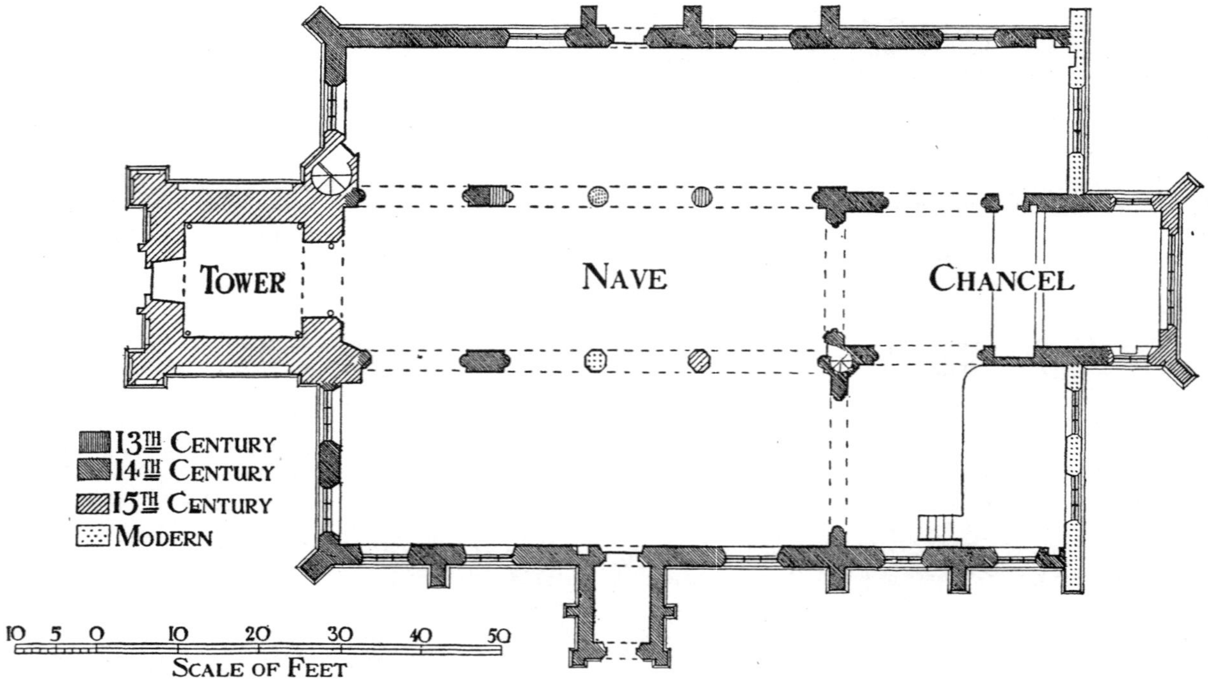

PLAN OF ST. MARY'S CHURCH

The nave has arcades of four bays. The three eastern bays on the north are of 13th-century date with two-centred arches of two orders, and round columns and semicircular responds having moulded caps and bases; the east respond and the second column have been rebuilt. There is a break between the third and fourth bays on both sides, which marks the 14th-century extension. The western arch on the north side and south is of two orders with semi-octagonal responds having moulded caps and bases. The three eastern bays of the south arcade were rebuilt in the 15th century with the exception of the 13th-century semicircular responds, which are similar to those on the north. The arches are of two orders and rest on octagonal columns with moulded caps and bases; the second column was rebuilt

[53] *Whittlesey Official Guide*, 23.

in 1927, and this date appears on it. There are four clerestory windows on either side, consisting of two trefoil-headed lights with a trefoil above. The newel stair to the former rood-loft is in the south-east angle of the nave; it has plain upper and lower doorways towards the nave and an upper doorway at a lower level in the aisle, an arrangement which indicates a difference in the height of the nave and aisle sections of the screen. The tower arch is two-centred and of two orders with a profusion of mouldings and a hood; the inner order springs from a rounded shaft with embattled cap and embattled band in the middle; the outer order has continuous mouldings which die into the responds half-way down, and below is a niche. The nave has plain coped parapets.

The north aisle has two lateral windows of two uncusped lights under an almost rounded head and a west window of three cinquefoiled lights with rectilinear tracery and a hood-mould. There are three lateral buttresses with one set-off and a diagonal gabled buttress at the north-west angle. The north doorway has a two-centred arch of two orders with continuous chamfers and a hood-mould terminating in masks. There is a string-course beneath the windows. The south aisle has a buttress with two set-offs at the junction with the lateral wall of the chapel and three others with one set-off. The three windows in the south wall are of three trefoiled lights with hood-mould, which in the case of the easternmost terminates in heads. There are two windows in the west wall, both of three trefoiled main lights with flowing tracery.

The 14th-century south porch is of two bays. The exterior has been completely refaced. There are clamped buttresses terminating in pinnacles and with a shallow niche on the face. The east and west walls have each a plain buttress with one set-off. The outer doorway has a two-centred arch with continuous mouldings and the inner a trefoiled head and continuous mouldings. The porch is covered with a plain quadripartite vault. There was originally a parvise, the blocked doorway to which, with a shouldered arch, remains in the aisle. The south gable of the porch is embattled.

The tower is of three stages and is a rich and beautiful example of 15th-century design. It shows distinctive Northamptonshire features. It is indeed tempting to suggest that the spire, with its tiers of lucarnes of 3, 2, and 1 lights, is a product of the Peterborough Abbey school of masons, who were almost certainly responsible for similar spires at Kettering and Oundle, or of a mason from the Thorney manor of Twywell near Kettering. The absence, however, of the finer nuances of design makes it more probable that Whittlesey spire was designed by a local man who may have seen the Northamptonshire work. Certain similarities with the spire at Yaxley (Hunts.), also a Thorney manor and only a few miles from Whittlesey, suggest a common origin.

The clamped buttresses of the tower are partially panelled, and run up into crocketed pinnacles, with brackets and canopies round the base. The lowest stage is panelled with an embattled cornice on the north, south, and west. The beautiful west doorway has an ogee crocketed head surmounted by a finial and continuous mouldings, and the arch, which is flanked by massive crocketed pinnacles, has a band of conventional flowers. The main west doorway window is of four lights with a large central mullion and good rectilinear tracery. There is a two-light window in the north and south walls of the ground stage with cinquefoiled heads and a quatrefoil above, and a crocketed hood-mould terminating in grotesques which support pinnacles. The second stage is panelled, with a two-light window pierced on the north and south. There are two belfry windows on each face with a double light above the transom and a single light below; the tracery of the north, east, and west windows is hidden by the clock-faces. The parapet is embattled and panelled. The crocketed spire has three sets of windows consisting of three lights, then two, and finally one light. Pierced flying buttresses spring from the angle pinnacles of the tower to the spire. The stair turret reaching to the second stage is of semi-octagonal form and is panelled and embattled with a conical cap; it is approached from the north aisle by a doorway with four-centred head and continuous mouldings. The ground stage has a good lierne vault with bosses carved with the Evangelistic symbols and a round bell-hole in the centre.

The chancel has a cambered beam roof with wall posts and brackets resting on embattled stone corbels; the principals and purlins are moulded and there are large carved bosses with traces of colour in the centre of the principals. The south aisle and chapel have modern roofs. The nave has a cambered beam roof with wall posts and braces resting on plain stone corbels; the purlins, principals, and intermediates are moulded, and there were formerly bosses at the intersections. The north aisle has a plain lean-to roof with wall posts and braces, those on the south resting on carved stone corbels; all the old roofs are probably of 15th-century date.

The font is modern and poor. Part of the base of a 15th-century screen is now at the west end of the nave; the base beam is missing and the southern portion of the wainscot is modern; the uprights and middle rail are well moulded and there are traces of a painted inscription on the latter; the northern portion of the wainscot has elaborate applied tracery with green and gold decoration. There is a plain oak chest probably of 17th-century date.

The altar was given in 1916 by Mrs. Saunders in memory of her son, Lieut. E. W. Saunders. There is a bust of Sir Harry Smith, by G. G. Adams; three of the stained-glass windows, and the reconversion of the east end of the south aisle from a school into a chapel, were carried out in his memory.[54] On the north wall of the chancel is a marble mural monument to Thomas Hake, 1590. Its two panels are now blank, but it is probable that they had figures. There are some good 18th-century headstones in the churchyard.

The plate includes two chalices, 1731, two patens and a flagon, 1765, and two alms-dishes.

The tower contains eight bells: 1st, 2nd, 5th, and 8th by Osborn and Dobson of Downham, 1803; 3rd, 4th, 6th, and 7th by Joseph Eayre of St. Neots, 1758. Some of the bells have interesting inscriptions. The 1st is inscribed 'The Lord to praise my voice I'll raise', the 3rd 'Peace and good neighbourhood', the 4th 'Do justice, love mercy, and walk humbly with thy God', the 6th 'The five old bells into six was run with additional metal near a tun', and the 7th 'Prosperity to the Establish'd Church of England and no encouragement to enthusiasm'.[55]

The registers begin in 1654, but are not complete before 1683.

[54] Ibid.

[55] J. J. Raven, *Church Bells of Cambs.* (1869), App.

The church of *ST. ANDREW* consists of chancel, north and south chapels, clerestoried nave, aisles, south porch, and west tower. The material is rubble and Barnack ashlar, and the roofs are covered with lead except those of the chapels, which are slated. The earliest surviving work is found in the west end of the south aisle, which dates from the middle of the 13th century. The chancel, chapels, and aisles belong to the second quarter of the 14th century, when the refashioning of the earlier church was begun. The nave arcades were rebuilt at the end of the 15th century and the tower and porch added about the same time. A vigorous 'restoration' in 1872 left the interior rather dull and depressing.

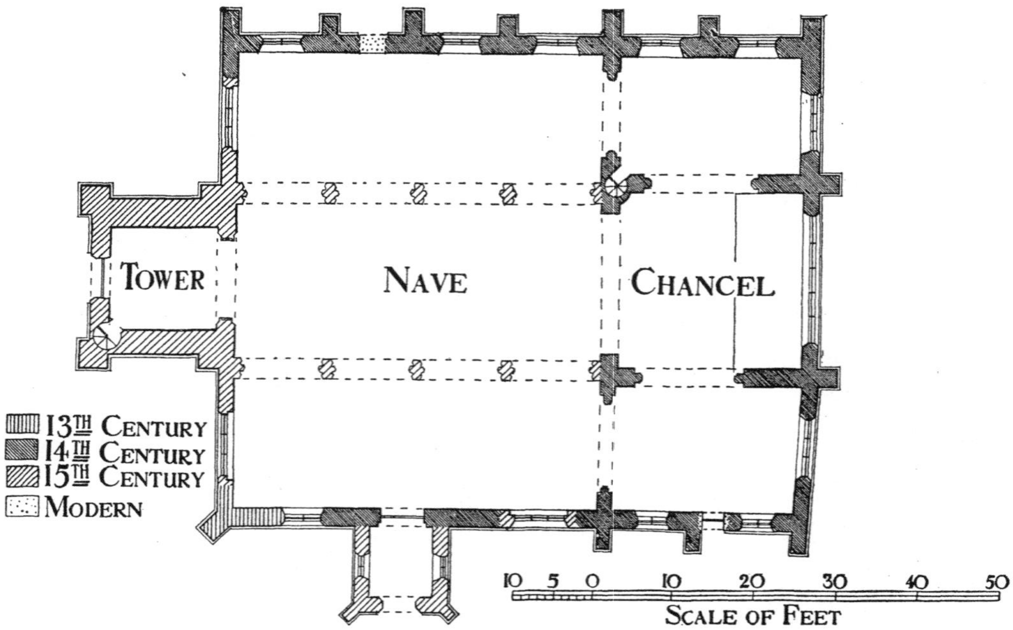

PLAN OF ST. ANDREW'S CHURCH

The chancel has an east window of five lights with flowing tracery of rather unusual design. The east gable is embattled. The chancel communicates with the chapels by a single arch on each side, of two orders with a hood-mould on the north and south faces, those on the south terminating in heads; there are semi-octagonal responds with embattled caps and moulded bases; the eastern respond on the north has been renewed. The lofty chancel arch is two-centred and of two orders, and springs from modern corbels. There is a modern piscina recess and sedilia. The south chapel has a three-light east window with reticulated tracery. There are two lateral windows of two trefoiled lights under a square head with a hood terminating in heads. Between the windows is a plain square-headed doorway, probably inserted early in the 19th century. There are two buttresses with one set-off. The chapel opens to the aisle by a two-centred arch of two orders springing from semi-octagonal responds with moulded caps and bases. The lateral parapet is embattled and there is an ancient foliated cross on the west gable. The north chapel, which like the chancel and south chapel dates from the second quarter of the 14th century, has an east window of three trefoiled lights and flowing tracery under a depressed head. The two lateral windows are similar to those in the south chapel. The arch to the aisle has semi-octagonal responds with embattled caps and moulded bases. There are two buttresses with two set-offs. The eastern gable is crocketed and the lateral parapet is embattled. In the south-west angle is the rood stair, with a plain rectangular doorway and above two openings to light the stair, one a quatrefoil and the other rectangular; the upper doorway is blocked. In the south wall there is a round-headed piscina recess and in the north wall an aumbry with pointed head.

The nave has arcades of four bays with two-centred arches of two orders, with a continuous hood, springing from clustered columns with moulded caps and bases, all of late 15th-century date. The clerestory consists of four two-light windows on each side, with cinquefoiled heads and a quatrefoil above. The tower arch, which is contemporary with the arcades, is panelled and without caps. The south aisle has varied fenestration and a plain coped parapet. The south-east window is of the 15th century, with three cinquefoiled lights under a depressed head and an external hood-mould terminating in heads. The south-west window dates from the middle of the 13th century and has two uncusped lights with a quatrefoil above and a hood terminating in masks. The west window, a 15th-century insertion, has three cinquefoiled lights under a depressed head and a hood-mould terminating in heads. There is a 14th-century buttress with one set-off at the junction of the aisle and chapel, and a 13th-century diagonal buttress with gabled top at the south-west angle.

The porch, a late-15th-century addition, has diagonal buttresses with one set-off and crowned with crocketed pinnacles. There is an embattled gable on the south with a crocketed pinnacle as a finial; the east and west walls are finished with plain coped parapets. The outer doorway is of two orders with a hood-mould; the outer order has continuous mouldings and the inner springs from shafts with embattled caps and moulded bases. There is a shallow niche on either side and above the arch, and a window of two trefoiled lights in the east and west walls. The inner doorway, which is of early 14th-century date, has continuous mouldings and a hood terminating in masks.

The north aisle has three lateral windows of two lights, similar to those in the chapels, and five buttresses with two set-offs, the second and fourth being smaller than the others. The west window is of three trefoiled lights with a trefoil above and a hood terminating in heads. The blocked 14th-century north doorway has a two-centred arch of two orders with mouldings dying into the chamfered jambs and a hood terminating in heads. The north parapet is embattled and with pinnacles, and on the west is a plain coped parapet.

The late-15th-century tower is of three stages, with an embattled parapet having crocketed pinnacles at the angles. It is of a type common in Northamptonshire and the adjacent districts, with clamped buttresses which are embattled at the top of the first stage. The newel stair is contrived in the south-west angle. The west doorway has a four-centred arch under a square label with carved spandrels displaying the Tudor rose and chamfered jambs. The west window is of three trefoiled lights with two quatrefoils above, beneath a four-centred arch with a hood-mould terminating in embattled caps. The second stage has a rectangular opening with a hood-mould on the north and south. The belfry windows are of two cinquefoiled lights with a quatrefoil and a transom.

The chancel has a plain cambered beam roof, probably of 15th-century date. The roof of the south chapel is of king-post type with wall posts and struts to the centre of the tiebeams; the principals, purlins, and intermediates are moulded. The north chapel has a plain tiebeam roof with wall posts resting on carved stone corbels. Both these roofs are ancient but of uncertain date; the former is probably of the 14th century. The nave has a cambered beam roof of late 15th-century date with wall posts resting on stone corbels carved with heads; the principals, purlins, and intermediates are moulded, and there are carved bosses at the intersections. The aisles have lean-to roofs; the south has been much renewed and the north, which has wall posts resting on stone corbels, is entirely modern. The porch has a cambered beam roof with wall posts resting on wooden corbels; the purlins and wall plates are moulded.

The font and other fittings are modern and poor, with the exception of an 18th-century chest, the oak pulpit (a memorial to Lieut. J. D. Smalley (1916)), and the altar erected in 1928 in memory of Mr. and Mrs. Ellis Staffurth.[56] The plate includes a silver chalice of 1569, beautifully chased but clumsily repaired and not now in regular use, a silver paten (possibly originally an alms-dish) of 1632, a bread stand (1701) of Britannia ware, a flagon, paten, and spoon of 1766, and another chalice of 1772.[57]

The tower contains six bells: 1st, 4th, and 5th by T. Mears of London, 1843; 2nd and 3rd by Joseph Eayre of St. Neots, 1759; 6th by Gillett and Johnston of Croydon (20th century, replacing an Eayre bell of 1760).

The registers begin in 1653. They are complete up to 1695, the entries for 1687–95 being copied from the Bishop's Transcripts. From 1695 to 1815, when the vicarages of the two churches were held in plurality, the St. Andrew's registers are combined with St. Mary's; since 1815 they have again been separate.[58]

The church of Holy Trinity, *COATES*, erected in 1840, is a brick structure in an effective adaptation of the Norman style. It consists of nave, aisles, and tower at the north-east corner with pyramidal roof. The building was designed by James Wild, a pupil of Basevi and decorative architect to the Great Exhibition. The plate includes a silver chalice, paten, and flagon inscribed 'Coates Chapel', and another small silver paten; there is one bell in the tower. The registers begin in 1850.

There are Anglican chapels at Angle Bridge (St. Stephen, 1877) and North Side (St. Guthlac), served by the clergy of St. Mary's and St. Andrew's respectively. The latter is situated in the portion of Whittlesey transferred to Thorney (q.v.) in 1933.

GUILDS In the Middle Ages there were many guilds in the town, but none of any great wealth or importance. In 1542 there were ten guilds in St. Mary's parish and two in St. Andrew's, rated to the lay subsidy at 30*s.* in two cases, 20*s.* in four, and 10*s.* in the remainder.[59] There was also a guild of St. Thomas the Martyr, a tenement called 'Thomas Beckett's guildhouse' in St. Mary's parish being sold by the Crown to William James and John Grey in 1572.[60] The hall of St. James's guild, in Arnold Street, came to the family of Thomas Dove, Bishop of Peterborough 1601–30, who was succeeded by his son Sir William (d. 1634) and grandson Thomas.[61]

NONCONFORMITY There were ninety-six Dissenters in Whittlesey in 1676. At the end of the 17th and beginning of the 18th centuries there was a regular Baptist congregation there, worshipping under J. Cooper and T. Shearman.[62] In 1715 Thomas Speechley, an 'Anabaptist' preacher, had a congregation of 160, including 3 forty-shilling freeholders, at Whittlesey and March.[63] There is no evidence that the Whittlesey Baptist congregation had a continuous existence throughout the 18th century, but early in the 19th century communities of both Particular and General Baptists are found. Zion Particular Baptist Church in Gracious Street dates from 1836 and replaces an earlier building. In 1851 it had an average congregation of 210, including 50 Sunday scholars.[64] Zion Church has a dependent chapel (Beulah) at the Turves, the congregation dating from 1862 and the chapel from 1866.[65] The General Baptist congregation, reorganized in 1821,[66] built a church in Windmill Street in 1830, which in 1851 had 180 worshippers.[67]

An Independent meeting began in Whittlesey in 1809. William Evenett, its first pastor, came to the town in 1811, and four years later a regular congregation was formed. In 1820 there was a Sunday school with 40 children and a weekday prayer meeting.[68] The Congregational Church (Ebenezer) in Broad Street was built in 1813 and enlarged in 1863.[69]

The Methodist churches were of later foundation than the foregoing and had, individually, smaller congregations in 1851.[70] These churches now (1951) are:

56 *Whittlesey Official Guide*, 25.
57 Infm. Revd. P. W. Dorman, Vicar.
58 Sweeting, *Parish Churches around Peterborough*, 108.
59 Palmer, *Village Gilds of Cambs.* 355, quoting E 179/82/190.
60 C 66/1083, m. 7. It was situated in Gracious Street.
61 C 142/478/44 and /476/98.
62 Adam Taylor, *Hist. Engl. Gen. Baptists* (1818), i, 356 n.
63 Evans MSS., f. 9*b*. Dr. Williams's Libr.
64 H.O. 129/7/192.
65 Inscription on building.
66 *Bapt. Handbk.*, 1950.
67 H.O. 129/7/192.
68 *Cong. Mag.* iii, 114 (1820).
69 Inscriptions on building.
70 H.O. 129/7/192.

Queen Street (Wesleyan, 1826), Coates (Wesleyan, 1831, enlarged 1866), Church Street (United, originally Primitive, 1841, rebuilt 1872), Eastrea (Wesleyan, 1845), and King's Dyke (United, *c.* 1898). There was formerly another Primitive Methodist church in Woolpack Lane, founded *c.* 1860. It was at one time used as a Masonic Hall.[71]

In about 1924 the Salvation Army opened a hall in premises now about a hundred years old in Church Street.[72]

SCHOOLS There is some evidence for the existence of schools in Whittlesey at the end of the 16th century,[73] but the first permanent primary school in the town dates from more than a hundred years later. Adam Kelfull, by his will dated 1735, gave 17 acres of pasture at Lipnea and 12 acres of fen at Wype, from the issues of which a sum of £15 was to be devoted to teaching 15 poor boys or girls to read, write, and say their catechism.[74] In 1798 the master of Kelfull's School, the Revd. G. Burgess, was teaching 15 boys in accordance with the above bequest, 12 more in consideration of £10 advanced by the parish authorities, and about 25 others who were presumably paying pupils. There were also several schools in the town 'occasionally kept'.[75] Kelfull's bequest was augmented by John Sudbury, who, by his will dated 1811, left a farm-house and 40 acres at the Turves, and 15 acres at Blackbush, to provide £40 to enable 10 of the 'more discreet' boys to learn reading, writing, and arithmetic at Kelfull's School or at a new one. When the Charity Commissioners reported (1837) the schoolmaster was drawing £15 15*s.* a year from Kelfull's charity and £20 from that of Sudbury; all the 15 places under the former charity were filled (by boys only), and 5 boys were being taught under Sudbury's bequest. The parish authorities had increased their contribution to the schoolmaster's salary to £12 for 12 boys. Recently, however, the adoption of the agricultural gang system in the neighbourhood had caused a drop in attendance, and shortly before 1837 the town contribution had been reduced to £6 6*s.* for teaching during the winter months only. The master was allowed to take paying pupils, and drew £17 15*s.* a year from this source. No house was provided.[76] The amounts allocated to the master's salary and the number of free place holders were the same in 1867 as they had been thirty years before, but by 1867 Kelfull's School had become merged in the boys' national school (see below). None of the boys was learning anything besides reading, writing, and arithmetic, but the schoolmaster was now certificated and had an assistant.[77]

By his will Sudbury had directed that the residue of the rents arising from his devise should be devoted to apprenticing. In 1837 £17 15*s.* per annum was being set aside out of the charity for this purpose, and by this date 52 boys had been so placed, with a premium of £12 each.[78] By 1867 the practice of apprenticing had lapsed.[79] Under a scheme made by the Charity Commissioners in 1868 it was, however, revived. The charity is now devoted to giving a commercial education to boys. It amounts to about £69.[80]

From 1816 there was a Sunday school in Whittlesey associated with the National Society. In 1830 it was attended by 86 boys and 66 girls.[81] In that year a new building was erected, providing accommodation for 160 boys and 100 girls, and whole-time instruction upon National Society principles began in Whittlesey.[82] This school, which was in the town, was supplemented in 1843 by another at Coates.[83] In 1846–7 it was stated that, out of a population of about 7,000, about 300 children attended day schools in the parish, and 300 more on Sundays only.[84] In 1849 a new girls' school, with places for 225, was built in Whitmore Street, and two years later the boys' school to accommodate 240 was rebuilt on a site between Station Road and Scaldgate. The site was given by J. W. Childers, lord of the manor. The National Society made grants of £60 and £35 respectively towards these schools.[85] About 1867 the average attendance was 115 boys and 97 girls during the winter, and in summer about half that number.[86]

Stimulated no doubt by the establishment of a School Board (1875),[87] the church authorities built (1876–7) another school at Angle Bridge in the south-east of the parish. The building served also as a mission room. The National Society made a grant of £40 towards this school,[88] which provided 130 places, scaled down after the 1910 reorganization to 80. This school became redundant after the opening (1910) of the Turves Council School (see below). At the end of 1931 there were only 39 children on the books, and it was then closed.[89]

The accommodation in the four church schools, which by the middle seventies amounted to 615 places, was not adequate, especially for the younger children,[90] and in 1877 the Board built two new schools. One, at North Side, is now in Thorney (q.v.); the other, in Broad Street, was capable of holding 420 children including 240 infants. These arrangements proved sufficient until the end of the 19th century, when the brickfields at King's Dyke began to be opened up and a fairly rapid increase of population took place. A. W. Itter, one of the brick manufacturers, built a school in 1904[91] on Peterborough Road near the level crossing, to take 60 children of his employees. This, however, was a private school, not subject to government inspection and with an uncertificated teacher, and by 1905 there was a total deficiency of 200 places in the town. The County Council planned an enlargement of the Broad Street school, into separate boys' and girls' departments with 536 places in all, and this was carried

71 The sources for this paragraph are the ecclesiastical returns of the 1851 Census (H.O. 129/7/192) and inscriptions on the buildings.

72 Kelly, *Dir. Cambs.* (1925).

73 *V.C.H. Cambs.* ii, 338 n., quoting Archdeacon's Visit. Bks. (Bp.'s Mun., Ely, B 2), which mention schoolmasters in St. Mary's parish in 1590 and in St. Andrew's parish in 1596.

74 *Rep. Com. Char.* (*Cambs.*), [103], 316 et seq., H.C. (1837–8), xxiv.

75 Bp.'s Mun., Ely, B 6. In the 1860's there were seven dame schools in Whittlesey. Night schools had been instituted, but 'have failed in this parish; the children are too tired to give the requisite attention' (evidence of the Revd. W. Waller, Vicar of St. Mary's, to the Commission on the Employment of Children, &c., in Agriculture, ([4068–I], 336, H.C. (1867–8), xvii).

76 *Rep. Com. Char.* (*Cambs.*), [103], 316 et seq., H.C. (1837–8), xxiv.

77 *Schs. Inq. Rep.* [3966–XII], 518, H.C. (1867–8), xxviii (10).

78 *Rep. Com. Char.* (*Cambs.*), 316 et seq.

79 *Schs. Inq. Rep.* [3966–XII], 518, H.C. (1867–8), xxviii (10).

80 *Whittlesey Official Guide*, 21.

81 Reports of Nat. Soc.

82 Archives of Nat. Soc.

83 Gardner, *Dir. Cambs.* (1851), 542. The school was on the site of a former burial ground of the Society of Friends.

84 Report of Nat. Soc. (privately printed).

85 Archives of Nat. Soc.

86 *Rep. Com. Empl. Children, &c., in Agric.*, [4068–I], 337, H.C. (1867–8), xvii.

87 *Lond. Gaz.* 1875, 5261.

88 Archives of Nat. Soc.

89 M. of Ed. File no. 12/55.

90 The population of Whittlesey declined from 1851 (7,687), but never fell below 6,334 (1891).

91 Inscription on building.

out in 1905–6 in spite of considerable local opposition. The position was also eased by: (*a*) the opening in 1910 of the Turves School, with 100 places, at Three Horseshoes Siding about half-way along the March Road, and (*b*) the adaptation of Itter's School at King's Dyke for 115 girls and infants. The Turves School, to which a teacher's house was added in 1914, belonged to the County Council from the first; that at King's Dyke was bought from Itter's executors in 1930 for £350.[92]

The Broad Street schools[93] were reunited as one department in 1927, and were again overcrowded in the late 1930's. Temporary accommodation for absorbing the overflow was provided in St. Andrew's Church Hall (60 places) in 1938 and in the Wesleyan Methodist Sunday school (50 places) in 1943. Some 36 boys who lived near the brickworks were transferred to King's Dyke in 1941, and a temporary extension has also been provided at the Turves School.

The Coates National School[94] was transferred to the School Board in 1877, when it was rebuilt at a cost of £1,135. After enlargement in 1900 it provided places for 120 mixed and 88 infants. It was again enlarged for an extra 42 in 1915, and completely overhauled in 1937–8, when £3,840 was spent on four new classrooms, a kitchen, dining-room, and staff room. Whilst the school was under reconstruction the children were taught in the old workhouse in Eastrea Road.

The National Schools in the town[95] were not transferred with the one at Coates, since the School Board and church authorities could not come to an agreement. Both schools were enlarged by an additional classroom in the late 1890's, and after further improvements had been made in 1911 the sixty-year-old buildings were recognized as capable of accommodating 173 children of each sex. In the early 1920's the numbers attending were lower than at the Broad Street Council Schools, although the recognized accommodation was about the same. Reorganization in Whittlesey was held up for some time by the fact that the town contained one provided and one non-provided school each for boys and girls, while the only infants' school was a provided one. In 1926 it was agreed to convert the two voluntary schools into senior schools, leaving the Broad Street schools for juniors and infants. Between 1928 and 1938 the number of school-children in Whittlesey rose from 715 to 838, and although theoretically 882 places were available, the two voluntary schools were very unsatisfactory. It was resolved therefore to build new senior and infants' schools. The site of the disused workhouse in Eastrea Road was obtained for the seniors, but the outbreak of war put a stop to the scheme.

In 1943–4 the Urban District Council, which had laid out a sewage system just before the war, threatened to close the two voluntary schools unless they were connected to the system. The managers were summoned for failure to abate a nuisance, and though the case was dismissed, great local feeling was aroused. It was pointed out that the Station Road and Whitmore Street schools, now nearly a century old, overcrowded, with poor lighting and ventilation and no proper washing or drying facilities, needed more than improved sanitation, and in 1945 they were taken over by the County Council under Section 38 of the 1921 Education Act. The following year the most urgent improvements were carried out at a cost of nearly £1,000. Some relief of the overcrowding was obtained in 1949 by the erection of a Horsa hut on the new site, and buildings for a senior school are now (1951) being prepared.

CHARITIES The charities of Whittlesey were very considerable in 1837.[96] The Town Lands, which at that date comprised 262 acres bringing in £309 9*s.* 6*d.* a year, were the accumulation of a number of bequests of money and land dating from the 17th century and earlier. They were managed by twelve trustees known as Governors, and were let annually by public auction in the market place. The proceeds were devoted mainly to paving, lighting, and policing the town, repairs to bridges, and other local purposes. There were also seven 'town houses' and nineteen 'Governors' houses' let rent free to the poor and the 'grasses' (see above) enjoyed by the parish officers.

A messuage and 16 acres of land were given by one Folliett at an unknown date to provide the Whittlesey contribution to the repair of Aldreth Causeway. The issues arising continued, at least nominally, to be used for this purpose until 1898, when the Charity Commissioners ordered that one-quarter of this charity should be devoted to church purposes and the remainder to education.[97]

Adam Kelfull, besides his educational charity (see Schools), left £150 to purchase land for the general benefit of the poor. The total revenue of his charities in 1837 was £120 a year, of which about £45 was given to the poor indiscriminately on New Year's Day and £35 on Easter Monday.

John Sudbury (see Schools) also devised 22 acres in Flegcroft and 10 acres in Glassmoor for the support of decayed tradesmen, farmers, and their widows, and other small pieces of land for general charitable purposes. His relatives disputed his will and claimed half his property, but in 1837 about £20 was given to decayed tradesmen in sums of 5*s.* to 10*s.*, and £9 to the poor indiscriminately in similar amounts.

The Commissioners of 1837 also recorded the existence of the charities of Anne Randal (1716), Richard Noble (1727), John Dow, and of persons called Bull and Hardley. Together these charities produced £45 12*s.* 6*d.* All were designed for the general benefit of the poor except Hardley's Charity of £5 5*s.*, which was allocated to poor widows of Coates hamlet.

The Commissioners spoke scathingly of the conduct of the Whittlesey charities. 'The various sums of money, from halfpence to silver, which are distributed in this parish at different periods of the year, are considered to be injurious to the poorer classes, inasmuch as a great portion of the money is frequently spent in beershops, and a day of distribution is generally a day of dissipation; moreover, the labouring poor frequently neglect their certain work to obtain a less sum of money from the charity. They demand money as their right, and abuse their superiors if it is withheld, or not paid to the amount of their expectations.' The Governors promised to mend their ways, especially in regard to Kelfull's non-educational charity which was distributed in sixpenny pieces, and resolved to give larger sums to the most deserving cases, with a preference to non-paupers.

The charities, with the exception of Sudbury's (see Schools), were consolidated under a scheme of 1868. Their annual value is about £80.[98]

92 M. of Ed. Files nos. 12/30 (Turves), 12/56A (King's Dyke).
93 M. of Ed. File no. 12/54.
94 M. of Ed. File no. 12/56.
95 M. of Ed. File no. 12/53.
96 *Rep. Com. Char. (Cambs.)* [103], 316 et seq., H.C. (1837–8), xxiv.
97 Kelly, *Dir. Cambs.* (1900).
98 *Whittlesey Official Guide*, 21.

SOUTH WITCHFORD HUNDRED

COVENEY WITH MANEA

Coveney is a secluded village lying near the centre of the administrative county of the Isle of Ely on a small 'island' rising to 43 ft. above sea level. It is only some 3½ miles west of Ely city as the crow flies, but nearly twice that distance by the main road. The village is situated on a by-road which leaves the main Ely–Chatteris road at Wentworth crossroads, about 2 miles south. This by-road, which has a branch to the hamlet of Wardy Hill, a mile west of Coveney village, used to be the only metalled road into the parish. But the droves across Ely West Fen, by Frogs Abbey, to Downham Hythe, and from Wardy Hill to Witcham have recently been made up for wheeled traffic. Coveney approaches the normal type of English village more nearly than most of those surveyed in this volume, since almost throughout its history the lordship of the manor was in the hands of intermittently resident laymen.

Manea, on another and lower fen 'island', had a history very similar to that of Coveney until the villages were effectually separated by the cutting of the Old and New Bedford Rivers in the 17th century. They are consequently now 16 miles apart by road, although the distance is only about 5 miles direct across the fen; in few parts of the county has drainage wrought such startling changes. Nor have the modern parishes of Coveney and Manea even a common boundary, for Welches Dam and Downham intervene. Since its separation from Coveney, Manea has become much the larger and more important village. Its road system is more convenient, linking it with Wimblington and March and with Welney and Christchurch, and more circuitously with Chatteris; its isolation is also lessened by the Ely–Peterborough line of British Railways (Eastern Region), which has a station about a mile north of the village. Manea lies roughly in the centre of the tract of fen land of which the drainage was taken over from the Earl of Bedford and his Adventurers by Charles I. The King is said to have planned a model town or village, to be known as Charlemont, near Manea.[1] The Civil War put a stop to this scheme, but the name remains as that of a small hill in the parish just south of the village. An area of 1,055 acres in Manea was inclosed in 1810[2] under an Act of Parliament of 1804.[3] Twenty-eight proprietors participated, besides the then lord of the manor, Lord Rokeby, and his two brothers. The chief beneficiaries were the trustees of John Clipson (78 acres), Thomas Martin (75½ acres), and the Revd. Thomas Crowe Munnings, in the right of his wife Margaret (63 acres). A further inclosure at Manea took place in 1830 under an Act of 1826,[4] when 115¾ acres were dealt with; William Whiting, with 30½ acres, was the chief of 30 allottees.[5] The trustees of the 'town estate' received 19½ acres under the first inclosure and the rector 10¾ acres under the second.

An attempt by the poor of Coveney in 1819 to take possession of the charity lands and divide them amongst themselves was frustrated by the magistrates arresting nine of the ringleaders.[6]

Coveney Mansion, a farm north of the church, is a timber framed thatched building of some architectural interest, dating from the 16th century and later. It is probably the successor of the manor house.

Conyers Middleton (1683–1750), Woodwardian Professor of Theology at Cambridge, was Rector of Coveney with Manea 1726–8 on the presentation of his wife Sarah, who had inherited a life interest in the manor and advowson from her former husband Robert Drake of Cambridge; her granddaughter Mrs. Elizabeth Montagu (1720–1800), authoress and leader of society, said to have been the first woman to be termed a 'bluestocking', spent her youth at Coveney.[7]

MANOR. Coveney is first mentioned in *c.* 1060, when it was assigned for life to Elswida, one of the daughters of Oswi and Leofleda, in return for her gift to Ely monastery of her life interest in Stetchworth; she retired here with her maidens to work at embroidery and weaving.[8] The vill is not mentioned in Domesday Book or the *Inquisitio Eliensis*, and it is quite possible that this small island 'in the bay'[9] between the Isle of Ely proper and the Doddington island was temporarily abandoned after Elswida's death as not worth occupying. In the troubles of Stephen's reign it was seized from the monks, but restored through the good offices of Bishop Niel (1133–69),[10] who granted it to Ralph his steward (*dapifer*) to be held of the prior and convent at 5*s.* yearly for all services. Prior Alexander (*c.* 1154–63) confirmed the grant.[11] The prior and convent remained overlords all through the Middle Ages. In 1297[12] and 1342[13] the manor was stated to be held of them in socage at 5*s.* yearly. The 5*s.* rent was still being paid in 1541, when the overlordship was transferred to the dean and chapter of the new foundation.[14] Owing no doubt to the long tenure of the Lisles and the Scropes (see below), the grant of 1541 gives the impression that Coveney was not regarded as an integral part of the Ely capitular estates in the same way as Sutton and other manors in the Isle proper. In 1563 the manor was stated to be held in socage of the queen;[15] this may not have been strictly correct, but it would seem that the dean and chapter took no steps to enforce payment of the 5*s.* rent, and no post-Reformation documents relating to Coveney survive in the capitular muniment room at Ely.

Ralph the steward already held Nedging (Suff.) of the prior and convent for 2 knight's fees. Nedging was

[1] B.M. Add. MS. 5810, f. 137*b*.
[2] 44 Geo. III, c. xlvi.
[3] Award at County Hall, March.
[4] 6 Geo. IV, c. lxv. Land in Doddington and Wimblington was also inclosed by this Act.
[5] Award at County Hall, March.
[6] *Fenland N. & Q.* ii, 9. No charity lands relating to Coveney were reported to the Commissioners in 1837.
[7] *D.N.B.*; *Compl. Peerage*, xi, 73.
[8] Kemble, *Cod. Dipl.* dccccxxxii.
[9] *P.N. Cambs.* (E.P.N.S.), 230.
[10] Wharton, *Anglia Sacra*, i, 619.
[11] E 210/7102; B.M. Add. MS. 5810, f. 135*b*.
[12] *Cal. Inq. p.m.* iii, p. 332.
[13] *Cal. Inq. p.m.* viii, p. 257. In 1356 Coveney was stated to be held by knight's service (ibid. x, p. 227). It is not mentioned, however, in the list of knight's fees granted by Robert de Lisle to the king in 1368 (*Cal. Close*, 1364–8, 494–7), so it is apparently correct to say that it was held in socage.
[14] *L. & P. Hen. VIII*, xvi, g. 1226 (12).
[15] C 142/150/188.

probably inadequate for the purpose, and Coveney may have been added to make up a holding capable of supporting 2 fees. Ralph can therefore be considered the ancestor of the Lisle family, which held 2 fees in 'Neding' in 1210–12,[16] and were the convent's tenants in Coveney for over 200 years. In 1275 Alice Lisle claimed £100 damages from Walter son of Roger and others for breaking into her manor house at Coveney.[17] In the following year the Bishop of Ely caused two of her men there to be arrested and imprisoned contrary to the king's mandate.[18] Warin de Lisle, Alice's grandson, died in 1296 holding Coveney and Manea in socage of the Prior of Ely at 5*s.* a year.[19] His son Robert, aged 6 on his father's death, was from 1311 summoned to Parliament as Lord Lisle (of Rougemont). Late in life he became a Franciscan.[20] In 1339 he bequeathed Coveney and his other manors to his daughters Alice (Seymour) and Elizabeth (Peverel) with remainder to his son John, who quitclaimed to his sisters but received the manor from them for life at a yearly rent of £20.[21] In 1344 John had licence of oratory for three years in his manors of Coveney and Manea.[22] His sons Robert and William were the last in the male line; in 1379 they quitclaimed Coveney to Richard Lord Scrope of Bolton (Yorks.),[23] who had licence (1393) to grant £10 from this manor to his chantry in Bolton Castle.[24] From 1403 his son Roger had licence of oratory during the bishop's pleasure.[25] Roger's grandson Henry made settlements of the manor between 1438 and 1446, dying seised thereof in 1459.[26] At his son John's death in 1498 Coveney and Manea were together reckoned to be worth £6, held of the Bishop of Ely by fealty and 7*s.* 11*d.* rent.[27] In 1541 the yearly rent was again 5*s.* and the manor said to 'belong' to Lord Scrope.[28] Nine years later a grant was made to Sir Edward North of an annuity of £100 from *inter alia* the manor of Coveney, as one of those in the Crown's hands owing to the minority of Henry Lord Scrope.[29] This Lord Scrope passed Coveney in 1562 to his kinsman Ralph Scrope of the Vine (Hants),[30] who the following year sold it to Symeon Steward, his son Robert, and their heirs. Robert (d. 1571) made a very complicated entail of his manors on about a dozen of his relatives,[31] of whom his brothers Mark, Thomas, and Nicholas were still alive at the death (1598) of their elder brother Edward, who had had the remainder next after Joan, Symeon's relict.[32] Each of the manors, including Coveney, seems to have been divided between the three brothers. Mark passed his third to Thomas Jermy and his wife Joan, Edward Steward's daughter and heir,[33] after a lawsuit regarding the entail and devolution of the manor.[34] It appears, however, that Coveney manor was afterwards reunited under Sir Simon, Mark's son, who was dealing with it in 1604.[35] Thomas Steward, Sir Simon's grandson and the last of the line, passed it to Thomas Allen in 1649.[36] By 1653 it was in the hands of John Childe;[37] another John Childe, probably his son, with his wife Elizabeth leased it to Joshua Gearing in 1677,[38] and the latter was dealing with it at the end of the century.[39] Another John Childe, probably the son of John and Elizabeth, was engaged in a lawsuit over the property with Sir Paul Whitchcote and others in 1703.[40] The subsequent descent is not very clear, but in the early 18th century Coveney was in the hands of Robert Drake of Cambridge, whose daughter and heir Elizabeth married Matthew Robinson of West Layton (Yorks.) (d. 1778);[41] their son, another Matthew, succeeded to the barony of Rokeby by special remainder in 1794. In 1810 his nephews Morris, 3rd Baron Rokeby, Matthew (later 4th Baron), and Charles Robinson shared the manor.[42] The 4th Baron Rokeby and his son were vouchees in a recovery of 1830,[43] and the lordship was stated to be in this family in 1851.[44] After the extinction of the barony in 1883 the estates were broken up, Coveney and its manorial rights being acquired by Walter Porter, who was lord in 1900.[45] Mr. Stanley Street-Porter was lord of the manor in 1933.[46]

MANEA is sometimes mentioned separately from Coveney,[47] but the descent has always followed the main manor. In the 16th century some of Lord Scrope's tenants here were paying their rent to the bishop as lord of Doddington, as is mentioned in Bishop Heton's alienation of that manor to the Crown in 1600.[48] A pamphlet issued by Thomas Neale, a former rector, in 1748,[49] gives some interesting information. He reckoned the parish to contain nearly 400 acres of high land and over 4,000 of fen and marsh.[50] In 1619 Sir William Ayloff and Anthony Thomas proposed to drain the neighbouring fens. They would thereby have received a third of those in Manea. Under the scheme, however, which Vermuyden actually executed for the Earl of Bedford, 700 acres in Manea, north of the Old Bedford River, were allotted to the Earl of Portland. This property in the early 18th century grew good crops of oats, coleseed, and wheat, and the store wethers raised there fetched £18 a score (ewes £15 a score). Rents at that time ran at 7*s.* an acre; in other parts of Manea fens they varied from 7*s.* 6*d.* to 4*s.* an acre. Drainage tolls of 2*s.* an acre were levied. Within the past generation, however, a great change for the worse had taken place. The average rental was reduced by a half, a third of the farms were in hand, tenants not

16 *Red Bk. of Exch.* (Rolls Ser.), 526.
17 K.B. 27/16.
18 *Rot. Hund.* (Rec. Com.), i, 149.
19 *Cal. Inq. p.m.* iii, no. 433, p. 332.
20 *Compl. Peerage*, viii, 72–73.
21 *Cal. Inq. p.m.* viii, no. 386, p. 257; *Cal. Close*, 1339–41, 332; 1341–3, 140.
22 *Ely Dioc. Remembrancer*, no. 58, p. 392.
23 *Cal. Close*, 1392–6, 123, 373; C.P. 40/476, m. 687.
24 *Cal. Pat.* 1391–6, 224.
25 *Ely Dioc. Remembrancer*, no. 185, p. 198.
26 C.P. 25 (1)/30/98; C 139/173.
27 *Cal. Inq. p.m. Hen. VII*, ii, no. 219, p. 140.
28 *L. & P. Hen. VIII*, xvi, g. 1226 (12).
29 *Cal. Pat.* 1549–51, m. 313.
30 C.P. 40/1199.
31 C 142/156/11.
32 C 142/252/62.
33 C.P. 25 (2)/93/864.
34 C 78/100/3.
35 C.P. 25 (2)/277/Trin. 2 Jas I. The appurtenances were then stated to include view of frankpledge. See also C 142/285/144, 467/89, and 474/74.
36 C.P. 25 (2)/538/Mich. 1649; C.P. 43/267, ro. 223.
37 C.P. 25 (2)/538/Trin. 1653.
38 C.P. 25 (2)/633/Mich. 29 Chas. II. See also C.P. 25 (2)/633/Trin. 27 Chas. II.
39 C.P. 25 (2)/815/East. 8 Wm. III and Trin. 13 Will. III.
40 C 5/231/13.
41 D. & S. Lysons, *Magna Brit.* ii (1), 172; *Compl. Peerage*, xi, 72. He was vouchee in a recovery of 1746 (C.P. 43/654, ro. 421).
42 They received respectively four, one, and one parts of the 54½ acres awarded in respect of the lordship of the manor in this year under the Inclosure Act of 1804.
43 C.P. 43/989, ro. 175.
44 Gardner, *Dir. Cambs.* (1851), 496.
45 Kelly, *Dir. Cambs.* (1900). The Rokeby estates in Cambridgeshire at the extinction of the barony in 1883 amounted to only 55 acres (*Compl. Peerage*, xi, 74).
46 Kelly, *Dir. Cambs.* (1933).
47 e.g. *Cal. Inq. p.m.* iii, no. 433, p. 332.
48 B.M. Add. MS. 5847, f. 89.
49 *The Ruinous State of the Parish of Manea in the Isle of Ely.* The references in the remainder of this paragraph are to this work, unless otherwise stated.
50 Actually 5,676 acres.

being prepared to take them even if they had only the drainage tolls to pay, and there was much emigration from the parish. The number of sheep kept was not much more than a tenth of what it had been, and the less fertile parts of the parish, which had formerly produced 15 or 16 coombs of oats an acre, now lay uncultivated.[51] The tithes, which had produced £100 a year from Manea alone, were now scarcely worth collecting when allowance had been made for the curate's stipend, dilapidations, and taxes. Neale attributed all this to several causes. First, pumping-mills, of which there were over a hundred in Whittlesey and Doddington alone, discharged their waters into Manea. Secondly, a tunnel had been dug in 1712 under the Forty Foot Bank and Drain 'to defend the lower parts of the level from the waters of the upper parts'. This, however, only resulted in the overflow from 10,000 acres pouring into the Twenty Foot Drain at Manea, which was too small to contain it. The third cause was the decay of Denver Sluice. Manea had thus become 'the sink and receptacle for the waters of a great part of the level'. Neale considered that fresh embanking and the erection of drainage mills was the only remedy, though he admitted that it would be difficult to get the many landowners to agree to concerted action. Haddenham, Ely, and Stretham, however, where this course of action had been successfully followed, furnished precedents.

A great breach in the north-west bank of the Old Bedford River between Manea and Welney occurred in November 1823.[52] In 1842 the drainage mills were replaced by a steam engine costing £8,000. In the same decade Manea was selected as the site of an Owenite experiment, 100 to 200 colonists cultivating 150 acres in common, and publishing a periodical called the *Working Bee*. 'But alas! for the mutability of human institutions!—the socialists have fled',[53] after a stay of only twelve months. The memory of the scheme, however, survives in the 'Colony' farm at Manea Fifties on the north-east edge of the parish by the Old Bedford River.

CHURCHES

The patronage of Coveney descended with the manor[54] until the break-up of the Rokeby estates in 1883, after which it was purchased by Athelstan Riley. He held it until his death (1945), latterly in conjunction with Lt.-Col. C. Riley, M.C., the present patron.[55] The church was valued at £4 in 1254,[56] and at £5 in 1291 and 1535.[57] In 1844 the tithes of Coveney were commuted for a rent-charge of £231 12*s.* 3*d.* and 31 acres of glebe and those of Manea for a rent charge of £850.[58]

In 1553, 2 acres of land in Coveney, held by Thomas Funden, which had been given for anniversaries in the church, were granted to John Butler and Thomas Chaworthe.[59]

The chapel of Manea is not mentioned in the *Valor Ecclesiasticus*.[60] The chapelry may date from 1646,[61] by which time the cutting of the Bedford Rivers had effectually separated Manea from its parent village and made separate spiritual provision necessary. Institutions of its curates are recorded from 1728[62] but it was still stated to be unconsecrated in 1763.[63] It did not become parochial until 1883 when a change occurred in the patronage of the mother church at Coveney (see above). The advowson of Manea was given to the Bishop of Ely.[64]

The church of *ST. PETER AD VINCULA, COVENEY*,[65] consists of chancel, north vestry, nave, south porch, and west tower. The material is rubble with stone dressings. The roofs are now tiled, but until 1896 the chancel and nave were thatched.[66] The nave and chancel belong to the first half of the 13th century. Early in the 14th century the chancel was extended, and the porch and the first two stages of the tower were added. In the 15th century another stage was added to the tower. The front of the porch was rebuilt in brick in the 18th century. In 1896 there was a thorough but careful restoration, when the front of the porch was reset and the vestry added. The interior was then, or subsequently, embellished with some handsome furniture of foreign origin.

The chancel has a three-light east window with flowing tracery having trefoil-headed main lights, and there is a contemporary hood-mould. There are diagonal buttresses with one set-off. The north-east and south-east windows have two cinquefoiled lights with a quatrefoil above and a hood-mould. To the west of these windows is a lancet of early 13th-century date. Finally, on the south is a large low-side window with two trefoiled lights and a hood-moulding of the 14th century, the tracery of which has been renewed though the mullion is original. There is no chancel arch. At the east end of the south wall is a double trefoil-headed piscina of the 14th century and next to it a lowered sill for the sedilia. West of the latter is another double piscina of 13th-century date. The chancel opens to the vestry by a plain modern doorway. The nave has varied fenestration. The south-east window has two trefoil-headed lights with a quatrefoil above and an external hood-moulding; it belongs to the 14th-century reconstruction. The next window on this side is a 15th-century insertion consisting of two cinquefoiled lights with a quatrefoil in the head. The westernmost window is a single light of 13th-century origin, but enlarged in the 15th century and provided with a cinquefoiled head. In the north wall are two unaltered 13th-century lancets. The plain north doorway is 13th-century work and has an arch of two orders with continuous chamfer; on the east side internally is a stoup recess. The south doorway of early 14th-century da[illegible] is recessed, and originally had jamb shafts, but only [illegible] moulded caps and bases remain. The plain double dc[illegible] is probably contemporary. There is a plain west door-

[51] Cole states that in his time (1747) all the land in Manea was pasture—'not an acre of arable' (B.M. Add. MS. 5847, f. 86).

[52] Diary of Revd. J. Jackson, 4 Nov. 1823 (Wisb. Mus.).

[53] Gardner, *Dir. Cambs.* (1851), 497.

[54] The bishop presented, presumably by lapse, in 1725, and in 1768 Thomas Topping had one turn (P.R.O. Inst. Bks.).

[55] *Crockford's Clerical Dir.* 1940.

[56] W. E. Lunt, *Valuation of Norw.* 216.

[57] *Tax. Eccl.* (Rec. Com.), 265. *Valor Eccl.* (Rec. Com.), iii, 499.

[58] Gardner, *Dir. Cambs.* (1851), 497–8.

[59] *Cal. Pat.* 1553, 290.

[60] The only chapels in the Isle to be mentioned are S. Mary by the Sea, Newton; Guyhirn and Murrow in Wisbech; Thetford in Stretham; and March in Doddington.

[61] A MS. copy of the marriage register from 1646 to 1837 is in the library of the Society of Genealogists.

[62] P.R.O. Inst. Bks.

[63] Ecton, *Thesaurus*, s.v. Manea.

[64] *Clergy List*, 1905, &c.

[65] The village fêast was in 1768 changed from 1 August (St. Peter ad Vincula) to the Tuesday before Midsummer Day, no doubt in order to avoid interference with the harvest (B.M. Add. MS. 5810, f. 137*b*, quoting advertisement in *Camb. Chron.* 14 May 1768).

[66] Coveney was the last church in the Isle to have this type of roof covering; Long Stanton St. Michael and Rampton in the county proper are still thatched (C. H. Evelyn-White, *Churches of Cambs.* (1912), xxix–xxx).

way with an arch of two orders and continuous chamfer. Above is a partially blocked lancet. Both these features are of early 13th-century date. There is an angle buttress with one set-off at the north-west corner. The porch has a plain brick outer doorway reset in 1896, but of 18th-century origin. In the east and west walls are plain rectangular openings of 14th-century date. The tower is built against the west end of the nave. It is of three stages, the two lower of the early 14th century and the third an addition of the 15th century. There is a passage through the base from north to south, a feature found in other churches of the district, e.g. March and Littleport. The north and south arches are two-centred and of two orders. In the second stage there is a recessed lancet on the south. The belfry windows have two cinquefoil-headed lights with a quatrefoil above. The tower is finished with an embattled parapet, which has been much renewed. There are diagonal buttresses on the west with three set-offs.

The nave and chancel have a continuous single-framed roof with collar beams and wind braces and three inserted tie-beams; the portion over the chancel has been much renewed and the section over the altar has modern panelling decorated in colour. The porch has a plastered ceiling. There are some old timbers in the tower.

The font has a plain octagonal bowl and is probably of the 14th century. There are three returned stalls in the chancel of 15th-century date but much renewed, with carved elbow-rests; the misericords are missing. The nave has much early 16th-century seating with quaintly carved poupée heads displaying the emblems of the Passion and other devices, including crossed keys. There is a good 16th-century carved reredos of German origin with the Crucifixion in the centre. The pulpit, dated 1706, is of Danish workmanship; on its panels are painted representations of Christ and the Evangelists with St. Peter and Moses; the base is modern, and there is an inscription in Danish. On the splays of the two lancets in the chancel are masonry patterns in red, and there is a portion of a consecration cross in a circle on the north wall of the nave; all these remains of mural paintings belong to the 13th century. There is a fine brass chandelier, probably of 18th-century date and Dutch origin.

The plate includes a communion cup of silver, 1570, inscribed 'FOR THE TOWN OF CHO-NE'.

The tower contains one bell by C. and G. Mears, 1847.

The registers begin in 1676 and are complete except for the marriage entries between 1812 and 1835.

The church of *ST. NICHOLAS, MANEA*, consists of chancel, north vestry, nave, north transept, and aisle. There is a bell-cote of timber covered with slates on the east end of the nave roof. The material is stone, faced with brick inside, and the roofs are covered with slates. The church was entirely rebuilt in 13th-century style in 1875. Its predecessor, which replaced an earlier structure in 1791, consisted of chancel and nave of brick with round-headed windows. The only visible relic of the old church is a 17th-century communion table of oak. In the churchyard are some good 18th-century headstones.

The plate consists of a chalice, 2 patens, 2 flagons, and an alms-dish.

The bell-cote contains one bell, unknown, probably modern.

The registers begin in 1646.

NONCONFORMITY Licence was granted in 1748 to Abraham Biggs for a Dissenters' meeting house at Coveney.[67] Its denomination is unknown, and it seems to have been abandoned by the end of the century. Methodism took an early hold in Manea, where the Primitive and Wesleyan chapels both date originally from 1814.[68] There is a Primitive Methodist chapel at Coveney village founded in 1845[69] and rebuilt towards the end of the 19th century, and a Wesleyan chapel, at Wardy Hill, founded 1837[70] and rebuilt 1903.[71] All these are still in existence. A Baptist congregation was founded in Coveney in 1833;[72] in 1851 it used a chapel which was the property of Waddelow Chambers, one of the principal landowners at that date;[73] it ceased to exist *c.* 1906.[74] Another congregation of this sect was started at Manea slightly later (1839).[75] It had died out by the late 1870's.

SCHOOLS At the end of the 18th century Mary Biggs, an Independent Baptist living at Wardy Hill, was teaching reading, writing, and arithmetic on undenominational lines.[76]

It was stated in 1831 that there were 85 children

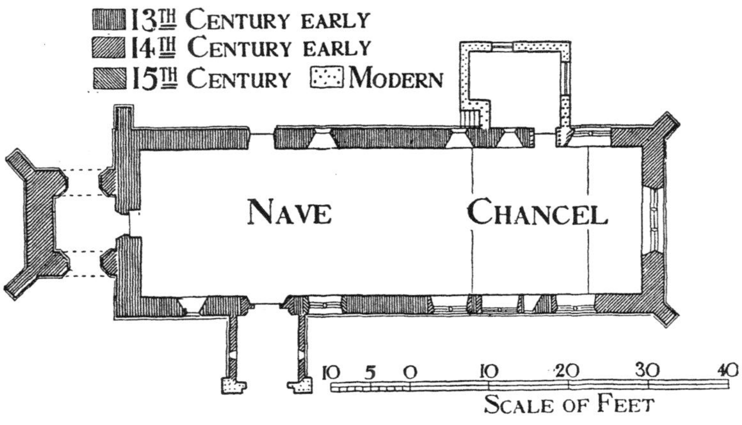

PLAN OF THE CHURCH OF ST. PETER AD VINCULA, COVENEY

67 Bp.'s Mun., Ely, A 6.
68 H.O. 129/7/190.
69 Ibid.
70 Ibid.
71 Inscription on building.
72 Ibid.
73 Gardner, *Dir. Cambs.* (1851), 497.
74 Kelly, *Dir. Cambs.* (1904) mentions it; the 1908 edition adds 'now disused'.
75 H.O. 129/7/190.
76 Bp.'s Mun., Ely, B 6.

in the parish requiring free education, besides some 30 born of parents living in Coveney but without settlement there. In that year a school was established in Coveney on a site given by Lord Rokeby. The cost of building and maintenance was defrayed by subscription, including £10 yearly from the rector. The school was in association with the National Society from an early date. The general inquiry made by this society in 1846–7 showed that there was a master with a salary of £25 a year with house. There were only 2 boys and 3 girls who attended on both weekdays and Sundays, but 28 more boys and 34 more girls attended on Sundays only.[77] The school was enlarged in 1872 to accommodate 112 children. The accommodation was reduced in 1910 to 90 (64 boys and girls, 26 infants). The numbers, however, have never reached even the lower figure; they were 75 in 1901, 60 in 1919, and 48 in 1938 and 1948. The children aged 13 and over were transferred in 1948–9 to the Cromwell School, Chatteris, pending the building of a secondary modern school for Ely Rural District.[78]

In 1753 Matthew Robinson, then lord of the manor, gave by copy of court roll, a commonable messuage and close, to the use of the poor of Manea. In 1796 a new trust was created, and the messuage converted into a schoolhouse at which 6 poor children were to be taught free. The rents of the lands were applied also to the apprenticing of the children and were allowed to accumulate. In 1804 an extra 22 acres were allotted under the Manea Inclosure Act. As a result of these augmentations and the increased value of land in the district, a very ambitious scheme was launched in 1816; this provided for a new school to take 200 children.[79] To finance this scheme the land was let at the high rent of £80. Even so, however, it was not possible to pay off the debt incurred, and in July 1834 the school (at which the average attendance had been only 30) was temporarily closed.[80] In 1851, by which time it had been reopened, the endowments amounted to £57 7*s.* 6*d.* a year and more than 50 children were being taught free.[81] In 1867 the endowments had increased to £71, of which £63 were applied to education and £8 to 'other benefit of scholars'.[82] No other information, however, was furnished about the school, and it may be inferred that it did not then flourish. In 1875 the trustees, stimulated no doubt by the 1870 Education Act and the fear that their activities might be superseded by a School Board, built a new school at a cost of £2,400. This provided 225 places. A School Board for Manea was established in 1876,[83] which took over the trustees' school. This had an average attendance of 195 in 1899,[84] and was enlarged in 1901–2 and in 1906 to accommodate 326; central heating was installed in 1930 for £318. The average attendance in 1938 was 165. The following year the school became one for junior mixed and infants, though its 'all ages' status was temporarily restored during the Second World War.[85]

CHARITIES No charities were reported to the Commissioners in 1837 in respect of Coveney. At Manea, in addition to the charity originated by Matthew Robinson and used for education (see above—Schools), there were six mud-and-stud houses, in tenantable repair, inhabited rent free by poor families. At an unknown date Thomas Neale placed a rent charge of 10*s.* on land in Manea for the poor of the parish. This had been paid within living memory, but the then owner of the land disclaimed any knowledge, though he was prepared to pay if the existence of the obligation could be legally proved.[86]

HADDENHAM

Haddenham is a large village in the south-west of the Isle, 7 miles south-west of Ely. It lies on the crest of a ridge rising to a height of 121 ft.[1] and in this countryside is therefore conspicuous from a distance. The hamlet of Aldreth lies 1½ mile south-west of the main village.

Haddenham is a road-centre of some importance. The north and east arms of the cross-roads round which it lies form part of a cross-country road from March and beyond. This road is often used by eastbound traffic to Newmarket and Bury in order to avoid awkward corners in Ely. The west arm of the cross forms one of the routes from Earith Bridge eastwards and is sometimes preferred to the main road through Sutton for the same reason. The railway to Ely was constructed in 1866, that to St. Ives in 1878.[2] The stations on this line, known as Haddenham and Earith Bridge, have, however, been closed since 1931.[3]

The village is spacious and dignified. The mansard roofs that cover several of its houses and the abundance of orchards and market gardens give it an appearance foreign to the Fens and more closely resembling that of a village in Huntingdonshire or upland Cambridgeshire. The best building in the village is Porch House, Hill Row, a fine brick structure of 17th-century date. The Limes, Station Road, is a building of 18th-century and later date perhaps with an earlier core. Other 18th-century buildings of note are: the Vine House; the pair of cottages 40 yds. west of the Limes, Hill Row; the pair of houses (originally one house) 120 yds. north-west of Porch House; the farm 40 yds. east of Highfield House; Linden House; and the Bell Inn. St. Ovin's, Highfield House, and the Limes, Hill Row, are early-19th-century buildings of note. The conspicuous windmill on the road between Haddenham and Aldreth was built in 1803.[4] In Cole's time Ovin's[5] stone, now in Ely Cathedral, stood on the east side of the road opposite the church. It was used as a mounting block.[6] Cole also mentions St. Audry's Well, ½ mile from the village on the Cambridge road, 'where the people of the neighbourhood do now resort to drink the waters of it, it being a sort of mineral water'.[7]

77 Archives of Nat. Soc.
78 M. of Ed. File no. 12/7.
79 The population of Manea at this date was only about 600.
80 *Rep. Com. Char.* (*Cambs.*), [103], 324, H.C. (1837–8), xxiv.
81 Gardner, *Dir. Cambs.* (1851), 498.
82 *Schs. Inq. Rep.* [3966–XII], p. 516, H.C. (1867–8), xxviii (10). Such information as was obtained came from the Charity Commission.
83 *Lond. Gaz.* 1876, 1853.
84 Kelly, *Dir. Cambs.* (1900).
85 M. of Ed. File no. 12/36.
86 *Rep. Com. Char.* (*Cambs.*), [103], 323, H.C. (1837–8), xxiv.

1 This is a considerable height for the Isle, the mean altitude of which is 6·34 ft. above sea level (R. W. L. Melbourne, *Isle of Ely*, Land of Britain, 71), 261.
2 *V.C.H. Cambs.* ii, 133.
3 Infm. Publ. Rel. Offr., Eastern Region, Brit. Rlys.
4 Inscr. on bldg.
5 St. Etheldreda's steward (d. 679).
6 B.M. Add. MS. 5803, f. 81*a*.
7 Ibid., f. 80*a*.

In the Middle Ages Haddenham and Aldreth commanded the principal land entrance to the Isle and derived a special importance from that fact. There was a crossing of the Ouse at Aldreth from early times. In 1169–70 a serjeant of the Bishop of Ely in Aldreth rendered 4*s*. 4*d*. to the Exchequer *sede vacante* for the farm of the fisheries and the passage of ships.[8] In 1172–3 the considerable sum of £6 10*s*. 9*d*. was spent on a 'brethasch' or causeway,[9] which suggests that at this time the original ferry to the mainland was being replaced by something more permanent. In 1279 a bridge is mentioned in addition to the causeway. Both had been out of repair for sixteen years, but had lately been restored by the bishop, who was responsible for their maintenance. His bailiff let the tolls for 20*s*. a year; horsemen paid ½*d*. each and foot passengers ¼*d*.[10] Though an indulgence was granted for the repair of the causeway by Bishop Fordham in 1406,[11] the duty of regular maintenance was in principle distributed among the customary tenants of the various episcopal manors. Even Outwell and Upwell, 20 miles away on the opposite side of the Liberty, were originally required to look after 44 perches of the causeway. In 1453, however, after attempts to disclaim their responsibility, their share was transferred to John Candeler, a Haddenham tenant who was in possession *inter alia* of part of the Queye property in Chewells manor (see below—Manors).[12] In 1548–9 the tolls of the bridge and causeway amounted to 17*s*. a year.[13] When Bishop Heton alienated Haddenham manor in 1600 he imposed the condition that he and his successors should be exempt from responsibility for Aldreth causeway. Thomas Lord Howard, on purchasing the manor in 1602, assumed the responsibility and was allowed a rebate of £300 for so doing.[14] Howard seems to have neglected his duties, for about 1613 the bridge collapsed and was replaced by a ferry, to the prejudice of the Bishop and Dean and Chapter of Ely and other landowners and their tenants. This ferry also was dangerous; by 1638 six or seven lives had been lost there.[15] By 1662 the bridge had perhaps been rebuilt.[16] It was certainly in existence again early in the 18th century, but in bad repair; William Cole, father of the antiquary, was in the habit, *c*. 1725, of dismounting from his horse before crossing.[17] Another partial collapse occurred in 1765, when the two middle piers fell down 'within a few minutes after two gentlemen had luckily got over' the bridge.[18] This disaster was probably caused by the succession of wet seasons and floods of the early 1760's. As early as 1676[19] the existence of an entry into the Isle through Stretham (q.v.) is noticed. This is the present approach to the Isle from the south. By the late 18th century it was brought into good repair, and with the alternative route formed by the road from Cottenham to Wilburton via Twenty Pence Bridge caused the main stream of traffic to be diverted from Aldreth causeway. The causeway, however, was still used fairly frequently until *c*. 1870, and the bridge was rebuilt in 1901–2.[20]

No doubt the causeway helped to bring prosperity to Haddenham which in 1562 was the most populous village in the county. Its householders then numbered 188.[21] In 1639–40 Haddenham was assessed for ship-money at £75 15*s*. Its rateable value at this time therefore was higher than that of the undivided parish of Doddington (£69 9*s*. 11*d*.) and not far below that of Ely (£85).[22] In Bishop Compton's census of 1676 Haddenham was credited with 700 persons of communicant age, which is a larger total than that assigned to any other village in the county.[23] Even in 1801, when Aldreth causeway was growing less frequented, the population of Haddenham was still larger than that of any other village in the Isle except Littleport and Thorney—both parishes with a greater acreage. The population increased with normal rapidity until 1851. It has subsequently declined greatly, so that at least six villages in the Isle now have a larger population. Moreover in density per acre Haddenham is now below the average for the Isle.

In 1612 Thomas Howard, Earl of Suffolk, as lord of the manor of Haddenham, was granted the right to hold a market in Aldreth on Thursdays and fairs there on the Tuesday after Trinity and at Michaelmas.[24] In 1858 Haddenham had 'a few shops, but on a small scale'.[25] A Haddenham Gas Company was formed between 1865 and 1869, and in 1879 the principal streets were lit with gas. The Company had ceased to exist by 1900.[26]

Haddenham was inclosed in 1847 under an Act of Parliament of 1843.[27] This inclosure dealt with 2,638 acres, including 35 acres of old inclosures exchanged, but not the roads and drains across the area allotted. A supplementary award of 1851 dealt with a further 39 acres. The open fields were Haddenham End, Linden End, Aldreth, and Hilrow Fields, whose position is shown by their names, and Hinton Closes in the northeast of the parish. There were 162 allottees under the award of 1847, the largest being the Earl of Hardwicke (355 acres, including 4 as lord of Haddenham and Hinton manors and 37 as lessee of the great tithes under the Archdeacon of Ely), Caius College (167½ acres, 102 leased to the Earl of Hardwicke and 65½ to Thomas Oscar Drage), Mary Cropley (146 acres), Edward Humphreys Greene (108 acres), John Read (107½ acres), and T. O. Drage (99 acres in his own right). Various members of the Camps family, who were principal beneficiaries at Wilburton a few years later, received 144 acres in Haddenham, 84 being allotted to John Camps.

On 18 March 1947 the whole western end of Haddenham parish, more than 2,000 acres in extent, was inundated to the eaves of the houses. The damage was more serious than in any other part of the Isle, and this was the one area where it was not possible to grow crops for the 1947 harvest.[28]

8 *Pipe R. 1169–70* (Pipe R. Soc. xv), 96.

9 *Pipe R. 1172–3* (Pipe R. Soc. xix), 162. For the history of the causeway see also above, p. 33, n. 71.

10 *Rot. Hund.* (Rec. Com.), ii, 452.

11 *Ely Dioc. Remembrancer*, no. 189, p. 39.

12 Ibid., no. 216, p. 111. The income of a charity at Whittlesey (q.v.) was applied to the repair of Aldreth causeway, at least in theory, until the end of the 19th century.

13 B.M. Add. Chart. 34274, f. 4.

14 *Cal. S.P. Dom.* 1638–9, 203.

15 Ibid.

16 W. Dugdale, *Hist. Imbanking* (ed. 1772), 355, mentions the bridge, but it is not quite clear whether the reference is contemporary.

17 B.M. Add. MS. 5803, f. 80*b*.

18 Ibid., quoting *Whitehall Evening Post*, 3 Sept. 1765.

19 *English Traveller's Companion.*

20 *Fenland N. & Q.* v, 64; vii, 179–80.

21 B.M. Harl. MS. 594, f. 198–200. Haddenham was nearly as large as Chatteris (206 households) and Doddington with March (201 households), both of which were urbanized. The only other Cambridgeshire villages with more than 100 households were Over (123), Cottenham (121), Fulbourn (105), and Bottisham (104). None of these was in the Isle.

22 *E. Anglian*, N.S. vi, 40.

23 Bp's. Mun., Ely, A 6.

24 C 66/1965, no. 14.

25 Kelly, *Dir. Cambs.* (1858).

26 Ibid. (1865, 1869, 1879, 1900).

27 6 & 7 Vic. c. viii. Incl. Award, County Hall, March.

28 M. of Agric. *Harvest Home* (1947), 46, 52, 53, 67, 70.

From about 1665 to 1680 Haddenham was the scene of operations of Christopher Graye, a son of the well-known bell-founder Miles Graye of Colchester. Christopher cast the three existing bells at Witchford and others in the county.[29]

'Gooding Day' (21 December), on which the widows of the parish went round collecting in money and kind for their Christmas festivities, was long celebrated in Haddenham.[30]

MANORS In 1086 Haddenham was held by the Abbot of Ely in three portions. The first of these was *LINDONE*, now represented by Linden End Field south-east of the village, which was purchased for the convent by Abbot Brithnoth.[31] The abbot held in Lindone 4 hides with land for 6 ploughs, of which 2½ hides (4 plough-lands) were in demesne. There were two sokemen who neither then nor formerly could depart from their land, 9 cottars, 1 bordar, and 10 serfs. The fisheries were important for a place on the landward side of the Isle, producing 3,333 eels and 4*s.* in tribute. The value of Lindone was £4 when purchased. It had risen to £9 T.R.E. and in 1086 was £8.[32] *HELLE*, now Hill Row, was a berewick of Lindone and rated at 2 hides (5 plough-lands), of which 1 hide 1 virgate and 10 acres with 3 ploughs were in demesne. Here there were 10 villeins, 4 cottars, and 5 serfs; a rather mysterious entry states that there were 'from the port (*portu*) three ploughshares'. This estate was worth 40*s.* when received, £6 T.R.E. and 100*s.* in 1086.[33] *HADDENHAM* itself was rated at 3 hides and was held by 7 sokemen under the abbot; like those at Lindon they were tied to the soil. They had land for 5 ploughs, and there were 8 villeins each with ½ virgate, 4 bordars each with 5 acres, and 6 cottars. The value when received was £8, and again £8 in 1086, but prior to the Conquest it had risen to £12.[34] In all cases there was sufficient meadow for the plough-teams, and pasture for all the cattle of the vills.[35]

Haddenham, like most of the valuable manors of the Isle except those in the immediate neighbourhood of Ely, was allotted to the bishop in 1109. In the early Middle Ages the manor was generally known as Lindon. This suggests that Linden End was the more ancient settlement and that we have here an example of an early settlement in the middle of a piece of territory (Lindon) supplanted by one at a cross roads (Haddenham). This process is common in many parts of England but rare in Cambridgeshire[36]. The theory is borne out by the absence of demesne in Haddenham in 1086, as opposed to Linden End and Hill Row.

The survey of Bishop de Fontibus (1221)[37] shows a demesne of 919 acres divided almost equally between the three fields of 'Northey', 'Middlefeld' and 'Scephey'. There were 68 acres of mowable meadow, but some land of this type had been lost owing to floods. There were two meres, 'Alkesmere' and 'Sothemere', and two fisheries. Besides the two tenants by military service (see below) there were 117 householders, divided into the six grades of *hundredarii* (owing suit of court at the hundred), customary tenants, holders of full and half lands, cottars and half-cottars. The Northwold survey of 1251[38] does not show so much progress as on many of the episcopal manors, but Haddenham on the landward side of the Isle had less room for expansion than, for example, Littleport, and even at Haddenham a few *novi feoffati* are recorded. The demesne had been slightly reduced to 834 acres, though assarts are mentioned in 'Sephey' and a small new field of 'Snota' is recorded. This demesne was rated at 5 carucates. The meadow land had increased to 134½ acres, besides 23 of 'great meadow' in the marsh. During the preceding generation the Ouse, now the Old West River, seems to have changed its course. Queenholm marsh, which formerly belonged to Haddenham, was now reckoned part of Willingham on the Cambridgeshire side.[39] The demesne was stocked with 30 cows and 2 bulls, 60 pigs and 2 boars, and 240 sheep. There was a windmill, let at 100*s.*, and 6 fisheries producing 48*s.* and 5,000 sticks of eels. There were now 3 holders of fees, and 136 other tenants. Fourteen holders of *plene terre* are specified as living at Hill Row: one of the similar tenants in Lindon and Aldreth, of whom there were 20,[40] was Cassandra relict of William son of Robert. Commutation of services for money rents had not gone far, the manor producing only £2 10*s.* 4*d.* on this score. During the vacancy of the see consequent on Bishop Balsham's death (1286) £4 16*s.* 9*d.* was received from commutation of 518 summer and 902 autumn works, and though the mill only realized £1 9*s.* 5*d.* the manor as a whole brought in £27 12*s.* 6*d.*[41] During the early years of the 14th century the value of the manor fluctuated; it was £51 6*s.* 11*d.* for the two years 1298 and 1299,[42] £35 0*s.* 8*d.* gross in 1302[43] and £34 4*s.* 7*d.* gross in 1316.[44] In 1299 the fisheries, one of which was at Aldreth and the other on the extreme edge of the manor and county at Earith, were rated for 4,900 eels, commuted for £2 9*s.* 6*d.*; in 1302 and 1316 they produced, with the mill, £4 3*s.* 1*d.* and £2 16*s.* respectively. In 1302 works to the number of 1,347 were commuted for £2 17*s.* 7*d.*; in 1316 only 549 were commuted for £1 2*s.* 10*d.* In 1345 the corresponding figure, for an unstated number of works, was £4 19*s.* 8*d.* out of a total revenue from 'Lyndonby' of £65 5*s.* 5*d.*[45] The survey of 1356,[46] made during Bishop Lisle's tenure of the see, shows retrogression. The demesne had shrunk to 380 acres, valued at £17 15*s.* 11*d.*, and 75 acres of meadow worth 6*d.* an acre. The immediate

29 J. J. Raven, *Church Bells of Cambs.* (1869), 48–49.

30 Kelly, *Dir. Cambs.* (1900) and (1933) mention that thirty-eight widows took part in 1895. This suggests that the custom lapsed after that date. It is however mentioned as recently as 1949 in Elizabeth M. Harland, *Norf. and Isle of Ely* (Penguin Guides), 224.

31 Jas. Bentham, *Ely*, 75. In the confirmation charter of Edgar (970) Lindon is stated to be bounded by Wilburton, Haddenham, Hill (Row) and Witcham (*Cart. Sax.*, ed. Birch, No. 1268). It is still so bounded.

32 *V.C.H. Cambs.* i, 368.

33 Ibid.

34 The decrease in value between 1066 and 1086 of this, the central portion of the vill, was much greater than in the eastern and western extremities. This suggests that Hereward's last stand may, as tradition asserts, have been at Aldreth causeway.

35 *V.C.H. Cambs.* i, 368.

36 The towns of Royston and Newmarket on the county boundary are other examples. Also perhaps the gradual extension of March from the top of its 'island' by the church down to the Nene.

37 B.M. Cott. MS. Tib. B. II, ff. 107*a*–10*b*.

38 B.M. Cott. MS. Claud. C. XI, ff. 53*a*–60*b*.

39 Queen Holme is still the name of the north-east corner of Willingham, and the former course of the river can be traced on the 1 in. Ordnance Survey map.

40 There were also 26 at Hengeton (Hinton) (Bps'. Mun., Ely, G 3 ('Old Coucher')).

41 S.C. 6/1132/9. These figures are from the audited account; another on the same roll shows profits of only £12 18*s.* 3*d.*

42 S.C. 6/1132/10.

43 S.C. 6/1132/11.

44 S.C. 6/1132/13.

45 Bps'. Mun., Ely, D. 5 (2).

46 B.M. Add. MS. 6165, f. 269, 285.

cause of this reduction was stated to be floods. The manor house, known as Lyndonbury,[47] was ruinous, like so many of the bishop's manor houses at this time, and had neither garden nor dovehouse from which profit could be extracted. The only sound portions were a small hall, chamber and kitchen, repairable for 2*s.* The outer gates were said to need complete rebuilding at a cost of 13*s.* 4*d.* The same was true of the hen house (*domus gallin'*) at half that cost. Other farm buildings were repairable at 8*s.* 8*d.* The windmill now brought in £2, and five several fisheries £5 8*s.* Rents came to £5 19*s.* 1*d.* and the movable property, including crops and stock, was worth £31 5*s.* 8*d.* At this late date 26 niefs were recorded in the four hamlets of Haddenham, Hill Row, Lyndon, and Aldreth. The Aldreth causeway needed considerable repairs, estimated at £2, the Earith causeway 3*s.* 4*d.* only. The warren of Haddenham is mentioned in 1468 and 1490, when its custody was entrusted to Elias Cliderow and his son Clement for 3*d.* a day and perquisites.[48] In 1612 it was let to Alice Davye for 10*s.* a year.[49]

Haddenham continued, however, to be one of the bishop's most valuable manors, bringing in £118 2*s.* 2*d.* gross in 1518, though the windmill was now only worth £1 13*s.* 4*d.* and the fisheries £2 17*s.* 8*d.*[50] In 1541, when the receipts totalled £91 5*s.* 7½*d.*, it ranked third, after the episcopal property in Wisbech and Ely.[51] Another account of 1548–9 shows a slight surplus. Rents of assize (£85 1*s.* 5*d.*) and the rent of the Delphs[52] (£10) accounted for most of the revenue.[53] A survey[54] made during the long vacancy after Bishop Cox's death (1581–1600), possibly on the occasion of a lease (1593) to Edward Skipwith of the manor house and demesne,[55] shows a gross value of only £38 8*s.* 10*d.* No rents, however, are mentioned other than rents of assize fixed at £20. The 'Berrysted' with its demesne of 77 acres was let at £3 10*s.* a year under an eighty-year lease made in 1529–30 to Robert Stannyowght.[56] A little reclamation of the fen seems to have been attempted; John James a copyholder had erected a windmill which, judging by its value (3*s.* 4*d.*), must have been one of the small 'draining engines' occasionally overthrown by conservative fenmen. A series of late-16th-century bailiffs' accounts,[57] recording gross receipts ranging from £82 15*s.* 1*d.* to £112 17*s.*, suggest that the estimate of this first survey may have been too low. Another survey of 1595[58] valued the pastures of the Delphs, the windmill, fisheries, causeway tolls, rents of land, and tenements and perquisites of court, at £50. In addition the sum of £40 was allowed yearly for the repair of Aldreth and Earith causeways and bridges.[59] Thomas Lord Howard (later Earl of Arundel) purchased the manor in 1602, at which date the rents of assize and customary rents amounted to £81 11*s.* 5½*d.*[60]

In 1612 the manor was transferred to the Earl of Arundel's cousin, Thomas Earl of Suffolk.[61] In 1642 and 1643 an order was made by the House of Lords for 'protection of certain lands of the Earl of Suffolk at Haddenham'.[62] Some time between these dates and 1659 James Earl of Suffolk, Thomas's grandson, conveyed the manor to Dr. Aylett and John Aylett of the Middle Temple, from whom Edward Eltonhead agreed to buy the reversion for £1,200. Eltonhead became financially embarrassed and only paid £600, but John Aylett agreed to lease the manor to John Eltonhead, presumably Edward's son, for £50 yearly exclusive of taxes.[63] The manor is next found (1674–5) in the hands of Frances, Edward Eltonhead's daughter, and her second husband Sir Joseph Douglas,[64] after which its descent is for a time obscure. In 1718 it was with the March family,[65] already owners of the subordinate manors of Grays and Hinton (see below). From 1742 when Pell Gatward and Sarah Rowland (March) his wife were dealing with them[66] the three manors descended together. Isaac Wollaston, Sarah March's second husband, obtained the baronetcy of Lawrence (created 1748) by special remainder, and was lord for a short time before his death (1750). His son died a minor in 1756.[67] In 1766 the manors were obtained by Philip Yorke, 2nd Earl of Hardwicke,[68] in whose family they descended for more than a century. At the beginning of the 19th century they were vested in Philip Yorke (lost at sea in 1808) and his cousin Charles Philip, both Viscounts Royston and heirs to the earldom.[69] A series of accounts from 1820 to 1828[70] shows ground rents from £52 18*s.* 7¼*d.* to £88 14*s.* 10¼*d.* Fines naturally varied greatly from year to year, but the balance due to the earl, after allowing for collector's expenses of £2 2*s.* and court expenses of about £10, averaged about £80 yearly. In the middle of the century the court proceedings were opened in the porch of the demolished manor house.[71] The manorial rights were still being exercised by the earl in 1900, but in 1933 the lords were Mr. James Spearing, and Mr. W. L. Raynes of Cambridge.[72]

In 1221 Simon de Insula held 3 carucates in *HINTON* (Henegeton) as 1 fee, and Philip de Insula held them in 1251.[73] In 1341 Sir William de Ruston received licence for an oratory in his house at Hinton.[74] Some time before 1459 Thomas Freeman, having settled the manor of Hinton on John Prysot, Chief Justice, and John Cheyne, sold it to John Colan, his cousin. Colan was slow in fulfilling his side of the contract, but Prysot expressed himself willing to give up his interest on completion of the transaction.[75] No connexion can be traced between these successive holders of the manor, who seem to have been merely tenants at will under the bishops. In 1544 Sir Edward

[47] An example of the manorial nomenclature common in the northern Home Counties but not generally found so far north. The term 'Burystead' was however applied frequently in the Isle, e.g. at Littleport and Sutton.
[48] *Ely Dioc. Remembrancer*, no. 244, p. 195; no. 282, p. 164.
[49] C 66/1965, no. 14.
[50] Bps'. Mun., Ely, D. 5 (4).
[51] A. Gibbons, *Ely Episc. Rec.* 12.
[52] Now Upper and Lower Delphs near Earith.
[53] B.M. Add. Chart. 34274, f. 4.
[54] S.C. 12/30/33.
[55] C 66/1403, m. 33–34.
[56] E 134/3 Jas. 1/E. 9.
[57] S.C. 6/Eliz./283; S.P. 12/268/100.
[58] S.P. 12/255/72.
[59] *Cal. S.P. Dom.* 1595–7, 155.
[60] C 66/1602, m. 22–5; *Cal. S.P. Dom.* 1638–9, 203.
[61] C 66/1965, no. 14. See also C.P. 25(2)/277/Trin. 7 Jas. I, and /278/ Trin. 10 Jas. I.
[62] *Lords Jnls.* iv, 640; vi, 48; Hist. MSS. Com., *5th Rep. App.*, 86.
[63] C.P. 25(2)/539/Hil. 1659; C 5/604/89.
[64] C.P. 25(2)/633/Hil. 26–27 Chas. II.
[65] C.P. 25(2)/1000/Mich. 5 Geo. I.
[66] C.P. 25(2)/1260/East. 15 Geo. II. D. & S. Lysons, *Magna Brit.* ii (1) 204.
[67] *Compl. Baronetage*, v. 98.
[68] Gardner, *Dir. Cambs.* (1851), 499.
[69] C.P. 43/889, ro. 176. C.P. 43/920, ro. 223. *Compl. Peerage*, vi, 308.
[70] B.M. Add. MS. 36231.
[71] Gardner, *Dir. Cambs.* (1851), 499. The last house on the site was built during the reign of Elizabeth by the Wren family, and was known as the 'Wren's Nest'. This family were lessees under the bishop of Hinton manor.
[72] Kelly, *Dir. Cambs.* (1900, 1933).
[73] B.M. Cott. MSS. Tib. B. II, f. 107*a*, Claud. C. XI, f. 53*a*.
[74] *Ely Dioc. Remembrancer*, no. 60, p. 416.
[75] *Cal. Close*, 1454–61, 390. C 1/26/481.

(later Baron) North had licence to alienate this manor to Thomas Wren of Haddenham,[76] who was already the tenant of the 'Delphs' under Bishop Goodrich.[77] His daughter Elizabeth later married Sir Edward's grandson, another Edward.[78] The estate pertaining to the manor extended into Wilburton, Wentworth, Witcham, and Sutton. Wren, who executed settlements of the manor on his wife Anne in 1565 and 1578,[79] died in the latter year leaving two daughters, Frances wife of John Wyllyams and Elizabeth (see above). He held the manor in chief by $\frac{1}{10}$ knight's fee, and land and a fishery of the bishop in socage and fealty as of his manor of Haddenham.[80] The two halves of the manor were conveyed by the respective heiresses and their husbands to Francis Brakyn in 1581 and 1588.[81] In 1602 Brakyn, in association with John Williams and Frances (Wren) his wife, passed the manor to Sir John Jolles.[82] Jolles, who was an alderman of London and died in 1621, made a complicated settlement of the manor, preserving the two moieties into which it had recently been divided. One was to go to his niece Alice (Towers) for life, with remainder to Danett Poyntell, a nephew; the other was settled on Poyntell and his sons in tail male, with remainder to his daughters and finally to John Towers, Alice's eldest son and her other sons successively in tail male.[83] Accordingly the whole manor ultimately came to the Towers family, of whom Thomas was in possession at the end of the century. Being financially encumbered he sold it to David Rowland in 1693.[84] Rowland died in 1717: by the marriage of his heiress it came to the March family[85] and thereafter descended with the main manor (q.v.).

The second military tenure in 1221 was that of Richard de Chevall, who held 2 carucates as a fee. His son Niel held similarly in 1251.[86] This fee became the manor of *CHEWELLS*. A manor of the same name in Hatfield (Herts.) passed before 1303 from Niel de Chevell to John de Queye, who in that year was joint holder of $\frac{1}{3}$ fee in Haddenham.[87] The de Queye family is recorded as holding by military service in Haddenham to 1428,[88] but had probably been extinct for some time, as Edmund de Suttone, the co-tenant in the 1346 inquiry, was in 1370 styled lord of 'Chiwalle' and had mortgaged his share, consisting of 2 messuages and 46$\frac{3}{4}$ acres of land, to Sir Robert Busteler. The heirs of the latter were John Hanchache a minor and four others of full age, un-named in the inquisition. These five enfeoffed Robert Parys of one messuage and 16 acres, and John Wroth, a lessee under Busteler, of the other messuage, named 'Thunderesplace', and 30$\frac{3}{4}$ acres.[89] Meanwhile the overlordship of this manor had been granted (1344) by Bishop Montacute to Peterhouse, Cambridge.[90] In 1594, when the college's tenant was the Betts family of Chatteris (q.v.), the manor was worth £10 a year.[91] In 1808 the tenant was the Revd. Samuel Hunt.[92] Court rolls exist for the period 1745 to 1937 in the strongroom of Messrs. Francis & Co., Cambridge.

In 1224 Hamon Passelewe was to be reseised of 1 carucate in 'Hedham' if Richard de Secheford, Beatrice his wife and Maud her sister had disseised him solely on account of the king's anger against Robert Passelewe, Hamon's brother. This anger seems to have been caused by Robert having taken the case to the court of Rome, and in 1225 he was reinstated in his carucate from which Richard and Beatrice de Secheford and Maud, with her husband Robert de Upwic had dispossessed him.[93] In 1277 Hamon Passelewe, perhaps Robert's nephew, held 25 acres in Haddenham of the manor of Rettendon (Essex)[94] for £1 a year.[95]

The co-tenant with John de Queye of a third fee in 1302–3[96] was Fulk Baynard or Barnard. He or his ancestor of the same name had in 1271 and 1272 received two portions of an estate of 2 messuages, 3 carucates and 7 acres of wood in Haddenham and Hatfield (Herts.) from John de Ditton and Philip Pertrick and their wives, presumably co-heiresses.[97] In 1346 Edmund de Suttone had succeeded to the Baynard portion of this $\frac{1}{3}$ fee, and his heirs are similarly recorded in 1428.[98] At the latter date John Grenelane and others held $\frac{1}{2}$ fee in Haddenham.[99]

The manor of *GRAYS* originated in an estate of a messuage and 320 acres of land, with 6 marks rent, held by William de Grey of the Bishop of Ely by unknown service. His son Thomas was a minor at his father's death (1495).[1] In 1541, when Edmund Grey, probably William's grandson, conveyed it to Thomas Waters, it included lands and rents in several Isle parishes, as far as Chettisham.[2] In 1580 Edward Cowper granted an eighty-year lease of it to Robert March,[3] whose son of the same name died seised of it in 1601. He held it of the chief lord (Thomas Lord Howard) by fealty and a rent of a pound of cummin or 3*d.* yearly.[4] During the minority of Humberston March, son of Robert March the younger, the issues of the manor were taken by his mother Anne and her second husband Thomas Castell. Humberston and his son William were dealing with this manor in 1652.[5] Since the 18th century this manor has descended with the main manor (see above).

Another fee is mentioned in 1303–5. In the former year a commission of oyer and terminer was issued regarding persons who broke into the estate of William de Tuddeham, king's yeoman, at Haddenham, and fished in his free fishery.[6] In the latter year William brought a suit against his tenants Nicholas de Cokayne and his son and daughter, who had forcibly secured certain cattle that William had distrained for arrears of services for lands held of his fee.[7]

[76] *L. & P. Hen. VIII*, xix, i, g. 141 (77), (p. 86). C.P. 25(2)/52/378, 379 (East & Trin. 36 Hen. VIII).
[77] E 134/37–8 Eliz./M. 46. A later lease of Bishop Cox assigned this land to Lord St. John of Blestoe. During the episcopal vacancy it was claimed by Charles Yelverton 'Her Majesty's Farmer' (E 134/42 Eliz./T. 10, E 134/44–5 Eliz./M. 2).
[78] *D.N.B.* sub. Sir Thomas North (1535?–1601?).
[79] C.P. 25(2)/93/842.
[80] C 142/184/17.
[81] C.P. 25(2)/93/845; C.P. 25(2)/93/852.
[82] C.P. 25(2)/93/867.
[83] C 142/396/147.
[84] C.P. 25(2)/814/Mich. & Hil. 5 Wm. & Mary.
[85] Lysons, *Magna Brit.* ii. (1), 204.
[86] B.M. Cott. MSS. Tib. II, f. 107*a*, Claud. C. XI, f. 53*a*.
[87] *V.C.H. Herts.* iii, 102. *Feud. Aids*, i, 151. Hatfield manor belonged in the Middle Ages to the bishops of Ely.
[88] *Feud. Aids*, i, 173, 193.
[89] *Cal. Inq. p.m.* xii (unpublished), no. 10 (pp. 7, 9).
[90] Bentham, *Ely*, 159.
[91] C 3/233/4.
[92] Lysons, *Magna Brit.* ii (1) 205.
[93] *Rot. Litt. Claus.* (Rec. Com.), ii, 6, 17.
[94] Also an Ely episcopal manor.
[95] B.M. Cott. MS. Claud. C. XI, f. 53*a*.
[96] *Feud. Aids*, i, 151.
[97] C.P. 25(2)/283/17; C.P. 25(2)/284/20.
[98] *Feud. Aids*, i, 173, 193.
[99] Ibid. i, 191.
[1] *Cal. Inq. p.m. Hen. VII*, i, 1173 (p. 520).
[2] C.P. 25(2)/52/376.
[3] C.P. 25(2)/93/843.
[4] C 142/280/12.
[5] *E. Anglian*, N.S. xiii, 293. The Marches sold the lordship to Martin Foulkes, curate in charge of Haddenham, for £2,000. This was probably for a term of years only as the manor, like the other ones in Haddenham, later came to the earls of Hardwicke.
[6] *Cal. Pat.* 1301–7, 185.
[7] K.B. 27/180, no. 53.

In 1537 William Buckenham, master of Gonville Hall, was licensed to alienate to his college 2 messuages called 'Madingleys' and 'Partriks'[8] (the latter in Hill Row), 2 tofts called 'Chevyns' and 'Codwyns', and 237 acres of land. This estate was worth £9 12*s.* a year and the purchase price was £200.[9] Minor adjustments during the following centuries reduced this estate to 171 acres, scattered in 283 pieces over the north and west of the parish. Under the Haddenham Inclosure Act (1843) the college received 167½ acres. When Grunty Fen was inclosed in 1861, it received a further 14 acres. These were near Wilburton railway station. Adjustments made with the Great Eastern Railway and others reduced the size of the estate slightly. The rents of the whole amounted to £332 in 1875, £346 in 1885, and £100 in 1895.[10] The reduction in rent in 1895 was due to the assumption by the college of responsibility for tithes and drainage taxes.

The antiquary William Cole also owned land in Haddenham, the history of which, as might be expected, is very well documented.[11] He traced his estate back to 1560, when Richard Tyrrell of London disposed of it by bargain and sale to Robert March of Ely for £160. It then consisted of a messuage and 72 acres, and had formerly been held by William Grace. In 1614 there were 4 messuages, including two called Graces and Hallyards and about 130 acres of land. In that year it changed hands within the March family for £1,000. Twenty years later it was sold by the Marches to Thomas Baron of Saffron Walden (Essex) for £920. The estate remained with the Baron family for nearly a century, gradually increasing in size and appreciating in value,[12] and was sold in 1728 by John Baron of Trumpington to William Cole of Babraham, father of the antiquary, for £2,205. The sale of 1728 included some 30 acres in Ewell Fen obtained by the Barons from the Mason, Phypers, and Worts families at the end of the preceding century. The total area sold in 1728 was 666 acres of arable, 26 of pasture, 95 of fen, and 10 of marsh (the Holmes), with 5 tenements including one with a 2-acre close at Aldreth. The then tenants of William Cole, the elder, were Francis Goodday and William Dunkin.

In 1754 Cole's tenant at Frog Hall, on the river bank near Aldreth High Bridge, was John Huckle. He showed the antiquary a document of 1728 rating the proprietors of lands in Ewell Fen at 4*s.* 3*d.* an acre for the repair of the fen bank. The 404 acres of the fen were divided amongst fourteen proprietors. The share of Cole's father amounted to 54 acres. This share was increased in 1733 by the antiquary's purchase of 2 closes from Joseph and Sarah Ogram. They totalled 24 acres and cost £525. The value of the property, and Cole's personal income, was much lowered by the serious floods of the 1760's.[13]

CHURCH The church of Haddenham, like others on the manors granted to the see of Ely in 1109, remained in the bishop's hands, and is so recorded in 1251.[14] It was worth £30 in 1217, £40 in 1254,[15] and £80 in 1291,[16] and may therefore be reckoned a valuable rectory. During the early Middle Ages there were frequent disputes between the bishops and archdeacons of Ely regarding the rights of the latter in the Isle.[17] To settle these disputes, Bishop Fordham was licensed in 1401 to unite the rectory of Haddenham to the archdeaconry.[18] This arrangement was confirmed in 1406,[19] and prevailed until the middle of the 19th century. The church was in effect appropriated to the archdeacon, though no vicarage was ordained and the church was served by a succession of curates-in-charge. In 1837 the rectorial tithes which had come to the earls of Hardwicke as lords of the capital manor were commuted for a rent-charge of £1,850, whereas those due to the incumbent were commuted for £250 only, although they included a portion of the great tithes.[20] The position was normalized in 1865, when Archdeacon Emery transferred the tithes to the Ecclesiastical Commissioners for £600 a year,[21] on condition that the living should be improved and a parsonage erected. This was done at various dates between 1866 and 1874,[22] and the living is now a vicarage in the gift of the Archdeacon of Ely.[23]

The rectory being a rich one, it is not surprising to find that the rectors and archdeacons treated it like a manor, and leased it. Thus in or before 1332 Geoffrey de Pakenham, rector, leased to Richard le Ferrour and his wife Margaret an acre of land in Haddenham for 6*s.* 6*d.* a year.[24] In 1601, when the rectory was leased to Richard Arkenstall for twenty-one years, the value of the rents was just over £8 a year; there were also 50 acres of glebe and the total value of the rectory was stated to be £100.[25] Arkenstall was shortly afterwards (1605) engaged in a lawsuit with Edward Phipers and other rectory tenants regarding his claim to the tithes.[26] From the latter half of the 17th century the rectory descended with the main manor,[27] but the tithes were sometimes leased out.[28]

The advowson lay with the bishops of Ely until 1401, and has since been held by the archdeacons, in accordance with the agreement of that year.[29] Before the transfer, a case of disputed presentation occurred, which dragged on for a great many years. In 1342 the rector in possession, John de Martham, was challenged by Henry de Harwedon, who during the vacancy of the see in 1337 had intruded himself by virtue of a papal provision. Harwedon continued his case against Bartholomew de Bourne, Martham's successor in 1344, and the advowson was for this reason retained in the king's hands after the other spiritualities were restored to Bishop Lisle (1345). Harwedon finally abandoned the suit in 1348 and was pardoned for contempt of court.[30] The rights

[8] Cf. Philip Pertrick mentioned as a landowner in Haddenham in 1272 (p. 144, above).
[9] *L. & P. Hen. VIII*, xii (2), g. 1008 (17). E. H. Gross, *Biog. Hist. Gonville & Caius Coll.* iv (2), 50.
[10] See Gross, op. cit. for a fuller history.
[11] B.M. Add. MSS. 5810, f. 200; 5841, f. 361.
[12] e.g. at £1,600 in a family transaction of 1688.
[13] B.M. Add. MS. 5824, f. 51*b*, &c.
[14] B.M. Cott. MS. Claud. C. XI, f. 53*a*.
[15] W. E. Lunt, *Valuation of Norw.*, 539.
[16] *Tax. Eccl.* (Rec. Com.), 265.
[17] *V.C.H. Cambs.* ii, 144.
[18] *Cal. Pat.*, 1401–5, 51. *Ely Dioc. Remembrancer*, no. 194, p. 143.
[19] *Cal. Pat.* 1405–8, 164.
[20] Gardner, *Dir. Cambs.* (1851), 502. Including Wilburton, where a similar arrangement had been in force.
[21] Kelly, *Dir. Cambs.* (1900).
[22] *Lond. Gaz.* 1866, 669; 1868, 6003; 1870, 3954 (augmentation of the living); ibid., 1867, 4367, 4751; 1874, 3770 (erection of parsonage).
[23] *Crockford's Clerical Dir.*
[24] *Cal. Pat.* 1330–4, 311. The date is that of confirmation.
[25] C 66/1561, m. 11, 12.
[26] St. Ch. 8/43/8.
[27] C.P. 25(2)/633/Hil. 26–27 Chas. II, C.P. 25(2)/1102/Trin. 11–12 Geo. II, C.P. 25(2)/1103/East. 15 Geo. II.
[28] e.g. to William Cole the antiquary, who in 1772 levied a fine upon them (C.P. 25(2)/1281/Mich. 12 Geo. III).
[29] Rymer, *Foedera*, ed. 1740, iv (1), 20
[30] *Cal. Pat.* 1340–3, 571, 593; 1343–5, 84, 399, 547–8; 1345–8, 229; 1348–50, 183.

of Martham and his successor Bourne were reaffirmed in 1363, but four years later Walter de Baketon was allowed to sue out possession of the church in ecclesiastical courts against Bourne by virtue of the papal provision of Harwedon, whose surrogate Baketon claimed to be.[31]

In 1386 the dedication feast of the church was changed from the vigil of the Nativity of the Virgin (7 September) to 20 October, as the former was a fast-day.[32] By 1841 the feast was celebrated on Trinity Sunday; 143 strangers attended it.[33]

ginning with the chancel. The transepts were refashioned in the 15th century, the chancel arch rebuilt in the same period, and another stage added to the tower. In 1876 a drastic 'restoration' was begun, under the supervision of R. R. Rowe.[40] The tower was completely rebuilt as was the greater part of the walling of the aisles and transepts, and the vestry and north porch added. The roofs were completely renewed and the 15th-century rood screen was thrown out. It reposed for thirty years in a builder's yard but was eventually restored to its rightful position.

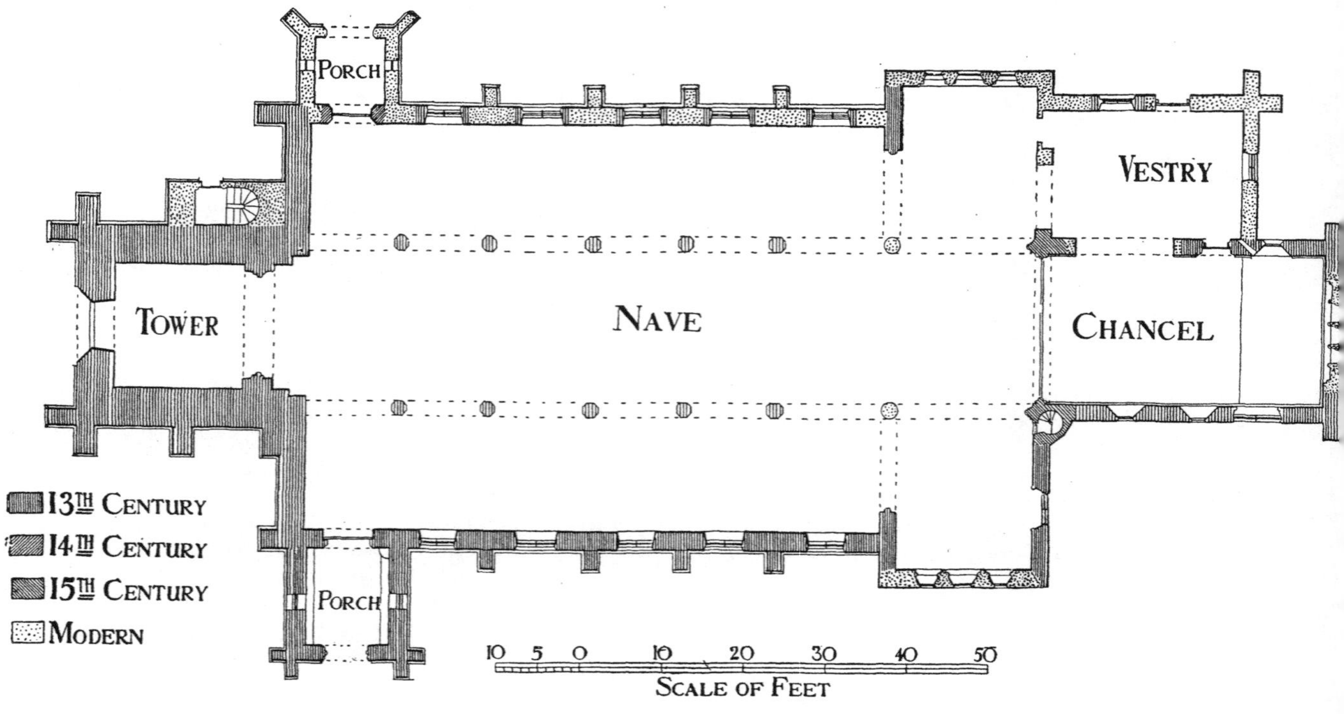

Plan of Holy Trinity Church

In 1335 the prior and convent confirmed a grant by Bishop Hotham to Robert de Orford, Agnes his wife, and Beatrice his daughter, of 30 acres in Westhall Field, reserving one good sheaf of corn per acre for the support of the lamp of St. Peter and St. Etheldreda in the cathedral.[34] In 1553 a ½ acre in tenure of Richard Wallis,[35] given for lights in Haddenham church, was granted to Sir John Butler and Thomas Chaworthe,[36] and 4 acres which had provided for a light in the rood loft were in 1571 granted to Richard Hill and Robert Don; they were then held by Edward Brome.[37]

In the early 16th century there were three guilds in Haddenham, of the Holy Trinity, All Saints and St. Katharine.[38] The hall of one of them was granted to Ralph Stretton and Edward Warner in 1561.[39]

The church of the *HOLY TRINITY* is built of ashlar and rubble with leaded roofs, and consists of chancel, north vestry, clerestoried nave, transepts, aisles, north and south porches, and west tower. The present building was constructed gradually during the 13th century, be-

The chancel has an east window consisting of three lancets, which are entirely modern and replace a 15th-century five-light window. There are 13th-century angle buttresses with one set-off. The south-east window, a late 13th-century insertion, is of two lights with trefoiled heads and a quatrefoil above; the jamb shafts have moulded caps and bases and there are external and internal hood-moulds terminating in heads. To the west are two lancets much renewed. The north-east window is a plain early 13th-century lancet. The doorway to the vestry is original 13th-century work with an arch of two orders, a continuous chamfer and a hood-mould. To the west is a modern arch of 13th-century character with corbelled responds. There is a much renewed 13th-century internal string-course. In the south wall is an aumbry and double piscina, the latter having a trefoiled head to its recess. The sill of the south-east window is lowered to serve as sedilia. In the north wall of the sanctuary is an aumbry and a tomb recess with trefoiled arch and a hood terminating

[31] *Cal. Pat.* 1361–4, 455; 1364–7, 417.
[32] *Ely Dioc. Remembrancer*, no. 132.
[33] *V.C.H. Cambs.* ii, 139 n.
[34] Hist. MSS. Com. *6th Rep., App.* 296*b*.
[35] In 1557 a Richard Walleys was pardoned for appropriating the skin and flesh of a neighbour's cow, which had strayed into his garden. Walleys set his dog upon the cow, which got its head fixed in the spokes of a cartwheel and broke its neck. (*Cal. Pat.* 1555–7, 454).
[36] *Cal. Pat.* 1553, 290.
[37] C 66/1074, m. 5.
[38] W. M. Palmer, *Village Gilds of Cambs.* 392.
[39] *Cal. Pat.* 1560–3, 106.
[40] *Cambs. Antiq. Soc. Procs.* xxxv, 84. A print of the church before the restoration hangs in the building. It corresponds closely with Cole's sketch, and shows that he was an accurate even if not a skilful draughtsman.

in heads of late 13th-century date; at the back of the recess is a squint to the vestry with a trefoil-headed opening. The 15th-century chancel arch is of two orders with semicircular shafts to the responds, having moulded caps and bases, and there is a hood-mould on the west terminating in heads. The modern vestry has an east window of two trefoiled lights and a larger two-light window in the north wall, both of late-13th-century origin but considerably renewed; they were probably removed from the north wall of the chancel when the present vestry was erected. There is a north doorway and a doorway and arch communicating with the transept, all modern.

The nave has arcades of seven bays, the easternmost arch on either side, which opens to the transepts, being larger than the others and of 15th-century date. The remaining arches on either side belong to the second half of the 13th century. The transept arches have an eastern respond consisting of a semicircular shaft with moulded cap and base; the western respond is modern with a brown marble shaft and a clumsy foliaged cap. The rest of the arches are of uniform design and are two-centred and of two orders with octagonal columns having moulded caps and bases. The west respond consists of a large and clumsy modern corbel. Several of the arches retain traces of red and yellow decoration. Above the arches is a modern string-course. The clerestory consists of six two-light windows on each side with trefoiled heads, except the first and fourth on the south and the second and sixth on the north which are uncusped; all appear to have had their tracery renewed but the rear arches are partly old and probably of late 13th-century date. There is a plain coped parapet. The tower arch is deeply recessed and acutely pointed, and springs from clustered shafts with moulded caps and bases; it belongs to the latter part of the 13th century but was rebuilt, largely with old materials, in 1876.

The south transept has an east window of two cinquefoil-headed lights with a spherical triangle above and jamb shafts with moulded caps and bases, and exterior and interior hood-moulds; it dates from the end of the 13th century. The rood stair turret is in the angle between the transept and the chancel with the entrance in the former and the upper doorway, now blocked, in the latter. There is a large shallow recess above the lower doorway with a cinquefoiled head. The south wall has been entirely rebuilt and has a window consisting of three lancets. The arch communicating with the aisle is probably of late 13th-century date and almost round; the south respond consists of clustered shafts with moulded caps and bases and the north respond, which is modern, consists of a marble column. There is a piscina recess in the east wall.

The north transept was completely rebuilt in 1876 and has a window consisting of three lancets in the north wall. The arch communicating with the aisle is similar to the corresponding one in the south transept. There is a canopied niche of 15th-century date in the north-east angle. Upon the outside wall of the former north transept there hung in Cole's time two dozen fire buckets.[41]

The south aisle has five two-light windows, the first and fifth cinquefoiled and the remainder trefoiled, the tracery being renewed except the two westernmost. There are four lateral buttresses and one at the west end, all with one set-off and renewed except the first and fourth. There is a plain coped parapet, mostly modern. The south porch is of early 13th-century origin, but rebuilt with old material farther to the west in the last century. It has an outer doorway with a two-centred arch of two orders having responds composed of clustered shafts with moulded caps and bases. There are angle buttresses with one set-off and a one-light trefoiled window under a square label in the east and west walls. There is a plain coped parapet and conical pinnacles. The inner doorway has continuous mouldings and a hood terminating in modern heads; above is a canopied niche. The plain bowl of a 13th-century font is fixed in the north-east angle of the porch.

The north aisle has been rebuilt, but it retains five old windows, similar to those in the south aisle, the tracery in which has been much renewed. The north porch is a modern addition with diagonal buttresses and a trefoil-headed single light in the east and west walls. The inner doorway is of early 14th-century date with continuous mouldings and a hood terminating in heads; the jambs have been renewed.

The fine tower was originally erected in the last quarter of the 13th century and consisted of three stages. In the 15th century another stage was added and the tower crowned with an embattled parapet and a lofty and slender leaded spire. In 1876 the whole structure was pulled down and the three lower stages rebuilt, largely with old materials; the top stage was not replaced. The old design has been followed more or less, but the stair turret, which was originally in the south-west angle, is now placed at the north-east. The west doorway is deeply recessed with an arch of five orders springing from jamb shafts with moulded caps and bases, all much renewed. The west window is of three lights with reticulated tracery and ball-flower ornament round the arch, and a hood-mould decorated with dogtooth ornament, which must be a later insertion. In the second stage there is a deeply recessed circular window on the north, south, and west filled with early geometrical tracery; those on the west and south have two circles of ball-flower in the splay while that on the north has in addition an inner circle of dog-tooth ornament. The belfry windows consist of recessed double lancets in each face with cinquefoiled heads and continuous mouldings. The tower is now finished with a corbel table. The modern stair turret in the north-east angle only reaches to the second stage and is crowned with a conical cap.

All the roofs are modern and poor, those of the chancel, nave and transepts being high-pitched and those of the aisles of lean-to type. Some of the stone corbels in the nave are ancient.

The font is octagonal and of 15th-century date with a shield bearing demi-angels and Tudor roses on the bowl and round the shaft seated lions and buttresses; it belongs to a type often found in Suffolk. The 15th-century chancel screen is of the arched type; it consists of eight bays, the two centre being occupied by the doorway. The vaulting is lost and the base beam, panels of the wainscot and top beam are modern. The uprights are buttressed and the middle rail well moulded; each bay has good tracery of rectilinear character. In the south transept is an elaborately carved communion table of early 17th-century date. In the chancel is a brass to William Noion, Rector, and Canon of York, Lincoln, and Chichester, 1405. The effigy in a cope is missing but the inscription and double

[41] B.M. Add. MS. 5803, f. 84.

canopy remain. In the tower, but formerly in the north transept, is a brass to John Godfrey and Margaret his wife, 1454, with effigies of a civilian and lady, and inscription. At the west end of the nave is a slab with indents of the effigies of a civilian and lady, and of two groups of children beneath.

The plate includes a silver chalice inscribed 'for the Town of Hadnam on the Hil' with cover paten, 1569, and a large silver flagon, 1701. These have been on loan to the Fitzwilliam Museum, Cambridge, since 1933.

The tower contains six bells, 1st, 2nd, and 3rd by Thomas Newman of Norwich, 1706, 4th by Joseph Eayre of St. Neots, 1741, 5th by T. Newman, 1725, 6th by George Oldfield of Nottingham, 1657, and inscribed in black letter 'All glory be to God for ever', with the churchwardens' names and founder's stamp.[42]

The registers begin in 1570 and are complete.

NONCONFORMITY In 1672 the house of John Woodbridge was licensed for Congregational worship.[43] In 1676 there were 30 Dissenters.[44] The houses of Thomas Clark and John Smethey were licensed for Congregational worship in 1745 and 1787 respectively.[45] John Berridge preached at Haddenham about 1768, and there was a congregation served by itinerant Independent ministers in the village all through the latter part of the 18th century. It was regularly organized as a Baptist church in 1812, and from 1816 had its own pastor, a local surgeon named Pinchard. There were about 50 members in 1819.[46] The chapel at Haddenham was built in 1817 and that at Aldreth in 1844. A Wesleyan chapel was built in 1800, rebuilt in 1843, and again in 1891.[47] All these three chapels are still in existence; the old Haddenham Baptist chapel was used as a Sunday-school after 1905, in which year a new building was erected at a cost of £2,000.[48] Its spire rivals the nearby church tower. The Aldreth chapel was also rebuilt, in 1908.[49]

SCHOOLS A school existed at Haddenham as early as 1463, when John Strother was appointed master,[50] and schoolmasters are mentioned in the Visitation Returns of 1590 and 1596.[51]

The early history of Arkenstall's School has already been narrated.[52] The buildings of 1863 provided 115 places; a new classroom was built in 1909, and the net result of this and the reorganization of the following year was to increase the accommodation to 129. The boys of 13 and over were in 1948–9 transferred to the Cromwell School at Chatteris.[53]

A British school was established by the Baptists in 1847.[54] It was still in existence in the 1860's,[55] but was closed by the School Board after the girls' and infants' school had been taken over and enlarged in 1874 (see below).

A school for children not old enough to attend Arkenstall's School was established under the will of Mrs. Elizabeth March (d. 1722), who left land at Oakington to provide five such schools at Haddenham, and at Brinkley, Fen Ditton, Fulbourn, and Histon in the county proper. Haddenham's share in this charity amounted to £20, and in the 1830's the churchwardens were in the habit of adding £14 from Arkenstall's charity and £6 from the extensive Town Lands to make up the schoolmaster's salary. In 1835 the master was teaching about 30 children.[56]

Haddenham presented a very favourable picture when the National Society made their inquiry in 1846–7. Education of some kind was provided for 468 children—almost the whole of the potential school population. There were 90 children in Mrs. March's infant school, 67 in a 'school of industry' for girls, and 197 in a National Sunday-school. There were also small primary schools at Aldreth and Hill Row.[57] The school of industry, the infant school and those in the hamlets employed 5 mistresses, and their total annual expenses were £83.[58]

In 1862 the girls' and infants' schools were rebuilt at a cost of £589. Of this sum £400 were subscribed locally. Support, however, was not sustained and in 1874 a School Board was established, which enlarged and improved the school. Further enlargements took place in 1879 and 1909–10 at a cost of £369 and £600 respectively. After the second enlargement the school provided for 122 girls and 98 infants. The two departments were amalgamated in 1927, and the senior girls transferred to Chatteris in 1948–9.[59]

In 1877 the Board rebuilt the Aldreth school as a mixed one for 80 children—a figure which was reduced to 63 in 1910. The chief interest of this school lies in the repeated attempts to close it. In 1907 there were only 14 children on the books as opposed to an average of 29 in 1900–4. The Board of Education, however, refused to sanction closure, as the number of children in Aldreth at that time was abnormally low, and the transfer of children from council to voluntary schools might have led to difficulties. The County Council, however, refused to appoint a new teacher and closed the school in August 1907. This led to a strike, the children assembling at the school and refusing to travel to Haddenham in the conveyance provided by the Council. The school was reopened in January 1908. Another attempt to close the school was made in 1929, when there were 40 on the books. This was abandoned owing to strong local opposition. The further proposal was made in 1940. It was held in abeyance for a time but was finally carried out in 1944.[60]

In 1846–7 the 42 pupils at the Hill Row school mostly attended on weekdays only. The school had probably been closed by 1851 and certainly by 1900.[61]

CHARITIES Besides the Arkenstall and March charities devoted to education, the following existed in Haddenham in 1837.

Sir John Jolles, lord of Hinton manor, by his will

42 J. J. Raven, *Church Bells of Cambs.* (1869), 48–9.
43 *Cal. S.P. Dom.* 1672, 63.
44 Bp's. Mun., Ely, A 6,
45 Ibid.
46 *Cong. Mag.* ii, 631 (1819).
47 H.O. 129/7/190.
48 Kelly, *Dir. Cambs.* (1933).
49 Inscr. on bldg.
50 *Ely Dioc. Remembrancer*, no. 238, p. 62.
51 *V.C.H. Cambs.* ii, 338 n.
52 Ibid. 334–5.
53 M. of Ed. File no. 12/27.
54 Gardner, *Dir. Cambs.* (1851), 500.
55 *Rep. Com. Empl. Children &c. in Agric.* [4068–I], 338, H.C. (1867–8), xvii.
56 *Rep. Com. Char.* (*Cambs.*), [103], 325, H.C. (1837–8), xxiv.
57 The National Society's estimate for the infant school is much higher than that of the Charity Commissioners some ten years earlier, so that the total of 468 may be somewhat overestimated. But there is no doubt that provision for schooling in Haddenham was far in advance of the average.
58 Archives of Nat. Soc.
59 M. of Ed. File no. 12/25.
60 Ibid., no. 12/26.
61 It is not mentioned in Gardner's Directory of 1851 or in Kelly's of 1900.

dated 1618 left £2 a year to the poor of Haddenham, which was paid through the Drapers' Company, of which Jolles was a member.[62] In 1837 the books of this charity were 'regularly and creditably kept'.

Two messuages called the Town Houses belonged to the parish from time immemorial. The Town Lands, originally 10½ acres copyhold, were increased in 1677 by 132 acres freehold in the North Fen awarded under the Bedford Level Act. In 1837 they were worth £185 15*s.* annually. After drainage taxes of £44 2*s.* 5*d.* and the diversions to educational purposes had been paid, £102 9*s.* was available for the poor. This was distributed to about 300 persons in sums of 3*s.* to 12*s.*[63]

MEPAL

Mepal is one of the smaller villages of the Isle, 8 miles from Ely and 4½ miles from Chatteris, and is situated on a loop in the main road which connects the two. The parish is crossed diagonally by the Old and New Bedford Rivers, which are spanned by bridges north-west of the village. These bridges carry the Ely–Chatteris road. They are the only important road bridges over these rivers in the Isle.

A great fire in the village between 1861 and 1871 was a contributory cause of the large decrease in population (510 to 397) that occurred in that decade.[1] In consequence of the fire there are few buildings of architectural interest in Mepal. Mention may however be made of Ash Cottage, a timber-framed thatched house dating from the 17th century and later, and of the early-19th-century Octagon House, whose unusual plan recalls the rather earlier 'Round House' at Little Thetford.

The open fields were inclosed in 1854 under the general Act of 1845. An area of 442 acres, including a few old inclosures, was distributed among 22 persons or institutions. The largest allotments were made to the Dean and Chapter of Ely (138 acres), Richard Bryan (72 acres), and William Cole (31 acres).[2] The Dean and Chapter's allotment was subdivided between their two tenants, Caius College and William Cattell, who received 89½ acres and 48½ acres respectively.

MANOR Mepal is not individually mentioned until the reign of John;[3] in early days it was perhaps reckoned as part of Sutton or Witcham, which lie close to it on the north-west edge of the high land. The first recorded landowner is Ralph de Saham, who held 1 fee under the Bishop of Ely at some uncertain date.[4] Subsequently Baldwin de Colne held ½ fee here of the Bishop of Ely. In 1321 Richard Frevyle, parson of Long Stanton St. Michael, who was presumably a trustee, conveyed the manor and advowson to William de Colne,[5] probably Baldwin's son. William held the manor in 1346.[6] In 1359 Joan, Sir William's second wife, held Mepal for her life, with reversion to John de Colne, their son; William Mochet was their tenant. Sir Henry Colvile, a member of the well-known family of that name at Newton, acting as agent for the prior and convent of Ely, obtained a conveyance of the manor from John de Colne, saving the interests of his mother and her tenant. In 1360 Joan de Colne released all claim in the manor except an annual pension of 20 marks.[7] The manor was alienated to the convent under a mortmain licence of 1309,[8] and was valued on alienation at £4. The purchase transactions cost the convent £254. About two-fifths of this sum was subscribed by the prior and twenty-two monks of the convent, and the residue raised by heavily mortgaging the manor itself, and by such surprising devices as the sale of kitchen utensils and old vestments.[9]

The manor was rated at ½ fee in 1428.[10] The receipts derived from it were £11 11*s.* 2*d.* in 1429, £5 in 1435, £5 13*s.* 6*d.* in 1437, £5 in 1440, and £9 15*s.* in 1467.[11] Thus its value in the 15th century was low. Just before the Dissolution it produced £11 14*s.*, less 8*d.* for the repair of Aldreth causeway (in Haddenham) and £1 in bailiff's fees.[12] The manor house was then on lease for £5 to Thomas Jetherell as heir to his father John, who had received a thirty-year lease in 1521–2.[13] Other rents realized £5 10*s.* 8*d.*, the sale of timber 16*s.* 8*d.*, and perquisites of court 6*s.* 8*d.* on an average.

In 1541 the manor was formally transferred to the dean and chapter of the new foundation.[14] In 1623 Sir Miles Sandys was tenant of the manor, and was in dispute with the dean and chapter over rights of common.[15] By an agreement between the dean and chapter and certain tenants in this year the commons, fens, and marshes had been inclosed and fines changed from arbitrary to certain (see Sutton).[16]

By 1649 the manor house with its demesnes had become so far separated from the manorial rights that it was necessary to value each independently. The demesne measured 231 acres. Of this area, 80 acres were arable, valued at £36 5*s.*, 51 acres pasture, valued at £37 18*s.*, and 100 acres fen, valued at £27. Some of the demesne had been encroached upon to make the New Bedford River. The timber was valued at £5, and there was a decayed fishery called Barr Load. To the manor proper was due the sum of £28 6*s.*, about equally divided between rents of assize and rents reserved on leases. It was stated that improvement (presumably by drainage) would raise the revenue to £106 3*s.*[17]

In 1650 Robert Croxton bought the whole estate from the Commissioners for the sale of Church lands. The manor was stated to be worth £15 16*s.*, a sum presumably representing pleas and perquisites of court and the like; Croxton paid £237 for it.[18] The price exacted for the manor house and demesne was £744 10*s.* The manor house when purchased by Croxton was a

[62] A. H. Johnson, *Hist. Drapers' Co.* iii, 469.
[63] *Rep. Com. Char.* (*Cambs.*), [103], 325–8, H.C. (1837–8), xxiv.
[1] *V.C.H. Cambs.* ii, 139.
[2] M.A.F. 1/662.
[3] *Pedes Finium* (Rec. Com.) 338.
[4] Bp's. Mun. Ely, G. 3 ('Old Coucher'), f. 229. This section of the book dates from c. 1330, but the information on this folio (a list of fees held of the bishop) probably relates to the earlier section of the book (1251).
[5] S. J. A. Evans, 'The purchase and mortification of Mepal by the prior and convent of Ely, 1361', *E.H.R.* li, 114.
[6] *Feudal Aids*, i, 151, 173.
[7] D. & C. Mun., Ely, Cartae 1133, and 1141 (John de Colne's release).
[8] *Cal. Pat.* 1307–13, 153.
[9] For further details see *E.H.R.* li, 113–20.
[10] *Feud. Aids.* i, 191.
[11] S.C. 6/1257/4–9.
[12] B.M. Add. MS. 5807, f. 129.
[13] L.R. 14/770.
[14] *L. & P. Hen. VIII*, xvi, g. 1226 (12). Confirmed in 1587 (C 66/1303, m. 1).
[15] C 2 Jas. I/E. 7/39.
[16] C 3/458/76 and /77.
[17] D. & C. Mun. Cupd. D.
[18] C 54/3442, no. 23.

timber framed building consisting of a hall, parlour, and buttery with three chambers over, a kitchen wing of two stories, outhouses, gardens and orchards amounting to 3 acres, and various closes, mostly contiguous, amounting to 132 acres. There was also a small cottage tenanted by Richard Lancaffe, and 140 acres of fen or 'new allotted' ground. All these premises had been let by the dean and chapter at Michaelmas 1649 for twenty-one years to Miles Carter for £5 and 12 quarters of wheat yearly. This was presumably a beneficial lease, as the improved value, due to the recently completed drainage and the construction of the Bedford River, was £111 13*s*.[19] Miles Carter was apparently undisturbed, and some time before 1661 sold his tenancy of the manor and demesnes, with a 'frish' called Mepal Holt, to John Lord and Elizabeth his wife, who in that year petitioned the dean and chapter to admit them as tenants.[20] A later lessee was James Halman, Master of Caius College, who in 1702 bequeathed his lease to the college for various benefactions. In the early 18th century the college were paying the dean and chapter an annual quit rent of £5 with 2 quarters of wheat, a calf and a boar. The fines payable for renewal were £70 in 1710 and 1717 and £87 in 1724. The estate then consisted of the manor house with outbuildings, a dovehouse, orchard, and 4 cottages, 80 acres of arable, 20 of meadow, 55 of old pasture, 80 of fen, and a sheepwalk for 300 sheep.[21] The Manor Farm was in 1702 sub-let for £85 a year. Halman's bequest was valuable to the college. The rent paid by the under-tenant rose steadily from £95 in 1710 to £432 10*s*. in 1866, and the fine on admission from £70 to £900. In 1871 the Ecclesiastical Commissioners, as successors to the dean and chapter in the lordship of the manor, bought out Caius College for £5,500.[22]

Mepal Holt, which originally included 60 acres, but some of which was 'cut away' by the Adventurers, was in the early 18th century on lease to a Mr. Carter, perhaps a descendant of the Commonwealth lessee of the manor, at £2 a year, with fines for renewal of £9 to £10. In 1716–7 the growing timber was sold to him for fencing for £30.[23]

CHURCH

The rectory, like the manor, was never a wealthy one; it was valued at £1 10*s*. in 1217, £2 6*s*. 8*d*. in 1254,[24] £3 6*s*. 8*d*. in 1291,[25] and £2 16*s*. 8*d*. in 1535.[26] In 1562 the value was stated to be the same as in 1535, subject to a pension of 6*s*. 8*d*. to the bishop.[27]

In 1553 an acre of land in the tenure of Thomas Smythe provided lights in the church.[28]

The advowson of the church was originally with the bishop. In the early 14th century it was in the hands of the Colne family as episcopal tenants.[29] Since the transfer of the manor in 1360 (see above) it has been continuously with the Prior and Convent and their successors the Dean and Chapter of Ely. From 1735 to 1844 the rectory was held with the vicarage of Sutton (q.v.), and since 1922 has been united with that of Witcham.[30]

The church of *ST. MARY* consists of chancel, nave, north vestry, south porch, and a double bell-cote on the west gable. The material is rubble and flint with Barnack stone dressings. The chancel and northern slope of the nave roof are covered with stone slates and the rest of the roofing is tiled. The nave and chancel date from the first half of the 13th century, and the fabric retained its original form until the 19th century. The present porch is modern, but it probably had a medieval predecessor.

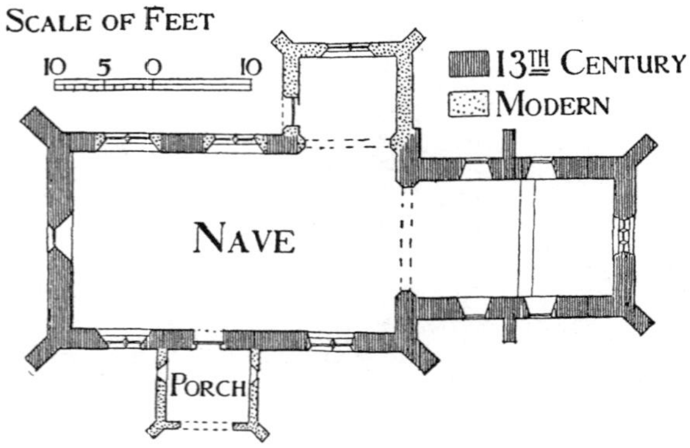

PLAN OF ST. MARY'S CHURCH

The church, which was described by Cole as a 'very mean and poor building',[31] is noteworthy as the only parochial one in the Isle without either tower or spire. An early restoration took place in 1849 at a cost of £600. It was executed under the supervision of James Hastie 'the talented clerk of the works' then 'going forward for the restoration of Ely Cathedral'. 'Much merit' was said to be 'due to the respected rector,[32] and to Mr. William Vipan (churchwarden) for this spirited restoration'.[33] There was also much restoration in 1876–84 and again in 1905 under W. D. Caroe.[34] The vestry was added in 1908. The fabric remains a good example of a small village church of the 13th century.

The chancel has an east window of three lights with plate tracery, which has been renewed on the exterior; there is a rather clumsy modern hood-mould of 13th-century character. There are diagonal buttresses with one set-off, and there is a similar buttress in the centre of the north and south walls. There are two plain lancets on the north and south, the latter having modern hood-moulds similar to that of the east window. There is a blocked low side window at the west end of the south wall. On either side of the east window on the interior is a 15th-century canopied niche. There is a piscina and credence recess with a plain rectangular aumbry. The chancel arch is two-centred and of two orders with moulded caps and bases to the responds.

The nave has two windows on the south side consisting in each case of a pair of plain lancets partly renewed, with a modern hood-mould. The south doorway has a plain two-centred arch with a continuous chamfer. There is a large single lancet in the centre of the west wall with a modern hood-mould and diagonal buttresses with one set-off. In the north wall is a modern two-centred arch opening to the vestry, and to the west of this two small windows each of two plain lights, which are modern insertions; there is no evi-

19 C 54/3449, no. 26.
20 B.M. Add. MS. 5807, f. 129.
21 D. & C. Mun. Bentham Transcripts.
22 E. H. Gross, *Biog. Hist. Gonville and Caius Coll.* iv (2), 63, 64.
23 D. & C. Mun. Bentham Transcripts.
24 W. E. Lunt, *Valuation of Norw.*, 539.
25 *Tax Eccl.* (Rec. Com.), 265.
26 *Valor Eccl.* (Rec. Com.), iii, 499.
27 *Cal. Pat.* 1560–3, 226.
28 Ibid. 1553, p. 290.
29 See note 5 above.
30 P.R.O. Inst. Bks.; *Crockford's Clerical Dir.*; *Clergy List.*
31 B.M. Add. MS. 5807, f. 129*a*.
32 Revd. F. Daubeny.
33 R. Gardner, *Dir. Cambs.* (1851), 563.
34 *Camb. Antiq. Soc. Proc.* xxxv, 84.

dence of ancient fenestration in the north wall. The vestry, which projects at right angles to the nave, has a plain window with two uncusped lights in the north wall and diagonal buttresses with one set-off. In the west wall of the vestry is a plain chamfered doorway. The modern porch has an outer doorway with continuous chamfer and diagonal buttresses with one set-off. In the east and west walls is a lancet. The walls are internally faced with brick.

The bell-cote, which is very effective, is corbelled out on the west and has a gabled top which has been rebuilt. All the roofs are modern and poor.

The font has a plain square bowl, probably of the 13th century, with slight traces of colour, and a modern shaft and base. Most of the windows are filled with clear glass.

The plate includes a silver chalice of 1569 and paten of 1570.[35]

The bell-cote contains one bell, uninscribed, cast in 1849.

The registers begin in 1659 and are complete.

NONCONFORMITY A congregation of 'United Protestant Dissenters' numbered about 60 in 1851. Their chapel dated from 1846.[36] This building is now used[37] as a Baptist chapel, but the congregation is not in association with the Baptist Union. The attached Hiam Memorial schoolroom dates from 1922.[38]

SCHOOLS In 1846–7 there was a Sunday school in Mepal, held in the church, and attended by about 20 boys and 30 girls. The master and mistress received £6 a year, the total annual cost being £8.[39] Such children as wished to receive daily education had to go to the endowed school over the hill at Sutton until 1875, when a Church school was built in Mepal village for 81 children. The National Society granted £25 towards the cost of £550, and the Ecclesiastical Commissioners gave the site. Proposals for this school had been made as early as 1872, but were not immediately realized since some of the inhabitants wanted a School Board.[40] The school was enlarged in 1914 by the addition of a new classroom, for which the National Society granted £9. It was improved in 1947 at a cost of £283. The school became one for junior mixed and infants in 1937, reverted to an all purposes school during 1939–45, and again became a school for junior mixed and infants in 1948. The average attendance was 45 in 1938.[41]

CHARITIES Alexander Aspland, by his will dated 1723, gave 6 acres for charitable purposes. In 1837 this was let by auction for £13 per annum, of which the churchwardens paid out £10 16*s*. to non-paupers in sums of about 3*s*. No books were kept.[42]

The Charity Commissioners reported that 81½ acres in Sutton 'supposed to have been the remainder of certain fenland after divers allotments had been made in pursuance of a decree of the Court of Chancery' (1624) were let for a sum varying between £122 and £160, according as the land was to be in pasture or arable. About £14 of this sum was given indiscriminately to the poor of Sutton and Mepal in money, and in providing coal at the rate of 25 bushels for each family.[43]

STRETHAM AND THETFORD

Stretham is a village and parish on the southern border of the Isle, 4½ miles south of Ely. It lies at the point where the Cambridge road (A 10) is crossed by that from Earith and Chatteris to Soham and Newmarket (A 1123 and B 1085). The 'street' from which the village derives its name is the Roman road leading northwards from Cambridge, which, though now the main road entry into the Isle from the south, was in the Middle Ages of little importance compared with that over Aldreth causeway. It began to supplant the Aldreth road (see Haddenham) towards the end of the 17th century, and is shown in the *English Traveller's Companion* (1676) and the 1695 edition of Camden's *Britannia*.[1] In 1763 the task of bringing it up to contemporary standards was begun, with a gift of £200 from Alderman Riste of Cambridge, and completed some twenty years later at a cost of upwards of £7,000.[2]

The parish is roughly bounded on the south and east by the Old West River and the Cam, but the line does not everywhere exactly coincide with the present watercourses. Half a mile north of the village a by-road leaves the main Ely road for Witchford. Stretham railway station (actually in Wilburton parish) is situated where this road crosses the Ely–St. Ives branch of the former L.N.E.R. The section on which the station lies dates from 1866.[3] The station has been closed for passengers since 1931,[4] but a considerable amount of goods traffic is still carried. The main line from Cambridge to Ely crosses the south-east of the parish, and has a signal box (Stretham Fen) and sidings, but no station.

In its present-day aspect Stretham is a typical village of the Isle. There is no 'great house', but there are several of medium size. There are also numerous 19th-century terrace cottages of the kind seen in industrial towns, and equally numerous council houses, mostly well designed, grouped in two separate blocks at opposite ends of the village.

Several fires have occurred in the village. In 1696 damage assessed at £2,170 was so caused and a general collection made by brief.[5] A more serious fire broke out on May Day 1844 in the blacksmith's shop. The whole of 'Bell Street' was destroyed and damage done to the extent of £20,000.[6] In the autumn of 1850 four fires, the work of incendiaries, occurred at Stretham in as many weeks, 'by which property to a large amount was

35 Infm. Revd. R. H. Gibbon, Rector.
36 H.O. 129/7/190.
37 Kelly, *Dir. Cambs.*
38 Inscr. on bldg.
39 Report of Nat. Soc. (priv. printed).
40 Archives of Nat. Soc.
41 M. of Ed. File no. 12/40.
42 *Rep. Com. Char.* (*Cambs.*) [103], 328–9, H.C. (1837–8), xxiv.
43 Ibid., pp. 329–31.
1 H. G. Fordham, *Cambs. Maps*, 127–9. It may be noted that the present Cambridge–Ely road makes two right-angle bends in Stretham village to aline with the now less important Earith–Soham road.
2 D. & S. Lysons, *Magna Brit.* ii (1), 41. 'Stretham Ferry' is first called 'Stretham Bridge' in the 7th (1786) edition of Paterson's Road Book.
3 *V.C.H. Cambs.* ii, 132–3. The line was opened from Ely to Sutton in 1866. It was extended to St. Ives in 1878.
4 Inf. Publ. Relations Offr. Eastern Region, British Railways.
5 *E. Anglian*, N.S., vii, 146; viii, 14–15.
6 *The Times*, 6 May 1844.

sacrificed'.[7] A detective from London made investigations, but the culprit was not discovered.[8]

In the Middle Ages the value of Stretham, in relation to that of other episcopal manors, was neither very high nor very low. It is worth noting, however, that it was one of the few villages in the Isle to indulge in the luxury of a stone spire for its church.[9] The erection of a fine cross and the presence of three guilds also point to some medieval affluence. After the Reformation Stretham seems to have advanced in size. In the 'censuses' of 1563[10] and 1676,[11] which credit Stretham-with-Thetford with 119 householders and 428 communicants respectively, the parish stands eighth in size among the towns and villages of the Isle; in the ship money assessment of 1639–40 (£50 *7s. 6d.*)[12] it stands sixth. Owing to its geographical position Stretham had little share in the agricultural development of the Fens in the 18th and 19th centuries, which caused such rapid increases at, e.g., Doddington and Littleport. The population density (25 per 100 acres), however, is now high for the southern part of the Isle.

A market and two annual fairs were established in 1634,[13] and were in existence forty years later, when Ogilby described Stretham as 'somewhat scattering'.[14] At the beginning of the 19th century they had long been in abeyance.[15]

In the main road to the east of the church is a fine stone cross of early-15th-century date, the most perfect surviving example in the county. It stands on a brick base, which was substituted some years ago for the original flight of stone steps. The shaft rests on an octagonal base ornamented with quatrefoils, and the square head has a shallow niche on each face with a bracket for a figure. In spite of the great fire of 1844, Stretham has some interesting old houses. The Rectory (see below—Churches) is of some architectural importance. The following 17th-century buildings are noteworthy: the house 10 yards east of the Red Lion Inn, now divided into two; the house next the grocer's shop in Front Street; the two cottages at the corner of Read's and Chapel Streets. Mention may also be made of: Manor Farm, 18th century but with portions possibly of earlier date; Orchard House, 18th century; and Hylton House, Top Street, a Regency villa. A fine tower windmill, now sailless, stands at the north end of the village.[15a] A parish room was erected in 1886.[16]

Mark Ridley (1560–1624), who became physician to the Czar of Russia, was born in Stretham. His father, who was rector, was a cousin of Bishop Ridley.[17] Another noteworthy rector was H. H. Baber (1775–1869), philologist and Keeper of the Printed Books at the British Museum. He resigned his keepership in 1837 to devote himself entirely to his parish. He was one of the chief sufferers in the crop of fires that occurred in the village in the middle of last century.[18] Dr. Ralph Perkins, rector 1696–1727, was a great benefactor to Ely Cathedral Library and to various ecclesiastical charities.[19] Charles Thomas, rector 1728–71, is described by Cole as 'a man of Parts, but of an odd and whimsical Disposition'—a reputation illustrated by his will, which contained complicated instructions about his burial.[20]

Thetford or Little Thetford is a hamlet of Stretham for ecclesiastical purposes, but a separate civil parish with its own Council. The small village is situated on a side turning from the Ely road about 2½ miles south of the city. It presents a picturesque appearance from the railway, the main line running close to its eastern end. The name, Theod-ford, 'ford of the nation (road)',[21] suggests that the main road into Ely may at one time have followed the lower ground past Braham Farm, where there is an ancient earthwork, and into the city by Cawdle Fen Drove and Potter's Lane to the waterside, the medieval focal point of Ely. An area of 233 acres was transferred from Ely Urban District to Thetford under the Isle of Ely Review Order, 1933, thus increasing the population by 29.[22] A parish room was erected in 1887.[23] The Three Horseshoes Inn at Thetford is an 18th-century building of some interest, built on a slight curve. The Round House, now two cottages, is even more remarkable in its plan. It also dates from the 18th century and is of brick with a thatched roof of conical shape with a finial.

The open fields of Stretham, measuring 2,390 acres, were inclosed in 1837.[24] This was about half the total area of the parish. The fields were named Crouch Field (east and west of the Ely road), Whitecross and Brook Fields (respectively north and south of the Wilburton road), and Meadow Field (east of the Cambridge road). There were also open commons and fens. Snoot Common, Middle Common, Elford Closes, Chear Fen, and Lazier Fen lay at the extreme south of the parish, and Starlock Hay Fen Common and Holt Fen in the south-east. Most of these names survive. There were 78 allottees, of whom the chief were William Edwards Read (228 acres), George Gould Morgan, lord of the manor (173 acres), Thomas Waddelow Granger (142 acres), and the Ely Feoffees (135 acres).

The Thetford fields, measuring 693 acres, were inclosed in 1844.[25] This was about half the total area. The fields were named Stone Bridge, Middle, and Goldsmere Fields. Goldsmere Common, and Hall, Holt, and Reed Fens were also affected. The land was rather more evenly divided amongst the 42 allottees than in Stretham. The devisees of Mary Hammond (60½ acres) and William Yarrow (45 acres) received the largest shares, and 7 other proprietors received more than 20 acres apiece.

Two thousand acres in Stretham and Thetford Fens were flooded in March 1947. The damage suffered was severe.[26]

MANORS

An estate in *STRETHAM* of 9 hides 24 acres, with two fisheries, was one of the benefactions of Ethelwold and Brithnoth to the church of Ely.[27] In 1086 Stretham was assessed at 5 hides. Three of these, comprising 4 ploughlands,

[7] R. Gardner, *Dir. Cambs.* (1851), 505.

[8] *Camb. Chron.* 5, 26 Oct. 1850. The property destroyed in the fires of 20 and 28 Sept. was insured for as much as £2,700.

[9] Only six churches in the Isle (Stretham, Ely St. Mary, Whittlesey St. Mary, Doddington, Leverington, March) have this feature, out of 29 parish churches and chapels of medieval date. The proportions in Huntingdonshire and the parts of Holland, neighbouring counties of similar area, are 34 out of 95 and 15 out of 37 respectively.

[10] B.M. Harl. MS. 594, ff. 198–200.

[11] Bp.'s Mun., Ely, A 6.

[12] *E. Anglian*, N.S., vi, 40, quoting S.P. 16/445/75.

[13] C 66/2643, no. 2.

[14] *Britannia* (1675), 86.

[15] D. & S. Lysons, *Magna Brit.* ii (1), 18.

[15a] See plate facing p. 91.

[16] Kelly, *Dir. Cambs.* (1920).

[17] *D.N.B.*

[18] Ibid.; Gardner, *Dir. Cambs.* (1851), 505.

[19] J. Bentham, *Ely*, 259.

[20] B.M. Add. MS. 5820. f. 72.

[21] *P.N. Cambs.* (E.P.N.S.), 242.

[22] M. of H., no. 76940.

[23] Kelly, *Dir. Cambs.* (1900).

[24] Incl. Awards, County Hall, March. The Stretham Act was 5 & 6 Wm. IV, c. *13*, the Thetford Act 7 & 8 Vic. c. xxix.

[25] Ibid.

[26] M. of Agric., *Harvest Home*, 44–45, 87.

[27] Jas. Bentham, *Ely*, 75.

were in demesne. Another 5 ploughlands were divided between 23 villeins, 12 of whom had 10 acres each; the remainder shared 1 hide. There were also 10 cottars and 2 serfs. Besides the usual meadows and pastures, there were fisheries producing 3,250 eels and 7*s.* 7*d.* in tribute. The estate, which had always been part of the demesne of the church of Ely, had been worth £6 when received, £12 T.R.E. and £9 in 1086.[28]

Stretham became an episcopal manor in 1109. It was one of the smaller ones, with no important tenants by military service, so that our main sources of information for its history during the Middle Ages are cartularies and bailiffs' accounts. In 1221[29] 338 acres were in demesne, and in 1251[30] 477½ acres, rated at 2 carucates. The arrangement of the fields is interesting. At both dates there were four comparatively small ones named from the points of the compass, representing perhaps a subdivision of an original two-field vill, but the two fields of 1221 known as 'Alvernercroft' and 'Brocfeld' do not reappear in 1251. It is possible that they were merged in 'Suthfeld', where the demesne increased from 60 to 158 acres. This field, however, was on the fen side of the vill and its increase may therefore have been partly due to reclamation. Another field, which appears in both cartularies, was called 'Hay', a precursor of the present Starlock Hay Fen Common south-east of the village. The lord's share of the meadow increased from 18 to 80 acres between 1221 and 1251. At the latter date the demesne was stocked with 20 cows and a bull, 20 sows and a boar, and 300 sheep. In addition more than 100 belonged to tenants. Stretham, being one of the manors on the landward side of the Isle, did not make much progress during the 13th century. Between 1221 and 1251 the free tenants decreased from 23 to 19, and the customaries holding a full virgate from 18 to 13. In each year, however, there were 7 half-virgaters, 9 cottars, and 1 'half-cottar'. In 1251 the marsh and fisheries produced £8 12*s.* yearly, and £1 14*s.* 4*d.* was due in rents. Stretham manor, like many others of the Bishop of Ely, seems to have reached its zenith as a source of profit about the beginning of the 14th century. The total receipts in 1286, 1298–9, 1302, and 1316 were £15 12*s.* 1*d.*, £37 5*s.* 8*d.* (two years), £24 19*s.* 3*d.*, and £17 19*s.* 7*d.* respectively.[31] Stretham seems at this time to have been about midway in the population scale among the episcopal manors in and around the Isle proper; in 1302 899 works were commuted.[32] The decline in value continued during the 14th century. In 1356[33] the total receipts fell to £13 18*s.* 8*d.*, of which £8 13*s.* 9*d.* came from a demesne of 347 acres. The fisheries, specified as those of Strethamere (held in severalty), 'Estewater', Hayfen, and 'Berlake', produced £3 19*s.* 8*d.*,[34] but the 20 acres of 'fennemedwe' were flooded and worthless, as was the ruined manor house of Strethambury. The bishop's movable property was valued at £19 18*s.* In the middle of the 16th century the value had risen again; it was £44 7*s.* 2½*d.* in 1540–1 and £61 14*s.* 1*d.* in 1548–9.[35] At the latter date the manor house was let for £9 13*s.* 4*d.*, the demesne for £2 1*s.* 6*d.*, and the fisheries for £15 7*s.* 4*d.* The average gross annual receipts at the end of the 16th century were £50.[36] Between 1581 and 1600 a fresh survey was made of the manor, preparatory to detaching it from the possessions of the see.[37] Rents of assize yielded £16 15*s.* 6*d.* a year, but it was not certain how much land was freehold and how much copyhold. The total capital value of the manor was set down as £1,656 at forty years' purchase, but it was considered that the value might be increased. For example, the fisheries were let to Richard Arkenstall, nephew of the late Bishop Cox, at £40 a year; they were therefore of £1,600 capital value; but it was estimated that they were worth £40 over and above this sum. The Berrystead or manor house was valued at forty times its annual rent of £9 13*s.* 4*d.* and £100 over and above that sum, i.e. at £486 13*s.* 4*d.* The demesne, or what was left of it, was still let at £2 1*s.* 6*d.* Early in the reign of Elizabeth the actual tenancy of the manor house was in dispute. Bishop Redman (1501–5) had leased it to Thomas Gallon or Gallant for a term which was still unexpired in the latter half of the century, when Henry Gallant of Wilburton, Thomas's grandson, claimed it against Robert March of Ely.[38] March based his claim on a lease made about 1529 by Bishop West (1515–33) to Richard Goode, and argued that this lease cancelled the previous one. Goode had bequeathed his tenancy to his sons Thomas and Richard jointly, and the sale of his moiety by Richard the younger to Thomas produced further complications.[39] Judgement was given (1567) in favour of Robert March.[40]

During the episcopal vacancy of 1581–1600 the lessee of the manor was Richard Drake,[41] who seems to have sub-let it to Henry Vernon. The latter was reported in 1601 to be desirous of purchasing Stretham manor, of which he was tenant.[42] The actual grantee, however, in the same year, was Sir Miles Sandys.[43] Sandys was soon in dispute with his tenants over common rights, and a series of lawsuits occurred between 1605 and 1610.[44] Several documents referring to these suits were preserved in the parish chest, and have been published.[45] It is noteworthy that even at this late date copyholders had to plough for the lord in respect of each yardland 2 acres in wheat and barley. Sir Miles, however, withdrew his right to any other kind of day works and to heriots, and agreed to demand only fines certain on admission to a copyhold.[46] Copyholders were also allowed to sell the timber on their holdings, and to sub-let them. By decree of the Exchequer in 1607 Sandys was empowered to inclose his demesne, and 1,600 acres were to be set aside as common pasture for the tenants of the manor. Hole, Lazier, Chair, Stallock, and Hay[47] Fens are specified as being included in this settlement, as part of the 1,600 acres of common pasture. Two years later complaints were made that the number of working horses which might be fed on

28 *V.C.H. Cambs.* i, 368.
29 B.M. Cott. MS. Tib. B. II, f. 102*b* et seq.
30 B.M. Cott. MS. Claud. C. XI, f. 43*b*–48.
31 S.C. 6/1132/9, /10, /11, and /13.
32 The largest number of works commuted then in this area was in Haddenham (1,347), the smallest in Littleport (477).
33 B.M. Add. MS. 6165, f. 270.
34 Stretham Mere fishery is assessed in the MS. at £20, presumably a scribal error for 20*s.*
35 A. Gibbons, *Ely Episc. Rec.*, 12.
36 S.C. 6/Eliz./282 and /283; S.P. 12/268/110.
37 S.C. 12/30/33, f. 11 et seq.
38 C 3/129/15.
39 C 3/209/48.
40 C 78/32, m. 32.
41 C 66/1394, m. 33.
42 Hist. MSS. Com., *11th Rep.*, App., 585 (Salisbury MSS.).
43 D. & S. Lysons, *Magna Brit.* ii (1), 260.
44 E 134/3 Jas. I/M. 30; E 178/3609.
45 By Archdeacon Cunningham, *Camden Misc.* xii, 253–89 (1910).
46 The change from fines arbitrary was made about the same time at Doddington by Sir John Peyton, another grantee of the Ely episcopal estates, and slightly later by the Dean and Chapter at Sutton in the Isle.
47 The two last named fens were adjacent, and reckoned as one in the 1622 document quoted below. Cf. the modern Starlock Hay Fen Common.

the commons was insufficient. It was then laid down that the tenant of every commonable house could pasture 5 such horses, with an extra 1 for every 15 acres, up to 60 acres, that he cultivated; the rector was allowed an additional 3 horses. A slightly lower scale was laid down in respect of Goldsmore Common in Thetford. In this case 3 horses were allowed for each tenement, with an extra one for every 15 acres up to 45 acres; the lord of the manor was allowed 5 horses in all. A further set of orders and by-laws made by the commoners of Stretham in 1622 defined their rights in greater detail. Each commoner was entitled to pasture 6 cows and 2 weanling calves, and an extra cow if he was in possession of a bull, in the 'heardwalke'. He might also pasture 8 working horses or mares with their foals in Starlock Hay Fen, which had become the horse pasture. Cattle might be hired to make up the permitted total, but joisting, or the pasturing of cattle from outside villages for money payments, was not allowed.[48] The ditch of Barlake, between the cow and horse pastures, was to be communally scoured each year. Cattle were not to be allowed on the horse pasture nor horses on the cattle pasture; pigs were not to be allowed on either. Chair, Hole, and Lazier Fens—apparently the commons set aside for cattle pasture—were not to be used for turf cutting. A scale of penalties was imposed. It varied from 2*d.* for letting swine into the pastures to 40*s.* for joisting and for trapping fish in 'netts or engines' in the breaches and gulls of the fen banks.

Sir Miles Sandys, with his son of the same name, levied a fine upon the manor in 1630.[49] The younger Sandys died without issue in 1654,[50] after which the manor passed, probably by purchase, to Sir Charles Sedley, the Restoration dramatist and wit. In conjunction with his wife Catherine (Savage), he was dealing with it in 1658.[51] After this date the descent becomes obscure. Cole states that Sir Walter Walker was joint lord with John Hampson in the following year.[52] Walker was a Judge of the Courts of Admiralty and of Probate;[53] his grandson of the same name was vouchee in a recovery of Stretham manor in 1694,[54] and died without issue before 1703.[55] In 1758 'Mr. Walker's estate' was purchased by the Earl of Hardwicke for £500.[56] During the 18th century the Lords Byron of Rochdale were in possession of a moiety of the manor.[57] William, the 5th baron, sold it in 1774 to (Sir) Charles Gould,[58] who inherited the other moiety which had passed through the Hampson and Harbord families.[59] He took the name Morgan on his elevation to the baronetcy, and the Stretham portion of his estates, comprising 1,213 of the 4,085 acres in the parish,[60] eventually devolved upon G. G. Morgan, sometime M.P. for Brecon. Morgan died in 1845, and his executors were recorded as holding manorial rights six years later.[61] In 1861 the lordship was divided between Lt.-Col. (afterwards Gen. Sir) G. H. S. Willis and the Revd. W. N. T. Marsh in the right of their wives, who were Morgan's two daughters.[62] Willis died without heirs in 1900, being then sole lord.[63] Manorial rights have since lapsed.[64]

LITTLE THETFORD was described as a berewick of Ely in Domesday Book. It was rated at 1 hide held in demesne; there were a villein with 6 acres and 4 cottars. The Thetford fisheries yielded 500 eels and 4½*d.* in tribute. The vill had been worth 20*s.* when received, 30*s.* T.R.E., and 40*s.* in 1086.[65]

In the 13th-century cartularies Thetford was treated as a part of Stretham, but the 'hayfen' of Thetford, a marsh lying between 'Bramemere'[66] and 'Averingmere', is expressly mentioned in 1251. Moreover one of the 1221 free tenants, and three of those of 1251, were described as 'de T(h)eford'.[67] Thetford first emerges as a separate manor in 1539, when Edmund Knyvett and his wife Joan (Burghden) made a settlement of it.[68] Thomas Knyvett was dealing with the manor in 1572,[69] and ten years later sold it to John and Oliver Thornton for £480. It then consisted of 20 messuages with tofts, gardens, and orchards, a dovecote, land estimated at 300 acres of arable, 100 acres of meadow, 100 acres of pasture, 200 acres of furze and heath and 2 acres of woodland, £16 in rents and foldage for 300 sheep.[70] Sir Roger Thornton died seised of the manor in 1631[71] and his son Samuel made a family settlement the following year.[72] At this time Thetford manor was still reckoned to be subsidiary to Stretham; orders were made for its common pasture of Goldsmore in the same manner as at Stretham (see above). By 1676 John Childe was in possession of Thetford.[73] In his will (1681) he directed the manor to be sold, the proceeds to go to his son John (6 parts), his daughters Penelope and Mary (3 and 2 parts), and his relict Elizabeth (1 part). Between 1703 and 1707 John Childe the younger held it.[74] Later in the 18th century it was held by Elizabeth Fisher.[75] She mortgaged it to William Sharp for £100,[76] and with her second husband William Aldred sold it to John Drage in 1768.[77] Drage's representatives, the Revd. Dr. Nasmith and James Merest, were lords in 1808.[78] In 1851 Henry Pigott was recorded as the owner,[79] and in 1861 Eliza Pigott, G. S. Hill, and his wife Hannah Mary held the lordship jointly.[80]

The convent of Ely obtained some interests in Stretham and Thetford although the manorial rights had been assigned to the bishops. For example, in 1150 Bishop Niel confirmed the gift made by Wigan the priest in frank almoin of 2 acres 1 rood of land and 3 roods of meadow; at an unknown date Richard of St. Albans granted 4 acres and 1 rood of land in the fields of Stretham at a yearly rent of 4*d.*, which rent was quit-

[48] The name Joist Fen survives in the neighbouring parish of Waterbeach, and at Lakenheath (Suff.).
[49] C.P. 25 (2)/400/East. 6 Chas. I.
[50] G. E. C., *Compl. Baronetage*, i, 90.
[51] C.P. 25 (2)/539/East. 1658.
[52] B.M. Add. MS. 5847, f. 11, quoting Rand's Coll. See also E 134/1659/M. 1, and /18 Chas. II/M. 8.
[53] B.M. Add. MS. 5820, f. 74*a*.
[54] C.P. 43/446, ro. 88.
[55] G. E. C., *Compl. Baronetage*, iv, 114.
[56] B.M. Add. MS. 5820, f. 76.
[57] C.P. 25(2)/1281/East. 9 Geo. III, and /1282/Hil. 14 Geo. III; C.P. 43/755, ro. 315. The poet Byron was a member of this family.
[58] C.P. 43/763, ro. 134.
[59] Gardner, *Dir. Cambs.* (1851), 504.
[60] Stretham Incl. Award, County Hall, March.
[61] Burke, *Peerage*, s.v. Tredegar; Gardner, *Dir. Cambs.* (1851), 504–5.
[62] Incl. Award for Grunty Fen (M.A.F. 1/1114).
[63] *D.N.B.*, Suppl.; Kelly, *Dir. Cambs.* (1900).
[64] Kelly, *Dir. Cambs.* (1933).
[65] *V.C.H. Cambs.* i, 368.
[66] cf. Braham Farm, south of Ely.
[67] B.M. Cott. MSS. Tib. B. II, f. 103*a*; Claud. C. XI, ff. 44*b*, 45*a*.
[68] C.P. 25(2)/52/373.
[69] C.P. 25(2)/260/Trin. 14 Eliz.
[70] C.P. 25(2)/846/Mich. 24–25 Eliz.
[71] C 142/464/24.
[72] C.P. 25(2)/400/East. 8 Chas. I.
[73] C.P. 25(2)/633/Mich. 28 Chas. II.
[74] C 5/231/13.
[75] C.P. 25(2)/1103/Hil. 22 Geo. II, and /1281/Hil. 7 Geo. III.
[76] E 134/10 Geo. III/H. 6.
[77] C.P. 25(2)/1281/Hil. 8 Geo. III.
[78] D. & S. Lysons, *Magna Brit.* ii (1), 260.
[79] Gardner, *Dir. Cambs.* (1851), 506.
[80] Incl. Award, Grunty Fen (M.A.F. 1/1114).

claimed to the convent by his son Henry.[81] A licence to acquire land in Thetford and elsewhere to the value of £10 was granted in 1387.[82] In 1541, when the conventual property was formally made over to the dean and chapter, the Stretham portion was worth £2 17*s.* 8*d.*[83]

Denny Abbey also held land in Stretham. In 1558 a sixty-year lease was granted to Sir Robert Chester of *inter alia* fisheries in Stretham known as 'Weremere', 'le Cote', and 'Fourdewere', with their appurtenant ferries and osier holts.[84]

In 1303 Peter de Champvent held 42 acres of land in Stretham, Thetford, and Haddenham of the Bishop of Ely by suit of court every three weeks.[85]

CHURCHES The advowson of Stretham was not, like the manor, alienated from the see of Ely in 1600, but has remained with the bishops continuously to the present day.[86] During the confiscation of the temporalities of the see at the end of Bishop Lisle's episcopate (1358), a dispute arose between royal and papal nominees to the living. This was finally settled in 1378 in favour of Robert de Stratton, the papal candidate.[87]

The church was valued at £16 in 1217, £20 in 1254,[88] and £26 14*s.* 4*d.* in 1291.[89] In 1555, however, the value, including the chapel of Thetford, had fallen to £22.[90] Shortly before 1851 the tithes of Stretham were commuted for land,[91] and at the end of the 19th century the glebe comprised 496 acres.[92]

In 1550 a cottage with a yard, and 4 acres in the common fields of Stretham, given for anniversaries, were granted to William Moryce and Edward Isaak. A rent of 6*d.* was saved to the dean and chapter; Robert Hytche was tenant at the time of the grant.[93] An acre of meadow devoted to the same purpose, of which the churchwardens were tenants, was in 1553 granted to John Butler and Thomas Chaworthe.[94]

In 1276 Thomas de Wymbych, rector, renounced all claims to tithes from the lands in the fields of Ely which were held by the men of Thetford.[95] The chapel of Thetford is not mentioned in the *Taxatio*, but the present building displays 14th-century work. Its dedication (St. George) suggests that it may have been newly built at that time.[96] An agreement between Richard Rysley, rector of Stretham, of the one part, and William Riplingham and eighteen other householders of Thetford, of the other, laid it down that the stipendiary in charge of Thetford chapel should have the tithes of the hamlet.[97] The inhabitants were, however, to attend the mother church at Stretham on St. James's Day (the dedication festival), and burial rights were to be reserved to Stretham.[98] Shortly before 1851 the Thetford tithes were commuted for a rent charge of £124.[99] Lands originally intended for the provision of lights and anniversaries in the chapel were sold to Richard Hill and Robert Don, the Elizabethan land speculators, in 1571.[1]

Another chapel called Haveringmere or Harrimere stood on the bank of the river south-east of Thetford. A presentation to this chapel is recorded in 1381.[2] It was included in a grant made in 1563 to Cecily Pickerell, widow, of Norwich, in satisfaction of a debt owed to her late husband by the Duke of Somerset.[3] In 1571 the chapel came into the hands of Richard Hill and Robert Don.[4] It is now commemorated in the names Chapel Hill and Harrimere Drain.

The church of *ST. JAMES, STRETHAM*, consists of chancel, north vestry, south organ chamber, clerestoried nave, transepts, aisles, south porch, and west tower and spire. The material is mainly rag and Barnack stone, and the roofs are tiled. The fabric, in so far as it is ancient, is mainly of early-14th-century date, except the east wall of the chancel which appears to belong to the end of the 12th century. In 1876 the church underwent a severe restoration. The chancel was practically rebuilt and the transepts, south aisle, and clerestory added. The work involved the destruction of the south porch and the provision of new roofs; the chancel was curtailed and the western portion thrown into the nave, and the window tracery renewed throughout in 14th-century style. A photograph of the church before restoration is preserved in the rectory. The cost of the work amounted to over £4,000.

The chancel has an east window of five lights with modern geometrical tracery, and there is a hood-mould terminating in heads. There are clamped buttresses which date from the end of the 12th century. The north-east and south-east windows are modern two-lights with hood-moulds terminating in heads. The chancel communicates with the vestry and organ chamber by modern arches of 14th-century character which die into the responds. The modern chancel arch is two-centred and of two orders, the inner resting on corbels and the outer with continuous mouldings, and there is a hood on the west face. The south-east window has a lowered sill for sedilia. In the north wall is an oblong rectangular aumbry and to the west of it a tomb recess with an acutely pointed arch.

The modern vestry has a two-light window on the east and north and opens to the transept by a modern two-centred arch of 14th-century character. The modern organ chamber has a two-light window on the east and south.

The nave has arcades of four bays with an additional arch of larger dimensions to the east, opening to the transepts. The modern transeptal arches are of two orders with moulded caps and clustered responds. The north arcade is of early-14th-century date with two-centred arches of two orders and hoods terminating in well-carved heads; the columns are octagonal and the responds semi-octagonal with moulded caps and bases. The modern south arcade has round columns with moulded caps and bases and hoods terminating in

81 B.M. Add. MS. 5847, f. 11, quoting Liber M., ff. 158, 533–4.
82 *Cal. Pat.* 1494–1509, 536 (*inspeximus*).
83 B.M. Add. MS. 5847, f. 11; *L. & P. Hen. VIII*, xvi, g. 1226 (12).
84 *Cal. Pat.* 1557–8, 436.
85 *Cal. Inq. p.m.* iv, p. 94.
86 P.R.O. Inst. Bks.; *Crockford's Clerical Dir.*
87 *Cal. Pat.* 1354–8, 604; 1358–61, 72; 1377–81, 314.
88 W. E. Lunt, *Valuation of Norw.* 539.
89 *Tax. Eccl.* (Rec. Com.), 265.
90 *Valor Eccl.* (Rec. Com.), iii, 499.
91 R. Gardner, *Dir. Cambs.* (1851), 505.
92 Kelly, *Dir. Cambs.* (1900).
93 *Cal. Pat.* 1549–51, 275–6.
94 Ibid. 1553, 290.
95 B.M. Add. MS. 5847, f. 16, quoting Bp.'s Mun., Liber M, f. 158.
96 St. George took the place of Edward the Confessor as the patron saint of England during the reign of Edward III; St. George was the patron saint of that king.
97 B.M. Add. MS. 5847, ff. 12–13.
98 The minor road and footpath leading south-west from Thetford village direct to Stretham is still known as the 'Burying Way'.
99 Gardner, *Dir. Cambs.* (1851), 506.
1 C 66/1074, m. 4–5.
2 *Ely Dioc. Remembrancer*, no. 123, p. 139. Harrimere as a locality is mentioned in the 1221 Ely cartulary (*P.N. Cambs.* (E.P.N.S.), 199).
3 *Cal. Pat.* 1560–3, 554–5.
4 C 66/1074, m. 5.

heads. The modern clerestory consists of two-light windows with trefoiled heads. The fine tower arch of 14th-century date is acutely pointed and of three orders with continuous mouldings.

The transepts are entirely modern with the exception of a re-used 14th-century doorway in the east wall with a hood-mould terminating in heads. The north and south windows are of three cinquefoiled lights, and modern arches dying into the responds communicate with the aisles. There are gabled angle buttresses with one set-off. The north transept probably occupies the site of the 'exceeding handsome and elegant Chapel' lights with flowing tracery. In the second stage there is a recessed lancet on the north, south, and west. The belfry stage has two-light windows with trefoiled heads and a quatrefoil above. The tower is finished with a plain parapet. The spire, which is of mid-14th-century date, has two tiers of windows on alternate faces, the lower of two lights and the upper of one, all with crocketed gables.

The roofs are modern, and high-pitched in the case of the chancel, nave, transepts, and porch; the aisle roofs are of lean-to type. The tower ceiling has some moulded beams of 15th-century date.

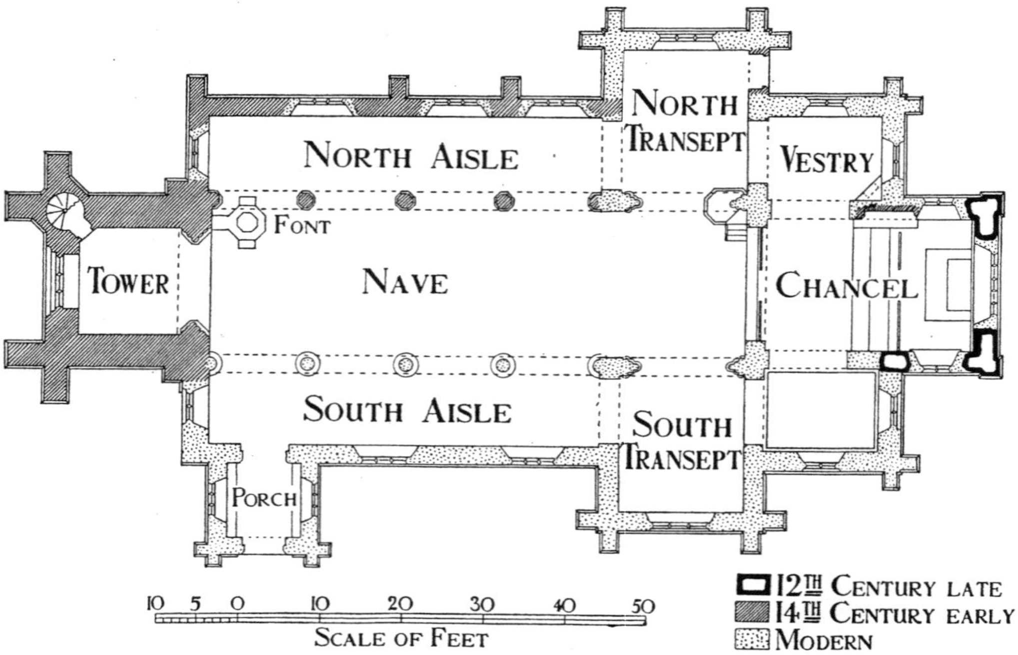

Plan of St. James's Church

noted by Cole in 1748 at the east end of the north aisle. It was said to have been a mausoleum of the Sandys family, and at that date was 'in great Squalidness and used for any Rubbish'.[5] After a period of use as the village school the chapel was demolished in 1770.[6]

The north aisle has three straight buttresses of considerable projection and a base course which is carried round the buttresses. The fenestration is entirely modern, the windows on the north being of three trefoiled lights under a square head and that in the west wall of two trefoiled lights with a trefoil above. The south aisle is entirely modern and the fenestration is similar to that of the north aisle, except that there are only two lateral windows instead of three. There is one buttress with one set-off.

The modern porch has inner and outer doorways with continuous mouldings and a hood; there are small straight-angle buttresses and three trefoil-headed lights on the east and west. Some fragments of 12th-century carved stones are built into the walls on the interior.

The tower is of three stages with angle buttresses having two set-offs, surmounted on the west by diagonal buttresses. The newel stair is contrived in the north-west angle and carried up to the parapet. The base course is carried round the buttresses. The west window has an acutely pointed head and is of three

The font is modern. The mid-15th-century chancel screen is of the rectangular type and of eight bays, with elaborate traceried heads. The two bays in the centre form the doorway, and there are two half doors. The wainscot has applied tracery, but the boarding is modern except in the case of the doors; the uprights are buttressed, and the base and top beams are modern. In the tomb recess on the north of the chancel is a slab with a French inscription in Lombardic lettering to Nicholas de Kingeston, rector at the end of the 13th century; there is the indent of a small brass demi-effigy. In the south transept, but formerly in the chancel, is a good brass effigy of a lady, which formerly had a triple canopy; the inscription is lost but it is known from Cole's account of the church in 1748 to commemorate Joan Swan, who died in 1497 and was the mother of two successive rectors of Stretham. There are two stone coffin lids with foliated crosses, now in the north transept. A handsome black marble slab commemorates Anne, wife of Dr. Brunsell, rector 1662–78, and sister of Sir Christopher Wren; it has a shield of arms with Brunsell impaling Wren. The other fittings are modern.

In Wilburton manor house are fragments of 15th-century painted glass which came from Stretham church. In the church and village hall of Wilburton is some panelling formerly in Stretham church.

[5] B.M. Add. MS. 5820, f. 73.

[6] W. M. Palmer, *M.I.... from Cambs.* 261, quoting Faculty Bks. i, 107.

The plate includes a chalice and paten of 1686, and a paten given by Ralph Perkins, LL.D. (rector, 1696–1727), in 1712.[7]

The tower contains five bells. The treble is a 20th-century addition; the 2nd and 5th are by J. Warner of London, 1876, replacing bells (1591 and 1615) by Thomas and John Draper of Thetford respectively; the 3rd is by Joseph Eayre of St. Neots, 1766, and the 4th by Henry Penn of Peterborough, 1727.[8]

The registers begin in 1558.

The church of *ST. GEORGE, THETFORD*, consists of chancel, nave, north porch, and south vestry. The material is rubble and the roofs are covered with tiles. The fabric is of 14th-century origin. A century later new windows were inserted in the chancel. There was a drastic restoration in 1863, when the nave was slightly extended and the porch added. The vestry was erected in 1885. Until the 1863 restoration the nave was separated from the chancel by a solid wall with a doorway in the centre and a square squint on either side; above were three brackets, probably for supporting the rood and its attendant figures. The roofs were thatched.

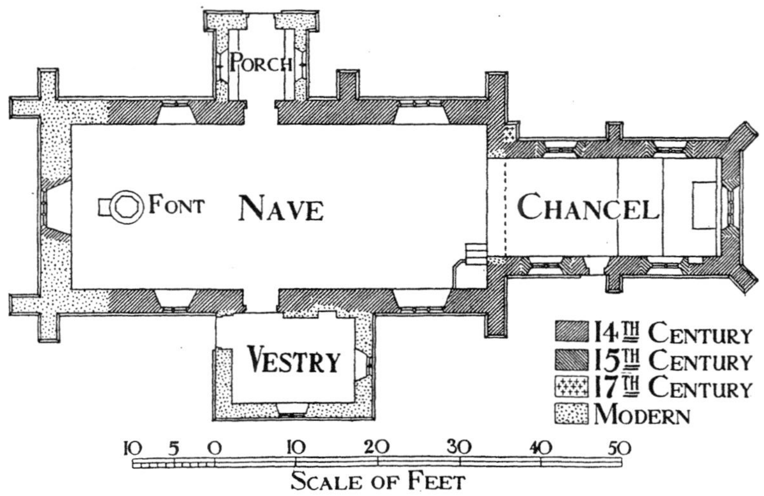

PLAN OF ST. GEORGE'S CHURCH

The chancel has an east window of three cinquefoiled lights under a depressed head, with a hood terminating in heads. There are diagonal buttresses with one set-off, and similar buttresses on the north and south between the windows. The lateral fenestration consists of two windows on each side, of two cinquefoiled lights with a quatrefoil above. There is an external base course and beneath the windows an internal string, which has been largely renewed. The two-centred chancel arch is modern. At the north-west angle there is a plain brick buttress, probably of 17th-century date. In the south wall is a plain piscina with a stone credence shelf, and a 14th-century door just west of the buttress.

The nave, which is wider than the chancel, has a large straight-angle buttress at the east end of the north and south walls and a similar one in the middle of the north wall. The east gable was rebuilt in brick in 1665, as appears from the date on the external face. The 14th-century north-east and south-east windows are of three cinquefoiled lights with a quatrefoil above, under a square head. The north-west and south-west windows are of similar date and design, but of two lights. The west wall, though rebuilt when the nave was extended, retains much old material and a good 14th-century window of three lights with flowing tracery and a hood terminating in masks. The tracery has been renewed. Above are two recesses for bells. There are pairs of straight-angle buttresses, which are modern. The plain north and south doorways are of the 14th century and have continuous mouldings. There is an internal string-course, which has been much renewed.

The porch, which is entirely modern, has an outer doorway with roll mouldings dying into the chamfered responds and a small two-light window on the east and west with trefoiled heads and a trefoil above. The modern vestry, which is entered from the nave by the original south doorway, is faced internally with brick and has a modern two-light window in the east and south walls. There is an external doorway on the west.

All the roofs are modern.

The 14th-century font has an octagonal bowl and shaft with a human head projecting from each alternate face of the bowl; there are tracery heads under the bowl.

The plate consists of a modern chalice and paten of silver and a modern brass alms-dish.[9]

There is one bell, dated 1769, which came from the chapel of Ely House, London.

The registers begin in 1654 and are complete.

The Rectory, to the south of the church, is a large structure of various dates. The material is brick and stone and the roofs are tiled. The oldest remaining portion is at the north end, and is of 14th-century date. It consists of a ground-floor room with a similar chamber above, and probably formed part of the medieval parsonage. The entrance consists of a stone doorway with a depressed head, and there are two small contemporary windows. The central portion of the house is probably of 16th-century date, but its

[7] Infm. Revd. J. E. Cowgill, late Rector of Stretham cum Thetford.

[8] J. J. Raven, *Church Bells of Cambs.* (1869), s.v. Stretham, attributes the 4th to T. Newman of Norwich.

[9] Infm. Revd. J. E. Cowgill, late Rector of Stretham cum Thetford.

appearance was greatly altered in the 18th century when new windows were inserted. Beneath the floor of the dining-room are some original oak joists. Early in the 19th century the house was extended towards the south and dormer windows inserted in the older portion. The interior has been completely modernized. In 1820 a faculty was granted for the demolition of the rectory dovecote, which 'is an injury and a nuisance'. The rectory barn was still standing at this date.[10]

Stretham had three guilds, of the Purification of the Virgin (1344), Corpus Christi (1350), and All Saints (1456). The two former made grants to disabled members of *3d.* a week in victuals and *2s.* a year in clothing 'as long as the goods of the gild last'.[11] A guild of St. George in Thetford is mentioned in a will of 1489.[12] Its hall was in 1571 sold to Richard Hill and Robert Don.[13] It was subsequently acquired by John Townsend, who at an unknown date gave it to the inhabitants of Thetford. Townsend desired that the chaplain of Thetford should have the use of a room in the building, but his wish was not realized. In 1837 the building was dilapidated and used as a workhouse and a school.[14]

NONCONFORMITY During the Interregnum the Baptists of Stretham, of whom John Tabram was the most prominent member, were visited by brethren from the Fen Stanton (Hunts.) congregation.[15] In 1669 Dissenters of unknown denomination met at Stretham 'by stealth and in the night'.[16] Bishop Compton's 'census' of 1676 shows that 29 of the 428 persons of communicant age in Stretham were Nonconformists.[17]

A congregation of Baptists under a Mr. Clark met in the early 18th century, but ceased to be an organized body after his death (*c.* 1725). A generation later (*c.* 1760) itinerant Baptist preachers began to visit Stretham, holding meetings in a barn and later in a disused malthouse. A proper meeting house was built in 1772 and enlarged by the addition of galleries in 1799. Under Joseph Howlett of Cambridge this congregation made rapid strides. It was organized on a regular basis in 1801, and the 14 original members were soon reinforced by 7 newcomers. Its doctrine was Calvinistic and it practised open communion. The meeting house of this body was rebuilt on a larger scale in 1818.[18] In 1851 there was a Sunday school.[19] The present Stretham Baptist Church dates from about 1935. A daughter church was built at Thetford in 1842 and rebuilt in 1867.[20] In 1672 the house of William Outlar in Stretham was licensed for Congregational worship,[21] but this branch of Nonconformity seems soon to have died out.[22] The Wesleyan Methodists built a chapel in 1814 by public subscription; it was rebuilt in 1888.[23] The London Evangelistic Mission built a hall at Stretham in 1884;[24] this still exists, but does not seem to be used.

SCHOOLS In 1579 Thomas Hitche was licensed to act as curate and schoolmaster in Stretham.[25] In the middle of the 18th century a school was held at the east end of the north aisle of Stretham Church, in a chapel which was pulled down in 1770 (see Churches). In 1789 there was a private school in the village, and nine years later there were two schools—one for infants, kept by John Sabbatin, a Dissenter with a 'harmless' reputation, and one for older children under Robert Hopkins, a Churchman 'of a fair and inoffensive character'.[26]

When the National Society made its report in 1846–7 there were said to be 244 children[27] receiving education. Most of them attended a Church school in Stretham held on weekdays, weekday evenings, and Sundays in hired premises. It would seem that many of the children attended part time, for the expenses amounted to only £67 *18s.* a year although a staff of eight was employed.[28] Lack of endowments[29] prevented any real progress; both in 1846–7 and in 1851[30] the rector is mentioned as the sole private contributor.

This state of affairs lasted for another generation. In 1870 the schools at both Stretham and Thetford were still in hired premises. The latter was held in a room lent by the Townsend Charity, which was in such bad condition as to be 'disqualified for government inspection'. With the aid of local subscriptions of about £1,000, a grant from the Diocesan Board, the gift of a site by the rector, the Revd. Hugh Pigot, and contributions from the charities,[31] schools were built in both villages at a total cost of about £1,850. The buildings were designed by the well-known architect J. P. St. Aubyn, and in 1872, their first full year, cost £220 *18s.* *9½d.* (Stretham) and £62 *14s.* *1d.* (Thetford). They provided places for 200 and 75 children respectively.[32]

By 1889 the school population of Stretham had outgrown the existing resources and a School Board was set up.[33] The church authorities leased their school to this Board, which in 1894–5 enlarged it to take 256 children and built another school for 110 infants, some little distance away from the existing school. In 1910 the accommodation of the two schools was scaled down to 198 (mixed) and 100 (infants). In 1948–9 the senior children were moved to the Cromwell School, Chatteris, pending the erection of a Secondary Modern School for Ely Rural District.[34]

At Thetford, the school remained non-provided. Under schemes made by the Charity Commissioners in 1871 and 1950 it was endowed with £25 annually from Townsend's charity. In 1895 the building was enlarged to take 90 children, and in 1950 the school was given Voluntary (Controlled) status.[35]

In 1870 there was a private school in Stretham, kept by Dissenters and catering mainly for the children of tradesmen and small farmers, and a dame school in Thetford.[36]

10 W. M. Palmer, 'Faculty Bks. in Dioc. Reg., Ely', *Camb. Antiq. Soc. Proc.* xxv, 70 (1933–4).

11 W. M. Palmer, *Village Gilds of Cambs.* 344, 398.

12 *Ely Dioc. Remembrancer*, Oct. 1908, p. 149.

13 C 66/1074, m. 4–5.

14 *Rep. Com. Char.* (*Cambs.*), [103], 331, H.C. (1837–8), xxiv.

15 Adam Taylor, *Hist. Engl. Gen. Baptists* (1818), i, 144.

16 *V.C.H. Cambs.* ii, 187.

17 Bp.'s Mun., Ely, A 6.

18 *Cong. Mag.* iii, 57–8 (1820).

19 H.O. 129/7/190.

20 Ibid. Inscr. on bldg.

21 *Cal. S.P. Dom.* 1672, 63, 90.

22 H.O. 129/7/190; Gardner, *Dir. Cambs.* (1851), 505. These authorities do not mention this denomination as existing in Stretham.

23 Inscr. on bldg.

24 Kelly, *Dir. Cambs.* (1900).

25 Bp.'s Mun., Ely, G 2.

26 Ibid., B 6.

27 The population of Stretham, including Thetford, was 1,357 in 1841.

28 Rep. of Nat. Soc. 1846–7 (priv. printed).

29 £7 a year was given by one of the charity trusts.

30 Gardner, *Dir. Cambs.* (1851), 506.

31 For this reason the National Society did not eventually make its promised grant of £70.

32 Archives of Nat. Soc.

33 Kelly, *Dir. Cambs.* (1900).

34 M. of Ed. File no. 12/44.

35 M. of Ed. File no. 12/46.

36 Archives of Nat. Soc.

CHARITIES In 1553 Richard Rysley, rector, gave a rent of £2 charged upon an estate at Upton (Hunts.), and 6*s.* 8*d.*, paid through Christ's College, to the poor of Stretham. By 1837 the value of this charity had increased to £18 5*s.* 8*d.*, which was given indiscriminately in money. Rysley also gave house property in Stretham, including the Chequers Inn and the blacksmith's shop, to the value of £14 10*s.* No documents relating to this part of the charity existed in 1837,[37] by which date the rents had risen to £81 7*s.* 6*d.*, including £31 from the Chequers. After payment of taxes (£10) the residue was given in amounts of 5*s.* to those who could produce tickets signed by three feoffees of the charity. Under the Stretham Inclosure Act[38] this part of Rysley's charity was replaced by the allotment of 30½ acres in Starlock Hay Fen Common for the benefit of the school (see above).

In 1715 Thomas King and Alice his wife in return for £61 surrendered a moiety of 11½ acres to the churchwardens for the use of the poor. Martha Digby, by her will dated 1717, gave 10 acres of copyhold land in Stretham to the poor, and 3¾ acres of copyhold land in Thetford, for the benefit of six poor widows. In 1837 these two bequests produced £27 15*s.* Of this total £5 were given to about 50 poor persons in sums of 1*s.* to 4*s.*, and the remainder to the widows.

In 1727 Lady Effingham Howard gave the interest on £50 stock to the poor of Stretham. In 1774 £70, secured on 6 acres in Stretham fields, had been lent to Edward Morden. In 1837 the interest on this loan was also devoted to indiscriminate poor relief.

At an unknown date John Townsend gave the guildhall (see above—Churches) and 3 roods of land in Thetford fields to the use of the inhabitants of the hamlet. In 1837 the common rights attached to the building brought in £12 and the 3 roods £2 yearly. The income after deductions was given in clothes, coals, beer, and money to non-paupers.[39] Since 1871 this charity has been devoted to educational purposes.[40]

SUTTON

Sutton is one of the larger villages in the county, and is situated at the western end of the Isle of Ely. It lies on the main road from Ely to St. Ives (Hunts.) (B 1381), about 6 miles from Ely. The approach to the Isle along this road from Earith Bridge carries the traveller through typical fenland scenery, the road running along the south-east bank of the New Bedford River, below its water level but above the fields on the landward side of the road. The two western promontories of the Isle enclose the valley leading to Grunty Fen, at the top of which the towers of Ely may be seen in clear weather. The village was formerly also on the main road from Ely to Chatteris (A 142), but during the Second World War a satellite airfield in connexion with Witchford R.A.F. Station was laid out at the point where this road crosses the ridge of the hill, and the road has not since been reopened. A by-road leaves the road to Earith Bridge at the west end of the village street, and crossing the Bedford Rivers at Sutton Gault,[1] serves the fens west of the rivers. It is not, however, a through road.

Sutton has perhaps the most attractive situation of any village in the Isle. It stands on a southern slope which, as at Haddenham, is much given over to fruit growing. In 1675 Ogilby noted the cherry gardens of Sutton as a feature of interest to travellers.[2] The slope is very steep for this part of the country, the ground falling about 60 ft. in half a mile. The village looks picturesque from a distance. The church is a fine one of distinctly East Anglian type. The other buildings in the village of architectural distinction are the Burystead (the former manor house—see below), Eastwood House, with shaped brick gables, on the south side of the mile-long street near the church, and the 18th-century house occupied by E. G. Backler and Lloyds Bank. The reputed Guildhall in the main street and one or two cottages in the Row, the lower village street, are timber-framed and plastered and may be medieval. For a thousand years a corporate body were lords of the manor, and the inevitable consequence of absenteeism goes far to explain a certain air of meanness about the village. This meanness is not entirely dispelled by the presence of some large houses in their own grounds, such as Sutton House at the east end of the village. In this respect Sutton may be contrasted with Thorney, a village of about the same size but with a very different history. It is significant that Cole, though he noted that the church was 'very grand and spatious', found hardly any monuments there; what there were had no coats of arms and only one legible inscription.[3] That labour relations in this very 'open' village were bad during the period of high farming is shown by the crop of destructive farm fires mentioned in the 1851 edition of Gardner's *Directory of Cambridgeshire*.[4]

In 1312 the Prior of Ely received the right to hold a weekly market at Sutton on Thursdays.[5] The village was the only conventual estate outside Ely itself[6] with such a privilege, but the market had disappeared before the middle of the 18th century, and probably much earlier.[7]

In 1623 a lawsuit occurred between the dean and chapter of the one part, and the tenants of their manors of Mepal (q.v.) and Sutton of the other, regarding rights of common in the fen. The defendants were Sir Miles Sandys of Mepal, Henry Jetherell of Sutton manor, Richard Wigmore, Vicar of Sutton, and twenty-seven others.[8] The dispute was settled by inclosure,[9] and in 1675 a parliamentary bill to confirm the decision obtained a second reading[10] but apparently was not enacted. It is uncertain how much land was affected by this 17th-century inclosure. Probably it amounted to the whole of the fen land reclaimed by that date; when the final inclosure of the open fields took place in the 19th century, the proportion of Sutton

37 *Rep. Com. Char.* (*Cambs.*) [103], 330, H.C. (1837–8), xxiv.
38 5 & 6 Wm. IV, c. xiii.
39 *Rep Com. Char.* (*Cambs.*) [103], 331, H.C. (1837–8), xxiv.
40 M. of Ed. File no. 12/46.
1 Probably from Middle English 'galle', marshy ground. There is no gault here (*P. N. Cambs.* (E.P.N.S.), 240).
2 *Britannia*, 145.
3 B.M. Add. MS. 5807, ff. 129–31. Since Cole's time even these monuments have disappeared.
4 p. 507.
5 *Cal. Chart. R.*, 1300–26, 185.
6 The market at Whittlesey (q.v.) was not established until 1715, long after the church of Ely had relinquished its interest in that town.
7 *V.C.H. Cambs.* ii, 87, quoting Carter's *Hist. Cambs.* (1753), 3. The lack of documentary evidence suggests that the market was obsolete by the time of the Interregnum, if not by the Dissolution.
8 C 2/Jas. I/E 7/39.
9 C 5/550/98; C 5/456/76 &/77.
10 *H. of C. Jnls.* ix, 380.

and Mepal that was affected amounted in each case to about one-quarter of the parish, an area roughly equivalent to the 'high lands'. A Sutton Inclosure Act was passed in 1838,[11] and an award[12] made two years later. The award mentions six fields:[13] Branghill Field west and Church Field east of the Chatteris road, Towns End Field north and south of the Ely road, Farm Hill and Claydon Fields in the north-east corner of the parish between Mepal and Witcham, and Every Year Field[14] north-west of the village; also the Meadlands and Upper and Lower Washes in the fenland in the west of the parish.[15] The total area affected was 1,786 acres, including 71 acres of ancient inclosures reallotted. Those of the 121 proprietors who received the largest amounts were John Vipan (93 acres), Thomas Vipan (91 acres), four other members of the Vipan family (207 acres between them), William Cole (100 acres), the trustees under the will of Anne Waddington (97 acres), and the Dean and Chapter of Ely (92 acres). The last allotment included 31½ acres which the dean and chapter received as impropriators of the rectory and 1 acre as lords of the manor.

A branch of the Great Eastern Railway was opened from Ely to Sutton in 1866 and extended to St. Ives in 1878;[16] the station was closed, except for goods traffic, in 1931.[17]

In 1618 the inhabitants of Ely and certain other places in the Isle petitioned the Council to direct the Commissioners of Sewers not to impose penalties for failure to open the drain called Sutton Lode by the appointed time. They claimed that 100,000 acres would be overflowed if the lode was opened before the main drain (the West Water) and its outfalls at Wisbech were completed. By way of answer the people of Sutton and Mepal petitioned that the stopping of Sutton Lode was highly prejudicial to them, because, though the Lode was a branch of the West Water, it had a separate outfall, which was the only good one for the Ouse as far as Lynn.[18] The floods of 1937 and 1947 were especially serious in the tract of fen between Sutton and Earith Bridge; in the latter year the damage had not been fully repaired five months after the floods.[19]

MANOR. Three hides in *SUTTON* were given to the church of Ely by Ulsius, his wife, and his son Alfius when the monastery was restored by Ethelwold and Brithnoth after the Danish invasions.[20] In 1086 Sutton was assessed at 5 hides. There was land for 10 ploughs. Two of the 5 hides were in demesne with land for 3 or possibly 4 ploughs. There were 9 sokemen, who neither then nor formerly could depart without leave from the abbot, 8 villeins each with 7½ acres, 15 cottars and 7 serfs. There was a small wood, providing for 5 pigs, and 44*s.* from fisheries. The vill was worth £12 when received and £16 in 1066. In 1086 its value was again stated to be £12.[21]

The manor remained with the convent in 1109, and was transferred to the reconstituted dean and chapter in 1541.[22] It is now held by the Church Commissioners, so that ownership has been continuous for nearly a thousand years, broken only during the Interregnum.[23] A grant of free warren was made in 1252[24] and confirmed with other manorial privileges in 1418.[25]

Philip de Insula appears as tenant of the manor of Sutton *c.* 1250, when the demesne consisted of 40 acres of arable land and 30 of meadow,[26] but for the rest of the Middle Ages it seems to have been kept in hand by the prior and convent. It was usually reckoned as the most valuable part of their property, other than the various estates in and around Ely itself. In 1291 the manor was valued at £64 15*s.* 5*d.*[27] In 1324–5 the value had increased to £67 0*s.* 5½*d.*[28] Prominent sources of revenue then were £6 6*s.* from fisheries, £8 12*s.* 8*d.* from mills and £3 1*s.* 6*d.* from the sale of stock. Wool and fells were also sold, though on a smaller scale than at Witchford and Wentworth: 71 skins yielded £1 6*s.* 7½*d.* In 1380[29] receipts had risen to £83 2*s.* 10*d.* Nine fisheries produced £4 18*s.* and the sale of stock and dairy produce £8 2*s.* 3*d.* Some of the demesne was arrented at £3 18*s.* This rent included 3*s.* 4*d.* from an osier bed in the 'Holt' let to Sir John Wake. Works were commuted for £4 4*s.* 6*d.* In this year corn and stock were purchased at Royston (Herts.) for £4 1*s.* 11*d.* and the prior paid a personal visit of inspection. During the 15th century the receipts diminished considerably, fluctuating between £27 17*s.* 8*d.* in 1429 and £54 13*s.* 4*d.* six years later; Sutton now ranked in value somewhat below the priory's manor in Whittlesey.[30] In 1522–3, however, the receipts had risen to £81 16*s.* 11½*d.*, including rents in kind valued at £1 7*s.* 9¾*d.*[31] In 1527–8 the receipts had again fallen to £41 4*s.*[32] In 1541 they had risen to £67 13*s.* 3¾*d.*[33] The general high degree of prosperity in Sutton is shown by the nickname 'Golden' Sutton, found in Chancery proceedings of 1599.[34]

Lying as it did on the border of the Isle, Sutton was the occasional scene of boundary disputes. Thus in 1292 Prior Salomon complained that Abbot Sawtry of Ramsey and some members of the de Insula family had broken into his Sutton property, mowed his meadows and reed bed and felled his trees.[35] In 1441 the successors of these disputants agreed that the boundary between their respective manors in Sutton and Chatteris should lie along the dike in the marshes of 'Hollode'[36] and 'Popilholt'.[37]

Many grants to the priory are recorded amongst the capitular archives, mostly small in amount. They are

[11] 1 & 2 Vic. c. ii.

[12] Award at County Hall, March.

[13] In 1794 there were said to be four fields, of 700 acres (W. E. Tate, 'Cambs. Field Systems', *Camb. Antiq. Soc. Procs.* xl, 81 (1939–42), quoting Vancouver, *Gen. View Agric. Cambs.*).

[14] Probably the name is an indication of superior fertility; this was the field lying nearest the fen portion of the parish, with a preponderance of peat over clay in its soil.

[15] None of these names, except the Meadlands and Washes, survives on modern maps.

[16] *V.C.H. Cambs.* ii, 133.

[17] Infm. Publ. Relations Offr. Eastern Region, Brit. Rlys.

[18] *Cal. S.P. Dom.* 1611–18, 552, 566.

[19] See *The Times*, 16 Mar. 1937, 17 Mar. 1947 and subsequent days; personal observation of present writer, Aug. 1947.

[20] Jas. Bentham, *Ely*, 75.

[21] *V.C.H. Cambs.* i. 370.

[22] *L. & P. Hen. VIII*, xvi, g. 1226 (12).

[23] C 54/3448, no. 22; C 54/3624, no. 13.

[24] *Cal. Chart. R.* 1226–57, 391.

[25] *Cal. Pat.* 1416–22, 184, 185.

[26] Bp.'s Mun., Ely, Liber M, ff. 374*a*, 374*b*.

[27] *Tax. Eccl.* (Rec. Com.), 270.

[28] Treas. R. transcribed by Archdeacon S. J. A. Evans.

[29] Bailiff's R. of 3–4 Ric. II (D. & C. Mun., Ely, Cupbd. B). A good many similar rolls are preserved, but mostly in very bad condition and in several cases incomplete.

[30] S.C. 6/1257/4 to /9.

[31] Transcript *penes* Archdeacon S. J. A. Evans. Of this sum, £22 16*s.* 8*d.* was appropriated to the pittancer.

[32] D. & C. Mun., Ely, Treas. R. no. 20.

[33] B.M. Add. MS. 5847, f. 120.

[34] *P. N. Cambs.* (E.P.N.S.), 239.

[35] *Cal. Pat.* 1281–92, 578.

[36] Now represented by Holwoods Farms (Chatteris) and Sutton Holwoods at the extreme west of the common boundary towards Huntingdonshire.

[37] D. & C. Mun., Ely, Cista II D, Carta. 1149. See also E 60/14335.

of little interest except to the student of place names; such names as 'Athelyngsbrygge', 'Waldhithewaye', the 'Haspehoue', and 'Fenhoue' occur.[38] The prior's vineyard is mentioned in 1312.[39] Most of these grants date from the early years of the 14th century, at which time, as in other of the conventual manors, the Fresyngfeld family made many gifts.[40] In 1371 it was reported that since the Statute of Mortmain (1279) the convent had acquired in Sutton, mainly without licence, no fewer than 68 properties, to the yearly value of £6 6*s.* 5*d.* These included a 100-acre estate called 'Pillerynestenements', worth £2, and marsh, worth 6*s.* 8*d.*, held of the bishop. The latter was acquired in order to enlarge the manor place.[41]

At the end of the 16th century the lessees of the manor under the dean and chapter were the Jetherell family.[42] A member of that family, Henry, was a party to the disputes that occurred between the dean and chapter and the villagers of Sutton in the 1620's (see above).[43] Concurrently with the inclosure of the fens the fines were changed from arbitrary to certain. The sum of £40 was to be collected from the tenants to compensate the dean and chapter for this concession. In 1654 some portions of the sum still remained unpaid.[44] The parliamentary survey of 1649 gave the 'improved' value of copyhold lands in the manor as £98 13*s.* 8*d.* a year,[45] amended to £101 3*s.* 8½*d.* when in the same year George Perrier, scrivener of London, bought the manor for £2,243 9*s.* 2½*d.*[46] The fines certain were now fixed at 4*d.* per acre a year. Perrier did not hold the manor long; in 1651 it was regranted, for £2,048 8*s.* 2*d.*, to Hamond Ward, a London merchant.[47] The demesne was in two parts, the larger, of 13 closes (140 acres), 33 acres of lammas ground in the North and South Meadlands and 105 acres of fen and marsh, was worth £132 17*s.* 4*d.* yearly and carried £50 of timber; the smaller, of 5 closes of 46 acres, and 24 acres of fen and marsh, was worth £42 19*s.*

At the Restoration the manor returned to the dean and chapter; in the early 18th century it was leased to the Ward family at a yearly rent of £8, 4 wethers, a brawn, and a calf; fines for renewal were £45 in 1710 and 1717 and £60 in 1724.[48] Courts baron were held until 1844.[49]

The manor house normally followed the descent of the manor. In 1541, however, it was let to Thomas Colin for £8 yearly.[50] In 1630 the tenancy was transferred by Richard Upcher and his wife Anne to David Dunbar and his wife Elizabeth.[51] The Dunbars made a settlement of their interest in 1637.[52] David Dunbar was engaged in a lawsuit with the dean and chapter about this time.[53] In 1851 John Vipan was the occupier of the Burystead.[54]

The Burystead Farm at the west end of the village incorporates a complete early-14th-century chapel which, though long desecrated and converted into a house, retains most of its original features in a remarkable state of preservation. In 1651 it comprised the main portion of the manor house, and was stated to consist of a block 39 by 24 ft., containing a hall and parlour on the ground floor and two rooms over. The south wing (rebuilt in 1742, see below) was 27 by 18 ft. and included another parlour, the kitchen, buttery, larder, and other offices downstairs and four small rooms above.[55] Judging from the position of the fine double piscina the actual chapel was raised on a low undercroft and was probably approached by an internal newel stair at the south-west angle. There was a large window in both the east and the west walls; that in the west, with plain intersecting tracery, survives complete though blocked. There were at least two lateral windows of two lights on the north and south which survive, though either wholly or partially blocked. There is a blocked two-centred doorway towards the east of the north wall. The present entrance on the south has a pointed arch, which however does not appear to be medieval. The materials are rubble with ashlar dressings, much of the surface being plastered, and there are angle buttresses with one set-off. On the west gable is the base of a cross. The roof, though retaining its original pitch, is a post-medieval reconstruction reusing some medieval material. The approximate internal length is 36 ft. and the walls are nearly 3 ft. thick. The fact that human remains have been discovered in the surrounding ground suggests that the building was something more than a private manorial chapel.

At right angles to the chapel and attached to it on the south is a brick two-story wing with sash windows and round-headed doorway with fanlight; on the front is the date 1742. Later additions have been made on the west.

In 1548 Sir William Fyldyng or Fielding held a capital messuage in Sutton, which he had settled in 1525 on his son Basil on his marriage to Goditha Wyllington.[56] Basil was in possession at his death (1586), having in turn settled it on his son William and his descendants.[57]

In the early 18th century 160 acres of fen in Sutton Holwoods were on lease to Mrs. Blatt at £30; fines for renewal ranged from £26 to £30. An area of 248 acres in Middlemore, which provided £20 a year for the schoolmaster, was let to feoffees at £3 11*s.* 6*d.*; fines for renewal were £65 in 1709 and 1716 and £68 in 1723. Other 'new rents' from inclosed and drained fen yielded £40 a year.[58]

CHURCH

The church was early appropriated to the hosteller of Ely monastery—an arrangement which Bishops Northwold (1254)[59] and Balsham (1276)[60] confirmed. It was valued at £13 6*s.* 8*d.* in 1217 and £12 in 1254.[61] A vicarage was ordained in 1254.[62] In 1291 the rectory was valued at £16 13*s.* 4*d.* of which £4 was set aside for the vicar.[63] The vicar's stipend had been increased to £10 by 1535.[64] Six years later the rectory was let to

38 D. & C. Mun., Cartae 294, 672, 221.
39 Ibid. Carta 228.
40 *Cal. Pat.* 1313–17, 121.
41 *Cal. Inq. Misc.* iii, no. 784 (pp. 293, 294).
42 C 3/243/25.
43 C 2/1/E. 7/39.
44 C 3/456/76 and /77.
45 D. & C. Mun., Cupbd. A.
46 C 54/3448, no. 22.
47 C 54/3624, no. 13.
48 D. & C. Mun., Bentham Transcripts.
49 Ct. r. exist up to this date in the capitular mun. at Ely (Cupbd. A).
50 B.M. Add. MS. 5847, f. 120.
51 C.P. 25(2)/400/East. 6 Chas. I.
52 Ibid. 13 Chas. I.
53 C 3/400/161.
54 R. Gardner, *Dir. Cambs.* (1851), 508.
55 C 54/3624, no. 13.
56 C 142/86/14.
57 C 142/211/194.
58 D. & C. Mun., Bentham Transcripts.
59 Bentham, *Ely*, 147 n.
60 *Ely Dioc. Remembrancer*, no. 193, p. 123.
61 W. E. Lunt, *Valuation of Norw.* 539.
62 Bentham, *Ely*, 147 n. The Norwich Valuation of 1254 mentions no vicarage, so that the ordination was presumably one of the last public acts of Bishop Northwold; he died 9 Aug. of that year.
63 *Tax. Eccl.* (Rec. Com.), 265.
64 *Valor Eccl.* (Rec. Com.), iii, 499.

John Bucke for £13 6*s.* 8*d.*[65] A pension of £1 was granted to the Bishop of Ely from Sutton vicarage in 1562.[66] There is less documentary evidence for the subsequent history of the rectory than for other churches impropriate to the dean and chapter, but in 1610 it passed from Benjamin Hodson and Cassandra his wife to Robert Storye.[67] From about 1684 the rectory was held by Jesus College, Cambridge, as trustees for Bishop Gunning's charity for the improvement of poor livings. In the early 18th century the college paid a yearly rent of £13 6*s.* 8*d.* and 2 quarters of wheat or the price thereof at Ely market on the Saturday before St. John Baptist's Day; fines for renewal amounted to £80 in 1710, £76 in 1717, and £100 in 1724. The rectory then consisted of a barn, dovecote, 39 acres of arable, 12 of old pasture, 40 of meadow and fen, and the great tithes.[68] The vicarial tithes were commuted in 1840 for £1,175 and the rectorial for £500.[69] There were then 44 acres of glebe, which by 1900 had been reduced to 20 acres.[70]

The advowson of Sutton has been continuously with the prior and convent and their successors the Dean and Chapter of Ely. From 1735 to 1844 the rectory of Mepal was consolidated with Sutton vicarage. Bentham told Cole, who visited Sutton in 1745, that this had been done by Dr. John Warren, formerly Vicar of Wisbech and Rector of Elm.[71] Warren alleged that the two livings had not been a 'maintenance' separately, but Cole believed that Warren 'took them solely to build up a new rectorial house and barns and then to resign them'.[72] A later incumbent, Dr. Samuel Peach, was appointed to the livings (then worth £200 per annum) in consideration of giving up the papers of Dr. James Bradley the astronomer, his father-in-law, which had been bequeathed to the public.[73] In 1781 a faculty was granted to Peach to pull down the rectory at Mepal and to use the materials to repair the vicarage at Sutton.[74]

An area of 12½ acres in the common fields of Sutton, held by Robert Auker, William Bucke, and Henry Gunton, which had been given to anniversaries and lights in the church, was in 1553 granted to Sir John Butler and Thomas Chaworthe.[75] Various small pieces of land in the fields of Sutton which had been given for anniversaries and lights in the church, came in 1568 into possession of Hugh Counsaille and Robert Baker.[76] Three years later Thomas Jennings and Edward Forthe, other Elizabethan land speculators, undertook the disposal of other obit land in 'Southmede'.[77]

The church of *ST. ANDREW* consists of chancel, north vestry, clerestoried nave, aisles, south porch, and west tower. The material is rubble and Barnack stone. The roofs of the chancel and vestry are tiled, and the others covered with lead. The fabric dates from the second half of the 14th century, and exhibits the transition from Decorated to Perpendicular. The arms of Bishops Barnet (1366–73) and Arundel (1374–88) occur on the vaulting of the porch, and also those of the see of York, to which Arundel was translated in 1388. The design is uniform throughout, but the chancel seems rather earlier than the nave, and the tower somewhat later. The lavish use of battlemented parapets and the unusual window tracery give the church a marked individuality.

The chancel, of 3 bays, has a large east window of 5 cinquefoil-headed lights with a transom and elaborate tracery. There are 8 angle buttresses each with 1 set-off, 2 pairs at the eastern corners, 2 on the north wall, and 2 on the south. There are 3 windows on the south and 2 on the north, all uniform in design, consisting of 3 cinquefoil-headed lights with tracery above. All the windows in the chancel and aisles are set under lofty arches with moulded shafts—a feature common in contemporary East Anglian work and found in the Isle also at March and Wilburton. A low stone bench extends along the walls. In the middle bay on the south side is a doorway with continuous mouldings and a hood-mould. The walls are surmounted by a plain coped parapet. In the east bay on the north a doorway with continuous mouldings opens to the vestry. The absence of a window in this bay is proof that a vestry formed part of the original design. There are graduated sedilia and a double piscina in the usual position. On either side of the east window inside is a large niche with cusped and crocketed canopy and a bracket above; they are occupied by modern figures of St. Andrew and St. Etheldreda. The lofty chancel arch is two-centred and of two orders, the outer with continuous mouldings and the inner springing from semi-circular shafts with moulded caps and bases; on the west face there is a hood-mould terminating in heads.

The vestry has been entirely rebuilt; it has a two-light east window, an external doorway on the west and a lean-to roof.

The nave has arcades of six bays with two-centred arches of two orders, the outer with continuous mouldings and the inner springing from semi-circular shafts with moulded caps and bases: there are hood-moulds terminating in heads and on the east respond there is a bracket. The clerestory consists of five windows on each side set over the columns and of two trefoil-headed lights with a quatrefoil above. The parapets are embattled. The rood loft newel staircase is situated in the north-east angle and has plain upper and lower doorways. The tower arch is two-centred and of two orders, the outer with continuous mouldings and the inner springing from semi-octagonal shafts with concave sides and embattled caps and moulded bases.

The aisles have uniform fenestration and buttresses with one set-off. The windows are of three cinquefoil-headed lights with tracery of late 14th-century character. The north doorway has continuous mouldings and a hood terminating in heads. The south aisle is embattled and the north has a plain coped parapet; both parapets have gargoyles. In the south-east angle of the south aisle is an elaborate piscina recess, above which is a large niche containing a seated stone figure almost life size and now headless; it is probably the Blessed Virgin and retains considerable traces of original colour, as does the niche. On either side of the east window of the north aisle is a stone bracket.

The porch is of two bays with parvise above, and an embattled parapet with gargoyles; there are angle buttresses with one set-off. The outer doorway has

[65] B.M. Add. MS. 5847, f. 120.
[66] *Cal. Pat.* 1560–3, 226.
[67] C.P. 25(2)/177/Trin. 7 Jas. I.
[68] D. & C. Mun., Bentham Transcripts.
[69] Gardner, *Dir. Cambs.* (1851), 507.
[70] Kelly, *Dir. Cambs.* (1900).
[71] Cole's statement that he became bishop of St. Davids in 1779 appears to be a mistake. That John Warren was born in 1730 and admitted to Caius College in 1747; he held various Cambridgeshire livings but never Sutton, Wisbech, or Elm (see *D.N.B.* and Venn, *Biog. Hist. Caius Coll.* ii, 60).
[72] B.M. Add. MS. 5807, f. 126*b*.
[73] Ibid.
[74] W. M. Palmer, 'Faculty Bks. in Dioc. Regy., Ely', *Camb. Antiq. Soc. Procs.* xxxv, 69 (1933–4).
[75] *Cal. Pat.* 1553, 290.
[76] C 66/1046, m. 2.
[77] C 66/1077, m. 39.

a two-centred arch of two orders, the outer with continuous mouldings and the inner springing from semicircular shafts with embattled caps and moulded bases; there is a band of conventional flowers beneath the caps and a hood-mould terminating in grotesque heads. The inner doorway has continuous mouldings. There is a lierne vault in two bays with moulded shafts rising from the lateral stone benches; the bosses at the intersection of the ribs are carved with the arms of Barnet, Arundel, and York, and various heads including a bishop. The parvise is approached by a stone newel stair contrived in the north-west angle and entered from the aisle.

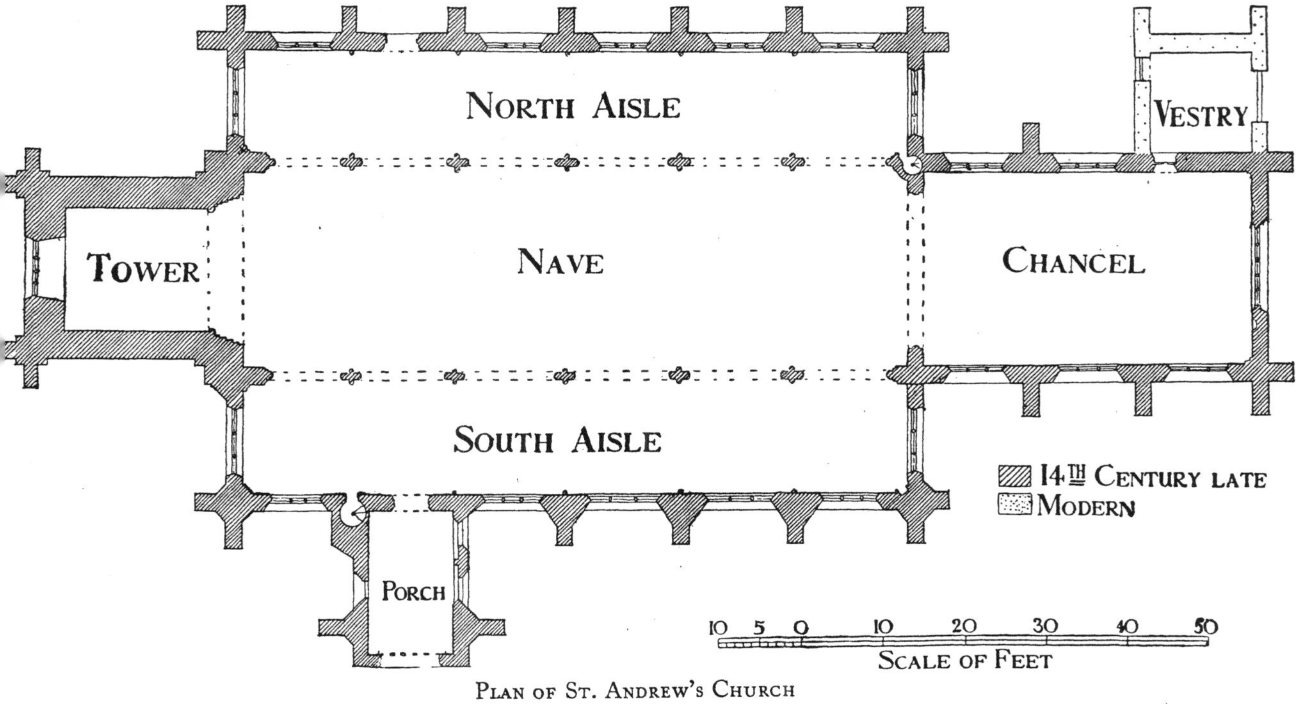

Plan of St. Andrew's Church

The tower consists of four stages surmounted by an octagonal stage which is crowned by a smaller octagon. It is probable that a spire was originally intended, the second octagon being substituted owing to lack of funds. The resulting effect is unique in English medieval church towers and produces a somewhat bizarre outline. The tower, which is lofty in itself and on high ground, can be seen for great distances, e.g. from near Welney Bridge across the Bedford Rivers, and the railway between Ely and Stretham. There are diagonal buttresses reaching to the top of the fourth stage and crowned with mutilated crocketed pinnacles, and smaller additional angle buttresses at the base. Round the base and above the west window are bands of quatrefoils. The large west doorway has a four-centred arch set in a square frame springing from jamb shafts having embattled caps and moulded bases; in the spandrels are quatrefoils enclosing shields with the arms of Ely and St. Andrew. The west window has four uncusped main lights and a transom with rectilinear tracery, all renewed. In the third stage there is a two-light uncusped window on the north, south, and west. The belfry windows in the fourth stage consist of three plain lights contained within a square hood with quatrefoils in the spandrels. The first octagon has an embattled parapet decorated with quatrefoils and remains of crocketed pinnacles on alternate sides; the upper octagon has an embattled parapet with crocketed pinnacles at each angle. There is a good lierne vault of wood with carved bosses and stone springers. The newel staircase is contrived in the south-west angle.

The roofs are modern; that of the chancel is of collar-braced construction with modern stone carved corbels, while the nave roof is of tie beam construction with the old stone corbels, embattled and carved with grotesque figures; the aisle roofs are of lean-to type with moulded stone corbels.

The plain octagonal font is probably of late 14th century date; beneath the bowl there are conventional flowers carved in relief. There are some old poupée heads in the nave seating.

The plate includes a communion cup, paten, and flagon of silver, 1784, inscribed 'The gift of Joseph Manwaring the younger of Islington in the County of Middlesex Esquire', and a plate of pewter.

The tower contains six bells: 1st by William Dobson of Downham, 1819, 2nd by Miles Graye of Colchester, 1654, 3rd by Charles Newman, 1691, 4th by the same, 1700, 5th and 6th by Henry Penn, 1722, inscribed respectively *Plebem voce congrego clerum* and *Defunctos plango vivos moneo.* The registers begin in 1558 but the entries between 1645 and 1655 and between 1686 and 1694 are missing.

In 1553 the church was more lavishly equipped with plate, ornaments, and vestments than any in the Isle except Ely St. Mary.[78] In 1686, however, it was reported that the church had lacked a surplice for more than twenty years.[79]

A guild of St. John is mentioned in the will of William Daynes of Ely (1457), who bequeathed 3*s.* 4*d.*[80] In 1568 the guildhall of Corpus Christi (*sic*),

[78] *Cambs. & Hunts. Arch. Soc. Trans.* ii, 152–3, printing the inventory of the 1553 commissioners (E 315/395).

[79] Ibid. 154.

[80] *Fenland N. & Q.* v, 112; W. M. Palmer, *Village Gilds of Cambs.* 398.

which had been given to pious uses by Richard Yongeman, sometime vicar, was sold to Hugh Counsaille and Robert Baker.[81] A timber-framed and plastered cottage opposite the church is still known as the guildhall.[82]

NONCONFORMITY Dissent was fairly strongly established in Sutton in 1676, where 52 out of 490 inhabitants were not members of the established church.[83] This may have been due to the influence of William Hunt, an ejected minister, who continued in Sutton as a schoolmaster, and from whom the Baptists (see below) traced a tenuous descent. The earliest dissenting congregation to be established seems to have been that of the Friends, whose existence is recorded in 1659 and 1664. In the former year there were four Quaker households,[84] and although there were few Quakers in the Isle generally 'who have any visible estate', it was stated that Henry Foster of Sutton, a member of one of those households, was 'fit for a justice'. Minute books of the Sutton Monthly Meeting survive for 1758–61,[85] and nearly a hundred years later there was an average congregation of 6 at the Friends' meeting house here. The meeting died out *c.* 1855.[86] In 1672 the house of William Hunt was licensed for Presbyterian worship.[87] A Particular Baptist chapel was built in 1791.[88] In 1820 it had a regular membership of nearly 60 and a congregation of 200 to 250.[89] In 1851 there was a Sunday-school attached to it.[90] A Wesleyan Methodist congregation was established in 1790. In 1851 a Sunday-school was attached to it.[91] The chapel was rebuilt in 1914.[92] A Primitive Methodist chapel was erected *c.* 1880 and closed *c.* 1935.[93] In 1900 the Salvation Army were holding meetings in the Jubilee Hall but their operations ceased after about ten years.[94]

SCHOOL William Heye was licensed as a schoolmaster at Sutton in 1579,[95] and a schoolmaster is recorded in the Visitation Return of 1596.[96] At the beginning of the 18th century £20 rents from Sutton Holwoods were appropriated to the salary of a schoolmaster,[97] and in 1798 an unspecified number of children were being taught reading, writing, arithmetic, and the Catechism.[98] The condition of this school was supposed to be good as there were no public complaints. The master may have been a certain Mr. Custnall, who kept a private school at Sutton and according to a report to the bishop of *c.* 1800–10 was a Methodist. This report also mentioned four dame-schools, two of them kept by Dissenters.[99] In 1837 there were about 20 boys at school in Sutton during the winter; an 'efficient' schoolmaster received £15 a year from the dean and chapter;[1] some time before 1851 his salary was increased to £20.[2] According to the National Society, however, 'the educational wants in this parish are very great. The respectable inhabitants are strongly impressed with the necessity of a National School, and the attention of the National Society will shortly be called more fully to the condition of this parish'. In 1859 the existing school was united with the National Society, and two years later the Society granted £20 towards rebuilding the premises.[3] About 1867, when owing to the operation of the agricultural gang system the state of education in Sutton was 'very deficient', the numbers on the books were: (winter) 70, (summer) 51, with average attendances of 58 and 37 respectively.[4] The school building was a large one, providing 230 places in 1900.[5] In 1912 a new classroom, for which the National Society granted £8 and the Ecclesiastical Commissioners £30, brought the total accommodation to 285. The school was reorganized as one department (Mixed and Infants) in 1930. It has always been a voluntary school, and in 1937 a proposal was made by the church authorities to rebuild the premises and include a senior department for the older children from Sutton and the neighbouring villages. This proposal was favourably considered by the County Education Committee but was abandoned in the end as it was found impossible voluntarily to combine sufficient villages to make a Senior School at Sutton practicable. The older children were transferred to Chatteris in 1939, brought back to Sutton during the Second World War, and removed to Chatteris again in 1948.[6]

In 1873, when abortive proposals were made for a separate infants' school at Sutton Gault, it was stated that as many as 60 children in Sutton were attending private schools.[7]

CHARITIES The Charity Commissioners in 1837 reported that 81½ acres in Sutton supposed to have been the remainder of certain fen land after various allotments had been made in pursuance of a Chancery decree of 1624 were let for a sum varying between £122 and £160, according as the land was to be in pasture or arable. About £14 of this rent was given indiscriminately to the poor of Sutton and Mepal in money, and in providing coal at the rate of 25 bushels per family.[8]

WELCHES DAM

This parish lies between Manea, Mepal, Witcham, and Coveney, and is the only one of the various extra-parochial areas in the Isle that survived as a civil parish after 1933.[1] Its name is derived from Edward Welsh, an employee of the Adventurers, who built a dam across the Bedford River.[2] It has little history, though Fortrey's Hall, the seat of the family of that name, lies just within its borders. The hall is a 17th-century

81 C 66/1046, m. 2.
82 *Cambs. & Hunts. Arch. Soc. Trans.* i, 151.
83 Bp.'s Mun., Ely, A 6.
84 *Cal. S.P. Dom.* 1658–9, 358; 1664–5, 142.
85 Infm. Librarian, Soc. of Friends.
86 It is last mentioned in Kelly's *Dir. Cambs.* of 1853.
87 *Cal. S. P. Dom.* 1672, 100, 119.
88 The meeting dated from 1749 (*Bapt. Handbk.*, 1950).
89 *Cong. Mag.* iii, 57 (1820).
90 H.O. 129/7/190. Kelly's *Dir. Cambs.* (1858 and 1865) mentions *two* Baptist chapels in Sutton.
91 H.O. 129/7/190.
92 Inscr. on bldg.
93 Kelly, *Dir. Cambs.* (1883 to 1933 editions).
94 Ibid. (1900); not mentioned in 1912 edition.
95 A. Gibbons, *Ely Episc. Rec.*, 176.
96 *V.C.H. Cambs.* ii, 338 n.
97 D. & C. Mun., Bentham Transcripts.
98 Bp.'s Mun., B 6.
99 Ibid.
1 *Rep. Com. Char.* (*Cambs.*), [103], 329–31, H.C. (1837–8), xxiv.
2 Gardner, *Dir. Cambs.* (1851), 508.
3 Archives of Nat. Soc.
4 *Rep. Com. Empl. Children &c. in Agric.*, [4068–I], 338, H.C. (1867–8), xvii.
5 Kelly, *Dir. Cambs.* (1900).
6 M. of Ed. File no. 12/45.
7 Archives of Nat. Soc.
8 *Rep. Com. Char.* (*Cambs.*), [103], 329–31, H.C. (1837–8), xxiv.
1 Isle of Ely Review Order, 1933 (M. of H. no. 76940).
2 *P.N. Cambs.* (E.P.N.S.), 242.

building roofed with old tiles of local origin. The Fortreys were refugees from Brabant in the 16th century, and became co-adventurers with the Earl of Bedford in the drainage of the Fens. The romantic career of Sir James Fortrey, who died in 1719, is recorded on a wall monument in Mepal church.[3] The area now comprised in Welches Dam civil parish was formerly common to the surrounding villages, and was much affected by Vermuyden's drainage schemes, the Old and New Bedford Rivers cutting right across it. In 1666 an award was made in a suit between the Dean and Chapter of Ely, as lords of the manor of Witcham, and the inhabitants of Coveney, Manea, Chatteris, Wentworth, Witchford, Downham, and Ely regarding 1,900 acres in Byall Fen. The dean and chapter received thereunder 150 acres between the Downham and Coveney portions of the Fen, 8 acres known as Pennyes Pingle on the north side of the Old Bedford River near Fortrey Hall, 41 acres in Widden Fen near Witcham Hithe, 17 acres in Old Fen near Wentworth Drove Way, 7 acres in Hale Fen between Coveney Causeway and the Adventurers' Grounds, 10 acres in Byall Fen Wash between the Bedford Rivers, 11 acres in Dam Head, and certain small pieces of waste ground totalling 14 acres. The value of this 258-acre allotment was £17 11*s.* 2*d.*[4]

The Anglican mission church of St. Eanswyth dates from 1909 and is served from Manea, where is the nearest school.[5]

A Wesleyan chapel was built in 1822. In 1851 a Sunday school had just been started.[6] The chapel was closed during the First World War.[7]

Welches Dam acquired an unhappy notoriety in 1849, when 11 fatal cases of cholera occurred in a parish that had had only 187 inhabitants at the time of the last census. Most of the fatalities were amongst the 81 occupiers of a group of 16 cottages huddled against the Old Bedford River bank and hence very damp.[8]

WENTWORTH

Wentworth, a very small village, is situated just south of the high road from Ely to Chatteris, 4½ miles west of the city. The parish, which is almost in the centre of the Isle proper, is mainly on the south side of this road, which forms part of the boundary. This boundary, however, encloses a 'tongue' of land about a mile long to the north-east, which consists of a section of the fen north of Witchford. Wentworth Sedge Fen (mainly in Witchford parish) shows by its name the purpose of this irregularity. There were formerly two detached parts of Wentworth, owing their origin to the same cause, which were amalgamated with Coveney and Grunty Fen in 1884 and 1886 respectively. The Domesday entry for Wentworth suggests that the village was then a fairly large one. Later, however, it declined in size. In 1428 it was recorded that there were only 9 resident parishioners.[1] The returns of 1563[2] and 1676[3] show only 12 householders and 78 communicants respectively. In 1639–40 Wentworth was assessed for ship-money at only £14.[4] The depopulation that had been occurring at least from the 15th century may have been partly due to inclosure, which was a feature in Wentworth earlier than in many Isle parishes, and to a long succession of lay tenants of the manor. An area of 177 acres of land and pasture round the manor house seems to have been inclosed by 1649,[5] and a further 1,004 acres by Act of Parliament in 1830.[6] By the later inclosure 32 proprietors benefited. The chief of these were Henry Francis (362 acres), the Dean and Chapter of Ely (118 acres in respect of the Rectory and Wold Farms), the Revd. H. J. Sperling (57 acres) and Anne Vipan (64 acres). The costs of the Act were £2,544 9*s.*, of which £1,424 2*s.* was borne by Francis.[7] The proportion inclosed (about 70 per cent. of the total area of the parish) was the highest in the Isle. Though so small, Wentworth has two public houses, but there is no post office. Burwell House is a late Georgian building of some merit: it was the rectory until the union of the benefice with Witchford in 1938.[8]

MANOR It is not known when the church of Ely obtained Wentworth. In 1086 the vill was rated at 3½ hides, one of which was in demesne. This demesne had 1 plough 'and there might be another'. There were 3 sokemen, 2 sharing 1 hide and the other holding a virgate, who neither then nor formerly could sell their land without the abbot's leave. There were 9 villeins with 10 acres each. These sokemen and villeins had 5 ploughs. There were also 17 cottars. There was woodland for 20 pigs. The value was, and had been when received, £10 10*s.*: in 1066 it was £12. The vill had always formed part of the demesne of the church of Ely.[9]

The manor was left with the convent in 1109, and has continued in possession of the chapter of Ely (now represented by the Church Commissioners) to the present day.

Bentham[10] mentions that after the death (1072) of Thurstan, the last Saxon Abbot of Ely, 'the King's officers took away a considerable sum of Gold and Silver that the monks had laid up' at Wentworth. This might imply that it was then either a very obscure or a very considerable place. In 1291 Wentworth was assessed at £30 10*s*[11]—a sum which corresponds with the average value of the convent's manors. From 1252 the priors had enjoyed free warren.[12] This right, with the other usual manorial privileges, and the wastes and marshes within the manor as parcel thereof, were confirmed in 1417 when the respective rights of the bishop and prior were more clearly defined.[13]

In 1381 an unsuccessful attempt was made by the men of Sutton to seize the Sacrist of Ely, who was at Wentworth when the Peasants' Revolt broke out.[14] He was perhaps supervising the collection of rents, which had long been appropriated to his office.[15]

[3] R. Gardner, *Dir. Cambs.* (1851), 503, giving the text of the inscription.
[4] D. & C. Mun., Ely, Cupbd. A.
[5] Kelly, *Dir. Cambs.* (1933).
[6] H.O. 129/7/191.
[7] Kelly, *Dir. Cambs.* (1916, 1922).
[8] *Camb. Chron.* 5 Oct. 1850.
[1] *Feud. Aids*, i, 193.
[2] B.M. Harl. MS. 594, ff. 198–200.
[3] Bp.'s Mun., Ely, A 6.
[4] *E. Anglian*, N.S. vi, 40; S.P. 16/445, no. 75.
[5] D. & C. Mun., Ely, Cupbd. A.
[6] 10 Geo. IV, c. xxiv.
[7] Award at County Hall, March.
[8] Infm. Revd. E. H. Yates, Vicar of Witchford and Rector of Wentworth.
[9] *V.C.H. Cambs.* i, 369.
[10] *Ely*, 105.
[11] *Tax. Eccl.* (Rec. Com.), 270.
[12] *Cal. Chart R.*, 1226–57, 391.
[13] *Cal. Pat.* 1416–22, 184, 185.
[14] E. Powell, *Rising in E. Anglia*, 47.
[15] Copy of rental of 1522–3 *penes* Archdeacon S. J. A. Evans.

Wentworth was let out to farm earlier than the other manors in the Isle belonging to the convent, and so does not appear in the set of account rolls of the mid-15th century.[16] The tenants were the Scrope family of Bolton (Yorks.), whose connexion with Wentworth is first detected in a family settlement of 1445.[17] John Lord Scrope of Bolton, who died in 1498, held Wentworth manor of the Prior of Ely by service of 5*s.* yearly; it was then extended as a capital messuage and 100 acres of land and 20 of meadow, and was valued at £6 with 13*s.* 4*d.* in rents.[18] A good deal was, however, kept in hand by the convent for the benefit of the sacrist, who received from Wentworth £18 12*s.* in 1522–3 and £19 16*s.* 2½*d.* in 1541.[19] In the latter year the chief lordship of the manor was formally transferred to the newly constituted Dean and Chapter of Ely.[20] During the reign of Elizabeth the manor appears in the tenancy of a junior branch of the Scrope family, of Hambleden (Bucks.).[21] The alternative name *SCROOPES* continued to be attached to Wentworth manor for a century later.[22]

A survey by the Parliamentary Commissioners in 1649[23] showed rents of £13 17*s.* 1½*d.* In 1650 the manor was sold to John Smyth of Wentworth for £1,069 11*s.* 8*d.*[24] It was then more fully surveyed. The estate appears divided into two parcels. The first of these consisted of the remnant of the demesne, containing 43½ acres of arable in 14 pieces in the fields, and 2 acres of meadow. With the usual manorial perquisites, it was valued at £21 11*s.* 8*d.* yearly. The other parcel consisted of the manor house itself, 133 acres of arable and 44 acres of pasture, and was worth, with foldage for 400 sheep, £97 0*s.* 8*d.* The manor house, which was said to have been formerly in the occupation of Michael Barnard of Coveney, had been let in 1621 to Jeremy Daveys of Cambridge for £8 and stock to the value of £2 18*s.* 8*d.* yearly. The distinction between demesne and manor-house lands is not drawn in a terrier made in 1663.[25] This terrier records 5 closes of pasture immediately surrounding the manor house and totalling 22 acres, and other lands totalling 194 acres, together with 14 poles of fen ground in Staple Fen. Briery, Broadwater, Long Hey, Maraway (cf. Mere Way in Witchford), and Staple Fields are mentioned in one or both these surveys. At the beginning of the 18th century Thomas Bacon held the manor under the dean and chapter at a beneficial rent of £8 a year together with a brawn, 4 wethers, and a calf (in kind, value £4 10*s.*). The fine for the renewal of the lease was £45 in 1709 and £60 in 1718. The estate comprised 172½ acres of arable, 26 of inclosed pasture, 8 of fen, 7 of meadow, a willow holt of 17 acres, and a sheepwalk for 400 sheep.[26] In Cole's time (1744) the principal tenants in Wentworth were Dr. Conyers Middleton, who is more closely associated with Coveney (q.v.), 'Dr. Goddard's wives sister' and Mr. Thompson of York. Dr. Middleton's farm had 'one of the largest barns I ever saw, except the Bishop's barn at Ely'.[27] In 1814 the tenant of the manor house under the dean and chapter was John Sanxter. The lease was renewed, still at the mid-17th-century rent of £8, to William Sanxter in 1822. In the same year he assigned his lease (for £1,900) to Mrs. Diana Brown, who reassigned to Henry Francis in 1825. The manorial estate was divided in 1843 when 166 acres were leased to (Sir) Samuel Bignold for £3 9*s.* 4*d.* yearly. The lease was renewed to him in 1850 and 1858; he surrendered it in 1863.[28]

Lands in Wentworth called *GRANSDENS* can be traced back to 1541,[29] when they accounted for £1 15*s.* of the conventual receipts from this manor. In 1649 they had increased in value and were divided into moieties, worth £3 15*s.* and £2 11*s.* 1*d.* respectively.[30] About 1730 they provided £2 17*s.* 6*d.* in rents.[31] The name presumably derives from the original tenant from Great or Little Gransden on the Huntingdonshire border. It is preserved in a corrupted form in Granny's End Road.[32]

Part of the Caius College Haddenham Estate extended into Wentworth, amounting to about 7 acres. It was leased to the Earl of Hardwicke until *c.* 1857, and after that direct to the Earl's sub-tenant at £11 10*s.* The tenant was responsible for payment of the tithe apportionment of £2 3*s.* 3*d.*[33]

In 1830 the Dean and Chapter of Ely were leasing Rectory Farm to William Sanxter and Wold Farm to Needham's Charity, Ely.[34]

CHURCH The church was valued at £10 in 1217.[35] It was appropriated by Bishop Northwold (1229–54) to the sacrist of the priory.[36] In 1291, however, it was still being described as an *ecclesia* and the value was still £10.[37] This suggests that by that time no vicarage had been ordained. Vicars are, however, mentioned in 1364[38] and 1390.[39] Bentham states, though without authority, that disappropriation took place between 1418 and 1446.[40] In 1535 the living was a rectory valued at £10.[41] The patronage has remained continuously with the prior and convent or dean and chapter. Since 1938 the living has been united with Witchford.[42]

The church of *ST. PETER* consists of chancel, nave, south porch, and west tower. The material is rubble with ashlar dressings. A drastic 'restoration' in 1868 involved the rebuilding of the nave and porch and a complete reroofing. The existence of a church on the site in the 12th century is proved by the reset north and south doorways. The chancel was rebuilt, probably on an extended scale, in the 13th century, and the tower added a century later. It appears that a chapel was added on the south side of the nave in the 14th century and subsequently destroyed; the blocked arch remained until the 'restoration'.

The chancel has an east window consisting of three plain lancets under a containing arch, all of 19th-century date. There are angle buttresses with one set-off. In the lateral walls are three 13th-century lancets, and there is a rectangular low-side window at the west

16 S.C. 6/1257/4–9.
17 C 139/173.
18 *Cal. Inq. p.m. Hen. VII*, ii, p. 140.
19 Copy of rental 1522–3 *penes* Archdeacon Evans; B.M. Add. MS. 5846, f. 156.
20 *L. & P. Hen. VIII*, xvi, g. 1226 (12).
21 C 142/164/121, /180/18, and /209/58.
22 C.P. 25 (2)/771/Trin. 3 Jas II and /814/Hil. 1 Wm. & Mary.
23 D. & C. Mun., Ely, Cupbd. A.
24 C 54/3515, no. 20.
25 D. & C. Mun., Ely, Cupbd. A.
26 Ibid., Bentham Transcripts.
27 B.M. Add. MS. 5805, f. 87.
28 D. & C. Mun., Ely, Cupbd. B.
29 B.M. Add. MS. 5846, f. 156.
30 D. & C. Mun., Ely, Cupbd. A.
31 Ibid., Bentham Transcripts.
32 *P. N. Cambs.* (E.P.N.S.), 243.
33 J. Venn, *Biog. Hist. Caius Coll.* iv (2), 93.
34 Incl. Award at County Hall, March.
35 W. E. Lunt, *Valuation of Norw.* 539.
36 Jas. Bentham, *Ely*, 127.
37 *Tax. Eccl.* (Rec. Com.), 265.
38 *Cal. Pat.* 1361–4, 14.
39 Ibid. 1388–92, 295.
40 *Ely*, 127.
41 *Valor Eccl.* (Rec. Com.), iii, 499.
42 P.R.O. Inst. Bks.; *Crockford's Clerical Dir.*

end of the south wall. There is a 13th-century doorway in the south wall with a continuous chamfer and a hood-mould terminating in masks. An original internal string-course runs round the north, south, and east walls. The chancel arch is two-centred and of two orders, and springs from large modern corbels clumsily carved. There is a double piscina with a central shaft having a moulded cap and base, and in the north wall is a modern aumbry.

The nave has been rebuilt, and all the fenestration is modern and of 14th-century character. On the north and south are two windows of two cinquefoiled lights with internal and external hood-moulds terminating in heads. At the west end of the north wall is a single

high relief and dating from the first half of the 12th century. It portrays a figure in mass vestments and pall, holding a book in the left hand and what appears to be an asperorium in the right; there is an architectural canopy and a twisted shaft on the left; the right side is missing and it is probable that there were originally more figures; when complete it may have served as a reredos. The surviving portion is in a remarkably good state of preservation.

The plate includes a silver chalice, paten and flagon of 1869–70, presented by Robert Raynbird, rector 1859–87.

The tower contains one bell by C. and G. Mears, 1845.

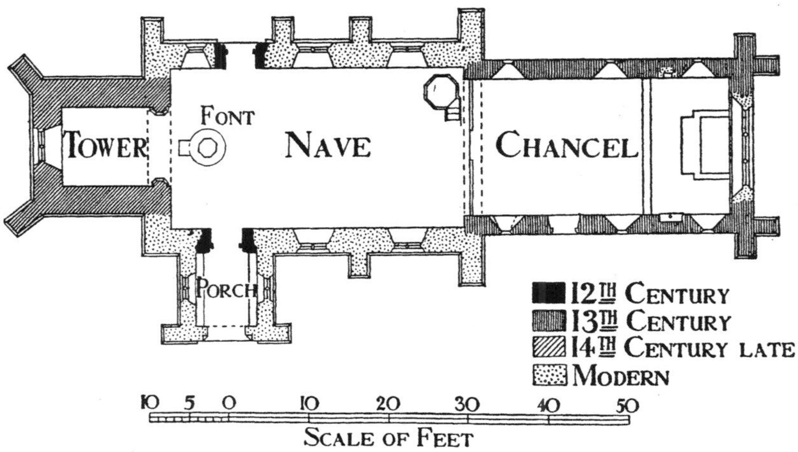

PLAN OF ST. PETER'S CHURCH

light with trefoiled head. The north and south doorways are of 12th-century date, though much restored; they have plain tympana and a roll moulding, and the south has cable angle-shafts and cushion caps. The tower arch is two-centred and of two orders with semi-octagonal responds having moulded caps and bases. There are four buttresses on the north and two on the south, all with one set-off and modern. The porch, which is entirely modern, has an outer doorway with continuous mouldings and a hood-mould terminating in heads. There is a two-light window with trefoiled head on the east and west, and small angle-buttresses.

The tower is of three stages with diagonal buttresses having four set-offs. The west window has two cinquefoil-headed lights which have been renewed, and an original hood-mould terminating in heads. In the second stage is a modern single light on the north, south, and west. The belfry windows have two trefoiled lights with a quatrefoil above and a hood-mould; the tracery has been renewed but the hoods are partly original. There is an embattled parapet, and the tower is crowned by a low pyramidal cap which is tiled.

All the roofs are modern.

The 13th-century font has a plain octagonal bowl resting on five shafts, and there is leaf foliage under the bowl. The chancel screen is of 14th-century character, the muntins consisting of turned shafts; it has been completely renewed with the exception of the tracery heads and part of the top beam. Attached to the north wall of the sanctuary is part of a stone panel carved in

The registers begin in 1684 (baptisms), 1685 (burials) and 1686 (marriages).

NONCONFORMITY In 1672 the house of William Birchall was licensed for Presbyterian worship,[43] but for at least a hundred years there has been no Nonconformist place of worship in the parish.

SCHOOL Wentworth had a small dame-school in 1789.[44] In 1846–7 it was reported that a Sunday-school, attended by 8 boys and 15 girls, was held in the chancel of the church. The parish clerk taught the pupils, and also held classes on week-day evenings in summer. Shortly afterwards the rector (George Peacock, Dean of Ely) built a school and had it conducted at his own expense. In 1872–3 a new school was built at a cost, with a teacher's house, of about £500, to which the National Society granted £30.[45] This school held about 50 children. In 1899 the average attendance was only 38 and in 1922 only 16. In the latter year the County Council, in the face of much local protest, closed the school. Children now attend school at Witchford.[46]

CHARITIES In 1837 Wentworth Town Lands consisted of 8¼ acres, which produced £10 a year. The rents were received and distributed by the poor of the parish themselves, in sums of about 6*s*.[47]

43 *Cal. S.P. Dom.* 1672, 100, 119.
44 Bp.'s Mun., Ely, B 6.
45 The old school building was in existence as late as 1922 (M. of Ed. File no. 12/52).
46 Ibid.
47 *Rep. Com. Char.* (*Cambs.*), [103], 332, H.C. (1837–8), xxiv.

WILBURTON

Wilburton lies on the southern border of the Isle, 6½ miles south-west from Ely. The village, which extends along the important cross-country road from Earith Bridge to Stretham, is rather small, but more substantial in appearance than most isle villages. It is still as true as it was a hundred years ago that the village 'is very neat and contains some excellent houses'.[1] Among the more noteworthy of these houses are the Burystead (the former manor house), the post office, one of the few surviving half-timbered houses in this part of the country, and Victoria Place, a substantial row of cottages. In the Pell family Wilburton is fortunate in having had for more than a century a succession of resident lords of the manor, many of whom have been keenly interested in agricultural and social questions. As at Haddenham there are many market-gardens and orchards, producing an attractive wooded appearance. Minor roads lead north and south from the village to Wentworth and to Witchford and to Cottenham *via* Twenty Pence Bridge, which, some 2½ miles downstream from the former Aldreth High Bridge, has to some extent supplanted it. There is a railway station, since 1931 open for goods traffic only, on the Ely–Sutton branch (1866) of the former L.N.E.R. St. Peter's Hall, a village assembly room, was built by O. C. Pell in 1891. It contains panelling formerly in Stretham Church.[2]

An area of 807 acres in Wilburton was inclosed under the general Act of 1845, the award being made in 1855. At that date there were six open fields in the parish: Little (north-west of the village), Flexon (north-east), Mill (east), Towns End (south-east), Dog House (south-west), and New Ditch (on the southern edge of the parish). In all sixty-five proprietors participated in the award, but over half of the inclosed land went to the Pell and Camps families. Lady Pell received 203½ acres, 2½ as lady of the manor, 131 as a freeholder and 70 as lessee of the Archdeacon of Ely. Four members of the Camps family—Edward, Read Tansley, Sarah, and William—accounted for 262½ acres between them. The only other allottees on a large scale were Alexander Simson (70 acres) and William Martin senior (39 acres). Two acres were allotted for the poor, 1 for a public recreation ground and 1 for a sandpit to provide road-mending material, all on the east side of the road to Twenty Pence Bridge.[3]

The Burystead, about ½ mile to the east of the church, is a late Elizabethan house of brick erected about 1600. In plan it consists of a central block with north and south wings projecting towards the east and large stair turrets in the angles. The front of the house is on the east. At the back are four gables in a line, but the two in the centre are probably later additions. The house is now divided into two residences. The principal gables have moulded brick copings with finials at the springs and apex. There is a string-course of moulded brick dividing the two stories, which is carried over some of the windows to form a hood-mould. Most of the windows have modern wooden casements, but several have late 17th-century casements. There are massive brick chimney-stacks with octagonal tops in groups of two and three. The roofs are tiled. It seems probable that the central portion was originally the hall and open to the roof, and this would explain the two staircases, since the upper stories in the wings would be isolated completely from each other. There is spacious attic accommodation in the roofs. While the exterior has been but little altered the interior has undergone great changes and the original size and appearance of the rooms is difficult to visualize. Even the main entrance, though probably occupying approximately its original position, is now quite modern. Some chamfered beams are visible and one original fireplace of clunch is exposed, with conventional flowers carved in relief on the lintel. Some of the original oak floors remain, but they are covered with modern deal boarding. The brick garden walls and a barn, also of brick, are of 17th-century date.

The Burystead was replaced as the manor house of Wilburton by 'a very gentlemanlike brick house' by Alice (Towers), wife of Sir John Thompson and lady of the manor *c.* 1632–50.[4] Wilburton Manor was again rebuilt in 1851 to the designs of Pugin.[5]

The Grange, to the north-west of the church, is a timber-framed house with brick filling and tiled roofs. The ground stage is of old brick, while the filling in the upper stage is modern, replacing reed and clay. The south wall and gable are entirely of brick. The original house was L-shaped and must have been erected about 1500. A century later a wing was added to the east, making the plan roughly T-shaped, and a large chimney-stack was inserted where the new addition joined the original work. This wing has an overhang in the upper story supported on brackets. In the ground-floor wing of the later wing the moulded uprights which supported the medieval front can be seen. The oak joists are exposed in the ceiling of this room and in the kitchen, which lies to the west in the original portion. There is a large open fireplace of early 17th-century date in the kitchen.

To the north of the church is Bell Gable House, a late 17th-century farmhouse of brick with curved gables and tiled roof. The windows and doorway are of later date, probably mid-18th century. The only internal feature of interest is a good staircase with turned balusters.

North-east of Wilburton parish, and amalgamated with it under the Isle of Ely Review Order of 1933,[6] lies the former extra-parochial tract of Grunty Fen—1,793 acres of low-lying ground that was formerly common pasture for the inhabitants of Ely, Witchford, Wentworth, Haddenham, Wilburton, Stretham, and Thetford.[7] This area was inclosed in 1861 under the general Inclosure Act of 1845. Eight acres were sold for £240 to defray costs, 24 acres were allotted as recreation gounds, and 249 allotted to the poor of the adjacent parishes. The several lords of the manors received 22½ acres. The remainder, in 670 pieces, was divided amongst the 396 persons or corporations who had held common rights in the Fen. The existing road system was fenced and stabilized at this time.[8]

MANOR. Five hides in *WILBURTON* were among the benefactions of Ethelwold and Brithnoth to the church of Ely.[9] In 1086 they contained 7 plough-lands, 3 of which, assessed at 3 hides 1 virgate, were in demesne. There were 4 soke-

[1] R. Gardner, *Dir. Cambs.* (1851), 510.
[2] Kelly, *Dir. Cambs.* (1900 and 1933).
[3] M.A.F. 1/382.
[4] B.M. Add. MS. 5819, f. 95*a*.
[5] Kelly, *Dir. Cambs.* (1900).
[6] M. of H. no. 76940.
[7] B.M. Cott. MS. Claud. C. XI, ff. 24*b*, 43*b*.
[8] M.A.F. 1/1114 and 1/85 (map).
[9] J. Bentham, *Ely*, 75.

National Buildings Record

UPWELL RECTORY

National Buildings Record

THE BURYSTEAD, WILBURTON IN 1948

(A second stair-turret is hidden by the gable on the right)

men (not mentioned in the *Inquisitio Eliensis*) bound to the land, 9 villeins, 9 cottars, and 8 serfs, and there was sufficient meadow and pasture for all the plough-teams and cattle. The sum of 1*s.* 4*d.* was received from the sale of reeds (*juncis*). The vill had been worth £4 when received, £10 T.R.E. and £7 only in 1086. It was and had always been part of the demesne of the church of Ely.[10]

Like its neighbours Haddenham and Stretham on the southern rim of the Isle, Wilburton was allotted to the bishops of Ely in 1109. In 1221 the demesne, in the three fields of Estfeld, Sudfeld, and Oswenhoue, amounted to 246½ acres, with 40 acres of meadow, 10 for hay.[11] The survey of 1251[12] shows a rather larger demesne of 276 acres, rated at 2 carucates: Oswenhoue was now merged in a larger field known as Northfeld *cum pertinenciis*. The meadow, in five pieces known as 'Brok,' 'Springuuelle', 'Lytlemed', 'Radeys' and 'le Hee', was about the same (38 a. 3 r.). The demesne was stocked with a bull and 10 cows, a boar and 16 pigs, and 200 sheep. The human population consisted of 9 *hundredarii* and free tenants, 21 customaries and virgaters and 11 cottars; they paid 31*s.* in rents, including 'witep'' and 'seggeselver'.

The gross income from this manor during the episcopal vacancies of 1286, 1298–9, 1302, and 1316 was £8 6*s.* 8*d.*, £9 18*s.* 9*d.*, £10 3*s.* 4*d.*, and £11 14*s.* 7*d.* respectively.[13] No specially interesting items of revenue are recorded. The two later accounts record the commutation of 834 (1302) and 423 (1316) 'works'. It is noteworthy that no mill is mentioned. The general upward trend in value shown by these accounts was in Wilburton as elsewhere reversed during the next generation, and the survey made in 1356,[14] during Bishop Lisle's troubled episcopate, shows a gloomy picture. The manor house was ruinous and of no value, and the demesne, producing £3 11*s.* 2*d.*, had shrunk to 214 acres. Twenty acres were worthless owing to floods, as were 8 acres of meadow: the unflooded meadow, of which there was also 8 acres, brought in 8*s.* The stock and other movables were worth £20 9*s.* 5*d.* 'Seggesilver' (8*s.* 6*d.*) and 'wychselver' (1*s.* 6*d.*), mentioned amongst the rents, recall the payments of 1251. The rents totalled £2 11*s.* 6*d.* There were no fisheries, Wilburton being on the landward side of the Isle, and again no mill. During the latter part of the 14th century conditions seem to have improved; fifteen accounts of the reign of Richard II show an average net profit of about £37.[15]

A long series of court and account rolls of the 14th and 15th centuries, in possession of the Pell family as lords of the manor at the end of last century, was examined by Maitland.[16] They show the extreme conservatism that prevailed in a manor held by an absentee ecclesiastical lord. Money commutation of services had hardly begun before about 1350, and was not completed for another two generations. In or about 1426 the demesne, consisting 246 acres of arable and 42 of meadow, was let for £8 a year to one of the customary tenants, then acting as reeve. This rent remained constant up to the alienation of the manor in 1600, but the rent charged to the customary or copyhold tenants gradually sank. In the early 15th century it had been 1*s.* an acre with the messuage thrown in, i.e. £1 4*s.* for a full virgate, and brought in about £22 a year; by the beginning of the 16th century, when some of the demesne was let to the customaries at £1 6*s.* 10*d.* a year, the gross rental had become fixed at £17 16*s.* 1*d.* The manor as a whole brought in between £25 and £30 a year. The profit gradually decreased at first, but at the end of the 16th century it remained fairly constant at about £29.[17]

Wilburton was left in Crown hands rather longer than the other manors alienated by Bishop Heton.[18] In 1609 it was granted in socage to Sir John Jolles, alderman of London, who paid the large sum of £1,261 18*s.* 4*d.*—about thirty-eight years' purchase at its then value.[19] Jolles, who died in 1621, also possessed the neighbouring manor of Hinton in Haddenham. By a settlement of 1616 he bequeathed Wilburton to his nephew Danett Poyntell, with remainder to his niece Alice Towers and her sons and daughters living at her decease. Poyntell was 59 years of age at Jolles's death.[20] Some litigation seems to have resulted from Jolles's settlement, as his nephew did not obtain livery of the manor until 1624.[21] He must have died soon after, for in 1650 we find Alice Towers, now the wife of Sir John Thompson, bt., of Husborne Crawley (Beds.), involved with her husband in a lawsuit against a tenant, William Patrick.[22] It was stated that the manor was settled on her for life in 1632.[23] Some time before 1656 Francis Towers her son sold the manor to Haynes Barlee of Clavering (Essex) for £1,000.[24] This family remained in possession for over a century until in 1778 the manor was passed by Charles Barlee to Catherine Buckle,[25] who was lady of the manor at the time that Lysons wrote (1808).[26] She in turn passed it to Sir Albert Pell in 1817,[27] in whose family the manor has since remained. Sir Albert died in 1832, and his relict the Hon. Lady Margaret Letitia Matilda Pell held it in dower until her death in 1868.[28] In 1900 Sir Albert's two surviving sons were joint lords. Albert Pell the elder, a noted agriculturalist and authority on the poor law,[29] died in 1907 and was succeeded by his nephew Albert Julian, who had been acting as steward. Oliver Claude Pell, the third of Sir Albert's sons, was chairman of the Isle of Ely County Council from 1889 until his death in 1891. On the death of Albert Julian Pell in 1916 his nephew, Beauchamp Stewart Pell, succeeded.[30]

CHURCH. The church was appropriated (*annexata*) to the archdeaconry of Ely before 1291, when it was valued at £14 13*s.* 4*d.*[31] Earlier valuations had been £10 (1217) and £13 6*s.* 8*d.*

10 *V.C.H. Cambs.* i, 368.
11 B.M. Cott. MS. Tib. B. II, f. 105.
12 Cott. MS. Claud. C. XI, ff. 49–52.
13 S.C. 6/1132/9, /10, /11, and /13.
14 B.M. Add. MS. 6165, ff. 277–8.
15 *E.H.R.* ix, 429.
16 Ibid. 417–39.
17 Ibid. 435; A. Gibbons, *Ely Episc. Recs.* 12; B.M. Add. Chart. 34274, f. 2*b*; S.C. 6 Eliz./783; S.P. 12/268/110.
18 For the alienation see Bentham, *Ely*, 194–5.
19 *E.H.R.* ix, 436; C 66/1800, no. 9. The annual value in 1609 was estimated at £33 8*s.* 9*d.*, included assessed rents of *c.* £20, farm of the demesnes £1 16*s.* 10*d.*, farm of the site of the manor house £8, and perquisites of court amounting to *c.* £3 10*s.*
20 C 142/396/147.
21 C 60/495, no. 8.
22 Cf. Partriks, a messuage in Haddenham.
23 C 5/382/123.
24 C 5/23/33; C.P. 25(2)/538 (Hil. 1650). Haynes Barlee was sheriff of Cambs. in 1669.
25 C.P. 25(2)/1282 (Trin. 19 Geo. III).
26 *Magna Brit.* ii (1), 282.
27 C.P. 25(2)/1283 (Hil. 57 Geo. III).
28 *D.N.B.* sub Albert Pell (1820–1907); M.A.F. 1/1114.
29 See *D.N.B.* He is rather to be associated with Northants, where he had an estate at Haselbech, and S. Leics. which he represented in Parliament from 1867 to 1885.
30 F. Boase, *Mod. Engl. Biog.*, ii, 1440; Burke, *Landed Gentry*, ed. 1937.
31 *Tax. Eccl.* (Rec. Com.), 265.

(1254).[32] No vicarage was ever ordained. In 1851 the perpetual curate received a stipend of £68 only and was obliged to augment his income by acting as headmaster of Huntingdon Grammar School.[33] With effect from 1864 his stipend was increased by £55 out of the Common Fund[34] and in 1874 by a further £71.[35]

In 1870, £1,400 was granted out of the Common Fund for the erection of a vicarage house.[36] A further £100 was granted in 1871.[37]

The rectorial tithes were from time to time leased by the archdeacons; to Sir Miles Sandys in 1608 for the lives of himself and his three sons at £24 yearly,[38] and later to the Malabar and Pell families. They were commuted in 1844 for a rent charge of £570[39] and transferred to the Ecclesiastical Commissioners in 1865.[40]

There was a guild of Our Lady in Wilburton in 1534. Its hall was sold in 1572 to Richard Hill and his heirs.[41] An acre of land held by John Tyllynghame, given to anniversaries in the church, was in 1553 sold to Sir John Butler and Thomas Chaworthe.[42]

The church of *ST. PETER* consists of chancel, north vestry and organ chamber, nave, north chapel, south porch, and west tower. The material is rubble and Barnack stone, partly plastered, and the roofs are covered with lead. The chancel arch and the tower are of the 13th century. There was an extensive rebuilding in the second half of the 15th century, and the chancel, vestry, nave, and porch are of this period. The organ chamber and north chapel were added in the second half of the 19th century.

The chancel has a large east window of five cinquefoiled main lights with small trefoiled lights above and a four-centred arch with an external hood-moulding terminating in heads. There are diagonal buttresses with two set-offs. The lateral windows, which are all of uniform design, are inserted under lofty blind arches with moulded caps and bases; they are of three cinquefoiled lights with rectilinear tracery and hood-moulds; the easternmost on both sides is curtailed to accommodate a doorway and the westernmost on the north now looks into the organ chamber. There is a string-course beneath the windows. The south doorway has continuous mouldings and a hood-mould. There is a coped parapet with gargoyles, and the uniform lateral buttresses have two set-offs. The chancel arch is of the 13th century and has deep mouldings and responds consisting of clustered shafts with moulded caps and bases. Internally there is a canopied niche on either side of the east window, in the angles, each with a shield, that on the north having the arms of Alcock, Bishop of Ely and lord of the manor, 1486–1500. A stone bench serves as a sedile. The doorway communicating with the vestry has continuous mouldings, and there is a good contemporary door with applied tracery. The vestry has a plain rectangular east window without a mullion; there is also a small rectangular opening through the north-east buttress of the chancel, and in the north wall is a small two-light window under a square head. There is an aumbry in the east wall. The organ chamber has a modern two-light window in the north wall and a diagonal buttress with two set-offs; it communicates with the nave by a plain modern doorway and modern two-light window above. It has a plain coped parapet.

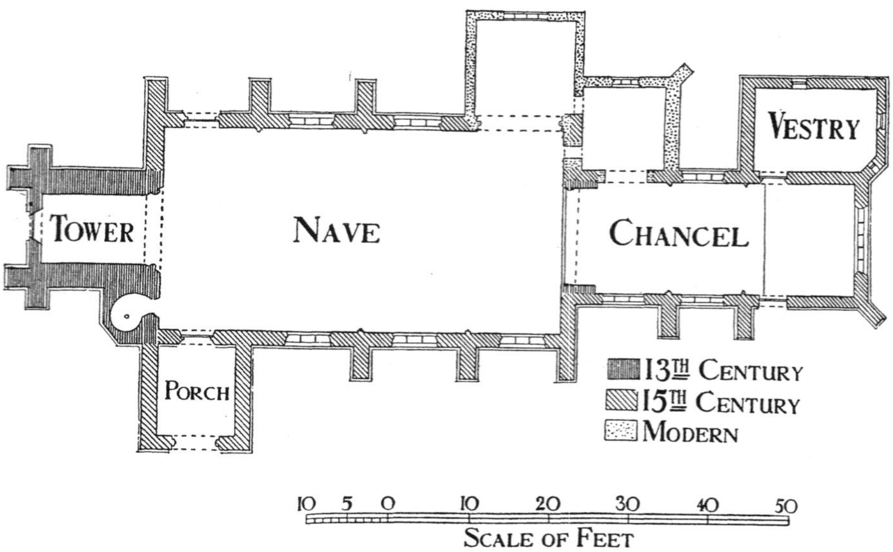

PLAN OF ST. PETER'S CHURCH

The nave has three windows on the south and three on the north, all uniform with those in the lateral walls of the chancel; the westernmost on the north is curtailed to accommodate a doorway, which has mouldings dying into the chamfered responds and an external hood-mould. All the tracery has been renewed. The transeptal north chapel opens to the nave by a modern arch of 15th-century character with clustered responds having moulded caps and bases. It has a 15th-century north window of three lights moved out from the north wall of the nave when the arch was made. There is a string-course beneath the windows of the nave and

[32] W. E. Lunt, *Valuation of Norw.* 539.
[33] H.O. 129/7/190.
[34] *Lond. Gaz.* 1866, 669.
[35] Ibid. 1874, 2358.
[36] Ibid. 1870, 1788.
[37] Ibid. 1871, 600.
[38] B.M. Add. MS. 5847, f. 21.
[39] Gardner, *Dir. Cambs.* 510.
[40] Kelly, *Dir. Cambs.* (1933).
[41] W. M. Palmer, *Village Gilds of Cambs.* 401.
[42] *Cal. Pat.* 1553, 290.

lateral buttresses with two set-offs, all uniform in design. There is a coped parapet with gargoyles. The 13th-century tower arch is two-centred and of two orders, the outer with continuous mouldings and the inner springing from shafts with moulded caps and bases. A portion of rough walling in the north-west angle of the nave and tower is probably a relic of the earlier church.

The porch has an outer doorway with a two-centred arch of two orders, the outer dying into the chamfered jambs and the inner springing from semi-octagonal shafts with embattled caps and moulded bases. Above is a cinquefoil-headed niche under a square label terminating in demi-figures. The inner doorway has an arch of two orders, which die into the chamfered jambs. There is a parvise above, which is approached from the tower stair turret; it has a two-light cinquefoiled window on the east and west under a square label, the former being now blocked. An early 13th-century shaft with carved caps and moulded base, and part of another cap, are preserved in the porch.

The tower is of three stages and dates from the middle of the 13th century. There are angle buttresses with two set-offs reaching to the top of the first stage, while the second stage has clamped buttresses rising from the string-course. The tower was refashioned in the second half of the 15th century. The west doorway has an arch of two orders with mouldings dying into the chamfered jambs and a hood terminating in heads. The west window is of three lights similar to those in the nave, with a hood terminating in heads. There is a blocked rectangular opening in the north and south walls with a slightly pointed inner arch. The second stage has a rectangular opening on the south. The belfry windows consist of two plain lancets under a plain containing arch. There is an embattled parapet and a small leaded spire, and a hexagonal stair turret to the belfry at the south-east corner with an embattled parapet.

The chancel has a cambered beam roof with wall posts resting on embattled stone corbels; the principals, purlins and intermediates are moulded and the intermediates have demi-angels at the feet holding shields. The vestry has a plain lean-to roof. The nave has a very fine cambered beam roof, almost flat, with wall posts and braces with pierced spandrels rising from wooden corbels, which consist of demi-figures, crowned and holding shields; the principals, purlins, intermediates, and wall plates are moulded and there is vertical boarding between the rafters; there are large bosses at the intersection of the principals and ridge purlins with the arms of Alcock, and cocks standing on globes are carved in the spandrels; the principals have a delicate cresting and a cock on a globe in carved wood is suspended from one of them. The porch roof is of cambered beam construction with wall posts and embattled wall plate; there is modern deal boarding but no rafters. The tower ceiling retains some plain old timbers. The roofs of the organ chamber and chapel are modern and plain.

The font is modern. The chancel screen of arched type is of six bays, the two in the centre being occupied by the doorway. The uprights are buttressed and terminate in shafts with embattled caps; the main openings have feathered cusping and tracery above, at the apex of the doorway is a demi-angel with three cocks on each side, all much renewed; the middle rail is moulded and the wainscot has tracery with carved spandrels; the base beam and the vaulting are modern, but part of the cornice on the east side is original. The altar rails are of early 18th-century date with turned balusters. There are four poupée heads in the chancel of 15th-century date. On the north wall of the nave are remains of 15th-century paintings, which represent St. Christopher, St. Blaise, and St. Leger. The two latter are fairly distinct but very little now remains of the St. Christopher, and all three have greatly deteriorated since their discovery in 1851. A portion of 17th-century panelling from Stretham church is fixed to the west wall of the nave.

Three panels of an altar tomb are now fixed against the north wall of the sanctuary; they have sunk quatrefoils in diamond-shaped frames enclosing shields, to which were formerly attached metal shields. The following brasses remain: (1) Richard Bole, Archdeacon of Ely, 1477, an effigy in a cope under a mutilated canopy, with a marginal inscription; this brass is attached to its matrix but the latter is now fixed to the north wall of the chancel; it is probable that originally it formed part of the above-mentioned altar tomb; (2) John Hyll and Margaret his wife, 1506, with five sons and two daughters beneath, and the Evangelistic symbols at the corners of the matrix; (3) William Byrd and Margaret his wife, 1516, with three sons and five daughters beneath. Both these brasses are now on the east wall of the nave to the south of the chancel arch. (4) Inscription to Robert Wetheringset, Archdeacon of Ely, 1444, now fixed to the south wall of the sanctuary.

The plate includes a Communion cup with cover of silver, 1569, inscribed 'For the Towne of Wilberton', a paten of silver, 1724, and a flagon of silver, 1714.

The tower contains five bells: 1st by Miles Graye III, 1651, 2nd and 5th by John Taylor of Loughborough, 1850, 3rd, no inscription, 4th, 1661 (probably by Christopher or Miles Graye). The original 2nd was by Charles Newman of Lynn, 1695, the original 5th (? by one of the Grayes) was the gift of Thomas Towers, 1661. They were re-cast at the expense of Margaret Letitia Matilda, Dame Pell.[43]

The registers begin in 1735 and are complete.

NONCONFORMITY. At the end of the 18th century prayer meetings were being conducted in Wilburton by Oliver Houlet. In 1802 a room was hired in Mr. Camps's house and converted into a chapel. From 1808 a Mr. Langford was pastor. A regular society of about 20 members, with a congregation of 100 on Sundays, existed about 1820.[44] This chapel and congregation may have been the Baptist chapel mentioned in 1831,[45] which was rebuilt in 1845[46] and still exists. In 1851 there was a larger average attendance at chapel than at church, though fewer Sunday scholars.[47]

SCHOOL. Bishop Yorke's inquirers found no school in Wilburton in 1789.[48] In 1846–7 there was a Sunday school for children of all ages, and infants' school opened in 1839, and a 'School of Industry' for girls, open during the six winter months. Apart from parents' fees of 1*d*. a week, these

[43] J. J. Raven, *Church Bells of Cambs.* App.
[44] *Cong. Mag.* iii, 114–15 (1820).
[45] Lewis, *Topog. Dict. Engl.* s.v. Wilburton.
[46] *Bapt. Handbk.* 1950.
[47] Baptists 148, Church 135. Sunday schools: Baptist 68, Church 80 (H.O. 129/7/190).
[48] Bp.'s Mun., Ely, B 6.

schools were supported by Lady Pell, who had endowed them with 15 acres of land producing £30 a year and by the Archdeacon of Ely as rector and impropriator. At that time 188 children were being taught (rather more than half on Sundays only), and a master and two mistresses were employed at salaries amounting to £36 15*s.*[49]

In 1855 the infants' school, which held 35, was enlarged to provide 90 places for children of all ages. The total cost was £687. Towards this sum the National Society granted £30 and £471 was raised locally.[50] The school was usually full or nearly so, and to avoid overcrowding after the new scale of accommodation of 1910 had been imposed, an extra classroom was added. Thus 111 places were provided. The children aged 13 and over were transferred to the Cromwell School, Chatteris (q.v.), in 1948–9.[51]

CHARITIES Thomas Buck, by his will proved 1566, left a rent charge of 40*s.* to be paid through the Cutlers' Company, of London, of which he had been warden, to the poor of Wilburton, with preference to persons called Buck. In 1692 Haynes Barlee, lord of the manor, bequeathed a messuage at Berden (Essex) to the use of the poor of six parishes, including Wilburton. The revenue, £10 yearly in 1837, was paid in rotation to the various parishes; Wilburton's latest turn had been in 1834, when it was used for premiums for three apprentices.

In 1714 Anne Wade left lands in Low Fen, the rents of which, subject to a life interest of her old servant Nathaniel Champions, were to be devoted to putting children to school. The land seems never to have come into possession of the parish, and was not identifiable in 1837.

In 1837 the Town Lands consisted of 24 acres in Wilburton, Haddenham, and Stretham, 15 acres being copyhold of the manor of Wilburton and paying 10*d.* ground rent, and two tenements in Town Yard inhabited rent free by labourers and their families. The rents of the Town Lands, then amounting to £29 6*s.*, had customarily been given indiscriminately to about 100 poor persons in sums of 1*s.* to 3*s.* It was ordered that in future they were to be used for the education of poor children. The books were kept 'in a slovenly and careless manner', and the treasurer owed £46 to the charities.[52]

WITCHAM

Witcham is a parish and village in the Isle, 5 miles west of Ely. The village stands on a hill and has a picturesque appearance from the Ely road, its houses grouped around the church with its low and massive tower. It was formerly situated on a by-road in the centre of its open fields, and provided a text-book example of the nucleated village settlement typical of the East Midlands. The construction of a satellite airfield at Sutton during the Second World War, in connexion with Witchford aerodrome, has, however, somewhat altered the layout. This airfield blocks the former high road to Chatteris, so that traffic now passes through Witcham village, along the road from the Horse and Gate cross-roads to Mepal. During the Second World War also, a hutted camp was built along the north side of the Sutton road. The camp was later used as a Polish Resettlement Centre and as a Women's Land Army hostel.

The men of Witcham formerly enjoyed common rights in the fens to the north (see Welches Dam). This fact gave the parish a confused boundary, which was considerably altered in the past century. In 1806 it was reported that Witcham and Coveney were separated by 'an ancient Ditch or Dike called the Division Dike', maintained by the occupiers of land on the Coveney side. The bounds of Witcham had last been perambulated about 1750.[1] In 1885 detached parts of Witcham were transferred to Coveney, Manea, and Mepal, and detached parts of Ely St. Mary, Ely Trinity, and Witchford were transferred to Witcham. This still left one part of Witcham parish, 124 acres in extent, detached. This fragment was joined to the main portion in 1933[2] when the intervening parish of Witcham Gravel, formed in 1896 by County Council order, was thrown into Witcham.

The fields of Witcham as specified in a 13th-century grant to Ely priory were north, east, and south fields, Shakelond and le Slade.[3] In course of time other fields are named. Burnt Hill in the north-east of the parish occurs as 'Burnewrthehil' in 1215–19.[4] It gave its name in the 16th century to Burn Hill Field.[5] The Market Way, now a by-road leading east from the village towards Ely, also occurs in 1215–19.[6] It gave its name to Market Way Field, which with Burn Hill Field is mentioned in the Parliamentary Survey of 1649.[7] In 1353 the men of Witcham bound themselves not to claim rights of common in Heathcroft,[8] now an unidentified area.

In 1649 the demesne included 45 and 55 acres respectively in Burn Hill and Market Way Fields; also 40 acres of pasture at Swarm Haugh Closes in the south-east corner of the parish by Wentworth cross-roads, and four closes comprising 15¼ acres near the site of the manor house.[9]

The final inclosure took place under an Act of 1838.[10] The award mentions Market Way Field south of the village and Burn Hill Field on the east of the parish, together with Wardy Way and West Fields north and west of the village and Littlebury Field between the village and Burn Hill Field. Wardy Way, now represented by Wardy Hill Road, first occurs in 1311.[11]

An area of 1,032 acres (excluding roads, drains, &c.) was inclosed and divided amongst 74 proprietors. The principal allottees were the Dean and Chapter of Ely who received 157 acres. These were divided amongst 3 lessees, of whom William Saberton obtained the largest share. Robert Poole received an allotment of 114 acres and Thomas Saberton 78 acres in his own right and 2½ acres jointly with Elizabeth his wife.[12]

49 Rep. of Nat. Soc. 1846–7 (privately printed).
50 Archives of Nat. Soc.
51 M. of Ed. file no. 12/58.
52 *Rep. Com. Char.* (*Cambs.*), [103], 332, H.C. (1837–8), xxiv.
1 E 134/46 Geo. III/M. 1.
2 Isle of Ely Review Order (M. of H. no. 76940).
3 B.M. Egerton MS. 3047, f. 67*b*.
4 D. & C. Mun., Carta 734.
5 *P.N. Cambs.* (E.P.N.S.), 245, quoting charters of Gonv. and Caius Coll.
6 D. & C. Mun., Carta 734.
7 Ibid., Cupbd. A.
8 Ibid., Carta 745.
9 Ibid., Cupbd. A.
10 1 & 2 Vic. c. viii.
11 *P.N. Cambs.* (E.P.N.S.), 245, quoting D. & C. Mun.
12 Award at County Hall, March.

Witcham possesses several interesting secular buildings. These include the Hall, an 18th-century building, 'gothicized' early in the 19th century, with casement windows and an embattled parapet. It has a gateway and barn. A farm-house 150 yards west-south-west of the Hall is probably of the late 16th century, with some mullioned and stone-framed windows.

MANOR The date at which *WITCHAM* was given to the church of Ely is not known. The Domesday commissioners found, however, that it had always formed part of the demesne of the church. When given and in 1086 it was worth £5. In 1066 it had been valued at £7. This reduction in value was due no doubt, here as elsewhere in the Isle, to Hereward's rebellion. In 1086 it was assessed at 4 hides 1 virgate. Two hides were in demesne, with 2 ploughs and enough land for another. There were 12 sokemen with 2 hides less 1 virgate, who could not grant away their land without the abbot's permission, 2 villeins with 10 acres and 2 bordars with 5 acres each. These 16 men had 4 ploughs. There were also 4 cottars and 5 serfs.[13]

Witcham remained with the convent after the formation of the see in 1109, and experienced the uneventful history of such villages. The only landmarks in its history are the usual ones in Ely conventual manors—the grant of free warren to the prior in 1252,[14] its confirmation together with that of other manorial perquisites in 1417,[15] and the formal transfer to the dean and chapter in 1541.[16] In 1302–3 Thomas de Lenne is stated to have held $\frac{1}{12}$ fee in Witcham of the Bishop of Ely.[17] The bishop was lord presumably in his capacity of titular head of the convent, for there is no other evidence of episcopal interests in Witcham. The same fraction of a fee was held by de Lenne's successors, Hervey Pelet and Thomas Laushille, in 1346.[18]

The convent seem to have kept the manor in hand all through the Middle Ages, and many grants to them in frank almoin of property in Witcham are recorded in documents in the Dean and Chapter muniment room at Ely.[19] Most of them are quite small in amount. In 1343 Edward Chesewyk granted 4 acres in return for half a gallon of convent ale daily.[20] More important grants, for which licences to acquire in mortmain have been preserved, were made in 1314 and 1364.[21] The former, comprising 7 messuages, 163 acres of land and 19 of meadow, and 21*s.* of rent, in Witcham and Ely, came from Geoffrey de Fresyngfeld, a relative of the then prior. In 1371 it was stated that the convent had recently acquired three parcels of land, 'Crispestenements', 'Fotestenements', and 'Huythecroft', value £1 annually, without licence.[22]

Witcham was not, however, a very valuable manor. In 1291 it was assessed at £27 13*s.* 10¼*d.*,[23] a figure slightly below that for Wentworth. The 'Long Roll' of 1324–5[24] shows a demesne of 2 carucates, producing £23 19*s.* 3½*d.* Expenses in that year amounted to £21 14*s.* 3½*d.* As at Witchford there was considerable trade in wool and fells, 85 skins being sold. A mid-14th-century extent[25] shows a total value of only £11 7*s.* 8*d.*, including a mill worth £2. The manorial dovecote needed repairs to the value of £2 6*s.* Later accounts show in some cases rather higher figures. Total receipts in 1367 were £23 10*s.* 11*d.*, including 1*s.* 4*d.* (with 3*s.* 4*d.* in arrears) from a small fishery at 'Pylgrymesee',[26] and in 1452 £17 11*s.* 2*d.*[27] On the other hand certain ministers' accounts at the Public Record Office, whose detachment from the Ely series has not been accounted for, return £9 16*s.* (1429), £5 17*s.* (1435), £8 3*s.* 4*d.* (1437 and 1466).[28] A roll of 1527–8[29] shows receipts of £13 19*s.* 10½*d.*, the lowest sum received from any of the convent's Cambridgeshire manors except Mepal and Wisbech Murrow. In 1541, when taken over by the dean and chapter, the net receipts were £12 16*s.* excluding the rectory.[30] Out of this were allowed 1*s.* 6*d.* as contribution to Aldreth causeway, £1 for bailiff's fees, and £2 13*s.* 4*d.* pension to a former vicar. The manor house was let to William Gibson for £5 6*s.* 8*d.* a year. The Parliamentary Survey of 1649[31] shows that the manor house had been pulled down: its site was a 4-acre close of pasture with a 'great barne' of timber, thatched. The fines were arbitrary, but usually paid on a yearly valuation. The tenants had the right to fell timber, and courts leet and baron were held. The yearly value was stated at £11 6*s.* 4½*d.* The £5 6*s.* 8*d.* yearly rent for the site of the manor was resumed in 1661, when Michael Holman of Chatteris (q.v.) had a 21-year lease. From the early 18th century the Wright family were tenants at the same rent. The manorial estate then comprised 67 acres of arable, 60 of pasture, and 10½ of old inclosure (including meadow). An area of 191 acres in Byal and Widden Fens was let to the same tenants at the additional rent of £3 5*s.* 6*d.* The fines for renewal were £45 in 1710 and 1717 and £60 in 1724.[32] The size of the estate was much reduced during the 18th century, terriers of 1745 and 1766 showing only 45 acres apart from the site of the manor house.[33] In 1785 Richard Wright, the then lessee of the manor, was drawing £123 7*s.* 6*d.* a year in rents from his sub-tenants.[34] Subsequent lessees include Mrs. Elizabeth Martin (1851)[35] and William Saberton (1863).[36] The latter, a member of a family still represented in the district, was a prominent farmer. The Church Commissioners, representing the Dean and Chapter of Ely, retain the lordship.[37]

CHURCH The church has always been in the gift of the Prior and Convent or Dean and Chapter of Ely. Its value was £10 in 1217 and £13 6*s.* 8*d.* in 1254.[38] It was given by one of the early bishops of Ely to the chamberlain to provide clothing for the monks, an arrangement confirmed by Bishop Northwold, who instituted a vicarage probably just before his death in 1254.[39] The appropriation was

13 *V.C.H. Cambs.* i, 369.
14 *Cal. Chart. R.* 1226–57, 391.
15 *Cal. Pat.* 1416–22, 184, 185.
16 *L. & P. Hen. VIII*, xvi, g. 1226(12).
17 *Feud. Aids*, i, 151.
18 Ibid. i, 173.
19 e.g. Cartae 160, 445, 726, 729, 733–5, 738, 743, 745, 750, 752, 764, 765, 1123.
20 Ibid., Carta 160.
21 *Cal. Pat.* 1313–17, 121; 1364–7, 14.
22 *Cal. Inq. Misc.* iii, no. 784 (pp. 293–5).
23 *Tax. Eccl.* (Rec. Com.), 270–1.
24 D. & C. Mun., Ely.
25 Ibid., Cupbd. A. Dated Weds. after St. Ambrose Edw. [II].
26 A small fen island belonging to the Pelrymes, a local family of some importance. It occurs in 1362 (D. & C. Mun., Carta 1127).
27 D. & C. Mun., Cupbd. A.
28 S.C. 6/1257/4, /5, /6, /9.
29 D. & C. Mun., Treas. R. no. 20.
30 B.M. Add. MS. 5847, f. 75.
31 D. & C. Mun., Ely, Cupbd. A.
32 Ibid., Bentham Transcript.
33 Ibid., Cupbd. A.
34 Ibid.
35 R. Gardner, *Dir. Cambs.* (1851), 511–12. She was a 'principal landowner' and the only 'private resident' in the parish apart from the vicar, but her address is not stated.
36 D. & C. Mun., Ely, Cupbd. B.
37 Kelly, *Dir. Cambs.* (1933).
38 W. E. Lunt, *Valuation of Norw.* 539.
39 Jas. Bentham, *Ely*, 147. The Norwich Valuation of 1254 does not mention a vicarage at Witcham. Bishop Northwold died on 9 Aug. of that year.

confirmed by Bishop Balsham in 1276.[40] In 1291 the rectory was returned at £13 6*s.* 8*d.* and the vicarage at £4 13*s.* 4*d.*[41] In 1535 the value of the vicarage had risen to £8 10*s.* 11*d.*:[42] the rectory was not separated from the other spiritualities of the convent but seems to have decreased in value, as it was let six years later to Nicholas Kyme for £7 10*s.* only.[43] A long series of beneficial leases of the rectory, at £5 18*s.* with 4 quarters of wheat, exists at Ely.[44] The first recorded lessee is John Thorogood, who was succeeded in 1618 by John Millett of Barley (Herts.). After the Restoration the lessees changed frequently until 1696. William Papworth then took the rectory and his family retained it until 1794. Terriers of 1745 and 1766 show 58 acres of glebe in 54 pieces. Another one of 1773 shows the same area of arable and 45 acres of fen. In 1808, when the rectory was leased to Thomas Maylin, the rent was raised to £6 13*s.* 8*d.* and 4 quarters of wheat.[45] The net value of the rectory seven years later was £386 5*s.* 6*d.*, the glebe (103½ acres) being worth £151 11*s.* 3*d.*, and tithes £310 9*s.* The vicar's stipend was then £39 18*s.*, raised in 1820 to £42 and before 1851 to £100 net.[46] The rectorial tithe had then been commuted for £595, the vicarial for £116. In 1863 the glebe comprised 106 acres.[47] In 1851 there was a Sunday school of 27 children.[48]

The church of *ST. MARTIN* consists of chancel, clerestoried nave, aisles, south porch, and west tower. The material is rubble with much repair in brick; the roofs are tiled; the chancel was formerly thatched. The earliest part of the fabric is the chancel, which belongs to the first half of the 13th century, and the tower is of approximately the same date. The aisles were added early in the 14th century, and about fifty years later a new east window was inserted in the chancel. In the 15th century the porch was added, the chancel arch rebuilt, and new windows inserted in the lateral walls of the chancel. In the latter part of the 17th century the west side of the tower was partly rebuilt in brick. The church escaped Victorian restoration and has been carefully repaired in recent years.

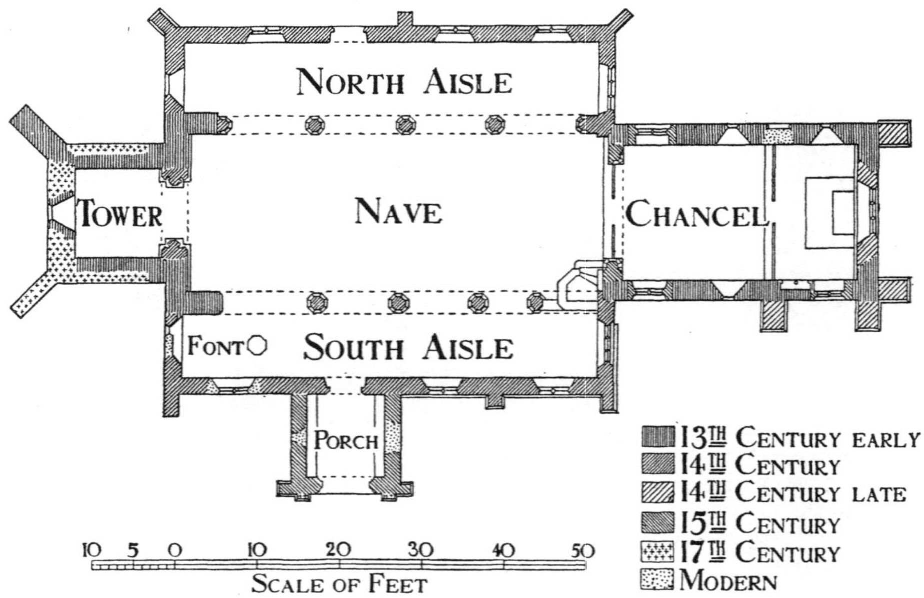

Plan of St. Martin's Church

The chancel has a three-light east window with flowing tracery much renewed and a hood-mould; on either side are traces of former lancets. There are angle buttresses with one set-off. In the north wall are two plain lancets and a 15th-century two-light window with cinquefoiled heads under a square label; also a blocked 13th-century doorway with continuous chamfer and a hood-mould. There is in the south wall a buttress with one set-off. The fenestration of this wall consists of a 15th-century two-light with cinquefoiled heads under a square label, which has been much renewed on the exterior, then a plain lancet, and finally a two-light window with cinquefoiled main lights and trefoiled tracery under a square label. There is an internal string-course of 13th-century date. The 15th-century chancel arch is two-centred and of two orders with semi-octagonal responds having moulded caps and bases. There is a double piscina with cinquefoiled heads and a plain octagonal central shaft; the central part of the heads has been renewed. There is a bracket on either side of the east window.

The nave has early 14th-century arcades of 5 bays on the south and 4 on the north; the first 2 bays on the south are on a smaller scale than the rest. All the arches are two-centred and of two orders with octagonal columns having moulded caps and bases; the columns on the north are more slender than those on the south and may be rather earlier; the eastern arch has been grooved for the insertion of a screen. The responds are semi-octagonal except the western on the south, which is in the form of a large corbel. The clerestory consists of three windows on each side, of three lights with trefoiled heads and a quatrefoil above under an external hood. The tower arch is of two orders with continuous chamfers and a hood-mould; it is prob-

[40] *Ely Dioc. Remembrancer*, no. 193, p. 123.
[41] *Tax. Eccl.* (Rec. Com.), 265.
[42] *Valor Eccl.* (Rec. Com.), iii, 499.
[43] B.M. Add. MS. 5847, f. 75.
[44] D. & C. Mun., Cupbd. A.
[45] Ibid.
[46] Gardner, *Dir. Cambs.* (1851), 511.
[47] D. & C. Mun., Cupbd. B.
[48] H.O. 129/7/190.

ably of 14th-century date and above are indications of a larger 13th-century arch. There is a foliated cross, probably of the 14th century, on the east gable of the nave.

The south aisle has been much repaired with brick. The east window is of three cinquefoil-headed lights under a square head. There is a south-east angle buttress with no set-off and a similar one on the south; at the south-west is a plain buttress with one set-off. The first two windows on the south are of early 14th-century date and consist of two plain lights; the remaining window on the south is a modern two-light with cinquefoiled heads under a square label; there is a plain lancet in the west wall, partially blocked in modern times. There is a plain brick parapet, probably of 17th-century date.

The porch has an outer doorway with a two-centred arch of two orders springing from semi-octagonal responds having moulded caps and bases; there is a plain hood-mould. There are straight angle buttresses largely of brick. In the east wall is a blocked two-light window under a square label, and in the west wall a single light, also blocked. The plain inner doorway of early 14th-century date has a continuous chamfer.

The north aisle has an east window of three irregular lights with trefoiled heads. There are diagonal buttresses at the north-east and north-west of one set-off and a similar one on the north, all much repaired with brick. The lateral fenestration consists of a trefoiled two-light with quatrefoil above, next a plain two-light, and finally a window similar to the first. The plain north doorway has a continuous chamfer. The west window is a plain single light. There is old coping on the east and west walls, but no lateral parapet. On the sill of the north-east window is a piscina and there is a bracket on the east wall.

The massive tower is of three stages. The west face and the return walling on the north and south were rebuilt, or at any rate refaced, in brick in 1691, as appears from the date cut on a brick beneath the west window. There is an original lancet on the west of the ground stage but no doorway. The second stage has a lancet on the north and south, the former now blocked, and the top stage a lancet on the north, south, and west. There are diagonal buttresses on the west with three set-offs, mainly of brick. The embattled parapet is of brick with stone dressings, and there is a small leaded cap.

The chancel has a modern braced rafter roof; the nave roof is of queenpost construction and modern except for the wall posts and brackets, which rest on stone corbels with demi-angels; the aisles have plain lean-to roofs, probably of 17th-century origin but much renewed; the porch roof is ancient and of collar-braced construction; there are some old timbers in the tower ceiling.

The 13th-century font has an octagonal bowl, the sides of which are carved with human heads alternating with grotesques and in one case a censing angel; it rests on five shafts with moulded caps and bases. The chancel screen is of very light construction and probably of early 16th-century date; the uprights do not run through in the ordinary manner. On the top is part of the parapet of the loft, with cinquefoil-headed openings; there are two bays on each side of the doorway with traceried heads. The wainscot has applied tracery which is largely modern, as is the base beam. The double doors, which do not extend above the level of the wainscot, are carved with arabesques above the transome. There is a most interesting stone pulpit of 15th-century date in the south-east angle of the nave, with its original base and stone steps; the three exposed sides have cinquefoiled heads, and the base is connected with the first column of the arcade by a contemporary dwarf wall. In the south-east angle of the nave, above the pulpit, are traces of wall paintings which are partly medieval and partly of the 17th century, the latter consisting of texts and borders in a fragmentary condition. At the west end of the nave is some 15th-century seating with plain poupée heads. There is a 17th-century communion table and good early 18th-century altar rails with turned balusters. In the base of the tower is a dilapidated oak chest, probably of 14th-century date. Most of the windows have clear glass, some of which is in rectangular leading.

The plate includes a silver chalice and paten, of 1669, and a silver salver of 1699, the gift of Dorothea Taylor.[49]

The tower contains one bell by C. and J. Mears, 1849.

The registers begin in 1663 and are complete.

NONCONFORMITY In 1851 the vicar (Revd. F. W. Packer) complained that 'this parish has been greatly neglected: and Dissent consequently has much prevailed, and laxity of principle'. A Wesleyan Methodist chapel had been established in 1813, and a Primitive Methodist chapel *post* 1840.[50] The former chapel was rebuilt in 1898:[51] the latter has not been otherwise encountered. In 1851 there were congregations of 90 and 19 at these chapels respectively. The former also had a Sunday school of 40 children.[50]

SCHOOLS John Fernley was licensed as schoolmaster for the parish of Witcham in 1582.[52] In 1798 Witcham parish contained two schools, kept by Mary Alison (Anglican) and Mary Remington (Nonconformist). Neither was licensed, and the latter seemed likely to close owing to lack of support.[53] The National Society report of 1846 described a school which was open both weekdays and Sundays, with 69 children in attendance, 32 on Sunday only. There were a master and a mistress, who received £5 and £1 19*s.* a year respectively, but 'a schoolroom is wanted and the funds to support it'.[54] This deficiency seems to have been fatal: no school is mentioned in Gardner's *Directory* (1851) and no existing provision was reported when the question of a National Society grant for a church school was taken up in 1870. A school providing 90 places, with a teacher's house, was built and opened in 1873. The National Society granted £40 towards the total cost of £537 13*s.*, of which about two-thirds was raised by subscription.[55] The land was bought from Clare College for £20. The school was reclassified for 73 children (53 mixed, 20 infants) in 1910, and again in 1939 for 35 juniors and infants, the reduction being due to the abandonment of the previous infants' classroom. There were 30 children on the roll in 1948.[56]

49 Infm. Revd. R. H. Gibbon, vicar.
50 H.O. 129/7/190.
51 Kelly, *Dir. Cambs.* (1900).
52 Bp.'s Mun., Ely, G 2.
53 Ibid., B 6.
54 Archives of Nat. Soc.
55 Ibid.
56 M. of Ed. File no. 12/71.

WITCHFORD

Witchford is a small and rather narrow parish near the centre of the Isle proper. 'The Town stands on a plesant Hill att 3 miles west of Ely,'[1] and the road between the two places crosses some of the highest ground in the Isle (about 80 ft. above sea-level), commanding fine views over the southern extremity of the Fens to the Suffolk hills beyond Newmarket. Witchford is at present one of the smaller villages in the county but was formerly of greater importance; Domesday Book refers to the two hundreds of Ely which meet at Wiceford,[2] and the name 'Witchford Hundred' is found from 1128.[3] North Witchford is still the name of one of the rural districts, and the village is one of the few in Ely Rural District which have increased in population since 1931.[4] The soil and subsoil are both clay, so that there is more permanent pasture around Witchford than in the low lands, but as at Downham there are some orchards,[5] which take advantage of the south-easterly slopes. The village straggles along the main road (A 142) from Ely to Chatteris and St. Ives. It contains no buildings of special architectural interest, though Ivy House and the farm at the east end of the village date from the 17th century. The general appearance is neat even if of no special distinction, but a hundred years ago Witchford 'in point of filth would bear comparison with the worst place in any part of the kingdom', with an open drain all along the main street.[6] At the eastern end of the village a by-road runs southwards across Grunty Fen to the Ely–Cambridge road near Stretham. An R.A.F. aerodrome was established at Witchford during the Second World War. The landing ground was on the plateau east of the village, in Ely; part of the buildings, on the north side of the village, have been converted into a Youth Hostel. An exchange of territory with Ely under the Isle of Ely Review Order, 1933,[7] resulted in a net gain to Witchford of 162 acres and 3 persons.

The proportion of land dealt with by inclosure awards in Witchford was 1,087 acres or about 50 per cent. of the total area of the parish, a proportion distinctly higher than anywhere else in the Isle except Stretham, Thetford, Wentworth, and Witcham. There were two Witchford awards. The first, made in 1813 under an Act of Parliament of 1806,[8] divided 460½ acres amongst 18 proprietors, the chief beneficiaries being Thomas Moxon (76 acres), Charles Cross senior (53 acres), John Cropley (52 acres in his own right and 32 as lessee under the Dean and Chapter of Ely), Clement Hitch (42 acres), Edward Cropley (38 acres), and Clare College (28 acres). Most of these families and institutions figure prominently in the second award of 1840,[9] which divided 626½ acres amongst 30 claimants. This time Clare College received the largest allotment (110 acres), George Cropley (95¼ acres in his own right and 74 with Luke Dench as devisee under the will of Clement Hitch) and Christopher Pemberton (79¾ acres) also taking large shares. The dean and chapter had by this time parted with most of their land and received only 1¼ acre; 8½ acres were awarded to the vicar as glebe. The open fields of Witchford numbered four—Belham's End and Marrow Way Fields north and north-west respectively of the village, Hale Field south-west, and Briery Field west of this latter on the Wentworth border. The 'built-up' area along the village street almost completely severs the farm lands.

Thurstan, the last Saxon Abbot of Ely (d. 1072), was a native of Witchford.[10] Samuel Bentham, father of James Bentham the historian of Ely, was Vicar of Witchford at the time of his son's birth (1708).[11]

MANORS. An area of 300 acres in *WITCHFORD* and Wold was amongst the endowments of Ethelwold and Abbot Brithnoth when they revived Ely monastery after the Danish invasions.[12] This estate was worth £8 when received, £12 in 1066, and £10 in 1086. The fall in value was no doubt due to the devastation caused by Hereward's exploits. It was rated at 3 hide or 7 ploughlands. The demesne was rated at 1½ hide and consisted of 2 or possibly 3 ploughlands. There were 5 sokemen with ½ hide who could not depart from the land, 17 villeins each with 7 acres, 7 cottars, and 8 serfs. There was meadow for 7 ploughs and pasture for all the cattle of the vill. Witchford had always formed part of the demesne of the church of Ely.[13] Its population and hidage as recorded in 1086 exceeded those of some places in the Isle now of greater importance.

Like most of the manors in the immediate neighbourhood of Ely, Witchford remained with the convent when the see was founded, and its subsequent history is uneventful. The prior acquired the right of free warren in 1252,[14] which was confirmed at a general definition of the episcopal and monastic rights in 1418.[15] The valuation of the *Taxatio* (£52 16*s.* 6½*d.*)[16] is much higher than any subsequently encountered, and higher than that of the monastic property in Ely itself. Perhaps the compilers of the *Taxatio* included in Witchford that tract of land west of Ely, which includes Emery Barn and Keten's manors. In 1324–5 the convent's receipts from Witchford amounted to £24 11*s.* 5*d.* only, and its expenditure to £24 19*s.* 10½*d.*[17] There was then a valuable mill, farmed at £4 8*s.* 11*d.*; the profits of the court leet came to £3 2*s.* 9*d.* There was a thriving trade in wool and fells; 108 fells were sold to John de Whatefeld and 53 to Bernard de Bradefeld, two East Anglian merchants who frequently occur in Ely rolls at this time. The demesne was of 4 carucates.[18]

Various small gifts of property in Witchford were made to the prior and convent during the 14th cen-

1 B.M. Add. MS. 5820, f. 83.
2 *V.C.H. Cambs.* i, 367.
3 *P. N. Cambs.* (E.P.N.S.), 229, quoting Red Bk. of Thorney.
4 Witchford, Littleport, Mepal, and Wilburton have increased, Stretham, Sutton, and Witcham remained stationary, and Coveney, Downham, Haddenham, and Wentworth decreased (Infm. Isle of Ely County Planning Officer).
5 These were noted as a feature of the village in 1851 (Gardner, *Dir. Cambs.* 512).
6 *Camb. Chron.*, 5 Oct. 1850. At this date 37 persons in Witchford were living in 5 houses of 2 rooms only.
7 M. of H., no. 76940.
8 46 Geo. III, c. vi. The award is at County Hall, March.
9 K.B. 122/1341, m. 3100.
10 *V.C.H. Cambs.* ii, 201.
11 *D.N.B.*
12 Jas. Bentham, *Ely*, 75.
13 *V.C.H. Cambs.* i, 369.
14 *Cal. Chart. R.* 1226–57, 391.
15 *Cal. Pat.* 1416–22, 184–5.
16 *Tax. Eccl.* (Rec. Com.), 270–1.
17 A deficit was at this time not uncommon in Ely manors, e.g. Wisbech Barton (Bailiff's R., 13–14 Edw. II, Bp.'s Mun., D 7).
18 D. & C. Mun., Ely.

tury,[19] but the manor seldom brought in much more than £10 a year in the 15th century[20] and the valuation of 1527–8 (£19 1*s.* 10*d.*) ranked Witchford as one of the less valuable of the monastic manors.[21] Another valuation of 1541, however, records net profits of £36 9*s.* 1*d.*, excluding the rectory.[22] In the same year the manor, whereof the rents of assize were appropriated to the *custos capelle*,[23] was formally transferred to the newly constituted Dean and Chapter of Ely.[24] This grant was confirmed in 1587.[25]

In 1650 the manor was sequestered to William Underwood and to Henry Whiston of London. It was then valued at £141 9*s.* 11½*d.* a year gross and the grantees paid £2,425 9*s.* 6*d.* for it.[26] In a survey made in 1649 the manor was said to comprise 245 acres held in demesne valued at £18 5*s.* and 224½ acres of rented land valued at £128 3*s.* 8*d.* The meadow land, which was rented, amounted to 100 acres. It had been let by the dean and chapter for £72 yearly and by the sequestrators for £100. Rents of assize were valued at £13 2*s.* 9*d.* and timber at £21 3*s.* 4*d.*[27]

A good deal of private inclosure had by this time been carried out. Various closes are mentioned in the deed of sale of 1650. Such are Dean's, Archdeacon's (north of Grunty Fen), Mr. Boys's close now belonging to Archdeacon Wigmore, a close occupied by George Coney in Alderforth and West Fen, Dr. Collins's, Dr. Cannon's, Dr. Tanner's, Dr. Cressener's, Dr. Knight's, Dr. Perkins's, Dr. Davies's, Dr. Ashton's, Mr. Rogers's or Beele's Close, Thistle Close, Dr. Goodridge's, and Mr. Buckeridge's. The names suggest considerable investment by cathedral dignitaries and their families in landed property in Witchford during the previous century.[28] The dean and chapter, as lords of the manor, had recently received 40 acres in Alderforth in the Vermuyden draining of the fens.[29]

At the Restoration the manor was returned to the dean and chapter, who, with their successors the Church Commissioners, have continued to be lords.[30] In the early 18th century the tenant was Sir Roger Jenyns. He was succeeded in 1725 by Soame Jenyns who paid in rent £31 10*s.* a year together with a brawn valued at £1 6*s.* 8*d.* He was also required to meet the expenses of an annual court dinner given to the dean and chapter and their officials.[31]

WOLD, which in Bentham's time survived only as 'some arable and pasture lands near Witchford',[32] and is now completely lost, was of some importance in the Middle Ages. In 1299 William Berewyk de Todeham and his daughter Alice, Ela relict of Simon de Insula and John and Agatha (le Despencer) his son and granddaughter were accused of ejecting John Deen and Amice (de Insula) his wife from property in Wold and Haddenham, which was settled by Simon de Insula on Amice on her marriage to John Deen.[33] The plaintiffs described Wold as a 'manor', but the phrase was not used in documents which they submitted in evidence. The suit was decided in favour of the plaintiffs, and in 1302–3 William de Todeham and his coparceners held a knight's fee in Wold, Sutton, and Chatteris of the Bishop of Ely for 40*s.* In 1346 this fee was shared between William de Ruston, the Prior of Ely, and John de Deen. The prior alone was in possession in 1428.[34] Wold was still a separate estate in 1649, when it comprised 120 acres (41 in pasture). About 1730 the tenant was Henry Stevens at £2 yearly.[35]

CHURCH Between 1219 and 1224 Bishop de Fontibus granted the church to the prior and convent and ordained a vicarage of 5 marks value,[36] Pope Honorius III confirming the grant in the latter year.[37] The vicarage was not actually instituted until after 1291, when the church, described simply as an *ecclesia*, was worth £10;[38] earlier valuations had been £10 13*s.* 4*d.* in 1217 and £10 in 1254.[39] In 1535 the vicarage was rated at £9 18*s.* 9*d.*, the rectory, which was appropriated to the hosteller, being returned with the spiritualities of the convent.[40] In 1541 the rectory was let to Thomas Rayne for £3 yearly.[41] In 1562 a pension of 19*s.* 10½*d.* was paid to the Bishop of Ely, as a tenth of the value of the vicarage.[42] The prior and convent, and since the Reformation the dean and chapter, have always been the patrons, though they have occasionally made grants of a turn, as in 1616 to Dr. Andrew Willett, prebendary of Ely and a well-known divine and controversialist.[43]

The rectorial and vicarial tithes were commuted for rent charges of £430 and £140, with 24 and 20 acres of glebe respectively, in 1839.[44]

The church of *ST. ANDREW* consists of chancel, nave, north porch, and west tower. The material is rubble and brick with stone dressings, and the roofs are tiled. The tower is of the 13th century and is the survival of an earlier church. In the third quarter of the 14th century the nave and chancel were rebuilt on an extended scale. The reconstructed building was consecrated in 1376.[45] In the 15th century the present porch was added. There were considerable repairs in the first half of the 19th century, which included the reconstruction of the porch and the partial rebuilding of the top stage of the tower. Later in the century there was much 'restoration' of the interior, with unhappy results.

The chancel has an east window of three lights with rather elaborate tracery of 14th-century date, the mullions of which have been much renewed. There are angle buttresses with one set-off and a similar buttress in the centre of the north and south walls. In the north wall there is a window of two cinquefoil-headed lights with a quatrefoil above. There are two windows of two trefoil-headed lights with a quatrefoil above in the south wall and to the west of these a low side window with cinquefoiled head. There is a piscina in a cinquefoil-headed recess in the usual position and a double aumbry in the north wall of the sanctuary. The chancel arch is two-centred and of two

[19] e.g. *Cal. Pat.* 1343–5, 478; 1364–7, 14, 172.
[20] £11 10*s.* in 1429, £8 in 1435, £12 10*s.* in 1467 (S.C. 6/1257/4–9).
[21] D. & C. Mun., Treas. R., no. 20.
[22] B.M. Add. MS. 5847, f. 77.
[23] D. & C. Mun., Treas. R., no. 20.
[24] *L. & P. Hen. VIII*, xvi, g. 1226 (12).
[25] C 66/1303, m. 1.
[26] C 54/3509, no. 37.
[27] D. & C. Mun., Cupbd. D.
[28] C 54/3509, no. 37.
[29] D. & C. Mon., Cupbd. C.
[30] Gardner, *Dir. Cambs.* (1851), 513; Kelly, *Dir. Cambs.* (1933).
[31] D. & C. Mun., Bentham Transcripts.
[32] Bentham, *Ely*, 75 n.
[33] J.I. 1/96, m. 65 d.
[34] *Feud. Aids*, 151, 173, 191.
[35] D. & C. Mun., Bentham Transcripts. Pasture called 'le Would' was held at the same rent in 1541 (B.M. Add. MS. 5847, f. 77).
[36] D. & C. Mun., Carta 67.
[37] B.M. Add. MS. 5819, f. 4. See also *Ely Dioc. Remembrancer*, no. 193, p. 123.
[38] *Tax. Eccl.* (Rec. Com.), 265.
[39] W. E. Lunt, *Valuation of Norw.* 539.
[40] *Valor Eccl.* (Rec. Com.), iii, 499.
[41] B.M. Add. MS. 5847, f. 77.
[42] *Cal. Pat.*, 1560–3, 226.
[43] Inst. Bks. (P.R.O.).
[44] Gardner, *Dir. Cambs.* (1851), 513.
[45] B.M. Add. MS. 5820, f. 81. *Ely Dioc. Remembrancer*, no. 119, p. 73.

orders with moulded caps and bases to the rounded responds. On either side of the arch in the east wall of the nave is a cinquefoil-headed niche.

The present nave is considerably wider than the tower, and the junction of the lateral walls with the latter is effected rather clumsily by means of a short section of slanting wall on either side, probably done with the object of economizing in quoins. The south-east window is of three lights with rather clumsy tracery of 14th-century date. The remaining two windows on this side and the three in the north wall are all similar and consist of two trefoil-headed lights with a quatrefoil above. Beneath the north-east window is a double aumbry. In the slanting curtain walls at the west end are two large and clumsy lancets of 19th-century date.[46] The south doorway is plain 14th-century work with continuous chamfers and hoodmould. The east gable of the nave has been rebuilt in brick, and the south buttresses have been repaired in brick. The large porch is of 15th-century origin, but the side walls and much of the front were rebuilt in brick in 1838, which date appears over the outer doorway. There is a stone seat along the east and west walls. The outer doorway is original and is of two orders with semi-octagonal responds having moulded caps and bases; there is a hood-mould terminating in heads. The inner doorway is of the 14th century with mouldings dying into the chamfered jambs.

The tower is of three stages. It opens to the nave by a low arch turned in brick, which is of uncertain date. The west window of the ground stage consists of coupled lancets, which have been renewed. There is no west door, but a small brick buttress, probably of the 16th century, is placed beneath the west window. In the second stage there is a plain lancet in the north and south walls. The west face of the top stage has been rebuilt in brick. The belfry windows consist of coupled lancets. The embattled parapet is a 19th-century restoration in brick. There is a gable cross built into the east face of the tower. The nave roof is ancient and of braced rafter construction, and has been much patched at various times; the chancel has a modern braced rafter roof and the porch a plastered ceiling.

The 14th-century font has a plain octagonal bowl with a round shaft and base. The rest of the fittings are modern and poor.

The plate includes a silver chalice and paten of 1694–5.

The tower contains three bells by Christopher Graye, 1671, probably cast at Haddenham.

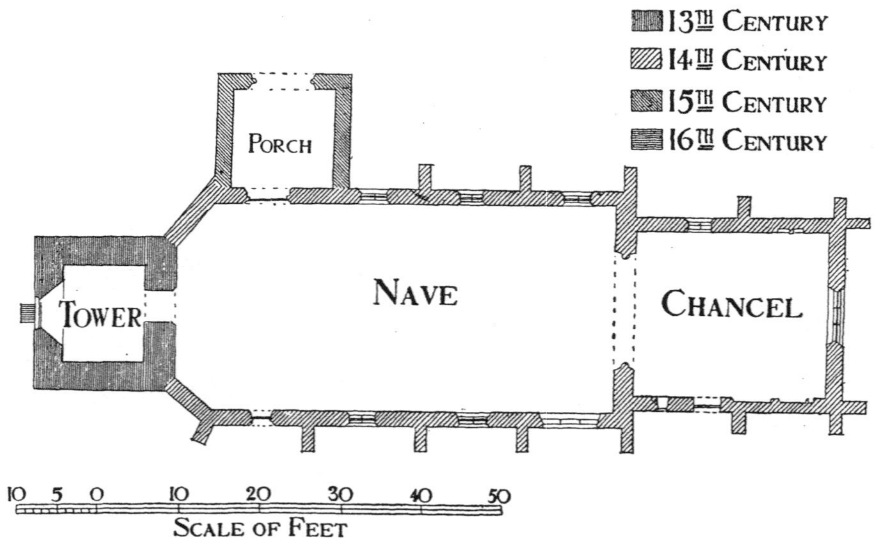

PLAN OF ST. ANDREW'S CHURCH

The registers of baptisms begin in 1788, marriages in 1754, and burials in 1788, and are complete.

NONCONFORMITY There were no Papists or Dissenters in Witchford in 1676.[47] A Primitive Methodist chapel was established in 1846 and one for Wesleyan Methodists in 1849.[48] The Wesleyan chapel ceased to be used about 1880; the Primitive Methodist chapel before the end of the century.[49] A Baptist chapel was founded in 1871; the existing building was erected four years later on land presented by James Cropley.[50]

SCHOOLS In 1789 there was a small dame school in Witchford. The older boys went to Ely for their schooling, if they received any.[51]

By the time the National Society made their inquiry (1846–7) Witchford had become one of the best-provided villages in the Isle. About 100 children, in a population of 561, were said to be attending school both weekdays and Sundays. The cost of running the school amounted to £30, of which the master received £20 a year in salary.[52] In 1851 the numbers attending were stated to be 74 on weekdays and 100 on Sundays, but the existing room could properly contain only about half that number. The parents were charged 1*d.* a week for each child, or 1*s.* a quarter; farmers were charged extra. The school seems to have been con-

[46] These windows are not shown on the accompanying plan.
[47] B.M. Add. MS. 5847, f. 76.
[48] H.O. 129/7/190.
[49] Kelly's *Directories*.
[50] *Bapt. Handbk.* 1950; Inscription on building.
[51] Bp.'s Mun., Ely, B 6.
[52] Rep. of Nat. Soc., 1846–7 (priv. printed).

ducted in hired premises, but in 1851 a site for a new building was given by Christopher Pemberton, of Cambridge, one of the local landowners; the National Society contributed £30 towards the building costs and the Dean and Chapter of Ely, lords of the manor, £15. The total cost, £164, was unusually low, since it was possible to use a converted cottage as master's house. The addition of an infant's classroom in 1869 brought the accommodation up to 129. The following year a new master's house was built at a cost of £204, of which the National Society contributed £12.[53] In 1910 the recognized accommodation was reduced to 106 (61 mixed, 45 infants), but the average attendance has never in recent times been as much even as the lower figure; it was 53 in 1899,[54] 69 in 1919, and 61 in 1938. Wentworth school was closed in 1922 and the children transferred to Witchford. Since the Second World War the school has been 'decapitated' by the removal of the older children to Ely,[55] but by 1951 the numbers had risen to 73.[56]

[53] Archives of Nat. Soc. [54] Kelly, *Dir. Cambs.* (1900). [55] M. of Ed. File no. 12/72. [56] Infm. Rev. E. H. Yates, Vicar.

WISBECH HUNDRED

ELM

Elm, on the Norfolk border between Wisbech and Outwell, is one of the larger parishes in the Isle. It was originally of the long, narrow shape usually found in Wisbech hundred, with the church and village on the firm silt land at the north-east end, just behind the 'Roman Bank'. The alterations made under the county Review Order of 1933 and the Ministry of Health Provisional Orders Confirmation (Ely, Holland, and Norfolk) Act, which came into force the following year, brought a net increase of 2,617 acres in area and a decrease in population of 67. The realinement of the county boundary transferred a small built-up area to Emneth (Norf.) which was not quite counterbalanced, as regards population, by the inclusion (from Wisbech Borough) of the hamlet of Ring's End, on the Nene opposite Guyhirn. Although it is questionable whether the name is derived from the common tree,[1] the country in the north-eastern part of the parish is undoubtedly more wooded and picturesque than is usual in the Fens. The northern end of the village practically adjoins Wisbech. It is divided by the Wisbech Canal, formerly the Well Stream, into Elm High and Elm Low Roads; the series of pools which is all that remain of the canal contribute largely to making Elm one of the prettiest villages in the Isle. The 'needle' spire of the church recalls the southern part of Cambridgeshire, while the fruit-growing hamlet of Begdale, about a mile west of the village, is reminiscent of the Vale of Evesham.

Among the few buildings of architectural interest, apart from the church, may be mentioned the Black Horse Inn, Elm House, and the Limes. The first has a tablet on the porch dated 1665; the others are 18th-century buildings of some merit with well-designed doorways. The stone-slated roof of the Limes is noteworthy, in a district where such roofs are not common. The village hall dates from 1921.[2]

The River Nene formerly crossed the south of the parish diagonally. After turning gradually through a right angle past Upwell and Outwell, it returned to Elm to form the parish and county boundary. Now, however, it takes a more direct course through Wisbech; since 1933–4 the new course has formed the boundary between Elm and Wisbech St. Mary. There are many subsidiary drains. Elm High Road forms part of the main road from Wisbech to Outwell, Downham Market, and Ely (A 1101). The Low Road (B 1101) forms the village street and continues in a zigzag course through Friday Bridge and Coldham, two hamlets of Elm, to March. Railway communication with Wisbech is provided by the Upwell tramway opened in 1884. Since 1928 it has been used for goods only. There are two passenger stations in the parish—Coldham for Pear Tree Hill, on the March–Wisbech line (1847), and Guyhirn on the March–Spalding line (1867). They are both, however, too remote to be of much local importance.

The northern part of the parish, on silt soil, is largely given over to orchards and market-gardens; in the south-west, on the peat, farms are larger and are devoted to cereals, potatoes, and sugar-beet. As stated above, there are fairly numerous isolated trees round the village, and a long shelter belt running north and south along the March road at Coldham Hall, the site of a former manor house.

At the beginning of the 19th century Elm had the reputation of a law-abiding village. The peace celebrations of 1814, when '738 persons dined abundantly, and capital sports succeeded' were specially commended by Jeremiah Jackson in his diary.[3] Twenty years later, although barn robberies were common, sheep stealing was not prevalent in Elm, and the state of the parish 'though unpleasant, is not such as to excite any apprehension'. 'Much cheerfulness' was shown at a parish meeting called to consider measures to alleviate the state of the poor. There was at this date (1834) an effective parish fire-engine.[4] Some of the ill feeling was probably due to inclosures, of which a good many were being carried out privately at this time.[5] In 1834 an Act[6] was passed to inclose the commons, droves, banks, and waste lands in the parish. As in most Marshland parishes, the amount of land to be so treated was small (195 acres), and the award made under this Act was not signed and put into force.[7]

Among the incumbents of Elm-cum-Emneth have been: (i) Sir Richard Swale (1545?–1608), Chancellor of Ely and a noted civil lawyer; he was appointed rector in 1588 by dispensation, being a layman;[8] (ii) Dr. Joseph Beaumont (1616–99), poet and Master of Peterhouse, rector in 1645;[9] (iii) Edmund Castle (1698–1750), Master of Corpus Christi, Cambridge, and later Dean of Hereford, vicar in 1729;[10] and (iv) Jeremiah Jackson (1775–1857), Headmaster of Wisbech Grammar School, 'whose name leaps to the eye from every local record of contemporary social activity',[11] vicar 1826–57. Mrs. E. B. Tanqueray, wife of a vicar of Coldham, was the author of *The Royal Quaker* and other novels that had a vogue at the beginning of the present century. Robert Tillotson, rector in 1711, was a nephew of Archbishop Tillotson.[12]

MANORS Like the other vills in the extreme north of the Isle, with the exception of Wisbech, *ELM* is not mentioned in Domesday Book. It is therefore conceivable that the district in which it lay was not reclaimed until *c.* 1200 and that the fine 13th-century church is the first on the site. At all events, the overlordship was at that time with the Bishop of Ely[13] and it has so continued, the Church Com-

[1] *P. N. Cambs.* (E.P.N.S.), 266–7.

[2] Inscription on building.

[3] 20 July 1814. Diary in Wisb. Mus. At Wisbech, although there was no 'gross intoxication', the sports were spoilt by the 'unmanageableness of the spectators'.

[4] H.O. 52/4 (Bp. Ely to Home Sec., 3 Feb. 1834, quoting a report made by the Revd. Jeremiah Jackson, Vicar of Elm).

[5] Ibid.

[6] 4 & 5 Wm. IV, c. xiv.

[7] W. E. Tate, 'Cambs. Field Systems', *Camb. Antiq. Soc. Procs.* xl (1939–42), 83.

[8] *D.N.B.*

[9] T. A. Walker, *Admissions to Peterhouse 1615–1911*, 43.

[10] *D.N.B.*

[11] *V.C.H. Cambs.* ii, 329. His manuscript diary in Wisb. Mus. is a valuable source of information on contemporary local history.

[12] *D.N.B.*; E 134/10 Anne/M. 1.

[13] *Bk. of Fees*, 628–9.

missioners being the present lords of the manor.[14] As at Wisbech and other episcopal manors in Wisbech hundred, John Thurloe was in possession during the Interregnum.[15]

In 1210–12 the bishop had six tenants at Elm who held by military service. Richard de Argentem or Argentan held 1 fee; Roger Mareis and Robert de Banstede (*recte* Haustede)[16] ½ fee each; Richard de Melkesham a ¼ fee in Elm and Wisbech; and Simon de Insula and Gerard de Vernun ⅙ fee each.[17] Most of these fees subsequently developed into subordinate manors (see below). They all reappear in the Ely cartularies of 1221 and 1251, but with changed assessments in some cases. In 1221 the Argentan fee was rated on 1,452 acres of land in Needham and 60 acres in 'Langebech'; in 1251 it was held by Giles de Wechesham. The Mareis tenement remained constant at ½ fee, as had that of Robert de Haustede. In 1221 Richard de Melkesham was assessed at a whole fee, on 380 acres of land, and he was stated to have made a purpresture of 60 acres in the marsh; in 1251 the rating of the Melkesham tenement is unstated. In 1221 Simon de Insula also was assessed at a whole fee, on 420 acres; in 1251 his tenement, now held by Philip de Insula, had been reduced to ⅙ fee again. The assessment of the de Vernun tenement had risen in 1221 and fallen in 1251 in a similar manner. These increases and reductions were no doubt due to early reclamation followed by the great sea flood of 1236. As at Wisbech, the 1251 cartulary showns the obligation of the customary tenants to repair Wisbech Castle and dig at 'Ramersdike' as well as the normal agricultural services. In spite of the flood, the numbers of free and customary tenants rose between 1221 and 1251 from 8 and 84 to 15 and 115 respectively. At the later date the total rents amounted to £45 5*s*. 6*d*.[18]

As usual in the 14th century, a decline in prosperity is evident. In 1391 the rents amounted to only £37 12*s*. 6*d*. By this time a good deal of reclamation had been undertaken. Fourteen fields of varying size are mentioned. Each embraced from 5 to 50 small tenements worth from 1*s*. to 2*s*.[19] Some of these fields survive in name to the present day, for example, Old, New, Town, Halfpenny, Redmoor, Needham, Needham Hall, and Coldham Fields.

In 1303 Richard Loveday held the Argentan–Wechesham tenement, again as 1 fee, of Giles Mumpenzon, who held of the Bishop of Ely.[20] Loveday became bound to Bishop Walter Langton of Lichfield in various recognizances, and in 1321 the inquisition *post mortem* on the bishop shows him as holding, *inter alia*, a messuage called Bealford or *BEAUFORD*, together with 776 acres and 7 marks rent. Of this property, 501 acres were stated to be held of Giles de Wechesham as 1 fee, and 80 of the Bishop of Ely for 26*s*. 8*d*. annual rent.[21] The fee was in 1346 held by Elizabeth Peverel, the eventual heiress of Bishop Langton;[22] in 1428 William Venour was the holder.[23] By 1536 the property had come to be styled a manor, and was held by Alexander Balam by service of ¼ fee and 25*s*. yearly. Balam was also tenant of 100 acres in the marsh at 10*s*.[24] The manor was situated in the Friday Bridge area,[25] and under the name Beauford or Bewford remained in the Balam family until 1699.[26] In that year Susan Balam, John Kelsall and Anne his wife, and William Jay, the last three probably brothers-in-law and sister of the first, settled it upon trustees.[27] Thomas Jenkinson, one of the trustees, was dealing with it in 1720.[28] Beauford House still exists as an 18th-century building with a doorway of good design.

In 1303 the Mareis tenement was again assessed at ½ fee and was held by a woman, Clemencia de Marisco.[29] She was probably the last of her line, for in 1321 Bishop Langton held the messuage of 'Mareystede' and 140 acres in chief of the Bishop of Ely, at ½ fee, and 863 acres of land and marsh at 7 marks rent.[30] In 1346 Elizabeth Peverel, Langton's heiress, held the ½ fee jointly with Sir John Colvile.[31] This estate is not subsequently recorded individually, and probably became merged with Langton's other property in Coldham (see below).

The Haustede estate is not entered in the lists of fees of 1303, 1346, or 1428. It was situated in *BROKENE*, on the north-west side of Elm towards Wisbech.[32] In 1470 it reappears, as a manor of Brokene, in the hands of the Colvile family of Newton (q.v.). It was again assessed at ½ fee and was valued at £5 6*s*. 8*d*.[33] It had been settled by Sir John Colvile, son of the founder of the Newton chantry, on himself and his wife Anne for their lives. Anne subsequently married Sir Robert Brandon, and outlived her son Francis Colvile. At her death in 1494 Brandon forcibly excluded Richard Colvile, Francis's son, then aged 14, from his lands.[34] Richard proved his age in due course and the manor, under the name *HALSTED*, remained in the family another hundred years, until it was sold (1591) by Francis Colvile to William Hynd.[35] Hynd died without issue in 1606; after which the manor seems to have come to William Steward of Ely, whose son-in-law, Francis Lord Aungier, died in possession in 1632.[36] He left his Cambridgeshire estates to his fourth son George, who was living in 1635.[37] In the early 18th century Halsted manor was held by three sisters, wives of Rupert Elcie, Richard Dent, and Thomas Briscoe; it is not recorded after 1720.[38]

In 1299 Richard de Melkesham conveyed his lands in Elm, Wisbech, and Upwell to Bishop Walter Langton by fine.[39] The following year the bishop was granted free warren in his demesnes in *COLDHAM* and elsewhere,[40] and in 1303 he held Coldham for ⅓ fee.[41] At his death (1321) he held, besides the Argentan–Wechesham–Loveday fee and the Mareis

[14] Kelly, *Dir. Cambs.* (1933).
[15] C.P. 25 (2)/616/Mich. 1650.
[16] So in *Bk. of Fees*; probably intended for 'Haustede' (see below).
[17] *Bk. of Fees*, 628–9; *Red Bk. of Exch.* (Rolls ser.), 526.
[18] B.M. Cott. MSS. Tib. B. II, ff. 153*b*, 156*b*; Claud. C. XI, ff. 89, 93*b*, 94*a*.
[19] Bp's Mun., Ely, A 6. This extent survives in a copy of 1600.
[20] *Feud. Aids*, i, 141, 151.
[21] *Cal. Inq. p.m.* vi, p. 196.
[22] *Feud. Aids*, i, 173. [23] Ibid. i, 191.
[24] B.M. Add. MS. 5837, f. 203.
[25] W. Watson, *Hist. Wisb.* (1827), 504. Beauford House is situated a few yards south of the cross-roads in the centre of Friday Bridge village.
[26] C.P. 25 (2)/277/Trin. 2 Jas I and /Mich. 8 Jas I.
[27] C.P. 25 (2)/815/Hil. 11 Wm. III.
[28] C.P. 25 (2)/1000/Trin. 6 Geo. I. Jenkinson was the founder of a clothing charity in Elm. [29] *Feud. Aids*, i, 151.
[30] *Cal. Inq. p.m.* vi, p. 196.
[31] *Feud. Aids*, i, 172. Sir Roger Colvile, Sir John's grandfather, married Desirée, daughter of Geoffrey de Mareis (Watson, *Hist. Wisb.* 482). The ½ fee was assessed on lands in Elm and Tydd St. Giles; it is probable that the Colvile interest was in the latter vill.
[32] B.M. Cott. MS. Tib. B. II, f. 156*b*.
[33] A. Gibbons, *Ely Episc. Rec.* 433 (Bp.'s Mun., D 6).
[34] *Cal. Inq. p.m. Hen. VII*, i, 1098.
[35] C.P. 25 (2)/93/855/Trin. 33 Eliz.
[36] C 142/487/39.
[37] *Compl. Peerage*, viii, 121.
[38] C.P. 25 (2)/910/Hil. 4 Anne, /911 /Hil. 10 Anne, /1000/Trin. 5 Geo. I, and /1000/Mich. 6 Geo. I.
[39] C.P. 25 (1)/26/43 and /47.
[40] *Cal. Chart. R.* 1257–1300, 481.
[41] *Feud. Aids*, i, 151.

fee (see above), the manor of Coldham with 420 acres of the Bishop of Ely for 1 fee. His heir was his nephew, Edmund Peverel, then aged 14.[42] He died in 1331, when he was recorded as holding, jointly with his wife Elizabeth, the three estates of Coldham, Bealford, and Mareystede, and the fishery of 'Estiworth', by knight's service and a yearly rent of £6 13*s.* 4*d.*[43] Elizabeth Peverel had licence of oratory in her manor house for five years from 1344.[44] Two years later the assessment of Coldham was again ⅓ fee.[45] The subsequent descent of this manor is obscure. In 1368 the custody of lands in Coldham, late of Hugh Gros, was granted to Alan Palmer; Gros held these lands by knight's service of the heir of Edward de Montacute, a ward of the king.[46] In 1402 William Moore, vintner, was tenant, and desired his feoffees to sell a moiety of his interest in the manor and devote the proceeds to pious uses.[47] There is no further record until 1536, when Thomas Clarke of Elm held the manor as 1 fee.[48] In 1578 Humphrey Mychell made a settlement of it.[49] The manor changed hands several times at the end of the 16th century;[50] in 1586 it was in possession of George Carleton, Constable of Wisbech Castle,[51] and in 1612 it came to Sir John Peyton of Doddington (q.v.).[52] Towards the end of his life Peyton was engaged in a lawsuit with Bishop Felton over the boundaries between his manor of Coldham and the bishop's manor of Waldersea alias Bewdesert in Wisbech,[53] and at Peyton's death in 1630 he held it of the Bishop of Ely, together with other land in Elm and Needham Fen.[54] Peyton's son, also Sir John, died in 1635,[55] and the manor continued in the family until the middle of the 19th century.[56] The manor house was replaced by a farm-house about 1793.[57]

In 1303 John Ode of Lynn held the de Insula tenement, again as ½ fee, of the heirs of Simon de Insula for life.[58] In 1346 this holding was divided, at the same assessment, between Sir John Dene, Sir John de Colvile, Sir William de Ruston, and Robert de Waldusee.[59] It has not been subsequently traced.

The Vernun property remained in possession of that family and was returned at its 1251 assessment of ⅛ fee in 1303[60] and 1346.[61] It may be identified with the manor of *VERNONS* which appears in 1593 in possession of William Steward, together with lands in Old Field and elsewhere in Elm, and in other places in the Isle.[62] His son Thomas, then a minor, had livery of his estates three years later.[63] Thomas died unmarried before 1605, when Edward Fincham, his cousin,[64] was dealing with the manor.[65] At his death in 1631 Fincham held it of the Bishop of Ely as of his manor of Wisbech Barton. He also held land called Vernons Fen, a messuage, and Needhams Field. The association of these tenements suggests that the property lay near Needham Hall in the east of the parish, south of Old Field. The property was settled on Edward Fincham's relict Frances (Richman) and son Thomas.[66] The latter died in 1666 leaving two daughters, Mary and Frances.[67] The Fincham estates in Norfolk and the Isle seem then to have been broken up, and this manor, with other property in Elm, Outwell, and Upwell, passed from Rachel Fincham and other members of the family to Baptist May.[68] In 1700 it came to Israel Wilkes of Clerkenwell, father of the politician,[69] and remained in his family until at least 1774.[70] The Wilkes family had long been settled in the neighbourhood. In 1592 William Wilkes held land bordering on the Crabhouse nunnery estate (see below).[71] In 1633 he or a namesake was seised of a messuage in Begdale and land called 'Gooles' in Haustedfield.[72] Needham Hall, 'formerly a most respectable residence', was at the end of the 18th century in possession of Dr. John Fountayne, Dean of York, who was succeeded in 1802 by his daughter, Catherine Judith. Two years later the Hall, being dilapidated, was pulled down and replaced by a farm. On Catherine's death in 1824 the property came to her nephew, Richard Fountayne Wilson, M.P. for Yorkshire 1826–30.[73] He died in 1847 and Needham Hall Farm passed, presumably by purchase, to F. Fryer.[74] In 1900 and 1933 the owner was (Sir) W. W. West, J.P.,[75] sometime Chairman of the Isle County Council.

The manor of *VAUX* seems to have originated in the lands of the Lords Vaux of Harrowden (Northants.), who were recorded as holding property in Elm worth £2 15*s.* 9*d.* in 1600.[76] These lands may have come to them through the marriage of Thomas, 2nd Baron, to Elizabeth, daughter and heir of Sir Thomas Cheyne of Fen Ditton, *c.* 1530–5.[77] In the latter half of the 17th century Vaux manor was in the Stukeley, Squire, and Swaine families.[78] In 1801 the manor house was purchased by the parish officials of Elm from John Swaine of Stretham for use as a workhouse. It had long been let as a farm-house.[79]

Lands in Elm were given to Nuneaton Priory (Warws.) by Robert, Earl of Leicester, about 1154.[80] In 1210–12 4½ acres of this estate was held by John son of Rainer for 2*s.* 3*d.* and 2 acres and a square perch (*perticatam*) of Hildebrand the clerk for 1*s.*[81] The lands do not figure in the *Taxatio*, but they remained in possession of the convent until the Dissolution, when they were valued, with land in Upwell and Emneth, at 79*s.* 7*d.*[82] In 1541 they were granted, as 'manors', to Thomas Manners, Earl of Rutland.[83]

William son of William de Wendling (Norf.) gave to the abbey of that name, which he founded in 1265, a messuage and lands in Elm.[84] These were worth £2 in 1291[85] and £2 5*s.* 7*d.* in 1391,[86] but only 18*s.* in 1535.[87] Presumably they formed the 'manor' of

42 *Cal. Inq. p.m.* vi, p. 196.
43 Ibid. vii, p. 257.
44 *Ely Dioc. Remembrancer*, no. 58, p. 392.
45 *Feud. Aids*, i. 173.
46 *Cal. Pat.* 1367–70, 79.
47 *Fenland N. & Q.* vii, 201.
48 B.M. Add. MS. 5837, f. 203.
49 C.P. 25 (2)/260/East. 21 Eliz.
50 C.P. 25 (2)/93/848/Hil. 26 Eliz., /262/Mich. 37–38 Eliz., /262/Hil. 41 Eliz., and /262/Hil. 42 Eliz.
51 E 134/29 Eliz. H. 8.
52 C.P. 25 (2)/385/Hil. 9 Jas. I.
53 C 2/Jas. I/E 2/53.
54 C 142/741/133.
55 C 142/573/62.
56 D. & S. Lysons, *Magna Brit.* ii (1), 182; R. Gardner, *Dir. Cambs.* (1851), 548.
57 Watson, *Hist. Wisb.* 503.
58 *Feud. Aids*, i, 151.
59 Ibid. 173.
60 Ibid. 141.
61 Ibid. 173.
62 C 142/278/103.
63 C 60/422, m. 35.
64 *Visit. Camb.* (Harl. Soc. xli), 10, 51.
65 C.P. 25 (2)/277 East. 3 Jas. I.
66 C 142/720/120.
67 F. Blomefield, *Hist. Norf.* vii, 462.
68 C.P. 25 (2)/633/Hil. 19–20 Chas. II.
69 C.P. 25 (2)/815/Trin. 12 Wm. III.
70 C.P. 25 (2)/1282/Mich. 14 Geo. III; C.P. 43/763, ro. 462.
71 C 66/1382, m. 23.
72 C 142/770/83.
73 Watson, *Hist. Wisb.* 504–5.
74 R. Gardner, *Dir. Cambs.* (1851), 550.
75 Kelly, *Dir. Cambs.* (1900, 1933), s.v. Friday Bridge.
76 D. & S. Lysons, *Magna Brit.* ii (1), 182; Gardner, *Dir. Cambs.* (1851), 548.
77 Watson, *Hist. Wisb.* 503.
78 C.P. 25(2)/910/Hil. 4 Anne, /911 /Hil. 10 Anne, /1000/Trin. 5, Mich. 6 Geo. I.
79 Watson, *Hist. Wisb.*, 504.
80 *V.C.H. Warws.* ii, 66.
81 *Bk. of Fees*, 629.
82 *Valor Eccl.* (Rec. Com.), iii, 76.
83 *L. & P. Hen. VIII*, xvi, g. 947 (11).
84 Dugd., *Mon.* vi, 890.
85 *Tax. Eccl.* (Rec. Com.), 270.
86 Bp.'s Mun., Ely, D 5 (2).
87 *Valor Eccl.* (Rec. Com.), iii, 328.

WENDLING which, with lands and rents in Emneth and Wisbech, was passed by Edward Story and Katherine his wife to James Dudley in 1575.[88] They, together with certain otherwise unrecorded property of Wormegay Priory (Norf.), had in 1571 passed through the hands of the land speculators Thomas Jennings and Edward Forth.[89]

The Abbot of Crowland (Lincs.) had two free tenants in Elm in 1210–12, John and Ralph de Croyland, each of whom held 30 acres and a fishery for 30*s.*[90] In 1291 this property was worth £2. In 1535 it was worth £2 18*s.* 8*d.*, less 2*s.* to the Bishop of Ely.[91] In 1572 the fisheries, then held by William Payne and William Froman, and 60 acres held of the Bishop of Ely by Lord St. John (of Bletsoe), probably also ex-Crowland property, were granted to Richard Hill and William James.[92]

The nunnery of Crabhouse (Norf.) possessed pasture lands in Elm and Emneth, valued in 1535 at £1 6*s.* 8*d.*[93] They were leased in 1539 to Henry Webbe[94] and in 1552 for twenty-one years to Thomas Sydney. In 1555 the reversion was granted to Sir John Gage, Chamberlain of the Queen's Household.[95] Twenty-eight acres of former Crabhouse land in Elm, which lay in two portions, was in 1592 granted to William Tipper and Robert Dawe at an annual rent of 4*s.*[96]

The possessions of the guild or chapel of the Holy Trinity next Stathedyke in Walsoken extended into Emneth and Elm. They were first granted (1545) to Ralph Stannowe, the late warden, then in 1552 to Mary, Duchess of Richmond and Somerset.[97]

In 1536 Sir Robert Brandon, lord of the manor of Halsted (see above), was recorded as having held, in the right of his wife Anne (Colvile), 36 acres in Elm and other land in Brokene which was formerly in the tenure of Robert Halsbeech, by service of ½ fee, 50*s.* rent, and suit at Wisbech hundred court.[98]

CHURCHES The church of Elm with its chapel of Emneth was valued at £40 in *c.* 1217 and 1254, and at £17 in 1291;[99] the last figure does not include £5 from the tithes payable to Lewes priory from their estate of Lewes Fee in Emneth. It is recorded as an *ecclesia*, but a vicarage had been ordained by 1275.[1] The church, however, was not appropriated, and appointments were made to the sinecure rectory until the early 19th century.[2] The vicarage with its appurtenances except the house and garden was in 1494 leased for seven years at £8 6*s.* 8*d.* yearly by Edmund Roray, vicar, to Thomas Aleyn, rector.[3]

In 1535 the rectory was worth £17 10*s.* and the vicarage £14 15*s.* 10*d.*;[4] three centuries later their comparative value is shown by the commutation of the rectorial tithes for £545 annually, and the vicarial for £215.[5] The patronage of both has been continuously with the see of Ely.

Emneth was made a separate vicarage in 1858, after the death of the Revd. Jeremiah Jackson. The parishes of St. Matthew, Friday Bridge, and of St. Etheldreda, Coldham, were formed out of Elm in 1860 and 1874 respectively. The three vicarages are also in the gift of the bishop; Friday Bridge and Coldham have been held in plurality since 1931.[6]

The Lewes tithes had risen in value to £6 10*s.* in 1339,[7] and to £8 13*s.* 4*d.* in 1535.[8] At the Dissolution (1537) the last prior conveyed them with the other possessions of the convent to the king by fine. In 1538 they were regranted to Thomas Cromwell.[9] They were resumed by the Crown after Cromwell's attainder. The next grantee was the Earl of Essex. They finally formed part of the miscellaneous spiritualities given to Bishop Heton in 1600 in return for the alienated episcopal manors.[10]

In 1370 William de Petworth, rector, made an agreement with Prior Cherlew of Lewes to pay the convent an annual pension of £20, on pain of £5. This was confirmed by the Bishop and Convent of Ely. Early in the reign of Henry VI Prior Nelond sued Thomas Reynold, Petworth's successor, for £65 representing this pension and arrears. Reynold denied that he had ever received tithes from Lewes priory lands in Emneth, and pleaded that the confirmation of Petworth's agreement at Ely was invalid. In 1430 judgement was given for the priory in Reynold's absence, but Reynold put in a writ of error. The case was still in progress in 1432, and its conclusion, if any, has not been traced.[11] Probably the priors managed to enforce payment of the pension, for its existence was still remembered in 1589. As it was then stated that payment had neither been demanded nor made for more than thirty years, the pension presumably lapsed at the Dissolution.[12] In 1608, however, it was revived and granted to Thomas Howard, Earl of Suffolk, and his heirs.[13]

Petworth was licensed in 1377 to receive 8 acres of land in Elm and 1 acre and a house in Emneth, to provide *inter alia* 7 lighted tapers on the altar at high mass on feast days.[14] In 1553 6 acres in Elm, in tenure of the churchwardens, were stated to have been given for anniversaries in the church.[15]

Petworth makes one further appearance in 1396, when he came to blows in the church with John Wetyng, his vicar. As no serious hurt was caused Bishop Fordham held that the church had not been polluted.[16]

The church of *ALL SAINTS, ELM*, consists of chancel, clerestoried nave, aisles, north porch, and west tower. The fabric belongs mainly to the first and last quarters of the 13th century, with windows and roofs inserted in the 14th and 15th centuries. The plan has

88 C.P. 25 (2)/93/840/Hil. 17 Eliz.
89 C 66/1077, m. 39.
90 *Bk. of Fees*, 629.
91 *Tax. Eccl.* (Rec. Com.), 270; *Valor Eccl.* (Rec. Com.), iv, 86, 87.
92 C 66/1083, m. 16.
93 *Valor Eccl.* (Rec. Com.), ii, 396. This estate must have been given to Crabhouse after 1291; it is not mentioned in the *Taxatio*.
94 *L. & P. Hen. VIII*, xiv (1), p. 611.
95 *Cal. Pat.* 1554–5, 183.
96 C 66/1382, m. 23.
97 *L. & P. Hen. VIII*, xx (1), g. 1211. *Cal. Pat.* 1550–3, 455.
98 B.M. Add. MS. 5837, f. 203.
99 W. E. Lunt, *Valuation of Norw.* 539.
1 In this year Hugh, Vicar of Elm, witnessed a document (*Bk. of Seals* (Northants. Rec. Soc.), no. 456). The same or another Hugh was a commissioner *de walliis et fossatis* in 1303 (*Cal. Pat.* 1301–7, 185). Blomefield (*Hist. Norf.* viii, 409) mentions a vicar named Peter in 1254, but does not give reference.
2 D. & S. Lysons, *Magna Brit.* ii (1), 182.
3 *Ely Dioc. Remembrancer*, 1909, p. 46.
4 *Valor Eccl.* (Rec. Com.), iii, 500.
5 Gardner, *Dir. Cambs.* (1851), 550.
6 *Clergy List*: *Crockford's Clerical Dir.*
7 J. H. Bullock & W. M. Palmer, *Cambs. portion of Chart. Prio. Lewes*, xi, quoting Reg. Montacute.
8 *Valor Eccl.* (Rec. Com.), i, 328.
9 *L. & P. Hen. VIII*, xiii (1), g. 384 (74).
10 Watson, *Hist. Wisb.* 496.
11 *Yr. Bk. 1 Hen. VI* (Selden Soc. no. 50), 119–30.
12 Hist. MSS. Com., *Var. Coll.* ii, 243.
13 C 66/1757, no. 7.
14 *Cal. Pat.* 1377–81, 55.
15 Ibid. 1553, 289.
16 *Ely Dioc. Remembrancer*, no. 180, p. 97.

remained unaltered and the structure is a magnificent example of early Gothic design. There was the usual restoration in the last century but it was less destructive than was often the case. Much of the tower was refaced at this time, the porch was rebuilt, and there was some renewal of tracery. The roofs are leaded except that of the chancel which is slated.

The chancel has a modern east window of five lights and there are angle buttresses of the 13th century with two set-offs. At the north-east corner there is a doorway with round arch and continuous mouldings. The unusual position of this doorway would suggest that the chancel was originally longer, but if this was the case it must have been curtailed soon after its erection. The lateral fenestration consists of long two-light windows with 14th-century tracery consisting of trefoiled heads with a quatrefoil above inserted in 13th-century openings, and external and internal hood-moulds. There is a 13th-century string-course round the interior, and an aumbry in the north and south walls, but no piscina. The roof is modern and of braced rafter construction. The chancel arch is two-centred and of two orders with moulded caps and bases, the latter partly covered by the raising of the floor.

The nave arcades are of six bays with piers alternately round and octagonal, and the two-centred arches are of two orders with moulded caps and bases and a hood-mould. Above the arcades is a string-course which probably indicates the original height of the walls. The clerestory is an addition of the last quarter of the 13th century and consists of ten lancets with angle-shafts having moulded caps and bases. There is an external corbel table of masks. Above the chancel arch is blind arcading contemporary with the clerestory and on the west wall the pitch of the original roof is clearly indicated with a blocked lancet near the apex. The nave roof is a superb example of 15th-century carpentry; it is of double hammer-beam construction with demi-angels having outspread wings at the termination of the hammer-beams, some of which are missing; all the principals are richly moulded and there are long moulded wall-posts and carved spandrels. The ends of the principals and rafters are plastered. There is a very wide tower arch of two orders, two-centred with rounded responds and recessed shafts, all having moulded caps; the bases are concealed. The east and west windows of the aisles are of three lights with cinquefoiled heads and tracery, and of 15th-century date; all with the exception of the west window of the south aisle are inserted in 13th-century openings with angle-shafts having moulded caps and bases. The north-east window in the north aisle is a 14th-century three-light, while the remainder on this side are of two lights with cinquefoiled heads and quatrefoils above, of 14th-century date. Much of the north wall was rebuilt in the 14th century, and the buttresses with two set-offs are of this period. The west portion of the wall, however, is partly of the 13th century and the internal string-course remains here. The south-east window in the south aisle is a 15th-century three-light with cinquefoiled heads and tracery under a depressed arch, and is inserted in a 13th-century opening with angle-shafts as in the east window. The remainder of the windows in this wall are 14th-century two-lights similar to those in the north aisle, but the first two are inserted in 13th-century openings with angle-shafts and have external and internal hood-moulds. The walls and buttresses of the south aisle are of 13th-century date. To the east of the north and south doorways are rounded responds with moulded caps, the bases of which are beneath the floor-level. The origin and purpose of these responds is obscure. Several of the windows have old clear glass in rectangular leading. The north doorway is excellent, with seven recessed

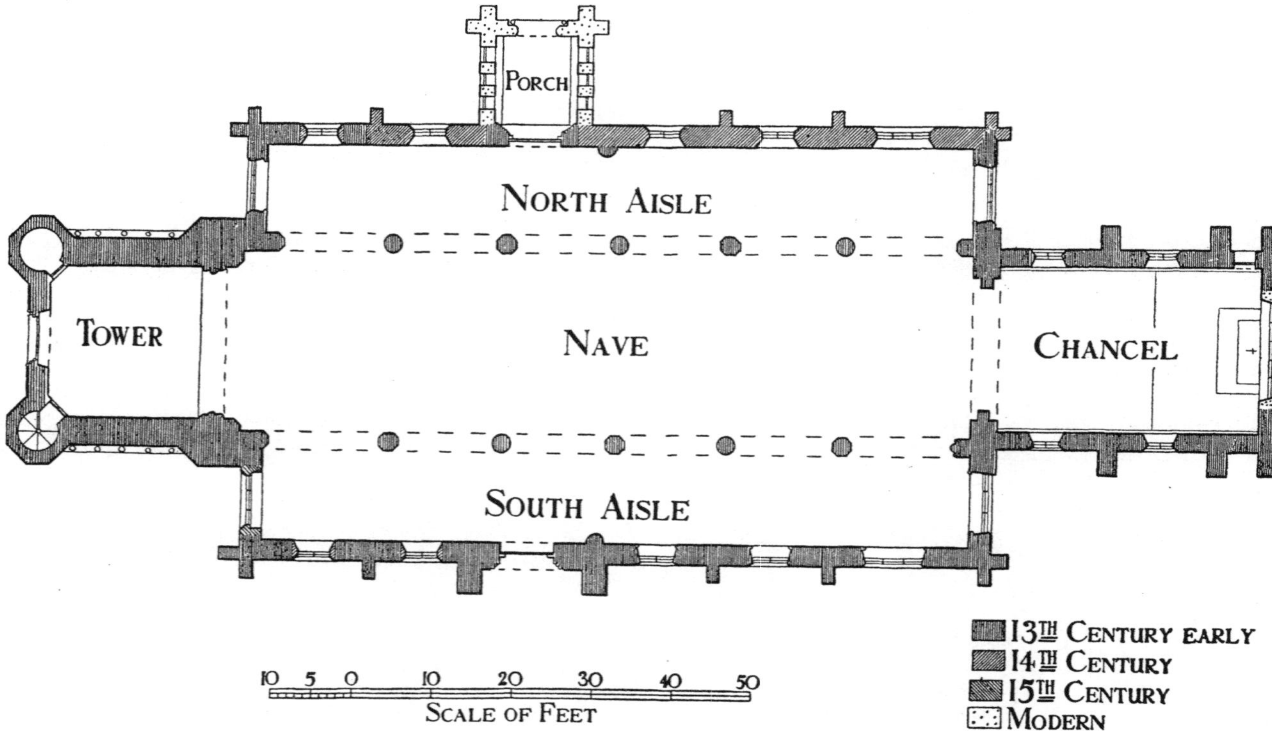

Plan of All Saints Church

shafts and good 13th-century mouldings to its two-centred arch. The south doorway is more simple but quite effective, with three recessed shafts and deep mouldings. Both the aisles have external corbel tables of masks. The aisles have lean-to roofs with wall-posts and struts, that on the north having an embattled wall-plate; they are probably of the 15th century.

The tower is of four stages, the three lower having external shafted arcading; the two lowest are open to the church and there are two lancets with banded shafts on the north and south and three on the west side; at the angles are octagonal turrets which are carried up above the parapet. The main newel staircase is in the south-west turret and there is a passage along the sill of the west windows to another stair in the north-west turret; beneath the latter on the ground floor is a small chamber with a domical ribbed vault having carved corbels; there is a quatrefoil opening on the west and a plain doorway communicating with the church on the east. The west doorway is semicircular with three recessed shafts and good 13th-century mouldings. In the third stage there is a lancet opening on the north, south, and west. The belfry windows are semicircular and of two lights. There is a corbel table of masks as in the nave and aisles. The embattled parapet has been rebuilt, and there is a small leaded spire.

All the fittings are modern and poor.

The plate includes a communion cup, 1753, a chalice, 1897, a paten on a foot, 1753, another paten, 1897, and a flagon, 1639, all of silver.

The tower contains five bells, the 1st and 2nd by Thomas Norris of Stamford, 1637, the 3rd 15th century but uninscribed, the 4th by G. Mears of London, 1865, the 5th by Thomas Gardiner of Sudbury, 1738.[17]

The registers begin in 1539 and are complete.

The church of *ST. MARK, FRIDAY BRIDGE*, was erected in 1860 and is a brick structure with stone dressings, in the Early Decorated style. It consists of chancel, nave, chapel, north-west tower, and spire. These last show serious settlements owing to inadequate foundations. There is one bell.

The church of *ST. ETHELDREDA*,[18] *COLDHAM*, is a stone building in the Early English style, dating from 1876. It consists of chancel, nave, south porch, vestry, and west turret containing two bells.

There were at various dates in the later Middle Ages seven guilds in Elm—those of St. Katherine, St. Mary, St. Giles, St. John Baptist, All Saints, Corpus Christi, and the Assumption. The first named was founded before 1389, and this and the next three were maintaining lights in the church in 1456. At the earlier date St. Katherine's Guild possessed twelve cows, given by members, which were let out and produced 2*s.* yearly for the guild.[19]

NONCONFORMITY. In 1835 a Primitive Methodist chapel was built in Elm village, and three years later a Wesleyan Methodist congregation was established at Friday Bridge;[20] the chapel of the latter dates from 1843.[21] Another Primitive Methodist chapel was built at Coldham in 1915.[22] All these chapels are still in existence.[23] The Rings End Methodist Chapel of the March Circuit was brought into Elm parish by the boundary changes of 1933. The congregation, originally Wesleyan, was formed in 1849;[24] the present chapel dates from 1866.[25]

SCHOOLS. In 1582 Thomas Brise was licensed as schoolmaster at Elm.[26]

Thomas Squire by his will (1689) gave a messuage and 22 acres in Redmore Field, and a cottage and ½ acre for the support of a master for the school he had already established in a 'low room or place in his house newly built'.[27] The school was reconstituted in 1692, when 111 acres of the Town Lands were added to the endowment. In 1730 this amounted to £20 a year. The school was then in association with the S.P.C.K.[28] About the beginning of the 19th century the school was rebuilt[29] to accommodate 120 children, and in the 1830's there were about 50 boys and 25 girls learning reading, writing, and arithmetic; the school population of Elm and Emneth was estimated to be 225. Attendance was irregular owing to the claims of farm work, and the farmers were said to be abusing the foundation by getting their children taught free, to the exclusion of the industrious poor. By this time the endowments amounted to over £53; all the proceeds were paid to the schoolmaster, who 'has filled his present situation for about six years, and is considered an efficient person'.[30] When the National Society made their inquiry (1846–7) the school contained about 30 boys who attended weekdays and Sundays, 20 who attended on weekdays only, and 30 girls attending on Sundays only.[31] In 1859 the school was again rebuilt, to provide 130 places for boys, and seven years later there were 2 masters and 84 boys on the books. Seven of the boys were learning 'mensuration and book keeping'.[32] The trustees and inhabitants of Elm and Emneth subscribed £200 to bring the school up to the requirements of the 1870 Education Act, and a new scheme for the regulation of the school was promulgated by the Charity Commissioners (23 June 1874). Under the 1910 reorganization the recognized accommodation was reduced to 105. The school still provided for boys from both Elm and Emneth, and by 1922 it was seriously overcrowded, with 134 on the books. The situation was eased in *c.* 1925 when a senior school was erected by the Norfolk County Council at Emneth. Since 1931 Squire's School has been reserved for infants, the junior boys attending the Elm National School, and the seniors the new school at Friday Bridge. In 1944 there were 47 on the roll.[33]

In 1860–1 a separate girls' school was built in association with the National Society. It cost £622 of which £26 was contributed by the Society and rather less than half raised locally. The school and house were designed by S. S. Teulon.[34] The original accommoda-

[17] The 1st is inscribed 'Alexander Balam Esq. gave part of this bell. 1673' (*sic*). The 4th replaces another Norris bell of 1637 (J. J. Raven, *Church Bells of Cambs.* (1869), App.).

[18] It is remarkable that no medieval church in the Isle was dedicated to the Isle's patron saint. The destroyed church at Histon in the county proper had this dedication, as have the modern Isle churches at Coldham and Queen Adelaide.

[19] W. M. Palmer, *Village Gilds of Cambs.* 339, 369.

[20] H.O. 129/7/193.

[21] Inscription on building.

[22] Ibid.

[23] *Meth. Ch. Bldgs. Stat. Return* (1947).

[24] Inscription on building.

[25] Ibid.

[26] Bp.'s Mun., Ely, G 2.

[27] *Rep. Com. Char.* (*Cambs.*), [103], 262, H.C. (1837–8), xxiv.

[28] *V.C.H. Cambs.* ii, 343. The association with the S.P.C.K. dated from 1710.

[29] 'Within these few years' (W. Watson, *Hist. Wisb.*, 509; published in 1827).

[30] Bp.'s Mun., Ely, B 6. *Rep. Com. Char.* (*Cambs.*), as above.

[31] Archives of Nat. Soc.

[32] *Sch. Inq. Rep.* [3966–XII], 514, H.C. (1867–8), xxviii (10).

[33] M. of Ed. File No. 12/13.

[34] Archives of Nat. Soc.

tion was about 75, increased by a new classroom in 1884 to 139 (97 girls, 42 infants),[35] and scaled down in 1910 to 103. In 1931 this school was reorganized for junior boys and girls, and in 1949 voluntary (controlled) status was adopted.[35]

During the 1860's the vicar and churchwardens of the new parish of Friday Bridge had devoted much time to the provision of mission churches and Sunday schools. In 1870 the question of establishing a Church day school was formally raised. The National Society granted £50 towards the £886 required; nearly £600 was subscribed locally and the site was given by Lord Overstone, a 'principal landowner' whose name is commemorated in the public house of Coldham hamlet.[37] The school was opened in 1871,[38] with accommodation for 127, reduced in 1910 to 106 (71 mixed, 35 infants), and increased by an extra classroom in 1913 to 144. The cramped site of this school has prevented any further enlargement, and since the building of the Friday Bridge Secondary Modern School the National School has been reserved for juniors and infants, of whom there are now about 120 on the books. In 1950 the school accepted voluntary (controlled) status.[39]

The Church school at Coldham was built in 1897–8 for 108 children. In 1910 the recognized accommodation was reduced to 91 (60 mixed, 31 infants), and again in 1934 to 78. At this date the older children were moved to Friday Bridge. There are now (1951) about 42 children on the roll.[40]

The Friday Bridge County Secondary Modern Mixed School was planned to relieve the great overcrowding then prevalent in the schools of Elm parish.[41] The school, in Maltmas Lane, near the primary schools, was opened in 1928, for 108 senior children.[42]

Two schools were transferred from Wisbech to Elm in 1933.

The South Brink School was built by the Wisbech School Board in 1879 for 49 boys and girls and 16 infants. It was closed in 1929, and the buildings, which stand near the main road from Wisbech to Guyhirn, were for a time used as a wayside café. They were sold in 1933 for £335.[43]

Rings End School was built in 1860 as a combined school and mission room. The classrooms were all on the upper floor. The school was not accepted as satisfactory until 1882, when it was leased to the Wisbech School Board. The building was enlarged in 1894–5 and again in 1905, and provided 78 places before and 65 after 1910. In 1927 the County Council bought the school from the Wisbech church authorities for £425. In 1935 the senior children were transferred to the Hereward School at March, the nearest town. There were only 29 children in attendance when the school was closed in 1949; 10 children from the County Council small holding estate at Goosetree went to Coates School (Whittlesey) and the remainder to Guyhirn School across the river.[44]

In 1846–7 there were four dame schools in Elm, teaching about 70 children.[45] In 1870 there were three such schools (one kept by a Dissenter) and three Church Sunday schools in Friday Bridge.[46]

CHARITIES[47] In 1597 William Payne gave 16 acres in Whales or Wales Field for the repair of the church and the relief of the poor. William Mayner, by his will dated 1638, gave a messuage and 6 acres in Oldfield. In 1837 the Mayner and Payne estates were let at £50 per annum, the proceeds being distributed in flour to about 250 families indiscriminately.

Thomas Coward, by his will dated 1674, gave 6 acres in Oldfield to provide clothing and coals for the poor. In 1837 4 chaldrons of coal were distributed indiscriminately in quantities of one bushel, and 5 coats and 5 petticoats given to non-paupers. 'The persons selected are out of the better class and best-conducted labourers.'

Thomas Adams, by his will dated 1697, gave 4 acres near New Common Bridge, on the north end of the parish adjoining Wisbech town, to provide 3 gowns or coats. Any surplus was to be given in bread to the poor. In 1837 the land was let at £13 15*s.* and provided for 8 men's coats and 2 women's gowns. Persons who had received parish relief during the previous three months were disqualified from this charity.

Thomas Jenkinson gave (1737–8) 2 acres in Whales Field. In 1837 this was let for £4 to provide gowns for women.

Daniel Swaine at an unknown date gave a rent-charge on 4 acres of land at the Gooles. In 1837 £1 was given to the vicar for a sermon on St. Thomas's Day, and £1 to the poor in 40 sixpenny pieces.

The poor of Elm were entitled to cut turf on a piece of land of about 40 acres which was inclosed about 1817. By 1837 the turf-cutting had been substituted by the letting of this land for £70 yearly. This sum was distributed in coals, at the rate of 6 bushels each, to families who paid less than £20 a year in rent.

LEVERINGTON

Leverington is a large parish and village on the north-west of Wisbech.[1] The lay-out—a frontage upon the Ouse (later the Nene) outfall, a village site protected by the 'Roman Bank', and a landward extension into the fen—was originally the same as in the other Marshland villages of the Isle, but in Leverington has been considerably modified. In the first place, the fen portion of the ancient parish, containing Parson Drove (q.v.), has been separated both for civil and ecclesiastical purposes, and the hamlet of Gorefield is also a separate parish ecclesiastically.[2] Secondly, the proximity of Wisbech and the extensive development of

[35] At a cost of £120 (leaflet appealing for funds, in possession of Mr. G. M. G. Woodgate, Wisbech).

[36] M. of Ed. File no. 12/12.

[37] Archives of Nat. Soc.

[38] Kelly, *Dir. Cambs.* (1900).

[39] M. of Ed. File no. 12/15.

[40] M. of Ed. File no. 12/14.

[41] There was an average attendance of 412, and only 352 places available in the two schools at Elm and the one at Friday Bridge.

[42] M. of Ed. File no. 12/73.

[43] M. of Ed. File no. 12/63.

[44] M. of Ed. File no. 12/62.

[45] Rep. of Nat. Soc. (priv. print).

[46] Archives of Nat. Soc.

[47] *Rep. Com. Char.* (*Cambs.*), [103,] 319 et seq., H.C. (1837–8), xxiv.

[1] By the M. of H. Provisional Orders Confirmation (Ely, Holland, and Norf.) Act, 1933, and the Isle of Ely Review Order (M. of H. No. 76940) of the same year, 16 acres, with a 1931 population of 81 were transferred from Leverington to Wisbech Borough, and 145 acres (21 population) from Walsoken (Norf.) to Leverington.

[2] By the Leverington Rectory Division Act, 1870 (33 & 34 Vic. c. clv). It may be noted, however, that Leverington is not divided into wards for parish council purposes, as is the case with some of the larger and more populous Isle parishes, e.g. Elm and Upwell.

G. F. Webb

MARCH, ST. WENDREDA'S CHURCH. NAVE ROOF (EARLY 16TH CENTURY)

F. R. P. Sumner

ELM, ALL SAINTS' CHURCH. WEST TOWER (13TH CENTURY)

F. H. Crossley

THORNEY ABBEY. 12TH-CENTURY NAVE

LEVERINGTON CHURCH FROM THE EAST
(From a photograph of 1853)

ELM, ALL SAINTS' CHURCH
(From a photograph of 1856)

small holdings have caused a rapid increase in population, particularly during the present century.[3] The main part of the village, around the church and Leverington Hall, stands on a ridge just inside the old sea wall. There is ribbon development along the new main road all through Leverington parish, and also along the Nene bank at its southern end. There is much scattered development along the minor roads, northwest from Leverington village towards Gorefield, and westwards along Leverington Common towards Parson Drove.

The Bedford Level Act of 1663[4] authorized the inclosure, division, and allotting of 339 acres of marsh land, between the Roman Bank and the Nene outfall, to the owners of commonable messuages or tenements in Leverington. Each allotment appears to have averaged 2 acres. Before this time various creeks, such as Carlton's Creek,[5] penetrated a long way inland and allowed the conveyance of seaborne goods right up to the village. An old boat has been dug up in the vicinity of this creek, just under the sea wall.

A great development of small holdings took place in Leverington in the early 20th century, and the population rose by about 75 per cent. between 1901 and 1931.

The village is an attractive one, with several large houses in well-timbered grounds.[6] The church spire, perhaps the most graceful in the county, is a prominent landmark.

The parish fields are as follows. In Leverington: the Marsh, Spitalfield, Thummins, Margerie's Croft, Farthing Field, Outnewlands, Fen Croft, Church Croft, Paps Field or Hill Croft, Seafield, Ives Dyke Field or Doole, Park Field, Woolcroft or Walcroft, Wratfield, and Snailcroft, otherwise Fendyke Field, West Meadow, Beaconhoe, or Maysfield. In Gorefield: Ox or Fitton Field, Gorefield, Harp, Hart or Harpley Field, Long Meadow, Cat Field, Richmond Field, New Field, Shire or Shear Field, Black Lane Field. In Parson Drove and Southea: Pock or Poke Field, Remers Field, Popefield, Canon Field, Old Eaufield, Elbow Field, Flain Field, North and South Inham, and Parson Drove Fen.

MANORS Leverington is not mentioned in Domesday Book, and the method and date of its acquisition by the church of Ely is unknown. As at Wisbech, both bishop and prior obtained interests in Leverington after 1109. About a century later (*c.* 1210–12) William de Longchamp held 400 acres for 2 marks of the Prior of Ely.[7] This is probably the origin of the manor of Graces (see below). But the prior was more concerned as a tenant of the bishop. Two of the episcopal tenancies mentioned in 1221[8] were held in 1251[9] at the same rents by the prior. These were the tenement of Walter son of Walter (120 acres) who held by a rent of 20*s.* with suit of court, and the tenement of Ralph de Tirington (3 virgates) who held by a rent of 5*s.* 4*d.*

The bishop's manor of Leverington was in the 13th century a fairly large one, but it did not make so much progress between the surveys of 1221 and 1251 as some of the others in the Isle. At the earlier date there were 5 free tenants, 64 customaries, and 9 cottars; at the later 7 free tenants (counting the Prior of Ely's two holdings separately), 60 customary tenants, 5 others who held land by customary services without messuages, and 5 cottars. In 1251 there was a demesne of 44½ acres, probably in open fields though not so specified, and two areas of marsh totalling 84½ acres. It was stated that some of the larger piece, Northale, was overflowed and laid waste by the sea.

The lord had the usual manorial perquisites, including, since Leverington was near the sea, the rights of wreck and of royal fish. To the finder of such a fish 4*d.* were payable. The total rents amounted to £15 14*s.* 6¾*d.*

The later Middle Ages saw the rise of a number of manors in Leverington, and it is not easy to associate any of them with the episcopal estate mentioned above.[10] The manor of *RICHMOND* has been generally regarded as the chief manor. It lay in Gorefield, and the homestead moat that surrounded the manor house still remains near the modern Richmond Hall Farm.[11] The first mention of this manor under its own name occurs in 1390, when Geoffrey Richmond was granted the right to hear mass in the oratory of his house provided he attended the parish church on Sundays and festivals.[12] It is possible that this manor is identifiable with the messuage and 160 acres in Leverington held in 1344 by Sir John de Shardelow of the prior of Ely for 13*s.* 4*d.* yearly,[13] for the next definite mention is in 1391, when Robert de Shardelow, great-grandson of Sir John,[14] settled it upon trustees.[15] He died eight years later,[16] when the surviving trustees granted it to his relict Ela for life, with remainder to Robert's son, another John.[17] Ela survived until 1457,[18] and between 1432 and 1434 seems to have made over the manor to her son and daughter-in-law. In the former year Beatrice (Everard) and Alice (Caius), coheirs of John Hiptoft, released all their rights in the Shardelow property to Ela;[19] in the latter a fine was levied concerning the manor to which John Shardelow and Margaret his wife were the deforciants.[20] Another fine of 1453 brought Richmond manor to Sir Gilbert Haultoft and his wife Margaret.[21] Their daughter Mary married Thomas Kervile of Wiggenhall St. Mary (Norf.), and the manor remained in this family

3 The 25 per cent. increase (1,124 to 1,424) between 1901 and 1911 was specifically attributed by the census authorities to the development of small holdings. A further rise of about 40 per cent. (to 1,981) took place between 1911 and 1931. (The 1921 figure of 2,484 was inflated by the temporary presence of fruit pickers.) Between 1931 and 1947 the population of Leverington itself continued to rise, while that of Gorefield remained stationary (*Survey of a Fen County* (I. of E.C.C. Planning Dept.), Map no. 52).

4 15 Chas. II, c. 17.

5 This creek crossed the Sutton Road nearly opposite Chamberlin's Farm, by the old Mill Drain, and turning southward reached the bend at Parson Drove at the point occupied by the Pits. The old name of this road was Crooked Drove (Wisb. MSS., deeds of 1767 and 1787).

6 Gardner, *Dir. Cambs.* (1851), 551, mentions that Leverington village was 'scattered and partly shaded by trees'.

7 *Bk. of Fees*, 630.

8 B.M. Cott. MS. Tib. B. II, ff. 148*a*–151*a*.

9 B.M. Cott. MS. Claud. C. XI, ff. 82*b*–86*a*.

10 Leverington is not mentioned in the survey of episcopal manors made in 1356 during the seizure of the Ely temporalities in Bishop Lisle's episcopate (B.M. Add. MS. 6165), unless it is included with Wisbech.

11 *V.C.H. Cambs.* ii, 34. Medieval pottery has frequently been dug up on this site.

12 Reg. Fordham, f. 175.

13 *Cal. Inq. p.m.* viii, no. 519 (p. 365).

14 J. Gage, *Hist. Suff.* (*Thingoe Hund.*) (1838), 60.

15 *Cal. Pat.* 1452–61, 206–7 (Exemplification of 1455).

16 C 137/8.

17 *Cal. Pat.* 1452–61, 206–7.

18 Gage, *Hist. Suff.* (*Thingoe Hund.*), 60.

19 *Cal. Pat.* 1452–61, 207.

20 *Cambs. Fines* 1196–1485, ed. W. Rye, 151.

21 Ibid. 154.

for more than a century.[22] Henry Kervile made a settlement thereof in 1592;[23] at his death (1615) it was stated to be held of the Bishop of Ely in socage as of his manor of Leverington.[24] His son, then aged 22, was a recusant; he suffered imprisonment for allowing meetings of a treasonable character at his house, but was released before his death in 1624.[25] His son died before attaining his majority, and this branch of the family thereupon died out. The manor eventually became vested in John Pell, of Holme next the Sea (Norf.). It is possible that he obtained it by descent, for his grandfather, John Pell of Dersingham (Norf.), married into the Gawsell family,[26] but the evidence is not conclusive.

In 1687 John Pell suffered a recovery of the manor and its appurtenances in Leverington and Wisbech.[27] The estate consisted of a messuage, 20 acres of arable, 5 of meadow, 200 of pasture, 20 of fresh marsh, and 100 acres of land covered with water.[28] He died in 1714, leaving three daughters as coheirs, Frances wife of Richard Mason of Necton (Norf.), Mary wife of Robert Goodricke, and Elizabeth Pell. In 1718 they passed the manor by fine to Frances's husband,[29] who held his first court in 1721. He died the following year, and his only son Richard also (under age) in 1735; the latter was succeeded by his cousin William Mason, who bequeathed the manor and other property to his nephew, another William Mason.[30] In 1762 a private Act of Parliament[31] was obtained, authorizing the sale of the Mason property, and under this the Leverington Estate was in 1800 auctioned in 13 lots.[32] The buyer of the manor house and some 200 acres was George Johnson, whose family had for some time been occupiers under the Mason trustees.[33] Shortly before his death (1830) he sold the estate to Jonathan Peckover of Wisbech. During the 19th century the old manor house was replaced by the present Richmond Hall Farm. The estate remained in possession of the Peckover family until 1920, when shortly after Lord Peckover's death it was sold in lots by his daughters. The house, with a considerable part of the land, has now been reunited in the ownership of Mr. Michael Newling, who lives at Richmond Hall.

The last court baron of this manor seems to have been held in 1777, when William Mason the younger was lord; after this date manorial rights reverted to the Bishop of Ely as superior lord. A rental of 1750 shows 57 tenants, paying £9 1*s.* 2½*d.* There were then 22 messuages or cottages.[34]

The holdings of the Prior of Ely, in his own right and as a tenant of the bishop, have already been detailed.[35] Little is known about the early history of the conventual and capitular manor of *GRACES*. In 1291 the convent's property in Leverington was valued at £11 16*s.*, and there was also a block extending into Wisbech and Elm which was worth £5 9*s.* 7*d.*[36] At the Dissolution the manor, so described, was worth £8 13*s.* 5*d.* net; the lessee was William Edwards.[37] As elsewhere, the estate was formally transferred to the newly constituted dean and chapter in 1541.[38]

The position of this manor in Leverington is shown by the fact that its demesne extended into the following fields: Farthing, Sea, Park, Oxfield, the west end of Gorefield, Catfield, Newfield, and South Inham. Little can be gathered as to its organization, as the court rolls are no longer extant, but it is known that lands were held at quit-rents.[39] As in the capitular manors in the Isle proper, lands in this manor were leased at beneficial rents, and during the 18th and 19th centuries the Colvile family were prominent tenants. The last Colvile lessee was W. E. A. Colvile of Horringer (Suff.), who died in 1860.[40] At this date the fines and fees levied at the seven-yearly renewals were £770.

In 1792 the demesnes amounted to 141¾ acres,[41] and there were seven sub-tenants. It seems probable that the manor house was in Newfield, on the site of the modern Bleak House. In 1860 there were five sub-tenants of this manor in Leverington and one in Parson Drove; they held 135½ acres for rents of £266 1*s.* 6*d.*[42]

In 1866 the dean and chapter let the manor for twenty-five years to the Revd. John Husband. Including a piece of pasture called Mill Hill and all the appurtenant lands the estate comprised 175½ acres.[43] Husband died in 1869, and the manor was sold for £4,850 to Emily Mary Colvile and C. B. Phillips, the trustees of W. E. A. Colvile, subject to the existing lease.

The manor of *FITTON* originated in lands held by Alan de Fitton of the Bishop of Ely. About 1210–12 these comprised 4½ virgates at a rent of 16*s.* 2*d.* and 100 acres of marsh at 10*s.*[44] In 1221 they are specified as 4½ virgates and 120 acres of newly reclaimed land (*nova terra*), held for 16*s.* 6*d.*[45] By 1251 Alan had been succeeded by Elias de Fitton, kt., who held as in 1221 with the addition of half a virgate for 3*s.* 8*d.* and the duty of suit of court. In 1327 Alice and Maud de Fitton are recorded as owning lands in Leverington; they were assessed at 7*s.* and 2*s.* 1*d.* respectively, and seem to be the last representatives of the family in this neighbourhood.[46]

In 1347 Sir Lawrence de Flete was granted free warren in his demesne of Fitton in Leverington.[47] His daughter Agnes married Sir Bartholomew Everard, into whose family the manor descended. In 1366 a chantry was founded by John Hode of Fleet and Simon son of Martin of Holbeach 'within the manor', and endowed with 41 acres of land in Leverington.[48] It was established for the good estate of Simon (Langham) Bishop of Ely[49] and his successors, and for the souls of Sir Lawrence de Flete, Isabel his wife and Margaret his sister, and Sir Philip Everard and Robert Braunch.

[22] C.P. 25 (1)/30/101/Mich. 10 Hen. VII.
[23] C.P. 25 (2)/262/Mich. 34–35 Eliz.
[24] C 142/348/141; Wards 7/52/55.
[25] M.I. in church of Wiggenhall St. Mary; Parkins, *Topog. Freebridge Hund.* (1762), 299.
[26] His wife Ursula was the sister of Sir Edmund Gawsell, the second husband of Winifred wife of Henry Kervile (d. 1615) (*Visit. Norf.* 1563, 1589 and 1613 (Harl. Soc. xxxii), 133).
[27] C.P. 43/719, ro. 287.
[28] Wisb. MSS.
[29] Ibid.
[30] Ibid.
[31] 2 Geo. III, c. 36.
[32] Wisb. MSS.
[33] John Johnson, George's father, was sheriff in 1781. He had six very tall sons, who attended him at the Assizes as javelin men.
[34] Wisb. MSS.
[35] See above, and notes 7 and 9.
[36] *Tax. Eccl.* (Rec. Com.), 270.
[37] Dugd., *Mon.* i, 494.
[38] *L. & P. Hen. VIII*, xvi, g. 1226 (12).
[39] e.g. a toft in Oxcroft late of John Newbrook and Richard Odam, in 1574 held of the Dean and Chapter of the manor of Graces at rent of 3*s.* 4*d.* (C 142/168/23; Wards 7/15/14).
[40] Wisb. MSS.
[41] Ibid.
[42] Ibid.
[43] Ibid.
[44] *Bk. of Fees*, 630.
[45] B.M. Cott. MS. Tib. B. II, f. 148*a*.
[46] *Lay Subs. R. Cambs. 1 Edw. III*, ed. C. H. Evelyn White.
[47] *Cal. Pat.* 1345–8, 480.
[48] C 143/358/15.
[49] Appointed Archbishop of Canterbury in 1366; died 1376.

The advowson of the chantry was granted to the last two.[50] Within a century the chapel of this chantry had become ruinous, and the income derived from the endowment so reduced—owing to flooding of the lands and other causes—that from 1449 onwards no chaplain had celebrated divine service there. Therefore John Everard and Henry Braunch, the holders of the advowson of the chantry, were in 1459 allowed by the bishop and convent of Ely to transfer the endowments to the chapel of Parson Drove (q.v.).[51] This transfer did not exempt them from the operation of the Act of 1545 for the suppression of chantries and diversion of their revenues to the Crown, and eight years later the original endowments of John and Simon Hode, with 64½ acres in Leverington and elsewhere which had been given since the foundation of the Fitton chantry, were sold for £1,539 15*s.* to Thomas Wren and Edward Slegge.[52]

Sir Bartholomew and Sir Lawrence Everard and their wives are represented in medieval stained glass in the church (see below). William Everard, Sir Lawrence's second son, succeeded his father in the manor, but at the date of his will (1419) had probably settled it on his own son John,[53] who resided at his other manor of White Hall in Wisbech, and at the time of the transfer of the chantry lands (1459) the Fitton Hall manor house was ruinous.[54] The manor continued in the Everard family for several generations. William Everard, grandson of John, died seised of it in 1534,[55] leaving a son, Bartholomew, aged 30. Bartholomew's only son died without issue and the manor devolved on his sister Grace, wife of Richard Buckworth of Wisbech.

It was probably under the Buckworths that the manor house was rebuilt. In 1827 it was described as a 'brick building' which still retained the name of Fitton Hall, though it appeared to be 'no longer known as a Manor'.[56] Anthony Buckworth, great-great-grandson of Richard and Grace Buckworth, made a settlement of the manor, on his marriage in 1660 to Anne (Fisher) relict of James Edwards of Wisbech.[57] His grandson Thomas, who succeeded to the property in 1692, conveyed it in 1696 to Humphrey Hyde. In 1728, when William Hyde of Long Sutton (Lincs.), clerk, made a settlement, the manor consisted of the house and 87½ acres, 5 of which were woodland surrounding the house, 11½ arable and the remainder pasture including 10 acres in Gorefield. The whole was in the occupation of Simon Hardy. In 1730 the property was sold to George Worrall of Long Sutton. Seven years later he settled it upon his second wife Elizabeth (Whitehall) and their issue if any. Worrall had no children by his second wife, and in 1761 she, with her stepson G. H. Worrall of Emneth (Norf.), barred the entail. Elizabeth Worrall survived her stepson, and in 1774 she devised the manor and estate to her nephew, W. H. Worrall of Spalding. In 1801 he made a settlement, giving his wife Ann a life interest, which she sold in 1818 to William Pass of Sheffield, her son-in-law. At this date the property comprised 92 acres, late in the occupation of John Musson, and then of Thomas and Humphrey Watts. In 1824 W. H. Worrall, the younger, with his sister Ann Pass, to whom the remainder in the estate had been granted by their father (d. 1806), sold it for £2,990 to Samuel Roberts of Sheffield, a silver plater. The younger Worrall was apparently a spendthrift, and enlisted in the 13th Dragoons; the sale may have been occasioned by his being drafted to the East Indies. The estate continued in the Roberts family until 1919, when Sir Samuel Roberts, M.P., sold it to Mr. Thomas Tansley. More recent owners have been Messrs. J. L. Cooper, E. L. B. Motley and Mrs. Ellen Wing. The present possessor is Mr. Aubrey Hammond.

The wood surrounding the house was cut down about 1750, and the manor house mentioned by Watson was replaced by the present farmhouse in the latter part of the 19th century.

LESSER ESTATES[58] The house at the south end of Church Lane, sometimes called *CROSSE HALL* but now *BEECHWOOD*, was probably built by Thomas Crosse who died there in 1633.[59] In 1624 Crosse was apparently living in Wisbech, which had been the home of his father and grandfather and other members of the family who had played a prominent part in the life of the town.[60] Thomas's second son John (d. 1639)[61] succeeded to Crosse Hall and was succeeded in turn by a grandson John II (d. 1666)[62] and a great-grandson John III (d. 1704).[63] The next owner was John IV (d. 1743), the eldest son of John III. He was succeeded by a son Edward, who in 1788 was appointed a D.L. for the Isle. In 1762 he inherited a further 200 acres under the will of Adam Hawkins of Leverington,[64] but he appears to have dissipated much of his fortune through horse racing. Several of his race horses are buried in the paddock, including the noted Alonzo whose grave was marked by two pines.[65] Much of Edward's property was sold at his death (1795),[66] but Samuel (1763–1847) his third son, Fellow of Peterhouse and Vicar of Hunstanton (Norf.), retained a part of it including Crosse Hall. His son Samuel Massey Crosse (d. 1861)[67] succeeded to Crosse Hall. In 1853 the remaining property (190 acres) was sold by auction[68] and S. C. Crosse went to live at Barton Lodge, Wisbech, and subsequently in London. Henry Sharpe subsequently lived for many years at Beechwood (or Crosse Hall) where he kept his shrievalty. On his death (*c.* 1930) there the place was sold to E. J. Newborn,

50 *Fenland N. & Q.* vi, 377.

51 Ibid.

52 *Cal. Pat.* 1553, 218, 221.

53 There is no mention of Fitton Hall in his will, in which he bequeathed his messuage called Durham's Place in Leverington to his son Lawrence, with 8 acres in the croft adjoining it called (Out)-newland, abutting on 'le mylhill' on the east and Kirkgate on the west. This seems identifiable with the site of the present Leverington Hall; when the neighbouring Cherry Tree Hill was excavated in 1933 it was conclusively established that it was an artificial mound—probably a mill mound as a large fragment of mill stone (now in Wisb. Mus.) was dug up. William Everard also bequeathed to the image of the Virgin Mary of Aldenhale a golden cross enclosing some particles of the Holy Cross.

54 *Fenland N. & Q.* vi, 377.

55 C 134/57/100; E 150/85/1.

56 W. Watson, *Hist. Wisb.* 472.

57 The owner retains the title deeds of this manor from 1660. They form the sources for the remainder of this section, together with the Wisbech MSS.

58 Except where otherwise stated the main source for this section is the title deeds which are the property of the respective owners.

59 M.I. in church in Cole's time.

60 It was in this year that his daughter Elizabeth married, at Leverington, William Steward of Barton Mills (Suff.). The children of Thomas Crosse were all baptized at Wisbech.

61 Leverington par. reg.

62 Ibid.

63 Ibid.

64 Wisb. MSS.

65 Infm. Mr. W. F. Crosse.

66 Ibid. The property was given to him for life, with remainder to the heirs of his body.

67 M.I.

68 Wisb. MSS.

upon whose death it devolved upon his daughter Mrs. Tansley, the present (1951) owner.

A rough drawing of the house in an earlier state is given in the map of the Leverington Hall estate[69] made in 1782. From this it can be seen that the house consisted of a central block with wings extending forward on each side, something like an H. Sharpe demolished much of it in 1892 and built a completely new front with large rooms behind. The walled garden and some of the farm buildings are probably as old as the original house. The dovecote is at least 300 years old,[70] and is a fine specimen of its kind. It is an octagonal brick building with ingress through the top. Within was an octagonal wooden funnel, some 7 ft. long. This acted as a kind of decoy, for a bird flying down would find difficulty in making its way out, and by the time that it had done so would usually become accustomed to its surroundings and would remain of its own accord. Around the inside of the building are over 800 brick-built nests, arranged in tiers. Access to these was provided by a revolving ladder, which could be swung round in any direction to provide approach to every nest in the building.

DECOY HOUSE in Gorefield Fen, next the roadway known as Turnover Bank, was formerly in the possession of the family of Samuel Clark, from whom it passed to the Revd. Jeremiah Jackson (d. 1857), Vicar of Elm. About the middle of the 19th century the roof and walls collapsed and the house was rebuilt by Jackson, and was occupied for many years by Joseph Ellard Griffin and subsequently by his nephew Joseph Scrimshaw. On Jackson's death the estate devolved upon his son the Revd. Frederick Jackson, Vicar of Parson Drove, on whose death (1904) it was sold to Walter Ward. The latter's son Mr. George Tom Ward succeeded on his father's death and is now (1951) the owner, the property being in the occupation of his son Mr. Stanley T. Ward. It is said that one of the rooms formerly contained ten oil-painted panels, probably three centuries old, two of which were supposed to represent the duck decoy, from which the house derives its name, with fanciful additions.[71] The panels were taken from the old house and were built into the present one.[72]

The decoy itself (about 6 acres) lies about ¼ mile behind the house, on the opposite side of Goredyke Bank, and actually just within the Newton boundary, in the old parish field known as Leets Lane End.[73] The property was conveyed, with other lands in Leverington, in all about 125 acres, by Robert Mears of Wisbech to Robert Mears his son in 1761, for a life annuity of £80; and on the death of Robert Mears the son in 1781[74] it descended to his only son, Noah Newton Tund Mears, who died in infancy. In 1786 Samuel Rolling of Leverington (cousin and next heir) owned the property and from him it passed to Thomas Smith[75] of Newton, the owner in 1833.

With the improvement of drainage in Newton and Tydd St. Giles fen, the decoy was eventually abandoned, as was also the similar decoy in Tydd St. Giles fen, and both decoys are now ordinary arable land. Aerial photographs of the site show no trace of them. These duck decoys, formerly a feature of the fenland, attracted the notice of Daniel Defoe when he visited Lynn and the Isle of Ely in 1722.[76]

FENCROFT, at the south end of Church Lane in Fencroft field, at the entrance to the village, is not much more than a century old, but stands on the site of, and incorporates some of the materials belonging to an older house. That this older house was of some importance is proved by the representation of it on the map of Leverington Hall, drawn in 1782.[77] A dovecote then lay to the west of the house.

In the 17th century the house with 12 acres belonging, extending from Church Lane to Paps Lane, belonged to Adam Denison of Leverington, grandson of Robert Denison of Leverington.[78] Adam Denison by will dated 1715 bequeathed all his lands in Leverington and Parson Drove to his grandson Richard Cumberland in tail.[79] In a deed of 1728 preliminary to barring the entail over the land devised to Richard the property is described as the capital messuage lately occupied by Adam Denison and then by Charles Gerbow, with 12 acres belonging called 'the Homestead' and other lands comprising 4 messuages and 274 acres in Leverington and Parson Drove.[80] After this the house with 12 acres was sold to William Hardy, who later sold it to John Swaine of the City of London, son of Thomas Swaine of Leverington Hall.[81] John Swaine occupied the house at his death (1753). John Swaine his son, in his will dated 1763, stated that the house was then in his own occupation, and gave it with much other land to his son Spelman Swaine, subject to his wife's life interest. Spelman Swaine bequeathed all his unsettled lands to trustees upon trust for sale; but the greater part of his property, including the house and 12 acres, had in 1790 been settled upon himself and his wife for life with remainder to his son Walter Swaine absolutely. He died in 1803 and apparently the house was purchased from the trustees by Captain Spelman Swaine, R.N. (1769–1848), the younger son.[82]

On his retirement from the Navy Swaine continued to reside at Fencroft, and became a J.P. and the last Chief Bailiff of the Isle of Ely. In 1846 he received the rank of rear-admiral. Prior to 1834 he sold his Leverington property. By 1834 Philip Godfrey had become the owner of the house with 12 acres. It was he who rebuilt the house. The property continued for many years in the Godfrey family. Early in the present century it was purchased by Mr. G. W. Whitehead, after whose death Mr. H. S. Littlechild purchased it and now resides there.

GLENDON, a square-built house on the west side of the Sutton Road, was built in 1831[83] by the Revd.

69 *Penes* Mr. G. C. Munday of Leverington Hall. It was drawn by John Watté, a local surveyor.

70 Infm. Mr. W. F. Crosse.

71 Infm. the late J. T. Marshall.

72 F. J. Gardiner, *Hist. Wisb.* (1898), 381.

73 Parish Map of Newton, revised to 1833, contains a drawing of the decoy. The map is in Wisb. Mus.

74 M.I. in church.

75 Wisb. MSS. The property belonging to Thomas Smith is numbered 237–40 on the Newton parish map, and contained 42 a. 2 r. 30 p.

76 *Fenland N. & Q.* i, 19 and vii, 36.

77 *Penes* Mr. G. C. Munday.

78 Wisb. MSS.

79 Ibid.

80 Lease and release of 1728 between Richard Cumberland Archdeacon of Northampton, Richard Reynolds Bishop of Lincoln and Charles Parker, Richard Cumberland, Francis Tregeagle, and Robert Harper.

81 It is not quite clear whether the purchaser was John Swaine or his father Thomas.

82 Wisb. Mus., Copy of Leverington Terrier. He purchased other lands from his father's trustees in Feb. and Apr. 1805, being described as of Leverington (Wisb. MSS.). He sailed round the world with Vancouver and in the course of the voyage made some valuable ethnological collections which he presented to the Wisbech Museum.

83 From particulars of auction sale, 1864.

Samuel Crosse (see above—Beechwood), Fellow of Peterhouse. Latterly he lived at Glendon and died there.[84] He devised it, with other property, to his younger son Edward Bailey Crosse, subject to the payment of £600. E. B. Crosse commanded the merchant ship *Violet* of Wisbech on her last voyage.[85] The vessel left Glasgow in 1851 and was captured by Barbary pirates in the Bay of Bologna. E. B. Crosse died a prisoner in their hands in 1851. Just before setting off on this voyage he gave a party, at which the row of walnut trees now standing was planted.

In 1864 the property was in the occupation of Francis Nicholas Taylor. It comprised 'two parlours, two kitchens, good cellar, five chambers, brick and tiled stable with loft over, chaise-house and piggeries, enclosed by a brick wall and neat iron fence with paddock of prime pasture land'. It was then sold to Thomas Steed Watson of Wisbech. It eventually passed into the ownership of the Houlden family from which it descended to Mrs. Cook who lives there.

LANCEWOOD, a pleasantly situated house in Outnewlands, opposite Beechwood, originally belonged to Anne Crosse (d. 1685),[86] relict of Thomas Crosse of Crosse Hall (see above—Beechwood). Anne devised it as 'a messuage, stable, chaise-house, orchard and land containing 2 a. 1 r. 25 p.' to her daughter Mary Amry,[87] from which family it passed to Adam Hawkins of Leverington and appears to have been occupied by him. On the death of his sister Conquest Mayhew in 1762 the house and much other property was inherited by her aunt Elizabeth Denison, wife of the Revd. Richard Cumberland, Archdeacon of Northampton. His grandson Richard Cumberland, the memoir writer, sold it in 1789 with other lands to Samuel Stanton of Leverington, whose son Stephen was living there in 1806.[88] The property was bought by Frances, Elizabeth, and Ann, daughters of John Johnson of Richmond Hall, High Sheriff of Cambs. and Hunts., and it was they who erected the present house. They were living there in 1834[89] and until the death of Ann Johnson in 1853, whereupon her two surviving sisters in the same year sold the place to James Gregory, a Wisbech merchant.[90] It was purchased from James Gregory in 1871 by Matthew Webster of Leverington Hall and was sold by his trustee in liquidation in 1879. After various changes in ownership it was purchased in 1931 by Mr. Francis Jermyn Smith, M.B.E., who has greatly improved it and now lives there.

Although not a manor house, *LEVERINGTON HALL*, standing in the middle of the village, near the church and school, is the principal house in the parish. It has been supposed that the original house on this site was Durham's Place, the property and residence of William, second son of Sir Lawrence Everard[91] (see above—Manors, Fitton Hall). The first authentic owner and occupier of the present house was Robert (d. 1705) son of Thomas Swaine of Wisbech (d. 1639). Robert was living there in 1641.[92] He was a J.P. and became High Sheriff in 1681. He must have succeeded to some of his father's property in 1639, and in the same year he became entitled to substantial benefits under the will of his elder brother Thomas Swaine of Wisbech.[93] His marriage in 1640 with Mary daughter of William Freeman, merchant, of London and Leigh (Surr.),[94] may have brought him additional wealth. All this strengthens the belief that it was Robert Swaine who, out of his then abundance, rebuilt Leverington Hall.[95] He also added considerably to his property at Leverington. He was succeeded by his only surviving son Thomas (1645–1728), a J.P. for the Isle. Thomas's second surviving son, Spelman, succeeded to Leverington Hall and died in 1761, devising the Hall to his nephew Daniel, a merchant of King's Lynn (d. 1782). Daniel thereupon took up residence at Leverington Hall, and became a J.P. and in 1775 high sheriff. By will dated 1780 he directed his trustees to sell all his lands, and in 1785 Leverington Hall with much other property was offered for sale by auction with the buildings, dovecote (situate near the house, at the south-east corner) and about 38 acres surrounding it. The house (which had been let to the Revd. Joseph Plumtree), 38 acres surrounding it, and 32 acres on the opposite side of Church Lane (constituting the entire block between the rectory and the Fencroft house and land belonging to Daniel's nephew Spelman Swaine), were bought by Edward Stone of Tydd St. Mary, who subsequently sold the property to John Johnson. In 1842 Johnson sold the Hall to Thomas Webster of Newton. The latter died in 1862, and by his will[96] gave Leverington Hall with 140 acres of land in Leverington to his son Matthew Webster. Matthew resided at Leverington Hall. He became involved in financial difficulties, however, as a result of which the property had to be sold in lots by the trustee in liquidation in 1879. The Hall with much land was bought by Henry Sharpe of Leverington, who never himself lived there. The Hall was let successively to Medland Newsham, George Carrick, a Wisbech solicitor, and Harold F. M. Peatling. The last eventually purchased it in 1927. Peatling was chairman of the Isle of Ely County Council and High Sheriff in 1925. He died in 1938 and in 1946 the property was sold by Mrs. Peatling to Mr. George Campbell Munday, M.C., tenant for ten years previously, who now (1951) resides there.

Leverington Hall[96a] is a building of mellowed brick with slated roofs.[97] It consists of two gabled wings connected by a slightly recessed central block. This plan suggests an Elizabethan or Jacobean origin—a surmise borne out by two chimney breasts, at the back and on the north side, which are certainly of late 16th-century date.[98] The rest of the house may at that time have been of timber; there are no other remains. The house seems to have been rebuilt in the 17th century, probably by Robert Swaine who held the property

84 M.I. in Leverington church.
85 M.I.; infm. Mr. W. F. Crosse.
86 Leverington par. reg.
87 Wisb. MSS.
88 Wisb. Mus., Copy of Leverington Terrier.
89 Ibid.
90 Within three months of each other, in 1858; they were buried in Leverington churchyard.
91 See above, note 53.
92 The descent is authenticated in the pedigree recorded at the Visitation of 1684 (Coll. of Arms). The arms given were: azure a chevron between three pheons or on a chief argent three maidens' heads proper crowned or, the crest: a maiden's head proper resting on wings displayed.
93 P.C.C. 69 Harvey.
94 London marriage licence, Apr. 1640.
95 In the Subs. R. for Cambs. 1640–1 (E 179/84/410) the six principal landowners in Leverington are: John Daye (£8), John Swaine (£4), Robert Swaine (£3 10*s*.), Thomas Crosse (£3), David Jones (£2 10*s*.), Nicholas Lumpkin (£2).
96 Will proved in P.P.R., 1863.
96a See plate facing p. 207.
97 The original stone slates—a rare feature in the Fens—survive at the back of the house; in front they have been replaced by modern Welsh slates.
98 Many very old bricks were discovered in the foundations when repairs and alterations were carried out *c*. 1920–30.

from 1639 to 1705, and to this date most of the walling and windows belong. At the back there is a rainwater head, with the Swaine crest, the initials TSE and the date 1716, which points to a restoration by Robert's son Thomas (d. 1728). To this period may also be ascribed the four fine chimney-stacks of three linked shafts each, two of which surmount the chimney breasts above mentioned, and the gables at the back, which are constructed in the Flemish manner with skew butts dovetailed into the horizontal courses. The front gables were probably rebuilt in the late 18th century; an estate map with drawing of the house shows curved 'Dutch' gables in this position. The porch is a 19th-century addition. The forecourt has a pair of fine brick gate piers topped by stone pineapples.

Inside, several rooms have early Georgian wainscoting and exposed ceiling beams. There is a well staircase with turned balusters and plain newels, an early 18th-century fireplace of marble in the hall and one with Adam decoration in the drawing room.[99]

LEVERINGTON HOUSE lies in the Marsh next Peatling's Lane on the seaward side of the old sea-bank upon which it abuts. It was built *c.* 1852 by Thomas Peatling of Wisbech (d. 1881), brewer and wine merchant. It embodies several interesting features of earlier date, including carved woodwork taken from the premises of the Wisbech Literary Society and portions of the lantern tower from the Octagon Church at Wisbech, when the lantern was removed (see Wisbech). Peatling died there and the property passed to his daughter Emily Peatling (d. 1909). She devised it to her nephew H. F. M. Peatling (see above—Leverington Hall). Since 1925 the house has been occupied by Mr. G. M. G. Woodgate, the present owner.

PARK HOUSE lies in Park Field on the north side of the Gorefield Road. The first recorded owner is John Lumpkin (1655–1743), a member of a family recorded in the Leverington marriage registers in 1612. John Lumpkin was succeeded by his son Nicholas (1685–1748) and he by Nicholas II (1728–94). The second Nicholas was D.L. for the Isle. In 1794 he owned 134 acres in Leverington and lands outside the parish also. He was succeeded as owner of Park House by his son Nicholas III (d. *ante* 1828), who in 1798 was obliged to mortgage the house to his brother-in-law Samuel Taylor for £2,300. He obtained a further £800 from this source two years later. By 1805 he found himself owing another £1,700, and decided to sell the property.[1] It was accordingly conveyed to trustees for sale for payment of creditors, the ultimate surplus to be returned to Lumpkin. At the auction in 1805 Samuel Taylor purchased the bulk of the property, but died before it could be conveyed to him, and accordingly, under directions in his will, Park House and about 40 acres were conveyed in trust for his son Nicholas, who went to reside there and by sundry purchases added considerably to it. Other parts of the Lumpkin property were conveyed to Samuel Taylor and Francis Taylor, two other sons. Nicholas Taylor died in 1869 leaving everything to his wife Sarah, who died three years later. In 1880 the property, about 150 acres, belonged to their son Francis Nicholas Taylor, being then in the occupation of his son Francis Nicholas Taylor. The greater part of the estate was sold in 1889 to Sir Samuel Roberts of Sheffield, who in 1919 sold it to Alfred Coates, the occupying tenant. After Coates's death it was sold in lots in 1944 when Park House with about 25 acres was purchased by Mr. G. A. Shippey of Gorefield, who subsequently purchased adjoining land to add to it. The house was empty on the outbreak of war in 1939 and some families of London evacuees who were installed in it did great damage to the interior. An attempt has since been made to repair some of the damage.

The front of the house is of 18th-century date and has been imposed upon an earlier structure. Oliver Goldsmith is reputed to have written *She stoops to conquer* under the shade of a fine old mulberry tree in the garden, for long pointed out to visitors but now removed. The pond in which Mrs. Hardcastle was 'draggled up to her waist like a mermaid' after her 'circumbendibus' is still pointed out. It is now within the paddock in front of the house, but when the play was written was part of the open common and therefore readily available on a dark night for such an accident as is related. It is also claimed that the play was written at Leverington rectory, where Goldsmith is reported to have stayed as a visitor, but there is no authentic evidence to support either tradition.

CHURCHES The advowson has always been held by the Bishop of Ely.

The living was a rich one in the early Middle Ages, being returned at £30 in 1217, £53 6*s.* 8*d.* in 1254, and £85 in 1291[2] and 1341.[3] In 1463, however, the revenues of the benefice had so much decreased that the taxed value was reduced from £40 to £20.[4] In 1535 the rectory was valued at £25 0*s.* 8*d.*[5] In 1851 the rectory, with the chapelry of Parson Drove, was returned at £2,099; the tithes had been recently commuted for a rent charge of £2,415.[6]

A rectory manor existed from the later Middle Ages until the beginning of the 18th century.[7] It was a small one, comprising land in some half-dozen of the fields of Leverington and in Sayers Field, Wisbech St. Mary.[8] The freehold quit-rents amounted to only 7*s.* 4*d.*, a pound of wax, and a capon, and copyhold rents in Leverington to 1*s.* 9*d.* By 1706 the steward's fees, which amounted to 12*s.* 8*d.*, had swallowed the income derived from the manor, and the rector seems to have allowed his rights to lapse.[9]

Under the Leverington Rectory Division Act, 1870[10] the chapelry of Parson Drove (q.v.) was erected into a separate ecclesiastical district, and the ecclesiastical district of Gorefield was formed; they were given the status of vicarages, in the patronage of the Bishop of Ely.

The church of *ST. LEONARD* consists of chancel, south chapel, clerestoried nave, aisles, south porch, and west tower with spire. The material is Barnack stone. A large church was erected in the middle of the 13th

[99] *Country Life*, ciii, 126–8 (16 Jan. 1948) (illustrated article).
[1] By deed dated 1805.
[2] W. E. Lunt, *Valuation of Norw.* 539.
[3] *Inq. Non.* (Rec. Com.), 203.
[4] B.M. Add. MS. 5805, f. 109.
[5] *Valor Eccl.* (Rec. Com.), iii, 500.
[6] Gardner, *Dir. Cambs.* 552.
[7] The earliest Ct. R. extant dates from 18 Edw. IV (1478–9), the latest from 1706.
[8] The Leverington fields were Woolcroft, Hillcroft, Fencroft, Spitalfield, Oxcroft, and Gorefield; place-names mentioned in the rolls include Mikelwellane (Muck Hill Lane), and Chapelgate south of Sayers Field.
[9] Ct. R. of Leverington Rectory Manor. See note 7 above.
[10] 33 & 34 Vic. c. clv. 'An Act for making better provision for the Cure of Souls within the limits of the Parish of Leverington and of certain adjacent Parishes. . . .'

century consisting of chancel, south chapel, nave, and west tower; the nave was probably aisled. The only trace of earlier work is a small 12th century cap preserved in the parvise. Early in the 14th century the east end of the chancel, the south chapel, and south aisle were rebuilt and the porch and spire added. In the second half of the 15th century there was a great reconstruction, which included the rebuilding of the chancel arch and the west arch of the chapel, the insertion of new windows in the lateral walls of the chancel and south aisle, a vestry on the north of the chancel, since destroyed, and the rebuilding of the is two-centred and of two orders with moulded caps and bases and marks on the soffit for the fixing of a wooden tympanum. There are three carved stone corbels on the north and three rough wooden ones on the south for the posts of the former roof; the present low-pitched roof is modern. There are triple and graduated sedilia with continuous mouldings of the 13th century. The chapel has a five-light east window with flowing tracery of the 14th century. There are three windows in the south wall, the second of which is now blocked; they are of three lights sharply pointed and cinquefoiled under a square head. The 14th-

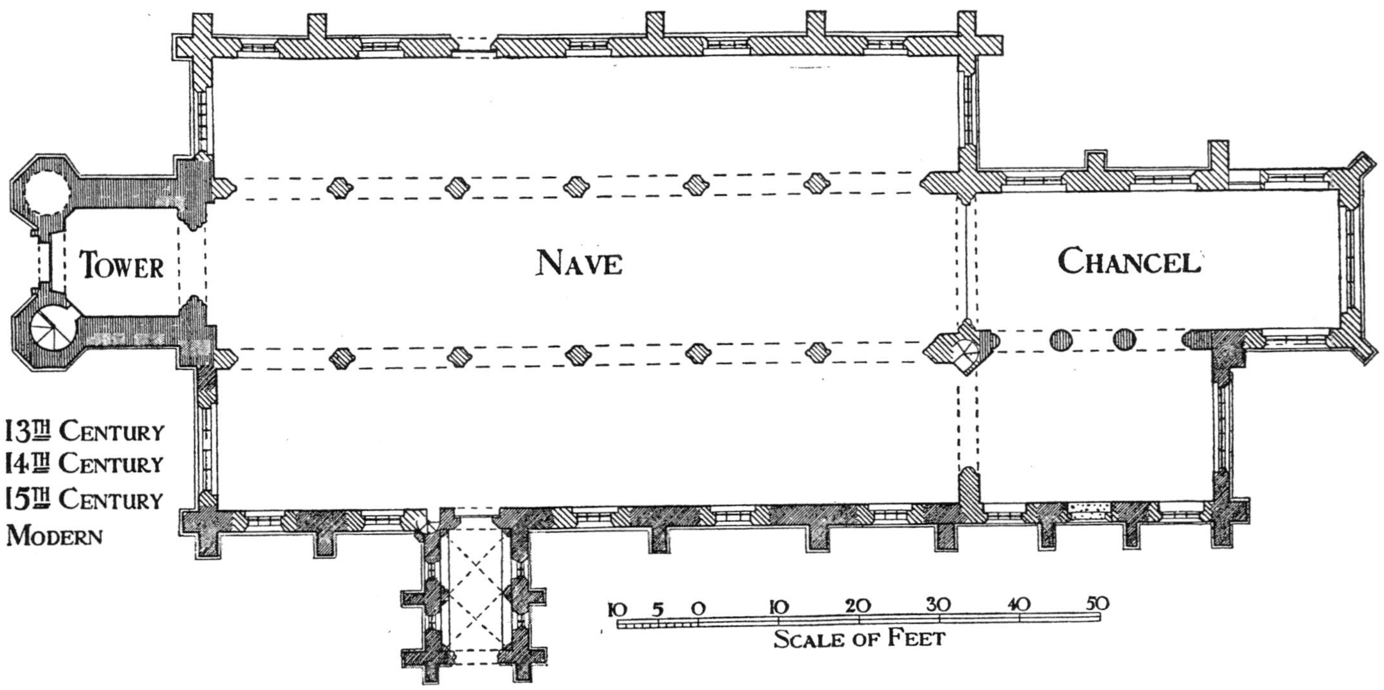

PLAN OF ST. LEONARD'S CHURCH

nave arcades and the north aisle. In the 19th century there was considerable restoration and the roofs were renewed, with the exception of that of the north aisle. The spire was rebuilt in 1901.

The east end of the chancel is most effective. There is a large window of four lights with a massive forked central mullion and geometrical tracery of elaborate design; above in the gable is a trefoiled opening set in an arch. The gable has pierced and crocketed coping and there are diagonal buttresses with crocketed gables. There was formerly a vestry at the east end of the north wall, the plain 15th-century doorway of which remains and part of the west wall incorporated in a buttress. There are three windows on the north and one on the south, all 15th-century insertions of three lights; the north-east is curtailed to clear the roof of the former vestry and has cinquefoil-headed lights, while the others have transoms, cinquefoiled main lights and rectilinear tracery. There is a 15th-century buttress on the north with two set-offs and another formed of part of the west wall of the destroyed vestry. The chancel opens to the south or Swaine chapel (see below—Guilds) by a mid-13th-century arcade of three bays with two centred arches with deep mouldings and clustered piers and responds having moulded caps and bases. The lofty 15th-century chancel arch

century buttresses have two set-offs. There is a plain brick parapet. The arch communicating with the aisle is similar to the chancel arch and of 15th-century date. There is a piscina with crocketed cinquefoiled head, two shields above and an embattled cornice of the first half of the 14th century. To the north of the east window is some 15th-century stone panelling with a mutilated canopy, which probably formed part of the reredos. The roof is modern.

The nave arcades are of six bays with wide and lofty two-centred arches of two orders, the outer of which continues to the ground, while the inner rests on shafts with moulded caps and bases. The clerestory windows are of three cinquefoiled lights under a square head. There are fine stone roof corbels carved with grotesques resting on semi-octagonal shafts which rise from the piers, but the roof itself, of the queen-post type, is modern. Three modern straining arches of wood span the nave beneath the clerestory. The rood stairs are contained in a turret at the south-east angle of the nave with upper and lower doorways, the latter approached from the aisle; the lower doorway has continuous mouldings and a hood terminating in the heads of a king and queen, and a plain contemporary door, all late 14th century. The turret is continued above the nave roof with a conical cap and finial. The tower arch is two-centred and of two orders with foliaged caps and moulded bases to the responds. The line of the high-pitched 13th-century roof is visible on the wall above the arch.

The north aisle appears to have been almost entirely rebuilt in the second half of the 15th century, and the windows and buttresses with two set-offs are all of this period; the east window has five cinquefoiled lights under a depressed head, and the west is similar; the five north windows have three cinquefoiled main lights. There is a plain contemporary parapet. The north doorway is of the 15th century with continuous mouldings and a hood; the original door with carved paterae round the frame remains. The roof is of lean-to type with braces and pierced spandrels, and is probably of the end of the 15th century. There is a piscina recess with continuous moulding. The south aisle has five windows in the lateral wall of three cinquefoiled lights under a square head, and a west window of four lights uncusped under a depressed head. All are late 15th-century insertions. The buttresses are of the 14th century with two set-offs similar to those of the south chapel, but much renewed. The roof is modern.

The porch, a splendid design of the first quarter of the 14th century, is of two bays with sexpartite vaulting springing from shafts with embattled caps and moulded bases. The outer arch is two-centred and of two orders with moulded caps and bases and an ogee hood-mould with flanking pinnacles; there are two windows of two lights on each side and stone benches within. Above is a parvise, which has a narrow rectangular opening on the east and west, and a two-light window with an ogee hood and flanking pinnacles on the south and over all an empty niche. The parvise is approached by a newel stair contained in a turret at the north-west angle and with a lower door opening to the aisle. There are angle buttresses with two set-offs and niches with crocketed canopies, pinnacles, and finials. The gable has crocketed coping, a finial, and angle pinnacles. There is a buttress on the east and west with gabled top, two set-offs, and a niche on the face. The parvise is covered by a pointed stone roof supported on stone ribs with an external roof of stone slates and a stone cresting pierced with a running pattern. The inner doorway is two-centred with deep mouldings and plain moulded caps and bases; the door is contemporary with plain hinges and an 18th-century wicket inserted.

The tower has four stages, the three lower of the 13th century and the top one of the early 14th. At the west angles are large octagonal buttresses, which terminate at the top of the third stage and support angle buttresses with gabled tops. The projecting west doorway has a two-centred arch with a gable above in which is a niche with the much-weathered figure of an ecclesiastic. The ribbed doors are probably late 13th century, but were badly handled in the 19th century and daubed with dark paint. The second stage has two lancets on the west with banded angle-shafts, and similar arcading of three bays on the north and south, the centre arch being pierced; the third stage has two lancets with clustered angle-shafts on the north, south, and west, those on the north being now bricked up; the top stage has two-light belfry windows with trefoiled heads and trefoils above, there is an embattled parapet and embattled octagonal turrets at the angles; the beautiful early 14th century spire has canopied windows of two lights. There are newel stairs in both the west angles, the approach being by a door in the south-west angle, and there is a connecting passage to the north stair along the sill of the west windows.

The 15th-century font, standing on risers, has an octagonal bowl and shaft; round the former are seated figures of saints under crocketed canopies and resting on foliated brackets, the panels being separated by pinnacled buttresses; the shaft has niches occupied by standing figures with folded hands with spreading foliage above, and the base is ornamented by paterae.

There is a 15th-century wooden eagle lectern with a hexagonal stem supported on lions couchant; the colouring has been restored. Inserted in the modern reading-desk are two 15th-century tracery heads, probably from stall desks. There is a plain oak chest probably of late-17th-century date.

The remains of 15th-century painted glass are most noteworthy. In the east window of the Swaine chapel are three panels, formerly in the south-east window of the chancel: in the centre is a Pieta beneath a flat canopy and on either side a kneeling knight and lady representing Sir Bartholomew Everard and Agnes (Flete) his wife and Sir Lawrence Everard and Margaret (Colvile) his wife; upon the mantle of the latter are the arms of Colvile, and beneath one of the ladies the word Margaret; over each group is a couplet:

Jesu fro Sine make us fre
For Jonn's love wot baptisyd ye

and

Lady lede us wele fro Harm
To hym that lay ded in your Barm;[11]

This glass was restored in 1908. In the central window on the north side of the chancel is good, but much restored, glass with figures of St. John Evangelist, St. Matthew, St. Philip, St. Andrew, St. Peter, and St. James the Greater; in the tracery are the arms of St. John's College, Cambridge, Ely, Drake, and Drake with a chief (or a mullet between two crosslets sable); the heraldry dates from the restoration of the glass in 1924. The east window of the north aisle is filled with glass, extensively restored in 1900, depicting the Tree of Jesse. The figures are placed in ovals within the loops of a vine; the kings bear scrolls and have attendant prophets; thirteen of the figures are original, seventeen partly restored, and thirty-one entirely modern. There are slight remains of mural paintings in the chancel consisting of 17th-century texts and borders, and in the north aisle is an 18th-century text and border. There is the base of a cross in the churchyard near the porch.

The Swaine chapel, the walls of which are adorned with fine monuments and hatchments, contains the family vault of the Swaines. Thomas Crosse (d. 1633) is buried in the chapel. Elsewhere in the church are monuments to Anne, relict of Thomas Crosse, John Crosse III (d. 1704), John Crosse IV (d. 1743). John Lumpkin (d. 1743), of Park House, his sons Anthony (d. 1747) and Nicholas (d. 1748) and Capt. Anthony Lumpkin (d. 1780) son of the last are also buried in the church. John's grave is before the chancel steps.

The plate includes a chalice, two patens, a large plate, and a flagon all of silver and inscribed: 'Leverington/Presented by J. H. Sparke, M.A./to replace the Ancient Service/The gift of Mrs. D. Spelman/Sacrilegiously stolen from the church/A.D. 1831.'

[11] As given by Cole (B.M. Add. MS. 5805, f. 112).

The tower contains six bells: 1st and 3rd by Taylor, 1871; 2nd by Osborn and Dobson of Downham, 1803; 4th and 6th by Lester and Pack, 1766.

The registers begin in 1558 and are complete. In a chest, kept in the north aisle, is a valuable accumulation of church records. Another accumulation, repaired in 1950, is kept in the church safe. It covers the period from Henry VII to Elizabeth.

The church of *ST. PAUL, GOREFIELD*, erected in 1870, is a stone structure in the 'Early English' style consisting of chancel, nave, vestry, south porch, and west turret containing one bell.

Leverington Rectory is a handsome Georgian building, with portions belonging to an earlier date. It is partly enclosed in a high wall of mellowed brickwork. A dovecote formerly belonged to it.

The following have been rectors: John Warkworth (d. 1500), Master of Peterhouse and reputed author of a chronicle of the reign of Edward IV, who became rector in 1473 and made several bequests to the church; Thomas Yale (? 1526–77), ecclesiastical lawyer; John Warren (1730–1800), Bishop of St. Davids and later of Bangor; James Nasmith (1740–1806), an antiquary who was also one of the clerical magistrates who have been so prominent in this county; J. B. Jenkinson (1781–1840), rector 1812–17 and later Bishop of St. David's and Dean of Durham. Richard Reynolds (1674–1744), Bishop of Bangor 1721–3 and of Lincoln 1723–44, was the son of a rector.[12]

GUILDS Leverington had three guilds, of St. Mary, St. John and the Holy Trinity. The first of these was founded in 1386.[13] The guildsmen used the chapel now known as the Swaine chapel in Leverington church. The guilds of St. John and Holy Trinity are mentioned in the will of Richard Adam of Leverington (1528).[14] Adam left a dozen young swans every year for four years, the profits to be applied at the discretion of the aldermen and brethren of the guilds, and 6 acres in Pope's Lane Field, Leverington, for the priest of St. John's Guild to pray for his soul.

The guildhall, which occupied the site of the present school (see below), belonged to St. John's Guild. In 1525 a loft was added to it, and in 1529 it was let to Richard Salter the parish clerk on a five-year lease.[15] In 1549, when the tenant was William Taymer, the building was bought by William Warde.[16] It eventually became vested in the Hawkins family; in 1713 William Hawkins bequeathed it to the Leverington Feoffees, subject to a life annuity of his wife Sarah. Other property of St. John's Guild was sold in 1568.[17]

It is possible that this guild may have had some connexion with the Hospital of St. John Baptist in Leverington. By 1686 the hospital had 'long ceased to exist' and all its endowments had been 'swallowed up.'[18] Its name, however, survives in Spital Field, and it is conjectured that the almshouses in Little Dowgate (formerly Poorhouse) Lane, which adjoins this field, may have been built on its site.

NONCONFORMITY In 1851 there were two Nonconformist chapels in the present parish, one at Gorefield, belonging to the Independents and dating from 1834, and the other at Leverington village belonging to the Primitive Methodists and dating from 1843.[19] Both chapels still stand. The Methodist chapel at Leverington was rebuilt in 1885.[20] The old chapel was then converted into a Sunday school; in 1904 a new Sunday school was built, but the original building is still in use for general purposes.

SCHOOLS Provision for education was made in Leverington earlier than in many parishes in the Isle, thanks to the extensive and valuable Town Lands. In 1789 the feoffees of these lands provided a school giving free education to the children of occupiers rated at less than £30 a year; the master's salary was £25.[21] The school was held in 'an ancient house called the *Town-hall*',[22] which had once been the headquarters of the guild of St. John and was later the parish workhouse.[23] When Dr. Jobson, Vicar of Wisbech, reported in 1814 on the schools of Wisbech hundred, the master was teaching reading, writing, arithmetic, and the catechism to about 17 boys and 8 girls. He was said to be competent, although Jobson described his £25 salary as 'miserable' and expressed doubts whether the charity estates, which at that time produced about £200 a year, were properly administered. At this date there were 134 children of school age in the parish, 84 of whom were of poor parents.[24] Shortly afterwards the Town Lands, 111½ acres in extent, were let for £285 yearly, but when the Charity Commissioners reported (1837) for £239 only. By this date the schoolmaster's salary had been increased to £40, and there were two assistant mistresses who received £8 each. Between them they staffed three day schools and one Sunday school. The original school provided free education for 15 pupils and a varying number of fee-payers; the two new day schools had 20 and 10 children, and expenses were equally divided between the parents and the feoffees. The Sunday school, attended by about 30 children, was entirely supported by the feoffees.[25] The National Society's inquiry of 1846–7 showed 58 children at school in Leverington.[26]

In 1860 the long and useful life of the guildhall was brought to an end, and a new school with teacher's house built by the feoffees. Alterations and enlargements in 1885 and 1891 brought the accommodation up to 134 (84 mixed, 50 infants). In 1906 the school was transferred to the County Council. In 1922 St. Leonard's parish hall was temporarily hired to cope with an overflow from the village school, and three years later two new classrooms were provided at a cost of £1,245, increasing the accommodation to 201. In 1948 the school was 'decapitated', the older children going to the Queen's School, Wisbech.[27]

The rent paid by the County Council for the master's house at Leverington school, with a further fund producing about £40 a year, is managed by a special body

[12] *D.N.B.*; Reynolds was described by Philip Doddridge as 'a valuable person' (*Dict. Engl. Ch. Hist.*). Nasmith 'distinguished himself by his edition of Bishop Tanner's *Notitia Monastica* and other works' (Gardner, *Dir. Cambs.* 552).

[13] W. M. Palmer, *Village Gilds of Cambs.* 360 (printing the 1389 certificate).

[14] P.C.C. 31 Porch.

[15] Leverington Par. Accts.

[16] *Cal. Pat.* 1548–9, 326.

[17] C 66/1046, no. 57.

[18] T. Tanner, *Notitia Mon.* (1744), 56.

[19] H.O. 129/7/193.

[20] Kelly, *Dir. Cambs.* (1900).

[21] Bp.'s Mun., Ely, B 6.

[22] Gardner, *Dir. Cambs.* (1851), 553.

[23] *Rep. Com. Char.* (*Cambs.*), [103], 265–9, H.C. (1837–8), xxiv.

[24] Bp.'s Mun., Ely, B 6.

[25] *Rep. Com. Char.* (*Cambs.*), s.v. Leverington.

[26] Archives of Nat. Soc.

[27] M. of Ed. File no. 12/28.

of trustees known as the Town Lands Educational Foundation, on behalf of the inhabitants of Leverington and Parson Drove. In the past the income was applied in various directions for which the state is now responsible, and it is now used in aiding parents and schoolchildren in cases where assistance is unobtainable from elsewhere.

The subsidiary schools mentioned by the Charity Commissioners seem to have lapsed quite early,[28] and a School Board was compulsorily formed in 1875[29] under the Education Act. This Board built a school at Gorefield in 1877. The accommodation (108 places) became inadequate, and in the latter part of the First World War the infants' classes were held in the parish room (erected in 1904).[30] In 1925 three new classrooms were built at a cost of £1,520, and the accommodation increased to 170. The school became one for junior mixed and infants in 1948.[31]

CHARITIES[32] Leverington is well endowed with charities, which are administered by trustees known as the Leverington Feoffees. The feoffees control both the ancient charities derived from the Town Lands and charities subsequently established by various benefactors.

The full number of appointed trustees is twelve who hold office for life, and fresh appointments are made when the number is reduced to six. The two Churchwardens serve ex-officio, and three representative trustees are appointed by the Parish Council. In former days the charities were managed in rotation by one of the feoffees known as the town bailiff, who held such office for one year only; but this practice has long been discontinued and the administrative duties are now carried out by the clerk.

In connexion with proceedings in Chancery in 1696 a search was made for early documents relating to the charities. Nothing, however, was traced except a deed of 1557 appointing new trustees and the Chancery Master was only able to report that there were 117 a. 1 r. of the ancient endowments, of which about 13 a. 3 r. had been given in Henry VIII's reign to the Rector of Leverington on condition that he should pray for the soul of the donor and his ancestors; and that the remaining 103 a. 2 r. had been held by feoffees upon various trusts including the repair of the parish church.[33]

We are hardly better able to trace the accumulation of the Town Lands than was the Chancery Master of 1696. A close study, however, of the wills of 18 Leverington inhabitants, proved in the Ely Consistory Court[34] between 1450 and 1460, shows that in nearly every case money or the proceeds of the sale of land was given either in alms, or in various ways for the benefit of the church or of those serving it. Moreover the origin of the charity land in Oxfield is now known; for in 1503 William Digby of Godmanchester (Hunts.), son of John Digby of Leverington, quitclaimed to the 'provost' of Leverington church and to the inhabitants his right in a messuage and land in 'le Oxfield' or Fytton Field.[35]

The charity monies so bequeathed in early days were not, it seems, in all cases immediately distributed, but were, here as elsewhere, used to provide a stock which was lent to husbandmen and others to help them in starting or carrying on their business. Interest on the stock was charged, and two bondsmen were required to ensure repayment. These loans were generally arranged on or about Plough Monday.

Another form of stock consisted of cows which were hired out for the year at an agreed rent; and for poor people unable to buy cows this must have been of great value in providing milk and calves. The uninclosed commons afforded grazing. For many years the accounts of the churchwardens, feoffees, and parish officials were intermingled. It was not until the 17th century that distinct accounts began to be kept. The acreage above referred to cannot be fully reconciled with the existing acreage, for the standard acre differed from that used at an earlier period; moreover, additions have been made on the inclosure of the marshes, of the commons, and of the land in Parson Drove Fen; furthermore, certain sales have taken place and the proceeds have been reinvested in the purchase of other lands or in stock.

There are or were the following charities distinct from the Town Lands:

The origin of Charlton's charity, or (more correctly) Charton's, is obscure, but the most probable founder was Philip Charton, buried at Leverington in 1597. The income from this charity is given annually in money to poor people on the Sunday before Christmas.

Hawkins's charity was established by the will of William Hawkins of Leverington in 1713. Hawkins gave to the feoffees the messuage known as the town hall, upon trust that they should yearly 'new-clothe two poor men and two poor women on Christmas Day at the Town Hall and should allow each of them half a pint of strong beer and a halfpenny loaf for their breakfasts and half a pint of strong beer for each after dinner and allow the landlord or landlady one shilling each for their dinners, and also allow the same to their clerk, who should dine with them and go before them into the church both at morning and evening service and place them decently, as was usually done in Wisbech Church on like occasions'. The beneficiaries were obliged to qualify by at least five years residence in Leverington, and for greater secrecy the election was to be held in the little parlour. He also gave an acre of land in Gorefield Fen containing three riggs to buy books for any person to read that might come to church before the beginning of morning or evening service, the books to be chained.

Criplin's charity was founded by the will proved 1718[36] of Thomas Criplin of Leverington, who charged his estate, subject to certain remainders, with a payment to two poor widows in Newton and Leverington, in alternate years, and an annual payment of 20*s*. to the rector and churchwardens of Leverington. The estate fell into possession of the feoffees in 1833, and consisted

[28] There is no mention of any school in Leverington, other than the endowed school in the guildhall, in Gardner, *Dir. Cambs.* (1851).

[29] *Lond. Gaz.* 1875, 5809.

[30] Kelly, *Dir. Cambs.* (1933).

[31] M. of Ed. File no. 12/29.

[32] Information concerning the charities is derived mainly from the records of the charities, the par. accts., and the report of the Charity Commissioners ([103], 265–9, H.C. (1837–8), xxiv).

[33] In 1702 Henry Perne, Rector of Leverington, quitclaimed to the feoffees his rights, as rector, to the 13 a. 2 r. of land. The date 1696 is derived from a deed of 21 Nov. 1827 which cites the report and decree. No copy of the decree has been traced in the Chanc. Entry Bk. of Dec. and Ord. *Rep. Com. Char.*, p. 266, gives 1796 in error.

[34] Now at Peterborough.

[35] Deed 7 May 18 Hen. VII (F. 74, D. and C. Wigorn docs.).

[36] Proved in Archdeaconry Court of Ely.

of 4 a. 3 r. 16 p. in Wratfield and 5 a. 1 r. 10 p. of pasture land in Park Field, Leverington.[37]

Anne Crosse's charity. Anne (d. *c.* 1681), relict of Thomas Crosse of Leverington, left £20 to be invested in land for the benefit of the poor of Leverington. It was lost by 1837.

Thomas Swaine's charity. Thomas Swaine of Leverington Hall by will dated 1725 charged 5½ acres of land in Allen's Drove, Leverington, with the payment of 40*s.* a year, at his father Robert's desire, for the provision of 2*d.* wheaten loaves of bread, to be given to poor people. It is still in force.

John Swaine's charity. In 1735 John Swaine of Leverington, son of the last, gave 4 acres in New Field, Leverington, called the Long Four Acres, and 5½ acres at the Horse-Shoe, in Leverington Marsh, to provide boots and clothing for poor people.[38] The election of beneficiaries was to be made on each first Sunday in September after evening service. The greater part of the land at the Horse-Shoe was required by the railway company when the railway line was being laid and was sold to good advantage, the proceeds being invested in the purchase from John Chamberlain of a house and two acres on the west side of the Sutton Road. This house, but not the land, was sold in 1950 and the proceeds invested in government stock. The remaining 1 a. 1 r. 20 p. of land at the Horse-Shoe has also been sold and the proceeds also invested in government stock.

Griffin's charity. Joseph Ellard Griffin of Gorefield, by will proved[39] 1895 gave to the feoffees the sum of £600. The money was invested in the purchase of 9 acres of land on Leverington Common and the annual income is applied according to the discretion of the feoffees. In exercising this discretion they have generally favoured such poor people as may have suffered unexpected misfortune through the loss of a horse or cow.

Almshouses. The feoffees administer twenty almshouses in Little Dowgate and Leverington Common. Three of these were purchased from Abraham Leahair, the village schoolmaster, in 1825. Six or eight cottages, or replacements thereof, were derived under the will of Thomas Crosse, proved 1633.[40] Crosse devised four newly erected tenements, with two others to be later erected upon the piece of ground called the Midd Feather to be used as dwellings for poor inhabitants of Leverington for ever. The original trustee and his successors were to keep these buildings in repair. The origin of the remaining almshouses is lost. The almshouses are let free of rent to poor people in Leverington.

CHAPELRY OF PARSON DROVE

Parson Drove, originally a township and chapelry of Leverington, became a separate ecclesiastical district under the Leverington Rectory Act (1870). It had long maintained its own poor,[1] and its separate existence for civil purposes was virtually recognized in 1874, when a separate Parson Drove School Board was established.[2] The parish consists in effect of the fen end of Leverington; it has no separate manorial history, and may be compared in status with Gedney Hill over the border in the Parts of Holland. About 1,000 acres in Inkerson Fen and its neighbourhood were transferred from Holland to the Isle and added to Parson Drove in 1934.[3] The added area is at the extreme west of the parish; it had formed a projecting salient into the Isle between Parson Drove and Thorney, and had been counted in with Holland as the only road communications were northward.

The scattered village is situated about 6½ miles west of Wisbech and 4½ miles south-west of Leverington. The network of roads is close, but few of them are of more than local importance. The March–Spalding line of British Railways, Eastern Region, opened in 1867,[4] crosses the western part of the parish diagonally. There are no stations in the parish, but that of French Drove and Gedney Hill is very near the north-west corner and those of Murrow (East and West) are just outside on the south. Some of the houses of Murrow village are in Parson Drove civil parish, and one of the local ecclesiastical districts (Southea cum Murrow) extends about equally into Parson Drove and Wisbech St. Mary.

Fendyke Bank, the great bank stretching from Cloughs Cross on the Lincolnshire border southwards to Guyhirn, is one of the most important in the district. For many hundreds of years it was the bastion of defence against the fresh waters coming down from the upland counties, and the landward counterpart to the old sea bank on the east side of Leverington. Fendyke protected the whole district on the north side of Wisbech which includes Tydd St. Giles, Newton, Leverington, Guyhirn, and Wisbech St. Mary. Its importance cannot therefore be exaggerated, and the most stringent measures were taken to ensure its safety. The obligation to maintain the bank was imposed on the landowners in the protected parishes.

A great breach was made in the bank in 1437, when 13,400 acres were flooded through the default of one Thomas Flower, the owner of 24 acres in Wisbech High Fen.[5] Further breaches occurred in 1570[6] and in 1770. At the latter date a gap 130 yards wide was made, probably at Abel's Gull. Parts of the country-side were flooded to a depth of 6 feet, and were not brought back into cultivation for three years. So sudden was the disaster that some fled for their lives to Thorney Abbey and the higher lands around.[7]

The Fendyke may be said to mark the boundary between the 'peat' and 'silt' portions of the parish. The former, comprising Parson Drove Fen, has always been less highly valued and was formerly used mainly as sheep pasturage; it is sparsely populated. The latter, which is the area of ancient settlement, forms part of the Wisbech fruit-growing and market-gardening district.

37 Thomas Criplin is buried near the south entrance to Leverington church. His gravestone has recently been restored by the Feoffees.

38 Swaine gave elaborate directions concerning the quality of the clothing to be provided, and the price to be paid for boots, overcoats, &c.

39 In the Peterborough District Probate Registry.

40 P.C.C. 91 Russell.

1 When the Charity Commissioners reported in 1837 there were 19 'town houses' in Parson Drove, distinct from those of Leverington: *Rep. Com. Char. (Cambs.)*, [103], 265, 269, H.C. (1837–8), xxiv.

2 *Lond. Gaz.* 1874, 6277.

3 Under the M. of H. Provisional Ord. Conf. (Ely, Holland & Norf.) Act, 1933.

4 *V.C.H. Cambs.* ii, 132–3.

5 W. Dugdale, *Hist. Imbanking*, 328. The lands flooded were in Wisbech (4,400 acres), Leverington (4,600 acres), Newton (1,400 acres), and Tydd (2,000 acres).

6 R. Holinshed, *Chron.*, ed. 1808, iv, 255.

7 W. Watson, *Hist. Wisb.* 388; *Fenland N. & Q.* ii, 235.

The 2-mile road called Parson Drove or Parson Drove Gate, along which the nucleus of the village is built, was formerly a green drove and wider than it is now. The inclosure of pieces of common land beside the road has brought it down to its present width; a small strip of common remains at the western end, and the former extent of the rest of the commons is still clearly visible. A fence used to stretch across the eastern end at Gates End Bridge, to prevent cattle straying upon Overdyke Bank.

Pepys, who visited Parson Drove on 17 and 18 September 1663, described it as a 'heathen place' where he had to sleep in a 'sad, cold, stony chamber in a miserable inn'. His visit was made in connexion with the affairs of his deceased aunt Beatrice, relict of John Day of Wisbech,[8] who held much property in Leverington, especially in Outnewlands Field and Fen Croft.[9] 'Uncle Perkins', mentioned by the diarist as then living in Parson Drove in a poor way, was the husband of Jane Pepys, the diarist's aunt.[10] The inn in which the diarist lodged was the Swan which in 1834 belonged to Charles Boucher, a brewer, who altered it drastically. Pepys, who was very susceptible to environment, reacted unfavourably to the Fens. The roads, houses, living conditions, even the gnats from undrained swamps, come in for severe criticism.

A woad mill at Parson Drove[11] near the vicarage was one of the last of its kind to remain at work in England. It continued in operation until about 1910, when the farm was sold to the Isle of Ely County Council for small holdings. The structure of the mill, as described in 1899,[12] resembles the earliest type of woad mill; doubtless the pattern was handed down. It was built with walls of sods, 3 ft. thick at the bottom and narrowing at the top, arranged in herringbone fashion. The roof was of circular cone-like shape, made of timbers and hurdles thatched with reeds. The grinding-wheel was in the middle of the building, and the baskets of woad were emptied into the hollow centre, where the plant was crushed by a horse-drawn wheel and made into a pulp. It was made up by hand into balls as large as a Dutch cheese and left exposed to the weather in open sheds until the balls shrank to the size of oranges. All around were the skeleton drying-sheds six or seven stories high, rudely made of poles and hurdles, and the balls were laid on twigs of twined hazel, called fleaks, until dried. The last owner of the Parson Drove mill was Fitzalan Howard of Holyrood House, Spalding (Lincs.).

THROCKENHOLT is a farm and cluster of houses on the Lincolnshire border about a mile north-west of Parson Drove village. It is first authentically referred to in *c.* 1133–51, when Bishop Niel of Ely granted a square mile (*milliarium*) of marsh, first called 'Everdewike' but afterwards 'Trokenholt', to Thorney Abbey.[13] This grant was confirmed by Bishop Longchamp (1189–97) and by Edward III in 1348.[14] The land was given for the service of God,[15] and it seems to have been hoped that some kind of hermitage[16] or cell would be established. At all events a chapel was built which survived until *c.* 1540, when it is shown on a map of Wisbech hundred.[17] This chapel was where Throckenholt farm house now stands; fragments of stone, bones, and other relics have from time to time been uncovered on the site. The map marks the eastern and southern boundaries of the site as heavily wooded, and shows in its cartouches that in 1274 and later Throckenholt was claimed by both Lincolnshire and Cambridgeshire.

In 1792 the farm belonged to Abraham Ulyat, who built the present house in 1806. John and Henry Ulyat are recorded as farmers in Parson Drove in 1851.[18] Throckenholt Farm contained 209 *a.* 2 *r.* 31 *p.*, and subsequently belonged to John Goodman.

In the middle of the 19th century Throckenholt was described as a 'wide bleak fen, productive indeed, but with no other recommendation to a civilized being.'[19]

CHURCHES. The chapel of St. John the Baptist, since 1870 (see Leverington) the parish church of Parson Drove, attained at a comparatively early date a more independent status than was usual with parochial chapelries. Burial rights were granted by a composition between the rector of Leverington and the inhabitants of Parson Drove in 1397, though the latter had to observe the dedication feast of the mother church, and present their chaplain to the rector for admission. The chaplain was required to live at his own expense, to undertake the repair of the chapel with its chancel, tower, and churchyard enclosure, and to go in procession to the mother church on Wednesday in Rogation Week.[20]

In 1459 the endowments of Fitton chantry (see Leverington—Manors) were transferred to Parson Drove and a chaplain, Adam Silk, licensed to celebrate in both chapels.[21] In 1487 the Bishop of Ely directed his clergy to receive kindly the proctor of the chapel whenever he should visit their parishes soliciting alms therefor, and granted an indulgence of forty days to all contributors.[22]

The following year another indulgence was granted for the repair of a chapel of St. John the Baptist in a hermitage in Leverington. It is possible that this was merely a repetition of the previous indulgence, but there is a persistent tradition that a hermitage existed opposite Parson Drove Church, on the site of the present vicarage, and the indulgence may refer to this other hermitage.[23]

The system of presenting the chaplain to the rector, at whose will he was removable, lasted until 1749. By a decree of this date the right of appointment was vested in the feoffees of the former Fitton Hall lands which provided the endowment.[24]

The Revd. Henry Pujolas, incumbent from 1692 to 1749, was an ejected Huguenot. Cole described him as

[8] John Day was buried at Wisbech in 1648. Beatrice's will is dated Nov. 1650.

[9] Wisbech MSS.

[10] So referred to in wil lof Robert Pepys of Brampton (1657).

[11] For a fuller description see *Fenland N. & Q.* ii, 93, and *Country Life*, 20 Apr. 1951.

[12] F. J. Gardiner, *Hist. Wisb.* 363, contains an account of the mill by Miss Peckover, late of Harecroft House, Wisbech. Relics of the Parson Drove woad industry, with paintings and a descriptive account, are exhibited in Wisb. Mus.

[13] Dugd. *Mon.* ii, 600–1, quoting Red Bk. of Thorney, f. 11. The authenticity of Wulfhere's charter to Peterborough (664), which mentions Throckenholt, is doubtful.

[14] *Cal. Chart. R.* 1341–1417, 82.

[15] Dugd. *Mon.* ii, 600.

[16] The Red Bk. of Thorney gives a history of Throckenholt which, while not accurate in detail, is probably based on genuine tradition; it is stated *inter alia* that Throckenholt had been used as a hermitage since 1107.

[17] In Wisb. Mus. The map is No. 44 in the list of Fenland maps given in *V.C.H. Hunts.* iii, 299, 306.

[18] Gardner, *Dir. Cambs.* (1851), 554.

[19] Ibid. quoting Walker & Craddock, *Hist. Wisb.* (1849).

[20] Ely Reg. Fordham f. 187.

[21] *Ely Dioc. Remembrancer*, No. 101, p. 260. Collation followed by induction; licence was granted for him to celebrate in the chapel and to minister to the inhabitants of Parson Drove at a competent salary.

[22] Ely Reg. Alcock, f. 13.

[23] Ibid., f. 31.

[24] J. Bacon, *Liber Regis* (1786), 241.

'a little black man' with an imperfect command of English even at the age of 83.[25] Like Pujolas, several families of French origin, mainly from Guisnes and Picardy, settled in Parson Drove. Some names of French origin survived in the parish to within the last hundred years.[26]

As late as 1837 it was reported to the Charity Commissioners that 137¼ acres, producing £301 13*s.* a year, were devoted to the endowment of the chapel. This sum constituted the incumbent's stipend, after the deduction of £25 to £30 a year for church repairs and 40*s.* given to four aged persons.[27]

probably belong to the first quarter of the 16th century. The clerestory consists of six three-light windows on each side set over the arches with cinquefoiled heads under a square label.

The south aisle has an east window of three cinquefoiled lights under a depressed head and hood-moulding of 15th-century date. The south wall was entirely rebuilt early in the 19th century and contains plain pointed windows with wooden frames. The west window is of two cinquefoiled lights with an octofoil in the head and a hood-mould terminating in heads.

The 15th-century porch has an outer doorway with

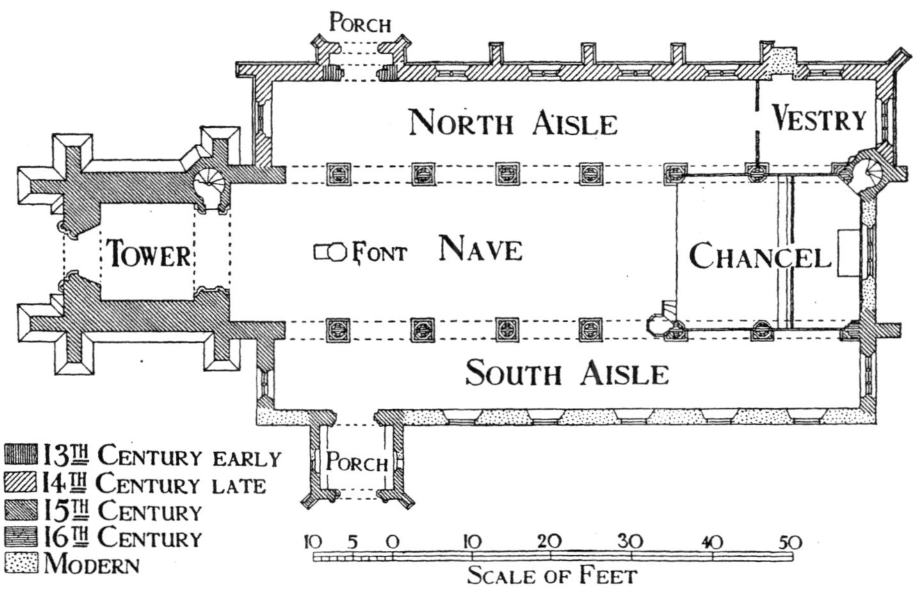

Plan of the Church of St. John the Baptist

The church of *ST. JOHN THE BAPTIST* consists of clerestoried nave, aisles, north and south porches and west tower. The chancel was destroyed by floods in 1613.[28] The material is mainly rubble and brick, but the tower is partly faced with Barnack stone. The nave roof is slated and the roofs of the aisles and south porch are leaded. The earliest portion of the present fabric is the north doorway, which belongs to the first half of the 13th century. The north aisle and porch are of 14th-century date and the rest of the fabric of the latter part of the 15th century. The arcades and clerestory may be even later. Early in the 19th century the lateral wall of the south aisle was rebuilt. In 1895 the nave roof was reconstructed and an east window inserted in the blocked chancel arch.

This arch is of two orders with moulded caps and bases to the responds and probably dates from the 15th-century reconstruction. The modern east window is of three lights with tracery of 14th-century character. To the north of the chancel arch is the stair to the rood loft; the upper doorway is blocked and the lower has a four-centred arch and continuous mouldings under a hood. The arcades are of seven bays with four-centred arches of two orders resting on clustered columns with moulded caps and bases. The west bay on each side consists of only a half arch returned against the buttresses of the tower. These arcades are very late in style and

a two-centred arch of two orders and jambs engaged shafts with moulded caps and bases. There are diagonal buttresses with one set-off and in the east and west walls a window of two cinquefoiled lights under a square head, partly blocked. The 15th-century inner doorway is of two orders, the outer with a continuous moulding and the inner with engaged jamb shafts having moulded caps and bases.

The north aisle belongs to the third quarter of the 14th century and has an east window of three cinquefoiled lights with quatrefoils above under a segmental head. The contemporary buttresses are uniform with one set-off. There are five windows in the north wall, all of two cinquefoiled lights with octofoils in the head and a hood-moulding terminating in masks. The west window is similar. The north doorway belongs to the first half of the 13th century and is two-centred and of two orders with deep mouldings and angle-shafts. There is a piscina recess with acutely pointed head. The shallow north porch is of the 14th century with a plain outer arch having a continuous chamfer.

The tower is a particularly fine example of 15th-century design. It is of three stages with angle buttresses having four set-offs. On the base of the north-east and south-east buttress is a small recumbent figure carved in relief. The west doorway has continuous mouldings and a hood springing from shafts with

25 B.M. Add. MS. 5808, f. 119*a*.
26 Gardner's *Dir. Cambs.* (1851), s.v. Parson Drove, contains the names Aubin, Goult, Jellis, Kingerley (perhaps originally Quimperlé), Marchant.
27 *Rep. Com. Char.* (*Cambs.*), [103], 265, 269, H.C. (1837–8), xxiv.
28 *Fenland N. & Q.* ii, 18.

moulded caps and bases. The plain door is contemporary. There is a three-light window with cinquefoiled heads and a sexfoil above on the north, south, and west. The second stage has a lancet on each side. The belfry windows consist of two cinquefoiled lights with quatrefoils above and a hood-moulding. The parapet is embattled and there is a newel stair turret on the north-east which extends to the parapet. The tower opens to the nave by a lofty panelled arch, a feature rarely found in this part of the country. The entrance to the newel stair is contrived in the north respond of the arch and has a plain contemporary door. There is a recess in the north wall of the ground stage. The ground stage has good octopartite vaulting with bell-hole in the centre and carved bosses displaying the Tudor rose and other devices; the springers terminate in grotesques.

The nave has a plain roof with king-posts, struts, and wall-posts, and retains notwithstanding its reconstruction some old timbers of uncertain but earlier date. The north aisle has a plain lean-to roof with wall posts and is of 17th- or 18th-century date; there is a series of stone corbels, all plain except two which have grotesque heads. The south aisle has a plain lean-to roof probably of early 19th-century date. The lowpitched roof of the south porch has old rafters.

The font has a panelled octagonal bowl and is of 15th-century date. In the windows of the north aisle are fragments of painted glass of the 15th century including the arms of Ely, a Trinity shield, and a shield with three chalices on a red ground. The oak pulpit, which is dated 1677, has shallow carving on the panels; the base is modern and of deal. There is an iron-bound chest in the vestry of 16th- or 17th-century date.

The plate includes a communion cup of silver, 1599, and a paten of silver, late 15th century, the only unaltered piece of medieval plate in a Cambridgeshire church. It is circular with a sexfoil depression. In the centre is a Vernicle with an aureole. The spandrels formed by the sexfoil are filled with leaves.

The tower contains four bells by T. Osborn of Downham, 1787.

The registers begin in 1657 and are complete.

The present vicarage house was built in 1760 by the Revd. John Dickinson.

The church of *EMMANUEL, SOUTHEA*, erected in 1872 by the Ecclesiastical Commissioners, is a brick structure with stone facings in the Early English style, and consists of apsidal chancel, nave, north aisle, and turret over the chancel arch containing one bell. There are four handsome brass chandeliers formerly in St. Paul's Cathedral.

NONCONFORMITY A Wesleyan Methodist congregation was established at the Southea end of the village in 1833.[29] Its chapel, built in 1838,[30] is still in use. A United Methodist chapel was built on land purchased in 1865 from Thomas Meadows Fisher.[31] The building was closed after 1937[32] and was sold in 1944. It thereupon became a village hall and institute.[33] In 1952 the trustees of the institute sold the building, which is now used as a store. The Salvation Army started work in Southea between 1900 and 1904.[34]

SCHOOL In 1814 there were said to be 35 children of school age in Parson Drove, 25 of poor parents. The school, which had at that time been in existence for about twenty years,[35] was endowed with £24 a year from the rent of 8 acres of the common; this was used for the salaries of a master and mistress, for whom a house was provided. They were teaching 12 children free, and could take paying pupils. The subjects taught were reading, writing, arithmetic and the Catechism.[36] A generation (1846–7) later the numbers attending this school had increased to 83, most of whom came both on Sundays and weekdays.[37] The school was rebuilt in 1850,[38] and its endowment regulated by a scheme of the Charity Commissioners of 1858. It was not in association with the National Society, and had been closed for some years before 1874. In this year 'the scholastic arrangements for the parish were in a most lamentable condition' and a School Board was set up. The Board rebuilt the school for £900, to afford 180 places, reduced in 1910 to 147. In 1930 the buildings, which had been erected on a disused drain, were found to be dangerous. The seniors were therefore temporarily transferred to Newton and the juniors to Murrow, leaving the infants in the one safe classroom. The new buildings, which were designed to take the seniors from Murrow as well as all the Parson Drove children, provide 160 places; the cost was £4,579 for the school itself and £895 for a teacher's house. On the opening (in 1933) the school was renamed the Payne Council Primary School after Alderman J. W. Payne, J.P., Chairman of the County Education Committee.[39]

CHARITIES John Bend, by will dated 1593, gave a messuage (now the Butcher's Arms Inn), with 15 acres in South Inham Field, the rents of which were to be devoted to 'making a stock to set the poor people to work' and to be distributed to poor persons born in Parson Drove or who had lived there for six years without having relief. The property was increased by inclosures of new-drained fen, and in 1837 brought in £80 a year.[40] On the other hand two portions were sold off, 1 a. 2 r. 29 p. when the New Wryde Drain was made, and another part in 1866 to the Great Northern and Great Eastern Railways. The house itself was sold about 1948.[41]

In 1725 Thomas Swaine gave a rent-charge of £1 a year for bread for the poor of Parson Drove, as well as his similar charity for Leverington (q.v.).[42]

The poor of Parson Drove in 1837 received 6*s*. 8*d*. yearly under the charity of Margaret Bende (see Wisbech St. Mary).

29 H.O. 129/7/193. 30 Inscr. on bldg.
31 Deeds *penes* Messrs. Fraser, Woodgate & Beall, Wisbech.
32 Meth. Ch. Bldgs. Stat. Returns as at 1 July 1940; Kelly, *Dir. Cambs.*
33 Deeds *penes* Messrs. Fraser, Woodgate & Beall, Wisbech.
34 Kelly, *Dir. Cambs.*
35 *Rep. Com. Char.* (*Cambs.*), [103], 269, H.C. (1837–8), xxiv.
36 Bp's Mun., Ely, B 6.
37 Archives of Nat. Soc.
38 Kelly, *Dir. Cambs.* (1900), s.v. Southea.
39 M. of Ed. File no. 12/43.
40 *Rep. Com. Char.* (*Cambs.*), [103], 269, H.C. (1837–8), xxiv.
41 For an account of the charity see *Fenland N. & Q.* ii, 232.
42 *Rep. Com. Char.* (*Cambs.*), [103], 269, H.C. (1837–8), xxiv.

NEWTON

Newton is a village about 4 miles north of Wisbech, lying just west of the 'Roman Bank'. The parish, which in 1934 was increased by 551 acres from West Walton (Norf.),[1] is of the elongated shape usual in the marshland areas of the Isle and Holland, stretching from the light silt soils by the Nene through heavier silts around the village to peat on clay at the extreme western end in the fen. With an area of about 6 square miles it is smaller than the majority of Isle parishes.

Until 1934 the old course of the Well Stream formed the eastern boundary, which is also the county boundary.[2] The North Level Main Drain, the principal artificial watercourse of the district, forms part of the boundary between Newton and Tydd on the north-west. The boundary with Leverington on the south is not so clearly defined, and the small hamlet of Fitton End lies across this boundary. The Wisbech–Long Sutton main road (A 1101) crosses the eastern side of the parish. A second-class road (B 1165)—an alternative route from Wisbech to Spalding—forms the main street of the village, and there are several minor roads. The Peterborough–Wisbech–Sutton Bridge branch of British Railways (formerly Midland and Great Northern Joint), which was opened in 1866,[3] runs along the eastern border parallel to the Nene outfall and has a station about a mile east of the village, known as Ferry from the former crossing for foot passengers to Walton Dam. The centre of the parish is crossed from south-west to north-east by the main electricity transmission line from Peterborough to King's Lynn.

In 1839, out of a total of 2,771 acres of land in the parish surveyed in connexion with the tithe award, there were 1,751 acres of arable, 781 of meadow and pasture, and 62 acres of inclosed common. By an inclosure award made in 1849, 166 acres were distributed in small lots. A further 73 acres were distributed by a supplementary award of 1851.[4] At the present day a large part of the land is devoted to fruit, potatoes, and market-gardening, as in all the northern parishes of the Isle.

THE MARSHES Like Leverington, Tydd, and other parishes on the 'North side', Newton was protected from the sea by the ancient sea bank.[4a] By inquisition taken in 1437, a jury found that this bank beginning at Tydd Gote and extending to Bevis Cross in Wisbech ought to be kept 15 ft. high and 6 ft. wide at the top, and to be maintained by every tenant of lands in Wisbech, Leverington, Tydd St. Giles, and Newton according to the proportion of their holdings.[5] The marshlands of Newton were inclosed by decree of Chancery in 1665. William Colvile, as lord of the manor, claimed the ownership of the soil of all the common marshes, and that the inhabitants of Newton had only a right of common over them. These contentions were upheld by the inclosure commissioners, who awarded Colvile 371 acres out of the marsh, free from all claims of the Bishop of Ely and the inhabitants or others claiming under them. The remaining part of the marshland was allotted at the rate of 5 acres for each commonable house. Every tenant had 1*s.* for each lot added to his quit-rent. The allotments comprised a total of 518 acres, at five score to the hundred. The following roads were laid out: Sutton or Lincolnshire Road (10 a. 2 r. 28 p.), Hundred Acre or Wesel Hill Drove (now known as the Little Ramper) (3a. 0 r. 8 p.), Lott or Chapel Drove (2 a. 3 r. 11 p.), Ferry House Drove (2 a. 3 r. 36 p.), and the Drove or roadway below the Bank from Fitton Gold Gate to the Chapel on the Sea (3 a. 0 r. 28 p.). The marsh was first ploughed in 1667 or 1668.

By the award the Bishop of Ely became entitled to 50 acres of the newly inclosed marshlands. This land lay on the south side of the Little Ramper and abutted upon the Leverington border.[6] It was let on 21-year beneficial leases, which were generally renewed by the tenants some time before their expiry. In 1871 the Bishop sold all his rights to the then tenants for £2,550.[7]

The earliest terrier or town book extant, so far as is known, was made in 1688 by Thomas Skepp.[8] It comprised only the high lands (some 948 acres), held by 32 owners,[9] together with the manor house, 34 messuages, 2 tenements, and 2 cottages. The principal owners were the lord of the manor, William Colvile (520 a.), the heirs of John Colvile (67 a.), the Master of the Chapel, i.e. the Rector (97 a.), Thomas Edwards (83 a.), John Snell (44 a.) and Thomas Davis (32 a.). The next surviving terrier was made in 1712. It shows that there were then 518 acres in the Marsh and 818 in the Fen held by 10 and 11 owners respectively.[10] In 1744 John Lumpkin made a terrier of Newton Fen 'as it lyes now, since it has been drain'd and exchanged'. The drainage can only have been that carried out by Henry Dereham in 1635 (see Tydd St. Giles) and Lumpkin's words suggest that there had taken place in Newton after the drainage a wholesale reallotment of lands by exchange in order to bring dispersed holdings together into more manageable units. According to the 1744 terrier the Fen consisted of 818 acres. The chief owners were: the lord (417 a.); the heirs of Henry Pitchford (86 a.); Nicholas Lumpkin (76 a.); the Rector (68 a.); John Pitchford (57 a.); Robert Mears (37 a.). Only two houses are mentioned—both in Earl's Doles.

The parish fields in Newton are as follows. In the high land: Meadow and West Fields, Fall Hirn, New Croft, Goole or Gull and Long Fields, Hard Croft, Crofts, Fitton Croft, Old Rowles and Gaul Fields, Whirler's Hill, Parrock Field, Short Lands, Fen Lands, Karrow or Seagate Field, Church Croft, Mud Croft, Dooles or Gotelands, Sterts or Sterdyke Field, and New Lands. In the fen: Leets Lane End, Earl's Doles, Hundred Acre, Black Lane, Bar, Midlinham, Richolme, Queen's Dolfe and Turnore Fields,[11] and Watery Pits.

1 By the M. of H. Prov. Orders Conf. (Ely, Holland and Norf.) Act, 1933.

2 The 551 acres transferred from West Walton in 1934 represent the area between the former and present courses of the Nene.

3 *V.C.H. Cambs.* ii, 132–3.

4 Awards at County Hall, March.

4a See plate facing p. 90.

5 W. Dugdale, *Hist. Imbanking*, 319.

6 The substance of the award and the information about ploughing the marsh are entered in the terrier of 1688—see note 8 below.

7 Wisb. MSS.

8 Copy *penes* Mr. G. M. G. Woodgate. There are earlier terriers of the glebe alone (Bp.'s Mun., H. 1).

9 The Court R. (in Wisb. Mus.) show the names of a number of copyholders or small freeholders whose names do not appear in the terrier, so that the latter cannot be regarded as completely exact, unless such owners as were entitled held their lands in the Fen.

10 Copy *penes* Mr. G. M. G. Woodgate.

11 Cf. Turnover Bank in this field.

MANORS Newton is not mentioned in Domesday Book. The first authentic reference occurs *c.* 1210,[12] and the name of the village suggests that it may have been a very late settlement, like Elm (q.v.). It is significant also that Newton is not mentioned at all in the earliest (1221) of the Ely episcopal cartularies,[13] and only in a cursory manner in that of 1251.[14]

As in all the manors of the Isle that were not explicitly reserved to the prior and convent in 1109, the Bishop of Ely was overlord, and he was by implication still so considered in 1851.[15] But for many centuries before he had held no immediate rights in Newton. At the beginning of the 13th century there was one tenant by military service, Stephen de Marisco, who held 1 fee of the bishop in Newton and Walsoken (Norf.).[16] It seems that during the next half century the manor was subinfeudated to his family, for in 1267 Geoffrey de Marisco, Stephen's son, was granted free warren in his demesnes at Newton.[17] The 1251 cartulary shows that the bishop held no demesne in Newton in severalty, nor are any of his customary tenants mentioned.[18] The rents payable to the bishop moreover amounted to no more than 12*d.* Stephen (*sic*) de Marisco,[19] on the other hand, held an unspecified amount of land by military service which had been commuted for 6*s.* 8*d.*

Geoffrey de Marisco was the last in the male line, and his daughter Desirée, by her marriage to Sir Roger Colvile of Carlton Colvile (Suff.), brought the manor to the family which was so prominently associated with Newton until the end of the 18th century.[20] Several of the Colviles were men of mark. Sir Geoffrey (d. 1317), son of Roger and Desirée, was knighted in company with Edward Prince of Wales (1306). His son Sir John I fought at Crécy and took part in the later French campaigns of Edward III. To him the grant of free warren, made to de Marisco in 1267, was confirmed in 1353.[21] Before his death (1361) he settled the manor for his own benefit should he return from France, or for that of his son, another John. In 1361 the manor was extended at 6*s.* 8*d.* for the site of the manor house, 200 acres of arable demesne at 1*s.* 6*d.* an acre, 40 acres of pasture (4*d.* an acre), a windmill worth 15*s.*, rents of £16 0*s.* 4*d.*, commutation of labour services 40*s.*, and perquisites of court 20*s.* It was stated to be held by socage.[22] John Colvile II, who succeeded his father,[23] was knighted in 1367 and died in 1394.[24] His eldest son, Sir John Colvile III, was perhaps the most remarkable member of the family. In 1409 he was sent by Henry IV on an important mission to Italy,[25] and at the end of the following year he was appointed Constable of Wisbech Castle, a position which he held until at least 1446.[26] In 1412 he accompanied an expedition to France against the Duke of Orleans, and the following year commanded a naval force which won a victory over the French in the English Channel. He fought at Agincourt, and for these and other services was granted a pension of £40 a year for life. He was the founder of the College of St. Mary by the Sea in Newton.[27] Sir John Colvile IV, his eldest son, was sheriff in 1459 and died ten years later.[28] His eldest surviving son Francis was then an infant, and the family estates were entrusted to the Bishop of Lincoln, who sold the custody to Anne, John's relict. Anne's second husband, Sir Robert Brandon, forcibly excluded Francis from his lands when he came of age, and he apparently never enjoyed them, as Anne outlived him by a few months. At her death in 1494 the manor passed to Richard Colvile, Francis's son; it comprised 12 messuages, 10 cottages, land, estimated at 100 acres of arable, meadow, and marsh severally, together worth £30, and £10 rents in Wisbech, Leverington, Newton, and Tydd.[29] At Richard's death in 1525 the manor, with lands in the village and in Leverington, Tydd St. Giles, and Wisbech, was valued at £30.[30]

Son succeeded father for another three generations,[31] until the death (1611) of Thomas Colvile (knighted 1611).[32] He left no son and no will, and his brother Richard succeeded. Richard's eldest son John was accidentally killed at the age of 7,[33] and on his death in 1650[34] he was succeeded by William, his second son, who was a devoted Royalist and sheriff in 1660. He died without surviving issue, and was succeeded by his brother Jeffrey (d. 1699), who lived at Walsoken. Richard, Jeffrey's surviving son, was a member of the Inner Temple and resided chiefly there or at Harrow. He died childless in 1723. He devised all his lands, subject to the life interest of his relict, to his nephew Robert Barker for life with remainder to his issue in tail. In default of issue the lands were to go to his nephew Colvile Lumpkin, son of Richard's sister Susannah by her second marriage to John Lumpkin, in like manner.[35] Portions of the estate had to be sold or assigned to meet debts and legacies,[36] and when Robert Barker (who assumed the name and arms of Colvile) succeeded his uncle the estate was much encumbered. He immediately began the task of restoring the fortunes of the family. He would rise early to see that the work on the estate was properly carried out, and gave his personal attention to every particular. In his time the estate was richly wooded, chiefly with oak and ash, and he would not allow any felling. Church services had to wait until he had taken his seat, and no one was allowed to walk

[12] Newton is mentioned in the (probably spurious) charter of Wulfhere to Peterborough, 664. The ref. in *P. N. Cambs.* (E.P.N.S.), 274, to A. J. Robertson, *Angl. Sax. Chart.* 192, 440, probably refers to Newton (Suff.).

[13] B.M. Cott. MS. Tib. B. II.

[14] Ibid. Claud. C. XI, f. 86*a*.

[15] R. Gardner, *Dir. Cambs*, 555.

[16] *Bk. of Fees*, 629.

[17] *Cal. Chart. R.*, 1257–1300, 99.

[18] B.M. Cott. MS. Claud. C. XI, f. 86*a*.

[19] The fact that a long-dead de Marisco tenant is still set down as holding by military service suggests that this section of the cartulary was compiled from earlier ones and not checked on the spot.

[20] A privately printed *Hist. of the Colvile Family*, by Sir C., C. R., and Z. Colvile, 1896, from which many of the following references are taken. It includes a detailed survey of the manor, 1395.

[21] *Cal. Pat.* 1350–4, 406.

[22] C 135/155/15.

[23] *Cal. Inq. p.m.* xii, no. 461 (p. 435).

[24] *Hist. of the Colvile Family.*

[25] Ibid.

[26] *Ely Dioc. Remembrancer*, no. 208, p. 181.

[27] For an account of this foundation, see *V.C.H. Cambs.* ii, 312–14.

[28] Will dated 5 Mar. 1469 (P.C.C. Godwyn 30).

[29] *Cal. Inq. p.m. Hen. VII*, i, no. 1098, p. 468.

[30] C 142/43/63.

[31] Richard Colvile, will dated 28 Aug. 1525 (P.C.C. Bodfield 39), Geoffrey Colvile, will dated 20 Oct. 1571 (P.C.C. Pyckering 58), John Colvile, will dated 1 Apr. 1590 (P.C.C. Harrington 4).

[32] To illustrate the rise in values then taking place, the one-third value of the estate payable as dower to Sir Thomas's relict amounted to £350 a year (Wisb. MSS.). She was also assigned, amongst other things, a coach with three horses, a road mare, one small hand basin, a ewer of silver, and other effects.

[33] 'John son of Richard Colvile Esq. of Newton a boy of 7 years, suddenly, by a fall from his horse and receiving a mortal wound from the horse's hoof was killed 7 September and buried the following day' (entry in Newton Parish Reg.).

[34] Will proved Dec. 1723 (P.C.C. Richmond 253).

[35] Ibid.

[36] Copy of conveyance *penes* Mr. G. M. G. Woodgate.

out of church before him. In spite of his severity and reserve he was greatly loved by all the villagers, and long after his death personal anecdotes were current in Newton.[37] He was a county magistrate, sheriff in 1739 and Town Bailiff of Wisbech in 1769; he possessed much property in the town in the right of his second wife. His only son Richard, who succeeded him in 1778, was also Town Bailiff of Wisbech, but his tenure of that office was unfortunate.[38] He had the reputation of a spendthrift, and at his death (1784) the estate subject to incumbrances devolved upon his son Robert, the last of the Colviles to enjoy the property. Robert Colvile was amiable but unbusinesslike, and in 1792 contracted to sell the whole to James Redin for £47,532. The estate comprised manorial rights in Newton, Walpole, and West Walton, the Hall, and nine farms comprising 1,682 acres.[39] Redin sold the estate in lots at a considerable profit and demolished the Hall,[40] replacing it by a farmhouse which still exists. His son, another James Redin, lived there until 1824. A subsequent owner (1864–71) was Richard Young, M.P., the Wisbech shipowner. His representatives retained possession until 1901, when the Hall was bought by John Wing (d. 1910). In 1920 it was sold by John Wing's trustees to G. F. Hoyles, who lived there until his death in 1951.

The manorial rights and certain other parts of the estate were bought by Steed Girdlestone, a Wisbech solicitor, in 1814 for £11,900. By 1851[41] they had come to the Jackson family of Wisbech, another well-known lawyer family, and were held by trustees of this family in 1933.[42]

In 1232–3 William de Lirling held, of the honour of Wormegay,[43] 1 fee in Larling, Rushworth (Norf.), and Newton, and in 1253 he had a grant of free warren in all these places.[44] Maud de Lyrling, the last of this family, in 1302–3[45] held ½ fee in Newton of the same honour, for 10*s*.[46] In 1346 this was held by her grandson John Cunvyle,[47] and in 1428 by Edmund Gonvile.[48] The estate descended twice through female lines to his great-granddaughter Anne, relict of John Scrope of Bolton, knight.[49] At her death in 1498 it is described as a manor, worth 6 marks and held of the Bishop of Ely, again at ½ fee.[50] Before Anne's death the manor had been settled upon trustees to the use of John son of Henry Scrope. Trustees were dealing with the manor in 1499,[51] and again in 1528, when the manor was estimated at 200 acres of land, 20 of meadow, 40 of pasture, 10 of wood, and 40*s*. rent.[52] John Scrope of Hambleden (Bucks.) and Spennithorne (Yorks.), to whom the manor had been devised in 1498, is found in possession *c*. 1538–44, when he was at issue with Robert Boswell over the tenancy of a messuage and 20 acres held of the manor of *LARLINGS HALL* by homage, fealty, and 12*s*. rent.[53] By 1592 this manor had become vested in John Colvile of Newton, who then entailed it;[54] after this it descended with the main manor.

In 1494 Francis Colvile died seised of the manor of *BODERS* or *BEDERS*, held of the Bishop of Ely by knight's service.[55] This manor is not again recorded, but it was perhaps the same as the manor of *BUSSEY'S* or *RUDHALL'S* with which John Colvile was dealing in 1584.[56] Rudhall's thereafter descended like Larlings, with which it is invariably associated.

THE LUMPKIN HOUSE Adjoining the west end of the churchyard is an old house, formerly the property of Robert Jackson, who succeeded to it on the death of his mother Margaret about the year 1659.[57] He in turn gave it by will to his daughter Ann, who in 1720 sold it to John Lumpkin of Newton.[58] The latter was born in 1681 and was the eldest son of John Lumpkin of Park House, Leverington (q.v.), but settled at Newton and in 1709 married Susan daughter and heiress of Jeffrey Colvile, owner of the Newton Hall estate. Jeffrey Colvile had died in 1699 leaving Susan, his only surviving child, who had married as her first husband Robert Barker. Robert Barker the father had died in 1708 leaving a son Robert who thus became the heir presumptive to the great Colvile property; and when his mother married John Lumpkin as her second husband and had issue by him an only son, Colvile Lumpkin, born in 1712, the boy was designated as next heir in the event of his half-brother's prior death.[59] According to tradition[60] young Robert Barker was sent to sea, not without the connivance (so it was said) of the Lumpkins, and was badly injured by a falling spar which crushed his foot and made him lame for life. Apart from this he returned unscathed to Newton in time to succeed to his uncle Richard Colvile's estate on the latter's death in 1723, subject to the life interest of Frances, Richard Colvile's relict. Robert Barker assumed the name and arms of Colvile in accordance with his uncle's will.

Meanwhile John Lumpkin had established himself in the house next the church, which he appears to have rebuilt, and took a very active part in the life of the village and church (see Church). He purchased much land in Newton, Tydd St. Giles, and West Walton. By will dated 2 September 1760[61] he gave his Newton properties to his wife for life and then to his daughter Elizabeth Harber and her two children.

Elizabeth, upon whom the Lumpkin House thus devolved, was John's only surviving daughter by his second marriage. She married Bartholomew Harber (or Arbor),[62] who died in 1753. John Harber, her son, was succeeded by his son Bartholomew, who died intestate in 1853 and the Lumpkin House passed to his daughter Elizabeth Lumpkin Harber. She died in 1893, when the house and 7½ acres of land were sold to Canon A. R. Evans, Rector of Newton, who subsequently gave

37 *Hist. of the Colvile Family*.
38 See p. 257.
39 Copy of deeds of Lease and Release 1793, *penes* Mr. G. W. G. Woodgate.
40 A view of the Hall prior to its demolition, showing the gardens overgrown into shrubberies, is in W. Watson, *Hist. Wisb.* (1827). It is supposed that the cellars under the present dwelling house are those of the old Hall.
41 Gardner, *Dir. Cambs.* 555.
42 Kelly, *Dir. Cambs.*
43 *Bk. Fees*, 1466.
44 *Cal. Chart. R.* 1226–57, 418.
45 F. Blomefield, *Norf.* i, 285.
46 *Feud. Aids*, i, 141, 151.
47 Ibid. 172.
48 Ibid. 191.
49 *Complete Peerage*, xi, 545.
50 *Cal. Inq. p.m. Hen. VII*, ii, no. 193 (p. 127).
51 C.P. 25(1)/30/101.
52 C.P. 25(2)/4/19.
53 C 1/1069/5.
54 C 142/233/98/3.
55 *Cal. Inq. p.m. Hen. VII*, i, no. 1098 (p. 468). It was, with the main manor of Newton, worth £40 yearly.
56 C.P. 25(2)/93/849.
57 He was admitted to the copyhold portion in 1659 (Wisb. Mus., Newton Ct. R.).
58 Ibid.
59 P.C.C. Richmond 253.
60 *Hist. of the Colvile Family*.
61 Office copy will among Wisbech MSS.
62 Marriage settlement dated 3 Dec. 1737 comprising Harber property in Leverington; followed by a settlement of about 32 acres of Lumpkin land in Newton dated 7 and 8 Mar. 1744, of which latter settlement the Rector of Leverington and the Vicar of Elm were the Trustees (Wisbech MSS.).

a portion of the land for enlarging the churchyard and sold the remainder to the late Thomas Eyre. The old house has now been converted into two cottages and belongs to Mr. Gee, but in earlier days it had a number of attractive features. A stone panel carved with the Lumpkin arms extended over the front door, with the date 1728 and the initials J. and E. L.[63] The door opened into a square entrance hall from which a handsome flight of stairs led to the first floor. There were two other staircases; and a fourth (outside) staircase of stonework used to exist but has long since been taken down to make room for a pantry. Some wood panelling lined what was probably the drawing-room, and there were until recently remains of old carving and Dutch tiles.

ABBEY GREEN, so called from John Abby who lived there, was originally part of the Colville estate and was included in the sale to James Redin in 1793. Redin resold to Francis Taylor of Newton, afterwards of Wisbech, who died in 1818, devising his Newton property to his wife Sarah. From the Taylors it passed to the Pecks, who added to the house the block of large rooms on the south side. Thomas Patrick of Lynn, who married into that family, sold the property early in the present century to the late Thomas Eyre, whose son Mr. T. E. Eyre now owns it and resides there.

THE SHRUBBERIES lies in Crofts, near the corner formed by High Green or the Newton High Street, and Mill Lane. The predecessor of the present house belonged to Thomas Edwards the elder, of Wisbech, who in 1702 settled it upon his son Thomas Edwards of Isleworth (Mdx.) on the latter's marriage with Joyce Sawyer of St. Martin in the Fields. The younger Thomas devised the property by will dated 1725 to trustees, the survivor of whom sold it in 1748 to Samuel Shepheard of Exning (Suff.). In 1793 Shepheard's trustees sold the house and about 145 acres to John Taylor, then tenant, for £5,000. He was succeeded by his only son John, who died in 1802 v.p., leaving the Revd. John Taylor his son and heir. It is probable that it was he who rebuilt the Shrubberies. He died in 1867, and his son Edward Whitty Taylor sold the whole property in lots, when the house and 62 acres were purchased by Matthew Webster. It was acquired eventually by John Wing, who died in 1910, and was sold by his Trustees in 1945 to Mr. G. J. Smith, whose son Mr. G. T. Smith is now (1951) the owner and occupier of it.

NEWTON HOUSE, in New Croft, belonged to the Colvile family on the break-up of the estate, and was ultimately purchased by Thomas Foster. The latter sold it for £4,100 to Robert West of Newton, who dying in 1803 was succeeded by his son Robert. The latter died in 1852, devising the property to two of his sons, Samuel and Stephen. Stephen, the sixth and youngest son, lived at Newton House and farmed extensively in the village. The farm was sold by his executors in 1898 to John Wing, who resided there until 1901, when he purchased Newton Hall (q.v.). In 1920 Newton House and lands were sold by John Wing's Trustees to Mr. John Mills Wilson, whose son Mr. Gordon Mills Wilson now resides there.

CHURCH. The advowson of the church has always been with the Bishop of Ely, and like most of the churches on episcopal manors, Newton was never appropriated. The value was £13 6*s.* 8*d.* in 1217, £23 6*s.* 8*d.* in 1254, and £40 in 1291.[64] The same amount was returned in 1341[65] but reduced to £16 13*s.* 4*d.* because 1,520 acres of land formerly arable were devastated and submerged by the waters.[66] In 1535 the value, including the chapel of St. Mary in the Marsh, was £18 14*s.* 8*d.*[67] This chapel, which had belonged to the Chantry or College of St. Mary by the Sea founded early in the 15th century,[68] was one of the few of its kind to remain after the suppression of chantries under Edward VI, being specifically excepted in the Act of Suppression of 1548.[69] The endowments of the college, of which the Rector of Newton was Master,[70] were transferred to the rectory of Newton, which explains the large amount of glebe—at one time some 400 acres in 11 neighbouring parishes—attached to the living. In 1839, when the tithes were commuted for a rent charge of £710, there were 176 acres of glebe.[71] In 1918 about 324 acres of the Newton glebe, comprising all the parts in other parishes and some in Newton itself, were sold in 46 lots.[72]

There were two guilds at Newton in 1452, those of St. Katherine and St. John. In that year each received a legacy of 20*d.* under the will of John Gowe of Newton.[73] The former at least was still in existence in 1525, when it was the contingent residuary legatee in respect of £240 left by Richard Colvile.[74] Lights of the Blessed Virgin, the Crucifix, St. Katherine, and St. John are mentioned in the will of Katherine Austin of Newton, 1455.[75] The third and fourth of these were doubtless maintained by the guilds of St. Katherine and St. John.

The church of *ST. JAMES*, formerly of St. Katherine,[76] consists of chancel, small north organ chamber, nave, aisles, south porch, and west tower. The walls are covered with cement and the roofs are leaded. The fabric is of late-12th-century origin, to which period belong the columns of the nave arcades. In the 14th century the aisles were rebuilt and widened, the south porch added, and the present tower erected. Early in the 15th century the eastern part of the north aisle was fitted up as a lady chapel by Sir John Colvile III (d. 1446); figures of him and his wife Alice (Wythe) existed in stained glass in the second window in the aisle in Cole's time (1744)[77] but have now disappeared. For many generations members of the Colvile family were buried in a vault at the east end of this aisle. During the 14th century the arches of the arcades were rebuilt and the piers raised, a clerestory added, and the chancel rebuilt. In 1718 John Lumpkin (d. 1762), churchwarden for 51 years,[78] carried out some decorations. 'The whole church', says Cole, 'was neatly painted and adorned with figures and sentences and the pillars painted like marble. The pulpit', he adds, which stood against the last pillar in the middle aisle, was 'curiously painted and gilt with a fine gilt canopy'.[79]

[63] No authority for their use has been found and it must be assumed that the user was arbitrary. The arms consisted of two bars gemel on a plain field with a crescent for difference.
[64] W. E. Lunt, *Valuation of Norw.*, 539.
[65] Ibid.
[66] *Inq. Non.* (Rec. Com.), 203.
[67] *Valor Eccl.* (Rec. Com.), iii, 500.
[68] *V.C.H. Cambs.* ii, 312–14.
[69] 1 Edw. VI, c. 14, s. 15. Printed in *Stat. of the Realm*, iv, 28.
[70] S.C. 12/6/19.
[71] 2,771 acres of land were surveyed.
[72] Much of the proceeds of the sale was devoted to the augmentation of benefices in the diocese.
[73] B.M. Add. MS. 5805, f. 104.
[74] P.C.C. 39 Bodfield.
[75] W. M. Palmer, *Village Gilds of Cambs.* 396.
[76] So called in will of Beatrice Edmonde of Leverington, proved 1454 in Ely Consistory Ct., and in Newton Terrier of 1688 (copy *penes* Mr. G. M. G. Woodgate).
[77] B.M. Add. MS. 5805, ff. 111–15.
[78] As recorded in the entry of his burial in the parish reg.
[79] B.M. Add. MS. 5805, f. 114.

Lumpkin and his three wives were buried just below the altar step of the north aisle but the grave slabs have gone.[80] The church was restored between 1853 and 1878, when new roofs were provided and the organ chamber built. Until 1879 two bays of each aisle were bricked off. That of the north aisle has been used as a vestry and a British school. The other had contained the staircase to the organ gallery, which occupied the west end of the nave. The brick walls and other partitions were then removed. At the same time the font was removed from the centre of the cross aisles to the south aisle, a new heating apparatus was installed, a part to its two-centred arch, and a hood-mould. The north aisle (the Colvile Chapel) has buttresses with two set-offs, and the south aisle buttresses are gabled with one set-off. The rood stair is contrived in the south-east angle of the north aisle, with plain upper and lower doorways both in the aisle, the upper being connected with an opening in the north-east angle of the nave wall by a small bridge of brick; the whole arrangement belongs to the latter part of the 15th century. There is a piscina in the north aisle with cinquefoiled head. The 14th-century porch has a two-centred outer arch with a hood-mould; the columns are of two orders with

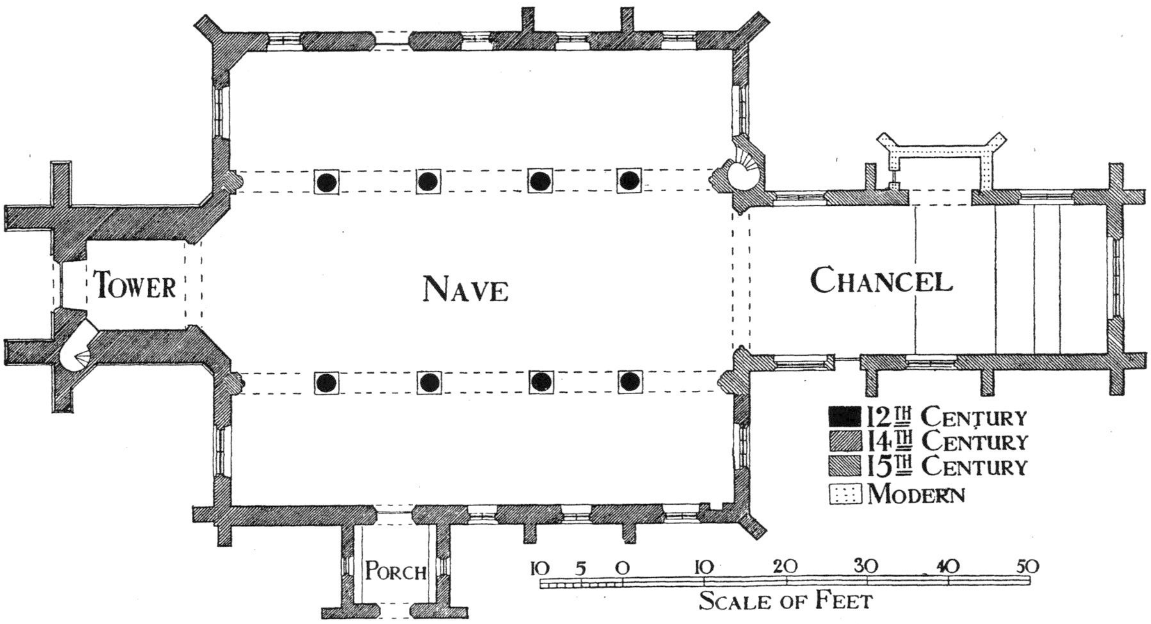

PLAN OF ST. JAMES'S CHURCH

of the north aisle was enclosed by a low screen to form a vestry, and the organ was placed on the floor towards the west end of the nave. In 1890–3 a new organ, choir stalls, altar table, rood screen, and two stained glass windows were provided at the cost of Canon A. R. Evans, rector 1885–1907.[81]

The chancel has an east window of four lights with renewed tracery of 15th-century character. There are angle buttresses and two windows in the north and south walls which are of two cinquefoiled lights with rectilinear tracery under a square head. The 15th-century chancel arch is two-centred and of two orders with moulded caps; the bases are concealed.

The nave arcades are of five bays with 15th-century two-centred arches of two orders resting on octagonal moulded caps; the piers are round with square bases and are of the end of the 12th century. The responds belong to the 15th-century reconstruction and have rounded shafts and moulded caps. The clerestory windows are of two cinquefoiled lights. The tower arch is two-centred and of two orders with moulded caps to the responds, the bases of which are concealed; it is of the 14th century. All the aisle windows are 14th-century work and of two lights, except the east and west windows of the north aisle and the east window of the south aisle, which are of three lights; all are trefoil-headed. The north doorway has continuous mouldings moulded caps and bases. There are two-light windows on the east and west. The inner doorway has continuous mouldings and a hood terminating in the heads of a king and queen.

The tower is of three stages, with gabled angle buttresses on the west and diagonal buttresses on the east, also gabled. The newel stair is contrived in the south-west angle. The west doorway has a two-centred arch and continuous mouldings, with a hood terminating in heads, now much decayed. Above is a window of three lights with mid-14th-century tracery and a hood terminating in heads. The middle stage has single trefoil-headed lights on the north, south, and west, and the belfry windows have two cinquefoiled lights which appear to be 15th-century insertions. There is an embattled parapet.

All the roofs are modern.

The 15th-century font has a hexagonal bowl, the sides of which have sunk quatrefoils containing blank shields; the shaft is panelled. There is a 13th-century stone coffin-lid in the chancel with a foliated cross, much weathered as a result of having long served as a part of the coping of the churchyard wall. Above the tower arch on the west wall of the nave is a blazon, the royal arms of George IV, on canvas of tympanic form, which was originally in the chancel arch.

The plate consists of a communion cup, paten, and

80 Ibid. f. 113.

81 Memo. at end of Newton parish reg., Vol. II, printed in W. Watson, *Hist. Wisb.*, 481.

flagon, all of 1663 and silver gilt, and an alms dish of silver, 1662. The first three items were presented by the executors of Bishop Laney in 1674.[82] In 1552 there were 3 large and 3 small bells and a sanctus bell.[83] Shortly before 1641–2, 3 of these bells were cast into 4 and new frames provided.[84] The tower now contains six bells by T. Osborn of Downham, 1786, which were rehung in a steel frame in 1912.[85]

The registers begin in 1653 and are complete.

In *c.* 1547 the rector lived in the College of St. Mary.[86] The present rectory is a comely late Georgian house of two stories with attics and a back addition of 3 stories.

The chapel of St. Mary in the Marsh stood in the grounds of the present rectory. The building, though recognizable in 1572[87] and in the later 18th century,[88] has long since disappeared, but incumbents of Newton are still inducted into the rectory of Newton with St. Mary in the Marsh.

NONCONFORMITY There were no recusants in Newton in 1676.[89] In 1869 both the Free and Primitive Methodists erected chapels. The latter chapel was closed before 1900 and was converted into cottages, which in that year were sold by James Smith to John Wing.[90] The cottages are at the western end of the village, nearly opposite the Woodman's Arms Inn. The Free Methodist chapel, which has a Sunday school, is a brick building seating 130 persons.[91] It is in the High Street of the village.

SCHOOL A schoolmaster, Roger Pinchbeck, was appointed at Newton in 1582.[92] At the end of the 18th century there were about 10 boys and 10 girls at the village school out of a total child population of 49, of which about two-thirds required free education. The schoolmistress received £8 a year from the parish, and £2 2*s.* as a donation from the rector; some of the inhabitants were paying 3*d.* a week for their children's education, and 3*d.* was also payable for girls learning needlework. The boys, as was generally the case, were removed from school as soon as capable of farm work.[93] This school was held in the vestry of the church, and in 1851 in the tower.[94] A better state of things was inaugurated in this year, when Dr. G. E. Corrie, Master of Jesus College, was appointed rector. In 1853 he built at his sole cost a permanent school for about 100 children,[95] which in 1876 was reported to be 'amply supplied with funds by private contributors'.[96] In 1910 the recognized accommodation, which had been 114, was reduced to 93 (71 mixed, 22 infants). During the early years of the 20th century Newton, like other marshland parishes, had an increasing population, and by 1916 the school was so crowded that the infants had to be taught in the parish room. The situation was aggravated by the deterioration of the buildings. In 1926 one classroom was declared unusable, but it was found impossible to raise money voluntarily to improve the buildings and in 1928 the school was transferred to the County Council. A new school was erected on a new site, and named on its opening (1930) the Colvile County Primary School, after the family so long associated with the lordship of the manor. This school, which provided 124 places, cost £4,400; it included a practical instruction room, and in 1938 additional land was bought for a school garden and playing-fields. As a new building, the school has not so far been 'decapitated', and since 1933 has taken the senior children from Tydd St. Giles. The resultant overcrowding (140 on the roll, 1949) was remedied by the erection of a two-classroom hut in 1950.[97] The old school has been purchased by the Newton Women's Institute.

OUTWELL AND UPWELL

These two villages lie astride the Well Stream. Each is therefore partly in Norfolk and partly in the Isle of Ely.[1] In the Middle Ages they were collectively known as 'Welle' or 'Welles', and today they are absolutely continuous, so that it is impossible for a stranger to tell where one ends and the other begins. Together they stretch about 4 miles along the old course of the Nene, and are locally considered, probably with truth, to be 'the longest village in England'. They also form a very large village community. In 1931 the population of the four constituent parishes was just under 5,000, though the area (some 20,000 acres) is also considerable.

In the Middle Ages the Nene dictated the lay-out of the villages, but the topography of the whole district has subsequently been much altered by several large drains which run through these and neighbouring parishes. Such are Popham's Eau, an early 17th-century cut to guide the waters of the old Nene into the Ouse at Salter's Lode; the Old Bedford River (part of the south-east boundary of Upwell), the main drainage channel, with the New Bedford River, of the southern fens; and the Middle Level Main Drain, which takes the outflow from the Chatteris and Doddington areas diagonally across Upwell and Outwell to the Ouse at the Wiggenhalls. The bursting of the bank of the Middle Level Drain in May 1862 inundated one-sixth of the parish of Outwell, which had not fully recovered three years later.[2] With these waterways may also be numbered the Wisbech Canal, now disused, which followed the course of the Well Stream as far as Outwell church and then struck across in a south-easterly direction to join Popham's Eau at Nordelph (formerly a hamlet of Upwell but now a separate Norfolk parish), about 3 miles east of Upwell and Outwell.

As befits a former market-town,[3] Upwell is something of a road centre. There is the main road from

[82] There is a note to this effect in Newton parish reg.
[83] E 315/495, p. 156.
[84] C 3/424/55.
[85] Kelly, *Dir. Cambs.* (1933).
[86] S.C. 12/6/19.
[87] *V.C.H. Cambs.* ii, 313.
[88] Parish Map of Newton in Wisb. Mus.
[89] Bp.'s Mun., Ely, A 6.
[90] Title Deeds.
[91] *Meth. Ch. Bldgs. Stat. Returns as at 1 July 1940* (priv. printed 1947).
[92] A. Gibbons, *Ely Episc. Rec.*, 187.
[93] Bp.'s Mun., Ely, B 6.
[94] Gardner, *Dir. Cambs.* 556.
[95] Kelly, *Dir. Cambs.* (1900).
[96] M. of Ed. File no. 12/41.
[97] M. of Ed. File no. 12/41.
[1] Upwell, which is much the larger parish (about 17,000 acres to 3,000 in Outwell), is fairly equally divided between the two counties; Outwell is largely in Norfolk. Both churches are just on the Norfolk side of the boundary and were formerly in Norwich diocese.
[2] Archives of Nat. Soc. Difficulty was experienced in 1865 in raising money for a National School.
[3] An Upwell halfpenny tradesman's token (1668) is known (S. H. Miller & S. B. J. Skertchly, *The Fenland Past & Present*, 478–80).

Copyright 'Country Life'

BEAUPRÉ HALL, OUTWELL. GATEHOUSE BLOCK IN 1923

Copyright 'Country Life'

LEVERINGTON HALL IN 1948

Copyright Raphael Tuck & Sons Ltd.

Upwell Village

Reece Winstone

Cottages and Windmill at Doddington

Wisbech to Downham Market (A 1122) following the canal; the Wisbech–Ely road (A 1101), which leaves this at Outwell church; a second-class road to Chatteris (B 1098), which follows the Sixteen Foot Drain (a continuation of the Middle Level Main) from the south end of Upwell village; and another from Nordelph (B 1094), which takes a very circuitous course through Christchurch, the southern hamlet of Upwell, to Manea and Wimblington. The Upwell steam tramway from Wisbech, a light railway of 4 ft. 8½ in. gauge, was opened by the G.E.R. in 1884;[4] it conveyed passengers until 1928, and is still used for goods traffic. It forms an unusual feature in the landscape, and running for much of its course along the high road, is to some extent responsible for the very pronounced ribbon development which has made the road to Wisbech (via Emneth) a built-up area almost continuous for the 7 or 8 miles of its course.

Except for the two churches, Beaupré Hall (the former chief manor house), the Upwell Baptist chapel (1844), James Lee's Almshouses (1819), and the modern Secondary and Beaupré Primary schools, the buildings of the villages have individually no great architectural distinction. They form, however, pleasant groups fronting the water-courses, and Upwell has been illustrated at least twice in books on the English countryside.[5] William Watson's description, written over a hundred years ago with reference to Marshland villages in general, is still applicable to these two in particular. 'The pale brick and tile', he says, 'manufactured in these parts, give a very neat appearance to the houses . . . and though the country cannot boast of rural beauty, there is an air of neatness and of greater comfort than . . . in many other agricultural districts.'[6] It was unfortunate that soon after he wrote, Upwell was one of the places most severely attacked by the 1832 cholera epidemic, 67 persons dying here and in Nordelph in less than two months.[7] The district was already known for its market-gardens in the 1830's—some time before they became general in Marshland.[8] As elsewhere in the north of the Isle, inclosure mostly took place gradually as the land was reclaimed and drained, but Outwell was affected by the Act of 1841[9] (see also Leverington and Tydd St. Giles), under which Cottons Common (98 acres) and 15 acres of waste land were inclosed. By this inclosure 2 acres were set aside for a recreation ground, the Bishop of Ely as lord of the manor was allotted 1¼ acre, 17 acres were sold to defray costs, and the remaining 93 were divided amongst 26 proprietors, of whom James Lee (14 acres) was the only one to receive a holding of any size.

A public hall was erected at Upwell by a limited liability company in 1868 and is still in use.[10] The Outwell Working Men's Institute at Outwell dates from 1891, though the building is not now used for its original purpose. A Gas Company was formed in 1867.[11] Outwell was probably the birthplace of Robert Wells or Steward (d. 1557), the last Prior and first Dean of Ely.[12]

Welney, about 6 miles south of Upwell on the road to Ely, was an inhabited site in Roman times, but in its present form is mainly a modern settlement. Its population increased from 265 in 1801 to 1,019 in 1921, a very large increase even for a fen village; there was a slight decrease in 1931. It is well known for the many distinguished skaters whom it has reared, and for the skating championships which have occasionally been held there. It was formerly a hamlet of Upwell, but was wholly transferred to Norfolk in 1895.[13] The distinction between the Cambridgeshire and Norfolk portions was preserved until 1935 in the division between West Welney (Cambs.) and Welney (Norf.) parishes. The important bridge over the New Bedford River was originally a suspension bridge. It was erected in 1824–6 by the Revd. W. G. Townley, Rector of Upwell and Welney, at a cost of £3,000. The engineer was a Captain Brown.[14] The road over the bridge was turnpiked in 1827.[15]

Beaupré Hall is a large 16th-century house mainly of brick, which was built by the Beauprés and enlarged by their successors the Bells. The oldest parts date from about 1500 and include much of the central block running south-west to north-east, with a long wing running north-west at an angle. The gatehouse was placed in front of the main block and probably dates from about 1525. Fifty years later, after Sir Robert Bell succeeded to the property in virtue of his marriage with the heiress of Edmund Beaupré, the north-east section was rebuilt from the screen of the hall, a porch with an upper story was added on both sides, and a bay added at the daïs on the front; about the same time a large wing was constructed at right angles to the south-east, and connected with a wall to the gatehouse to form a court. This wing and part of the main block were destroyed a century ago, though part of the south-east end wall still remains. Before the end of the 16th century another court was formed to the south-west by a wing projecting from the main block and abutting upon the south-west side of the gatehouse. Considerable alterations, mainly internal, were made about 1750.

The gatehouse is of brick with stone dressings, the upper part being mainly of ashlar. The arches of the passage are four-centred. Above is a room lighted back and front by a square-headed window with stone mullions and transom. There are embattled angle-turrets rising well above the parapet. The room contains a late-16th-century fireplace. The spacious chapel occupied the extreme end of the north-west wing. It is now roofless and the walls have partly collapsed. It is not correctly orientated and the altar was at the north-west end. Here there is a six-light window with uncusped heads; the mullions are missing. There is a similar window, now blocked, in the wall on the left. There are some good 16th-century chimney-stacks. The main door of the house has 16th-century linenfold panelling. Several rooms on the first floor retain late-16th-century panelling; another room has early 18th-century panelling and yet another Georgian wainscoting. The drawing-room, formerly part of

4 F. J. Gardiner, *Hist. Wisbech*, 85.

5 D. K. Wickham, *Villages of England* (1933); T. Sharp, *Anatomy of the Village* (1946).

6 *Hist. Wisb.* 600–1.

7 There is a memorial brass in Upwell Church.

8 Bell's *Gazetteer of Engl. & Wales*, 1837.

9 4 & 5 Vic., c. xvi. Award at County Hall, March.

10 Kelly, *Dir. Cambs.* (1900).

11 White, *Dir. Norf.* (1883), s.v. Upwell.

12 'Wells in Norfolk' (*D.N.B.*). His father Nicholas was of Outwell and his brother Richard of Upwell (Blomefield, *Hist. Norf.* vii, 384, viii, 222; *Genealogist*, N.S. ii, 36–37). The Steward family were prominent landowners in the Isle in post-Reformation times. Robert must have taken the name of his birthplace as his name in religion on entering Ely Priory.

13 Kelly, *Dir. Cambs.* (1900).

14 R. Gardner, *Dir. Cambs.* (1851), 565; Diary of Revd. J. Jackson, 12, 16 Aug. 1826 (Wisb. Mus.).

15 Jackson's Diary, 15 Feb. 1827 (Wisb. Mus.).

the hall, has an early 17th-century chimney-piece and a deep wooden cornice, the sole remnant of panelling now no longer in existence. The back of the house was somewhat altered in the 19th century and suffered greatly in the process. Of late years a number of windows which had been modernized in the main block have been restored to their original form with stone mullions and transoms. The building at the southwest angle retains its characteristic flanking finials, which were also formerly found on the porch and other parts. The roofs are covered with stone tiles except some portions which have been repaired with blue slates. To the south are some fine contemporary farm buildings with stepped gables, moulded brick string-courses, and massive timbers. The two windows of the present entrance hall are filled with fine heraldic glass dating from 1570–80. The heraldry displays the matrimonial alliances of the various owners before 1580; the mantling is particularly fine. The following coats occur: (1) Beaupré impaling Meeres, (2) Beaupré impaling Fodryngaye, (3) Beaupré impaling Mountford, (4) Beaupré quartering Fodryngaye, (5) Beaupré impaling Fodryngaye, (6) Thomas Fodryngaye, (7) Coggeshall, (8) Fodryngaye, (9) Bell, (10) Sir Robert Bell, (11) Bell impaling Harrington, (12) Beaupré quartering Fodryngaye, (13) Bell quartering Beaupré, (14) Beaupré impaling Fodryngaye. The Beaupré panels are larger than the Bell panels and slightly earlier in date.

Beaupré Hall Farm is an early-17th-century building, much modernized but with stepped gables characteristic of the Eastern Counties.

A house adjoining the school, 180 yards south-west of Upwell church and on the east side of the street, dates from the early 18th century but has been much altered. It is of two stories with attics. The roof is partly thatched. A stone-walled house on the west side of the street, just north of Lloyds Bank, has a blocked doorway with a four-centred head and semi-octagonal columns with chamfered bases. This feature is probably late 15th-century and may have been reset. Farther south-west on the same side of the street is a three-story house with a late 18th-century front of some merit; the doorway is recessed and has Ionic columns. There are also some diamond-shaped chimney-stacks, which suggest that the building has an earlier date.

MANORS

King Edgar in 974 confirmed the gift of Aylwin his foster-brother of 60,000 eels yearly from 20 fishermen at *WELLES* for the support of his newly founded monastery at Ramsey (Hunts.).[16] Domesday Book records 16 bordars with lands worth 5*s.* as belonging to this abbey.[17] The later 12th century seems to have been a period of extensive reclamation here, as at Elm; in 1202 the abbot was granted by charter view of frankpledge and a market at 'Welles',[18] and free warren in 1251.[19] The market (on Wednesdays) and a yearly fair on the feast of St. Peter and St. Paul were renewed to the Abbot and the Bishop of Ely as joint lords in 1486.[20] The market had become 'trivial' by 1845 and had disappeared altogether by 1864, when the fair, on 29 and 30 June, was merely a pleasure mart.[21]

Two extents of the manor, made about the time that the market was first granted, are preserved. The earlier, made by order of Abbot Eudo (1200–2),[22] records a small demesne, consisting of land, formerly held by Seman, whereon the new hall of the abbot had been built. The annual tribute of eels had risen slightly above the 60,000 ordained by Aylwin, and many tenants of land and messuages, as well as of fisheries, paid their rent in this form. A few years later (1206–7) a new scale of rents was drawn up, sticks of eels were reckoned by the old and the new count, and many fish-rents had been commuted for money.[23] On this occasion 28 free tenancies are recorded and the number was increasing. New purprestures were charged at 1*d.* an acre. In 1291 the total value of the Ramsey property in Outwell and Upwell from rents and fisheries was £19 13*s.* 4*d.*[24] Three years later the abbot made an agreement with the Bishop of Ely to avoid such incidents as that which occurred in 1200–2, when Ailbrith son of Walter, a nief of the abbot, claimed to be a bishop's man.[25] The court leet of Welles was to be held by the abbot's steward, the bishop's bailiff being entitled to attend if he liked. The bishop was to hold a separate court for his own tenants.[26] The boundary between the two blocks of property in this neighbourhood was defined as being in the centre of the (Well) stream, the bishop having the land north and west thereof, the abbot that south and east.[27] The latter was thus in Norfolk in the hundred of Clackclose, of which the abbot was lord.[28] The value of the Ramsey Abbey possessions in 1538–9 was stated as only £13 9*s.* 9*d.*[29] This, however, may be an underestimate, as eight years later Edmund Beaupré, to whom the manor with all appurtenances was granted in fee, paid £282 15*s.* for it.[30] Edmund Beaupré was the last of his family in the male line, his daughter bringing the manor by marriage to Sir Robert Bell, Speaker of the House of Commons in 1572 and later a judge, who died at the Black Assize at Oxford (1577).[31] The marriage seems to have taken place shortly before 1561, when Beaupré settled the manor on his heirs male with remainder to Bell as his son-in-law. It was held of the queen in chief by $\frac{1}{20}$ knight's fee.[32] The Bell family continued to hold the manor for nearly 200 years. Beaupré Bell senior, the last but one of the male line, was 'one of the strangest of mortals, letting his wild colts and cattle of 20 or 30 years old come into the very house, which was quite uncovered and every other way suitable, in a very ruinous condition'.[33] He seems in his youth to have run away from school or been kidnapped, £100 reward being offered for his return by his mother and guardian Dorothy Bell.[34] His son Beaupré was an antiquary, making a collection of coins and medals which he pre-

16 Vide *Cart. Sax.* ed. Birch, iii, no. 1310. This charter is preserved only in the form of an *inspeximus* of Edward III, but its content is genuine.

17 *V.C.H. Norf.* ii, 138. The 8 fishermen, who rendered 5,260 eels to Ramsey Abbey from Wisbech (*V.C.H. Cambs.* i, 371) should also probably be reckoned as belonging to Outwell or Upwell.

18 *Pipe R. 1202* (Pipe R. Soc. N.S. xv), 136.

19 *Cal. Chart. R.* 1226–57, 366.

20 *Cal. Pat.* 1485–94, 87.

21 White, *Dir. Norf.* (1845), 627 and (1864), 847.

22 *Cart. Monast. Rames.* (Rolls Ser.), iii, 296–9.

23 Ibid. ii, 318–20.

24 *Tax. Eccl.* (Rec. Com.), 101.

25 *Cart. Monast. Rames.* (Rolls Ser.), iii, 299.

26 Ibid. i, 275. In 1275 Henry le Noble had been holding an unlicensed court in 'Welle' to the prejudice of the abbot (*Rot. Hund.* (Rec. Com.), i, 459). The bishop's bailiff also had tried to usurp the functions of the court leet (ibid. i, 519).

27 Dugd., *Mon.* ii, 577.

28 *Feud. Aids*, iii, 450.

29 Dugd., *Mon.* ii, 590.

30 *L. & P. Hen. VIII*, xxi (1), g. 1383 (81).

31 Blomefield, *Hist. Norf.* vii, 459–60.

32 C 142/182/28.

33 B.M. Add. MS. 5809, f. 114.

34 B.M. Add. MS. 5804, f. 55*b*, quoting an advertisement in the *London Gazette* of 26 April 1688.

sented to Trinity College, Cambridge.[35] The younger Beaupré Bell's sister and heir Elizabeth, to whom he bequeathed the manor, married William Greaves of Fulbourn.[36] Their daughter Jane brought it by marriage to the Townley family, in which it has since continued, Mr. Charles E. Townley being lord in 1937, though the ownership of land is much divided.[37]

In 970 King Edgar confirmed the grant by Aethelwold, Bishop of Winchester, of 10,000 eels annually from 'Wyllan' to the monastery of Ely.[38] Neither Outwell nor Upwell is individually mentioned in the Cambridgeshire section of Domesday Book, but it is reasonable to suppose that some of the 15,500 eels due to the monastery from Wisbech actually came from this neighbourhood.[39] The Ely property here mostly came to the bishop on the foundation of the see, but never compared in importance with the episcopal property in other vills in the Isle. There were no tenants by military service, and only 4 free tenants in 1210–12[40] and again in 1221, the most important being Adam (succeeded by Henry) de Sancto Edmundo, who held 200 acres in the marsh for 20*s.* At the latter date the fisheries were worth £14 6*s.*[41] Bishop Northwold's survey (1251) shows a very small demesne of 8 acres called 'Hallecroft', rent from land and messuages of £2 13*s.* 7*d.*, and from 7 fisheries of £15 10*s.* 4*d.* The bishop was entitled to a wind- or a horse-mill, though his tenants were not obliged to grind at it; he also had faldage once a year of beasts pasturing in the marsh, worth £1 4*s.* 6*d.* more or less, and the right of wrecks, royal fish (4*d.* payable to finder) and other manorial perquisites.[42] The responsibility of the bishop's tenants in Outwell and Upwell for the upkeep of 44 perches of Aldreth causeway, which was disputed by them, was in 1453 transferred to John Candeler, one of the bishop's tenants in Haddenham.[43] As late as 1490 many of the bishop's tenants earned their living on water rather than land. Richard Purviaunce of Outwell 'husbandman alias waterman' received exemption from fair and market tolls in that year.[44] On the eve of the Dissolution (1536) 200 acres of marsh, specified as being in 'Beryallfield', were held by Antony Cotton in socage. There was also a military tenure, Thomas Fincham holding 240 acres known as 'Warnamy' formerly of Everard de Vernun, by service of ⅛ fee. The total rents were lower than in the 13th century, viz. £2 3*s.* 2*d.*[45] A long series of bailiff's accounts, mainly of the reign of Edward IV, shows rents of assize usually of £3 6*s.* 8*d.* and perquisites of court ranging between 10*s.* and £1 a year. The total net receipts were as much as £7 14*s.* 2*d.* in 1465. They tended, however, to decrease, for they were only £5 19*s.* 10*d.* in 1481, £4 16*s.* 4*d.* in 1492, and £3 19*s.* 1*d.* in 1521. Arrears of rent were exceptionally high in the earlier rolls, amounting in several years to more than £10.[46] In 1548–9 the receipts had risen to £6 6*s.* 7*d.* owing partly to an increase in court perquisites to £4.[47] At the end of the 16th century the receipts averaged £4, with a tendency to rise.[48] This manor was not one of those alienated to the Crown by Bishops Cox and Heton,[49] and in the late 18th century the bishops were still granting leases in Euximoor and Laddas Fens.[50] By the mid-19th century, however, the bishops of Ely were no longer 'principal landowners' in either Outwell or Upwell.

The cathedral priory retained certain rights in the vills after 1109. Rakenold of Well was responsible for providing 2,000 turves yearly for heating the guest house.[51] At the Dissolution these turves were still due each year to the hosteller of the monastery.[52] One of the fisheries, 'Livermere' on the borders of Elm, Upwell, and March, was retained by the monks; it was worth £1 2*s.* in 1291.[53] At the Dissolution the priory estates in Upwell were valued at £2 16*s.* 9*d.*;[54] they were formally made over to the dean and chapter in 1541.[55]

A ploughland in 'Wella' held in 1086 by Rainald son of Ivo has been ascribed to Upwell, but is more likely to refer to Beechamwell in Norfolk.[56]

In the reign of Edward I, John son of Gilbert de Beaupré held a manor in 'Welles'.[57] His wife was Christian, heiress of the St. Omer family, which had held common of pasture in Upwell marsh at the end of the reign of Henry III and numbered several public officials among its members.[58] The Beauprés continued in possession for more than 200 years. Edmund, the ninth of this line, was dealing with the manor of *BEAUPRÉ HALL* in 1541[59] and received the Ramsey abbey manor shortly afterwards.

Another lay tenant in Upwell in 1086 was Hermer de Ferrers. His Saxon predecessor Wihenoc had seized (*occupavit*) all the six subtenants, i.e. had made them customary instead of free tenants, although only three were rightly of that status. There were also 9 bordars. The extent of Hermer's tenement was 2 ploughlands and 15 acres, with 2 ploughs, worth £1 6*s.* 8*d.*[60] Hermer's chief seat was at Wormegay, his descendant William de Wormegay holding a number of fees in Norfolk.[61] William's daughter Alice married Reynold de Warenne, and their granddaughter Beatrice married Doun Bardolf,[62] in whose family these possessions descended. Upwell first figures individually in the Honour of Wormegay at the death of Hugh Bardolf in 1304, when with Stow Bardolph it was extended at a capital messuage, 265 acres of land, 5 of them woodland, a marsh, a windmill, 2 fisheries producing 57 sticks of eels, 2*s.* rents and pleas of court.[63] This was held in chief by commuted service to Norwich castle guard and to 'waytefe'. Stow Bardolph, perhaps including Upwell, is mentioned in other inquisitions on the holders of the barony,[64] but before the end of

35 W. Watson, *Hist. Wisbech*, 539.
36 Blomefield, op. cit. vii, 459. Greaves refitted the derelict Hall 'in a very elegant manner' (B.M. Add. MS. 5809, f. 114).
37 Gardner, *Dir. Cambs.* (1851), 556, 563; Kelly, *Dir. Cambs.* (1933); Burke, *Landed Gentry* (1937), s.v. Townley.
38 *Cart. Sax.*, ed. Birch, iii, no. 1266.
39 *V.C.H. Cambs.* i, 369.
40 *Bk. of Fees*, 629.
41 B.M. Cott. MS. Tib. B. II, ff. 158–9.
42 B.M. Cott. MS. Claud. C. XI, ff. 94–96.
43 *Ely Dioc. Remembrancer*, no. 216, p. 111.
44 Ibid. no. 282, p. 164.
45 B.M. Add. MS. 5837, f. 205.
46 Bp.'s Mun., Ely, D 5 (3).
47 B.M. Add. Chart. 34274, f. 19.
48 S.C. 6 Eliz./282 & /283.
49 Blomefield, *Hist. Norf.* vii, 456, states that it was, but gives no reference.
50 A. Gibbons, *Ely Episc. Rec.* 25.
51 B.M. Add. MS. 5847, f. 150*b* (Rand's Cambs. Collect, no date stated).
52 D. & C. Mun., Ely.
53 *Fenland N. & Q.* iii, 142–4; *Tax. Eccl.* (Rec. Com.), 101.
54 Dugd., *Mon.* i, 494. This does not include the site of Molycourt Priory (see below, note 83).
55 *L. & P. Hen. VIII*, xvi, g. 1226 (12).
56 *V.C.H. Norf.* ii, 155; Blomefield, op. cit. vii, 286–7. The descent is more convincing if this estate is ascribed to Beechamwell.
57 Blomefield, op. cit. vii, 457, quoting MSS. of Bell family.
58 Ibid. vii, 456; *Cal. Pat.* 1247–58 to 1272–81.
59 C.P. 25(2)/52/376/Trin. 33 Hen. VIII.
60 *V.C.H. Norf.* ii, 199.
61 *Red Bk. of Exch.* (Rolls Ser.), 392, 395, 399.
62 Blomefield, op. cit. vii, 493–4.
63 *Cal. Inq. p.m.* iv, no. 236 (p. 169).
64 Ibid. xi, 573; C 139/103.

the 13th century at least some of the Upwell property had been granted to William de Cryketot. He died in 1299 holding 4 cottages, 4 acres of land, and 27 of marsh of Sir Hugh Bardolf, the abbot of Ramsey and others, mainly by money rents.[65] The last male Cryketot proved his age in 1368[66] but died childless, and the estate passed to his sister Joan. She with her husband Richard de Pakenham made a settlement of the manors of Outwell and Upwell in 1373.[67] Four years later they passed these manors, and the fourth part of a fishery, to Nicholas Goddard of Terrington.[68] The next recorded owner was Gilbert Haultoft, baron of the Exchequer, who was buried in Outwell church (1458). From him it passed twice through female lines to the Finchams, of whom John (d. 1528)[69] was the first possessor. This family still held the lordship when Blomefield was writing (*c.* 1730–40); the manor was then known as *LORD BARDOLPH'S* alias *CRIKTOTS*.[70] About 1800 it was in the possession of Heaton Wilkes, youngest brother of the politician, who being financially embarrassed sold it to James Lee of Upwell, and to Messrs. Boyce and Allenby, the last of whom soon afterwards resold his portion in small lots.[71]

William de Warenne had also six bordars in Outwell in 1086.[72] Nearly 200 years later William de Goldinham, the steward of his descendant, aided by Hugh de Blicwic the under-bailiff, was reported to be holding a court at Outwell by unknown authority, compelling attendance from outside (*forincecos*) tenants and levying, to the king's prejudice, fines amounting to about £40.[73] Soon afterwards the Warenne holding disappears. It was probably granted to Lewes or to Castle Acre priories, each of which had had a free tenant in 'Welles' in 1210–12,[74] and whose possessions there in rents were respectively valued at £3 1*s.* and 8*s.* in 1291.[75]

The Prior of Lewes, as lord of West Walton, had rights in the common pasture of Upwell marsh, which brought him into conflict with other commoners in that village towards the end of the reign of Henry III.[76] These two priories (Lewes and Castle Acre) seem to have disposed of their estates in Upwell before the Dissolution, as they are not mentioned in the fines conveying the possessions of these monasteries to the Crown.[77]

A great part of Outwell and Upwell passed at various times into monastic ownership; in 1291 no fewer than 16 religious houses together held lands, rents, and fisheries there valued at £64.[78] There were two small priories in the locality itself. In 1291 Molycourt, on the Norfolk side, where the Well Stream is now crossed by the Middle Level Main Drain, held in Welles lands valued at £5—a sum representing rather more than half the total value of its estates.[79] In 1306 John son of Gilbert leased 3 acres in Outwell called 'Milleresland' to the priory for 12*d.* a year for 2,000 years. This is interesting as an early example of a term so long as to amount almost to a perpetuity.[80] Molycourt Priory suffered much from floods in the next hundred years,[81] and was amalgamated with Ely cathedral priory in 1446, when the endowment included 7 houses and some land in Outwell and Upwell.[82] At the Dissolution it yielded a net return (including the actual site of the priory) of £6 19*s.* 2*d.*[83]

The estates of this cell spread over Marshland, in Downham, the Wiggenhalls, Elm, Emneth, Wisbech, Walsoken, and Leverington as well as Outwell and Upwell. They were kept together after the Reformation and let out on twenty-one-year beneficial leases by the Dean and Chapter of Ely for £20 5*s.* 11*d.*, as the manor of *MULLICOURT HOUSE*. The first recorded tenant (1638) was Oliver Cromwell. After the Restoration this 'manor' was leased to successive generations of the Legg, Gibson, and Burton families, all non-resident, until 1803 when a twenty-one-year lease was granted at the customary rent to John Miller and William Stevens, both of London, trustees of the executor of Philip Burton. Their successors were (1827) 'the representatives of the late amiable Bishop Horne, of Norwich'.[84] After this date the estate was divided into three parts and tenancies were sometimes acquired by local persons, e.g. William Lee (1839) and Hanslip Palmer (1847), both of Upwell.

In a survey of 1649 quit-rents were valued at £1 12*s.* 4*d.* and perquisites of court at 3*s.* 4*d.* The total property then comprised 259 acres of an annual value of £187 12*s.* 8*d.* Of this area 88 acres—or rather more than one-third—was in Outwell and Upwell; it was valued at £85 2*s.* 8*d.* The following year the manor was sold for £1,296 10*s.* 11*d.* to Blunt and Edward Sadleir, merchants of London, who resold in 1651 at the same price to Samuel Calverley of Outwell. Calverley was allowed to remain as tenant when the dean and chapter returned as lords in 1660, but his twenty-one-year lease made that year was not renewed beyond 1668.[85] Fines for the renewal of the lease were £110 in 1704, £300 in 1717, and £185 in 1724.[86]

In 1251 the priory of Marmont,[87] now represented by Marmont Priory Farm on the Cambridgeshire side of Upwell village, held 100 acres as a free tenant of the bishop at 2*s.* yearly.[88] In 1291 the priory drew £11 0*s.* 8*d.* from rents and fisheries in Upwell.[89] In 1535 it derived £7 9*s.* 10*d.* from Upwell—9*s.* 9*d.* from rents on the Norfolk side and £7 0*s.* 1*d.* from a 'manor' on the Cambridgeshire side.[90] The site was granted in 1568 to Percival Bowes and John Mosyer;[91] it retained its identity as a reputed manor for another hundred years,[92] and early in the 19th century belonged to a Mr. Bacon, who had purchased it from Mr. Thomas Audley of Lynn.[93]

Besides Ramsey other monasteries not locally situated

65 *Cal. Inq. p.m.* iii, no. 537.
66 Ibid. xii, no. 254.
67 *Cal. Close*, 1369–74, 576, 582.
68 W. Rye, *Norf. Feet of F.*, 369.
69 C 142/49/62.
70 *Hist. Norf.* vii, 462. See also C 3 /377/18, /383/95, and /402/54.
71 Watson, *Hist. Wisb.* 531.
72 *V.C.H. Norf.* ii, 83.
73 *Rot. Hund.* (Rec. Com.), i, 462, 519, 543.
74 *Bk. of Fees*, 629.
75 *Tax. Eccl.* (Rec. Com.), 101.
76 *Norf. Portion of Chart. Lewes* (Norf. Rec. Soc. xii), no. 198. This cartulary records a number of small grants here to Lewes Priory, mainly 13th century.
77 Dugd., *Mon.* v, 20, 55.
78 *Tax. Eccl.* (Rec. Com.), 101.
79 Ibid. The prior owed suit of court to the Bishop of Ely in 1251 for a croft, rent 4*d.* (B.M. Cott. MS. Claud. C. XI, f. 95).
80 D. & C. Mun., Ely, Carta 856.
81 For general history of this priory, see *V.C.H. Norf.* ii, 349–50.
82 *Cal. Pat.* 1441–6, 456.
83 Dugd., *Mon.* v, 20, 55.
84 Watson, *Hist. Wisb.*, 532.
85 D. & C. Mun., Ely, Cupbd. B.
86 Ibid. Bentham Transcrs.
87 See *V.C.H. Cambs.* ii, 258–9.
88 Bp.'s Mun., G 3 ('Old Coucher'), f. 49. This piece of land had formerly been held by Henry de Hauteville.
89 *Tax. Eccl.* (Rec. Com.), 101. The prior was also himself a tenant of the Bishop of Ely in respect of 40 acres (B.M. Cott. MS. Claud. C. XI, f. 95).
90 *Valor Eccl.* (Rec. Com.), iii, 379.
91 B.M. Add. MS. 5805, f. 103.
92 C.P. 25(2)/400/Mich. 2 Chas. I; C.P. 25(2)/632/Mich. 12 Chas. II.
93 Watson, *Hist. Wisb.* 543.

in the district held property in Upwell and Outwell. Of these the abbey of West Dereham (Norf.) was the most extensively endowed. During the 13th century several local persons made grants to it. The most important of these benefactors was William le Curteis, who gave 23 acres, 7*s.* 4*d.* in rents and the homage of five tenants in 'Welle',[94] and Thomas de Burch, who gave a messuage and 50 acres in Upwell.[95] In 1291 the West Dereham estates were reckoned as a manor, worth £10 13*s.* 4*d.*[96] At the Dissolution the value had increased to £15 13*s.* 3*d.* net.[97] In 1554 the manor was granted to Edmund Beaupré,[98] and thenceforth became merged in the main Ramsey abbey manor.[99]

The priory of Walsingham (Norf.) also received grants from the de Burghs and other local families;[1] its lands were valued at £1 10*s.* in 1291 but at only 16*s.* in 1535.[2] A 'great close' in Outwell called 'Wadingstowe', named in the cartulary of this monastery, was in the tenure of Thomas Calleye in 1544, when it was granted to John Eyer. At the same time Eyer received a messuage and croft in Upwell, formerly the property of Bury abbey, which had been leased to him the previous year.[3] This messuage and croft had been appropriated to the cellarer of the abbey in 1291, and were then worth 12*s.*[4]

Ixworth Priory (Suff.) held a fair amount of property in Upwell, of which unusually full documentary evidence survives. It lay in the west and south of the Cambridgeshire side of the parish. The estate was generally known as Thirling, and can be identified with Thurland's Drove on the western boundary of Upwell near Laddus Bank. Appurtenances of the manor, as the property was called after the Reformation, lay in Welney. Thirling itself seems to have been merely a farmhouse or grange.[5] After the Dissolution a church with a rectory and an advowson is mentioned at intervals in connexion with the manor.[6] The words may recall a chapel or oratory once attached to Thirling grange; on the other hand they may have been inserted into conveyances *pro majori cautela*.

The canons of Thirling held 80 acres in frank almoin in 'Well' in 1221,[7] and seventy years later the priory drew £2 8*s.* from rents and fisheries there.[8] In 1535 the value of the Ixworth property in Upwell had risen to £5 9*s.* 10*d.*[9] Some of it lay next the West Dereham lands, and was in 1355 leased in succession for three lives to Adam Borewyne, Maud his wife, and Henry his son.[10] After the Dissolution the Ixworth property was granted in fee to Thomas Megges.[11] Nicholas Megges, who succeeded Thomas in 1542,[12] and Joan his wife passed it to (Sir) Thomas Gresham and his wife Anne in 1551.[13] The manor of *THIRLING* was valued at £9 yearly at Sir Thomas's death in 1579,[14] and was held in chief as $\frac{1}{10}$ fee in 1596.[15] Gresham's son Richard died unmarried before his parents, and the manor passed at Lady Gresham's death (1596), in consequence of a former marriage of hers, to the Reade family.[16] Sir William Reade made a marriage settlement in 1606,[17] and at his death in 1621 the manor passed to his three granddaughters, Jane Viscountess Fitzwalter, Elizabeth wife of Sir George Berkeley, and Bridget Stanhope.[18] The last named was a minor at this date, and her third share was united with that of the Berkeleys in family settlements made in 1640 and 1653.[19] Bridget married George Fielding 17th Earl of Desmond[20] and as Countess of Desmond is recorded in 1649 as holding 200 acres in 'Thurlands', worth £15 a year.[21] Jane and her second husband Sir William Withypoll were dealing with their third in 1629, and their daughter Elizabeth with her husband Leicester Devereux in 1647.[22] The latter, as Viscount Hereford, conveyed his share in 1666 to John Berridge,[23] whose family had by 1681 obtained the other two-thirds. Berridges were still in possession in 1705.[24]

A manor of *BUDBECK* originated in a messuage and 12 acres of that name, of which Reynold son of Thomas son of William Andrewe made a settlement in 1442.[25] This family also held the manor of Vernons in Elm (q.v.), with which Budbeck descended. In 1649, when Thomas Fincham was lord, the quit-rents were £2 5*s.* 1*d.*[26]

The manor of *COURTEYS FEE* presumably originated in the lands of the Curteis family, tenants under Ramsey Abbey, and benefactors of that foundation and West Dereham Abbey[27] (see above). In 1592 Richard Buckeworth died seised of it, and it is recorded for nearly a century afterwards.[28]

CHURCHES. The Norwich Valuation of 1254 assessed the churches of Outwell and Upwell at £13 6*s.* 8*d.* and £6 13*s.* 4*d.* respectively.[29] Contrary to expectation, the *Taxatio* of 1291 put these churches at lower values, namely £6 13*s.* 4*d.* for Outwell and £4 for Upwell.[30] They were never appropriated, but in 1291 £2, and in 1535 £1 6*s.* 8*d.* were payable from Upwell rectory as portions to Ramsey abbey, and at the latter date 6*s.* 8*d.* was diverted from Outwell rectory to the church[31] of Elm.[32] The total values in 1535 were £16 18*s.* 3*d.* (Outwell) and £17 17*s.* 3*d.* (Upwell).[33] With the drainage of the fens, the value of Upwell rectory

94 B.M. Add. MS. 5842, f. 215. Blomefield (*Hist. Norf.* vii, 333) states that he bequeathed his body to be buried there.
95 B.M. Add. MS. 5805, f. 98*b*.
96 *Tax. Eccl.* (Rec. Com.), 109.
97 *Valor. Eccl.* (Rec. Com.), iii, 376–7.
98 *Cal. Pat.* 1553–4, 189–90.
99 C 142/182/28.
1 B.M. Cott. MS. Nero E. VII, ff. 132–5. The total area is put at 380 acres, but the 20 grants or so copied into the cartulary do not amount to anything like so much.
2 *Valor Eccl.* (Rec. Com.), iii, 385–6.
3 *L. & P. Hen. VIII*, xviii (1), p. 556; xix (2), g. 340 (22).
4 *Tax. Eccl.* (Rec. Com.), 101.
5 For the 'monastery' of Thirling see D. Knowles, *Relig. Houses of Medieval Engl.* 90–91.
6 e.g. *Cal. Pat.* 1550–3, 46; C.P. 25(2)/259/Mich. 12 & 13 Eliz. The Thirling Chapel may have been that of St. Botolph (see below—Churches).
7 B.M. Cott. MS. Tib. B. II, ff. 158–9.
8 *Tax. Eccl.* (Rec. Com.), 108.
9 *Valor Eccl.* (Rec. Com.), iii, 482.
10 B.M. Add. Chart. 18650. Other leases by Ixworth Priory are Add. Charts. 9191, 18761.
11 *L. & P. Hen. VIII*, xiii (1), g. 1309 (40).
12 C 142/69/95.
13 *Cal. Pat.* 1550–3, 46; C.P. 25(2)/66/549.
14 B.M. Add. MS. 5805, f. 103; C 142/192/11.
15 C 142/249/57.
16 Ibid.
17 C 142/396/137.
18 Ibid.
19 C.P. 25(2)/528/Trin. 16 Chas. I; C.P. 25(2)/538/East. 1653.
20 *Compl. Peerage*, iv, 258.
21 D. & C. Mun., Ely, Cupbd. B.
22 C.P. 25(2)/526/Hil. 4 Chas. I; C.P. 25(2)/528/East. 23 Chas. I.
23 C.P. 25(2)/653/East. 18 Chas. II.
24 C.P. 25(2)/634/Trin. 33 Chas. II; C.P. 43/489, ro. 195.
25 *Cal. Close*, 1441–7, 77, 119.
26 D. & C. Mun., Ely, Cupbd. B.
27 *Cart. Monast. Rames.* (Rolls Ser.), i, 99, 100; ii, 375, 376; iii, 296–9.
28 C 142/233/106; C.P. 25(2)/634/Hil. 32–33 Chas. II.
29 W. E. Lunt, *Valuation of Norw.* 408.
30 *Tax. Eccl.* (Rec. Com.), 88.
31 *Ecclesia*. It is not certain whether the pension went to the sinecure rectory or the vicarage of Elm, of which only the gross values are entered in the Cambs. Portion of the *Valor*. The Ramsey portion was appropriated to the sacrist. (Dugd., *Mon.* ii, 574).
32 *Valor Eccl.* (Rec. Com.), iii, 380.
33 Ibid.

increased very greatly. In 1851 it was stated to be worth £6,000 a year.[34] This is probably an exaggeration, but an authoritative return eight years later put the value at £3,855.[35] The parish, which at that time included Welney, Christchurch, and Nordelph, comprised some 25,000 acres of the most fertile soil in the country.

The chapel of Welney, which is not mentioned in the *Valor*, was in existence probably before the middle of the 17th century and certainly by 1763.[36] In 1862 it became parochial, under the Upwell-cum-Welney Division Act of sixteen years earlier.[37] The ecclesiastical districts of Christchurch and Nordelph were formed out of Upwell in 1866[38] and 1909[39] respectively. The latter is entirely in Norfolk and comprises portions of Downham Market and Stow Bardolph as well as Upwell.

The advowsons of the churches of Outwell and Upwell have followed the descent of the Ely and Ramsey manors respectively (q.v.). That of Welney rectory was originally held by the Townley family, patrons of the mother church, but since 1938 presentations have been made by the Bishop of Ely. Those of Christchurch rectory and Nordelph vicarage have been retained by the Townley family.[40]

A chapel or hermitage of St. Christopher was established in the high street of Outwell in 1348.[41] No trace remains. In 1571 its site, then occupied by Edward Jenkyns, was granted to Richard Hill and Robert Don.[42] According to Blomefield there was also a chapel in Upwell in the Middle Ages, dedicated to St. Botolph,[43] which was also granted to Hill and Don in 1571, when its ½-acre site was occupied by William Christen.[44] The following year it was regranted to Hill and William James.[45] Watson mentions an old house near the bridge, the last in the village on the road to Outwell, which had the appearance of formerly having been a chapel, and a roadway nearby called Pious Drove (now Pious Lane).[46] This may be the chapel which, with two cottages, all in the tenure of the churchwardens, was in 1549 granted to William Warde.[47]

A guild of St. John in Upwell was dissolved in 1547; its hall was sold to Richard Hill and Robert Don in 1571 and regranted to Hill and William James the following year.[48] This guildhall was the meeting-place of the courts leet of the Bishop of Ely in 1648.[49] Hill also received a close of pasture which had belonged to the chantry of Marmons in Upwell.[50] A 'Guyldehall' in Welney was in 1550 sold to William Place and Nicholas Spakeman,[51] and in 1568 to Hugh Counsaille and his heirs.[52]

The church of *ST. CLEMENT, OUTWELL*, consists of chancel, north and south chapels, clerestoried nave, aisles, north chapel in the form of a transept, south porch, and west tower. There was a church on the site in the 13th century, of which the only visible remains are the first three stages of the tower. In the 14th century the present arcades were built and the aisles assumed their existing form. In the second half of the 15th century there was an extensive scheme of rebuilding and enlargement which was not completed until about 1520, the north chancel chapel being approximately of this date. In 1863 the church was severely restored, the east end of the chancel being practically rebuilt, largely with the old materials, and new roofs provided for the chancel and north aisle. The walls were replastered, and this must have involved the destruction of at least one mural painting which was noted by Blomefield in the middle of the 18th century. The materials are Barnack and rag stone plastered, except the north nave chapel which is of brick with stone dressings. The roofs of the chancel, nave, south aisle, north nave chapel, and porch are slated, and the rest leaded. There are embattled parapets throughout, except to the north aisle.

The chancel has a fine five-light east window with embattled transom, with cinquefoiled heads above and below the transom and cinquefoiled tracery. There are angle buttresses with three set-offs. The plinth has a band of quatrefoils enclosing blank shields. The chancel extends beyond the chapels and has a north and south window of three lights with rectilinear tracery; the latter has a transom in the main lights and also in the tracery. There are crocketed angle pinnacles which have been renewed. The south window has a lowered sill to serve as sedilia. There is a four-centred arch opening to each chapel, the south with continuous mouldings and the north with demi-angels as caps. The lofty chancel arch is two-centred and of two orders with moulded caps. The chancel roof is modern, of hammerbeam construction; the 15th-century stone corbels remain and consist of demi-angels bearing shields.

The north chapel, which from the evidence of the heraldry of its roof corbels must have been erected in the first quarter of the 16th century, has an east window of six lights with an embattled transom and a four-centred head with hood-mould, on either side of which are much-worn figures under canopies. There is a diagonal buttress with three set-offs at the north-east angle, and in the north wall are two three-light windows with embattled transoms and four-centred heads. The chapel opens to the aisle by a truncated four-centred arch with a moulded cap having a shield on the north, but no respond. Near this arch on the south side are remains of the stairs which formerly led to the rood loft. There is a four-centred doorway leading to the north nave chapel with square hood-mould terminating in carved heads; it is set at an angle in the north-west corner. There is a piscina with cinquefoiled head in the south wall. The roof is of cambered beam construction with moulded principals and purlins, and demi-angels bearing shields. The stone roof corbels have shields charged with the arms of Fincham, Haultoft, and Derham, which indicates that the chapel was built by John Fincham, who married Elizabeth daughter and heiress of Thomas Derham, and Alice his wife, who was daughter and coheiress of Gilbert Haultoft. John Fincham died in 1527.

[34] Gardner, *Dir. Cambs.* 564.
[35] *Clergy List.*
[36] The register of baptisms dates from 1642; it is officially mentioned in Ecton's *Thesaurus* (1763). There seems to be no foundation for the statement in J. C. Cox, *Norf. County Churches*, ii, 106, that the chapel was chiefly of Norman date.
[37] 9 & 10 Vic. c. 7. The delay in implementing the Act was due to the fact that no vacancy in the undivided rectory occurred until 1862.
[38] Ibid.
[39] Kelly, *Dir. Norf.* (1937).
[40] *Crockford's Clerical Dir.*
[41] *Cal. Pat.* 1348–50, 92.
[42] C 66/1074, m. 5.
[43] Blomefield, *Hist. Norf.* vii, 470.
[44] C 66/1074, m. 5.
[45] C 66/1083, m. 16.
[46] *Hist. Wisb.* 548.
[47] *Cal. Pat.* 1548–9, 326.
[48] Ibid. 1548–9, 116; W. M. Palmer, *Village Gilds of Cambs.* 399; C 66/1074, m. 5; C 66/1083, m. 16.
[49] D. & C. Mun., Ely, Cupbd. B.
[50] C 66/1083, m. 16.
[51] *Cal. Pat.* 1549–51, 410, 414
[52] C 66/1046, m. 2.

The south chapel has a five-light east window with embattled transom and cinquefoiled main lights and three rows of tracery lights. There are angle buttresses with three set-offs having quatrefoils enclosing blank shields on the faces. The south window has six lights with an embattled transom and four-centred head. There is a tomb recess of the 14th century in the south wall with elaborate cusping to the arch and angle shafts. The roof is of hammerbeam construction with full-length angels as hammers on the south and demi-angels on the south wall-plate; the hammers on the north are moulded and formerly terminated in shields. The wall-plate on this side has an elaborate carved trail. There are unusually short wall-posts.

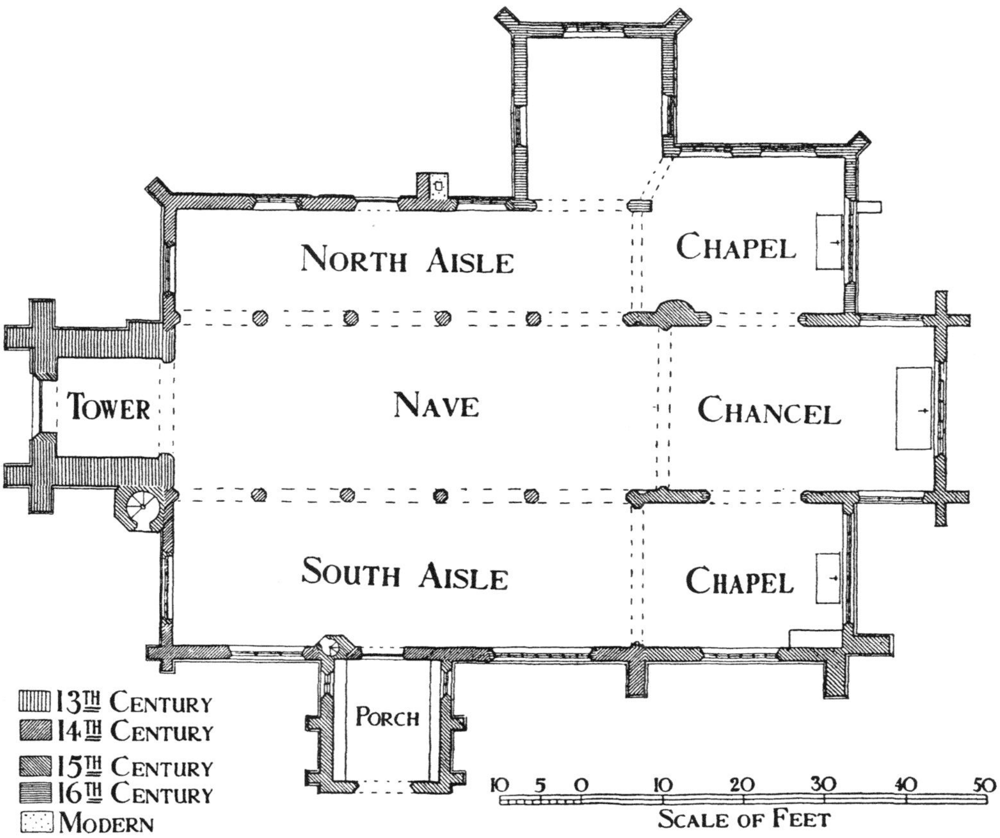

Plan of St. Clement's Church

The nave has arcades of five bays with two-centred arches of two orders and octagonal piers with moulded caps and bases; it belongs to the first half of the 14th century. There are five clerestory windows on each side, of three lights with cinquefoiled heads and hood-moulds terminating in human heads. There are gargoyles on either side. Over the chancel arch is a three-light window with cinquefoiled heads. The clerestory is a 15th-century addition. The lofty tower arch is of two orders, the outer with continuous mouldings and the inner with demi-angels as caps. The roof is a fine example of the Marshland type with alternate hammerbeams and tiebeams having queen-posts, the spandrels have pierced tracery, and there are demi-angels on the tiebeams, hammerbeams, and wall-plates. The wall-posts have full-length figures under canopies and there are carved stone corbels.

The north aisle has two lateral windows of three lights with cinquefoiled heads, of the 15th century, and a similar window at the west end. There is a four-centred arch from the aisle to the north nave chapel, with moulded caps and bases. The north doorway has a continuous chamfer and is of the 15th century. There are two buttresses with two set-offs, of the same period. The roof is modern.

The north chapel is set at right angles to the aisle. It has two diagonal buttresses with four set-offs. The east window is of three lights, the centre light trefoil-headed and the others uncusped. The north window has four uncusped lights under a depressed head, and the west window is of three lights with cinquefoiled heads. The roof is of hammerbeam construction with shield-bearing angels on the hammers and wall-posts.

The south aisle opens to the chapel by a late-15th-century arch of two orders. To the east of the porch is a seven-light window with embattled transom and depressed head, probably early 16th century. West of the porch is a three-light 14th-century window, cinquefoiled with a quatrefoil above. The west window is of the 15th century and has four cinquefoiled main lights with rectilinear tracery in the head. The plinth has a band of quatrefoils enclosing blank shields, and the buttresses have three set-offs. The roof is similar to that of the chapel, but it retains the shields on the north hammerbeams, and only the western portion of the wall-plate has a carved trail.

The porch, which belongs to the 15th-century reconstruction, is of two bays with two-light windows on each side in the inner bay, and with a parvise. It has an octopartite vault with brick web and carved bosses. The outer arch is four-centred with moulded caps and chamfered bases to the responds. The inner doorway has a continuous chamfer. The parvise is approached by a newel stair at the north-west angle with a doorway in the aisle having a four-centred head. The parvise is lighted by windows on the east, west, and south, of two lights.

The tower is of four stages, the first three being of the second half of the 13th century and the belfry stage a century later. The west doorway and window are 15th-century insertions, the former having a four-centred arch and continuous mouldings, and the latter being of four lights cinquefoiled with rectilinear tracery in the head. There are two blocked lancets on the north and south in the second stage. The third stage has windows of two lights with foliated circles in the head and shafts with moulded caps and bases as mullions and at the angles. The top stage is a 14th-century addition having two-light windows with trefoiled heads. There is a plain parapet and crocketed pinnacles at the angles, and the tower is crowned by a leaded pyramidal cap, which replaces a lofty spire removed in 1753. There are angle buttresses with four set-offs. At the south-east corner is an octagonal stair turret reaching to the third stage; this is a 15th-century addition and was formerly reached by a doorway in the south aisle, now blocked, with four-centred head and shields in the spandrels; the modern entrance is from the churchyard.

The fittings are important. The 15th-century font has a hexagonal bowl with two shallow cinquefoiled niches on each face and one on each face of the shaft. There is a fine eagle lectern of latten, also of the 15th century, and a massive iron-bound chest with seven locks. In the south aisle is a remarkable wooden alms box standing on a tall shaft, which dates from the beginning of the 17th century. Of the same period is the communion table now in the south chapel. There is a brass with inscription to Richard Quadryng, 1511, now on the wall of the north aisle; he is in armour of the period. In the same aisle is a mid-15th-century brass inscription to Margaret Haultoft, and in the south aisle is the matrix of a knight and lady. In the south-east corner of the south chapel is a Purbeck marble altar tomb with back-piece and small projecting canopy; on the north and west sides are quatrefoils containing shields, and on the back-piece is an alabaster tablet of later date commemorating Nicholas Beaupré and Margaret his wife, 1512. There is a shield with the arms of Beaupré impaling Fodryngaye. The tomb is apparently of this date, but the tablet is fifty years later and was probably inserted by his son Edmund, who is himself commemorated by an alabaster mural monument supported on the canopy of the altar tomb with his shield of arms. The south chapel is paved with black and white marble and in the west arch is an iron screen on a stone base, all 18th-century work. There is a considerable amount of early-16th-century painted glass. In the tracery of the east window of the chancel are portions of canopies, heads, and inscriptions. In the tracery of the east window of the south chapel are the following subjects: top row—(1) chalice and Host, (2) Deity seated and holding orb, (3) St. Faith, (4) Five Wounds; middle row—(1) fragments of canopies, (2) St. Olaf, (3) St. Edward the Martyr, (4) St. Walstan, (5) St. Martin, (6) St. Lawrence, (7) St. Stephen, (8) fragments; bottom row—(1) scroll, (2) hand holding cup, (3) shield of Ely, (4) Virgin and Child and two other figures, (5) St. Edward the Confessor, (6) fragments, (7) fragments, including shield of St. George, (8) part of female saint, (9) St. Ursula, (10) scroll. In the north-west window of the north chancel chapel is a large figure holding a covered cup, part of the Adoration of the Magi, which was almost complete when Blomefield visited the church in the middle of the 18th century; above is an angel holding a shield with Fincham quartering Haultoft.

The plate includes (i) a communion cup and cover paten without assay marks, but the former is inscribed 'the perishe of Outwell anno domini 1593'; (ii) a paten of silver without assay marks inscribed 'Outwell St. Clement'; (iii) a flagon of silver, 1652, inscribed 'This flagon belongs to the Parish of Outwell', and above these arms: per chevron argent and azure a crescent counterchanged, and for a crest an arm embowed mailed, the hand holding a tilting spear broken enfiled with a chaplet.

The tower contains six bells by Osborn and Arnold of St. Neots, 1778, the 4th recast by Dobson of Downham, 1827.

The registers begin in 1559 and are complete.

The church of *ST. PETER, UPWELL*, consists of chancel, north vestry, clerestoried nave, aisles, north porch, and north-west tower. The material is Barnack stone and rag. The earliest portions of the present fabric are the west wall of the nave and the two lower stages of the tower, which are of the middle of the 13th century. The fact that the east arch of the tower is contemporary proves that there was a north aisle to the nave at that time. A third stage, octagonal in plan, was added to the tower in the 14th century. In the second half of the 15th century the church was almost entirely rebuilt on an enlarged plan. In the 19th century, but rather earlier than usual, there was an extensive restoration and refitting, which greatly altered the appearance of the interior. There are large galleries in the north aisle and at the west end of the nave. The vestry, which is of brick covered with stucco, is probably of this period, though it is likely that it is on the site of an earlier structure. There is an embattled parapet on all parts of the building, but it is modern or at any rate greatly renewed.

The chancel has a four-light east window with renewed tracery of 15th-century type. There are angle buttresses with three set-offs, and this type is found in all parts of the church except at the west end, where a 13th-century example remains. There is a continuous string-course carried round the buttresses. The lateral windows of the chancel are of three lights with cinquefoiled heads and rectilinear tracery. The east window on the north is curtailed to accommodate the vestry, a proof that a vestry was contemplated from the first. The middle window on the south is similarly curtailed to accommodate a doorway with continuous chamfer and a hood-moulding terminating in masks; this doorway has been renewed with the exception of the hood. The doorway to the vestry has a four-centred head and continuous chamfer, of 15th-century character, but it has been so daubed with modern paint that it is difficult to decide how far it is old work. The wide chancel arch is four-centred with moulded caps and bases to the responds. There is a piscina with trefoiled head and ogee hood-moulding, and good canopied sedilia with panelled back and soffit but no divisions. The roof is of hammerbeam type with angels having outspread wings at the end of the beams and demi-angels on the wall-plates. The vestry has a three-light window on the north with cinquefoils under a square head, and a flat plastered ceiling.

The nave arcades consist of six arches on the south and five on the north; they are two-centred and of two orders; the outer order has continuous mouldings while the inner has moulded caps and bases. The composite piers are oblong in plan. Rounded shafts rise to the stone roof corbels. The clerestory consists of two-light windows with cinquefoiled heads under a square label. In the south-east angle of the nave is the rood stair turret, which rises above the roof and is crowned by a sanctus bell-cote of ashlar, the upper part of the turret continuous mouldings and a hood-mould terminating in masks. The 15th-century door remains, with swans carved on the frame. There is a lierne vault, which is almost flat. Above is a parvise with a three-light window on the north with cinquefoiled main lights and cinquefoiled tracery, all contained under a square label with a male and female head as stops; on either side is an empty canopied niche. There is a plain low-pitched beam roof. The embattled parapet is of yellow brick and probably dates from the first half of the 19th century.

The tower is of three stages, the two lower of which

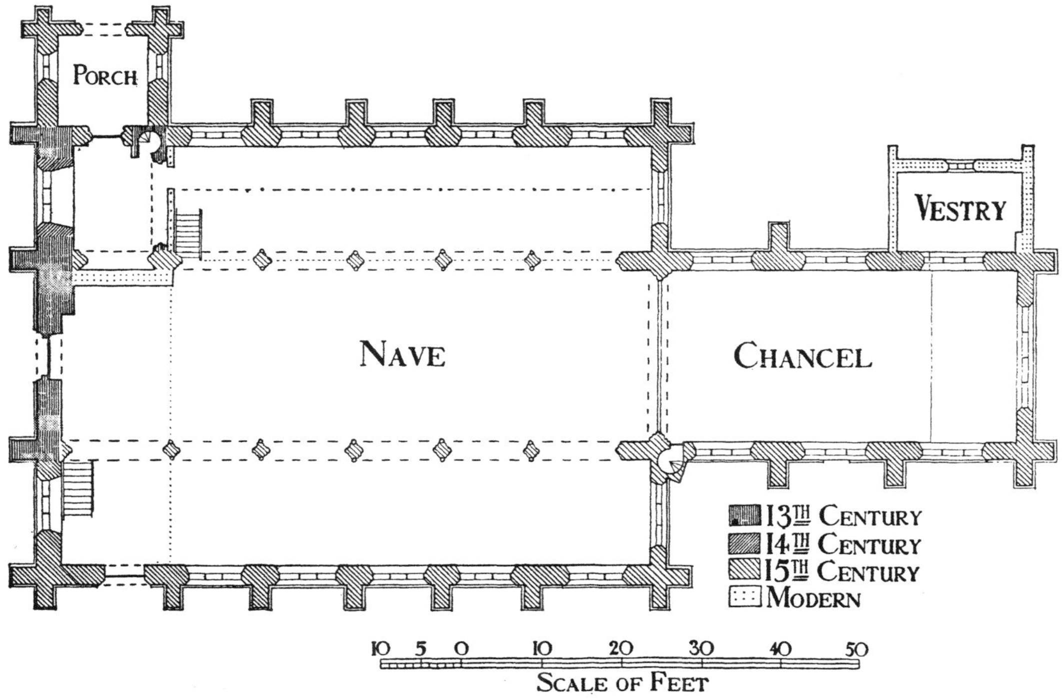

PLAN OF ST. PETER'S CHURCH

being of brick with stone dressings; the plain lower doorway with four-centred head remains but is partially blocked; it opens into the chancel. The west wall of the nave is of the 13th century with a contemporary buttress at the south-west. The doorway, also contemporary, has a two-centred arch of two orders with a continuous chamfer and a hood-mould. The window above is of four lights with intersecting cusped tracery, all modern. The roof is of Marshland type with alternate tiebeams and hammerbeams, the former having queen-posts while the latter are formed of angels with outspread wings; there are demi-angels with outspread wings on the wall-plate and the spandrels have pierced tracery.

The aisles have windows of uniform type with three cinquefoiled lights and trefoiled rectilinear tracery. There is a good south doorway with continuous mouldings and a hood terminating in grotesque heads. The roofs are of the lean-to type with hammerbeams at the outer ends and demi-angels with outspread wings on the wall-plates. The porch has an outer arch of two orders with moulded caps and bases and a hood-mould terminating in masks. There is a two-light window on the east and west with trefoiled heads under a square label. The inner doorway has a four-centred arch with are 13th century and the top stage 14th century. The west window in the lowest stage is of three trefoiled lights and two sexfoils above, and is a 14th-century insertion. The middle stage has a window of two lights on the east, north, and west with a shaft having a moulded cap and base in lieu of a mullion and a trefoil above, the whole contained within a chamfered arch with angle shafts having moulded caps and bases and a hood-mould. The top stage is octagonal with two-light windows on alternate sides with cinquefoiled heads and trefoils above, and a hood-mould terminating in carved stops. The embattled parapet is of brick and probably dates from the end of the 15th century; there are gargoyles at the angles. The base of the tower has arches on the south and east, now blocked. The former is of the 15th century and the latter of the 13th; both are of two orders, and two-centred with moulded caps and bases.

The 15th-century font has an octagonal bowl with demi-angels holding blank shields on the sides, and the shaft has crocketed canopies with flanking pinnacles and plain shields beneath. In the chancel is a fine brass with triple canopy and the figure of a priest in alb, stole, and cope; the inscription is missing, but it probably commemorates William Mowbray, rector, 1428. There is another brass of a priest similarly vested, but without a canopy; the inscription is now missing, but it is known to commemorate Henry Martyn, Rector of Yaxham (Norf.), 1435. There is a brass plate engraved with a man and woman kneeling at a desk, with seven sons and four daughters, commemorating Jane, wife of Sinolphus Bell, who died in 1621. There is an eagle lectern of latten of early-15th-century date with a round moulded base supported on three lions.

The plate includes a communion cup and paten of silver, 1629, an alms dish of silver, 1770, inscribed 'Cristo Crucifixo D.D.D. Thomas Audley', a flagon of silver, 1639, inscribed 'This flagon belongs to the Parish of Upwell', and another alms dish of 1766, inscribed 'the gift of Francis Dixon, 1767'.

The tower contains six bells, 1st and 2nd by Joseph Mallows of East Dereham, 1760; 3rd and 4th by John Draper of Thetford, 1613 and 1627; 5th by Thomas Norris of Stamford, 1634; 6th by C. and J. Mears of London, 1856.

The registers begin in 1650 and are complete except for the years 1671 to 1703.

The churchyard gates were erected when the churchyard was enlarged about 1840. They came from Peterborough Cathedral and the vases on top of the stone piers from Wanstead House (Essex).[53]

CHRISTCHURCH, at the hamlet of the same name, formerly called Brimstone Hill,[54] is a red brick structure with stone facings and tiled roofs. It dates from 1862, and is in the style of the mid-13th century with some foreign modifications. It consists of apsidal chancel, north and south chapels used respectively as vestry and organ chamber, transepts, nave, south porch, and boiler-room in the angle between the nave and the north transept. There is a wooden turret at the east end of the nave containing one bell; the original design included a tower on the north side, but this was demolished shortly before 1883 owing to the insecurity of its foundations.[55] There are oil paintings in the nave representing Christ crowned with thorns, and the Descent from the Cross. The former is a copy of a painting by Holman Hunt. The latter is a copy of a painting by Giuseppe Ribera (Spagnoletto) and was presented by Lt.-Col. E. R. Pratt of Ryston Hall (Norfolk); it was originally brought from Italy by Sir Roger Pratt the architect, a member of this family.[56]

The church of *ST. MARY THE VIRGIN, WELNEY*, is a stone building with a slate roof. It dates from 1848 and replaces an earlier building, the costs of reconstruction being met from 'the overplus of the charities of the township'.[57] The present church is in the 'Geometrical' style and consists of chancel with north vestry, nave, and bell-cote over the chancel arch. There is a chalice dated 1848, and a paten probably of the same date. There are two bells. The registers date from 1642 (baptisms), 1650 (marriages), and 1653 (burials).

The church of the *HOLY TRINITY, NORDELPH*, was erected in 1865 with money bequeathed by the Revd. W. G. Townley. It is of brick with slate roof, in the early 'Gothic' style, and consists of chancel, nave, vestry, south porch, and turret containing one bell.[58]

The Rectory of Upwell[58a] is a late-15th-century brick structure, considerably altered at the end of the 17th century, with a large wing added to the east in the 19th century. The house is a notable example of East Anglian brick construction, with some interesting and unusual features. It is situated to the south-west of the church and was originally approached through a fore-court of which the only remains are two small octagonal towers with newel staircases, which formed the outer angles of the court. The front of the house has three stepped gables, that on the right being over the porch, and there are two rows of blind arcading in brick. There are four stone-framed windows on the first floor on this side with hood-mouldings, which have lost their mullions and have had late-17th-century casements inserted. Below are three more windows, formerly of a similar nature, which have been mutilated for the insertion of late-17th-century leaded casements. The porch has an outer doorway with a two-centred arch of stone with moulded caps to the responds. The inner doorway has a moulded square head and the door itself is panelled with applied tracery. In the angles of the porch are stone roof corbels carved with the Evangelistic symbols. The roof itself is modern. The porch now opens into the modern wing, and on the opposite side of the hall there is another door, which is not *in situ*, with applied tracery and very similar to that on the north. The fact that the porch is now at the extreme east end of the ancient portion of the front suggests that the house originally extended farther in that direction on ground now occupied by the 19th-century addition. This supposition is further supported by the position of the right angle tower of the forecourt. At the back of the house there is a semi-octagonal turret for the stairs, and farther west a blocked doorway and a small window above. There are two ancient gables on this side. Two original windows remain on the west side, a two-light above and a single-light below. There is also on this side the lower part of a turret, which probably contained a newel stair. The chimney-stacks are modern as regards their upper portions. The roofs are slated except that at the west end, which is stone tiled. The interior has been greatly altered and the original plan obscured by later partitions and rearrangement of the rooms. Part of a fine late-15th-century fireplace with four-centred arch is visible on the first floor. The dining-room has early-18th-century panelling.

NONCONFORMITY Nonconformity was very strongly established in Outwell and Upwell by the middle of the 19th century. The congregations at the parish churches represented only about one-third (at Outwell) and one-quarter (at Upwell) of the total number of church-goers. The proportion of Anglican to Nonconformist Sunday school children in Upwell was even lower.[59] None of the dissenting congregations was of especially early foundation, though the Wesleyan Methodists in Upwell dated from 1814 and the Methodist Reformers at Nordelph worshipped in a building that was erected before 1800. The present Wesleyan churches at Upwell and Nordelph date from 1888 and 1861 respectively. Primitive Methodism obtained a very large following. The chapels of this

[53] White, *Dir. Norf.* (1845), 627.
[54] *P. N. Cambs.* (E.P.N.S.), 291; White, *Dir. Norf.* (1864), 848.
[55] White, *Dir. Norf.* (1883), s.v. Upwell.
[56] Infm. Archdeacon S. J. A. Evans, sometime rector.
[57] Gardner, *Dir. Cambs.* (1851), 565. One-third of Marshall's charity (1661) was devoted to the repair of Welney chapel and the bridge leading to it.
[58] Kelly, *Dir. Norf.* (1937).
[58a] See plate facing p. 168.
[59] The sources for this section are the ecclesiastical returns of the 1851 Census (H.O. 129/7/193), the year books of the various denominations, county directories, inscriptions on buildings.

denomination, with the dates of their foundation, were as follows: Outwell (1844), Upwell (two chapels, 1829 and 1835), Euximoor (1838), Three Holes (1839), Lakes End (originally Wesleyan, *ante* 1800), Welney (1832), and Suspension Bridge (1843). The chapel at Welney was built in 1839 and rebuilt in 1890. The Chapels at Lakes End and Suspension Bridge were rebuilt in 1914 and 1872 respectively. The United Methodists had chapels at Outwell (Bethel, 1851) and Christchurch; the latter congregation dates from 1833, but the present chapel was built in 1872. There are Baptist chapels at Upwell and Tips End. The former congregation began to meet in 1840; its church was built in 1844. The latter chapel (Zion) dates from 1878.[60] The Tips End chapel and the Methodist chapel at Christchurch were built by the same architect and builder (J. Kerridge and E. Everson), and are respectively of 'Gothic' and Classic design. The Upwell Salvation Army hall dates from 1883, but is not now used for its original purpose.

SCHOOLS The division of the villages between two counties has produced some anomalies; for example, the Outwell National School had its infants' department in the Isle and its boys' and girls' departments in Norfolk, and its successor the Beaupré School, though erected and controlled by the Isle of Ely County Council, is about ¼ mile over the county boundary in Norfolk.

In 1798 the sole educational provision in the two villages was a Sunday school in Upwell. In consequence of this the villages were described as a 'dark corner'.[61] In Outwell an attempt to establish a day school supported from the parish charity estates was made in the 1830's.[62] This failed, and in 1846 Outwell had still no school except a Sunday school, attended by 10 boys and 20 girls.[63]

The Revd. W. G. Townley, Rector of Upwell, who as brother of the lord of the manor and occupier of Beaupré Hall was almost as powerful in Outwell as in his own parish, was instrumental in establishing the Outwell National School in 1848. The school buildings were erected in 1854 for £779 including a teacher's house, towards which the National Society made a grant of £110.[64] After enlargement in 1873 the school could accommodate 199 children, a figure which was scaled down to 156 at the reassessment of 1910. The school was closed in 1878 owing to a controversy over the organization of the Outwell charities. It was reopened in 1881 when by a scheme made by the Charity Commissioners in that year about half of the £69 arising annually from the consolidated charities was assigned to education. As elsewhere in Marshland, the development of market-gardens and small-holdings led to an increase in population and great pressure on the schools. The situation at Outwell was only relieved when after the First World War an army hut, accommodating 77, was acquired. The school was closed on the opening of the Beaupré School in 1939.[65]

A National School for infants was built in Outwell in 1865. The site was given by C. W. Townley, lord of the manor, and the National Society made a grant of £14.[66] The accommodation was for 80.[67]

The Beaupré School, named after the former lords of the manor, was opened in 1939. It takes (1951) all the infants and junior children who formerly attended the two voluntary schools in Outwell. The accommodation is for 250.[68]

Some time after 1831[69] the Revd. W. G. Townley opened a school in Upwell, associated with the National Society. This school was attended by 217 children in 1846[70] and 250 in 1864.[71] Too much reliance seems to have been placed on the patronage of the founder, and though some attempt was made to provide for the out-parish by the erection of voluntary schools at Christchurch and Nordelph (see below), conditions deteriorated in Upwell after Townley's death in 1862. It was necessary to set up a School Board for the combined parishes of Upwell Isle and Upwell Norfolk in 1874.[72] The Board considered Townley's building unsatisfactory and in 1877–8 built a large new school for 304 children, at a cost, including two teachers' houses, of about £4,000.[73] The old building was left with the Church authorities, who established an Institute and Literary Association in it; it is now the parish hall.[74] In 1909 the Upwell St. Peter's Board School was enlarged by the Norfolk County Council, at a cost of about £1,500, to take 400 children.[75] In the 1930's this school was practically full; there were 154 boys on the roll in 1935, in a department with 156 places. The situation, however, was eased when the secondary school was completed, and from 1939 St. Peter's School has been reorganized for 120 'junior mixed' and 120 infants.

A proposal for a secondary school at Upwell, to be provided by both County Councils, was made in 1931. Owing to the economic crisis of that year the Norfolk County Council withdrew, leaving the Isle to provide for its older children in the building planned for the Christchurch school (q.v.). The scheme of 1931 was revived in 1934, but in a form that placed responsibility for provision solely upon Norfolk. A site was obtained in the following year. The school was planned for 140 senior children of each sex, and was opened in 1939. It cost nearly £23,000, including equipment.[76] This school now (1951) takes the older children from Emneth, Welney, and Nordelph as well as from Outwell and Upwell, and since the Second World War has had over 350 on the books. Additional accommodation in the shape of a Horsa hut, for 80 children, was provided in 1948.[77]

In 1865 a school, towards which the National Society granted £110, was built at Christchurch.[78] From 1875 it was leased to the newly formed Upwell School Board.[79] In 1884 additional accommodation was provided for 37 infants, raising the total capacity to 150. Though the school was situated in a well-populated area it was on the edge of the district which it served. In 1893 the School Board decided by a narrow majority to replace the school by one more centrally situated on the Sixteen Foot Bank. This

60 *Bapt. Handbk.* 1949. The chapel at Tips End is served from March. It seems to have replaced a Baptist chapel in Welney dating from 1853 or earlier (*P.O. Dir. Cambs.* (1853)).
61 Bp.'s Mun., Ely, B 6.
62 See below—Charities.
63 Rep. of Nat. Soc. (priv. printed).
64 Archives of Nat. Soc. Only £160—a rather low proportion of the cost—was subscribed locally.
65 M. of Ed. File no. 27/302.
66 Archives of Nat. Soc.
67 M. of Ed. File no. 12/42.
68 M. of Ed. File no. 12/80.
69 This school is not mentioned in Lewis's *Topog. Dir. Engl.* of that year.
70 Rep. of Nat. Soc. (priv. printed). At this date there were 4 dame schools in Upwell.
71 White, *Dir. Norf.* (1864), 848.
72 *Lond. Gaz.* 1874, 5407.
73 M. of Ed. File no. 27/433.
74 White, *Dir. Norf.* (1883); Kelly, *Dir. Norf.* (1933).
75 M. of Ed. File no. 27/433.
76 *Education*, 28 July 1939. Mr. F. G. Limmer was the architect.
77 M. of Ed. File no. 27/510.
78 Archives of Nat. Soc.
79 M. of Ed. File no. 12/51.

scheme, though it was revived by the County Council in 1914 and 1925, came to nothing, and the school was not moved until 1932, when a new site in the village, about ½ mile from the former one, was found. Here a new school called the Townley School was built and opened.[80] Owing to the fact that the plan for building a secondary school at Upwell in 1931 was not achieved, the County Council of the Isle found it necessary to use the hall of the new Townley School as classrooms, and to retain the old school building for the infants. By this means 180 places in all were provided. There were 125 children in attendance in 1949.

From about 1908 there was much local agitation on account of the hardships faced by small children in their long muddy walk from the end of Euximoor Drove to Christchurch. The persistent demand for a school on the Sixteen Foot Bank was mainly for the benefit of children living in Euximoor and other fens on the opposite side of the river to Christchurch. In 1910 the County Council opened a school for 30 children in a temporary building about half-way along Euximoor Drove. This building, which is said to have been originally used as a gymnasium at March Grammar School, was at first recognized for ten years only. It was in use, however, until the opening of the Townley School in 1932.[81]

The school at Lakes End was built by the School Board in 1876 for £2,000 and provided for the south-east side of the parish. Enlargements in 1893 and 1910 brought the accommodation up to 145—91 mixed and 54 infants. The isolated position of the school led to difficulties in staffing, especially in 1913–15. The numbers attending have gradually decreased. They were 91 in 1928 and 71 in 1939 after 18 senior pupils had been transferred to the new Upwell Secondary School.[82]

The date of erection and affiliations of the original school at Nordelph are uncertain.[83] The Upwell School Board did not consider its buildings suitable. The buildings, however, remained in use until the completion of a Board school in 1879. This school, which cost £2,053 and provided 95 places in the first instance, was enlarged in 1893 to take 119 children. After its transfer to the County Council it became overcrowded, the average attendance in 1908 being 118. A new classroom was provided in 1909–10 at a cost of £543. The accommodation was thereafter reckoned at 154 (106 mixed, 48 infants). Numbers remained fairly high for some time; they were 120 in 1930 and 121 in 1935. In 1939, however, when 12 senior children were removed to Upwell Secondary School, only 71 pupils were left behind in Nordelph. The remaining seniors were transferred to Upwell in 1946, and three years later the attendance was only 33.[84]

In 1661 William Marshall left nine pieces of land, amounting to about 470 acres, in Welney, Upwell, and Littleport. The rents were to be devoted in equal amounts to (i) the repair of Welney chapel and the bridge leading to it; (ii) the poor widows of the hamlet, and the apprenticing of poor children therein; and (iii) the repair of the highways. In 1819 a new scheme for the disposal of the charity moneys was introduced. A third of the rents were to go to the widows. The remaining two-thirds were to be devoted to the other original objects of the charity, and also to the establishment of a free elementary school on Church principles.

The greater part of the rental of about £650 was spent on the testamentary objects of the relief of widows (£130),[85] repairs to highways, bridges, and the chapel (£185), and drainage taxes, and it was not possible to appoint a schoolmaster until 1827. Even then he received only £20 a year instead of the £60 proposed in 1819, but by 1835 his salary had been increased to £28.[86] By 1847 sufficient funds had been accumulated to build the school, and also some almshouses, and to rebuild the Anglican chapel.[87]

In 1866 the school received £77 a year from Marshall's charity, and there were 135 children, taught by an uncertificated master and mistress. The school was not subject to government inspection.[88] By 1893 £100 a year was being devoted to the school, but complaints were made that this was inadequate, and that the school was entitled to a full third of the charity. The buildings accommodated 163 children before and 130 after 1910. The school was 'decapitated' in 1939, when the older children were moved to the new school at Upwell (q.v.).[89]

Another school, which served also as a mission room, was opened in 1874. It stood near the suspension bridge, and was intended for the children living on the far side of the Bedford Rivers.[90] It was closed in 1927, when its 43 pupils were transferred to Marshall's School in the village. This arrangement proved unsatisfactory owing to the almost annual flooding of the Washes between the Bedford Rivers, and shortly before 1938 the children were again transferred, but this time to Hilgay (Norf.).[91]

CHARITIES[92] In 1628 a rent of 10*s.* derived from the Outwell Town Lands, then first mentioned, was being paid annually to the poor of each 'side' of the village, i.e. Norfolk and the Isle. Subsequently the Town Lands must have increased greatly in area, for in 1795–6 the rental was £68 10*s.* This was applied in roughly equal proportions to the following objects: to the clerk and the sexton for salaries; to the churchwardens, and the overseers of each side (£10 each) to be spent at their discretion; to the poor in bread; to the poor in a general dole. In 1835 the property amounted to a messuage and 33 acres in Outwell, Upwell, and Emneth, and produced £117 10*s.* The parochial officials still received their £30, and £18 1*s.* was devoted to a school supported by voluntary subscriptions, which seems to have been short lived. The remainder of the proceeds were spent on incidentals and on indiscriminate gifts of bread and money to the poor, and the trustees were advised by the Commissioners to mend their ways. In 1864 £50 were devoted to the national school.[93]

A 10-acre piece known as the Town Ten Lands brought in £24 in 1835, and £36 in 1864, when its area was estimated at 12 a. 35 p.[94] This was applied by

[80] M. of Ed. File no. 12/75.
[81] M. of Ed. File no. 12/51A.
[82] M. of Ed. File no. 27/431.
[83] It is mentioned for the first time in White's *Directory* of 1864 (p. 848).
[84] M. of Ed. File no. 27/432.
[85] The widows received 2*s.* 6*d.* each per week, and free firing and rent.
[86] *Rep. Com. Char.* (*Norf.*), H.C. 216, pp. 528–33 (1835), xxi(1).
[87] Gardner, *Dir. Cambs.* (1851), 565.
[88] *Schs. Inq Rep.* [3966–XII], pp. 426–7, H.C. (1867–8), xxviii(10).
[89] M. of Ed. File no. 27/450.
[90] White, *Dir. Norf.* (1883).
[91] M. of Ed. File no. 27/450.
[92] The Commissioners' reports on the charities of the two villages are: *Rep. Com. Char.* (*Cambs.*), [103], 272–5, H.C. (1837–8), xxiv, and *Rep. Com. Char.* (*Norf.*), H.C. 216, 518–20, 526 et seq. (1835), xxi.
[93] White, *Dir. Norf.* (1864), pp. 840 (Outwell), 848 (Upwell).
[94] White, *Dir. Norf.* (1864), pp. 840 (Outwell), 848 (Upwell).

the overseers on the Norfolk side to the poor rate, and to the repair of a bridge in the Norfolk portion of the parish.

The churchwardens of Outwell had a house and 10 acres of land in the fen, which were let in 1835 for £19 10*s*. The money was applied to the church rate. By 1864 the rent was £29 5*s*.[95] Accounts existing in 1835 showed that the house had been in possession of the churchwardens since 1617 and the land since 1670.[96]

In 1616 William Lynne bequeathed 9 acres in Elm for the benefit of the poor of Outwell. In 1835 the rent arising was £20. This was distributed on Christmas Eve in sums varying from 2*s*. to 13*s*. according to the size of the recipient's family. The intended recipients were designated beforehand at a public meeting. The rent of this charity had in 1864 decreased to £17 7*s*.[97]

Several charities were to some extent consolidated in this fund. At an unknown date unknown donors gave in charity a house, at one time the blacksmith's forge, and 3 acres in Adcock's Hill, High Fen. In 1837 the rents arising from these two properties amounted to £12 15*s*. At an unknown date Matthew Bateman gave a rent charge of £5 to provide clothing for the poor at Christmas. By 1837 these three charities, with a portion of Foxe's charity (see below), had been merged by the parish officials into a general fund, amounting to about £120 a year. The interest on this sum was given to about 200 poor persons on the Isle side in sums of 2*s*. 6*d*. to £1 according to the size of the recipient's family, with a preference to those not on the rates. The Charity Commissioners did not approve of the absorption of Bateman's clothing charity into this general fund.

In 1626 John Foxe bequeathed 3 houses to charitable purposes. The rents of two of the houses were to go respectively to the poor of the two sides of the parish of Upwell, and the rent of the third to the repair of the church. As a result of drainage and inclosure in the fens, 15 acres in Neatmoor were allotted to the 'church' house.[98] This allotment had been increased in 1837 to 17 acres and then brought in £50. The land had formerly been let for £45 10*s*. The house itself was divided into two tenements and brought in £5. An attempt was being made in 1837 to divert some of the income to church rates, but the rector was against this, as the church itself was then (1837) in bad repair. By 1864 the rental had decreased to £40, all applied to the upkeep of the church.[99] The 'Isle' house was in 1665 awarded 16½ acres in Euximoor and Shrewsness Green. In 1837 the house itself was let to the overseers at £5 5*s*. as a workhouse. The land was let at £31 10*s*.—a sum which was carried into the general fund. The letting of the various lands appertaining to Foxe's charity was in 1837 carried out by public auction.

In 1562 Thomas Lamb bequeathed two small estates: a messuage and 3 acres in Plaw Field, and another house, the rent of which was to be applied for the benefit of the poor of the Norfolk side. In 1837 the former house was used as a workhouse for the Norfolk side and the land was let for £10. In 1665 14 acres in Neatmoor were awarded to this house under the inclosure of drained fen land. In 1837 this allotment produced £36 a year. In 1837 the rent of the second house and of another 14 acres obtained under the 1665 award was £49 10*s*. All this was applied to the Norfolk poor except for one-third of the rent of the house then used as the 'Norfolk' workhouse and its 3 acres in Plaw Field, although the Isle poor were also entitled to one-third of the £36 derived from the 14 acres in Neatmoor.

In 1864 the Norfolk share in the Foxe and Lamb charities consisted of two of the various houses and about 30 acres of land, producing £100 a year.[100]

The two small charities of John Boss and Thomas Dixon were in 1837 worth £10 and £20 respectively. The interest on the former was devoted to bread and on the latter to general purposes. The Isle poor received 6*s*. 8*d*. of the £1 interest of Dixon's charity, but all the rest went to Norfolk. These charities were formerly reserved for non-paupers, but in 1835 were distributed indiscriminately. By 1864 they were merged with the Foxe and Lamb charities and were distributed at Christmas.[101]

The balance of population in Upwell has always been slightly in favour of Norfolk, but the main reason for the apparently unfair treatment of the Isle poor in the above charities was probably the existence of one important charity entirely devoted to the Isle side of Upwell and Welney. In 1765 Bishop Mawson of Ely demised 35 acres of land in Welney Runns, known as the Bishop's Lands,[102] on a perpetually renewable lease to the rector and churchwardens of Upwell for the poor of the main village and the hamlet. Up to 1837 this land was let at £39 10*s*., after that date at £68 10*s*.; two-thirds of the income was given to the poor of Upwell Isle, and one-third to Welney Isle. The moneys were distributed annually, with a preference to non-paupers.

Details of Marshall's charity, the only one applicable solely to Welney, have been given under schools.

THORNEY

This large parish[1] is of a different type from the others in Wisbech hundred. Its nucleus is an island of gravel, not very perceptible except from the west, but rising to 25 ft. above sea-level, or some 15 to 20 ft. above the surrounding fen. Reclamation has proceeded outwards from this nucleus, resulting in an oblong area some 6 miles east to west by 4 miles north to south. The parishes near Wisbech, whose nuclei are on sites just behind the 'Roman Bank', are quite differently constituted, with narrow frontages to the Bank and long extensions into the fen. Since, moreover, Thorney is on peat and not on silt fen, the orchards and market gardens which are such a feature of the Wisbech area are largely absent,[2] potatoes, cereals, and sugar-beet

95 Ibid.

96 *Rep. Com. Char.* (*Norf.*), H.C. 216 (1835), xxi.

97 White, *Dir. Norf.* (1864), pp. 840 (Outwell), 848 (Upwell).

98 By Act of Parl., 15 Chas. II, c. 17. Sec. 38.

99 White, *Dir. Norf.* (1864), pp. 840 (Outwell), 848 (Upwell).

100 Ibid.

101 Ibid.

102 This piece of land may represent the remnant of the demesne of the small Ely bishopric manor in Welles (see above—Manors).

1 3,936 acres, with a 1931 population of 330, were transferred from Whittlesey under the Isle of Ely Review Order, 1933 (M. of H. No. 76940).

2 A large orchard on the south side of the Wisbech road, about two miles east of Thorney village, is an exception.

taking their place. There is also a larger area of pasture.[3]

Thorney is the only village in the Isle to have been for a long period the sole property of an occasionally resident landlord (the successive earls and dukes of Bedford), and the conspicuously neat aspect of the planned 'estate village' contrasts abruptly with that of its neighbours. The Bedford family has earned a reputation for enlightened estate management, especially in the matter of housing, and at the present time the limits of Thorney, though undefined by any natural landmark, are sufficiently discernible from the superior quality of the buildings within them. By the end of the 19th century over 300 cottages, of a total of some 450 in the parish, had been provided by the dukes, who made a practice of leasing them on direct weekly tenancies so as to avoid the drawback of the tied cottage.[4] The first village schools, the water-supply, and the fire station were also provided at the sole cost of the landlords. There is probably no village in England with a more extensive display of well-designed Victorian cottage architecture. The work of house-building has been energetically continued by the Thorney Rural District Council. Although this body controls one of the smallest areas in England, as regards both population and rateable value,[5] it had up to 1939 provided 34 houses, which with 14 built by private enterprise represented 8·1 per cent. of the total. This proportion was far above the average for rural districts of the size of Thorney, and slightly above the average for the country at large.[6]

The munificence of a great landowner has not hindered the rise of a strong community spirit; it is interesting to note that two of the five Community Centres sponsored by the County Council have been established at Thorney village and Wryde Croft.[7] The Rose and Crown Inn was acquired by the People's Refreshment House Association in 1899 and was one of their earliest properties.[8]

Francis, 4th Earl of Bedford, was granted the right to a market and two fairs at Thorney in 1634,[9] a privilege which was renewed to Wriothesley, 2nd Duke, in 1716.[10] The market was discontinued in 1830, but the fairs, for horses and cattle, were still much patronized in 1879;[11] they were held on 1 July and 21 September.

The village is situated where the main road from Peterborough to Wisbech (A 47) is crossed by that from Whittlesey to Crowland (B 1040), somewhat west of the centre of the parish. The former road is now one of the main cross-country routes of England (Leicester–Norwich); it was turnpiked from Peterborough to Thorney in 1792 and on to Wisbech in 1810.[12] In the 14th century there was a monastic grange at Wryde,[13] in the east centre of the parish, and there is a certain amount of diffused settlement, known generically as Wryde Croft, along the road leading north from the main Wisbech road towards Gedney Hill; also at Knarr Fen, along the droves south of the Wisbech road; and at Thorney Toll on the eastern boundary. All these places have modern chapels and schools.[14] The railway from Peterborough to Sutton Bridge, opened in 1866,[15] crosses the centre of the parish, with stations at Thorney and Wryde. The line from March to Spalding (1867)[16] crosses the north-east corner and has a station (French Drove and Gedney Hill) just across the Lincolnshire border.

French Drove and French Farm in the extreme north of the parish recall the settlement in the 17th century of French and Walloon refugees, who helped in the reclamation and long remained an important element in Thorney. The refugees were first admitted on condition of being allowed to sell at any market, and of being exempt from service overseas for 40 years, and from subsidies and fifteenths.[17] They held services in the church under their own pastors,[18] who kept a separate register from 1654 to 1727.[19] The settlement was fairly large; the register shows an average of 29 births a year in 1654–63, 37·9 in 1664–73, and 26 in 1764–83. After this date numbers began to diminish, partly no doubt owing to intermarriage and absorption in the general community. In the 10 years 1714–23 there was an average of only 7·9 births per year, and 7 in the last 4 years to 1727.[20] Cole, writing in 1744, mentions that French was still spoken by the descendants of the refugees in private conversation,[21] and a court roll of 1748 shows 7 French names among the 28 jurors.[22] Gardner's *Directory* of 1851[23] shows a few names, such as Barron, Bellamy, Charity, and Provost, which may be of French origin, but these have now died out.

Thorney Mill is mentioned in a 1470 commission *de walliis et fossatis*,[24] and the manorial appurtenances in 1787 were stated to include as many as ten mills, though some of them may have been for drainage purposes only.[25] A windmill still stands in the village on the Peterborough Road, but is now derelict. In the 18th and 19th centuries the Wing family, descendants of Vincent Wing the astronomer, were agents to the dukes of Bedford. John Wing (1752–1812) was in 1788 the subject of scurrilous attacks in connexion with a proposed new tax on the North Level.[26] Two companies of volunteer infantry, numbering in all 145 men, were raised between 1803 and 1805.[27]

The manor house, commonly called Abbey House,[28] is situated to the west of the church and is a large stone structure of various dates. The east wing is of the second half of the 16th century and was probably

3 R. W. L. Melbourne, *Isle of Ely* (Land of Britain, 71), 279, 297.

4 In 1895 nearly 20 per cent. of the cottages were occupied by non-estate workers. By that date the capital outlay on housing was about £57,000, and the net annual loss over £2,000 (Herbrand Duke of Bedford, *A Great Agricultural Estate*, 95–96).

5 In 1943–4 a 1*d*. rate produced only £25, a smaller sum than for any English rural district except Broadwood Widger (Devon).

6 M. of H., *Rural Housing* (1944), 67.

7 I. of E.C.C. Educ. Comm., *Scheme for Further Educ.* (1950), 12, 13. The other three centres so far established are at Aldreth, Black Horse Drove and Stonea; the demand seems to be greatest in the small isolated hamlets.

8 Ex inf. Sec. P.R.H.A.

9 C 66/2643, no. 2.

10 C 66/3515, no. 24.

11 R. H. Warner, *Hist. Thorney Abbey* (1879), 228.

12 Ibid. 261.

13 Ibid. 156–7.

14 For Thorney Toll chapel and school see Wisbech St. Mary, in which parish the building is locally situate.

15 *V.C.H. Cambs.* ii, 132–3.

16 Ibid.

17 B.M. Lansd. MS. 110, f. 4.

18 Miss Scott Thomson, in *Family Background* (1949), 177–8, establishes clearly that the restoration of the abbey was in the first instance for the especial benefit of the foreign settlers.

19 Printed as an app. to Warner, *Hist. Thorney Abbey*.

20 *Fenland N. & Q.* iii, 357–61.

21 B.M. Add. MS. 5805, f. 118.

22 Quoted by Warner, op. cit. 250–1.

23 Gardner, *Dir. Cambs.* (1851), 561.

24 *Cal. Pat.* 1467–77, 170.

25 C.P. 43/816, ro. 325.

26 *D.N.B.*

27 *Fenland N. & Q.* i, 128; ii, 123–4 (giving names).

28 For a fuller illustrated description see *Country Life*, xlvi, 392. The house has been National Trust property since 1943.

erected soon after the site of the monastery was granted (1550) to the 1st Earl of Bedford. It retains some original stone-mullioned windows, including a pleasant two-storied oriel with gabled top, on the south side. The east front was altered in the 18th century, and the brick chimney-stacks are probably of the same period. The main portion of the house is to the west, and is a particularly good example of mid-17th-century work. There is reason to believe that it was designed by John Webb, the nephew and pupil of Inigo Jones, as it has a marked similarity to his work at Thorpe Hall, Peterborough.[29] There is a fine staircase rising to the garrets, with turned balusters and ball and acanthus cup finials. Several of the rooms, particularly the dining-room, retain their panelling, and there are some good wooden-framed doorways. The house was enlarged towards the north in the 19th century. The roofs are covered with stone slates. To the north of the house is a 16th-century barn.

Thorney, as an 'estate village' near the stone belt, contains more examples of good domestic architecture than most villages in the Isle. The following houses, besides those above-mentioned, date from the 18th century. In Whittlesey Road: Nos. 1 and 2, and J. W. Harrison's premises. In Wisbech Road: Nos. 5, 6, 7 and 7A, and 14 to 17. Also Park House, and the premises occupied by Hugh Cure, and (outside the village), Willow Hall and Prior's Farm. St. Vincent's Cross, in the north-west corner of the parish on the road to Crowland, dates from the 15th century; it is now headless.

MANOR. During the Middle Ages what is now the parish of *THORNEY* was of little or no account except as the site and home farm or demesne of the abbey,[30] and references to it apart from the abbey are very scanty. The boundary with the Ely Cathedral priory property in Wisbech Murrow was adjusted in 1248. The Prior and Convent of Ely recognized 200 acres between the dike dividing Thorney and Wisbech marshes on the east, the alder wood of 'Estdelf', 'Bingebrigge', 'Midfentre' and 'Eldethorneyweye' as Thorney property. Thorney in return quitclaimed 400 acres of marsh called 'Eldegore' to Ely.[31] A manor of Thorney is mentioned *c.* 1458,[32] and the abbot had the right of free warren in his demesnes, although the recorded breaches of this may equally well refer to his share of Whittlesey Mere or other possessions in the neighbourhood as to Thorney itself.[33]

In 1291, when the abbey was at the height of its prosperity, its temporalities in Thorney were valued at £46 19*s.*[34] In 1539–40 the value had fallen to £34 13*s.* 4*d.*[35] This does not suggest an important estate, when allowance is made for the monastic buildings themselves. It is significant that no part of those buildings was converted into a church until long after the Dissolution, and that no grant of a market at Thorney was ever made to the abbot. The surrender was signed by the abbot and twenty monks, who may perhaps have had some 100 persons dependent upon them as servants and farm staff; these with their families would not have amounted to more than a fair-sized village.

Robert Moulton or Blyth, the last abbot, like many of his contemporaries demised portions of the abbey lands shortly before the Dissolution. The site of the monastery itself was leased for eighty years from 1538 to Walter Williams or Crumwell,[36] of Chatteris, the lease being converted into a twenty-one-year term by the Crown in 1541.[37] In 1550 the site of the monastery and its possessions in Thorney were granted outright to John Russell, 1st Earl of Bedford,[38] whose family held them until 1910, when the 11th Duke sold the Thorney Estate to the tenant farmers.[39] In 1545 the keepership of Thorney Chase was granted to Edward Elryngton.[40]

A survey of 1574[41] shows 240 acres of upland ground, 160 acres of wood with 1,000 oaks worth £500, and an unspecified amount of pasture worth 6*s.* 8*d.* an acre with £100 worth of timber. The total annual value of this part of the estate was £200. There were also 16,000 acres of fen, formerly dry but at the time of the survey waterlogged. The ground of the fen was worth 6*d.* an acre and the reeds and willows growing on it £200. The timber and what remained of the monastic buildings were valued at £1,000.[42] The total area of the estate, upland and fen, was put at 17,760 acres. This estimate coincides almost exactly with the area of Thorney parish before the recent changes of boundary which was 17,802 acres of land and 40 of water. The Elizabethan and modern statutory acres, however, are not necessarily equivalent. If, as Browne Willis says, the moat surrounding the precincts was a mile long and 20 feet broad,[43] the precincts themselves must have contained about 40 acres.

The first proposal to redrain the Thorney fens was made in 1626 by Robert Tipper and John Gason,[44] and from 1630[45] Francis 4th Earl, the first of his family to take an active interest in the Thorney estate, and his fellow Adventurers, accomplished a good deal with the professional help of Vermuyden.[46] In the early days of the 4th Earl's tenure (1629) only some 300 to 400 acres around the site of the abbey, let at £300 a year, were cultivable.[47] The market grant of 1634, and the re-edification of the church between 1636 and 1638, show how rapid was the progress of draining. The earl is said to have spent £100,000 on the work, making himself practically bankrupt. The area was declared to be drained at a Session of Sewers held at St. Ives in 1637,[48] but the following year the excuse offered for not collecting the £20 ship money due from Thorney was that the place was inaccessible in winter.[49] More-

29 The actual builder was John Lovin, mason, of Peterborough. His contract with William, 5th Earl and 1st Duke, is printed in Scott Thomson, *Family Background*, 180–4.

30 For history of the abbey see *V.C.H. Cambs.* ii, 210–17.

31 D. & C. Mun., Ely, Carta 184*a*.

32 B.M. Add. MS. 5805, f. 121*b*, in a copy of a letter from Maurice Johnson of Spalding to Browne Willis, quoting a 15th-century cartulary which the former had recently bought.

33 e.g. the takers of ducks' eggs and herons' eyries in 1275 (*Cal. Pat.* 1272–81, 118). See also *Cal. Pat.* 1301–7, 357.

34 *Tax. Eccl.* (Rec. Com.), 268.

35 Dugd., *Mon.* ii, 612.

36 L.R. 14/892/4/11 & /18.

37 *L. & P. Hen. VIII*, xvi, p. 725; E 326/7217.

38 *Cal. Pat.* 1549–51, 43. To be held in chief, at $\frac{1}{40}$ knight's fee with other property.

39 Kelly, *Dir. Cambs.* (1933).

40 *L. & P. Hen. VIII*, xx (1), p. 674.

41 S.P. 12/99/38.

42 The remains then consisted of the roofless nave, half the steeple, and a building 60 by 20 ft. used as a dwelling house.

43 B.M. Add. MS. 5805, f. 119*b*.

44 Hist. MSS. Com., *12th Rep.*, App. pt. i, 277.

45 In that year the earl wrote to Sir Henry Vane, Ambassador at The Hague, suggesting financial assistance since 'the Fen business is so feasible' (*Cal. S.P. Dom.* 1629–31, 311).

46 See *V.C.H. Cambs.* ii, 76 et seq.

47 Warner, *Hist. Thorney Abbey*, 220.

48 *V.C.H. Cambs.* ii, 76.

49 *Cal. S.P. Dom.* 1638–9, 343.

over, floods continued to cause great damage in some years, notably at the end of 1672, when a west wind backing south caused the upland waters to overflow all the country between Crowland, Spalding, Wisbech, and Ely. Many cattle were then drowned and houses flooded, and the colza crop, planted to provide lamp oil and also used in wool manufacture, totally lost.[50] Damage on a similar scale occurred in 1770, when the north bank of Moreton's Leam burst at Abel's Gull. After this a farm of 407 acres at English Drove, Thorney, several miles from the breach, had to be let at £20 for the first and £200 for the next six years, the tenancy to be determinable if there was a fresh breach.[51]

Nevertheless, under Francis's son William, 5th Earl and 1st Duke, Thorney had risen sufficiently to be considered a town by the traveller Gregorio Leti, who mentions that the earl had tamed six stags in his park here 'with a continual noise of drums, harps and other musical instruments', and presented them to Charles I to draw the royal coach.[52] At the outbreak of the Civil War the earl[53] was made General of the Horse in the Parliamentary army, but after the failure of an attempt at mediation he joined the Royalists and his estates were sequestrated.[54] There is no record that the estates were granted out, and they were in any case restored in July 1644 after the earl had returned to the Parliamentary cause.

After the Restoration the drainage and improvement of the Thorney estate began to justify itself; most of the corn used in the ducal household at Woburn, and many of the bullocks, came from Thorney.[55] In the middle of the 18th century the manor was for a short time leased to the Markham family.[56] About this time Cole remarked that 'within these 20 years [1724–44] Thorney is quite another thing from what it was, being handsomely adorned with many good buildings, and particularly with a very noble Inn as big as a College'.[57]

Between 1583 and 1787 the manor of Thorney was on many occasions the subject of family settlements among the Russells.[58]

The later history of Thorney is uneventful, but of some interest as that of a compact agricultural estate of 20,000 acres carefully administered.[59] At the beginning of the 19th century the estate showed a profit. Its average income was then nearly £10,000. The agricultural depression of the last quarter of the century was, however, severely felt. Deficits were incurred in five years between 1879 and 1895,[60] and the proportion of net to gross income fell from 42 per cent. between 1816 and 1835 to 19 per cent. between 1876 and 1895. At the end of the century the average income was only about £5,000 a year, representing only $2\frac{1}{7}$ per cent. on new capital outlay with no allowance for death duties. But the duke could claim that he had never evicted a tenant, and that Thorney was a healthy village with a minimum of crime and no pauperism. Thorney in fact came as near as any in the country to the ideal of 'a wealthy landowner understanding the economics of agriculture, a farmer master of its practice, a village not over-populated,[61] with pure water, decent houses, allotments and a school' instanced by a present-day writer as 'the most successful experiment in social organization that England had so far seen'.[62] The various earls and dukes of Bedford are said to have spent nearly £2 million in all on their Thorney estate.[63]

CHURCHES It is not known how the spiritual needs of the abbey tenants in Thorney were provided for in the Middle Ages. Neither rectory nor vicarage is mentioned in the *Taxatio* or the *Valor Ecclesiasticus*.[64] In consequence of the terms of the grant of the abbey estates to the 1st Earl of Bedford, the abbey church when it was restored to religious use in the 17th century was treated as a donative of the Russells. It continued to be so treated until *c.* 1894.[65] The advowson was with the Dukes of Bedford until the sale of the estate, and has subsequently been with the Bishop of Ely.[66]

There are Anglican chapels at Wryde Croft (1877) and Knarr Fen (1890).[67] There is also a mission church at North Side, served by the clergy of St. Andrew's, Whittlesey. It is an iron building erected *c.* 1902[68] at the cost of the 11th Duke of Bedford and dedicated to St. Guthlac. There is another Anglican place of worship at Willow Hall (*c.* 1930).

The church of *ST. MARY THE VIRGIN AND ST. BOTOLPH* consists of north and south transepts and nave. The former are modern, but the latter consists of the five western bays of the great conventual church, which had a total length of nearly 300 feet. It was dismantled after the suppression in 1539 and the site granted to the Earl of Bedford. In 1638 the western portion of the nave was patched up to serve as a parish chapel. The aisles were destroyed and the arcades walled up, the windows from the aisles being inserted in this walling. The clerestory was also removed and the triforium openings filled with 15th-century tracery from some other part of the building, possibly the clerestory. The whole was covered with a coved plaster ceiling. In 1841 the transept was added in tolerable imitation of 12th-century work. There was further restoration in 1888 when the galleries were removed and the church reseated. The old portion of the fabric is a splendid example of Romanesque design, resembling Ely and Peterborough, and dates from the beginning of the 12th century. In the 15th century the west front was considerably altered, a large window being inserted and the flanking turrets crowned with octagonal tops.

The chancel is arranged in the crossing, and there is a large pointed single-light east window with internal jamb shafts having moulded caps and bases. The tran-

[50] *Cal. S.P. Dom.* 1672–3, 380.

[51] *Fenland N. & Q.* ii, 235–6. North and Morris Fens in Thorney were again flooded after the breach at Great Postland in the floods of March 1947 (M. of Agric., *Harvest Home*, 51, 54).

[52] R. H. Warner, *Hist. Thorney Abbey*, 232.

[53] He was not created Duke until 1694 (*Compl. Peerage*, ii, 79).

[54] *D.N.B.*

[55] Gladys Scott Thomson, *Life in a Noble Household* (1937), 152–6, &c.

[56] C.P. 43/700, ro. 45, 46; C.P. 25 (2)/1281/Mich. 2 Geo. III.

[57] B.M. Add. MS. 5806, f. 118.

[58] C.P. 25 (2)/260/Trin. 25 Eliz.; C.P. 25 (2)/93/857/East. 35 Eliz.; C.P. 25 (2)/527/East. 9 Chas. I; C.P. 43/307, ro. 106 and/323, ro. 43; C.P. 25 (2)/761/Trin. 21 Chas. II; C.P. 43/598, ro. 290 and/816, ro. 323.

[59] The 11th Duke of Bedford's *Story of a Great Agricultural Estate* (1897) gives a full account, from which the following details are taken.

[60] During these years an average of 26 per cent. of rent was annually remitted to the tenants on account of bad harvests, &c.

[61] Its population density (12 persons per 100 acres) is very much below the average for the Isle.

[62] G. M. Young, *Victorian Engl.* (1937), 145.

[63] E. Conybeare, *Hist. Cambs.* (1897), 252.

[64] Incomplete for Cambs.

[65] W. D. Sweeting, *Hist. & Archit. Notes on Parish Churches ... Peterb.* (1868), 116–17; Kelly, *Dir. Cambs.* (1892, 1896).

[66] *Clergy List*; *Crockford's Clerical Dir.*

[67] Kelly, *Dir. Cambs.* (1933).

[68] Ibid. (1900, 1904).

septs have plain round-headed windows of Romanesque character, and there are lofty round-headed arches at the crossing. In the south transept is a large organ gallery added in 1888, the space beneath which is utilized as a vestry. The transepts are covered by a plain barrel vaulting and the crossing by a groined vault.

The nave[68a] has arcades of five bays of early 12th-century date with rounded arches of two orders, the outer with a roll moulding and the inner plain. The massive columns are alternately round and clustered with cushion caps and moulded bases; the responds are semicircular. The triforium has rounded arches of two orders with roll mouldings, springing from shafts with cushion caps and moulded bases. There are circular vaulting shafts running up the face of the columns. A small portion of the clerestory remains on both sides at the west end, but it is now entirely external; on the south it consists of a rounded arch with continuous roll moulding, but the northern section has billet moulding. Both openings are now blocked. The triforium is surmounted externally by an embattled parapet dating from the 17th-century restoration. The arcades and triforium are fitted with uniform windows of 15th-century date with three cinquefoiled lights. Both nave and transepts have high-pitched external roofs covered with stone slates. The west front is a very interesting composition and exhibits work of three distinct periods. It is flanked by large square turrets of the 12th century, which are surmounted by 15th-century octagonal tops richly panelled and with embattled parapets. The rest of the façade is mainly of 15th-century date. The doorway is deeply recessed and of four orders with jamb shafts having moulded caps and bases. The wall on either side is panelled in two tiers, and above is a string-course with faces alternating with conventional flowers. In the spandrels is the date 1638, which must refer to the window above. There is an inner doorway with ogee hood-mould terminating in a finial. The large five-light west window is set within the arch of one much larger; it has an embattled transom with trefoiled lights below and cinquefoiled above and was inserted in 1638; there is an ogee hood-mould with finial at the apex. Above the blocked larger window is a string-course, and the spandrels are panelled. A stone screen connects the octagonal turrets, with nine canopied niches on its western face containing contemporary figures in good preservation; it is uncertain whom they are intended to portray.

The east window contains stained glass inserted nearly a hundred years ago which is remarkably good for its period. It depicts the miracles of St. Thomas of Canterbury and is a copy of some 13th-century glass in Canterbury Cathedral. In the lower easternmost window on both sides of the nave is late 15th-century Flemish glass; on the north is the Mocking, the Women on their way to the Sepulchre, and the Harrowing of Hell; on the south the Denial of St. Peter, the Supper at Emmaus, and the Pieta. The second, third, and fourth windows on the north have the following shields of arms respectively: Russell, early 19th century; John of Gaunt, late 14th century; and Henry VIII impaling Katharine of Aragon, first half of the 16th century. The second window on the south has the royal arms, of early 16th-century date. On the north wall of the nave is a marble memorial tablet to Ezekiel Danois, 1674. Danois was the first minister of the French Protestant refugees who settled at Thorney in the middle of the 17th century. The other fittings, including the font, are modern.

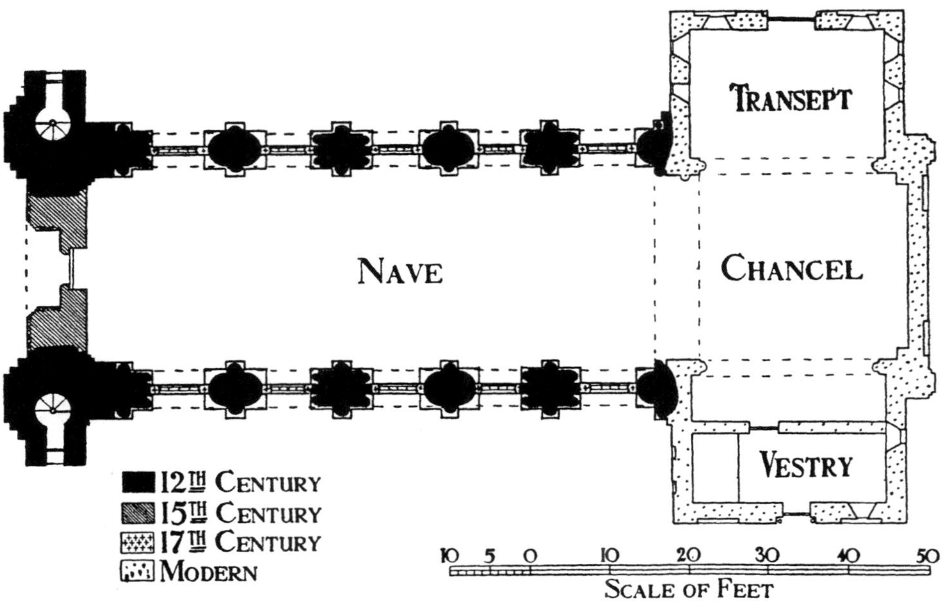

Plan of the Church of St. Mary the Virgin and St. Botolph

The plate includes a communion cup and paten of silver, 1709, inscribed 'Ex Dono Tho: Brecknock Minister Evangelii Thorney Abby',[69] and a shield of arms, a chevron between three lions the paws erased; a flagon of silver, 1876; an alms-dish of silver, 1750.

The north-west turret contains one bell dated 1720, by a Nottingham founder.

The registers begin in 1654 and are complete.

Virtually nothing remains above ground of the monastic buildings, but a 17th-century house about

68a See plate facing p. 186.

69 The last word occurs only on the cup.

150 yards to the south of the church has the remains of a moulded window of medieval date in the cellar; it probably formed part of the southern range. The vicarage possibly incorporates some of the eastern range. It is a good example of late 17th-century domestic architecture, and resembles in character many Northamptonshire buildings. There are brick chimney-stacks of later date and the roof is covered with stone slates.

NONCONFORMITY An Independent 'mission' was conducted by William Evenett, the Whittlesey minister, between 1815 and 1820. Mr. Wing, the Duke of Bedford's steward, was hostile, and threatened Mr. Harker, in whose house the mission was held, with eviction. The duke, however, was tolerant and Congregationalism for a time gained a foothold in Thorney.[70] It had died out by the middle of the 19th century. A Primitive Methodist chapel was erected in the village in *c.* 1880.

SCHOOLS The village school, like most of the public institutions in the village, owes its origin to the dukes of Bedford. Its date of foundation is uncertain. Early in the 19th century a tradition existed that 'a schoolhouse was erected by a member of the illustrious house of Russell',[71] and it is possible that this was Wriothesley, 3rd Duke, who in 1727–9 paid several visits to this part of his estates and might, had he lived, have substituted Thorney for Woburn as the principal residence of his family.[72] Thorney as a ducal donative does not appear in the various inquiries concerning education made by the bishops of Ely at the end of the 18th century, but early in the next century the dukes were making an allowance of £20 a year for a schoolmaster's salary.[73]

By 1846–7 the Duke of Bedford had provided three schoolrooms, for boys, girls, and infants.[74] In all, 284 children (in a population of 2,159) were being taught, about a quarter of them attended both weekdays and Sundays, and the staffs and their salaries were on a generous scale. The staff consisted of two masters, three mistresses, and two paid 'monitors', who received £225 between the five of them. In 1850–1 the girls' school was rebuilt with an infants' department.[75] This building still exists, but the boys' school was rebuilt in 1875, in which year a Thorney School Board was formed.[76] The Board took over the management of the schools but the buildings remained the property of the dukes of Bedford until 1912, by which time the Thorney Estate had been broken up. The County Council then bought the village schools, and those at Knarr Fen and Wryde Croft (see below), for £800 including improvements.[77] The village schools, in separate buildings on non-adjacent sites, were run as separate departments; before 1910 the total recognized accommodation was 328, afterwards 273 (106 boys, 89 girls, 78 infants). By 1925 both school buildings were overcrowded.[78] In 1935 a new site on Wisbech Road was obtained and an all-ages school built to accommodate 160 senior and 150 junior children of both sexes. This was opened in 1940, and named the Duke of Bedford School.[79]

Other schools were built by the dukes of Bedford in the north-east and south-east of the parish at Wryde Croft (1866) and Knarr Fen (1880) respectively. Like the village schools, they were managed by the School Board, and later by the County Council, though they remained the property of the dukes until 1912. The original accommodation was 80 in each case; both were enlarged in the early years of the 20th century to enable the Wryde Croft School to take 156 and the Knarr Fen school 137. These figures were scaled down in 1910 to 140 and 112 respectively. The senior children from the Wryde Croft School were transferred to the Duke of Bedford School in 1944 and the remainder in 1945 when Wryde Croft School was closed. The buildings, though of iron and timber construction, have lasted well and are now used as a community centre. Knarr Fen School lost its senior children to the Duke of Bedford School in 1944 and was closed at the beginning of 1948. The remaining children were then moved to the North Side School.[80]

North Side School, transferred from Whittlesey to Thorney parish in 1933, was built by the Whittlesey School Board in 1877; the Duke of Bedford, who owned the site, contributed £200. The building originally accommodated 90 children, and, after enlargement and reassessment, 124. The average attendance, however, seldom if ever reached the latter figure. In 1936 the senior children were transferred to the Thorney village schools pending the completion of the new Duke of Bedford School, and only 42 remained on the books. These have since been reinforced by the children, some 35 in number, from Knarr Fen.[81]

TYDD ST. GILES

Tydd St. Giles, the northernmost parish in Cambridgeshire, is situated 6 miles north of Wisbech. The Shire Drain divides it on the north and west from the sister parish of Tydd St. Mary in Lincolnshire. This stream has shrunk to a shadow of its former self but is an important boundary, separating, as it has done, two counties, two dioceses, and in all probability two Anglo-Saxon kingdoms. The upper part of the Shire Drain, forming the western boundary of Tydd, has been known since the 17th century as Lady Nunn's Old Eau, from the wife of a 14th-century Tydd landowner.[1] The eastern boundary of the parish is also a county boundary, and was in 1934 adjusted to conform with the modern course of the River Nene.[2] The neighbouring parish on the south is Newton-in-the-Isle. The most important watercourse in the parish, after

70 *Cong. Mag.* iii, 113 (1820).
71 Lewis, *Topog. Dict.* (1831), iv, 293.
72 Gladys Scott Thomson, *Family Background* (1949), 186–94.
73 D. & S. Lysons, *Magna Brit.* ii (1), 267.
74 Rep. of Nat. Soc. (priv. printed).
75 Gardner, *Dir. Cambs.* (1851), 560.
76 *Lond. Gaz.* 1875, 303.
77 M. of Ed. File no. 12/47, pt. i.
78 Ibid.
79 Ibid.; infm. Chief Ed. Offr., Isle of Ely C.C.
80 M. of Ed. Files nos. 12/49 (Wryde Croft) and 12/48 (Knarr Fen).
81 M. of Ed. File no. 12/57.
1 See *P.N. Cambs.* (E.P.N.S.), 7.
2 By the M. of H. Prov. Ord. Conf. (Ely, Holland, and Norf.) Act, 1933, which took effect 1 Apr. 1934, mutual exchanges were made between the Isle and Holland and the Isle and Norfolk, involving the parishes of Tydd St. Giles, Tydd St. Mary, Walpole St. Peter, and West Walton. This resulted in a net loss to Tydd St. Giles of 230 acres and 17 population (1931 Census). Previously, owing to the straightening of the course of the Nene in the 18th and 19th centuries, the eastern end of the parish round Oldroyd's Farm was inaccessible by road from the main part except by a 12-mile detour through Wisbech or Sutton Bridge.

the Nene, is the North Level Main Drain (1831–4), the successor to the Shire Drain, which cuts straight across from south-west to north-east and has considerably altered the local topography. The parish is crossed by the main road from Wisbech to Long Sutton (A 1101) and the Peterborough to Sutton Bridge branch of the former Midland and Great Northern Joint Railway (1866), which has a station (Tydd) about 3 miles east of the village just outside the county boundary. The soil is highly fertile and much given over to orchards and market gardens. As in most Marshland parishes, as opposed to those in the fens proper, there is a network of by-roads and a more scattered type of settlement. At Tydd, besides the main village round the church, which is roughly in the centre of the parish, there are hamlets at Four Gotes and Tydd Gote,[3] where the high road enters and leaves the parish, and at Foul Anchor in the extreme north-east corner, near the outfall of the North Level Main Drain into the Nene.

The name 'Tydd' defies explanation. Tydd St. Giles is not mentioned in Domesday Book, since it was within the liberty of the Bishop of Ely, but Tydd St. Mary appears there in the form of Stith.[4] The writers of Domesday may have identified the name with the Saxon word 'staith', meaning such parts of the banks of a watercourse or sea beach as were used for a regular landing-place for men or goods. Staithes were quite common in East Anglia, e.g. the Staithe in Wisbech (now called Staithe Road) on the old sea bank. Here, at Tydd, was the outlet to the sea of the old Nene or Sluice Drain or Elloe River—the last of which alternatives gave its name to the whole wapentake; and so Stith (or Staithe) St. Mary on the Lincolnshire side of the river would have its counterpart on the Cambridgeshire side. This argument depends upon the correctness of the Domesday 'Stith'. Alternative suggestions are that the name is derived from (*a*) Tidi, an Old English personal name used in the 9th century from which a number of other place-names are derived—though the use of such a name uncompounded would be most unusual; (*b*) from O.E. *titt*, used in the transferred sense of a slight hill.[5] The facile 'tide' is unwarranted. Although the name is of a single syllable it possesses nearly a score of variants. 'St. Mary's Tyd', which constantly recurs, might perhaps account for the appearance of the initial 'S' in Stith, which is a rare form; this would negative the 'staith' explanation. But the chain of manor houses along both banks of the river at Tydd points clearly to the use of the river as a highway, with its concomitant landing-places.

Probably the oldest house in the parish is Kirkgate House, lying opposite the present manor house. It is built on the site of the home of the Fisher family who lived at Tydd St. Giles from 1320 for several generations. Its south gable at one time bore the date 1587, and the north end was stated to be 100 years older.[6] The house was badly damaged a few years ago by the fall of a tree, which crashed through the centre of the building. The Fishers, as their name implies, were predominantly a seafaring family, some of them being engaged in commercial enterprises both at home and overseas.[7] Two at least of them took orders, apart from the John Fisher whose sepulchral monument in the church is referred to below.

Hannath Hall is a handsome Elizabethan residence probably somewhat older than the manor house. The property belonged to Richard Sparrow, whose family is referred to in early manorial records as 'Passur'. It was sold to John Laverock of Upwell in 1573.[8] John Laverock devised by will[9] to his son Edmund 'all free lands and my house that I have in Tyde Seyntt Jeyles and Tyde Seyntt Marye's that I lately purchased of Richard Sparrow', which suggests that the house was already in existence at the time of purchase. It was known as the 'Sparrows Nest', until in later years it became vested in the Hannaths. It does not appear that the Laverocks ever lived there. Edmund Laverock died *c.* 1618, leaving two daughters and coheirs, of whom Dorothy married Hamond L'Estrange of Hunstanton, brother of the first baronet. The property was sold by the L'Estrange family to the Traffords and was comprised in Lady Trafford's marriage settlement of 1791. It was then occupied by Thomas Watkinson and afterwards by John Hannath. In 1812 it was sold to Joseph Hannath of Tydd St. Giles, son of John Hannath, for £7,948 10*s.* under the description of 245 acres in Tydd St. Giles and Tydd St. Mary, which included lands other than of former Laverock ownership. Joseph Hannath died in 1868 after devising his property to his nephews and nieces, children of his late sister Jane Marshall, the wife of Joseph Marshall of Waldersea House, Elm. The devisees sold Hannath Hall and a portion of the land in 1870 to William Kilham of Tydd Gote, who devised the property to his son Frederick Kilham. From him it passed to George Williams, who was living at Hannath Hall in 1899,[10] and it still continues in possession of his family. A portion of the house appears to have been taken down but substantially it is intact, and capable of restoration.

The manor house dates from the early 17th century, with later alterations. It has good octagonal and diamond-shaped chimneys.

The rectory is a brick building erected for Canon John Scott, the rector, from the design of his brother Sir Gilbert Scott in 1868. It occupies the site of the former rectory which was a plain square ivy-covered house in a state of dilapidation. The present house is built on a raised terrace and is supported on solid blocks of reinforced concrete. Part of the field in front is an intake from the common. Close to the old boundary of the field, on the east side of the drive, was the tithe barn, and traces and fragments of the foundation have been discovered from time to time. The old name of the road passing the rectory, Barn Gate, is derived from this circumstance. The old house in New Field, nearly opposite the church, is said by persistent local tradition to have been the old rectory, but there is no evidence in support of this. It is of considerable antiquity and when first built must have been imposing. It has been

3 Gote, a corruption of go-out, the outfall of a drain. Four Gotes was where the Wisbech Drain, Leverington High Lode, Newton Lode, and Tydd Drain discharged into the Nene (*P.N. Cambs.* (E.P.N.S.), 285).

4 Stith is mentioned in Domesday (*Lincs. Domesday*, ed. Foster and Longley (Lincs. Rec. Soc. xix), 184) as a socland of Holbeach and Whaplode, and is identified tentatively as Tydd. There is no other holding in Elloe Wapentake with a name at all similar, and this identification may therefore be accepted.

5 See *P.N. Cambs.* (E.P.N.S.), 285–6, for a full discussion of the various possibilities.

6 On the authority of the late J. T. Marshall of Tydd Gote.

7 See *De Stemmate Piscatoris* by Mr. Frank Owen Fisher their descendant, containing an exhaustive and scholarly history of the family from the earliest times (privately printed *c.* 1910).

8 C.P. 25(2)/93/838.

9 P.C.C. 23 Pyckering. The Laverock arms are ermine a fesse engrailed gules.

10 Kelly, *Dir. Cambs.* (1900).

spoilt by modern alterations and is now divided into two cottages. Originally the property of Francis Dusgate, *c.* 1640, it was for a number of years owned by Timothy Hall and his family, afterwards by Joseph Hicks and Robert Hicks, and later by John Broughton. It was purchased from the representatives of the latter by E. B. Howlett, whose relict now possesses it.

Park House, in the Fen, was built by the Traffords and, together with a considerable area of land, was sold *c.* 1789 to Charles Littlewood. On Littlewood's death in 1796 it was again sold and eventually became vested in Lord Peckover of Wisbech, upon whose death (1919)[11] it was purchased by Mr. Abraham Licquorice. It has been considerably added to and altered.

In Fold Lane, off Kirkgate, are to be found the remains of the village pound, as well as the remaining portion of the old parish workhouse, now consisting of two cottages vested in the Tydd St. Giles Charity Trustees. Part of the parish workhouse was destroyed by fire in the last century.

PARISH FIELDS There are about thirty fields, or *campi*, extending from east to west: Hallcroft (the Angle); Hallcroft (South); Doddins; Gardyke or Laysoken; Furlong, Headacres or Spadeholme; Blowhead; Great Broad East; Little Broad East; Gores or Cotefield; Southcrofts or Dike Field Crofts; Long Pricks; Summer Leazure or Church Field; Edyke Field; Bladderwick; New Field or High South; Low South Field; Hornfield; Ryeland Field; Carrow Field; Bottle Lane or Cockley Field; Eagate Field; Ealeet Field; North Lane Field; Quaney Field; Fengate Field; Thridding or Treading Field; Fen Lane or Gitt Field; Shoffendyke or Shoddyk Field.[12] East of the Roman Bank are High Marsh and Low Marsh, representing the land inclosed from the sea. These marshes were gradually silted up as the sea receded, and became ripe for inclosure temp. Charles II. They were 'improved, set out and divided' under the Bedford Level Act of 1663.[13] The land was allotted to the lord of the manor and the owners of commonable messuages in the parish on the basis of 2 a. 1 r. 20 p. in the High Marsh and 2 acres in the Low Marsh in respect of each commonable messuage. By this means some 500 acres of marsh were brought into cultivation, and now provide some of the finest silt farms in the county. The final parliamentary inclosure took place in 1841.[14] Only 159 acres remained to be dealt with at this date, of which 2 were allocated for a recreation ground, 21 sold to defray costs, 17 given to the Bishop of Ely as lord of the manor, and the remainder divided, in very small allotments, among 52 proprietors.

DRAINAGE Tydd St. Giles, Newton, and Tydd St. Mary are associated with a particularly successful piece of land drainage. On 29 August 1632 Richard Colvile of Newton, Simon Wood, and other proprietors and commoners of the 3,000 or so acres of fen land in these parishes determined upon their recovery. The lands, it was said, were so waterlogged as to be of little value and were soon likely to be of none, 'by reason of their ancient drain the river of Wisbech, the outfall to the sea by the fower gotes, being diverted and drained'. The proprietors accordingly agreed with Henry Dereham of London that he would, within two years at his own cost, drain such lands to the outfall near Catsmere in such manner that the lands should always be kept dry one foot at least under the soil, and would maintain the drains and sluices he erected. In return he was to receive in fee simple two-fifths of the lands in question. This agreement was ratified at a Session of Sewers held at Wisbech on 20 September 1632. The Commissioners enacted that the fens in question should be charged with a rate of 40*s.* an acre, payable to Henry Dereham. In default of payment by any owner two-fifths of his lands should be set out and allotted to Henry Dereham absolutely. One moiety of Dereham's two-fifths was to remain liable for maintenance works, and every owner was to divide his land from that of his neighbours by such dikes as should be sufficient to carry away their waters to Dereham's main drains. All owners of high grounds in the three parishes were to be permitted to send their waters into Dereham's drain, for which an acre shot of 1*d.* an acre was to be paid.

This isolated drainage scheme, like that at Elm, must be considered to have been a success. Only a few years later, when the number of landowners in the parish of Tydd St. Giles was just under a hundred, Henry Dereham is shown to have been the owner of 13 acres in the Fen, which suggests that practically all the owners paid their 40*s.* per acre rather than hand over two-fifths of their land.

The drainage is now carried out by the Tydd and Newton Drainage Commissioners, appointed under an Act of 1773,[15] whose powers were amplified by a further local Act (1808).[16]

The Marsh lands, inclosed from the sea and lying on the east side of the old Sea Bank (now popularly known as the Roman Bank) possess natural drainage and did not in consequence in ancient times fall under any special drainage commission. The Marsh now falls within the area of the Westside Marshes Internal Drainage Board, a body which in 1939 was incorporated and re-formed under the Land Drainage Act (1930).

MANORS The manor of *TYDD ST. GILES* with the advowson of the church belonged to the bishops of Ely, and the sub-manors of Hockholds (Colvile's), Rickards, and De la Launds were held of this manor. In the inquisition upon the manor taken in 1222[17] Robert, son of Walter, was found to hold 8 virgates of arable land, each virgate containing 32 acres, by the fourth part of a knight's fee and suit to the Hundred Court, which 8 virgates in the inquisition of 1251[18] were found to be held by William de Weston, knight, by the same service. This is presumably Rickards manor. Similarly in 1222 Stephen de Marisco was found to hold 3½ virgates by the service of ½ knight's fee, presumably Hockholds manor. By the inquisition of 1222 Innocencia, widow, was found to hold 2½ virgates for 3*s.* 6*d.* in equal payments, which in 1251 were held by John de Lytbery and Innocencia his mother, presumably De la Launds manor. This last manor extended also into Tydd St. Mary, in which parish was the manor of Littlebury.

There were in 1251 94 separate customary tenants, holding 69 messuages and 9 cottages, and 32 virgates were accounted for besides 3 crofts, a virgate in Newton, certain other holdings and the demesne lands. The

[11] G. E. C., *Compl. Peerage*, 2nd ed., xiii, 84.

[12] These names are taken from the Tithe Award, the Tydd St. Giles terrier, deeds, and wills.

[13] 15 Chas. II, c. 17.

[14] 4 & 5 Vic. c. xvi. Award at County Hall, March.

[15] 13 Geo. III, c. 60.

[16] 48 Geo. III, c. xxiii.

[17] B.M. Cott. MS. Tib. B. II, f. 151*a*.

[18] B.M. Cott. MS. Claud. C. XI, f. 86*b*.

demesne lands detailed in the inquisition of 1251 were: 5 acres in Hevedakeres (Headacres, now Furlong Field); 2 acres in Suth Meadow Field (South Crofts); 3 acres in Bradherst Field (Broad East); 3 acres in Laysoken (Gardike); 5 acres in Hassecroft (Hall Croft); 4 acres in 'Thystlecroft'; 7 acres in 'Sumersleswe'; 2 acres in 'Stackstede'; 13 acres in 'Chapeleland'; making a total of 44 acres of profit-yielding (*lucrabilis*) land, which could and ought to yield profit according to the customs of the town. There also belonged to the manor a pasture which the lord held in severalty called Edych (Eadike), worth 10*s.*, and with other herbages of stubble worth 20*s.* more or less. There was also a salt pit which then returned half a load[19] of salt yearly; it used to yield more but had been almost entirely destroyed by the sea (*periit per mare*).[20]

Of the customary tenants, Oky son of John held a messuage and found a hen at Christmas and 10 eggs at Easter. He would plough with 5 oxen 2 days in winter and 2 days in Lent. He would go for seed to the lord's granary in Tydd St. Giles, would sow and harrow what he had ploughed, and would hoe for one day, receiving the lord's food. Among other services he would dig and embank about Wisbech Castle a portion in common with his mates (*paribus*), like the men of Wisbech. He should go with the lord's bailiff to distrain whenever necessary, pay *gersuma* or fine on his daughter's marriage, and the lord should have for relief at his death 20*s.* at most. He was liable for service as beadle or dike-reeve, if appointed. For every day that he mowed he should have 2 loaves, 4 herrings or 5 eels, or ½*d.* worth of cheese. The other messuages were held on similar terms. Ralph, son of Geoffrey del Wer, held a cottage for which he had to collect hens or eggs in Tydd St. Giles and carry them to (Wisbech) castle, for which he was to have his food.

Tenants of virgates could and ought to reclaim land towards the sea and marsh without increase of rent, and if any man's land or messuage should be destroyed by the sea, yet he must answer for his rent and customs. Elgere, son of Lenmere, held a cottage at the rent of 11*d.*, but nothing was then paid because it was destroyed by the sea. If any man could get a profit from the messuages and lands which had been for a long time drowned (presumably fen lands), the lord must have his due. Other tenements are mentioned as being destroyed by the sea. Sea wreck and royal fish, if found, should belong to the bishop as lord, but the finder should have 4*d.*

The whole yearly rental was £4 3*s.* 7*d.*, with 24 hens at Christmas, 800 eggs at Easter, and boon days.

Richard del Mers (Marsh), Symon Bryd (atte Bridge), and Richard de Fenna are names of significance. William de Stabulo is mentioned, but though the lord in later times possessed a mill, rented at the ancient rent of 8*d.*, no reference is made to it in the 1251 inquisition except inferentially, e.g. the reference to the lord's granary given above. Subsequently also reference is made to a chapel in Eadike, rented at 6*d.* This is doubtless the chapel near Tritton Bridge, by the 13 acres called 'chapel land', and until recently known as the Chapel Field, though the chapel has long since disappeared. The bishop's, or Barton manor appears to have been seized during the Civil War, for in March 1650 Lawrence Bromfield levied a fine in which Denis Taylor and Katherine his wife were deforciants, and in another fine of 1657[21] John Thurloe, Cromwell's Secretary of State, who had acquired other episcopal property, including Wisbech Castle, was plaintiff against Lawrence Bromfield and Sarah his wife, over the manor of Tydd St. Giles, with the rectory, all tithes, and the advowson of the church. At the Restoration the manor reverted to the bishops of Ely, who held it until it became vested in the Church Commissioners.

The manor of *HOCKHOLDS* was held of the Bishop of Ely, as above stated, by Stephen de Marisco as a knight's fee, being described as 3½ virgates in 1251; de Marisco obtained a grant of free warren there in 1268.[22] He was succeeded by his son Geoffrey, whose daughter Desirée carried it in marriage to Roger Colveil of Newton.[23] The ownership remained in this family until John Colvile of Newton, fifth in descent from Roger, died possessed of it and was succeeded by his son Francis, who died in 1509[24] without surviving issue, whereupon the manor devolved upon Richard, a younger son. Richard Colvile, who died in 1525, appears to have disposed of the manor,[25] which eventually became vested in the old Lincolnshire family of Welby.

The manor house belonging to this manor was originally in the north-eastern corner of Horn Field,[26] now called Willoughby's Corner, on a piece of land containing 4 acres, where now stands a small farmhouse. It had evidently fallen into decay, for a new manor house was erected by the Welby family on the north side of Kirkgate, being the house now known as the 'Manor House'.

It is probable that the manor was acquired by Thomas Welby of Long Sutton, whose relict Agnes (subsequently Agnes Deynes) and their eldest son Richard Welby possessed it. The manor was settled in 1579 upon Agnes for life, with remainder to Richard Welby and his heirs and remainder in default upon the right heirs of Mary, relict of Henry Adams of Tydd St. Mary, a daughter of Thomas and Agnes Welby.[27]

Richard Welby was probably the first occupant of the new manor house, which is a handsome Elizabethan building of red brick. On the east side a fleur-de-lis, the Welby crest, is carved in a panel in the brickwork. Originally the roof was thatched, and there were two upper stories, but one was taken down in the latter part of the 19th century by John Richard Tindall, the then owner, who roofed it with slate as it now appears. Richard Welby was living at Tydd St. Giles as early as 1574,[28] and his children were baptized there, the earliest in 1577. He married Audrey, daughter of William Callow of Holbeach, who survived her husband. He died in 1584, having by his will dated 3 April in the same year made provision for his younger

19 *Summa*, a measure containing 6 or 8 bushels.

20 It was estimated that 400 acres were devastated by 'sweet', i.e. fresh, waters temp. Edw. III (*Inq. Non.* (Rec. Com.), 203).

21 C.P. 25(2)/539/Hil. 1657.

22 Charter dated 20 June 52 Hen. III (1268) as given in *inspeximus* of 1466 (*Cal. Pat.* 1461–7, 498).

23 Sir C., C. R., and Z. Colvile, *Hist. of Colvile Family* (privately printed, 1896).

24 W. Watson, *Hist. Wisb.* (1827), 482.

25 For further details of this family see under Newton.

26 Tydd St. Giles parish terrier, *c.* 1647, in which the 4 acres of land is described as 'the site of the Manor of Hockholds'.

27 C 142/204/141. The settlement of 1579 affected 118 acres, part of the manor lands. See also C.P. 25(2)/93/843, a fine of East. 21 Eliz. between John Walpole and John Callow, and Anne Deans and Richard Welby.

28 When he attested the will of Robert Akers (P.C.C. 12 Pyckering).

children out of his other lands there. He was succeeded at Tydd St. Giles by his eldest son John Welby, who in 1601 married Jane, daughter of Sir Edward Osborne, Sheriff of London and Lord Mayor in 1583. In 1613 Welby entered his pedigree at the College of Heralds.[29] His elder children were baptized at Tydd St. Giles, but he eventually removed to Lynn, where from 1608 onwards his younger children were baptized. William, the second son, baptized in 1604, held the manor, containing 3½ virgates, of the Bishop of Ely, and sold it in 1629 to Thomasine Owfield or Oldfield, widow, who in turn sold it to Simon Wood of London.[30] Considerable property at Tydd was held by Richard Welby, younger brother of William, and according to a note in the parish terrier[31] 'Mr. Robert Girling bought all R. Welby's land and sold it to Mr. Wood'. Simon Wood[32] was a citizen and Merchant Taylor and evidently a man of considerable means, having apparently succeeded his father Robert Wood in the ownership of the manors of Dunton Hall and Tilney in Tydd St. Mary, though in 1643 he was at pains to disclaim any idea of wealth.[33] If he ever lived in the locality it would probably have been at Dunton Hall, though in the manor house at Tydd St. Giles, in a painted chamber window, was a shield bearing the arms of Robert Wood, i.e. Wood and two other coats quarterly, impaling the arms of Montague. From the bottom of the shield issued eleven sprigs with an escutcheon for each showing the differences for his eleven children.[34]

Simon Wood appears as owner of the manor in 1636, when the quit-rents of Hockholds amounted to about £2 per annum, and the rental due to him out of the two manors of Hockholds and Rickards amounted to £206 14*s.* His only daughter and heiress Margaret Wood carried the properties on her marriage to John Trafford of Low Leyton (Essex), descendant of a very ancient family and a firm adherent of the Royalist cause. He resided at Dunton Hall, where he died in 1666. His second daughter and eventual coheir, Anne, married Fisher Dilke of London, and their only daughter, another Anne, married (1692) Clement Boehm, a Director of the Bank of England, whose son Sigismund assumed the surname of Trafford upon succeeding to the estate. Sigismund Boehm Trafford made considerable additions to the family property in the Tydds, and rebuilt Dunton Hall at great expense. A few years before his death in 1723 he, like the Welby family, went to reside at Lynn. He was succeeded by his nephew Clement Boehm, and the latter, who died in 1741, by his son, another Clement, who was knighted on the occasion of his carrying an address to the king in 1761.

In 1759 Clement Trafford suffered a recovery upon the manors of Hockholds and Rickards and other properties, which passed to him under the will of Sigismund Trafford, in order to vest the same in himself absolutely, and part of it was then settled by him in trust for his wife and their issue. Matrimonial difficulties arose, which led to the break-up of the establishment at Dunton Hall; the building was demolished, and by deed dated 18 August 1779 the unsettled properties were conveyed to trustees upon trust for sale. The greater part of the property was disposed of in lots during the next few years, though some parts were still unsold at Sir Clement's death in 1786.

Lady Trafford obtained a private Act of Parliament in 1791[35] authorizing her to use her maiden name of Southwell alone. She died in 1809, and in 1812 the settled property was sold for £49,800, in lots, but although the two manors of Hockholds and Rickards were included in the recovery of 1759, they do not appear to have been separately disposed of subsequently, and must have lapsed. Included in the sales of 1812 was the manor house with 90 acres of land, purchased for £8,025 by Barnabas Coe of Terrington St. Clement, who two years later re-sold to John Cartwright of Holbeach Marsh. It does not appear that the latter ever lived in the house, and some years after his death, which occurred in 1845, the property was sold by his trustees to Richard Joseph Tindall, son of John Tindall of Little Hale (Lincs.), then resident at Tritton manor, Tydd St. Mary. Like some of his predecessors he spent the last years of his life at Lynn, dying there in 1890, when the property passed to his elder son J. R. Tindall, J.P. The latter died in 1946, leaving the property in trust for his daughter Edith Mary, the wife of Mr. A. H. Carter.

The manor of *RICKARDS*, like that of Hockholds, was held of the Bishop of Ely, and in 1529 belonged to William Depdale, citizen and painter-stainer of London. He directed by will that his manor called Ricards manor in Tydd St. Giles and his lands there held by Richard Ogle, and his co-feoffees, who were seised thereof to his use, should be sold and the money divided among his children.[36] The following year Thomas Tyd of Tydd St. Mary, yeoman, appears to have been possessed of a moiety, for he directed by will[37] that his feoffees should stand seised of a moiety of a manor in Tydd St. Giles called Ryckards manor and a messuage and 24 acres of land in Tydd Hern formerly belonging to his father Thomas Tyd. In 1550 the manor was held by Richard and Thomas Tydde.[38] By 1614 it had become vested in John Welby and thereafter was held in similar fashion to Hockholds (q.v.) until it lapsed. No trace of any manor house exists, but the site was identified in the Tydd St. Giles parish terrier, *c.* 1647, as being on 2 roods of land in Newfield next Kirkgate on the north, and land of William Watkinson on the east, being the property now and for many years occupied by the village blacksmith's shop, and not much more than a hundred yards from the site of Hockholds Manor.

In 1636 the quit-rents of the manor were given as £1 17*s.*

DE LA LAUND'S MANOR

A manor was held by Sir Thomas de la Laund of Gosberton (Lincs.) which, with many other manors in Lincolnshire and elsewhere, on the rebellion of Sir Thomas were forfeited to the Crown. In 1471 it was granted by Edward IV to his brother Richard, Duke of Gloucester,[39] and four

[29] Visitation of Norfolk of that year (*Visit. Norf.* (Harl. Soc. xxxii), 307). The arms of Welby, viz. sable a fesse between three fleurs-de-lis argent, were formerly in a window in Tydd St. Giles church and recorded by Layer (Bodl. MS. Rawl. B. 275, f. 182).

[30] B.M. Add. MS. 5838, f. 195*a*.

[31] *c.* 1647.

[32] Robert Wood of Langley (Notts.) married, *c.* 1539, Margaret daughter of Sir Edward Montague, C.J. of K.B.

[33] See his petition to Parliament in that year.

[34] B.M. Add. MS. 5809, ff. 121–4, where a drawing of the arms is given. This shield has disappeared since 1745 when Cole saw it, but in recent years Mr. A. H. Carter has put new glass of the same general design in one of the dining-room windows.

[35] 31 Geo. III, c. xxxix.

[36] P.C.C. F 13 Jankyn.

[37] P.C.C. 17 & 18 Jankyn.

[38] *Cal. Pat.* 1549–51, 414.

[39] Ibid. 1467–77, 297, 560.

years later was regranted (probably in trust) to Queen Elizabeth (Woodville), Richard (Beauchamp) Bishop of Salisbury, and William Dudley, Dean of the Chapel of the Household.[40] Richard Laund's heirs held property in Tydd St. Giles in 1435,[41] and a number of parcels in Tydd St. Giles were held of 'Dallalands Manor' (as the name came to be known in the passage of time) but it is evident that the manor was actually sited in Tydd St. Mary. On this point the will of Henry Adams, esquire, of Tydd St. Mary[42] is decisive. In it he devised to his son Robert Adam 'twelve acres of pasture at Clements Cross being the scite of the Manor of Dallalands'. Except that the 12 acres were in Tydd St. Mary, the exact location has not been established. In the reign of Henry VIII this manor was in possession of Thomas Denton, who made settlements of it in 1535[43] and 1538.[44]

Robert Adam after his father's death (*c.* 1586) wasted his inheritance, and sold this manor to his kinsman William Welby. In 1629 it was settled by fine by another William Welby; by this date it had acquired the alternative title of *DENTONS*, and the manors of Rickards and Hockholds were also included in this settlement.[45] In the following year the same three manors were acquired by Philip Culme and another from Thomasine Owfield, widow, John Janson, esquire, Joyce (*Jocosum*) Glover, clerk, Joseph Owfield, and John Owfield,[46] who were acting as trustees for Welby under the 1629 settlement. The ownership of this manor ultimately became vested in Thomas Pell, from whom it passed to his son Edmund, lord of the manor of Littlebury in Tydd St. Mary, of which he was in possession in 1677.[47]

References occur to another manor in Tydd St. Giles, held by the Thimbleby family and subsequently that of Cony, which may perhaps be identified with Rickards or De la Launds. About 1540,[48] William Walgrave, illegitimate son of Richard Thimbleby, esquire, the grandson and heir of the said Richard, was plaintiff in respect of lands at Tydd St. Giles set out to him in lieu of an annuity charged on manors in the two Tydds and elsewhere; and in 1558–9 Sir William Cordell, Master of the Rolls, obtained this manor, with that of Dunton Hall in Tydd St. Mary and other property in Lincolnshire, from various members of the Thimbleby family for £1,080,[49] followed by another fine of 1559 between the same parties[50] in which the manors of Dunton (Tydd St. Mary), Tydd St. Mary, and Tydd St. Giles are specifically mentioned. Ten years later William Cony senior, esquire, bought the manor from Sir William Cordell and Mary his wife. A third of the area at this date was stated to be marshland.[51]

CHURCH The advowson has always followed the descent of the main manor, and has been in the gift of the Bishop of Ely. The earliest recorded rector is Nicholas de Houton, 1320. The church, like most of those on the episcopal manors, was never appropriated, and in the early Middle Ages was a valuable one. It was assessed at £20 in 1217, £33 13*s.* 4*d.* in 1254, and £42 in 1291.[52] At the last date it was one of the richest in the Isle. In 1535, however, the value was only £21 13*s.*[53]

The church of *ST. GILES* consists of clerestoried nave, aisles, north porch, and detached south-east tower. It is of early 13th-century origin, to which period the chancel arch, nave arcades, and two lower stages of the tower belong. The aisles were rebuilt in the first half of the 14th century and the porch added, and at the end of the 15th century the clerestory was rebuilt and raised and another stage added to the tower. The chancel, originally 50 ft. long by 20 ft. wide, was completely destroyed in a gale in 1741. It was rebuilt in a shortened form and debased style the following year, but the 18th-century building was taken down in the 19th-century restoration and not replaced. This restoration was begun in 1868 by the rector, Canon John Scott, under the direction of his brother, Sir Gilbert Scott the architect. The whole church was refloored and re-seated, the nave re-roofed, screens provided for the vestries, and the musicians' gallery at the west end taken down. The tower was strengthened in 1888, and the organ installed in place of a harmonium about the same time.

The position of the tower, about 50 ft. to the south-east of the church, is probably due to the necessity of obtaining a firm ground for the foundations. It may be paralleled at Terrington St. Clement and West Walton (Norf.), and Fleet and Long Sutton (Lincs.) in the marshland. As at West Walton and Long Sutton, and also at certain churches in the Isle, e.g. Littleport and March, the tower rested on open arches which have now been filled in.

The blocked chancel arch is two-centred and has deep mouldings, while the responds have engaged shafts and caps with early foliage. A modern window of three lights has been inserted in the arch. There is a blocked opening in the wall on either side of the arch. The arcades consist of six bays with two-centred arches of two orders and hood-moulds; the piers and caps are round and the bases square; several of the caps have stiff foliage and the west responds have three engaged shafts and foliaged caps. The blocked upper doorway of the rood stair exists in the north-east angle of the arcade. On one of the piers is a 14th-century inscription reading 'Cest Piler Comencat Ricard le Prestre Primer Preyez Pur Luy'. The westernmost bay is of slightly later date than the rest, suggesting that there was formerly a tower in the normal position, which was replaced, when the present one was built, by an extra bay to the nave.

The clerestory consists of cinquefoiled two-lights under square heads. The blocked heads of the original clerestory lights exist between the present windows and at a lower level; they are plain lancets. The blocked rood stair is in the south-east angle of the north aisle and is corbelled out on the exterior. The large west window of the nave is of mid-14th-century date and has five lights with flowing tracery. It is flanked on the outside by large 14th-century buttresses with three set-offs and a canopied niche on the face, with gabled tops and angle pinnacles. The west ends of the aisles also have angle pinnacles and crocketed coping, all of the 14th century. The west doorway has a wide ogee arch with three niches above.

40 Ibid. 543, 567.
41 Wisbech MSS.
42 P.C.C. 65 Windsor.
43 C.P. 25(2)/4/20.
44 Ibid.
45 C.P. 25(2)/526/East. 4 Chas. I. Welby also included his Lynn property in this settlement.
46 C.P. 25(2)/400/Mich. 5 Chas. I. This fine is defaced, and the name of Culme's partner and the consideration paid are illegible. The three manors included a dove-house, orchards, and 240 acres of arable land and 450 of marsh.
47 E 134/29 Chas. II/T. 4.
48 C 1/1086/2.
49 C.P. 25(2)/83/713.
50 C.P. 25(2)/259/Hil. 1 Eliz.
51 C.P. 25(2)/93/835.
52 W. E. Lunt, *Valuation of Norw.* 539.
53 *Valor Eccl.* (Rec. Com.), iii, 499.

The east window of the north aisle is late 15th century and of four main lights cinquefoiled, with eight tracery lights, all under a square head. The first three windows in the north wall are of three trefoiled lights with mid-14th-century tracery; the remaining two windows in this wall are also of the 14th century and of two trefoiled lights with quatrefoils above. The west window is a 15th-century insertion of three lights cinquefoiled, with rectilinear tracery in the two-centred head. There is a large pinnacle at the north-east angle. offs with gabled tops. The north aisle has a plain plastered lean-to roof which is probably 17th century; the other roofs are modern and poor.

The font has a hexagonal bowl with the arms of Ely, the emblems of the Passion, the shield of St. George, an angel and two grotesque faces, representing sloth and gluttony, growing out of leaves. The shaft is traceried and the font stands on two stone risers. In the modern screens in the east bays of the arcade are some old tracery heads.

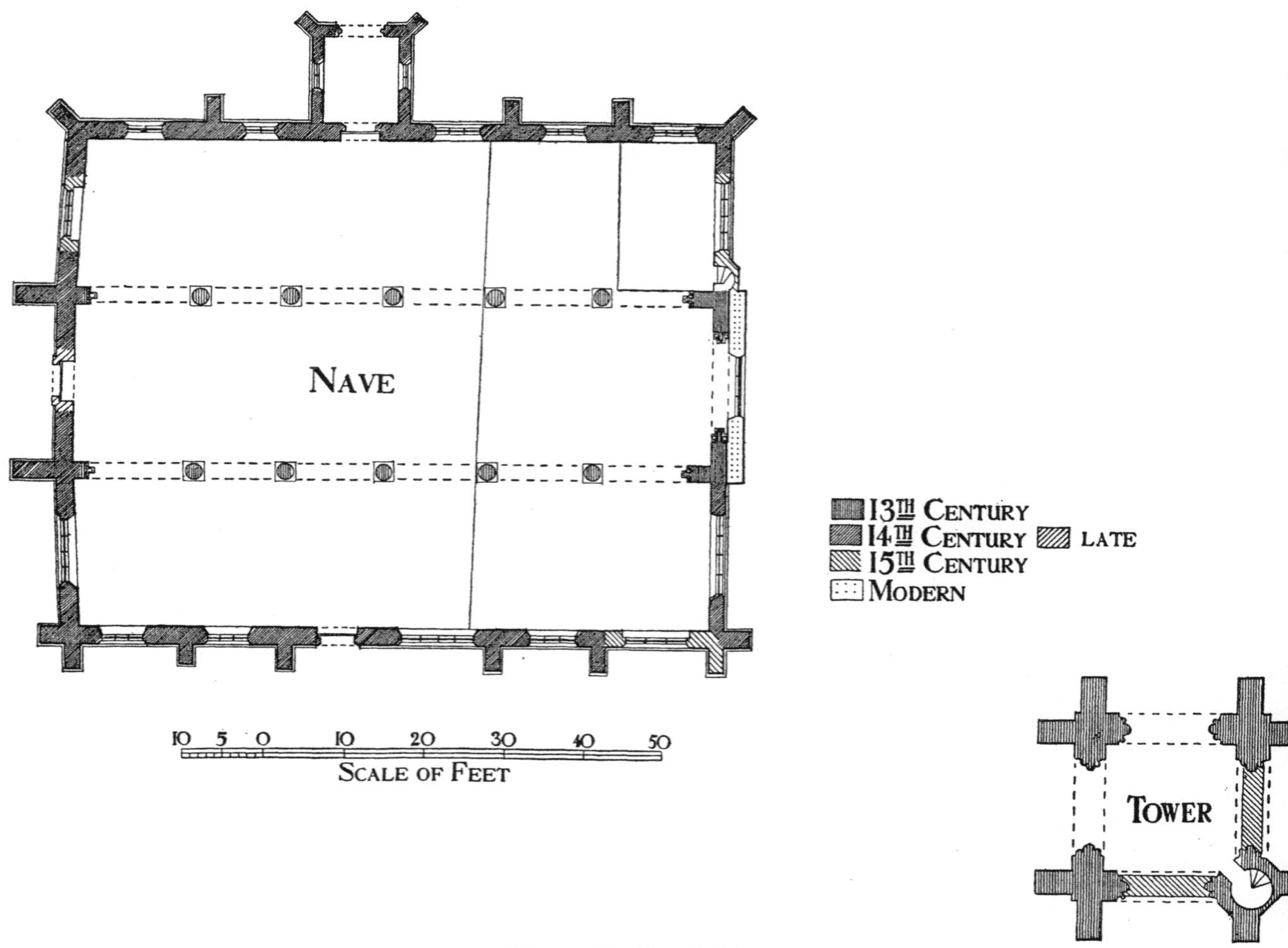

PLAN OF ST. GILES'S CHURCH

The south aisle has an east window of four lights with rather clumsy 14th-century tracery. The east portion of the south wall has been rebuilt in brick with stone dressings, probably at the end of the 15th century, and contains a contemporary window of three lights. This portion of the aisle, now the vestry, was in the Middle Ages used as a Lady Chapel[54] and contains a piscina. The second, fourth, and fifth windows are 14th-century three-lights and the third a four-light, with trefoiled heads and tracery of the period; the west window is also of four lights and of the 14th century. There is a niche with trefoiled head in the south wall.

The north porch has an outer arch of two orders with moulded caps, and a two-light trefoil-headed window on the east and west. The inner doorway has continuous mouldings. The buttresses throughout the building are almost entirely 14th century, of two set-

The detached tower is of three stages, the two lower being of the 13th century and of stone, and the top stage a late 15th-century addition in brick with stone dressings. It stands on four two-centred arches of three orders with moulded caps and bases, the east and south being blocked with a brick filling when the tower was raised. There are large and massive angle buttresses, which are square below and semi-octagonal above. The newel stair is contrived in the south-east buttress, which was raised in brick and stone in the 15th century so as to reach the top stage. There is a large 13th-century window on all four sides in the second stage. Each is of three uncusped lights now blocked with brick, with a quatrefoil pierced in the solid head, the whole contained within a two-centred arch of two orders with engaged shafts to the jambs—an early example of plate tracery. The top stage has a window of two cinque-

[54] Will of Robert Galyon, rector, 1435 (P.C.C. 20 Luffenham).

foiled lights on each face, and there is an embattled parapet.

The most interesting sepulchral inscription remaining is a 13th-century cross slab of grey marble with an inscription commemorating Sir John Fysuer. The stone was originally at the east end of the nave, but was moved to its present position behind the organ in 1868. According to John Layer, the antiquary (1585–1640), there were formerly inscriptions in the stained-glass windows commemorating (i) Robert Michell, rector, collated 1349 (east window), (ii) Thomas Howson, who 'built two faire wyndowes', (iii) Simone Howson, his wife, and John Howson, 1483, and (iv) Henry at the Bridge and his wife ('in the fair ancient window' in (?) north aisle). The following coats of arms were apparently also displayed (i) argent, a lion, over all a label of five points gules, (ii) or, two bars gules, (iii) or, a chief, and (iv) or, on a fesse gules, three roundels argent.[55]

The plate consists of a communion cup and cover paten of silver, 1569, the latter inscribed on the foot 'FOR TED SENT GYLS 1570', and a paten and flagon both of silver, dated 1869.

In 1552 there were three bells in the bell turret, another bell on the ground, and the sanctus bell.[56] The tower now contains six bells by Mears and Stainbank of London, 1887, replacing a ring of five. The inscriptions on the latter were: treble, 'The gift of Sigismund Trafford of Dunton Hall in Tidd St. Maries 1710. Henry Penn Fusor', with the Trafford arms and crest; 2nd, 'Omnia fiant ad gloriam Dei. Abell Hodges, Rector. Tobie Norris cast me, 1625'; 3rd, 'In aure Dei non clamor sed amor cantat. Tobie Norris cast me, 1625'; 4th, 'Coelorum Christe placeat tybe vox sonus iste. Jesus spede us. Johannes Welbe generosus et Clement Martin, Rector, 1603'; 5th, 'James Scribo, Adam Cook, Churchwardens, 1726'.[57] The cost of the sixth bell was met by public subscription.

The registers begin in 1559, with a gap from 1659 to 1689.

About 1865 Richard Young built a Church mission house at Foul Anchor at his own expense. He stipulated in doing so that the work of restoration of the parish church should be pressed forward with all convenient speed. The mission house was opened for use in 1866. On Young's death (1871) the ownership vested in his son the Revd. Belton Young, who several years later handed it over to the ecclesiastical authorities. It was closed for some years during the 20th century, but in 1950 was reopened for its original purpose and as a Sunday school.

GUILDS There were three guilds at Tydd St. Giles, those of Our Lady (founded in 1350), Holy Cross (1385), and St. Giles (1386).

Our Lady's Guild was remarkable in that its membership was strictly limited to twelve brethren, and every vacancy which occurred was to be promptly filled up. It was founded by William Everswell, chaplain, and Nicholas Clerk, who, wishing to do something for the improvement of their lives and for the benefit of their souls, and because they two alone were insufficient to carry out the purpose in mind, founded a chaplaincy and got others to join them. In 1535 the guild's properties were valued at £4 13*s.* 1*d.*[58] It possessed a hall of its own, with a house for the chantry priest, and certain lands. Two torches at festivals and one candle before the image of St. Mary were provided by the members.[59]

St. Giles' Guild maintained three candles at masses, and subscriptions were paid in the form of one, or two, bushels of barley. The members were required to attend vespers on the vigil of the feast-day, high mass on the day itself, and vespers following. All members were obliged to attend the funerals of dead brethren unless excused for illness or infirmity. The proctor was to suffer the bellman to go round the town ringing his bell, to make all men pray for the soul of the deceased, and to summon all the guild members to mass, at which each brother was to give 1*d.*, and the brotherhood 1*s.* 8*d.* in bread to the poor. The brethren each paid ½*d.* in soul arms.[60]

The Guild of Holy Cross possessed its own chapel at Sea Gate, at the Tydd Gote end of the village next the outfall of the Shire Drain, where the members were accustomed to meet. This guild did not possess a hall.[61]

There was a further Guild of St. John Baptist,[62] in respect of which no guild certificate exists. It is possible that the chapel at Tritton Bridge, previously mentioned, belonged to this guild.

NONCONFORMITY A congregation of Baptists was established at Tydd St. Giles by 1782. John Smith, who kept the village shop in the house next the rectory field, founded the chapel in Broadgate, adjoining a house belonging to him, with an open-air baptistry for immersion. Smith was ordained in 1795, and two deacons were appointed at the same time. In 1796 the congregation numbered 15. It had increased to 28 in 1800. In 1817 there were 65 Baptists at Tydd St. Giles, Sutton St. James, and Fleet. Smith died in 1807, and was buried beside his chapel. He was succeeded by Mr. Pocklinton and James Smith.[63] In 1851 there was a larger average attendance at the chapel than at the parish church, where the rector was non-resident.[64] The chapel was discontinued in 1922 and the building sold.

Two Methodist chapels, Wesleyan and Primitive, in the village and fen respectively, were established in Tydd in the latter part of the 19th century.[65] They are both still in use.[66]

SCHOOL In 1789 Tydd St. Giles was one of the sixty-eight parishes in the diocese without a school.[67] Dr. Jobson, reporting on the state of the schools in Wisbech hundred in 1814,[68] stated that there were about 50 children of school age in the parish, including many Dissenters. By this date a school had been established in the workhouse, where 30 to 35 children learned their catechism under a master who was allowed £12 a year by the parish. Jobson mentioned the possibility of applying £70 a year

55 Bodl. Lib. MS. Rawl. B. 275, f. 182.
56 E 315/495, p. 165.
57 J. J. Raven, *Church Bells of Cambs.* (1881), 172.
58 *Valor Eccl.* (Rec. Com.), iii, 499.
59 C 47/36.
60 C 47/35.
61 C 47/34; E 179/82/197 and /247.
62 See will of Richard Farrour, 1537, bequeathing 5*s.* for the repair of St. John's guildhall (P.C.C. 14 Dingley); also rental of Hockholds manor, 1636, showing land formerly of St. John Baptist held at a yearly rent of 4½*d.*
63 A. Taylor, *Hist. Engl. Gen. Baptists* (1818), ii, 300–1, 406–7.
64 H.O. 129/7/193.
65 These chapels are not mentioned in the 1851 Census ecclesiastical returns.
66 *Meth. Ch. Bldgs. Stat. Return as at 1 July 1940* (privately printed, 1947).
67 Bp.'s Mun., Ely, B 6.
68 Ibid.

from charity lands to education, but nothing seems to have been done until after the foundation of Marshalls' charity.

In 1834 Thomas Marshall purchased 4 a. 2 r. 29 p. to meet the cost of school fees for children of deserving families. In 1837 this land was let in eleven allotments at an average rent of 50*s.* an acre, to the sober and industrious poor. On the establishment of the School Board (1879) the charity became superfluous and was wound up by J. T. Marshall, the settlor's son, who had complete discretionary power for that purpose. Thomas Marshall, although born deformed with only one arm, by his diligence and industry amassed a considerable fortune. He attributed this largely to the good start he got by attending school at Leverington, and is said never to have missed the 5-mile daily journey except in flood-time. In his childhood, early in the 19th century, the land was mostly in pasture and it was possible to walk from Tydd to Leverington almost in a straight line.

The school thus endowed was held in the south aisle of the church. This was partitioned off and a door cut in the west wall, an arrangement which lasted until the restoration of the church in the 1860's. Soon after its establishment there were 35 boys, all over 6 years of age, receiving free education in reading, writing, and arithmetic, besides 16 boys whose parents paid 5*s.* a quarter for reading and 2*s.* a quarter extra for writing and arithmetic.[69] The schoolmaster was Samuel Clarke; the girls received their instruction from Mrs. Parker in a cottage near the church, where the village post office now is.

The first actual school building was erected opposite the church in 1866[70] at the instance of Canon John Scott, the rector, who collected moneys for the purpose and himself contributed £150; the National Society made a grant of £70.[71] A. J. Burling was the last master to teach in the church, and when the new school was opened he and Mrs. Burling found themselves with over a hundred children. He and his wife were still teaching in 1899, when there were 140 places and an average attendance of 110.[72] The school was originally a Church school, but failure to raise the funds for its necessary enlargement led to the compulsory formation of a School Board in 1879,[73] to which the Church authorities leased their school at a nominal rent. Early in the 20th century the school was again enlarged by the provision of a porch, and in 1927 a new head teacher's house was built on a fresh site in Church Lane, the old one being occupied by the school cleaner. Six years later the senior children, 30 in number, were transferred to the Colvile School at Newton, leaving 85 at Tydd. This number had decreased to 72 in 1949, when the school buildings were described as obsolescent.[74]

CHARITIES[75] The Brigstock and Wren's Charities' trustees, acting under a scheme established by the Charity Commissioners in 1910, administer about 29 acres of land, with 3 cottages, in Tydd St. Giles, and about 12 acres in Sutton St. Edmunds (Lincs.). Wren's charity originated when the English fleet attacked the Dutch in Sole Bay in 1672. One of the naval officers on board the *Royal Prince* was Matthew Wren, son of the Bishop of Ely, who had a presentiment that he would not survive the engagement. He made his will on board on 19 May 1672, in sight of the Dutch fleet and in expectation of fighting with them in a few hours' time, and thereby gave 15 acres in the Low Marsh for the use of the poor of Tydd St. Giles, subject to a rent charge. The 15 acres were let in 1835–7 to Robert Mills at £29 yearly.

The Charity Commissioners of 1837 declared that John Brigstock in 1667 had given the following lands to 9 trustees for the poor of Tydd St. Giles: a messuage and 3 roods in Sumer Leazure, the messuage in 1837 being used as the parish workhouse and very old and in bad repair; a messuage and 1¼ acre in Gardyke; two other parcels each of 1¼ acre in South Crofts and Carraway Fields; 2 acres in Low South Field; 2 acres in Furlong and Spade Holme Fields; 2 acres in Low Marsh; 2¼ acres in High Marsh. In point of fact none of the property ever belonged to Brigstock; he was the last surviving trustee of an older charity and the deed was simply an appointment of new trustees for the continuation of the trust. It is difficult to establish the origin of these charity lands, but (mainly with the aid of the Tydd Terrier) it may be stated as follows. The 2 acres in Low South Field were given by Robert Wilzey of Tydd St. Giles for providing, out of the rent, a supply of coals for the poor. The lands in Gardyke were purchased by the town from Coyton. The 2 acres in Furlong and 1 acre in Great Broad East were taken in exchange for the water mill. The land in the Marsh was allotted to the trustees on the inclosure of the marshes. The remaining lands are described as 'Town Lands' and are of uncertain origin, except that some of them were derived from one Coxon. The income of all of them, so far as is known, has always been, and now is, distributed to or for the benefit of the poor. Certain odd portions of lands were sold when the North Level Drain was cut in 1834, and separated from the remainder, and the proceeds of the sale were employed in the purchase of 12 acres at Sutton St. Edmunds (Lincs.), which are held upon the same trusts.

WISBECH ST. MARY

Wisbech St. Mary, one of the larger parishes of the Isle, covers an area of some 10,000 acres[1] on the left (north-west) bank of the Nene. It appears to have emerged from Wisbech (St. Peter) in 1109 when Wisbech vill was divided between the Bishop of Ely and the prior and convent. That this separation had not taken place in early times is suggested by the fact that the parish boundary stops more than a mile short of the 'Roman Bank' which bounds all the other marshland parishes of the Isle. The tradition that St. Mary's church was the mother church of Wisbech is rendered improbable by the line of this boundary and also by the fact that until the last century Wisbech St. Mary was for ecclesiastical purposes a chapelry of Wisbech (St. Peter).

[69] *Rep. Com. Char. (Cambs.)*, [103], 270–1, H.C. (1837–8), xxiv.
[70] Kelly, *Dir. Cambs.* (1900).
[71] Archives of Nat. Soc.
[72] Kelly, *Dir. Cambs.* (1900).
[73] *Lond. Gaz.* 1879, 736.
[74] M. of Ed. File no. 12/50.
[75] *Rep. Com. Char.* (*Cambs.*), [103], 270–1, H.C. (1837–8), xxiv. Infm. Messrs. Fraser, Woodgate, and Beall, of Wisbech.
[1] Before 1933, 10,187 acres of land and inland water. The Isle of Ely Review Order of that year (M. of H. no. 76940) reduced it to 9,793 acres.

Wisbech St. Mary is the least nucleated of all the Isle parishes. The silt soil which covers the north-eastern part provided a firmer foundation than peat, and though all the Marshland villages are more scattered than those farther south, Wisbech St. Mary is unusually so. The largest settlement is round the church, in the north-east corner of the parish about 3 miles south-west of Wisbech town. But this accounts for less than half the population, and there are other considerable villages at Murrow (which gave its name to the chief manor) on the northern and Guyhirn on the southern edge of the parish, and smaller ones at Tholomas Drove on the road between St. Mary's and Guyhirn and at Thorney Toll on the extreme western edge, 7 miles west of the parish church. There are also several scattered houses along the Nene bank and the droves across Wisbech High Fen in the west. In spite of the dispersed settlement pattern, the density of population (26 per 100 acres) is high; it is surpassed only by Outwell (77) and Leverington (46), where somewhat exceptional conditions prevail, and Littleport (27), among the rural parishes of the Isle.

A large section of the parish falls within the Wisbech market-gardening area, and near Hiptoft Farm east of Murrow village is one of the largest areas of orchard, covering more than a square mile. There is also a good deal of bulb-growing.

The Nene forms the southern boundary of the parish. The main road from Peterborough to Wisbech (A 47) crosses the fen in its course from Thorney to Guyhirn, and leaves the parish at its junction with the March road at Guyhirn Bridge. A second-class road (B 1441) keeps to the left side of the Nene and runs from Guyhirn to Wisbech through Tholomas Drove and St. Mary's. This was formerly the main road from Peterborough and Thorney to Wisbech. The section between Tholomas Drove and Bunkers Hill was originally straight, and is still discernible; it was diverted to a more northerly course in the 18th century.[2] Another road (B 1187) runs from Guyhirn into Lincolnshire, passing the west end of Murrow village, where it is joined by B 1169 from Leverington which runs along the north edge of the parish. The Peterborough–Sutton Bridge branch of the former M. & G.N. Joint Railway, opened in 1866,[3] has stations in the parish at Murrow (East) and Wisbech St. Mary. About ¼ mile west of the former station it is crossed[4] by the March–Spalding line, opened the following year, which has a station for Murrow (West); that for Guyhirn is just over the parish boundary.

The Manor House at Wisbech St. Mary is a late 18th-century building with a stone-slated roof. Inham Hall and a farmhouse in St. Mary's village about 120 yards west of the school are also noteworthy, though late in date (early 19th century). All these three houses have fine doorways; that at Inham Hall is protected by a portico porch.

MANORS On the division of the Ely estates between the bishop and the convent, half the vill of Wisbech was reserved to the latter. This half was more or less co-extensive with the present parish of Wisbech St. Mary, and became known as the manor of *WISBECH MURROW*. This manor remained continuously with the prior and convent or their successors the dean and chapter, but documentary evidence is slight. Pope Gregory IX, *c.* 1231, confirmed 160 acres in the marsh (*mora*) of Wisbech to the monastery.[5] John de Watford held the manor on a twenty-two-year lease at £16 a year from 1277.[6] The *Taxatio* of 1291 values the conventual property in Wisbech, with portions in Elm and Leverington, at £19 9*s.* 7*d.*[7] In 1324–5 it was let to farm at £51 12*s.* 6*s.*,[8] a figure so much higher than any previous or subsequent one as to suggest that it included all the convent's estates in the marshland vills. A mill was let at 30*s.* a year in 1385.[9] In 1448 Wisbech Murrow manor produced rents of £16 16*s.* a year, and perquisites of court of 13*s.* 5*d.*[10] By 1522–3 the value had risen to £20 7*s.* 7½*d.*, and the manor was appropriated to the pittancer of the monastery; in 1527–8, however, the manor was returned at only £13 and was the least valuable of all the conventual manors in Cambridgeshire except Mepal.[11] In 1541 it was formally transferred to the newly constituted Dean and Chapter of Ely.[12] In the early part of the 17th century the manor was farmed by the Pepys family, lessees under the dean and chapter, to Richard Bendish.[13] The survey of the parliamentary commissioners (1650) shows rents of £4 10*s.* 8*d.* and manorial perquisites valued at £5 18*s.* 10*d.* The only properties in demesne were two tenements, a thatched timber barn 66 ft. long and a small piece of arable ground; each of these was valued at 10*s.*[14] The whole had been leased in 1639 for twenty-one years to Samuel Pepys.[15] In 1641 a twenty-one-year lease was granted at £17 to Edward Crosse of Wisbech, whose daughter Katherine, a minor, was in possession in 1650.[16] The Commonwealth grantees of Wisbech Murrow were John Pepys and Percival Aunger. They paid £898 2*s.* 8*d.* for it—a lesser sum than was paid for most of the Ely capitular manors, showing that its relative importance had not increased in the past century. The appurtenant lands included 271 acres reclaimed from the marsh, almost all of which was in pasture.[17] At the Restoration the manor was restored to the dean and chapter. About 1730, when the demesne consisted of 20 acres of old inclosure and 300 acres of fen, Mrs. Susan Pepys paid the beneficial rent of £14 13*s.* 4*d.*, and one quarter of wheat or £1 6*s.* 8*d.*, and was obliged to provide the annual court dinner. The fines for renewal of the lease were rising sharply, from £85 in 1713 to £170 in 1720 and £190 in 1727. The sum of £7 15*s.* 1*d.* was received in quit rents and £3 3*s.* from profits of the court, at which the fines were certain.[18] Among subsequent tenants were Mr. Chettoe of Chesterton, a friend of Cole the antiquary (*c.* 1773),[19] and Cheeseman Williamson (1851).[20] In 1900 the lordship was vested in the Ecclesiastical (now Church) Commissioners as representing the dean and chapter.[21]

During the later Middle Ages various subordinate

2 Infm. Mr. G. M. G. Woodgate.
3 *V.C.H. Cambs.* ii, 132–3.
4 The only example of railway crossing railway on the level in the Eastern Counties, where it might be expected to be common. Apart from mineral lines and tramways there are only about a dozen such crossings in the whole of England.
5 B.M. Add. MS. 5819, f. 4.
6 Hist. MSS. Com., *6th Rep., App.* 291.
7 *Tax. Eccl.* (Rec. Com.), 270.
8 D. & C. Mun., Ely, Treas. R. 1324–5.
9 D. & C. Mun., Cupbd. A. 10 Ibid.
11 Rentals transcribed by Archdeacon S. J. A. Evans.
12 *L. & P. Hen. VIII*, xvi, g. 1226 (12).
13 Court R. (ex. infm. Mr. G. M. G. Woodgate).
14 MS. Collect. of Mr. G. M. G. Woodgate: Wisbech, i, 47 (from a manuscript at Lambeth). 15 C 54/3513, no. 15.
16 Woodgate MSS. Wisb. i, 47.
17 C 54/3513, no. 15.
18 D. & C. Mun., Bentham Transcripts.
19 B.M. Add. MS. 5847, f. 3.
20 R. Gardner, *Dir. Cambs.* (1851), 567.
21 Kelly, *Dir. Cambs.* (1900).

manors emerged in the Wisbech area, of which Hiptofts, Jacketts, Tuddenham Hall, and Bevis Hall were within the boundaries of St. Mary's parish.

In 1388 and 1406 Edward and John Hiptoft had licence for an oratory in their house, which must have been on or near the site of Hiptoft Farm, about a mile east of Murrow and a mile west of St. Mary's village.[22] In 1476 Isabel, wife of Sir William Norreys and formerly of John Nevile, Marquess of Montagu, held 100*s.* rents of assize in Wisbech, known as 'Hiptoftes rent'.[23] A manor of *HIPTOFTS* was in 1525–6 settled by Christopher Coote and Elizabeth his wife on John Hudleston and others, together with a fishery and 40*s.* rent in Wisbech and Leverington.[24] In 1620 it was held by Humphrey Gardiner of the Bishop of Ely as of his manor of Wisbech, but was no longer itself described as a manor.[25]

The manor of *JACKETTS* may be traced back to the lands of that name for which John Cave owed suit at Wisbech Hundred court in 1492–3.[26] By 1542 it was in possession of the Megges family, who were tenants of a good deal of Ely property including the bishop's manor of Wisbech Barton; in that year Thomas Megges died seised of it, his son Nicholas being heir.[27] The next recorded owner is Henry Adams of Tydd St. Mary, who bequeathed it to his brother Thomas, of Duxford, in 1587.[28] No connexion can be traced between this family and the Meggeses, or with the Steward family, into whose possession it had come in 1635. In this year William Steward died seised of it, bequeathing it in default of male heirs to his brother Thomas. It was then held of the bishop as of his manor of Wisbech Barton.[29] Thomas and his son of the same name were dealing with it 12 years later.[30] The manor is again recorded in 1668.[31] The name survived in 1777 as that of a piece of land of about 19 acres in Sayers Field, on the north side of the village street of St. Mary's near the old Primitive Methodist chapel.[32]

The manor of *TUDDENHAM HALL* must also have been in Wisbech St. Mary, as in 1392 it was held by Sir John de Tuddenham of the prior of Ely.[33] The first recorded lord was Sir Robert (1281), and it continued in the Tuddenham family until the execution of Sir Thomas at the accession of Edward IV, when his estates were forfeited.[34] The name of the family suggests that they were settlers from the high lands of mid-Norfolk, or possibly from one of the Tuddenham villages in Suffolk. In 1620 Tuddenham Hall manor, with 200 acres of arable land, was held by Humphrey Gardiner, to whom it had been bequeathed in tail male by his father Thomas (d. 1566).[35] His grandson and heir, another Humphrey, was then aged 12, and had livery of this manor and Hiptofts in 1636.[36] In 1677 it was conveyed by him, his wife Helen, and son Humphrey to John Willys,[37] and after being for a short time in the Penhall family[38] was in 1703 conveyed by John Penhall and Cecily his wife to Richard Reynolds,[39] whose family were in possession up to 1800.[40]

The reputed manor of *BEVIS HALL*, now represented by the farm of that name on the North Brink at the south-east corner of the parish, was settled in 1624 by William Reve of London, haberdasher, on his daughter Margaret, wife of George Bromley.[41] In the 18th century it passed to the Drury and the Southwell families, the latter being also the lessees of Wisbech Castle, and in 1746 it was bequeathed by Edward Southwell to his wife Jane, and by her to Sir Clement Trafford of Dunton Hall, Tydd St. Mary. In 1773, when the tenant was William Storey, it was mortgaged with Sir Clement's other property for £12,600, and sold after his death to Francis Saunders of Parson Drove. After being in the Culy family it was purchased in 1851 by Joseph Peck, whose son John sold it to W. G. Jackson. Mr. N. G. Jackson, grandson of the latter, sold it about 1910 to Mr. Frank Britain. The property was re-sold about 1933.[42]

CHURCHES

Bishop Northwold, when ordaining the vicarage of Wisbech in 1252, specified that the priest appointed should conduct services in the chapel of Kilhus in Wisbech St. Mary.[43] He assigned to the vicar's support the tithe of mills, land bought from the Prior of Spinney for the endowment of the chapel, and the tithe of faldage, fishery, lambs, cheese, butter, geese, calves, and pigs.[44] In 1275, when Wisbech rectory was appropriated to Ely priory, a house for the priest at Kilhus is mentioned as distinct from the rectory house.[45] Nevertheless the chapel is not mentioned in the *Taxatio*. The 14th-century reconstruction of the chapel produced a building of considerable size, and in 1535 it was officially mentioned as *capella Beate Marie*, though still not separately valued.[46] The Parliamentary Survey of 1650 showed the tithes as separate from Wisbech St. Peter and worth as much as £60,[47] but St. Mary's did not finally achieve its independence until 1854.[48] The living is now a vicarage in the patronage of the Bishop of Ely. About this time the Revd. Henry Fardell, the last vicar of the undivided parish of Wisbech, deplored the lack of church accommodation in the outlying areas of the parish. 'I employed a waggon to convey my poor parishioners of St. Mary's to church, but it soon became abused and I was compelled to abandon the scheme.'[49]

The chapel of Guyhirn originated in a chantry founded in 1337 by John de Reddik, who endowed it with a messuage and land worth £8 13*s.* 4*d.*, which at that time was the dower of his mother Florence.[50] This chapel was licensed for public worship by Bishop Fordham in 1398.[51] At the Reformation its endowments, amounting to 2 messuages (one of them the original Reddik 'mansion') and 118 acres of land, were worth £5 5*s.*, with stock worth £6 19*s.* 8*d.* It also possessed plate and vestments, valued at £2 5*s.* 8*d.*, including a

22 *Ely Dioc. Remembrancer*, no. 143, p. 78; no. 188, p. 20.
23 *Cal. Inq. p.m. Hen. VII*, i, 210 (p. 95).
24 C.P. 25(2)/4/19/65.
25 C 142/383/84.
26 Peckover MSS. (Terrier of Bp. Alcock) (Wisb. Mus.).
27 C 142/69/95. He was in possession of 24 acres called 'Jacketts' in 1536 (B.M. Add. MS. 5837, f. 200).
28 C 142/214/240; P.C.C. 65 Windsor.
29 C 142/524/90.
30 C.P. 25(2)/401/Mich. 23 Chas. I.
31 C.P. 25(2)/633/East. 20 Chas. II.
32 Infm. Mr. G. M. G. Woodgate.
33 C 136/78/3.
34 Woodgate MSS.
35 C 142/383/84.
36 C 60/531, no. 13.
37 C.P. 25(2)/633/Mich. 29 Chas. II.
38 C.P. 25(2)/634/Mich. 36 Chas. II.
39 C.P. 25(2)/910/Hil. and Trin. 2 Anne.
40 C.P. 25(2)/1000/Mich. 6 Geo. I; Woodgate MSS.
41 C 142/448/114.
42 Woodgate MSS.
43 The chapel stood in or near Kiln House Lane near the present church.
44 Document copied into Reg. Fordham (*Ely Dioc. Remembrancer*, no. 192, p. 97).
45 *Ely Dioc. Remembrancer*, no. 192, p. 98.
46 *Valor Eccl.* (Rec. Com.), iii, 500.
47 Woodgate MSS. Wisb. i, 47.
48 Order in Council, 11 Aug. 1854 (on death of the Revd. Henry Fardell, Vicar of Wisbech).
49 H.O. 129/7/193.
50 *Ely Dioc. Remembrancer*, no. 322, p. 55.
51 Gibbons, *Ely Episc. Rec.* 400.

9 oz. parcel-gilt chalice (36*s.*).[52] The lands, with others in the neighbourhood producing in all £111 6*s.* 3*d.*, were sold in 1549 to William Warde.[53] They descended through female lines to Margaret Morrice, who in 1640 sold them to John Jaggard of Haverhill (Suffolk). In 1668 they passed to his brother Daniel, a linen draper of Colchester, who re-sold them two years later in small lots.[54] By this time a small unconsecrated chapel (the still existing Old Church) had been built and endowed with £30 a year,[55] which by the middle of the 19th century had increased to about £100.[56] This chapel was served by a curate appointed by the Vicar of Wisbech until 1878, when Guyhirn was made a separate ecclesiastical district and the patronage vested in the Bishop of Ely.[57]

There was in 1517 a guild of St. Mary Magdalen (the patron saint of the chapel) at Guyhirn. In this year Thomas Kersey of March bequeathed 3*s.* 4*d.* to the guild.[58]

The chapel of Corpus Christi at Murrow was founded in or before 1376, when John Fowler of Wisbech, with his household and neighbours, were licensed to attend divine service in it for one year, except on festivals and high days. The licence was justified on the ground that Murrow was far from any other chapel and that the roads from it were dangerous.[59] The plate and vestments at Murrow were valued at £4 5*s.* 8*d.* at the Reformation, and included a 16 oz. gilt chalice valued at £3 4*s.*[60] In 1553 26 acres of land in Wisbech which had belonged to the chapel were granted to John Butler and Thomas Chaworthe.[61] The memory of this chapel is still preserved in the name Chapel Hill. A new chapel was built at Murrow in 1857, which became the church of an ecclesiastical district thirteen years later. The parish includes Southea in Parson Drove. The living is a vicarage in the gift of the bishop.[62]

Dr. Abraham Jobson, Vicar of Wisbech 1802–28, bequeathed £100 to support a Sunday school at Guyhirn and £200 to establish another at Murrow.[63]

The church of *ST. MARY* consists of chancel, clerestoried nave, aisles, south porch, and west tower with a pyramidal cap covered with lead. The exterior, with the exception of the chancel and tower, is plastered. The fabric is of 14th-century origin. The chancel was rebuilt in the 15th century and the clerestory was added about the same period; shortly afterwards the parapet of the tower was added or rebuilt and several windows inserted. The south aisle was repaired with brick in 1758, as recorded on a tablet in the south wall. In 1840 the porch was considerably repaired and in the second half of the 19th century the whole building was drastically restored and re-roofed.

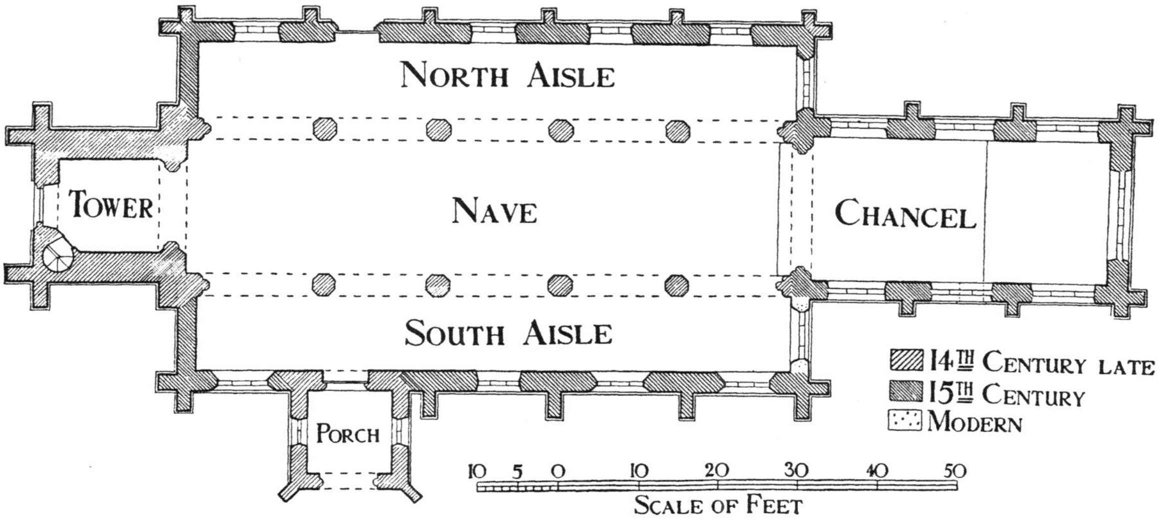

Plan of St. Mary's Church

The east window of the chancel is of five lights under a four-centred arch. There are angle buttresses with two set-offs. In the lateral walls are three windows, all of three lights under a four-centred arch; the middle window on the north side has been curtailed at some period. On the south is a priest's doorway of the 15th century with continuous mouldings. There is a plain piscina with a two-centred arch having a continuous chamfer. The chancel arch, originally of the 15th century, has been rebuilt but the responds are old, with moulded caps and bases. Above the arch is a 15-century window of three lights originally external but now opening into the chancel owing to the raising of the roof in the last century. On the east gable of the nave is a 15th-century sanctus bell-cote.

The nave arcades consist of five wide arches of rather clumsy appearance; they are of two orders with moulded caps and bases. In the east respond of the north arcade there is a shallow niche, and the caps of the first bay have been mutilated for the fixing of a screen, now destroyed. There is an image bracket on a column of the south arcade, and a curious carved relief in one of the spandrels, portraying a man and woman

[52] S.C. 12/6/22.
[53] *Cal. Pat.* 1548–9, 326–7.
[54] Woodgate MSS.
[55] Bp.'s Mun., Ely, A 5.
[56] Gardner, *Dir. Cambs.* (1851), 567.
[57] Kelly, *Dir. Cambs.* (1900).
[58] Gibbons, *Ely Episc. Rec.* 213.
[59] *Ely Dioc. Remembrancer*, no. 117, p. 35. Mr. G. M. G. Woodgate also suggests that risks would have been run if the Fen Dyke at Murrow had been left unguarded in times of flood.
[60] S.C. 12/6/22. This survey is incomplete, not mentioning the landed endowment of the chapel.
[61] *Cal. Pat.* 1553, 290.
[62] Kelly, *Dir. Cambs.* (1900).
[63] Gardner, *Dir. Cambs.* (1851), 568.

standing on either side of a large vessel. There are some stone roof corbels carved with grotesque heads. The clerestory consists of three-light windows with cinquefoiled heads. The east window of the south aisle is modern and bad, but the remainder of the fenestration of the aisles is late 15th century with square heads, except the north-east window of the north aisle; the windows are of three lights with the exception of the east window of the north aisle which is of four lights and has a hood-mould with grotesque heads as stops. The north aisle has a good 14th-century doorway with round arch and continuous mouldings and a blank shield above. The buttresses and string-course are also of the 14th century, the former with two set-offs. Over the north doorway on the inside is a demi-angel holding a blank shield. There are no windows in the west walls of the aisles. The porch, though it is stated to have been rebuilt in 1840, is obviously mainly original 14th-century work, even the stone coping on the gable being ancient. The arch is two-centred and of two orders with engaged shafts to the jambs. The side windows are of two lights with square cusped heads. The inner doorway is modern and mostly of plaster, and probably dates from the 1840 restoration. There is a mutilated stoup and a niche in the wall above it.

The tower, which is of four stages, dates from the second half of the 14th century and is well proportioned. It opens to the nave by a lofty two-centred arch with embattled caps and moulded bases to the responds. The west doorway has continuous mouldings and a hood terminating in carved heads. There are angle buttresses. In the second stage is a three-light 15th-century window with trefoiled heads to the main and tracery lights. The third stage has plain square-headed openings to the north, south, and west, while the belfry is lighted by windows of two lights with trefoiled heads and a trefoil above. The parapet is of brick and probably dates from the end of the 15th century. There is a newel stair to the belfry in the south-west angle. The roofs are entirely modern and poor with the exception of that of the south aisle, which contains some old timbers of uncertain date.

The font is of the 14th century with an octagonal bowl and shaft; the sides of the bowl have arcading except in one case, where there is a quatrefoil contained in a circle. Portions of the 17th-century altar rails are now at the east end of the south aisle in front of the side altar, which is composed of an early 16th-century sideboard. Some early 17th-century panelling serves as a reredos for this altar, and there is a mutilated image bracket in the north-east corner. In the chancel are some handsome gilt chairs and a settee, all of 18th-century date.

The plate includes a paten of silver, 1761, a modern chalice, paten and flagon of plated metal by Cox & Son of Southampton.

There are five bells in the tower by Osborn and Dobson of Downham, 1803.

The registers begin in 1570 and are complete.

The old church of *ST. MARY MAGDALENE, GUYHIRN*, consists of a simple parallelogram without any structural division and an open timber bell-cote on the west end of the roof. It is built of ashlar except the west wall and the upper part of the north wall, which are of brick. The roof is now covered with blue slates, but formerly with stone slates. It was erected in 1660 and this date appears over the south doorway. The present building replaced a medieval building on a different site. Since 1878, when a new church was erected about half a mile to the west, the old structure has served as a mortuary chapel. It was repaired in 1918, and is an interesting example of 17th-century Gothic.

There are east and west windows and one on the north and two on the south, which are all uniform in design; they consist of four uncusped lights under a square head and are glazed with old clear glass in rectangular leading. There are angle buttresses with one set-off; the north-west buttress has been rebuilt in brick. The doorway, which is placed at the west end of the south wall, has a four-centred arch and continuous chamfer; above is a shield with the date 1660 and the initials R.P. There is a base course. The open timber bell-cote has simple tracery of Gothic character in the sides and a low pyramidal top. Internally, there is a flat plaster ceiling. The plain wooden pulpit and the seating are probably late 17th-century work. There is a small modern stone font on a wooden base. The old brick floor remains. In the bell-cote is a bell dated 1637.

The modern church of St. Mary Magdalene consists of chancel, north vestry, south chapel, nave, south porch, and west bell-cote. The material is brick and the roofs are tiled. It was erected in 1878 from designs by Sir Gilbert Scott and is a particularly bad example of Victorian Gothic. It stands on, or near, the site of the medieval chapel and fragments of carved and moulded stones, now in the vicarage garden, were discovered when digging the foundations. All the fittings are modern.

The plate consists of a communion cup, standing paten, flagon, and dish, all plated, and a large alms-dish of Sheffield plate and parcel gilt with the Raising of Lazarus in high relief in the centre and repoussé work on the rim.

The bell-cote contains three bells, probably contemporary with the church, by Warner of London.

The registers begin in 1871 and are complete.

The church of *CORPUS CHRISTI, MURROW*, erected in 1857, is a brick structure consisting of apsidal chancel, nave, south porch, and west turret containing one bell. The registers begin in 1858 and are complete.

An Anglican mission room, formerly used also as a school, was erected at Thorney Toll in 1872.

NONCONFORMITY. Owing no doubt to the scattered nature of the parish and the inadequate provision of Church accommodation,[64] Nonconformity obtained a firm hold in Wisbech St. Mary.

Guyhirn was the birthplace of David Culy, the son of a French Protestant refugee. In 1695 he founded a sect, holding extreme Anabaptist doctrines, which was named after him. At the time of his death (*c.* 1725) there were 700 to 800 Culimites over the northern part of the Isle,[65] and in the middle of the 18th century there were 15 families of adherents, mainly drawn from Wisbech St. Mary and Guyhirn.[66] No Culimite

[64] A return to the House of Lords in 1818 reported a deficiency of 3,409 in the seating capacity of churches to the population of Wisbech borough (H.L. 93 (1818), xviii). The position in St. Mary's parish, though not officially set out, was as bad; St. Mary's and Guyhirn chapels (seating capacity approximately 360 and 100 respectively) served a population of 991 (1811 Census).

[65] *D.N.B.*

[66] Gardner, *Dir. Cambs.* (1851), 607.

chapel is known with certainty to have existed in Wisbech St. Mary itself, but the Baptist chapel in Guyhirn which was founded before 1800[67] may have taken over its meeting place. The Baptist chapel was closed between 1869 and 1875.[68]

In 1851 there were Wesleyan Methodist chapels at Guyhirn and in Wisbech St. Mary itself.[69] The latter is now demolished. The Guyhirn chapel, which dated from 1849, was rebuilt by the Primitive Methodists in 1868. The Primitive Methodist chapel at Murrow dates from 1835 and was rebuilt in 1875,[70] and other chapels of this denomination were subsequently built at Wisbech St. Mary village (1891, rebuilt 1926), Tholomas Drove (1845—now closed) and 'The Alley' or Hook's Drove (1841—Wesleyan Methodist Reformers, rebuilt *c.* 1900) in Wisbech High Fen.[71]

The Congregational chapel in Wisbech St. Mary, described in 1851 as a 'neat' building, dated from 1843.[72]

A Roman Catholic congregation existed at Thorney Toll from 1862; its present church dates from 1904.[73]

SCHOOLS Francis Hardy, by will dated 1726, bequeathed 4 acres of pasture in Richey Field and 8 acres of fresh marsh in High Fen to support a school in this parish. In 1774 8 acres, recently inclosed, were added by Hardy's last surviving trustee, John Sumpter.[74]

In 1814 these lands were bringing in £60 a year, and there was a good schoolroom and teacher's house at Tholomas Drove. The principal farmers appointed the master and succeeded in getting their sons taught to the exclusion of free place holders. Of these there was supposed to be one for every pound of rent, but in actual fact they seldom numbered more than a dozen. Dr. Jobson, who as vicar of the parish was especially interested, persuaded the farmers to allow one free scholar for every 30*s.* of rent, and considered the master should be capable of teaching the resulting number of 40 boys as well as the fee-paying farmers' sons. He also proposed to appropriate £20 from John Bend's charity and £10 from the vicarage of Wisbech to establish primary schools at St. Mary's, Murrow and Guyhirn, from which children could go on to the Tholomas Drove School (Hardy's) at 8 or 9 years of age. The teachers at these primary schools were to receive 2*d.* a week for each child from the appropriated charities, and 1*d.* a week from the parents. An extra 2*d.* a week was to be paid for girls learning sewing. There were at this date 194 children between 6 and 12 years of age in the parish, of whom only 74 were getting any kind of education.[75] It is not certain how far Jobson's intentions were carried out, but in 1837 Hardy's School was reported as taking in 21 'free' children.[76] In 1851 the number was 25.[77] Hardy's School was rebuilt on a new site in St. Mary's village in 1859,[78] but had lapsed into a stagnant condition a few years later. In 1866–7 the Schools Inquiry Commission found only 35 pupils and an uncertificated master. None of the children learnt anything beyond the elements or took up apprenticed employment on leaving.[79] Just after this inquiry the school was enlarged, to provide 140 places, the National Society making a grant of £15, and another grant of the same amount towards a further enlargement in 1912.[80] After this second enlargement the school was recognized for 147, but the numbers on the books have greatly increased, reaching 177 in 1925 and 201 in 1931. Another classroom, to hold 44, was therefore added in the former year, and a practical instruction room in 1933, bringing the accommodation up to 191. A kitchen, dining-hall, and another classroom, in the form of huts, were added in 1949. The average attendance in 1938 was 143.[81]

The Murrow Church of England School was built in 1868 for 75 children.[82] By 1913 this school was so full that children were being refused admission. It was not found possible to enlarge it from voluntary sources, so that the County Council erected another school in a different part of the village. This school, which accommodated 100 children, was opened in 1915, and for some years there were two primary schools in the village. In 1933, however, the old (voluntary) school was closed and its 29 remaining pupils divided between the Council school, which took the juniors, and the new Payne School at Parson Drove, for the seniors from both Murrow schools.[83]

Another Church school was erected in 1872 at Thorney Toll in the extreme west of the parish. It was a combined Mission room and school, and originally accommodated 70 (later 105) children, but was more than half empty shortly before the First World War. In 1937 the school was 'decapitated' by the transfer of the senior children to the Hereward School, March, and reorganized as one department for 78 juniors and infants. The foundations of the building were inadequate and in 1945 it was decided that it would be impossible to repair them properly. There were then about 45 on the books. The school was closed in 1948.[84]

A third Church school is the still existing one at Guyhirn. This was built in 1875 for 120 children, the National Society making a grant of £80 towards the cost of £718.[85] Since the First World War the numbers on the books have greatly increased, from 91 in 1927 to 149 in 1934, and the school was therefore remodelled and extended in 1937 at a cost of £1,450, two new classrooms, a staff room, dining-hall and kitchen, and central heating being provided. Since 1935 this school has been for juniors and infants only, in one department; there were 70 on the books in 1939. The senior children attend the Hereward School at March.[86]

In 1846–7 the National Society reported the existence of two schools in Wisbech St. Mary parish, staffed by a master and a mistress, open daily, and teaching 25 boys and 40 girls.[87] The boys' school was presumably Hardy's (q.v.); the girls' school was probably a dame school and short-lived. No such school is mentioned in the 1851 edition of Gardner's *Directory*.

CHARITIES Besides the purely educational charities of Hardy and Jobson, the following existed in Wisbech St. Mary in 1837.[88]

John Bend, by his will dated 1593, left a cottage and 3 acres in Chapel Field, and 58 acres in Richey

67 H.O. 129/7/193.
68 Kelly's Directories.
69 H.O. 129/7/193.
70 Kelly's Directories.
71 Kelly, *Dir. Cambs.* (1933).
72 Gardner, *Dir. Cambs.* (1851), 567; H.O. 129/7/193. This building is now demolished.
73 *Catholic Year Bk.* 1950.
74 *Rep. Com. Char.* (*Cambs.*) [103], H.C. (1837–8), xxiv, s.v. Wisb. St. Mary.
75 Archives of Nat. Soc.
76 *Rep. Com. Char.* (*Cambs.*), as above.
77 Gardner, *Dir. Cambs.* (1851), 568.
78 Archives of Nat. Soc.
79 *Schs. Inq. Rep.* [3966–XII], H.C. (1867–8), xxviii (10).
80 Archives of Nat. Soc.
81 M. of Ed. File no. 12/69.
82 Kelly, *Dir. Cambs.* (1900).
83 M. of Ed. File no. 12/68A.
84 Ibid. 12/70.
85 Archives of Nat. Soc.
86 M. of Ed. File no. 12/67.
87 Rep. of Nat. Soc. (privately printed).
88 *Rep. Com. Char.* (*Cambs.*) [103], 275 et seq., H.C. (1837–8), xxiv.

Field, to provide stock to set the poor to work. The proceeds from sales were to go to poor persons who had lived in the parish six years without receiving relief. A further 8 acres were awarded under the Bedford Level Act of 1677, and the total yearly rents derived from these sources in 1837[89] were between £140 and £150. By that date those wishing to benefit under this charity made personal application to the trustees, and if approved received 2*s.* 6*d.* each. After six months they were entitled to further sums, if they had been of good behaviour. In 1837 about ninety-eight persons received an average of 25*s.* each through this charity. In 1851 there were 130 beneficiaries.[90]

Margaret Bende, by will dated 1605, bequeathed £50 to the Corporation of Wisbech, to provide 6*s.* 8*d.* yearly for the poor of Parson Drove, the remainder of the interest to those of Wisbech St. Mary. In 1837 £2 3*s.* 4*d.* was distributed in small sums to the poor of this parish.

The parish possessed fifteen 'town houses', all of brick and two of them thatched, which were let to poor families at 1*s.* per annum. The 26 acres of poor's allotment under the Wisbech St. Mary Inclosure Act (1833) were let out in allotments of 1 or 2 roods at 2*s.* to 2*s.* 6*d.* per annum. In 1851 there were sixty-five allotment holders.[91]

WISBECH

Wisbech is the largest and only corporate town in the Isle, and the nodal point for the whole marshland area comprised by south-east Holland and Norfolk west of the Ouse. It is situated on the Nene about 12 miles from its present outfall, 94 miles north of London and 40 from Cambridge by road. It is one of a series of settlements spaced at two- to three-mile intervals along the landward side of the 'Roman Bank'[1] which runs around the edge of the Wash from Lynn to Boston and beyond. Like all these settlements the parish was originally long and narrow, stretching some 9 miles from the old course of the Well Stream, which divided Wisbech from Walsoken in Norfolk, to the far end of Wisbech High Fen beyond Guyhirn, where the boundary marched with Whittlesey for a short distance.

From at least 1610 the area was divided into three 'watches'—Town and South Brink (Wisbech St. Peter) and Murrow or North Brink (Wisbech St. Mary). The fact that the three obtained £1, 15*s.*, and 10*s.* respectively in a division of certain charitable funds in that year suggests their relative importance.[2] The borough has always been co-extensive with the parish of Wisbech St. Peter, and was by 1585 divided into ten wards (Timber Market, Church, Castle, South, Barton, White Cross, Old Market, Ship Lane, New Market, Mill), of which Barton, White Cross, and Old Market were on the left bank of the Nene.[3] These divisions have been obsolete since the reform of the corporation in the 1830's, but the names Castle, Timber Market, and South Brink have been preserved in three of the County Electoral Divisions which are comprised in the Borough.

Proposals were made in 1837 to reduce the borough to some 1,200 acres by detaching the south-western 'tail' of the parish and incorporating the portion of Walsoken already built over. These proposals, however, though supported by some of the Corporation, were not carried out, probably because the trial and disposal of offenders from the New Walsoken area, which was in Norfolk, presented complicated problems.[4] The old wards and the 'watches' were then replaced by the North and South Wards which lasted until the boundary alterations of the 1930's.[5]

By the Ministry of Health Provisional Orders Confirmation (Ely, Holland and Norfolk) Act, 1933[6] considerable alterations were made. Portions of Leverington in the Isle and of Emneth and Walsoken in Norfolk, already much urbanized, were transferred to Wisbech Borough, and the more rural parts of the latter were transferred to Elm and to March Urban District. The effect of these Orders was to reduce the acreage of the borough from 6,477 to 4,617. The Borough is now divided into East, North, South, and West Wards.

Rights of intercommoning of stock existed (i) on Norwood Common, with the tenants of March in the Bishop of Ely's manor of Doddington, and (ii) in Sutton Fen with the tenants of the royal manor of Long Sutton (Lincs.). Disputes, especially in the former case, sometimes led to bloodshed and slaughtering of livestock. Three letters dated 1561 and 1562 from Bishop Cox to his March and Doddington tenants, regretting their 'stowteness' and ordering them to restore the right of common to Wisbech, and a similar letter from Queen Elizabeth under the seal of the Duchy of Lancaster, in favour of Long Sutton and with a penalty of £500 for non-compliance, are copied into the concluding sheets of the Holy Trinity Gild Book.[7] The dispute with Long Sutton had been in progress as far back as 1487, when a commission was appointed to inquire as to the bounds and pasturage rights.[8] That with March and Doddington was not settled by Bishop Cox's letters; four tenants of Doddington manor refused to accept it and were sued in Chancery by various tenants on the Wisbech side of the fen. The Court finally (1569) upheld Bishop Cox's decision in favour of Wisbech.[9]

The town stands at the crossing of two Class A roads: from Peterborough to King's Lynn (A 47) and from Ely to Long Sutton (A 1101). The former is part of the important cross-country route from Leicester to Norwich and carries much traffic, particularly in summer to and from the east coast resorts. There are secondary roads to Parson Drove and Crowland (B 1169), March via Elm and Friday Bridge (B 1101), and Peterborough via Wisbech St. Mary (B. 1441). Another route to March, which leaves the main (South

89 The cottage had been pulled down 'many years ago'.

90 Gardner, *Dir. Cambs.* (1851), 568.

91 Ibid. 569.

1 Definitely not Roman; probably early medieval in date (*V.C.H. Hunts.* iii, 250 n.).

2 Corp. Rec. iii, 84. This division was observed in a Particular of all Buildings and Lands in Wisbech (1792) (MS. in Wisb. Mus.). It became obsolete in 1837.

3 Corp. Rec. ii, 177. Presumably each ward was represented by one of the 'ten men' or capital burgesses.

4 H.O. 52/33 (letter from Wisb. Corp. to Home Sec., 17 June 1837).

5 *Rep. Com. Municipal Corp. Boundaries*, H.C. 238, s.v. Wisbech (1837), xxviii.

6 23 and 24 Geo. V, c. lxv.

7 Corp. Rec. i, 114, 115, 118, 119.

8 *Cal. Pat.* 1485–94, 215.

9 C 78/36/39.

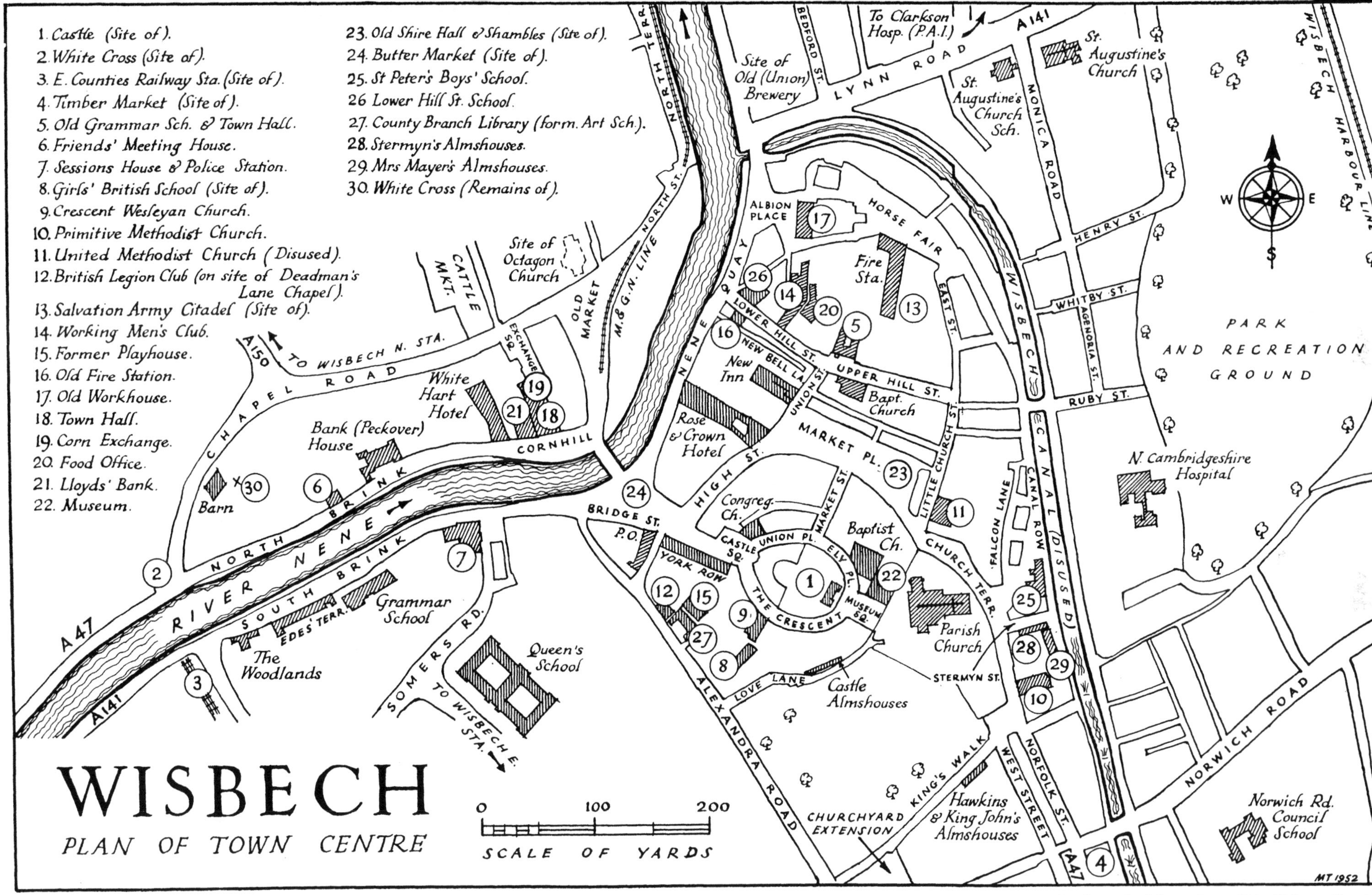

1. Castle (Site of).
2. White Cross (Site of).
3. E. Counties Railway Sta. (Site of).
4. Timber Market (Site of).
5. Old Grammar Sch. & Town Hall.
6. Friends' Meeting House.
7. Sessions House & Police Station.
8. Girls' British School (Site of).
9. Crescent Wesleyan Church.
10. Primitive Methodist Church.
11. United Methodist Church (Disused).
12. British Legion Club (on site of Deadman's Lane Chapel).
13. Salvation Army Citadel (Site of).
14. Working Men's Club.
15. Former Playhouse.
16. Old Fire Station.
17. Old Workhouse.
18. Town Hall.
19. Corn Exchange.
20. Food Office.
21. Lloyds' Bank.
22. Museum.
23. Old Shire Hall & Shambles (Site of).
24. Butter Market (Site of).
25. St Peter's Boys' School.
26 Lower Hill St. School.
27. County Branch Library (form. Art Sch.).
28. Stermyn's Almshouses.
29. Mrs Mayer's Almshouses.
30. White Cross (Remains of).
WISBECH
PLAN OF TOWN CENTRE
0
100
200
SCALE OF YARDS
MT 1952
W
E
S
WISBECH HARBOUR LINE
PARK AND RECREATION GROUND
N. Cambridgeshire Hospital
Norwich Rd. Council School
NORWICH ROAD
St. Augustine's Church
St. Augustine's Church Sch.
MONICA ROAD
HENRY ST.
WHITBY ST.
AGENORIA ST.
RUBY ST.
A141
To Clarkson Hosp. (P.A.I.)
LYNN ROAD
BEDFORD ST.
Site of Old (Union) Brewery
WISBECH CANAL (DISUSED)
CANAL ROW
FALCON LANE
CHURCH TERR.
Parish Church
STERMYN ST.
NORFOLK ST.
WEST STREET
A47
Hawkins & King John's Almshouses
KING'S WALK
CHURCHYARD EXTENSION
Castle Almshouses
LOVE LANE
ALEXANDRA ROAD
HORSE FAIR
EAST ST.
LITTLE CHURCH ST.
Fire Sta.
UPPER HILL ST.
Bapt. Church
Baptist Ch.
MARKET PL.
MARKET ST.
ELY PL.
MUSEUM SQ.
UNION PL.
THE CRESCENT
CASTLE SQ.
Congreg. Ch.
ALBION PLACE
LOWER HILL ST.
NEW BELL LA.
UNION ST.
New Inn
HIGH ST.
Rose & Crown Hotel
YORK ROW
BRIDGE ST.
P.O.
NENE QUAY
NORTH TERR.
NORTH ST.
M.&G.N. LINE
OLD MARKET
Site of Octagon Church
CATTLE MKT.
EXCHANGE SQ.
CORNHILL
White Hart Hotel
TO WISBECH N. STA.
A150
CHAPEL ROAD
Bank (Peckover) House
Barn
NORTH BRINK
RIVER NENE
SOUTH BRINK
EDES TERR.
Grammar School
The Woodlands
SOMERS RD.
TO WISBECH E. STA.
Queen's School
A47
A141

Brink) road to Peterborough[10] at Guyhirn, was turnpiked in 1730.[11] The road through Wisbech St. Mary to Thorney and Peterborough was turnpiked in 1810 and the Sutton road (from Tydd Gote through Wisbech to Downham Market) as late as 1849,[12] when it was diverted from its old course through Leverington village to its present one bisecting Spital Field. The repair of one of the roads to March, known as Lionell and later as Southewold causeway, was the subject of an indulgence in 1494.[13]

In 1796 there was a daily mail coach service to London, leaving at 4 p.m. and returning at 10 a.m., for which the fares were £1 5*s.* inside and 12*s.* 6*d.* outside, and there was also a stage coach three times weekly at 18*s.* and 9*s.* Until 1831, when the bank at Sutton Wash, by-passing Wisbech, was built, another coach ran on Tuesdays, Thursdays, and Saturdays from Lynn through Wisbech and Spalding to Boston.[14]

The first railway line to reach Wisbech, that from St. Ives and March, was opened in May 1847.[15] The original station was a terminus at the present sidings on the South Brink near the Grammar School, but in the following year the line was prolonged to join the East Anglian Railway at Watlington (now Magdalen Road) and a new station was opened on the site of the present East Station.[16] The two stations were both in use up to at least 1851, when they were 'temporary erections . . . conveniently situated at the south-east end of the town'.[17] They were used by the Eastern Counties and East Anglian Railways respectively. About the same time (1845–50) rival schemes for lines from Peterborough via Wisbech to Sutton Bridge were promoted by the Great Northern and Eastern Counties Railways. The former company proposed a new cut to the river and improved docks, but partly owing to the opposition of the Corporation to another bridge over the Nene below Wisbech which might hinder shipping, the Eastern Counties scheme was approved. The line, however, was not actually built until August 1866.[18] Each railway laid down branches to the quays for goods traffic, but the former Midland and Great Northern Line, terminating in the Old Market, is not now used,[19] and the Wisbech Harbour branch of the former Great Eastern Railway is used very little.[20] In 1884 the latter Company opened a 'steam tramway' of 4ft. 8½in. gauge from Wisbech to Upwell. Its receipts, about £2,500 a year in the 1890's, were even then not sufficient to maintain it, but the line is still much used for goods traffic. Passengers were carried until 1928, when a motor bus service was substituted.[21]

Wisbech owed its existence to its situation at the point where the Well Stream joined the 'Wysbeck', the stream from which the town derived its name. The Well Stream, now represented by the Wisbech Canal,[22] formerly carried the main outfall of the Great Ouse. The diversion of the Ouse, however, to its present course in *c.* 1300 and the cutting of Moreton's Leam in 1478 have reduced its importance. The 'Wysbeck' on the other hand has now swollen into the main River Nene.

The central part of the town, between the two water-courses, is a good example of a manorial borough, with the site of the castle midway in the peninsula and dominating the Market Place.

The Old Market, so called as early as 1221,[23] suggests that some kind of trading centre existed in Wisbech in very early times. It is on the left (west) bank of the river, and is not very far from the large moat at the north end of Barton Field, which is the probable site of the Barton manor house.[24] This house must have been the administrative centre of the Ely estates in Wisbech prior to the erection of the castle and the division between bishop and convent in 1109.

From the 12th century onwards the New Market by the castle, which was the headquarters of the richer of the two Wisbech manors, has outstripped the Old Market as a centre of trade. By 1492 there were 115 tenants of the bishop in the New Market as compared with 32 in the Old.[25] It should be observed, however, that although at the present time most of the best shops are in the Market Place and its continuations High Street and Bridge Street, the Old Market still has a few open-air stalls on Saturdays, and is 'the important part of the town to which farmers from the locality gravitate'.[26] It contains many of the banks, solicitors' and auctioneers' offices, and seed and implement shops; and the cattle market and the former corn exchange are in the immediate vicinity.

A third market, for timber, was at the south-east end of the town. The site, close to the Well Stream, was convenient for unloading heavy goods, and the narrow blocks of buildings between the canal, Norfolk Street,[27] and West Street represent the sites of the stalls and booths of this market.

The 'Roman Bank' and the two streams meeting at Wisbech have determined the line of the main roads leading from the town, those to Long Sutton and Lynn following the former[28] and those to Guyhirn and to Elm, Upwell, and Ely the latter. In 1830 the town consisted mainly of a 'hard core' between the Market Place, the river and the canal, ribbon development along the Nene above the bridge in the shape of the North and South Brinks,[29] and the houses round the Old Market and along North Street. Lower down the river, Bedford, Russell and Chase Streets had been laid out and partly built up, and the suburb of New

10 'Cromwell Road' was constructed as a relief to the narrow road along the South Brink in 1933 (Infm. the late E. J. Rudsdale, Curator Wisb. Mus.).

11 F. J. Gardiner, *Hist. Wisb.* (1898), 92.

12 MS. Books at Wisb. Mus.

13 *Ely Dioc. Remembrancer*, 1909, 47. Previous indulgences had been issued in respect of this road in 1457 and 1488 (ibid. no. 231, p. 169; no. 279, p. 120).

14 *Universal Brit. Dir.* (1796), iv, 800–5.

15 *V.C.H. Cambs.* ii, 133. Some of the trains to and from Cambridge and London still take this route and not the more direct one via Ely.

16 Gardiner, *Hist. Wisb.* 78–79.

17 Map dated 1850 (Wisb. Mus.); R. Gardner, *Dir. Cambs.* (1851), 624.

18 *V.C.H. Cambs.* ii, 133; Gardiner, *Hist. Wisb.* 80.

19 Most of the track has been removed.

20 The Wisbech Harbour line was opened in November 1862 (Gardiner, *Hist. Wisb.* 129).

21 Ibid. 85–86.

22 The actual course of the Well Stream was shown by the former boundary between Wisbech and Walsoken, which included a good deal of ground on the east (Walsoken) side of the Canal.

23 B.M. Cott. MS. Tib. B. II, f. 146*a*.

24 The use of 'Barton' as a suffix meaning manor house or grange is not usually found so far north-east. 'Burystead' is commoner.

25 Bp. Alcock's Terrier (Peckover MSS., Wisb. Mus.).

26 R. W. L. Melbourne, 'The Function of Wisbech in relation to its Geographical Setting' (unpublished M.A. thesis, Univ. Lond., 1939), p. 115.

27 Formerly known as Timber Market, and so shown in a town plan of 1830 in Wisb. Mus.

28 Part of the bank on the Lynn Road side was used in 1586 to level up the (New) Market Place (Corp. Rec. i).

29 So named in a Ct. R. of 28 Eliz. (1586). Houses were built along the Brinks early enough to be shown on the 1657 map.

Walsoken had begun to develop along Norwich (formerly Marshland) Road.[30] By 1837 this suburb had some 280 houses, 'mostly dwellings of an inferior description', occupied by about 1,200 'persons of the labouring classes' employed in Wisbech.[31]

Another development east of the canal occurred in the 1840's, when Agenoria, Henry, Ruby, and Whitby Streets[32] were built. The second and fourth of these are named after Henry Whitby Pooley, a former owner of the land.[33] By 1867, when an official map was prepared by the Borough Surveyor,[34] other important developments had occurred. Such were (*a*) the streets near the East Station, around Albert Street, Victoria Road,[35] and Milner Road; (*b*) the first of the houses in and near Leverington Road and Harecroft Road (formerly Pickard's Lane), where the railway—opened in 1866—was again a factor; (*c*) the development along Lynn Road and in the streets leading off it. This continued rather later, and was encouraged by the creation of the park (1869); it has resulted in the coalescence of Wisbech and Old and New Walsoken into a single built-up area. Development was less rapid after about 1875, for the town was experiencing a temporary decline in population. In the 1880's, a writer referred to the difficulty she experienced in finding a small house in Wisbech 'because for years no new ones had been built'.[36] Towards the end of the 19th century King's and Queen's Roads and the adjacent area were developed for housing purposes out of an estate that had belonged to the Colvile family. Between 1919 and 1939 the main trends were the filling up of vacant spaces in this district, e.g. Kingsley Avenue and Hillburn Road, and intensified development along and off Elm Road, making a continuous stretch of houses from Wisbech through the northern end of Elm village to Emneth. Since 1945 a new suburb at Mount Pleasant Bank, north of the town, has arisen. The land surrounding the town is of great fertility, which has prevented its reckless use for housing purposes; the numerous Corporation housing estates e.g. in Weasenham Lane are small, and private enterprise has concentrated more on filling empty spaces in roads already laid out than on opening up new roads.

Although Wisbech became a corporate town in the middle of the 16th century, its population at the time seems to have been small, even compared with neighbouring places. The earliest return with any degree of accuracy is one made by the bishops of the various dioceses in 1563,[37] which credits Wisbech St. Peter with 242 householders and Wisbech St. Mary with 71. The town seems to have been about half the size of Cambridge (517 householders), and ranked below Ely with its hamlets (400) and Whittlesey (355 householders in the two parishes). Another ecclesiastical return, made by order of Bishop Compton in 1676,[38] showed 1,424 persons of communicant age in St. Peter's parish and 284 in St. Mary's; the comparable figures for Ely and Whittlesey were 2,371 and 2,021 respectively. It may perhaps be inferred that the total population of the borough (St. Peter's parish) was about 1,000 and 2,000 at these dates. The tendency of the town to double in size in about a century continued during the 18th century, and the census of 1801 showed 4,710 persons. The town was again about half the size of Cambridge (9,276), and had outstripped Ely and Whittlesey, both of which were just below the 4,000 mark. The first half of the 19th century was a very prosperous time for Wisbech, and in 1851 the population of the borough was 10,594. The increase of about 120 per cent. was similar to that of the Isle as a whole, but below that of Cambridge, where the population had just trebled to 27,803. A sudden decrease occurred in Wisbech between 1851 and 1861, of over 10 per cent. to 9,276. This was about twice as much as in Cambridgeshire generally, and was due no doubt to the damage done by railways to the trade of the port. For the next generation the population of Wisbech remained virtually stationary (9,395 in 1891). Since the 1890's the emergence of the Wisbech neighbourhood as the chief market-gardening area in England, and the development of ancillary industries (e.g. fruit canning) in the town, have led to a steady increase of about 30 per cent. The 1931 Census showed a population of 12,006 on the then existing area and of 15,129 on the present area of the borough. It seems that this increase, on a slightly lower scale, is continuing; the Registrar-General's estimates for 1938 and 1947 have been 15,620 and 15,910, and the preliminary figure of the 1951 Census is 17,430.[39] The figure at the National Registration of 1939 showed 17,599 persons in Wisbech.[40]

Various travellers have recorded their impressions of Wisbech. Pepys, who was visiting relations in the neighbourhood in 1663, thought it 'a pretty town', and commented upon its 'fine church and library, where sundry very old abbey manuscripts', and its 'fine house built on the church ground by Secretary Thurlow'.[41] William Cole, the Cambridge antiquary, who passed through in 1772, mentions that 'the buildings were in general handsome, the inn we stopped at [the Rose and Crown] uncommonly so . . .'. 'But the Bridge,' he added 'stretching Rialto-like over this straight and considerable stream, with a good row of houses extending from it, and fronting the water, to a considerable distance, beats all, and exhibits something of a Venetian appearance'.[42] John Byng in 1790, however, found Wisbech 'an ugly dull town, with a large canal through it in the Dutch taste' and his inn was bad and expensive; 'my horses could not eat the hay, and I could not drink the wine'.[43] Cobbett, who 'speechified' to about 220 people in the Playhouse in April 1830, called it 'a good solid town, though not handsome' and re-

[30] The earliest dated house still standing in this road is dated 1793. N. Walker and T. Craddock in their *Hist. of Wisbech* (1849), p. 528, attribute the development here to a 'pleasant country aspect immediately adjoining the town' and also a 'certain leniency of rates'.

[31] *Rep. Com. Municipal Corp. Boundaries*, s.v. Wisbech.

[32] They are shown for the first time on the 1850 town plan (Wisb. Mus.).

[33] Wisb. Mus., Auction Bill of 1850 (Infm. the late E. J. Rudsdale).

[34] Copy in Wisb. Mus.

[35] Houses in these streets are dated 1840 and 1855 respectively.

[36] Mrs. Charles Roberts, *Recollections of a Long Life*, 132.

[37] B.M. Harl. MS. 594, ff. 198–200.

[38] A transcript of this return (as regards Ely diocese) was made by Bishop Fleetwood, and is preserved in the muns. at Ely (Ref. A 6).

[39] The 1947 and 1951 figures should have about 1 per cent. added to make them comparable with that of 1938, to allow for the non-inclusion of men called up for military service.

[40] This figure includes evacuees, but not persons already embodied in H.M. Forces on 28 Sept. 1939.

[41] *Diary*, ed. H. B. Wheatley, iii, 264. He was visiting relatives in the neighbourhood; John Pepys, of the senior (Cottenham) branch of the family (d. 1604), held 40 acres of marsh in Wisbech Fen in a field called Willick, and a tenement called Tyballs with 18 acres in Wisbech (C 142/288/123).

[42] B.M. Add. MS. 5804, f. 366.

[43] *Torrington Diaries*, ed. J. Beresford, ii, 239–40.

marked the export of corn.[44] In 1837 the Commissioners on the Boundaries of Municipal Corporations noted that 'a spirit of improvement' was visible in every part of Wisbech, though they did not consider that the town was likely to increase much; the rural part of the borough contained only about ninety houses, many 'of such an inferior description that they are not rated at all' and the rest not much better. In the town, the best houses at this date were in the North Brink, the Crescent, Ely Place, and Yorke Row; the average annual value in the last three streets was £28 10*s.*; labourers' cottages in Horse Fair and the adjoining streets were worth about £6 per annum.[45]

The town has few examples of really ancient domestic architecture. The following buildings, however, contain features which may be dated to 1714 or earlier.[46] No. 8 Bridge Street; Nos. 6 to 8 Yorke Row (Messrs. Ream, Ollard, and Rawlinson's premises), which probably form the house occupied by Secretary Thurloe's sons, and which were used for the administration of the manor from the second half of the 17th century;[47] Nos. 28 and 29 High Street, and the New Inn, Union Street, both dating from about 1500 and some of the rare examples of timber-framed buildings remaining in Wisbech; the Old Grammar School, Hill Street; Nos. 49 to 51 Horse Fair, early 18th century and retaining the small windows at end which lighted the spiral brick staircase formerly found in this type of house; Ely House, Lynn Road, an early-17th-century country house; the Rose and Crown Hotel, 17th century with a fine early-18th-century staircase and Regency front, and a two-story outbuilding dated 1601; the vault below and outbuilding behind No. 29 Market Place, late medieval; Nos. 33 and 34, 38 and 39, and the George Inn (No. 35) Market Place; Nos. 3 to 5 Nene Quay; No. 97 Norfolk Street, dated 1701, with a 'Dutch' gable; the barn and remains (not *in situ*) of the White Cross in the grounds of Bank House; 'Woodlands', South Brink. Thurloe's rebuilding of the castle is represented by the entrance gateway and certain fittings in the present buildings and elsewhere, and part of the garden wall in Love Lane.

Wisbech is noteworthy for its fine street architecture of the Georgian and Regency periods.[48] The group of buildings along the North Brink, from the Town Hall to No. 25, is of exceptional merit, and the streets laid out between 1793 and 1816 on the site of the castle—The Crescent, Castle Square, Ely Place, Market Street, Museum Square, and Union Place—and the block of warehouses between the Old Market and the river, have architectural distinction not often found in towns the size of Wisbech. The general lay-out of the Castle Estate, with the 'circus' round the present castle expanding into squares at each end, with Market Street bringing the whole into relation with the town centre and street plan, is a good example of formal town planning. The buildings, which have not suffered much from later alterations, have no great architectural distinction in themselves, but are of just the right scale for their setting. An interesting reference to the development of this estate is found in Pigot's *Universal British Directory* (1792–6);[49] 'several houses are begun, and, if carried on upon the same scale as begun, will be a great ornament as well as addition and accommodation, and particularly as houses are now so scarce in this town.'

The row of buildings along the North Brink forms one of the finest pieces of Georgian street architecture in the country. Its effect is greatly enhanced by the river and its banks. The most noteworthy individual buildings in this group are the Town Hall and Bank (now Peckover) House. The former, built in 1810–11, under the Improvement Act of 1810, is distinguished by its stone front. The lower story is rusticated, with arched doorways and windows. The three centre bays project slightly and carry a pediment supported by four graceful Ionic pilasters; in the pediment is a round shield with the badge of the borough, St. Peter's Keys. Bank House, about 200 yards south-west, is a fine early 18th-century[50] town house, of three stories and five 'bays'. It is of light-brown brick with darker brick dressings, and the entrance doorway is of stone with a semi-circular pediment. At the top of the façade is a cornice and panelled parapet, supported at the corners by brick pilasters. The railings along the street front are original. Inside there are some fine rooms with mid-18th-century carving similar to that at Houghton Hall (Norf.). In 1943 the house, with 48 acres of land, was given to the National Trust by the Hon. Alexandrina Peckover, the last member of the great Wisbech banking family to live there. The intervening buildings are not individually so distinguished, but all are good architecture and contribute their share to the general effect. The White Hart Hotel and No. 12 North Brink are perhaps the most attractive individually. No. 14 (which with No. 19 was transferred to the National Trust in 1950) has a fine staircase. No. 6 (next the White Hart on the south-west) is interesting as being of two periods; the two lower stories are of the early 18th century, while the upper two were added between 1820 and 1840 in an exactly similar style. The modern building of Lloyds Bank fits into the terrace very well in scale and style. Beyond Bank House, the Friends' Meeting House, the gabled terrace (Nos. 22 to 25), and Sibald's Holme beyond the corner of Barton Road are interesting as having been designed by Algernon Peckover, another member of the banking family.[51]

The South Brink, though not so remarkable as the North Brink, has also several buildings worthy of attention. Nos. 7 and 8 date from the early 18th century, with a pediment added after 1756; they were originally one house. Inside there are some good fireplaces, a fine staircase, a panelled room and one with a good late 18th-century plaster ceiling, and a clock on the staircase, dated 1720 and with the weights and pendulum running in a channel cut in the thickness of the wall. No. 7 was the birthplace of Octavia Hill. The early 19th-century Sessions House and Police

[44] *Rural Rides*, ed. Pitt Cobbett, ii, 308, 313.

[45] *Rep. Com. Municipal Corp. Boundaries*, s.v. Wisbech.

[46] This date has been chosen to facilitate comparison with the inventories of buildings in towns already surveyed by the Historical Monuments Commission. So far these inventories have been confined to pre-1714 buildings.

[47] In 1732 the house belonged to James Marshall and Emma his wife, and was commonly called the 'Moot Hall'. Emma surrendered it nine years later to Henry Banyer (see Worthies), subject to the obligation of providing accommodation for the stewards of the manor and a room for the juries on court days (Infm. Mr. G. M. G. Woodgate).

[48] Detailed accounts of Wisbech architecture are contained in the *Archit. Surv. of Wisb.*, reprinted from the *Wisb. Advertiser*, 10 Aug. 1938, the *Archit. Rev.* lxxxvi, 517 (Dec. 1939), and *Country Life*, 23 May, 6 June 1947.

[49] iv, 803.

[50] It is said to have been built in 1722 (*Archit. Rev.* lxxxvi, 517).

[51] The Peckover family lived at Harecroft House, now the High School for Girls.

Copyright 'Country Life'

SOUTH BRINK, WISBECH, IN 1947

Copyright 'Country Life'

BANK (NOW PECKOVER) HOUSE, NORTH BRINK, WISBECH, IN 1947

BOROUGH of WISBECH CAMBRIDGESHIRE

Station, on the opposite corner of Somers Road, are also good examples of their period. Ede's Terrace, Nos. 17 to 23, also early 19th century, is of somewhat unusual design, as also is No. 5 North Terrace, early 18th century with some Palladian windows with stone mullions and jambs.

There are many other 18th- and early-19th-century houses in Wisbech of more than average merit. No. 17 Hill Street, now the Food Office, is mid-18th century, with some rooms with original panelling. Nos. 19, 27, 30, 50, 52, and 54 Hill Street are similar to No. 17 but mostly rather later in date; No. 27 has a fine doorway, and the design of Nos. 50 to 54, with a high central archway running through two stories, is unusual—the building may have been erected as an inn. 'Flint House' (No. 67 Lynn Road) is an interesting little Regency Gothic house. Nos. 27 and 28 Market Place are some of the best buildings in a square where almost every one has some interest. The North Brink Brewery, *c.* 1790, is a good specimen of industrial architecture, forming a punctuation mark to the Brink terraces at the bend of the river. Until recently a somewhat similar building, the Old Brewery, stood at the corner of Lynn Road and Nene Parade. The Old Market has several good houses, notably Nos. 1 and 2, 9, 28 and 29, all late 18th century, and 'Sulehay House', *c.* 1730. 'Glan Dyfi', next door, is a successful imitation (*c.* 1890) of the Georgian style, harmonizing well with the older house. The Museum, though much later in date (1846–7) than the other buildings round the castle, fits into the scheme very well. It is interesting as one of the first buildings in the country designed as a museum; the architect was George Buckler.[52]

The early 19th-century development east of the canal, in New Walsoken and around Victoria Road, is mostly of a poor type. A later block, Nos. 1 to 5 Monica Road, *c.* 1875, is interesting on account of the 'ridge and furrow' type of roof, seldom found outside London. Its resemblance to contemporary terraces in such London suburbs as Finsbury Park and Shepherds Bush suggests that a London builder may have been responsible.[53]

MANORS. The first authentic reference to Wisbech occurs *c.* 1000, when Oswy and Leoflede, on the admission of their son Aelfwin as a monk, gave the vill to the monastery of Ely.[54] In 1086 Wisbech was held by the abbot, and rated at 10 hides, of which 1 hide and 1 virgate with 2 ploughs were in demesne. It was by no means one of the wealthiest of the Ely manors, its value (£6 T.R.E., 100*s.* in 1086) being below that of Ely itself and of more than half of the manors in the two hundreds of Ely. The fisheries were important. One belonged to the manor. It rendered 1,500 eels, and two fishermen in the vill paid the abbot 13*s.* 4*d.* and 14,000 eels. Another fisherman produced 5,000 eels for the abbot of Bury St. Edmunds, 8 fishermen rendered 5,260 eels to the Abbot of Ramsey, 3 rendered 4,000 to the Abbot of Crowland and 6 rendered 3,500 and 5*s.* to William de Warenne. The Abbot of Ely had soke of all the men of the vill. They comprised, besides the fishermen already mentioned, 15 villeins with 10 acres each, 13 sokemen, who had always been attached to the land, sharing 2½ hides, 17 cottars and 2 serfs.[55] Some of these men may have been members of the same household, and some households may not have come within the scope of the Domesday survey, but it seems that in 1086 there may have been some 65 to 70 families, or about 300 to 350 persons, in Wisbech manor. It must be remembered, however, that Wisbech, which is the only one of the Marshland vills of the Isle to be mentioned in Domesday, probably comprised the whole area from Tydd Gote down to the far end of Upwell at Welney.

At the formation of the see of Ely in 1109 the vill was divided. The manor later known as that of *WISBECH BARTON* went towards the endowment of the bishopric, while that later known as *WISBECH MURROW*[56] was retained by the convent.[57]

The manor of Wisbech Barton remained uninterruptedly in the hands of the bishops, except during vacancies in the see,[58] until 1652. It was then granted to Nathaniel Andrews and John Bancks,[59] and later purchased by Secretary Thurloe.[60] At the Restoration it was returned to the bishops, and the Church Commissioners are now lords.

In a survey of the manor made in the second year of the episcopate of Bishop de Fontibus (1221), the demesnes, including 140 acres of *nova purprestura*, amounted to 794 acres.[61] In 1251 during Bishop Northwold's episcopate there were 719¼, rated at 5 carucates.[62] By the latter date some reclamation had been carried out; we hear of 47 *novi feoffati* at Waldersea and 3 at Rumere[63] in the upper fen,[64] and there was pasture estimated at 20 acres. This pasture was situated towards the sea and was held by the bishop in severalty; its value depended on whether it had been recently flooded. The results of the great marine flood of 1236, when hundreds of lives were lost in Wisbech and its neighbourhood,[65] are evident in the numerous references to messuages, especially in the New Market, totally destroyed by the sea. In 1251 there were four fisheries, and a water mill, a horse mill, at which the customary tenants were obliged to grind their corn, and a newly constructed windmill. South-west, from the

52 Gardiner, *Hist. Wisb.* 199. The building cost £2,405. Owing to its situation over the former castle moat the foundations have sunk, causing serious settlements.

53 The roofs are arranged in short stretches at right angles to the street instead of in one long line parallel to it, and show a series of shallow gables at the back of the houses, generally masked in front by a cornice. This type of roof uses rather less timber than the normal one, but there is great risk of dampness in the top rooms under the valleys of the roofs. In the Wisbech example these valleys come on the top of the party walls instead of in the middle of the houses as is usual in London.

54 J. Bentham, *Hist. Ely*, 87. The authenticity of Wulfere's charter of 664, mentioning the river going to Elm and to Wisbech as the boundary of the Peterborough abbey lands, is doubtful.

55 *V.C.H. Cambs.* i, 369–71, 381.

56 For history of this manor, and its subordinate manors, see under Wisbech St. Mary.

57 The names are preserved in Barton Road and Farm, west of the town, and the hamlet of Murrow in Wisbech St. Mary. The manors coincided roughly with the parishes of St. Peter and St. Mary respectively.

58 The long vacancy (1581–1600) in the reign of Elizabeth is shown by the inscription under Barton manor *modo in manu Dom. Reginae* in the map of Marshland preserved in Wisb. Mus. It is dated 1657 but stated thereon to be copied from a map of 1597, which in its turn may be a corrected copy of an earlier one, as it shows Thorney Abbey standing complete.

59 C 54/3678, m. 29.

60 Gardiner, *Hist. Wisb.* (1898), p. 14. Thurloe took part in a recovery of the manor in 1658 (C.P. 43/301, ro. 18).

61 B.M. Cott. MS. Tib. B. II, f. 143*a*.

62 B.M. Cott. MS. Claud. C. XI, f. 72*b*.

63 This name is preserved in Rummer's Farm on the left bank of the Nene 3 miles above the town. The field of 'Rummere' was exempted from a commission *de walliis et fossatis* in 1336 (*Cal. Pat.* 1334–8, 291).

64 B.M. Cott. MS. Claud. C. XI, ff. 74*a*–75*a*.

65 Matt. Paris, *Chron. Majora.* (Rolls Ser.), iii, 379.

Horseshoe as far as Wryde in Thorney, stretched the great Heye or High Fen, common to the vills of Wisbech, Leverington, Newton, Tydd St. Giles, Elm, Outwell, and Upwell. In 1251 there was faldage worth 21*s.* 7*d.*, and stock consisting of 24 cows and 2 bulls, 360 ewes and 240 wethers and hoggets.[66] In spite of the floods, the number of tenants increased between 1221 and 1251. While the six holders of knight's fees or fractions thereof remained constant, the 25 free and customary tenants of 1221 had grown by 1251 to 36 free and 92 customary tenants. The total rents at the latter date were £37 15*s.* 3¼*d.* Some abnormal customary services are recorded. Thus at both dates the customaries might be required to find a boatman to convey the bishop by fresh water to his next manor, the lord providing the man's food. This might involve a journey to Somersham (Hunts.), Ely, or Brandon (Suffolk). Efforts to repair the flood damage are shown by the half day's labour required in 1251 *ad desiccandum campos*, the day's digging on 'Ramesinedik' and inside and outside around the castle, and the day's labour for repairing the park of Barton. The customaries in 1251 had also to provide guards for prisoners in the castle jail, or if that was unserviceable to keep them in their own houses. They were to provide bedding (*pannos scilicet chalones*) for guests arriving at the castle or else to take such guests into their own houses.[67]

At another survey made in 1356[68] the rents were only £28 10*s.* a year and further damage from floods, in some cases from upland waters, is recorded. The sites of the castle and of Barton manor house were valued at only £2, and there were many ruinous houses which would cost more to repair than they were worth. Seven hundred acres of land and pasture in Waldersea also were left unvalued, *quod superfluitas aquarum fere omnibus annis totaliter superfluit*, and the repair of Waldersea Dyke was considered to cost more than the land was worth.[69] At this date there were two windmills, worth £1 each, and the four fisheries, specified as those of 'Upstaven', 'Lathermere', 'Wellenee', and 'Hedmere', all let to farm, produced £3 6*s.* 8*d.* The livestock and produce were valued at £52 13*s.*

The long series of 14th- and 15th-century bailiffs' account rolls of Wisbech Barton manor[70] give many interesting details. As might be expected, the half-drained fenland which at that time made up so much of the estate was excellent grazing. The net receipts on stock usually amounted to some £10 a year, and in the early 16th century the sale of wool and fells often reached £7 or more. More surprising is the export of corn, which, especially in the early part of this period, sometimes reached extraordinary proportions. A net profit of £115 18*s.* 4*d.* on grain was made in 1322. There are violent fluctuations in this item of revenue, due no doubt to bad harvests, but the general tendency is downwards, the net profit being £36 11*s.* 6*d.* in 1419 and only £9 19*s.* 10*d.* in 1492. The manor was occasionally a liability to the bishop, as in 1320. Normally, however, it was an asset. The profit fluctuated from £15 in 1331 to £82 5*s.* 3*d.* in 1517. When let, the rent was usually £35, which presumably shows the average estimated profit.[71] The amount kept in hand as demesne varied considerably—198 acres in 1408 but only 15 acres eleven years later, and as much as 250 acres in 1492. The expenses of the dyke keeper (*custos fossatorum*) occur regularly but vary greatly in amount, occasionally approaching or exceeding £5.[72] Scouring old dykes or digging new ones cost 2*d.* or 3*d.* a perch, the labourers' wages being 4*d.* a day. A 'drainage tax' of ¼*d.* or ½*d.* an acre was levied to pay for the work.[73].

The survey made by order of Bishop Alcock in 1492–3[74] sets out in great detail the legal privileges attaching to the manor. The lord had the right of sac and soc, utfangthef, infangthef, tumbril, frankpledge, escheats, chattels of fugitives, deodands, wrecks, and treasure trove. It was stated that these had been held without contradiction since the reign of Edgar.[75] There was a Saturday market, assize of bread and ale and a fair in connexion with the feast of St. Peter. The bishop levied 4*d.* toll on all ships, not of the liberty of Ely, which entered or left the port.

The demesnes amounted to 1,137 acres. The four fisheries, one of which is specified as a sea fishery,[76] had increased in value to £4 16*s.* 8*d.* There is no reference to any lands or buildings laid waste by the sea, and the greater part of the manor had been parcelled out into large fields each divided into a multitude of small tenancies. These tenancies varied in number from 5 in Wisbech Holmes Field (rents 6*s.* 8*d.*) and in Herne Field (rents 3*s.* 5*d.*) to 81 in Oldfield (rents 35*s.* 2*d.*) and 77 in Wheatmaths (rents 46*s.* 9*d.*). Many of the field names survived until the 17th century, and some to the present day. The inquisition *post mortem* (1641) on William Edwards records that he held a pasture called Snelling, in Harecroft Field, land in Oldfield abutting on Bramble Lane on the west and Deadman's Lane on the east. He also held a messuage, cottage, and lands in Sandilands, near Core Corner, and property in Tydd St. Giles.[77]

The annual value of the bishop's manor in Wisbech in 1540–1 was £115 13*s.* 9*d.*[78]

The 1492 system of land tenure closely approximates to that of another survey, taken in 1792, when there were about 600 proprietors.[79] This survey, accounting for about 15,100 acres,[80] recorded rentals of £993 17*s.* 7¾*d.* on a total property value of £13,550 10*s.*, including £2,980 mercantile stock in the port and the proceeds of the customs and excise. After 1792

66 Bp.'s Mun., Ely, G 3 (Liber R or 'Old Coucher').

67 B.M. Cott. MSS. Tib. B. II, ff. 143*a*–147*a*; Claud. C. XI, ff. 73*b*–81*b*.

68 B.M. Add. MS. 6165, pp. 272–5.

69 The depressed state of Wisbech revealed by this survey, as compared with the prosperity of 1251, is partly due to the blow to trade caused by the diversion of the Ouse to the Lynn outfall, and to the Black Death a few years back. There seems, however, to have been a general falling-off in the Isle manors during the first half of the 14th century.

70 Bp.'s Mun., Ely, D 8.

71 As in 1378 and 1480.

72 As in 25–26 Hen. VI and 9 Hen. VIII.

73 See H. C. Darby, *The Medieval Fenland* (1940), 161.

74 17th-century copy at Wisb. Mus. (Peckover MSS.).

75 The view of frankpledge was stated to have been held 'from time immemorial' in 1327, when John de Staynton, Bishop Hotham's bailiff, was forcibly prevented from holding the court, of which the profits were valued at £10 a year (*Cal. Pat.* 1327–30, 154).

76 Probably the several fishery known as Depewere, which extended north-eastwards downstream from the bishop's fishery of Upstaven, and had been granted in mortmain to the bishop by the warden of the chantry of St. Mary, Newton, in 1411 (*Cal. Pat.* 1408–13, 345). It had been granted to the chantry by Sir John Colvile.

77 C 142/536/133.

78 A. Gibbons, *Ely Episc. Rec.* 12.

79 MS. book at Wisb. Mus. The survey may have been taken in connexion with the forthcoming sale of Wisbech Castle by Bishop Yorke.

80 The combined area of Wisbech St. Peter and St. Mary, before the recent alterations, was 16,664 acres.

values rose rapidly, the real property in the borough being estimated at £28,333 in 1815.[81] Inclosure was gradual and piecemeal, as reclamation proceeded, and no formal act or award for Wisbech St. Peter is known. The final inclosure of the commons, droves, and wastes in Wisbech St. Mary took place in 1833,[82] when only 181 acres, out of more than 10,000 in the parish as then constituted, remained to be so treated. About 5 acres each went to the bishop and to the dean and chapter of Ely as lord of the manor and impropriators respectively, 14 to the vicar of Wisbech in lieu of tithes, 26 to the overseers for the benefit of the poor, 44 were divided amongst the owners, numbering about 30, of common rights, and 86 were sold to defray the costs of the Act.

At the beginning of the 19th century a distinction was still made, in the Waldersea area, between 'high lands', which had been much improved by stocking and pasturing and were worth up to 40*s.* an acre, and the fen proper, which though drained fetched 15*s.* an acre only. In the 1730's, however, the fen had been worth only 2*s.* to 3*s.* an acre, and within living memory had been very wet, 'belly deep of a horse', until midsummer.[83] A 'manor' of Waldersea was leased to Lord Saye and Sele in 1776 for £30 a year;[84] its value had been returned separately from Wisbech as early as 1591, when it was £8 16*s.* 4*d.*[85]

In 1801 the arable acreage of the two parishes was 5,989, oats (3,340 acres) being by far the most prominent crop. There was an unusual proportion (1,164 acres) under rape and turnips.[86] At the tithe apportionment in Wisbech St. Mary (1840), out of a total of 9,599 acres, 5,615 acres were recorded as arable, 3,748 as meadow and pasture, and 55 as marsh and common.[87]

Two rentals of 1632 and 1778–9 are of some interest as showing the manorial organization then in force. At the former date a demesne let at £16 a year is recorded, and two fisheries worth £5 16*s.* Various other lands, mainly in Waldersea, produced £17 5*s.* 7*d.*, and there were quit rents of £42 17*s.* 8*d.* payable twice yearly. At the latter date these rents, held by Mr. Mayer as executor to Mr. Wensley, had gone up to £94.[88] Barton Farm, the descendant of the demesne land, was let at £32. Peter Thompson, a Wisbech rope-maker, held a windmill at the nominal rent of 1*s.* 9*d.* The Revd. Charles Gretton of Springfield (Essex) held 600 acres in the High Fen for £11 5*s.*, the low rent probably being due to the poor drainage.[89]

Manorial courts were held until 1922.[90]

FEES In 1221 the tenants by military service were as follows: William de Longchamp held 80 acres as 1 fee, with a messuage in the New Market belonging thereto, and also 240 acres at a money rent (20*s.*). Henry de Walpol held 80 acres with 4 messuages, as 1 fee, and 120 acres for 10*s.* 4*d.* Robert de Haustede held 94 acres and 3 cottages as 1 fee. John de Litlebury and Richard de Melkesham held 80 and 120 respectively, each as 1 fee.[91] These can all be identified in 1251, though the assessments had in most cases been reduced owing to flood damage. The Longchamp fee had been reduced to $\frac{1}{12}$ and equated with 40 acres only. The messuage had been destroyed by the sea. The 240 acres, however, were still rented at 20*s.* The Walpol fee, now held by another Henry, son of Osbert de Walpol, had also been reduced to $\frac{1}{12}$, though the rent of the tenant's 120 acres had in 1251 been increased to 12*s.* 4*d.* The Haustede holding, specified as being partly in 'Brokene', was assessed as $\frac{1}{2}$ fee, and the Melkesham 120 acres at $\frac{1}{8}$.[92] In 1302–3 Osbert de Longchamp was associated with John son of Osbert de Denever in holding $\frac{1}{6}$ of a fee in Wisbech of the Bishop of Ely,[93] which by 1346 had passed to the heirs of Sir John de Shardelowe.[94] The Haustede $\frac{1}{2}$ fee had by 1428 passed to the Colvile family,[95] who had in 1302–3 held another $\frac{1}{2}$ fee which in 1346 was in the hands of Reynold son of Thomas son of Nicholas de Waltone.[96] A $\frac{1}{12}$ fee, probably that which in 1251 was held by Henry son of Osbert de Walpol, had in 1302–3 come to William son of Geoffrey and in 1346 to his son Nicholas.[97] Walter de Langton, Bishop of Lichfield, held a $\frac{1}{3}$ fee in Coldham (in Elm and Wisbech), both in 1302–3 and at his death in 1321.[98] This descended to his nephew Edmund Peverel, in whose family it can be traced to 1428.[99] In the Alcock terrier of 1492–3 the Longchamp-Shardelowe holding, rated at a $\frac{1}{12}$ fee, is shown as divided between Thomas Rowsyll, who held 40 acres in Bryggesfeld on the left bank of the river and had several subtenants,[1] and Thomas Hunston, who paid 20*s.* rent for 240 acres in Sayer's Field.[2] The Colvile $\frac{1}{2}$ fee, rated on 94 acres in Brokene and owing suit of court, had come to Sir Robert Brandon through his marriage with John Colvile's relict. The 80 acres of the Litelbury family, assessed at a $\frac{1}{12}$ fee, after having been held by Andrew Reynolds, were now in the hands of John Mabyll. The 80 acres of Henry son of Osbert de Walpol are also mentioned, but the tenant was stated to be unknown and the land divided amongst several subtenants.[3]

Stephen de Segrave, justiciar of Henry III, held lands in Wisbech, and in 1243–4 his son Gilbert acknowledged the right of his (Gilbert's) stepmother Ida to hold them in dower.[4] This estate was increased by Bishop Balsham's grant of certain new purprestures for $\frac{1}{2}$ mark rent to the pittancer of Ely monastery,[5] and of a tenement and lands in Guyhirn from Reynold Serewynd of Cambridge.[6] At the death of John de Segrave in 1353 the property comprised $38\frac{1}{2}$ acres, held by 19 tenants for £4 18*s.* 5*d.*[7] Rents in Wisbech descended through Elizabeth (Mowbray), John's daughter, to the Dukes of Norfolk, who held them in 1433.[8]

81 *Rep. Com. Municipal Corpns.* H.C. 116, p. 2563 (1835), xxvi.
82 3 Wm. IV c. ix. The award (1835) is preserved in the office of Messrs. Fraser, Woodgate and Beall, 29 Old Market.
83 E 134/45 Geo. III/M. 1.
84 Gibbons, *Ely Episc. Rec.* 25. Bp.'s Mun., Ely, D 6.
85 S.C. 6 Eliz. 283. In 1592 the lessee was Richard Drake (C 66/1394, m. 33).
86 H.O. 67/9.
87 MS. at Wisb. Mus.
88 Including those from Tydd St. Giles.
89 Bp.'s Mun., Ely, D 6.
90 Infm. Mr. G. M. G. Woodgate.
91 B.M. Cott. MS. Tib. B. II, ff. 143*a*, *b*.
92 B.M. Cott. MS. Claud. C. XI, f. 73*b*. The $\frac{1}{8}$ (Melkesham) fee is entered in a later hand, and the assessment of the Litelbury property is left blank.
93 *Feud. Aids*, i, 141, 152.
94 Ibid. i, 172. Sir John (d. 1342) had held three shops in Wisbech of the Bishop of Ely for 8*d.* yearly, his heir being his nephew, another John (*Cal. Inq. p.m.* viii, p. 365). Sir John was a commissioner *de walliis et fossatis* in 1336 (*Cal. Pat.* 1334–8, 283, 291).
95 Ibid. i, 191.
96 Ibid. i, 141, 151, 173, 193.
97 Ibid. i, 141, 152, 173.
98 Ibid. i, 141, 151; *Cal. Inq. p.m.* vi, 330 (p. 196).
99 *Feud. Aids*, i, 193.
1 In 1536 these 40 acres (in Brigfeld) were held by Humfrey Carvyle for $\frac{1}{12}$ of a fee (B.M. Add. MS. 5837, f. 200*a*).
2 This piece was also held in 1536 by Carvyle by socage for 20*s.* (ibid. f. 201*a*).
3 Peckover MSS., Wisb. Mus.
4 C.P. 25(1)/283/11.
5 B.M. Campb. Chart. ii, 3.
6 B.M. Add. MS. 37671, f. 111*b*.
7 Ibid., ff. 96*a*, *b*, 97*a*, *b*.
8 C 139/60.

In 1235–6 Bishop Northwold granted to the Abbey of Thorney 1,000 acres in the marsh of Wisbech.[9] The Thorney property in Wisbech was worth only 8*s.* in 1291,[10] perhaps owing to floods, and there is no further mention of it until 1548, when a 21-year lease at 10*s.* a year was granted to Thomas Barrett and John Wrighte *alias* Goldwell. In 1560 the reversion in fee simple was given to John Norden of London and Clement Robertes of Little Braxted (Essex).[11] The priory of Spinney held 40 acres in Wisbech Fen, worth 10*s.* in 1291,[12] and known as Spineylond in 1347.[13] In 1544 it was granted in fee to George Carleton.[14] About 1275 Robert son of Edmund of Wisbech granted 10*s.* rent to Anglesey Abbey, from a messuage and 14 acres in Sayer's Field.[15]

The practice of leasing the manor or portions of it dates from at least 1330, when Bishop Hotham was licensed to approve or inclose waste land in Wisbech and his other manors to the value of £20 a year and to let the same to farm.[16] Three years later he granted a piece of land called 'Inlyk' (now Inlay Field) to Sir Simon de Drayton for life.[17] Bishop Alcock, who died at Wisbech Castle in 1500, was the last to be normally resident at the manor, and his successor (West) in 1529–30 leased the site of Barton manor and the demesnes to Thomas Megges for ninety-nine years at a yearly rent of £30.[18] Megges died in 1542, and three years later his relict Agnes sublet an 80-acre pasture called 'Cowleylond' for forty years. Nicholas Megges, his eldest son,[19] on whom the manor was settled after Agnes's death, sold his interest to Mark Steward and Thomas Wyseman. They resold it for 200 marks to Henry, the third son, but the transaction was declared invalid in Chancery.[20] Before her death, however, Agnes had conveyed her rights in the manor and demesnes to Henry for £40 a year.[21] Disputes later arose owing to a sublease executed by Nicholas Megges.[22] For 105 years after the Restoration the Southwell family leased the castle and manor,[23] until in 1793 the site of the castle was sold to Joseph Medworth and the estate finally broken up.[24] The Barton manor and its perquisites were sometimes leased separately, as in 1712 to Samuel Shepheard and in 1771 to Robert Wensley.[25]

RELATIONS BETWEEN THE BISHOPS OF ELY AND THE TOWN OF WISBECH

There is but little upon which to base an estimate of the relations between the bishops and the men of Wisbech during the Middle Ages. By the early 16th century, however, it became the custom for the townsmen to present the bishop with a substantial sum as 'recognition money' on his first coming to the town. In 1534 Bishop Goodrich received £7 6*s.* 8*d.*,[26] increased for Bishop Cox in 1560 to £7 16*s.*,[27] and £9 9*s.* 4*d.*, with 20 stone of best beef, for Bishop Felton in 1619.[28] In 1676, when Bishop Gunning visited the town, the sum was increased to 20 guineas, 'not doubting but his Lordshipp will have and continue his favour and kindness to this Townshipp and his Tenants'.[29] The payment was waived by Bishop Keene in 1775 and received for the last time by Bishop Sparke in 1813.[30] The recognition money did not preclude occasional presents in kind; for example, £3 6*s.* worth of fowl in 1617, a buck from Exton Park (Rutland), worth 20*s.*, in 1669,[31] and a £23 piece of plate the previous year.[32]

In actual fact the bishops, who had three other palaces in the Isle and one just outside at Somersham (Hunts.), do not seem to have visited Wisbech at all frequently, judging by the fewness of ordinations held in the church or the castle chapel. Of those who recorded the places of ordinations in their registers, Bishop Lisle (1345–61) held 4 in Wisbech of a total of 31, Bishop Arundel (1347–88) 2 out of 72, Bishop Fordham (1388–1426) 3 out of 142, Bishop Alcock (1486–1500) 5 out of 24, and Bishops Gray and West none out of 19 and 14 respectively.[33]

It was, however, at the instance of Bishop Goodrich that the town obtained its first charter in 1549,[34] and it was probably with a view to improving the silted river that Bishop Morton (1479–86) caused his Leam to be cut to Guyhirn and so through the town rather than by a more direct route to the sea.[35]

SUBSIDIARY MANORS[36]

The manor or capital messuage of *CORNERS* or *CORINS HOLT*, which lay in the north-west corner of Wheatmaths Field,[37] first appears in 1512. It had been settled on Sir Edward Ferrers of Baddesley Clinton (Warws.),[38] who bequeathed it to his relict Constance (d. 1551), with remainder to his daughter Alice for life, his grandson Edward in tail male, and to the right heirs of his younger son Nicholas.[39] In 1581 it was passed by Thomas Gray and Joan his wife to Henry Stapleton of Rempstone (Notts.),[40] who two years later settled it on his wife Elizabeth for life, with remainder to his daughter Faith, wife of Matthew Babbington. Faith predeceased her mother, after whose death (1593) the estate passed to John Stapleton, Henry's nephew.[41] In the last few years of Elizabeth's reign he was engaged in lawsuits over the property with his tenant Gabriel Armstrong.[42] Stapleton passed the manor in 1600 to Francis Lodge,

9 Hist. MSS. Com., *6th Rep.* (Leconfield MSS.), 292; C.P. 25(1)/24/17/51.

10 *Tax. Eccl.* (Rec. Com.), 268.

11 *Cal. Pat.* 1558–60, 396–7.

12 *Tax. Eccl.* (Rec. Com.), 270.

13 B.M. Harl. Chart. 50, 55.

14 *L. & P. Hen. VIII*, xix, i, g. 610 (54).

15 Undated MS. at Wisb. Mus.

16 *Cal. Pat.* 1327–30, 549.

17 Hist. MSS. Com., *6th Rep. App.* 296.

18 C 3/214/23. In 1542 the rent was £32, and the commutation of labour services brought in £10 (Bp.'s Mun., Ely, D 5 (4)).

19 *Misc. Gen. and Heraldica*, 2nd ser. ii, 378. The pedigree calls Thomas Megges's wife Ann.

20 C 3/129/20.

21 The whole manor was valued in 1588 at £142 4*s.* 10*d.* a year (W. Watson, *Hist. Wisb.* (1827), 116 n.).

22 C 3/126/21.

23 Edward Southwell paid £20 rent for the castle in 1778–9 (Bp.'s Mun., Ely, D 6).

24 Gardiner, *Hist. Wisb.* (1898), 11. Permission to sell was granted by Act of Parliament (33–34 Geo. III c. liii). Medworth was a Wisbech charity boy who went to London as a bricklayer and prospered there.

25 Gibbons, *Ely Episc. Rec.* 25.

26 Corp. Rec. i, 93.

27 Ibid. i, 111–14, with a schedule of the amounts levied on individual tenants.

28 Ibid. iv, 17*b*, 20*a*.

29 Corp. Rec. iv. (The later folios of this and subsequent volumes are unnumbered.)

30 N. Walker and T. Craddock, *Hist. Wisb.* (1849), p. 326.

31 Corp. Rec. v, 8*b*.

32 Ibid. iv, 161*b*.

33 *Ely Dioc. Remembrancer*, passim (Calendars of Bp.'s Reg. by Canon J. H. Crosby).

34 *Cal. Pat.* 1548–9, 339.

35 This suggestion was made by the late E. J. Rudsdale.

36 This section deals with the manors situated within the parish of Wisbech St. Peter. The other subsidiary manors of Wisbech are described under Wisbech St. Mary.

37 Infm. Mr. G. M. G. Woodgate.

38 C.P. 25(2)/4/18 (East. 3 and Hil. 7 Hen. VIII).

39 C 142/95/9.

40 C.P. 25(2)/93/844 (Mich. 23 Eliz.).

41 C 142/248/10.

42 MS. Collections of Mr. G. M. G. Woodgate at Wisb. Mus.

and the latter in 1606 to William Goodday.[43] In 1629 Thomas Holmes appears as tenant of 27 acres of pasture in Corners Holt.[44] Corners Holt was later vested in Thomas Swaine of Leverington. His grandson Spelman Swaine settled the freehold portion of 36 acres on his marriage. It passed to his son Walter, who died seised of it in 1803.[45]

The manor of *WHITE HALL*, now represented by the farm of that name on the North Brink about 1½ mile south-west of the town, is first mentioned in 1479–80, when Richard Cause or Caius was in possession of a messuage and 500 acres.[46] This came by marriage to William Everard, whose granddaughter Grace married Richard Buckworth of Wisbech. The manor remained in possession of the latter family[47] until 1703, when it was sold to Robert Harris of London.[48] His descendant Thomas Harris of Chelsea in 1779 bequeathed it to William Wills of East Ham in trust for his wife Mary for her life, then to Christopher Hand of Fleet Street, ironmonger.[49] The latter died childless and the property, after the life interest of his wife Fanny, passed to his nephew Philip Bennett and his heirs.[50] Philip Bennett junior made a settlement of the estate on his marriage (1822) to Ann Pilkington, when the house with 248 acres was let to Thomas Moore for £400 a year. William Moore, the latter's son, bought the property in 1842 for £11,000, and on his death in 1882 it was sold in small lots.[51] The manor house had been pulled down in 1807, and the farmhouse which replaced it was severely damaged by enemy action in 1940.[52]

A manor of *REPPES* was in possession of John Reppes in 1542.[53] In the later 16th century it belonged to the Coney family,[54] of whom Henry and Martha his wife conveyed it in 1622 to Edward Cason, who died seised of it four years later having settled it on his son John.[55] The latter with his wife Elizabeth passed it to Francis Polsted in 1634,[56] and it changed hands several times subsequently. The existence of this manor is last recorded in 1697.[57]

CHURCH A vicarage was instituted in or before 1252, when Bishop Northwold presented William de Northwold to it. The vicar was to receive the tithes except those of flax and wool. Land and other tithes were assigned to him for conducting the services at the chapel of Kilhus in Wisbech St. Mary (q.v.).[58] In 1275 Bishop Balsham appropriated the church of Wisbech to the prior and convent of Ely, with the rectory house and a house for the priest at Kilhus. The profits of the rectory were devoted to the refectory of the monastery, which was inadequately endowed.[59] The vicar was to be resident and was to bear the ordinary burdens of the church; special expenses and the repair of the chancel were to be borne by the convent and the vicar jointly. If the lands from which the convent received the tithes of sheaves were flooded, they were to receive the tithes of all the lands of the vicarage.[60] In 1298 the prior and convent paid 500 marks to retain this appropriation, which had been made without licence.[61] In 1338 they received licence to say mass in the oratory of the rectory.[62] The right to appoint the vicar remained, as it has since done, with the bishop.[63]

The church was valued in 1217 at £33 6*s*. 8*d*., and in 1254 the rectory and vicarage were returned at £20 and £6 13*s*. 4*d*. respectively.[64] The vicarage was unchanged in 1291, when the rectory had increased to £26 13*s*. 4*d*.[65] In 1535, the vicarage, with the chapelry of Wisbech St. Mary, was returned at this latter sum.[66] The rectory is not distinguished from the other spiritualities of Ely monastery in the *Valor*, but in 1538–9 it was farmed at £18 13*s*. 4*d*.[67] In the early 18th century it was on beneficial lease to Arthur Jermy at £17 a year, at fines of £96 (1708), £100 (1715), and £160 (1712). A pension of 6*s*. 8*d*. a year was also payable to the dean and chapter from the vicarage.[68] An earlier pension of £2 13*s*. 4*d*. (one-tenth of the 1535 value) had been awarded from the vicarage to Bishop Cox.[69]

The uncertainty of the boundary between Wisbech on the one hand and March and Doddington on the other caused as much trouble over the tithes as over rights of common.[70] In 1347 there was a cause in the Court of Arches between John de Herle, Rector of Doddington, and the prior and convent of Ely regarding the tithes of Henlonde. They were allotted to Ely.[71]

The patronage of the church of St. Augustine (1869) is with the Bishop of Ely. The Octagon Chapel in the Old Market, opened in 1831 as a chapel of ease to the parish church, was a perpetual curacy in the patronage of trustees,[72] but has been closed since 1946.

The church of *ST. PETER AND ST. PAUL* consists of chancel, south vestry, south chapel, clerestoried nave, two south aisles, south porch, north aisle and north-west tower. The history of the fabric is involved and of unusual interest; the plan, too, is extraordinary. The oldest part of the existing structure dates from the third quarter of the 12th century. At that time there was a chancel, clerestoried nave, aisles, and engaged west tower. At the beginning of the 14th century the chancel was rebuilt on an enlarged scale and a large chapel added on the south side, the south aisle was widened and provided with a clerestory and another aisle added to the south; the north aisle was also widened at this time. A little later in the century the south porch was erected. Early in the 16th century the west tower seems to have partially collapsed, and this necessitated a considerable reconstruction. The south arcade of the nave was rebuilt and the north clerestory reconstructed. Both the nave and inner south aisle were provided with ceilings of beams and plaster and a single high-pitched outer roof was placed over

43 C.P. 25(2)/93/864; C.P. 25(2)/277.
44 B.M. Add. MS. 5838, f. 196*a*.
45 Infm. Mr. G. M. G. Woodgate.
46 C 140/71.
47 C.P. 25(2)/526 (Hil. 4 Chas. I); C.P. 25(2)/632 (East. 16 Chas. II).
48 Woodgate MSS.; C.P. 25(2)/910 (Hil. 1 Anne).
49 Woodgate MSS. The will was proved in 1782.
50 Ibid. 6 Dec. 1797.
51 Woodgate MSS.
52 *Wisb. Advertiser*, 6 Nov. 1940.
53 C.P. 25(2)/52/376.
54 C.P. 25(2)/93/837; C.P. 25(2)/386 (Hil. 11 Jas. I).
55 C.P. 25(2)/260 (East. 20 Eliz.); C 142/748/4.
56 C.P. 25(2)/527 (Mich. 10 Chas. I).
57 C.P. 25(2)/617 (Hil. 1657); C.P. 25(2)/815 (9 Wm. III).
58 Doc. copied into reg. of Bishop Fordham (*Ely Dioc. Remembrancer*, No. 192, p. 97.
59 D. & C. Mun., Ely, Carta 33.
60 *Ely Dioc. Remembrancer*, No. 192, p. 98.
61 *Cal. Pat.* 1292–1301, 363.
62 *Ely Dioc. Remembrancer*, 1889, 344.
63 During the Commonwealth John Thurloe as lord of the manor presented William Goldwell (Woodgate MSS.).
64 W. E. Lunt, *Valuation of Norw.* 539.
65 *Tax. Eccl.* (Rec. Com.), 265.
66 *Valor Eccl.* (Rec. Com.), iii, 500.
67 Dugd., *Mon.* i, 493.
68 D. & C. Mun., Ely (Bentham Transcripts).
69 *Cal. Pat.* 1560–3, 226.
70 See above, p. 238.
71 D. & C. Mun., Ely, Carta 712. This suit had been started in or before 1341, as Prior Crauden, who died in that year, was cited in the first instance.
72 *Crockford's Clerical Dir.* 1940.

them. The east arch of the tower was removed and a large window was inserted in the west wall. The south-west stair turret of the tower was retained and covered with a bell-cote. About this time a vestry was added to the south-east of the chancel. Finally, about 1525, a fine tower was erected at the north-west of the north bays with 2-centred arches of 2 orders, chamfered and resting on clustered piers of three-quarter rounded shafts with moulded caps; the west respond has a foliaged cap. There is a plain square-headed piscina of the 14th century. The chancel arch is 2-centred with 14th-century mouldings and a hood terminating in

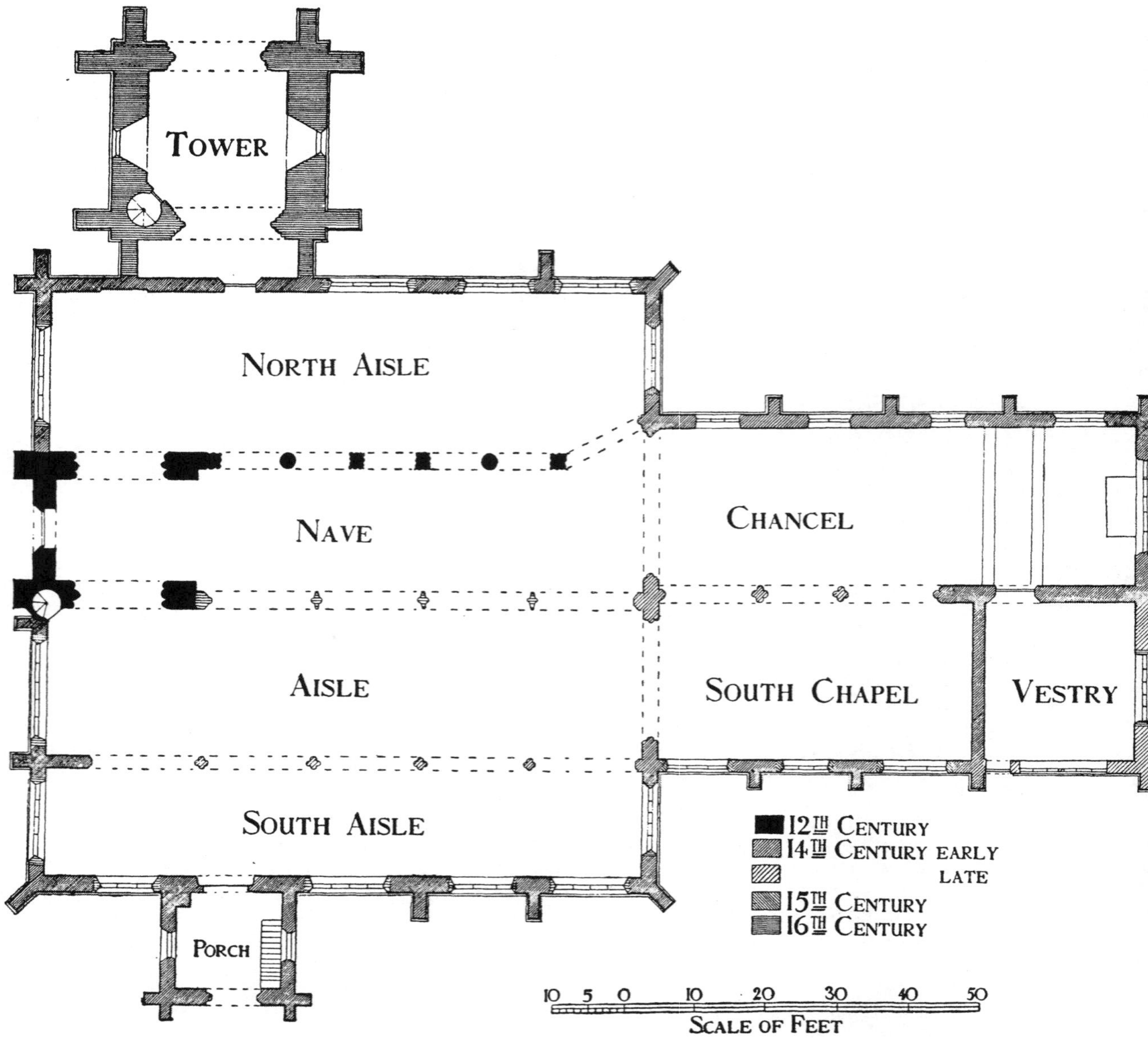

Plan of the Church of St. Peter and St. Paul

aisle, the base of which serves as a porch. An extensive restoration was begun in 1856. The chancel was re-roofed and the early 16th-century window at the west end of the nave was replaced by one of 14th-century character.

The chancel has an east window of 5 lights with geometrical tracery, which has been renewed. There are angle buttresses with 2 set-offs. In the north wall are 4 windows with flowing tracery, the east and west of 3 and the middle two of 2 lights. On the south side above the vestry is an early 16th-century window of 9 lights with trefoiled heads under a square label. The chancel opens to the south chapel by an arcade of 3 masks which have been renewed. The roof is modern. The vestry communicates with the chancel by a door with continuous mouldings and hood. The vestry has an east window of 5 trefoiled lights under a segmental head. The south window is divided by a heavy central mullion into 2 portions of 3 lights each. The embattled parapet has an elaborate band of decoration divided into panels each displaying a carved device; these include the emblems of St. Peter and St. Paul, the arms of Ely, a chalice and host, the Tudor rose and the monogram and rebus, four times repeated, of Thomas Burwell, who was an official of the Holy Trinity Guild at the beginning of the 16th century. The east side of the parapet rises in the centre to form a low gable. There is a doorway in the south wall with continuous chamfer.

The south chapel has a 15th-century east window of 5 lights, and 3 windows in the south wall each of 3 lights; the 2 eastern are of the 15th century with cinquefoiled heads and rectilinear tracery above, and the western of the 14th century with trefoiled heads and segmental arch. There is a plain doorway in the south wall with continuous chamfer. The arch communicating with the inner south aisle is similar to the chancel arch. There is a piscina with traceried head and a hood terminating in male and female heads. The high-pitched roof appears to be ancient; it is canted and boarded.

The nave has a 12th-century arcade of five bays on the north and an early 16th-century arcade of four bays on the south. The former has round arches which are plain except for one with cheveron ornament; the caps are square and chamfered with acanthus decoration. Two of the piers are round and the others, and the west respond, are clustered. In two of the spandrels are sunk quatrefoils and in another a circle. The chancel when rebuilt was widened on the north, as a result of which it was necessary to throw a skew semi-arch across from the east pier of the arcade, which was strengthened when the clerestory was rebuilt by the insertion of a supporting arch. The clerestory consists of seven three-light windows with cinquefoiled heads under a square label. The south arcade dates from the 16th-century reconstruction. The wide obtuse arches of two orders rest on lofty and rather slender piers with engaged shafts and moulded caps on their inner faces. In 1856 the bases of the original arcade were discovered beneath the floor. The west end of the nave was originally the base of the 12th-century tower. In the 16th century the tower arch and wall above were completely removed, but the side walls were retained. They are pierced by plain and massive pointed arches of Transitional character, which prove that the tower was engaged by the aisles. The large west window of four lights with tracery of 14th-century type was inserted at the restoration, and replaced an early 16th-century window. The west doorway is late 12th century and deeply recessed. The octagonal turret, formerly the stair turret of the tower, is crowned by a pyramidal stone cap and has an opening with a pointed head in each face of the top stage. The nave has a flat beam roof of 16th-century date, which from its plain character would seem to have been intended as a temporary covering.

The inner south aisle has an arcade of five bays similar to the chancel arcade, but the arches are less acutely pointed. There is a clerestory, the windows of which are alternately of two and three lights, the former with trefoiled heads and a trefoil above and the latter cinquefoiled under a segmental head. The west window is of five lights with flowing tracery. The roof is similar to that over the nave. The outer south aisle was provided with new windows early in the 16th century. The windows are of uniform design with cinquefoiled lights under a square head, the east, the third from the east and the west being of four lights and the remainder of three. In the east wall is a recess, possibly intended for relics. The buttresses are 14th century, but some of them were partly remade in the 16th century and the date 1586 appears on one of them. The roof is of lean-to character with moulded principals and rafters and brackets with pierced quatrefoils. The porch has an outer arch of two orders, two-centred with moulded caps and bases to the responds. There is a single trefoiled light on the east and west. The inner doorway has continuous mouldings. There is a flat plastered ceiling. Above is a parvise with a window of two cinquefoiled lights on the east, west, and south. There are angle buttresses with two set-offs. The plain low-pitched roof of the parvise is of uncertain date.

The north aisle, which was rebuilt in the first half of the 14th century, was provided with new windows early in the 16th century similar in design to those in the outer south aisle. There is a blocked 14th-century window at the west end of the north wall, and a piscina in a rectangular recess in the same wall. The lean-to roof is similar to that of the outer south aisle and probably dates from the beginning of the 16th century. The north doorway is an excellent piece of 14th-century work; it has a two-centred arch of several orders enriched with carvings in low relief of birds, animals, foliage, and grotesques on the exterior.

The tower consists of three stages, and is constructed throughout of ashlar; it is enriched with bands of carved panels. The lofty north arch is two-centred and of two orders with moulded caps and bases to the responds, while the spandrels have shields with the emblems of St. Peter and St. Paul. There is a two-light window on the east and west of the ground stage with cinquefoiled lights. The south arch, which gives on to a narrow space between the tower and the aisle, is two-centred and of two orders, similar to the outer arch. The ground stage has a plain wooden ceiling with massive beams. The lower string-course is ornamented with a band of quatrefoils, while the upper has a series of shields, which are blank except on the north where they are charged with the arms of England, Canterbury, and Ely and the emblems of St. Peter and St. Paul. The second stage has on each face a recessed two-light window with an embattled transom; the belfry windows are similar but are not recessed. Above on the south and east are angels holding religious emblems; similar figures on the north and west were destroyed to make room for the clock dials; over these are canopied niches and various devices including the emblems of St. Peter and St. Paul, the arms of Canterbury and Ely, the wheel of St. Catherine and the chalice and host. There is an embattled parapet pierced and stepped, with lofty angle pinnacles and lower pinnacles in the centre of each side. Below the parapet is a band of ornament with shields, floral designs and the emblems of St. Peter and St. Paul. The tower is crowned by a leaded pyramidal cap. The newel stair is in the south-west angle, and the lower doorway has a four-centred arch with hood-mould and traceried spandrels; the plain door and its hinges are original. All the roofs are leaded.

The 14th-century font has a plain bowl resting on clustered shafts. In the south-east window of the chancel and in the tracery of the east window of the south chapel are fragments of 15th-century painted glass, and several windows have clear glass in rectangular leading. There are three stalls now in the sanctuary with misericords carved with human heads and grotesques on the elbow rests. The communion table is early 17th century, and the altar rails and paving in the sanctuary are a century later. There is a particularly fine royal arms of James I, carved and coloured.

On the chancel floor is a magnificent brass of a knight in armour commemorating Thomas de Braunstone, Constable of Wisbech Castle, 1401. The canopy shields and part of the inscription in old French are missing. The dimensions including the border are 9 ft. by 3 ft. 10 in., and the effigy is one of the largest in England. In the south chapel is a brass inscription with shield of arms to Nicholas Sandford, 1638. On the north wall of the chancel are 2 marble monuments, each with kneeling figures of a man and wife. One commemorates Thomas Parke, 1628, and Etheldreda his wife, and the other Matthias Taylor, 1633. In the south chapel is a mural monument to Edmund Southwell, 1787, with allegorical sculpture by Nollekens. Bishop Watson[73] and Abbot Feckenham[74] (see Recusants in the Castle) are buried in the church without monuments.

The plate consists of two chalices, two patens and a flagon, all silver, 1862, a chalice of metal, and an almsdish of brass.

The tower contains ten bells by Dobson of Downham, 1823. There were formerly eight, of which the 5th was by John Brend of Norwich, 1566, the 6th and 7th by John Draper of Thetford, 1608 and 1640. The present 5th and 6th were the gift of Dr. Abraham Jobson, vicar 1802–28.[75] The present 4th is inscribed 'Long live King George the Fourth' and the 7th 'Prosperity to the town of Wisbech St. Peter, 1823'.[76]

The registers begin in 1558 and are complete.

The *OCTAGON CHURCH*[76a] is a brick building, faced with stone and embattled. It was erected under an Act of Parliament of 1827[77] as a chapel of ease to the parish church, and consists of chancel, octagonal nave, porch at the south side and a turret containing one bell. The building was designed by William Swansborough in imitation of the Octagon of Ely Cathedral, but the lantern became unsafe owing to defective foundations, and was replaced in 1846 by a battlement with pinnacles.[78]

The plate consists of two communion cups, three patens, and a flagon, all silver, 1829.

The church of *ST. AUGUSTINE*, erected in 1868–9, is of brick with stone dressings and consists of chancel, clerestoried nave, aisles and a turret containing one bell. The registers begin in 1870 and are complete. The ecclesiastical parish was formed in that year from Wisbech St. Peter and Leverington.

ROMAN CATHOLICISM

From 1840 Roman Catholics worshipped in what had been a carpenter's shop, a priest coming over from Peterborough for monthly services.[79] A church, in Queen's Road, dedicated to St. Mary and St. Charles Borromeo, was built in 1854.[80]

PROTESTANT NONCONFORMITY

In 1638 unlawful meetings and conventicles were held in Wisbech at the house of Thomas Bouth.[81] The Baptists are the first sect to be mentioned by name. In 1655 two Baptists from Fen Stanton (Hunts.) visited their brethren at Wisbech at the invitation of John Milles, a member of the Wisbech congregation, of which at the same time Edmund Smith and Israel Cave were elected elders.[82] Wisbech does not figure as a prominent centre of Nonconformity in Bishop Compton's 'census' of 1676, which enumerated only 12 Dissenters here as compared with 139 in March, 43 in Doddington, 96 in Whittlesey, and 67 in Littleport.[83] No licences were issued in respect of Wisbech under the Declaration of Indulgence (1672).

By the end of the 17th century there were already separate congregations of Particular and General Baptists in Wisbech. In 1692 the former built the first Nonconformist place of worship in the town—in Deadman's Lane (now Alexandra Road).[84] William Rix was appointed as resident pastor.[85] Rix was still in his ministry in 1715, when his congregation numbered 60, including ten 40*s*. freeholders.[86] In the middle of the 18th century the numbers began to decline, and in 1776 there were only 20 adherents, too few for a resident minister.[85] Under the Revd. Samuel Fisher of Norwich (minister 1781–1803) the congregation revived, but a schism occurred and some members, under Richard Wright, adopted Unitarian views and claimed the Deadman's Lane chapel.

In the mid-18th century a congregation of Johnsonian Baptists established itself in Wisbech. Deprived of his chapel Fisher allied himself with this community, and in 1794 built a new chapel in Ship Lane (now Hill Street).[87] This was rebuilt in 1859, and is the only modern building in Wisbech constructed entirely of stone.[88] An offshoot of this chapel, in Artillery Street, was closed in 1937.[89]

The General Baptists absorbed the congregation of Walpole (Norf.) towards the end of the 17th century. In 1697 they built a permanent chapel off Falcon Lane, on land belonging to Henry Place, a wealthy woollen draper and a member of the sect. The chapel measured 35 by 32 ft., and cost £120.[90] The graveyard and some fragments of the building remain.[91] About 1700 the General Baptists had 26 members but no resident pastor.[92] Later ministers of this congregation included Joseph Proud (1745–1836), who became a prominent Swedenborgian,[93] 'the excellent Mr. [Joseph] Freeston of Hinckley' (minister 1783–99)[94] and Mr. Jarrow, who was in office in 1820.[95] In 1803 a new chapel was built in Ely Place, which in turn was replaced by the present building seventy years later.[96] In 1820 the society maintained a flourishing Sunday school and

73 *D.N.B.*
74 Infm. Mr. G. M. G. Woodgate.
75 Gardner, *Dir. Cambs.* (1851), 599.
76 J. J. Raven, *Church Bells of Cambs.* (1869), App.
76a Demolished 1952.
77 7 & 8 Geo. IV c. xcii.
78 Gardner, *Dir. Cambs.* (1851), 605–6. A scathing denunciation of the style, planning, and construction of the chapel.
79 Gardner, *Dir. Cambs.* (1851), 608.
80 Gardiner, *Hist. Wisb.* (1898), p. 272.
81 *Trans. Cambs. and Hunts. Arch. S.* iv, 382; Bp.'s Mun., Ely, D 8.
82 A. Taylor, *Hist. Engl. Gen. Baptists* (1818), i, 138–41; ii, 301–2.
83 Bp.'s Mun., Ely, A 6.
84 Walker and Craddock, *Hist. Wisb.* (1849), 395. This chapel was on the site of the present British Legion premises; its graveyard still exists (infm. the late E. J. Rudsdale).
85 Wilson MSS. I 5, 283–6 (Dr. Williams's Libr.).
86 Evans MSS. f. 9*b* (Dr. Williams's Libr.).
87 *Cong. Mag.* iii, 169–70 (1820).
88 Gardiner, *Hist. Wisb.* (1898), 269–71. The materials were largely taken from the Stone Bridge and Butter Market, demolished in 1856.
89 *Bapt. Handbk.* 1950.
90 Taylor, *Hist. Engl. Gen. Baptists*, i, 138–41.
91 Infm. the late E. J. Rudsdale.
92 Taylor, *Hist. Engl. Gen. Baptists*, ii, 301–2.
93 *D.N.B.*
94 Author of *Encouragement for Travellers to Zion.*
95 *Cong. Mag.* iii, 169–70.
96 Gardiner, *Hist. Wisb.* (1898), 269–71. The site of the Ely Place chapel was 'very unwillingly' conveyed to the Baptists by Joseph Medworth (*Ely Place Chapel Souvenir* (1905)).

Church of St. Peter and St. Paul Wisbech, from the South-east

(From a photograph of 1857)

Octagon Church, Wisbech (Demolished 1952), from the North

(From a photograph of c. 1855)

Copyright 'Country Life'

'Wisbech Castle' in 1947, from Museum Square

Copyright 'Country Life'

Thurloe's House on the Castle Site, Wisbech (Demolished 1816)

(From a drawing in Wisbech Museum)

also the General Baptist Academy.[97] Three years later a plot of land in Crescent Passage was purchased for a burial ground, and on it a school was built in 1836. This was used by the Sunday school and by the girls' British school during the week.[98] The General Baptists were also instrumental in establishing (1818) the Christian Fund, a friendly society open to all church-goers, Anglican as well as Nonconformist.[99]

Another Baptist chapel (Zion) was built in Victoria Road in 1856;[1] it has a good classical façade.

In 1751 licence was granted to Isaac Culy for an Independent meeting-house in Wisbech.[2] This was undoubtedly for a congregation of Culimites, followers of David Culy of Guyhirn in Wisbech St. Mary (q.v.). The sectaries met in the Wool Hall, Exchange Square.[3] In 1851 the remnants of this sect were worshipping in the Calvinist Chapel at New Walsoken, 'a small mean building, erected in 1837'.[4]

The Presbyterians in Wisbech were organized as a society some time before 1694, when Ishmael Burroughs, their pastor, published a pamphlet on the conversion of T. Mackenesse, a condemned robber. Burroughs moved to London in 1724, and was succeeded by John Ford, who was paid £4 a year and moved to Sudbury (Suff.) in 1729.[5] Philip Doddridge preached at the ordination of William Johnston as minister on 8 June 1737.[6] The original chapel of this denomination is said to have been in Ship Lane (Hill Street). During the 18th century this congregation, like many of the English Presbyterians, seems to have become Unitarian. Its two most famous ministers almost certainly were of that persuasion.[7] They were John Godwin (minister 1748–58), father of the author of *Political Justice*, and William Hazlitt senior (minister 1764–6), who married Grace Loftus, daughter of a Wisbech ironmonger, and was the father of the essayist.[8] From the fact that the section of Particular Baptists under Richard Wright who seceded into Unitarianism about 1790 (see above) claimed the Baptist chapel in Deadman's Lane and ousted the orthodox section, it seems that the Presbyterian-Unitarian congregation may have for some time been worshipping there rather than in Ship Lane. The Unitarians were influential in Wisbech early in the 19th century.[9] Between 1851 and 1858 they moved to Great Church Street, but their chapel here was closed early in the 1870's.[10]

In 1816 the Congregationalists started public worship in the Wool Hall in Exchange Square.[11] The foundation stone of a chapel in Castle Square was laid in 1818 by the Rev. William Holmes, the pastor, and the congregation moved thither in the same year.[12] An addition made in 1911 comprised a new Sunday school. There are two subordinate chapels, at South Brink (1848), about 2 miles south-west of the town, and at Gorefield (1836) in Leverington.[13]

There are said to have been 8 members of the Society of Friends in Wisbech in 1660.[14] They were prominent in Wisbech in the 18th century,[15] and had a meeting-house, originally two cottages, on the North Brink on the site of the present building as early as 1711.[16] In 1851 the meeting had 14 adherents;[17] the present meeting-house was built three years later.[18]

Methodism did not immediately obtain a foothold in Wisbech; Charles Kyte and Ashmead of Lynn, its first missioners, were pelted with stones and mud when they preached in the Market Place in 1786.[19] By 1793, however, its adherents were numerous enough to hire a barn in Deadman's Lane, moving in 1803 to the present chapel (enlarged in 1835) in the Crescent.[20] Primitive Methodism began in Providence Chapel, New Walsoken, in 1838, and the present chapel in Church Terrace dates from 1868.[21] United Methodist Free Church meetings were held from 1850, at first in the public hall and later (1862) in the existing chapel (now sold) in Little Church Street.[22] A Seamen's Chapel ('Bethel') was erected in Russell Street in 1828. It was closed between 1879 and 1883.[23] The Salvation Army citadel (now demolished) in East Street dated from 1885. Its foundation stone was laid by General Booth.[24]

In the late 18th and early 19th centuries a considerable amount of co-operation occurred between the established and dissenting churches. The original Committee (1781) of the Literary Society included both Anglicans and Friends. The Committee of the Wisbech branch of the British and Foreign Bible Society included Anglicans, Quakers and Baptists.[25]

In 1824 open air services were held in the Market Place on Sunday evenings by the Independents, Methodists, General and Particular Baptists. 'The Church was silent, but perhaps had service been celebrated there at the same hour much good would arise from it, and at any rate some of the thousand loungers, riotous persons and squallers and squeezers in the Market Place would for a time at least have had their feet, and tongues and hands in a state of greater quietude and not less insincere'.[26]

CASTLE The castle, not mentioned in Domesday Book, is traditionally said to have been first built by William the Conqueror after the submission of Hereward.[27] It may have been granted

97 *Cong. Mag.* iii, 169–70.
98 Gardner, *Dir. Cambs.* (1851), 607.
99 Ibid. 625–6.
1 *Bapt. Handbk.* 1950.
2 Bp.'s Mun., Ely, A 6.
3 Gardiner, *Hist. Wisb.* 260.
4 Gardner, *Dir. Cambs.* (1851), 607.
5 Evans MSS. f. 9*a* (Dr. Williams's Libr.).
6 *Cong. Mag.* iii, 169–70.
7 *D.N.B.* states that William Hazlitt senior was a Presbyterian minister who became Unitarian, and that John Godwin was a 'dissenting minister'.
8 William Godwin the elder was born in Wisbech in a cottage attached to the Deadman's Lane Chapel, but Hazlitt had moved to Maidstone by the time of his son's birth.
9 James Hill, who was prominent in local affairs *c.* 1800–25 and was the father of Octavia Hill, was a Unitarian. The sect is mentioned several times in Jackson's Diary (Wisb. Mus.) as a power in Wisbech.
10 Gardner, *Dir. Cambs.* (1851), 607; Kelly, *Dir. Cambs.* (1858, 1869, and 1875). The Unitarians and their chapel are not mentioned in the last-named edition.
11 Gardiner, *Hist. Wisb.* 265.
12 *Cong. Mag.* iii, 169–70 (1820); Walker and Craddock, *Hist. Wisb.* 393; Gardiner, *Hist. Wisb.* 265. A photograph and memoir of Holmes appear in the *Evang. Mag.* 1853.
13 *Cong. Year Bk.* 1949.
14 Susanna Peckover and J. Brown, *The Friends in Cambs.* (1888). There were sixty-nine in Cambridge and fifty-one in Ely at this date.
15 See p. 257 below.
16 Walker and Craddock, *Hist. Wisb.* 389.
17 H.O. 129/7/193.
18 Gardiner, *Hist. Wisb.* 267–8.
19 Ibid.
20 Ibid.
21 Ibid.
22 Ibid.
23 Kelly, *Dir. Cambs.* (1879, 1883).
24 Notes supplied by the late E. J. Rudsdale.
25 Ibid.
26 Diary of Revd. J. Jackson, 29 Aug. 1824 (Wisb. Mus.). 'Whether the Socinians had anything of the kind [i.e. an open-air service] I am sorry I did not ascertain.'
27 Matt. Paris, *Chron. Majora* (Rolls Ser.), ii, 7; Walker and Craddock, *Hist. Wisb.* (1850), p. 209; *V.C.H. Cambs.* ii, 47.

to the see of Ely at its formation, but the original headquarters of the episcopal estate in Wisbech, from which the manor ultimately took its name, must have been on the site of Barton Farm on the opposite side of the river. The earliest dated evidence of episcopal tenure of the castle is in the vacancy of 1215–19, when it was entrusted in turn to Ralph de Normanville and Robert de Cantia, and to Richard (Poore), Bishop of Salisbury.[28] King John stopped there on 12 October 1216 on his last journey.[29]

Edward I visited it in 1292, 1298, 1300, and 1305.[30] From the late 13th century the building was mainly used as a prison and as a place for holding the bishop's courts. John de Lacy, a clerk, was pardoned after breaking out of the prison in 1291, and in 1315 Richard Lambert of Lynn, illegally imprisoned in Wisbech Castle, was so inhumanly gnawed by toads and other venomous vermin that his life was despaired of.[31] In 1350 a commission of oyer and terminer was issued when a prisoner, John de Wilton, was rescued by Christine and John de Wilton, his mother and son, and others; as a result of this the older John was outlawed. He was stated to have assaulted the bishop's bailiffs, threatened the constable and besieged the castle, so that a large garrison had to be maintained and the business of rent collecting and management of the bishop's property was at a standstill.[32] Another commission was issued in 1360 when many evildoers escaped to Walsoken and took sanctuary in the church there, from which they were rescued by other felons.[33]

The duties of the constable included the custody of the castle and its prisoners, the conduct of the courts, and in some cases at least the stewardship of Barton manor.[34] In the latter half of the 14th century the constable was entitled to perquisites as follows: a hall by the gates, chambers on both sides of the hall and over the gates, the easement of a kitchen and stabling for three horses. He had a salary of 20 marks yearly and robes for himself and his deputy, or 40*s.* in lieu, 40,000 turves, 12 cartloads of hay, and straw and a bushel of oats daily for the three horses.[35] The 20 marks had been paid as early as 1298–9, when the robes were valued at 53*s.* 4*d.*[36]

A long series of account rolls of the castle bailiff exists side by side with those of Wisbech Barton manor. The division of duties between the two officials seems to have been one of function, the castle bailiff being responsible for the legal, as the Barton bailiff was for the economic side of the manor. The castle accounts show the profits from manorial perquisites—assized rents (over £100 a year to the end of Henry VI's reign, then reduced to a fixed sum of £68 11*s.* 11*d.*), fisheries (which produced £10 to £15 a year), mills, markets, and fees of court. These last, serving as the court did for the whole of Wisbech hundred, were a valuable source of revenue at about £30 a year on average. The profits of the castle bailiff were considerably greater than on the Barton manor, rarely less than £100 a year and sometimes more than double that amount.[37] In the 15th century the castle fell into ruin, and was rebuilt during the episcopate of Bishop Morton (1479–86).[38]

During the brief reign of Lady Jane Grey, Robert Dudley, later Earl of Leicester, seems to have made Wisbech his headquarters, while attempting to win over the town of Lynn to her cause.[39] On 14[40] and 24[41] July 1553, however, Thomas Carwell, Edmund Beaupre of Outwell, and John Dethick were ordered to take possession of the castle for Queen Mary and make an inventory of its contents. The neighbourhood seems to have been somewhat disturbed at this time, for on 22 July 1553 the Privy Council dispatched a letter to 'Mr. Repps' 'for the staye and appeasynge of the inhabitants of the Ile of Elye'[42] and ten days later Beaupre warned the Council of a proposed assembly of 5,000 persons on Tilney Smeeth (Norf.), who intended to 'take all gentelmen into their rule and custody until redress were had of their wrongs done at the Queen's Majesty's hands'.[43]

During the reign of Elizabeth the castle was used as a prison for Roman Catholic recusants (see below), after which it again fell into disrepair. Bishop Andrewes (1609–19) spent £2,000 on it,[44] and Bishop White (1631–8) a smaller sum,[45] and it was used as a residence by most bishops until the Civil War. In return for £550 Andrewes made an agreement with Matthias Taylor and Rowland Bradford, constables of the castle, to accept a fixed salary of 40 marks a year, without any perquisites in kind. Bradford later claimed the full fees and perquisites, amounting to £94 a year, but got no satisfaction from Bishop White and only £10 a year from Bishop Wren his successor. Bradford alleged that Wren had failed to keep the castle in good repair, thereby enabling four prisoners to escape.[46]

RECUSANTS IN THE CASTLE

The idea of concentrating Roman Catholic prisoners from various jails into one place dates from 1572, when the Privy Council asked the Bishop of Ely (Cox) to report on the suitability of Wisbech Castle for this purpose. It was 'not ment that they shalbe burdenouse to any man, but to lyve upon theyr own charges, with consideracion of all such as shalbe nedefull to attend and guard them'.[47] Eight years later the deprived ecclesiastical Papists were ordered to be sent to castles such as Wisbech and Banbury; the laymen were to be placed

28 *Rot. Litt. Pat.* (Rec. Com.), i, 128; *Cal. Pat.* 1216–25, 122.

29 *Rot. Litt. Pat.* (Rec. Com.), i, Introd. It is possible that the loss of his baggage occurred between Walsoken and Wisbech, where the road from Lynn along the sea bank crossed the Well Stream. The writer is indebted to the late E. J. Rudsdale for this suggestion.

30 P.R.O. Round Room, Itin. of Edw. I (TS).

31 *Cal. Pat.* 1281–92, 434; 1313–17, 318.

32 C 47/50/6/97; *Cal. Pat.* 1348–50, 383–5. It is permissible to connect these sudden outbreaks of violence with the contemporary feud between Bishop Lisle and Lady Wake (*D.N.B.* xi, 1223), and with the depressed state of Wisbech in the middle of the 14th century.

33 *Cal. Pat.* 1358–61, 485; 1361–4, 533.

34 e.g. between 1538 and 1544 Thomas Megges held both offices (C 1/983/44–48).

35 *Cal. Pat.* 1358–61, 43; 1385–9, 460; 1399–1401, 534. The conditions were the same in 1446 (*Ely Dioc. Remembrancer*, Nos. 208, p. 181; 229, p. 131).

36 S.C. 6/1132/10 and /13.

37 Bp.'s Mun., Ely, D 7.

38 Bentham, *Ely*, 181. No trace of this building survives; it is said to have been of brick, like Morton's work at Hatfield and later at Lambeth, and the contemporary rebuilding of Buckden Palace by Bishops Rotherham and Russell.

39 H. J. Hillen, *Hist. Lynn* (1907), i, 279. The writer is indebted to the late E. J. Rudsdale for drawing attention to this episode, and for most of the references.

40 *Acts of P.C.* 1552–4, 302–3.

41 Ibid. 415.

42 Ibid. 301.

43 Hillen, *Hist. Lynn*, i, 280.

44 Gardiner, *Hist. Wisb.* (1898), 13.

45 *Cal. S.P. Dom.* 1638–9, 211. It seems to have been the custom of the bishops at this period to demand £400 from the executors of their predecessors for dilapidations.

46 *Cal. S.P. Dom.* 1635–6, 69; Hist. MSS. Com., *4th Rep. App.* 67.

47 *Acts of P.C.* 1571–5, 73.

under guard in London.[48] In August 1580 the bishop was enjoined to put Wisbech Castle 'in order and strength' to receive Papists[49] and by October the first eight priests, of whom Bishop Watson of Lincoln and Abbot Feckenham of Westminster were the best-known, had arrived.[50] In 1584 it was suggested that the number confined at Wisbech should be limited to twenty.[51] At first the discipline was strict, the recusants being locked in separate rooms except for meals and for half an hour's exercise before dinner and supper. After the Armada, however, at which time there were about thirty-five prisoners,[52] there was a relaxation. Servants were allowed to be kept, visitors to be admitted, friends to send in food. Priests were allowed to go into the town. They paid 12*s.* a month for their rooms. The castle became a kind of ecclesiastical college, with exhaustive theological debates, and a place of pilgrimage for Roman Catholic laity.[53] In 1590 allegations of slackness were made against Thomas Gray, the keeper of the castle, and a commission was sent to investigate.[54] It was recommended that the castle should be strengthened, and certain trustworthy townsmen appointed to assist the keeper, who was not to be absent from the castle without special dispensation. Visitors were only to be allowed by special permission, incoming letters were to be censored, communication between the prisoners was to be stopped except at meals, and they were not to be allowed outside the castle grounds. The diet of the Fleet prison was to be adopted. This was defined as 16*s.* a week for knight's commons and 10*s.* for gentleman's, with 6*s.* 8*d.* and 2*s.* 4*d.* respectively for their lodgings.[55]

During this second period of stricter discipline quarrels between Jesuits and seminary priests confined in the castle broke out. The leader of the former party was William Weston, a fanatical and ascetic Jesuit who arrived at Wisbech in 1587, of the latter Christopher Bagshaw, a comparatively recent (1582) convert to Roman Catholicism. Weston began to be active about 1593, he and his followers supporting the stricter discipline so as to make their imprisonment as conducive to sanctity as possible. As the more zealous party they received most of the alms of the faithful; they were suspected by Bagshaw and his followers as traitors, and the latter in return were accused of loose living and immorality.[56] Matters first came to a head with the introduction of a hobby horse for the Christmas festivities of 1594, and the following year a commission was appointed 'to take verie particular notice of all the prisoners there, of what condicion, qualitie and estate they are and how manye of them stand condemned'. It was considered that too many servants and outside persons had access to the prisoners, and it was proposed to send the latter overseas in due course.[57] Various persons who had had dealings with the prisoners were ordered to appear in the Star Chamber.[58] In 1597 the custody of the prisoners was entrusted to William Medley and William Brewster, who were ordered 'to see the prisoners more strightly looked unto than they have bin of late'.[59] In the same year some form of epidemic broke out, some of the prisoners being temporarily moved to Sir John Higham's house at Barrow (Suff.), and others allowed 'to walk abroad to take the ayre at suche tymes and so far as you [Higham] shall thyncke convenyent'.[60] Towards the end of 1598 various priests, including Weston, were moved from Wisbech to the Tower of London, and others who 'have conformed themselves and go orderly to church' were released to friends 'known to be well affected'.[61] In 1602 thirty-six priests were transferred to Framlingham (Suff.), guarded and manacled,[62] but some remained at Wisbech four or five years later. William Chester, constable of the castle, received £75 12*s.* for their diet in November 1607.[63] In February 1606 Bishop Heton reported that a priest, George Smith, had been converted and was 'an excellent musician, especially on the organs'.[64] In 1615 twenty-four priests were brought to Wisbech from London, and the regulations of 1590 for safe custody brought into force again.[65] Three of this fresh batch escaped, and the following year the hundred of Wisbech as well as the town was called on to provide trustworthy assistants to the keeper.[66] Nevertheless a further escape of five prisoners took place.[67] In 1618 nine of the priests were deported,[68] and eight years later the keeper of the castle was summoned before the Council to give a full report.[69] This marks the end of the use of Wisbech Castle for this purpose, though the idea of removing Jesuit and other priests from London prisons to stricter confinement at Wisbech was still being mooted in 1628.[70]

WISBECH AND THE CIVIL WAR

The ship money controversy affected Wisbech to the extent that two journeys were made by members of the Corporation 'concerning the King's Majesties Shippinge' and 'about the Towne business to get abatement for the shipp-money', though the results are not recorded.[71] The town as a whole sided with the Parliament, though there were a few Royalists; after the war John Fisher was fined £80 for having been in arms against the Parliament, and Edward Buckworth and Thomas Wragg pardoned.[72] Bishop Wren as a staunch Royalist was unpopular in the town, and in 1640 was dissuaded from attending a Commission of Sewers in the castle. A crowd of soldiers however clamoured for him, crying: 'give us Wren, that damned Bishopp Ely, wee will have him or else fyer the town and cutt the throats of all the Commissioners'. On his absence being discovered the shops of some of his supporters were

48 *Cal. S.P. Dom.* 1580–1625, 8.

49 *Acts of P.C.* 1580–1, 142. George Carleton, the constable, had to cut down his retinue to make room (ibid. 157–8).

50 Ibid. 1547–80, 681.

51 *Cal. S.P. Dom.* 1581–90, 157.

52 £12 was paid to George Cobham, one of the Queen's messengers, and to Robert Awfeild, for conveying twelve priests from London to Wisbech in March 1588, and six more were admitted in October (*Acts of P.C.* 1588, 4, 317).

53 T. G. Law, *Jesuits and Seculars in the Reign of Eliz.* (1889), pp. xl–xlii.

54 *Acts of P.C.* 1590, 170, 410, 420–2. Gray's daughter Ursula was said to have been perverted to Roman Catholicism (Law, *Jesuits and Seculars*, p. xlii).

55 *Acts of P.C.* 1592, 392–8.

56 For a full narrative see Law, *Jesuits and Seculars*, which prints many original documents.

57 *Acts of P.C.* 1595–6, 154–5, 174–5.

58 Ibid. 252, 418–20.

59 Ibid. 1597, 335–6. Several escapes occurred about this time; see e.g. *Cal. S.P. Dom.* 1595–7, 194; 1598–1601, 456. *Acts of P.C.* 1599–1600, 200, 462. Medley and Brewster later fell out over the profits of keeping the prisoners (*Acts of P.C.* 1597–8, 299).

60 *Acts of P.C.* 1597–8, 6, 115.

61 Ibid. 1598–9, 373–4, 381; 1599–1600, 201; *D.N.B.* xx, 1290.

62 *Cal. S.P. Dom.* 1601–3, 167–8.

63 Ibid. 1603–10, 382.

64 Ibid. 285.

65 *Acts of P.C.* 1615–16, 108–17.

66 Ibid. 205–6, 383.

67 *Acts of P.C.* 1616–17, 73–74.

68 Ibid. 1618–19, 202–3.

69 Ibid. June–Dec. 1626, 264–5.

70 *Cal. S.P. Dom.* 1628–9, 53.

71 Corp. Rec. iv, 79*a*, 89*a* (12 Dec. 1634, 31 Oct. 1638).

72 W. Watson, *Hist. Wisb.* (1827), 223. For an account of Buckworth's case see *Cal. S.P. Dom.* 1645–7, 37–38.

plundered.[73] During 1643 the castle and town were put into a state of defence. £11 12*s.* 6*d.* was spent on ironwork for the castle drawbridge. £7 17*s.* 6*d.* was spent on other fortifications, including an outpost at the Horseshoe, the object of which was to prevent the Royalists of South Lincolnshire from penetrating into the Eastern Counties.[74] The Parliamentary forces garrisoned locally took part in the siege of Crowland (1643), for which the capital burgesses granted them two barrels of beer.[75] The following year a mutiny occurred, for the suppression of which Captain Richard Le Hunt received a gratuity of £5.[76] In April 1643 Captain Dodson borrowed £100 from the burgesses to raise the Crowland force, which he commanded. This sum was still outstanding more than three years later and apparently was never repaid.[77] The effect of this and other loans is shown in 1656, when the burgesses presented a petition to the House of Lords. The town had raised a troop of horse for the Parliamentary army,[78] and it was claimed that Cambridgeshire had been overtaxed in comparison with neighbouring counties. The burgesses prayed to be exempted from arrears, especially since in the previous winter 14,000 acres in Wisbech, Elm, Outwell, and Upwell had been drowned by floods.[79] In answer to this petition the town received £170 from delinquents' estates in return for moneys lent during the war.[80]

The Restoration festivities cost £3, including £1 to the waits of Lynn, and 40*s.* was distributed in bread to the poor at the coronation of Charles II.[81]

LATER HISTORY OF THE CASTLE

Secretary John Thurloe, who purchased the manor during the Interregnum, replaced Morton's palace by a house in the style of the time, said to have been designed by Inigo Jones, but almost certainly by his pupil John Webb. It had a resemblance to Thorpe Hall, Peterborough, which is certainly the work of the latter architect.[82] At the Restoration the property reverted to the bishops. Most of them, however, spent little, if any, time at Wisbech, and the practice arose of leasing the castle to local families. Henry Pierson was the first post-Restoration tenant; later the Southwell family were lessees for more than a century. In 1778–9 Edward Southwell was paying £30 a year rent.[83] In 1793 statutory powers were acquired[84] by Bishop Yorke to sell the castle and grounds. The Bill passed despite the opposition of Sir James Eyre, Chief Justice of the Common Pleas,[85] and the premises were sold by auction in the same year to Joseph Medworth of Bermondsey[86] for £2,305.[87] In 1811 Medworth offered the house to the capital burgesses for the Grammar School, but they had hopes of obtaining possession after Medworth's death at a reduced price and the offer fell through.[88] Medworth, who had been developing the 5-acre grounds as a residential estate since 1793,[89] pulled down Thurloe's house and erected the present 'Wisbech Castle' in 1816.

GUILDS AND CHANTRIES

In 1389 there were six guilds in Wisbech. St. Peter's, dating from 1327, was the oldest, the remainder, with the possible exception of the Guild of Holy Trinity, having been founded during the reign of Richard II. Some of them were quite small. The Guild of St. John Baptist, however, though founded only in 1384, must have already collected a considerable membership. Its expenses during the five years of its existence amounted to £80 6*s.*[90] All these guilds (Holy Trinity, St. Peter, St. John Baptist, Our Lady, St. Thomas, Corpus Christi) were still in existence in 1462, by which date guilds of St. George, St. Lawrence, and Holy Cross had also been founded. In this year the Holy Trinity Guild allowed its hall to be hired by the others for payments ranging from 12*d.* (St. Peter, St. Thomas, Our Lady) to 4*d.* (Corpus Christi, Holy Cross).[91] The guilds of St. Mary, St. Peter, and St. John Baptist are mentioned in wills of 1495.[92]

St. Martin's Chantry lay on the North Brink adjoining White Hall.[93] The rental made at the time of its suppression shows it as owning 2 messuages, 19 acres of arable, 32 of pasture and 9 of reed ground, valued at £7 4*s.* 8*d.* Amongst its tenants was Robert Pygot, one of the Wisbech martyrs. Its goods and chattels were worth £3 6*s.*, the most precious item being a parcel-gilt chalice valued at £1 6*s.*[94] Two cottages and yards (*ortos*), which had been occupied by the chantry priest, were in 1551 granted to William Place and Nicholas Spakeman, haberdasher of London.[95] The rest of the chantry lands were leased between 1558 and 1565 to Edmund Downing, who in the latter year assigned his lease to Richard Everard of Wisbech.[96] A twenty-one-year lease was granted to Richard Buckworthe in 1582,[97] and a forty-year lease to Lancelot Johnson in 1606, the reversion being granted six years later to Francis Morrice and Francis Phellipps.[98]

73 Hist. MSS. Com., *12th Rep. App.* iv, 522.
74 *Cal. S.P. Dom.* 1644–5, 325–6; *V.C.H. Cambs.* ii, 407; *Brit. Arch. Assoc. Proc.* xxxv, 65; Corp. Rec. iv, 106*b*, 112*b*.
75 Corp. Rec. iv, 107*a*.
76 Ibid. 107*b*.
77 Ibid. 103*b*, 115*b*.
78 £250 was borrowed at 8 per cent. for its upkeep in 1648 (Corp. Rec. iv, 119*a*). This troop was still in existence in 1690, when £4 was voted for the expenses of a horseman to serve therein (ibid. 246*b*).
79 Walker and Craddock, *Hist. Wisb.* (1849), 223, quoting L. J.
80 *Cal. S.P. Dom.* 1655–6, 120.
81 Corp. Rec. iv, 142*b*, 145*a*.
82 A model of Thurloe's house is preserved in the Museum. For Webb and Thorpe Hall see *V.C.H. Northants.* ii, 458.
83 Bp.'s Mun., Ely, D 6.
84 By 33–34 Geo. III c. liii.
85 In a letter (written by Dr. John Warren, Bishop of Bangor) preserved at Ely (Bp.'s Mun., D 9), Eyre considered that it would be improper for so rich a see to alienate property, and imputed dishonesty to Bishop Yorke as the castle was said to be in bad repair. Warren and Eyre were sons-in-law of Henry Southwell of Bank House, Wisbech, a member of the family tenanting the castle, and to that extent interested parties.
86 Formerly of Wisbech. See p. 246, note 24.
87 *Fenland N. & Q.* iv, 362 (containing a site plan).
88 Gardiner, *Hist. Wisb.* (1898), 234–6; *V.C.H. Cambs.* ii, 328–9. The house was in good repair in 1791, the chimney-stacks and cornices having been repointed five years earlier, but there were holes in the stable roof and the gardens were neglected (MS. in Wisb. Mus.). The preamble to the Act authorizing the sale states that the castle had for many years been in a ruinous state.
89 York (*recte* Yorke) Place commemorates the bishop under whom the sale took place. A proposal to prolong Market Street (laid out by Medworth) on the other side of the Market Place, to form a new approach to Lynn Road, was not carried out. This proposal was still under debate in 1824, when it was suggested that the Grammar School should purchase Dr. Whitsed's house (formerly belonging to Dr. Fraser) and give up part of its premises in Hill Street to allow the new street to be made (Jackson Diary, 30 July 1824) (Wisb. Mus.).
90 W. M. Palmer, *Village Gilds of Cambs.* 377–9 (printing the gild certificates).
91 Corp. Rec. i, 30.
92 *Fenland N. & Q.* ii, 195–6.
93 Infm. Mr. G. M. G. Woodgate.
94 S.C. 12/6/22.
95 *Cal. Pat.* 1549–51, 410, 414.
96 B.M. Harl. Chart. 77 F 53.
97 L.R. 14/753.
98 C 66/1710 m. 5; C 66/1964 m. 6.

GUILD OF THE HOLY TRINITY

The Guild of the Holy Trinity was much the most important of these institutions. It became indeed the main feature of Wisbech life in the later Middle Ages and was the direct ancestor of the Corporation. It is reputed to have originated in 1379, the year of the earliest surviving account.[99] Certain entries in the account, however, make it plain that the Guild had already existed for some time before.

The guild was incorporated in 1453 at the petition of its alderman, chaplain, four brethren and four sisters. The guildsmen were then licensed to found a chantry of chaplains to celebrate at the altar of Holy Trinity in Wisbech church and to grant to the chaplains lands in mortmain to the value of £40.[1]

In 1397 there were 67 guildsmen, paying 1*s.* 8*d.* each, and 14 novices, paying 3*s.* 8*d.* each.[2] The novices' subscriptions no doubt comprised entrance fees; in the next century (1489) there is one instance of the payment of such fees by instalments.[3] The total number of brethren is not entered regularly, but the membership, of men and women, seems to have remained fairly constant. Thus in 1453 33 'couples' each paid 18*d.*, and in 1460 28 paid 16*d.*[4]

The officers of the guild, elected annually by a body of from 12 to 18 jurors or inquisitors (their title and numbers vary), were at first 6 in number. At the head of the guild was the alderman, empowered to prosecute for debts, to make leases of guild property with the consent of 12 of the 'most honest' members, and chief of the party in the church services on the annual feast day and at memorial services for dead members.[5] He had one of the three keys of the guild's chest, the other two being in possession of the chamberlains (*hostiliarii*), who preceded the alderman to and from church 'with their garlandds and typpe-stavyes in their handds'.[6] After the appointment of a bailiff they audited his accounts. The dean was responsible for warning the brethren of the death of one of their number, so that they could make offering for his soul, or in default 8 lb. of wax.[7] He led prayers at the guild services and was responsible for the four tapers at the Trinity altar on the feast day; he was entitled to demand wax of newly admitted members for this purpose.[8] The clerk, though presumably an official from the time of the first recorded proceedings, is not mentioned by name until 1456, when he was ordered to repair the hall.[9] In 1513 he was required to translate the statutes into English, and in 1531 to read them out at the dinner on Trinity Sunday. He was also required to write up the bailiff's accounts, under penalty of 20*s.*[10] There were also two wardens (*scabini*), who received legacies, paid out benefits and bought the wax for the altar lights. In 1519 the 'skyven' had to provide a 'competent' breakfast for the alderman and brethren on the guild day, 'every one to have drynke if they wyll', and in 1525 to keep the revel and bring the money therefrom to the alderman the following Sunday.[11]

The number of officers increased with the importance of the guild. By 1472 there was a full-time priest, whose salary was £5 6*s.* 8*d.*[12] In 1514 it was ordained that five masses were to be said for each departed brother or sister. Three years later the guild employed at least three priests whose respective duties were set out in detail. The priests were expected to help in the ordinary services of the parish church, and were not to be absent from the town at night without permission from the alderman. A special morrow mass priest is mentioned in 1524.[13] In 1549 the priests or chaplains numbered four, one being schoolmaster.[14] Besides the fully ordained priests there was a chaplain, whose office in 1459 and 1461 was combined with that of sacrist and custodian of the altar; he was probably in minor orders only. In the latter year he was required to say vespers at the guild altar after the principal vespers in the quire of the church.[15] In 1469 he assisted the priest in praying for the souls of the departed, and received a salary of 6*s.* 4*d.*, probably in addition to the 20*d.* he received from each member of the guild.[16] In 1497 it was ordained that eight chaplains should celebrate on certain days at the recently erected altar of the Holy Virgin.[17] A chantry was founded by the Holy Trinity Guild on the north side of the church in 1453, and endowed with land to the value of £40 a year for the support of its staff.

A bailiff is first mentioned in 1463, his duty being to manage the guild's real estate.[18] In 1470 he had a salary of 20*s.*, altered in 1531 to 12*d.* 'on every charge that he schall doo'. In 1472 he was allowed to call in the alderman to help in distraining upon the property of debtors.[19] In 1483 £10 14*s.* 4*d.* was set aside out of issues and fines from the guild estates to pay the priests' salaries.[20] In 1524 it was laid down that the bailiff was not to make leases of more than one year's duration without the consent of the alderman and chamberlains.[21]

The hall of the guild is first mentioned in 1423,[22] but it was probably in existence before then, as it was repaired in that year. Its site is not certainly known.[23] By 1453 there were two stewards (*senescalli*) of the hall, probably identical with the marshals of the hall mentioned seven years later.[24] In 1506 the steward and server was to entertain 'all the pore people then ther present' to what was left over from the Trinity Sunday banquet. In 1515 it was enacted that no member of the guild in good health could have his portion of the banquet sent to his house, but he could send a deputy to the feast if unavoidably absent.[25]

At its dissolution the guild's officers comprised alderman, bailiff, two chamberlains, a steward and a server in the hall, two cupbearers, two almoners, a 'skyvener', a clerk, and four priests or chaplains.[26]

[99] The guild records are very incomplete.
[1] *Cal. Pat.* 1452–61, 140.
[2] Corp. Rec. i, 1. The entrance fee was sometimes 3*s.* 4*d.* as in 1379 and in 1502 (3 admissions), but more usually 6*s.* 8*d.* as with 7 other admissions in 1502 and 4 in 1460 (Corp. Rec. i, 27, 59).
[3] Corp. Rec. i, 54.
[4] Ibid. 24, 27.
[5] Walker and Craddock, *Hist. Wisb.* 282.
[6] Ibid.
[7] Corp. Rec. i, 27.
[8] Ibid. 35, 42, 43.
[9] Ibid. i. 25.
[10] Ibid. i, 70, 89.
[11] Ibid. 80, 86.
[12] Ibid. 37. £4 6*s.* 8*d.* had been paid to a celebrant in 1379.
[13] Ibid. 71, 78, 85.
[14] Ibid. 106.
[15] Corp. Rec. i, 26, 28.
[16] Ibid. 22.
[17] Ibid. 34, 58.
[18] Ibid. 31, 33, 89.
[19] Ibid. 35, 37.
[20] Ibid. 48.
[21] Ibid. 71. The chamberlain was at the same time ordered to make a personal inspection of the estates every year.
[22] Ibid. 2.
[23] It may have stood in Hill Street (formerly Ship Lane), where was afterwards the first Town Hall and Grammar School; this building still exists. The late E. J. Rudsdale pointed out that underneath Oldham's chemist's shop, at the north-west corner of the Market Place next the New Inn, is a fine basement with a groined vault in three bays, dating from *c.* 1450. It must have served as the cellar of an important building, perhaps the Trinity Guild Hall.
[24] Corp. Rec. i, 24, 27.
[25] Ibid. 63, 72.
[26] Ibid. 103.

The annual festival began on the Friday before Trinity Sunday with a 'mornspeche' in the guildhall before 7 o'clock, after which the bailiff submitted his accounts and the ordinances of the guild were read out.[27] The first dinner on record, that of 1379, cost £1 8*s.* 8½*d.* in food and drink. On that occasion there was also paid 10*s.* to five minstrels, 5*s.* 8*d.* for apparel for ten dancers and 6*s.* 8*d.* to the brethren at Lynn for their care and labour.[28] The expenses were £2 5*s.* 3*d.* and £2 3*s.* 8*d.* in 1464 and 1465 respectively.[29] About this time it was laid down that the meal should consist of potage and a haunch of venison for the first course, a stroke of veal 'and no more' for the second, with pullets and chitterlings for supper, and 7 bushels of frumenty and 6 dozen of good beer.[30]

The guild possessed an image of the Holy Trinity at least as early as 1379, when £1 6*s.* 8*d.* was levied from members for its repair.[31] The first list of the guild's jewellery and plate is undated and unvalued;[32] it comprised a silver-gilt crown, a stag with a gilt head, a gold figure of Jesus, two *Agnus Dei* in silver, four crystals, a beryl and a branch of coral, and various crucifixes, rings, brooches, and beads. At the dissolution in 1549 the guild owned 192 oz. of plate, valued (at 4*s.* 8*d.* an ounce) at £44 16*s.* This included a silver-gilt cross, a pyx (silver parcel-gilt), a pair of silver saucers, a chalice (bought in 1460 for £3 12*s.* 4*d.*) and a pipe of silver for the cross-staff.[33]

The estates of the guild are important, as they were taken over by the Corporation and remained more or less intact until the early 19th century.[34] In 1524 they amounted to 187 acres, let for £17 5*s.* a year. The most valuable block was 40 acres in Meadowgate Lane, Emneth, for which £3 a year was received. This was still Corporation property in Watson's time.[35] The total annual value of the guild's estates in 1549 was put at £28 3*s.* 2½*d.*,[36] which is almost certainly an underestimate. A survey of about the same date shows 4 messuages and 466 acres in Wisbech and 2 messuages and 281 acres elsewhere, producing in all £55 3*s.* 8*d.*[37] —a rental comparable with that of some of the most valuable manors in the Isle, such as Doddington, Sutton, and Whittlesey St. Andrew. By 1614, when a schedule of Corporation property was prepared, some fresh purchases had been made and apparently also some old property relinquished. There were then 3 tenements and 341 acres in Wisbech and 2 and 277½ elsewhere.[38]

The guild was perhaps unpopular and its dissolution predictable as early as 1535, when John Zeugge, priest, and Robert Martyn were admitted members, provided they had no 'approbrious words' with the alderman for the time being and did not try to introduce 'any person spekyng ageyn us'.[39] The answers to the articles of the Commissioners for the dissolution show the guild in a favourable light; it supported a grammar school and many charitable objects in the town.[40] The chapel at Murrow is referred to at some length as one of these, but its connexion with the guild is not obvious.

SEALS OF THE GUILD AND THE CORPORATION

A 15th-century seal of the Holy Trinity Guild exists.[41] The Trinity is treated in an unusual manner, with the three Persons standing half length holding up the hand in benediction, on a castle with water underneath.

The common seal adopted on the incorporation of the town[42] shows full-length figures of St. Peter and St. Paul, the patrons of the parish church, under canopies in the Renaissance style. This design, with more Gothic canopies, has been adopted for the arms granted to the borough in 1929, which have a crest showing a ship with St. Peter's Keys on one of the sails.[43] This badge (keys) is used on the seals of the mayor and town clerk.[44]

CHARTERS OF INCORPORATION

The charter of 1549,[45] making Wisbech a corporate borough for the first time, cost £260 10*s.* 10*d.*[46] The burgesses were given the right of perpetual succession and the use of a common seal.[47] They were to hold the guild property in socage, and were licensed to acquire further property up to £100 in value, and to lease (up to twenty years), grant, sell and exchange their lands. Householders were to meet in the common hall annually from 1 November 1550 to choose ten of the better men among them to manage the property, who were allowed, but not enjoined, to take advice from their fellows. A schoolmaster was to be provided at a salary of £12 8*s.*, and an annual distribution of £3 15*s.* was to be made to the poor. The bishop was made Visitor of the Corporation. The sea shores, banks and streams were to be maintained by the Corporation, as they had been by the guardians and brethren of the guild. The last provision had important consequences. It enabled the Corporation to keep control of the port, in connexion with which they adopted, particularly in the late 18th and early 19th centuries, many well-intentioned but not always beneficial measures. The port of Wisbech still remains a Corporation enterprise, and has not, as is commonly the case, been entrusted to an *ad hoc* body.

BOROUGH OF WISBECH. *Azure representations of St. Peter and St. Paul or standing within a double canopy or.* (Granted 1929.)

No mention is made of any mayor or chief burgess in the charter. At the first recorded election of the 'ten

[27] Corp. Rec. i, 33. The bailiff also had to prepare an inventory of the jewels and ornaments of the guild, under the supervision of the alderman, to be rendered by Lammas each year.

[28] Ibid. 1.

[29] Ibid. 32, 33.

[30] Ibid. 27, 31.

[31] Ibid. 1. The repairs actually cost £1 5*s.* 3*d.*

[32] Ibid. 2.

[33] Ibid. 109–10. The proceeds helped to pay for the equipment of thirty-seven men for the suppression of Ket's rebellion (£28 9*s.* 8*d.*), various drainage improvements (£20 5*s.* 4*d.*), repairs to the church (£4 5*s.*), to the market cross (£4 3*s.* 4*d.*), and the paving of the Market Place (£17 2*s.* 2*d.*).

[34] Walker and Craddock, *Hist. Wisb.* 341–3.

[35] Watson, *Hist. Wisb.* 159.

[36] *Cal. Pat.* 1548–9, 339.

[37] S.C. 12/6/23. The 40 acres at Emneth were then let for £4 yearly.

[38] Copied into the back of Corp. Rec. v. The rentals are not stated for every holding.

[39] Corp. Rec. i, 94.

[40] Ibid. 106–8.

[41] B.M. *Cat. Seals*, i, 845.

[42] B.M. *Cat. Seals*, ii, 213; Lewis, *Topog. Dict.* iv, 523.

[43] C. W. Scott-Giles, *Civic Heraldry of Engl. and Wales*, 278–9.

[44] B.M. *Cat. Seals*, ii, 214.

[45] *Cal. Pat.* 1548–9, 339–40. The full text is printed in Watson, *Hist. Wisb.* (1827), 174–87.

[46] Rather more than nine times the official valuation of the guild property made over to the Corporation. But see above.

[47] See note 45.

men' (1564) Richard Best was appointed 'baley for the receipt of all the lands and tenements . . . for this yere to come',[48] but the office of town bailiff, as it came to be called, was never legally recognized.

In 1610, apparently as a result of a suit against the Crown regarding feudal dues on newly purchased estates[49], application was made for a renewal of the charter. The chief amendments asked for were authority to purchase lands and tenements up to £3 or to £5 annually, and the restriction of the franchise to £10 freeholders.[50] Neither of these amendments was included in the charter granted the following year,[51] which retained the £100 limit on the Corporation estates and restricted the franchise to 40*s.* freeholders only. The date of the election of the governing 'ten men', now styled capital burgesses, was changed from 1 to 2 November; if no meeting of the freeholders took place the outgoing capital burgesses were empowered to choose their successors.[52] The other clauses of the charter repeat those of 1549 in greater detail.[53]

After the Restoration there was a movement amongst the capital burgesses to get the charter amended so as to allow them a life tenure. Four of them, claiming to be the 'major part', i.e. presumably representing public opinion in the town, signed a resolution of protest.[54] These four were not re-elected in the following November (1668), but application was sought and granted for the renewal of the charter a second time, after a 'cavett and petition put in to stope our proceedings'.[55] The charter as renewed[56] does not differ in any material respect from that of 1611. The schoolmaster was to be appointed by the ten capital burgesses with the assent of ten other burgesses.

Wisbech thus enjoyed a reasonably democratic constitution before 1835, and there is little evidence of corruption or negligence in the Corporation's management of the town's affairs. In 1752 John Garland, the retiring town bailiff, refused to hand over his books, causing a long and expensive Chancery suit,[57] at the end of which (1756) he admitted his liabilities and agreed to hand over all deeds and writings in his possession.[58] In 1778 Richard Colvile as town bailiff induced several tradesmen to accept promissory notes from him in discharge of the sums owed them by the Corporation. Colvile's subsequent bankruptcy caused considerable distress, and a resolution of 21 October 1778 asserts 'that this mode is disgraceful, and very detrimental to the credit of the Corporation, and ought not to be followed by any of our Town Bailiffs for the time to come'.[59]

Some trouble was occasionally caused by the operation of the Test and Corporation Acts. Thus in 1819, James Hill,[60] an uncompromising Unitarian, obtained nomination to stand for capital burgess, and was rejected on show of hands by 12 to 5. He demanded a poll and got in, but refused to answer whether he had received the sacrament according to law. Thereupon the Revd. Jeremiah Jackson, Headmaster of the Grammar School and later vicar of Elm, who was eleventh on the list, demanded Hill's seat and was declared elected. Hill made similar fruitless attempts to become a capital burgess in 1820 and 1821.[61] In the later 18th century, however, the Quakers Jonathan Peckover (1758) and Joseph Hancock (1773) were elected to the Corporation; the latter was town bailiff in 1774.

More generally, the Corporation took seriously its responsibilities for poor relief, the paving and lighting of the streets, and the improvement of the port. During the 18th century, however, despite its ample endowments in land, it accumulated a considerable debt on account of these services. Matters came to a head in 1774–5, when an investigating committee reported that the bonded debt, which had stood at £1,750 in 1760, had been reduced in fifteen years by only £338 7*s.* 10*d.*, while £846 9*s.* had been paid in interest. It was recommended that £1,000 be raised by annuities on lives at £10 per cent. for ages of 60 and £11 per cent. for 65 years and upward.[62] Insolvency was staved off for another fifty years, during which time the building of the custom house, corn exchange, and cattle market severely taxed the resources of the Corporation. In 1822–3 it was reported[63] that the annual revenue was £1,919 6*s.* 9*d.* Of this £1,481 14*s.* 9*d.* came from 754 acres of landed property, and the remainder from profits of the port, fairs, and markets. Expenditure amounted to £2,399 1*s.* 11½*d.*, and bonded debt to £1,980. To redress the balance 'wise frugality' was proposed, together with a reduction in the number of watchmen and scavengers, and the elimination of over 200 of the 237 street lamps then in use.[64] But even in 1832 income stood at £1,779 18*s.* and expenditure at £1,954 14*s.* 8*d.*[65]

During the latter part of its unreformed existence the Corporation became obsessed with the idea[66] that any improvement of the drainage of the neighbourhood would react unfavourably on the port and Nene Outfall. They opposed without success several schemes of improvement—in Sutton Marsh (1719) and in Thorney Fen and the Bedford level generally (1749 and 1775–6).[67] Their efforts in 1720 and 1723 to preserve navigational facilities by a statutory prohibition upon any further embanking along the Nene

[48] Corp. Rec. i, 121. The title 'bailiff' and the fact that he received a gratuity suggest that he was regarded as the agent of the Corporation rather than its master. Wisbech was a manorial borough; the Bishop of Ely was Visitor of the Corporation as well as lord of the manor. On the other hand the town bailiff undoubtedly presided over meetings of the 'ten men' and was invested with executive powers on their behalf. In this sense his position was analogous to that of the chairman of an elected body.

[49] Walker and Craddock, *Hist. Wisb.* 318; Corp. Rec. iii, 53*b*. These estates were at Walpole, where the manor was in Crown hands owing to a vacancy in the see of Ely. It seems probable that the monetary equivalent of military service was demanded contrary to the terms of the charter, but the lawsuit has not been traced.

[50] Corp. Rec. iii, 91*b*.

[51] C 66/1877, No. 10.

[52] This occurred several times in the 17th century, e.g. in 1621 and 1623–6 (Corp. Rec. iv, 27*b* et seq.).

[53] Corp. Rec. iii, 94*b*.

[54] Ibid. iv, 163*a*.

[55] Ibid. 165*b*.

[56] C 66/3108, m. 1–7.

[57] C 11/2153/31 (Mitford).

[58] Walker and Craddock, *Hist. Wisb.* 332; Corp. Rec. vi, 10.

[59] Corp. Rec. vi.

[60] Father of Octavia Hill.

[61] Walker and Craddock, *Hist. Wisb.* 335. Jackson in his Diary (2 Nov. 1819) states that in that year Hill was 11th on the poll, Jackson himself being 9th and a certain John Johnson 10th. The incident referred to was in 1820. In the following year some windows in the Grammar School were broken after Hill's failure to get elected.

[62] Walker and Craddock, *Hist. Wisb.* 333–4; Corp. Rec. vi.

[63] The report was made by Jeremiah Jackson, and adopted by the capital burgesses on 13 Oct. 1823 (Jackson Diary, Wisb. Mus.).

[64] Corp. Rec. xi, 9–15.

[65] *Rep. Com. Municipal Corpns.*, H.C. 116, p. 2555 (1835), xxvi.

[66] Not confined to Wisbech. See H. C. Darby, *Draining of the Fens* (1940), 101–3.

[67] Corp. Rec. v, also vi (26 Dec. 1749, 29 Dec. 1775, 1 Jan. 1776); Walker and Craddock, *Hist. Wisb.* 332.

Outfall were also unsuccessful.[68] On the other hand, they combined with the local landowners in erecting an engine to drain Wisbech South Field.[69]

The Corporation can also be credited with a prompt and patriotic reaction to the wars with revolutionary France. On 26 November 1793 it was resolved to convene a public meeting 'to take into consideration the propriety of opening a subscription for supplying His Majesty's Forces serving in Flanders with comfortable articles of extra clothing'.[70] Next year £50 was voted towards 'the internal defence of the kingdom'.[71] In the famine year 1795 the capital burgesses combined with the 'respectable inhabitants' to buy up a supply of wheat against the coming scarcity (in July) and gave £105 (September) to a fund for reducing the price of bread to the poor.[72] A company of Volunteers, about 60 strong, was raised from 1797 to 1801, and was revived in 1807 when Wisbech, with Whittlesey and Thorney, produced 300, and later 600 men.[73] The peace celebrations of 1814 were on a grand scale. 'The *coup d'oeil* of the Market Place presented a most beautiful spectacle', and about 3,600 (two-thirds of the population of the town) sat down to dinner. 'It was a celebration which will long be remembered by the inhabitants and neighbourhood with gratitude and delight. The sports, though but indifferently managed from the unmanageableness of the spectators, served at least to prevent gross intoxication.'[74] The news of Waterloo was received in Wisbech on 23 June 1815.[75]

An Act of Parliament of 1810[76] vested in the Corporation power to improve the markets and the port and to pave, cleanse, light, and watch the streets. The chief consequences of this Act were the rebuilding of the Shire Hall, the building of the Town Hall, improvements in the markets and the widening of what is now Union Street. Minor provisions of some interest were the enforcement of brick or stone party walls of not less than 9 in. thickness for new buildings (cl. 31),[77] the prohibition of gates and doors opening outwards upon the street (cl. 34), and of street 'parking' for excessive lengths of time (cl. 35). The financing of the improvements was to be carried out by an annual rate not exceeding 1*s.* 6*d.*, and the borrowing of not more than £3,000. The paving was to be undertaken by special Commissioners, the remainder of the improvements under the direction of the capital burgesses.

The Commissioners of 1834 reported very favourably on the Corporation,[78] which, as has been shown, was a reasonably public spirited and democratically elected body. The Municipal Corporations Act of 1835, therefore, caused less of an upheaval in Wisbech than in many towns. As the burgesses stated in an address of gratitude to King William IV for the renewal of the charter, 'we, as a Town Corporate, do not derive from the measure advantages to the same extent [as other boroughs] in consequence of a peculiarity in our old Charter, which had already embrased to a considerable extent, the principle of popular controul recognized by the new law'.[79] The number of freeholders who voted in the later years of the unreformed Corporation, e.g. 166 in 1815, 172 in 1818, 160 in 1821, 129 in 1822,[80] compared favourably with the number voting at parliamentary elections,[81] and was substantial for a town with about 230 entitled to vote, and a population of some 6,000. The principal changes introduced by the 1835 Act were as follows. (1) The practice of taking fines on the granting of leases of Corporation property was prohibited. (2) The administration of the charities was transferred to trustees.[82] (3) The constitution was normalized by the transformation of the town bailiff into a mayor, and the substitution of aldermen and councillors for capital burgesses. (4) The Corporation was empowered to appoint a Coroner and a Recorder, and to hold a separate commission of the peace. (5) All clergy, both Anglican and Nonconformist, were disqualified from membership of the Corporation.[83] (6) The publication of abstracts of accounts, which had been done voluntarily since 1818, was made compulsory.[84] The franchise was given to occupiers of houses and shops rated for three years to the relief of the poor, if residing within 7 miles of Wisbech.

The 'establishment' of mayor, 6 aldermen and 18 councillors continued until 1933, when on the absorption of New Walsoken and the rearrangement of the borough into three wards, it was increased to 8 aldermen and 24 councillors. In 1940 the borough was reorganized in four wards.[85]

THE CORPORATION AS A LAW COURT

In its early days the Corporation was mainly concerned with the settlement of small disputes between the burgesses. Such powers of arbitration had not been provided for in the original charter, and no authority had been given to the 'ten men' to enforce their decisions,[86] which seem to have been accepted without question. For example, it was ordered in 1574 that a suit regarding the recovery of goods held in pawn was to be settled by the defendant giving the plaintiff a 'ryssyt gowne to make his wyff a cassok' and 30*s.*[87] Other typical disputes dealt with (1569–72) concerned the 'worowing of sertayne shepe', an action of trespass for the 'estray of a mare into a stack of hay', the recovery of small debts, or the hiring of a house in the New Market.[88] The occasional absence of this type of business at meetings of the 'ten' was specially noted, e.g. in November 1568 'there was no matter or cause of controversy dependyng betwene any person or persons of our neighbours . . . thanks be gyven to Almyghty God for the same'.[89]

CONSTITUTIONAL DEVELOPMENT OF THE CORPORATION

At the same meeting in 1568 it was laid down that the 'ten men' were to meet on the first Tuesday in every month, and once a year were to dine together 'at such place as shalbe agreed upon by consent'. The cost of the dinner was

68 Corp. Rec. v.

69 Ibid. iv, 145*b*.

70 Ibid. viii.

71 Ibid.

72 Ibid.

73 Gardiner, *Hist. Wisb.* (1898), 335.

74 Diary of Rev. J. Jackson, 4 Aug. 1814 (Wisb. Mus.).

75 Ibid. 23 June, 21 Aug. 1815.

76 50 Geo. III c. 206.

77 This was a usual provision in contemporary Acts of this type.

78 *Rep. Com. Municipal Corpns.*, H.C. 116, p. 2564 (1835), xxvi.

79 Walker and Craddock, *Hist. Wisb.* 338.

80 Figures given in Jackson's Diary.

81 The local government and parliamentary franchises were identical, but the numbers polling at municipal elections seem generally to have been slightly higher.

82 The Trustees were, however, nominated by the Corporation.

83 Up to 1835 seven vicars of Wisbech and two other clergymen had held the office of town bailiff.

84 Walker and Craddock, *Hist. Wisb.* 338.

85 Infm. the late E. J. Rudsdale.

86 Their proceedings were not authenticated by their signatures until 1666–7.

87 Walker and Craddock, *Hist. Wisb.* 309.

88 Corp. Rec. ii, 66, 71, 85.

89 Ibid. 45.

to be 4*d*. per head 'and no more', payable whether the burgess was present or not.[90] This resolution was signed by the thirty-two burgesses who had just elected the 'ten' for their year of office. There seems to have been some intention of making this dinner a kind of annual general meeting, intermediate between the meetings of the 'ten' and the election meeting on 1 November at which every householder was entitled by the charter to be present. There may also have been an intention to revive the annual feast of the Holy Trinity Guild, which had been dissolved only twenty years before, and of which some of the thirty-two burgesses may have been members. The following year (1569) it was enacted that any absentee refusing to pay his 4*d*. for the dinner should in future not be admitted. This was repeated in 1583 with the addition of a 2*d*. fine for a late arrival.[91] During the period up to about 1625 there were usually about twelve absentees each year—a number sufficient to prevent the dinner from developing into a business meeting in fulfilment of the apparent intention of 1568.

CORPORATION ELECTIONS

The annual election of the capital burgesses was made an excuse for festivities. At first they were modest. In 1617, for example, an expenditure of only 3*s*. 3*d*., which included bread, beer, and a fire in the common hall, was incurred.[92] After the Restoration they became more elaborate.[93] In 1671 £2 13*s*. 6*d*. was spent on wine and 3*s*. 10*d*. on beer, as well as 37*s*. 6*d*. for eatables and 12*s*. 4*d*. for tobacco and pipes 'for one whole year.[94] It is therefore not surprising to learn that this feast had by 1679 proved 'of inconveniancye to the Townshipp by retardin the election . . . and sometimes produced quarrellings and undecent language in the neighbourhood'. For this reason, and because the election that year was to fall on a Sunday, the feast was ordered to be discontinued unless and until future capital burgesses decided otherwise.[95] In the event, it was revived the following year (1680) at a maximum cost of £8, and in 1757 the provision of wine was limited to 25 gallons.[96] The free distribution of food and drink to electors was finally abolished in 1767,[97] but the prevalence of 'treating' in connexion with the annual elections still provoked controversy in 1812. The following year a Mr. Marshall put an anti-treating resolution before the candidates for election as capital burgess. Fourteen of the twenty candidates signed, but five of the burgesses actually elected did not do so, and the controversy smouldered on for several more years.[98]

OFFICIALS OF THE CORPORATION

The town bailiff, as stated above, first appears as an agent of the Corporation rather than as its presiding member. He received a 'gratuity' of £5 5*s*. a year, but by 1834 this sum was by established custom awarded to charity. This led the Commissioners to regard the office as 'burdensome without being profitable'.[1] The early town bailiffs, in this capacity of paid official, were not always members of the Corporation, but by 1800, when they were in fact though not in name mayors of Wisbech, the office was usually taken by the capital burgesses in rotation. In accordance with the terms of the charters, the capital burgesses were elected year by year. Often however they kept their seats for long periods—in the case of a certain Dr. Hardwick (*fl. c.* 1780–1830), for as much as fifty years.[2]

A regular town clerk was first employed in 1680; previously 2*s*. had been paid to a clerk when a lease or bond was drawn up, under the benefaction of William Holmes. Robert Cockson was the first town clerk, but his salary is not recorded.[3] In 1825 John Wing was appointed to the office at a salary of £35, and a new post of chamberlain was created.[4] In 1849 the town clerk's salary was £45 a year, and he also drew £15 a year as clerk to the Sub-Commissioners of Pilotage and £25 as clerk to the Charity Trustees.[5]

A beadle is mentioned in 1621. His wages were 12*d*. a week, and he also acted as scavenger.[6] In 1832 he was paid £6 6*s*. a year with livery.[7] References to the crier or bellman are numerous. In 1576 his wages were 2*s*. a quarter, and in 1609 'old father Parker' received a coat in virtue of this office.[8] The duties were laid down in 1678 as patrolling the streets continually from 9 p.m. to daybreak from October to May Day. Two years later the bellman was provided with a coat and a pair of boots, and a dog 'as cheape as one can bee procured'.[9] James Peddar, a new bellman, was appointed later in the same year (1680). His duties seem not to have included continuous night watching, for he was given 10*s*. towards bed and bedding in the church 'that he may lye in the steeple and with more conveniency ring the bell' at 4 a.m.[10] In 1688 he was paid off and the office abolished,[11] but it was in existence again in 1775, when the bellman was required to give notice of the arrival of fishing vessels.[12] In 1849 the salary attached to the combined offices of town crier, beadle, and hall keeper was £43 15*s*.[13]

Watchmen were sometimes appointed separately. In 1618 William Woodes, a tailor, was imprisoned for 'a strooke which he gave in the performance of his office, being a watchman, the party stricken being in danger of deathe'. Woodes received 6*s*. 8*d*. compensation from the Corporation.[14] Two watchmen were appointed in 1774 at 6*s*. weekly, to patrol from 11 p.m. to 4 a.m., ringing their bells and calling the hour, wind, and weather every five minutes at least.[15] In 1812 a 'noisy and riotous crowd' in the Market Place and the 'inevitable overthrow of all modesty and good morale among both sexes in the lower walks of society' brought caustic remarks on the inefficiency of the town's police and watching system from the Revd. Jeremiah Jackson.[16] By 1822 the sum paid in wages to watchmen was cut from £92 8*s*. to £15 for economy's sake.[17] In 1828

90 Ibid. ii, 45.
91 Ibid. 54, 157.
92 Ibid. iv, 9*a*.
93 It was perhaps this increasing expenditure that led six of the capital burgesses in 1668 to seek to amend the charter so as to convert the Corporation into a self-perpetuating body (see above, p. 257).
94 Corp. Rec. iv, 176*a*.
95 Ibid. 207*b*.
96 Ibid. vi.
97 Ibid.; Walker and Craddock, *Hist. Wisb.* 325.
98 Diary of Rev. J. Jackson, 7–8 Nov. 1812, 2 Nov. 1813 and subsequent years (Wisb. Mus.).
1 *Rep. Com. Municipal Corpns.*, H.C. 116, p. 2553 (1835), xxvi.
2 Ibid.
3 Corp. Rec. iv, 221.
4 Jackson Diary, 18 Nov. 1825 (Wisb. Mus.).
5 Walker and Craddock, *Hist. Wisb.* 339.
6 Corp. Rec. iv, 28.
7 *Rep. Com. Municipal Corpns.*, H.C. 116, p. 2554 (1835), xxvi.
8 Corp. Rec. ii, 103; iii, 84.
9 Ibid. iv, 217*a*.
10 Ibid. 219*b*.
11 Ibid. 240*b*.
12 Ibid. vi.
13 Walker and Craddock, *Hist. Wisb.* 340.
14 Corp. Rec. iv, 16*b*.
15 Ibid. vi.
16 Diary, 13 Sept. 1812 (Wisb. Mus.).
17 Corp. Rec. xi, 9–15. On the 1774

two superintendents of watchmen were appointed, to carry cutlass, pistol, and dark lantern; the equipment of the watchmen was lantern, rattle, and bludgeon.[18] A local police force of superintendent, sergeant, and 6 constables was established in 1836.[19] In 1900 the force consisted of a superintendent, 2 sergeants, and 7 constables; the first named officer was also deputy chief constable of the Isle.[20] It is now merged in the Isle force.

A town inspector was appointed under the Wisbech Improvement Act of 1810 (50 Geo. III c. ccvi), to keep the streets clear of encroachments and to prevent nuisances. In 1832 his salary was £26 a year with livery.[21] The duties of the town inspector were merged in those of the police in 1849.[22]

POOR RELIEF AND APPRENTICESHIP

As in some other towns the relief of the poor fell not to the churchwardens but to the Corporation, who had in any case inherited the function from the guild. The charter had enjoined the 'ten men' to distribute £3 15*s.* a year in charity.[23] This sum was supplemented in the latter part of the 16th century by weekly collections from householders, certain benefactions,[24] and the interest on 'town stock'. In the period between 24 August 1591 and 1 May 1592 there were two collectors for the poor, who received £21 0*s.* 11*d.* and disbursed £21 17*s.* 7*d.*[25] In 1591 William Skortred sold a gelding to William Styrman to pay during his life 'the weekly payment for the collection for the pore of Wisbich'.[26] The town stock consisted of the capital of well-to-do orphans invested on their behalf by the Corporation as trustees. In 1584 it produced £11 17*s.*[27] The expense of relief was considerable, and before the end of the century the Corporation began to take steps to set the poor to work.[28] This soon justified itself; the accounts of the Corporation are difficult to disentangle from its proceedings at this period, but a fairly clear picture can be drawn for the years 1623 and 1624. In the former year the raw materials, mainly hemp and tow, bought for the poor to work up, cost £10 2*s.* 7*d.* Wages to those employed amounted to £4. The finished goods were sold in 1624 for £36 8*s.*[29]

In 1720 the capital burgesses resolved to build a municipal workhouse on a 3-acre site between the Horse Fair and the Quay. They ordered 260,000 bricks, and Robert Twells, then tenant of the land, was induced to surrender his lease for 2 guineas and the first year's production of hemp from the workhouse. In all, £2,000 was borrowed for the completion of the building,[30] which still exists as Nos. 1 to 6 Albion Terrace. It is a good example of early Georgian architecture, though badly repaired. The Corporation kept the management of the house in their own hands. Those incapable of work, the sick, aged, and very young were not admitted. The inmates averaged about 80. Children were admitted when 'big enough to read' and were taught to write and to spin. Raw wool was bought direct from graziers by expert managers, and stockings and very good yarn were made for the Norwich market, as well as all the inmates' clothing.[31] Brewing and baking were also done on the premises. The master and mistress of the house were paid a fixed salary and were under the control of a board of 'four or more of the most prudent persons of the parish'. By the time the house was in full working order it was reckoned that the rates had been reduced from 3*s.* 6*d.* to 2*s.* in the £.[32]

About 1815 the master, a Mr. Matthews, was said to be doing well out of the business. The inmates consumed half a stone of flour a head weekly. When this commodity was at 4*s.* a stone an agreement had been made allowing Matthews a similar sum weekly for each pauper. The price of flour had, however, declined, and while the master of the Leverington workhouse was getting only 2*s.* 9*d.* a week for each of his charges, Matthews still received 3*s.* 10*d.* His net profits were said to be £182 a year. Matthews claimed that the earnings of the inmates, who averaged about 70 at this time, were 20*s.* a week.[33]

In the early 1830's the Poor Law Commissioners reported that the house 'was not ill-regulated', and that the rates had greatly diminished since 1817 although the population had increased by 1,500. There was a professional overseer.[34] The workhouse served for a few years after 1834 for the whole of Wisbech Union, which included the Rural Districts of Wisbech and Marshland (Norfolk) as well as the borough. In the early 1840's the present Public Assistance Institution (now Clarkson Hospital) was erected in Lynn Road. This building, which was planned for 560 inmates at a cost of £6,900, is of most elaborate design, and 'has not only an air of comfort but of grandeur'.[35] In 1850–1 the average number of inmates was about 300, supported at a cost of 2*s.* 7½*d.* per head per week.

So far as is known the 'ten men' bound their first apprentice in 1576.[36] During the 17th century the number of apprenticeships varied between 5 and 10. At the beginning of the 19th century the capital burgesses seem to have apprenticed 3 boys regularly each year.[37] The premium paid to employers for taking pauper apprentices was raised in 1822 from 50*s.* to sums varying from £3 10*s.* to £7 10*s.*[38]

The badging of paupers began in 1681, when the town bailiff was authorized to purchase 180 badges,

scale, the force of six watchmen was to be cut down to one watchman only.

[18] There were four watchmen at this date, their hours and wages being 11 p.m. to 5 a.m. in winter for 9*s.* per week, and to 4 a.m. in summer for 6*s.* per week (MS. in Wisb. Mus.).

[19] Walker and Craddock, *Hist. Wisb.* 341.

[20] Kelly, *Dir. Cambs.* (1900).

[21] *Rep. Com. Municipal Corpns.*, p. 2554.

[22] Walker and Craddock, *Hist. Wisb.* 339.

[23] Mann Hutcheson in his MS. History of Wisbech (p. 2) states that this sum represented an annual payment by the Holy Trinity Guild, but the records are not now sufficiently complete to prove this.

[24] e.g. in 1596–7 there were: the gift of Robert Beste, £3 6*s.* 8*d.*; £2 interest on the £20 gift of Bartholomew Edwards and his wife; and £1 13*s.* 4*d.* and £1 given by Robert Girden and John Strigett respectively (Corp. Rec. ii, 373).

[25] Corp. Rec. ii, 258.

[26] Ibid. 247.

[27] W. Watson, *Hist. Wisb.* 164. The 'orphan stock' in 1594 amounted to £280 9*s.* Some was lent out in sums of £4, £5 &c., on bonded security; there were fifty-nine such bonds in 1604 and seventy-one in 1609 (Corp. Rec. iii, 35–36).

[28] *V.C.H. Cambs.* ii, 93.

[29] Corp. Rec. iv, 35*b*, 43*b*.

[30] Ibid. v (14 Nov. 1720, 15 Nov. 1721, 26 Oct. 1722).

[31] In 1686–7 the clothing of the 'town children' had been estimated to cost up to £40 (Corp. Rec. iv).

[32] Ethel M. Hampson, *Treatment of Poverty in Cambs. 1597–1834* (1934), 74–75.

[33] Diary of Rev. J. Jackson, 5 May 1815 (Wisb. Mus.).

[34] *1st Rep. Comrs. Poor Law*, H.C. 44, 596–7 (1834), xxviii.

[35] 1851 ed., p. 623.

[36] Corp. Rec. ii, 103.

[37] Diary of Rev. J. Jackson, 28 Nov. 1812 (Wisb. Mus.).

[38] Hampson, *Treatment of Poverty in Cambs. 1597–1834*, 159 n., 161.

and a stamp.[39] Newcomers to the town were required to enter into a bond that they would not become chargeable to the rates;[40] those who did become so chargeable were given fourteen days' notice to remove.[41]

Thus to a great extent the Corporation undertook the duties of parochial officers. It is not surprising that it extended its activities to the audit of their accounts. For example, the surveyors of menworks (highways) had in 1588 collected £7 4*s.* and laid out £3 14*s.* 8*d.* Of the balance, 16*s.* 4*d.* was appropriated for paving the streets.[42] In 1603 £7 9*s.*, and in 1604 £14, was allowed to the churchwardens of Wisbech St. Mary for repairs to their church 'by the consent onely of the Tenne men, as a benevolence, at the earnest sute of the said churchwardens'.[43] In 1615 the Corporation paid for three trees, at 6*s.*, for the repair of the chancel of the parish church.[44]

At this time, in spite of the heavy demands of poor relief, the Corporation seems to have accumulated a surplus, which was used to buy more land.[45] It was the difficulties involved in these purchases that led to the first renewal of the charter.

EPIDEMICS, SANITATION The origin of the weekly collection for the poor, mentioned above, seems to have been the outbreak of plague. In December 1584 a balance of 11*s.* 6*d.* was allowed to John Robynson 'collector' for his pains in collecting money during the previous six weeks 'to be payd to the poore at Guyhurn vissyted with sycknes of the plage as supposed'.[46] For a time the epidemic was confined to the outparish, but in 1587 it broke out in the town itself. A pesthouse was set up at Barton, and, though many cases were not fatal,[47] William Skortred was co-opted one of the Ten to replace Robert Skortred, a plague victim, as representative of Barton ward. The total number of burials during the year was 206, including 42 in September and 20 in October, as opposed to an average of about 60 a year at this time.[48] The town seems to have escaped the 17th-century outbreaks of plague, and the first attack of cholera (1832), though causing some 20 to 30 deaths, was not so severe as in some neighbouring villages or in the country at large. The 1849 cholera epidemic, however, affected Wisbech badly. There were 140 cases, 66 fatal, in the town, and the government inspector sent the following year declared the water supply to be worse than in any other town he had visited.[49] The inspector's proposals included the establishment of a Local Board of Health for Walsoken,[50] which was to be combined with Wisbech for main sewerage. A further outbreak of cholera occurred in 1854, when the Wisbech death-rate (49 per 10,000) was the fourth highest in the country, and the following year the central Board of Health sanctioned the expenditure of £8,000 on sewerage and £13,400 on water supply.[51] By 1864 the general death-rate was down to 25 per 1,000, but water, sold at ½*d.* a bucket, was still being drawn from brick-pits. The Wisbech Waterworks Act of that year (27 & 28 Vic. c. clxxx) empowered the Wisbech Waterworks Company to obtain supplies from Marham (Norf.), 18 miles east of the town, and the works were completed in 1865 at a cost of £24,800.[52] They supply the whole of the borough and rural district, and a wide area around. The Water Company has now been transformed by statute into the Wisbech and District Water Board.[53] Sewage works at Foul Anchor, in Walpole (Norf.), were completed in 1877 for £11,000.[54]

FIRES Timber is scarce in the Fens and brick earth plentiful. Timber-framed construction was therefore superseded by brickwork at an early date. This perhaps explains why it is that Wisbech has not suffered severely from fires.[55] In 1558 William Styrmin was paid 40*s.* for his charges laid out after a fire in the Old Market.[56] Another fire in the same quarter occurred in 1607. Amongst other payees, John Norton received £10 compensation for his house 'pulled down for the preservation of the rest'.[57] After this it was resolved that the town's fire equipment should consist of 24 leather buckets, 3 large ladders, and 3 'cromes' with 14-yard ropes fastened to the chains. Another serious fire occurred in 1652, when some almshouses were destroyed. A general collection throughout the Isle yielded £14 0*s.* 10*d.*, including £4 16*s.* 5*d.* from Wisbech itself, which was distributed in amounts varying from 5*s.* to £1 12*s.* 6*d.* amongst sixteen beneficiaries.[58] Four years later (1656) the Corporation obtained the use of a fire engine for 20*s.* yearly; it was stored first in the church, and from 1679 in an engine house by the neighbouring almshouses.[59] A new house was built near the oil mills in the Old Market in 1776,[60] but as late as 1781 the town relied at least in part on amateur assistance in extinguishing fires, 10 guineas being paid to those who had helped to put out a fire in Bell Lane.[61] At the end of the 19th century the Wisbech Fire Brigade consisted of a captain, 3 officers, 4 subordinate officers, and 20 men, with 2 steam and 1 manual fire engines.[62] The fire station in Lower Hill Street was replaced by the present one in Horse Fair in 1932.

WISBECH AS A COMMERCIAL CENTRE Wisbech is now the commercial centre of the southern marshland, and the title 'Capital of the Fens', sometimes given to the town, is not without justification. This situation has long prevailed. No record of the grant of a market has been found, and this, taken together with the evidence adduced above,[63] suggests that the market may have been established before the Conquest. A definite indication of the growth of commercial interests in Wisbech is the grant (1190) by Richard I to the tenants of the manor of freedom

39 Corp. Rec. iv, 222*a*.
40 e.g. ibid. ii, 364 (1596); iv, 44*b* (1624).
41 e.g. Corp. Rec. ii, 244 (1591).
42 Corp. Rec. ii, 221.
43 Ibid. iii, 30*a*, 32*b*.
44 Ibid. 129*b*.
45 e.g. Corp. Rec. ii, 352 (1595); see also Walker and Craddock, *Hist. Wisb.* 318.
46 Corp. Rec. ii, 173.
47 For example, some patients were to be released after 21 November, and on 10 December they were to be placed 'in stools convenyent by them selves in the churche' (Corp. Rec. ii, 202).
48 Walker and Craddock, *Hist. Wisb.* 229.
49 Gardiner, *Hist. Wisb.* 65–67. Wisbech was an unhealthy town at this period; the death-rate was 30 per 1,000 for 1841–9 (average) as compared with 20 per 1,000 in England and Wales as a whole. For the cholera year (1849) the Wisbech rate was 37·88 per 1,000.
50 Wisbech as a borough was already a statutory authority for health purposes.
51 Gardiner, *Hist. Wisb.* 168.
52 Ibid. 169–76. Another water shortage occurred in 1877 owing to a dry winter and spring.
53 Infm. Mr. P. G. Laws, County Planning Officer.
54 Gardiner, *Hist. Wisb.* 179–80.
55 Walker and Craddock, *Hist. Wisb.* 238.
56 Corp. Rec. ii, 209.
57 At this time house property of the Corporation let for about £7 a year (e.g. Corp. Rec. iii, 91).
58 Corp. Rec. iv, 124–5.
59 Ibid. 133*b*, 146*b*, 210*b*.
60 Ibid. vi.
61 Anonymous late 18th-century diary in Wisb. Mus.
62 Kelly, *Dir. Cambs.* (1900).
63 See p. 240 above.

from toll in all fairs and markets throughout England.[64] A generation later, one of the free tenants of Bishop de Fontibus (1220–5) in Wisbech had the significant name of Alexander the merchant (*mercator*).[65] In both 1221 and 1251 several persons were holding messuages by money rents only—in effect by burgage tenure though not so described. Nevertheless the general impression derived from these 13th-century surveys is that Wisbech was still only a large and slightly urbanized village.

The growth of Wisbech was hampered by the sea-flood of 1236, the evil effects of which were still apparent in 1251 notwithstanding creditable efforts to repair them. The flood was followed (*c.* 1300) by the diversion of the Ouse. These factors taken together prevented the town from ever becoming one of the great English ports of the Middle Ages.

The Alcock survey of 1492, like preceding surveys, still gives an impression that Wisbech was an urbanized village rather than a town. It is true that it records as a privilege of the lord of the manor the power to levy 4*d.* toll on foreign ships entering or leaving the port. Since the blocking of the Well Stream, however, the only step taken towards the improvement of the port was the straightening of the course of the Nene by Bishop Morton. Thus the port was still largely inaccessible and the privilege unreal.

Though an elaborate tower was added to the church in the 16th century it is significant that the whole church was not rebuilt on a lavish scale as happened at Boston or Louth. It was only with the scientific draining of the fens in the 17th century that Wisbech rose to real importance as a commercial centre.

From at least 1662 the town had its own weights and measures. The Corporation still possesses a brass pint measure inscribed 'WISBECH ANO DOM 1662. IC. GM'.[66] Several tradesmen's tokens of this period are preserved in the Museum.[67] The Corporation also seems to have issued its own currency; in 1668 the town bailiff was authorized to spend £5 to £10 on farthings with the town arms, and in 1670 £20 on 'Wisbech halfpennies'.[68]

Evidence of increasing prosperity is to be found in the fine houses, dating from about 1720 onwards, on the Brinks and in such streets as Hill Street. It is significant that at the end of the 18th century Joseph Medworth, the speculative builder, thought it worth while to build some forty to fifty medium-sized houses on the former castle grounds.

By 1796 the town had become preponderantly commercial. There were 'hardly any people in the town', it was said, 'but what are in trade; even Sir Philip Vavasor[69] is a merchant, and trades in coal, firkin-butter &c.' The number and range of trades, though smaller than in Lynn or Boston, were larger than those practised anywhere else in the Isle, and a most rosy picture of the future of Wisbech was painted. 'The fields seem to laugh and sing with their superabundant produce . . . and, by the revival of its commerce, a true spirit of emulation has burst forth amongst its merchants and tradesmen, so that few places have the prospect of becoming so rich and prosperous.' To some extent these optimistic prophecies were fulfilled in the next half century.

At the present day the sphere of influence of Wisbech is the most extensive of any town in the Isle, embracing as it does all the county north of a line from Thorney through the northern outskirts of March to Welney.[70] A more factual index of the relative importance of the various Fenland towns is the proportion of persons engaged in commercial and financial occupations. Here again Wisbech is pre-eminent, with a percentage of 16·4 as compared with Chatteris (11·8), Ely (11·5), March (11·1), and Whittlesey (7·6); the Wisbech percentage is also higher than that of the three towns whose 'occasional' sphere of influence covers the Isle—Cambridge (13·3), King's Lynn (15·6), and Peterborough (13·7), and of the country as a whole (10·9). The proportion employed in the 'service' industries[71] is 53 per cent. This proportion is also above the national average (41 per cent.), but not so markedly, and it is below the proportion of such towns as Oxford (68 per cent.) and Worcester (59 per cent.).[72]

MARKETS AND FAIRS The market is held by prescription. In 1657 John Thurloe, as lord of the manor, was granted a fortnightly market at Wisbech by letters patent.[73] In the 1840's Wisbech was one of the largest if not actually the largest corn market in the country. In 1844–5 250,000 quarters were sold, the quantity changing hands in each week of that period sometimes being as much as 9,000 quarters.[74] About 1850, in spite of the recent revival of the markets at Chatteris, March, and Whittlesey, Wisbech ranked second only to Wakefield among the English corn markets.[75] There are now weekly markets for cattle on Mondays and for general produce on Saturdays.[76]

The right to hold a fair on the vigil, feast, and morrow of Trinity, and the nineteen days following, was granted to Bishop Hotham in 1327.[77] This fair had changed slightly in date by the end of the 15th century, when it was associated with the festival of St. Peter and St. Paul, to whom the parish church is dedicated. At the end of the 18th century there were five fairs, on the Saturday and Monday before Palm Sunday, the Monday and the Saturday before Whitsun, 25 July, and 1 and 2 August. Hemp, flax, cattle, and horses were the principal goods exchanged.[78] In 1778–9 Francis Bampton leased the fairs for £5 7*s.* a year, probably a beneficial rent; as far back as 1632 their annual value had been put at £5.[79] Until about 1930 fairs were held on the Saturday after 14 February, the second Thursday in May, 25 July, the first Thursday in August, and the third Wednesday in September. The last was for hiring.[80]

[64] Walker and Craddock, *Hist. Wisb.* 209–10; M. Hutcheson, Hist. Wisb. (MS. in Wisb. Mus.), 98.

[65] B.M. Cott. MS. Tib. B. II, f. 143*b*.

[66] Infm. the late E. J. Rudsdale.

[67] Incl. tokens of Henry Tunard, baker (¼*d.*, 1657), Richard Harrison, haberdasher (¼*d.*, 1664), John Bellamy, grocer (½*d.*, 1665, 1667), and Henry Coldwell, haberdasher (1668).

[68] Corp. Rec. iv, 165, 170.

[69] High Sheriff of Cambs. 1760. Created Knight Bachelor, 24 Jan. 1761 (W. A. Shaw, *Knights of Engl.* ii, 290). He lived at Nos. 7 and 8 South Brink.

[70] *Survey of a Fen County* (Draft Report of the County Planning Dept.), Map No. 50.

[71] viz. building, transport, commercial and financial, public administration, professions, entertainments, personal service, clerical work.

[72] These figures are derived from the social surveys of Worcester (*County Town*, by Janet Glaisyer and others, 1946) and Oxford (*Oxford Replanned*, by T. Sharp, 1948). The proportions for Cambridge, Lynn, and Peterborough were not worked out on account of their less representative industrial structures.

[73] A.O. 3/411.

[74] Cassey, *Dir. Cambs.* 209.

[75] Kelly, *Dir. Cambs.* (1933).

[76] Gardner, *Dir. Cambs.* (1851), 593.

[77] *Cal. Chart. R.* 1326–41, 11.

[78] *Univ. Brit. Dir.* (1796), iv, 802.

[79] Bp.'s Mun., Ely, D 6.

[80] Kelly, *Dir. Cambs.* (1933). The horse fairs were held in Lynn Road (infm. the late E. J. Rudsdale).

Market Place, Wisbech, from the East

(From a photograph of 1857)

Bridge Street, Wisbech, Showing Butter Market (Built 1801, Demolished 1856)

(From a photograph of 1856)

Shipping in Wisbech Port
(From a photograph of 1857)

The Bridge, Wisbech
(Built 1758, Replaced by Iron Bridge 1855–7)
(From a photograph of 1853)

Since at least 1851 travelling showmen and hawkers, on their way between Lynn to Stamford Fairs, have gathered in the town. This concourse is called the Mart. 'This saturnalia' it was said in 1851, 'sometimes continues five weeks, and during that period all the noise that can be tortured out of drums, dulcimers, lungs, trumpets, organs, and cymbals serves to enliven, if not to gratify. The utility of marts and fairs is now almost wholly superseded; and those of Lynn and Wisbech have degenerated into a mere gathering of freaks of nature, "harlotry players", dirty exhibitions, conjurers, wild beast [*sic*], and ragamuffin life in all its gipsyism.'[81]

During the Middle Ages the profits from the market and fairs were a manorial perquisite. They long remained so. Seventy years after incorporation an attempt to buy out the episcopal claim in the butchers' shambles failed.[82] The bishop's rights were not finally extinguished until 1811, when £20 was paid to the bishop and £300 to trustees for the holders of permanent stalls, whose annual rent had been 8*s*. 8*d*.[83] In the early 19th century the Corporation held the market and fairs at a fee farm rent of £5 6*s*. 3*d*., the tolls being let by the Corporation for £85 in 1819 and £153 in 1845.[84] The tolls produced £111 10*s*. in 1822–3.[85]

A market cross existed in the Middle Ages. In 1549 £4 3*s*. 4*d*. was spent on its repair.[86] A market house, probably a wooden shelter around the base and shaft of the cross, was repaired in 1614.[87] It was rebuilt in 1667 on a more substantial scale near the river.[88] Two years later, however, it was in a dangerous state from defective workmanship, and John Lewin, the builder, received £425 in final settlement.[89] Four 'dyalls' were set up on it in 1671.[90] Its successor, the butter market, a building of the common market-house type with an upper story carried on open arches, was erected at the foot of Bridge Street in 1801. In 1849 it was also known as the custom-house.[91] It was removed in 1856 when the bridge was again rebuilt.[92] An obelisk was erected in the Market Place in 1765, on which an effigy with an abusive inscription was placed after a robbery in 1770.[93] The Shambles at the east end of the Market Place were built in 1595 for £104 7*s*.,[94] and were let yearly for £7 10*s*. This rent was increased in 1610 to £8 10*s*.[95] The Shambles and the obelisk were removed when the Market Place was repaved under the Improvement Act of 1810.[96]

The Improvement Act of 1810 (see Corporation) resulted in the repairing of the Market Place, the erection of a town hall and the construction of a cattle market. The lower story of the Town Hall, of which the arches were originally open, was intended for a corn exchange. It was, however, little used for this purpose, merchants preferring to trade in the street. From 1831 the building was let out for public meetings, bazaars and concerts.[97] In 1858 the present Corn Exchange behind the Town Hall was built by a private company.[98] The first cattle market was enlarged in 1827 and 1838.[99] A new one was opened in 1869.[1] In the 1840's the hiring fees from the Corn Exchange and tolls of the cattle market amounted to about one-tenth of the Corporation revenue.[2]

PORT, NENE OUTFALL, CANAL

The original combined outfall of the Ouse and Nene, by Wisbech, had been diverted along Well Creek into the present Ouse outfall by 1300, though it was temporarily diverted back again soon afterwards by the blocking of the Creek.[3] No attempt seems to have been made to reopen the Nene channel until about 1480, when the flow of water was improved by cutting Moreton's Leam. Bishop West spent £200 on a new sluice, which was broken by a sea flood in or before 1523,[4] but by 1566 there was a certain amount of activity in Wisbech port. In that year there traded from Wisbech 13 keels and other vessels of 3 to 12 tons. They were used for the corn and coal trade by 7 owners employing 12 watermen.[5] In 1576 there was a proposal, supported by the Corporation, for a new sluice to be erected at Leverington at the cost of Wisbech hundred, but it was rejected by the Privy Council and local landowners as too expensive.[6] An undated 'breife description of the notable or most famus rivers fallinge into the Fennes' in a volume of collections towards a history of the Fens made by Maurice Johnson in 1710 sums up the situation admirably: 'Wisbech river . . . hath an indifferent descent and the outfall reasonable good but may be muche bettered . . . [by] an apt and straight course to reteyne the waters descending from Peterborough by Thorney, Whittlesey and Wisbech'.[7] The great flood of 1613, probably the worst since that of 1236,[8] brought matters to a head, and a special commission was sent to the Fenland by the Privy Council to investigate the problem. The commissioners found that the level of the river bed had been raised 6 ft. or more by silting, so that the inland water was no longer able to flow down the channel from Guyhirn, but rather the other way. The town of Wisbech asked that the county should provide a sluice at the Horseshoe, and that the river should be widened from 40 to 60 ft.[9] The commissioners, however, thought that if the outfall was sufficiently improved to carry away the upland waters, the channel would be so deepened as to undermine the proposed sluice. They recalled the fate of a previous one which was 'blowne up by the tyde' after a life of seven days.[10] Orders were, however, issued by

81 Gardner, *Dir. Cambs.* (1851), 595.

82 Corp. Rec. iv, 18*b* (1619). For Bishop Patrick's clothing charity relating to the Shambles, see p. 270.

83 Walker and Craddock, *Hist. Wisb.* 424.

84 Ibid. 438.

85 Corp. Rec. xi, 9–15.

86 See above, p. 256, note 33.

87 Corp. Rec. iii.

88 Close to the bridge, where the Butter Cross later stood.

89 Corp. Rec. iv, 160*a*, 163*b*, 166*b*, 170*a*.

90 Ibid. 175.

91 Walker and Craddock, *Hist. Wisb.* 427.

92 Gardiner, *Hist. Wisb.* 111–12. The materials of the building, which had cost £2,500, sold for £219.

93 Anonymous late 18th-century diary in Wisb. Mus.

94 Corp. Rec. ii, 334.

95 Ibid. iii, 87*a*.

96 Gardiner, *Hist. Wisb.* 25.

97 Gardner, *Dir. Cambs.* (1851), 612.

98 Gardiner, *Hist. Wisb.* 284–5.

99 Walker and Craddock, *Hist. Wisb.* 438.

1 Gardiner, *Hist. Wisb.* 190.

2 £438 out of £4,473 average for the years 1837–46 (Walker and Craddock, *Hist. Wisb.* 438).

3 H. C. Darby, *Medieval Fenland*, 96, 98–99; *Cal. Pat.* 1321–4, 386.

4 *L. & P. Hen. VIII*, ii, no. 1733; iii, no. 3476.

5 *Fenland N. & Q.* vii, 95.

6 H. C. Darby, *Draining of the Fens*, 15; *Acts of P.C.* 1575–7, 134–5; 1578–80, 190–1; E 134/21–22 Eliz./M. 33.

7 B.M. Add. MS. 35171, f. 193*b* (modern numeration).

8 *Acts of P.C.* 1613–14, 265–7, 299. This flood was commemorated by an inscription on the wall of the church. The text is given in Walker and Craddock, *Hist. Wisb.* 372–3. The inscription was largely destroyed in 1750, but some remains of it have been revealed during recent alterations (Wisb. Soc. *Ann. Rep.* (1951), 9).

9 *Cal. S.P. Dom.* 1611–18, 538–41.

10 *Acts of P.C.* 1618–19, 295.

the Privy Council for a sluice at the Horseshoe at the cost of occupiers whose lands would be thereby drained.[11] This sluice was finally constructed by Vermuyden in 1631 at a cost of £8,000, by a 'little Army of Artificers venting, contriving and acting outlandish devises'.[12] Seven years later the first attempt to improve the flow by straightening the Nene Outfall was made, by constructing a 2½-mile cut.[13]

In 1680 the trade of Wisbech had increased sufficiently for the port to be reckoned independent and no longer a member of Lynn.[14] In 1710 the Corporation was licensed to buoy the channel for the first time. By 1718–19 the customs amounted to £4,000 a year. The exports consisted of 1,000 tons of oil[15] and 40,000 quarters of oats, carried in 30 ships 'of burden' and 30 smaller vessels of 20 to 30 tons. In this year 241 ships were cleared.[16] About this time Charles Kinderley proposed to embank the Nene for 7 miles downstream from Wisbech to Peter's Point. This would have given a 5-foot fall, but the Corporation opposed the scheme on the ground that it would make navigation difficult, filed a petition in Chancery to stop it, and spent £150 in demolishing a bank that was part of Kinderley's improvements (1726).[17] By 1751 it was possible in a dry year to walk across the river bed under Wisbech Bridge[18] and vessels of any size had to unload at Foul Anchor or Cross Keys, 6 or 8 miles downstream respectively. Spring tides produced only a 4-foot rise and neaps did not reach the town at all. The silting of the river also resulted in disastrous floods, the upland water not being able to get away to sea. Such floods occurred in 1763, 1765, 1767, and 1770.[19] In the face of these disasters the Corporation became more accommodating and withdrew its opposition to the straightening and embanking, provided navigational interests were safeguarded.[20] 'Kinderley's Cut' was finally made in 1770–2 at a cost of £10,000, straightening the river down to Gunthorpe Sluice.[21] Further straightenings and improvements in the channel were effected in the early 19th century. Among these may be mentioned the Paupers' Cut of ¾ mile at Woodhouse Marsh,[22] and another cut from the end of Kinderley's Cut to Crab Hole. The latter cost £130,000 towards which the Corporation contributed £30,000.[23] The cut, with the contemporary embankment to carry the Lynn-Spalding road over Cross Keys Wash, caused much local uneasiness. The Corporation, always apprehensive about the effects on navigation of bridges below Wisbech, decided in 1820 to oppose the embankment if the bridge were to be at Gunthorpe Sluice, but to offer no objection if it were at the Washway (Sutton Bridge).[24] John Wing, writing to Lord Lansdowne on behalf of the Commissioners for the Nene Outfall and the local Justices, stated (20 September 1827) that the countryside was at the mercy of the navvies employed on the schemes. Of these 200 had already arrived and 800 more were expected. He recommended the dispatch of a military force.[25]

In spite of the forebodings of the Corporation, these improvements brought new life to the port of Wisbech. A custom-house was built in 1801.[26] In 1825–6 1,209 ships, all but 45 of them coastwise, cleared the port. The gross tonnage amounted to 70,320 and the customs to £29,531 15*s.* 9*d.*[27] The tonnage increased fairly steadily to a maximum of 167,442 in 1847.[28] Railway competition caused a rapid decline to 88,082 in 1854.[29] After this the tonnage remained fairly constant. Thus in 1910 it stood at 89,333, which represented 92 'foreign' and 502 coasting vessels.[30] This was a slight increase on the tonnage for 1909.

The exports in the early 19th century were mainly corn and rape-seed. Coal was imported and, from 1824, timber. In the period of high farming Wisbech was the biggest corn-shipping port in the country. A high level of 283,094 quarters was achieved in 1845, and though the opening of the railway shortly afterwards was a severe blow to this trade,[31] another 'peak' export of 67,782 quarters was handled in the autumn, winter, and spring of 1870–1.[32] After the First World War, however, the export of corn fell to an average of about 16,000 quarters a year, and was replaced by that of potatoes, of which 18,000 tons were shipped in 1931.[33] The imports of timber have also dropped from 76,968 loads in 1897 to 23,316 loads in 1932.[34] The expansion of the brickmaking industry at Peterborough and elsewhere in the East Midlands has led to an outward trade in bricks.[35] The total value of imports and exports was respectively £144,072 and £4,049 in 1894,[36] when the port was perhaps at the lowest ebb of its fortunes. A revival in the present century brought the figures up to £236,798 (imports) and £80,640 (exports) in 1937. At this date the chief imports were timber, seed potatoes, grain, cattle, foodstuffs, and petroleum, while potatoes, scrap iron, and wheat were exported.[37]

A certain amount of shipbuilding was carried on at Wisbech in the late 18th and early 19th centuries. Between 1778 and 1850 21 ships were launched, 12 of them in the last 10 years of the period. The yards were on the right bank of the river near the present

[11] B.M. Add. Chart. 33106, 33117.
[12] Darby, *Draining of Fens*, 42, quoting B.M. Lansd. MS. 213, f. 316*a*.
[13] Ibid. 60 n.; W. Dugdale, *Hist. Imbanking*, 414.
[14] A. A. Oldham, *Hist. Wisb. River* (1933), 3. There is an extra illustrated manuscript copy in Wisb. Mus. The merchandise carried in ships entering and leaving Wisbech is first entered separately in the Lynn Port Books in 1678 (E 190/438).
[15] Oldham, *Hist. Wisb. River*, 3.
[16] There were seven mills crushing oil from seed in Wisbech in 1735; one had been established as early as 1553–5 (C 1/1365/47).
[17] Corp. Rec. v, various dates between 1719 and 1729; Darby, *Draining of Fens*, 137–9.
[18] N. Kinderley, *Anct. and Present State of the Navig. of . . . Wisbech* (1751), 70.
[19] Oldham, *Hist. Wisb. River*, 112–13.
[20] Darby, *Draining of Fens*, 142.
[21] Walker and Craddock, *Hist. Wisb.* 174.
[22] Constructed in 1830 at a cost of £5,000 to relieve unemployment (Oldham, *Hist. Wisb. River*, 6). A gravel walk along East Field Bank was laid out in 1824 (Diary of Rev. J. Jackson, 22 Oct. 1824 (Wisb. Mus.)).
[23] Watson, *Hist. Wisb.* (1833), p. 74.
[24] Diary of Rev. J. Jackson, 8 Sept. 1820 (Wisb. Mus.). By 1825 £12,000 of the £35,000 needed for the scheme had been subscribed. (Ibid. 28 Nov. 1825.)
[25] H.O. 52/4. The former winding course of the river below Wisbech is still shown on many maps, since until 1934 it formed the boundary between the Isle of Ely and Norfolk.
[26] Walker and Craddock, *Hist. Wisb.* 426.
[27] Oldham, *Hist. Wisb. River*, 5.
[28] Walker and Craddock, *Hist. Wisb.* 427. The customs in this year were £25,000.
[29] In 1862 George Dawbarn, a Wisbech merchant, stated that 'the railways have almost annihilated the coal trade' (*Fenland N. & Q.* i, 330).
[30] Oldham, *Hist. Wisb. River*, 6–7, 11–12.
[31] Gardner, *Dir. Cambs.* (1851), p. 593.
[32] Oldham, *Hist. Wisb. River*, 14.
[33] Ibid. 12, 22.
[34] Oldham, *Hist. Wisb. River*, 19, 21. The 1932 figure included 3,879 loads of aspen logs for the local 'chip' fruit-basket industry.
[35] Personal observation. Four vessels were cleared during the middle fortnight of August 1948.
[36] Gardiner, *Hist. Wisb.* 152.
[37] Sir D. J. Owen, *Ports of the U.K.* (1939), 200.

petroleum depot.[38] In the period of greatest activity, between 1831 and 1856, 213 vessels were registered at Wisbech and Sutton Bridge, but by 1909 there were only 2 steamboats and 12 fishing-vessels on the books.[39] The greatest Wisbech shipowner was Richard Young (1809–71), who had at various times 43 boats operating from the port. Young was a man of more than local celebrity, and was M.P. for South Cambridgeshire from 1865 and Sheriff of London in 1871.[40] He was Mayor of Wisbech from 1858 to 1862.[41]

Various proposals were made during the 19th century for the construction of docks. The only outcome was a dock at Sutton Bridge, which was undermined by a subsidence shortly after its opening (1881) and has never been repaired.[42] Under the Wisbech Corporation Act (1889), however, new quays and a swinging berth were constructed at a cost of nearly £30,000 and the port was made available to ships of 245 ft. length, gross tonnage of 1,250 and draught of 17 ft.[43] Since 1940 the reconstruction of the quays has been accomplished for £103,000,[44] and a petroleum depot has been established. Wisbech is now accessible to ships carrying 1,700 tons cargo.[45] 'It is somewhat refreshing to observe that, while many of the smaller ports of the Kingdom are finding it difficult to keep up their business, Wisbech has succeeded in increasing its trade during the past few years.'[46]

There was formerly some passenger traffic upstream, the fare from Wisbech to Peterborough in 1756 being 2*s.* 6*d.*[47] In 1820 there was a service three times a week each way in the summer and twice a week in the winter.[48] As recently as 1938 a sea-going vessel, the *Constance H*, sailed as far as Peterborough, clearing Guyhirn Bridge by 1 inch.[49] In 1951 some sea-going barges, laden with corn, sailed from the Thames to Peterborough.[50] In 1850, 1851 and 1887 the river was the scene of regattas for sailing, rowing, and swimming.[51]

The former Well Stream was canalized under an Act of 1794 (34 Geo. III c. xcii). In the early 19th century it carried some traffic in coal; 2,508 tons in 1838 and 29,562 ten years later after the tolls had been reduced.[52] By the end of the 19th century the canal had become unremunerative owing to the competition of the Upwell tramway,[53] and it is now disused. The ownership of the canal was vested in the Corporation in 1944.[54]

BRIDGES The first mention of the bridge over the Well Stream is in 1317, when in the settlement of a dispute between Bishop Hotham and Sir Geoffrey de Coleville over their manors of Wisbech and Walsoken, certain houses erected by the latter at his end of the bridge are alluded to.[55] Grants of pontage were made to the bishop in 1327, 1328, and 1331.[56] References to bridges are frequent in the Corporation records, mostly to that across the Nene on the site of the present bridge. In 1586–7 and 1618 Walsoken bridge, the former 'great bridge' of Wisbech, was out of repair.[57] The new 'great bridge', the present Wisbech bridge, was built or perhaps rebuilt in 1583.[58] It was of timber and fairly often in need of repair, notably in 1651 and 1683 when over £100 was spent on it.[59] The repair of the original (Walsoken) bridge had been a responsibility of the hundred, and attempts were made in 1668 and 1679 to get contributions from Elm, Outwell, and Upwell for the repair of Wisbech bridge.[60] Finally, in 1739, John Addy, carpenter, undertook its upkeep for the Corporation for £18 a year exclusive of any damage caused by floods and shipping accidents.[61] Eighteen years later it was decided to replace this wooden bridge by a stone one. The result was a one-arch bridge which excited Cole's admiration; it was designed by a Mr. Shearman of Peterborough with the assistance of Sir James Burrough, and was completed in 1758 at a cost of £2,250.[62] This bridge, with its high camber and steep approaches, proved dangerous to road traffic,[63] and as early as 1819 an abortive proposal was made to replace it.[64] In 1855–7 it was succeeded by an iron structure designed as a swing bridge but never actually so used except in testing.[65] This bridge in turn became inadequate for traffic and dangerous owing to its high parapets and was replaced by the present concrete bridge in 1931.

POLITICAL ACTIVITY Except in 1659, when John Thurloe was elected to represent Wisbech in the parliament of Richard Cromwell,[66] the borough has never had separate representation. It was not even a polling station until after 1832, and there is little evidence of marked political activity in the 18th and early 19th centuries. The parliamentary electorate, which was identical with that for local government, numbered 188 in 1780, 141 votes going in that year to Lord Robert Manners, 101 to Sir Sampson Gideon, and 97 to Philip Yorke. In the first election of 1802, however, the town supported the Hardwicke candidate, Sir Henry Peyton, who received 160 votes to Lord Charles Manners's 53. This change of allegiance may have been due to the fact that the then Bishop of Ely (the Hon. James Yorke) was a member of the Hardwicke family. The same sympathies were shown, though less markedly, in the general election of the same year, when, out of

38 Oldham, *Hist. Wisb. River*, 55–56.
39 Ibid. 61.
40 He was the son of a Norfolk farmer, and acquired his mercantile interests through his appointment as Keeper of the North Level Sluice (1830) and Superintendent of the Nene Outfall Works (1833).
41 *Fenland N. & Q.* i, 327.
42 Gardiner, *Hist. Wisb.* 135–6.
43 Ibid. 138–44.
44 *Wisb. Advertiser*, 10 Apr. 1940 (describing the start of the scheme).
45 R. W. L. Melbourne, 'The Function of Wisbech in relation to its geographical setting', 1 (unpublished M.A. Thesis, Univ. Lond., 1939).
46 Owen, *Ports of U.K.* (1939), 200.
47 Oldham, *Hist. Wisb. River*, 19.
48 *Fenland N. & Q.* v, 115.
49 *Wisb. Advertiser*, 19 Jan. 1938.
50 Infm. the late E. J. Rudsdale.
51 Oldham, MS. Hist. Wisb. River, 170.
52 Walker and Craddock, *Hist. Wisb.* 498.
53 Gardiner, *Hist. Wisb.* 163.
54 By the Wisb. Corp. Act of that year (*Municipal Yr. Bk. 1950*).
55 *Cal. Pat.* 1317–21, 36; Hist. MSS. Com. *6th Rep. App.* 317.
56 *Cal. Pat.* 1327–30, 95, 227; 1330–4, 183.
57 Corp. Rec. ii, 194; iv, 12*a*.
58 Ibid. ii, 153. The Commissioners of Sewers ordered the Bishop of Ely to defray one-third of the cost of its upkeep (M. Hutcheson, MS. Hist. Wisb. (Wisb. Mus.)).
59 Corp. Rec. iv, 123*b*, 230*a*.
60 Ibid. 165*a*.
61 Corp. Rec. v (10 Jan. 1738–9).
62 Ibid. vi (21 Oct. 1757, &c.).
63 A traffic count in May 1852 showed that 73,760 foot passengers, 2,436 carriages, 3,669 horses, and 3,250 beasts crossed the bridge in one week, between a quarter and a third of the traffic being on Saturday (market day) (Oldham, MS. Hist. Wisb. River, 180 (Wisb. Mus.)).
64 The Revd. Jeremiah Jackson attended a meeting on this subject on 28 August of that year (Diary (Wisb. Mus.)).
65 Gardiner, *Hist. Wisb.* 111–14.
66 Walker and Craddock, *Hist. Wisb.* 246. Thurloe was also elected for Cambridge borough, for which he preferred to sit.

an electorate of 224, the Rt. Hon. Charles Yorke, the bishop's nephew, received 82 Wisbech votes to 94 for Manners.[67] In 1830 the town reproduced the general situation in the county;[68] of the 248 votes cast in Wisbech 208 were Osborne, 129 for Adeane and 112 for Manners. Wisbech continued to support 'reform' candidates; in 1832 Captain Yorke, nephew of the Earl of Hardwicke, was placed at the bottom of the Wisbech poll, though he was returned by the county at large; he was placed third in the Wisbech poll in 1835. At the elections of 1832 and 1835 365 and 352 votes were respectively cast.[67]

SCHOOLS

A reference in the Archdeacon's Visitation Book of 1581–3[69] to a certain Mr. Belson of Wisbech St. Peter, who had become negligent in his duties as schoolmaster, probably indicates the presence of a primary school. The continuous history of primary education in the town, however, does not begin until a century later. A school for boys was opened in 1707 and one for girls in 1710; both schools were supported by private subscriptions.[70] In 1709 there were 50 boys at school, who had 'caps and bands &c. given them, to distinguish them from other children', and in 1710, when a permanent school building with a master's house had been built and the girls' school started, there were 63 boys and 40 girls attending. The latter were taught to read, spin, and 'cypher'. There was an unspecified number of other children in the town 'more privately taught by dames'.[71] Elizabeth Wright, by her will dated 1729, left house property, the rental of which at the end of the century was £70, to the boys' and land worth £12 yearly to the girls' school.[72] In 1798, when Dr. Abraham Jobson, later vicar of Wisbech and a keen educationist, reported to the bishop on the schools of Wisbech hundred,[73] there were 250 boys and 30 girls being taught at a cost of about £117 a year, of which rather less than half came from the Wright charity. The boys learnt reading, writing, arithmetic, and the catechism. The girls omitted writing and arithmetic but learnt sewing and knitting instead. In consequence of the maladministration of the charity, a Chancery suit was in progress when Jobson reported. It was settled in 1811, when it was decreed that a further £20 should be allotted to the boys' and £28 to the girls' school from the charity.[72] In 1847, when the total income was £326, £125 was applied to education, including a gratuity of £5 yearly to one boy leaving the National School. The residue was divided, roughly in accordance with the wishes of the foundress, as follows: £2 2*s*. for sermons on St. Paul's and St. Barnabas' Days, £85 to 100 aged churchgoing widows, £10 to 15 aged men, and £105 to 344 poor families.[74]

The schools were successful in attracting many legacies and subscriptions,[75] and during Jobson's vicariate (1802–28) both were rebuilt and united with the National Society. The boys' school (1811) was in Church Terrace, that for girls (1814) in Lower Hill Street. The cost of the latter building was defrayed by the Corporation, and Jobson and John Edes each gave £400 for the endowment and £50 for the equipment.[76] The national schools were well attended in the first half of the 19th century. In 1827 120 boys and 104 girls attended. In 1830 the numbers were 130 and 90, in 1846 215 and 160. In 1846 there were also 18 boys and 25 girls attending on Sundays only.[77]

In 1868–9 the building of St. Augustine's Church made it necessary to consider a Church school for the new parish. In 1874 it was decided to take the town as the unit of Anglican education and not to provide for the education of each sex in each parish. The result was that (St. Peter's) boys' school was rebuilt, at a cost of £1,910, to accommodate 262 boys, and a school for 70 girls and 130 infants was constructed in St. Augustine's parish at a cost of £1,205. Towards these separate costs over £2,000 was subscribed locally and £190 was granted by the National Society.[78] The buildings of St. Peter's School were improved in 1908–9 and can now accommodate 288 boys from 7 to 11. In 1943 218 were on the roll.[79] St. Augustine's School had by 1913 become overcrowded. There were 204 on the books but places for only 196. Very young children had to be refused admission. The opening of the Queen's School (q.v.) in 1928 afforded some relief, but St. Augustine's was again over-full in 1939.[80] The Lower Hill Street school continued after 1874 to be used for girls and infants. It became a Council school in 1924 and was closed in 1928. It was badly sited and throughout its existence its buildings were not altered.[81] The buildings are now used as auction rooms.

Another charity school had been opened in Deadman's Lane in 1803. Its original promoter was James Hill the Unitarian, who intended 'to carry out a system of training similar to that proposed by Robert Owen, of socialist notoriety, but failing in his [i.e. Hill's] object, it was purchased for its present purpose',[82] and conducted on British and Foreign Schools Society principles. James Smith, the first master, received a salary of 80 guineas, and the school became popular, with an average attendance of about 240 boys. In 1814 31 were admitted and 32 left. In the same year 230 were 'learning their letters', 100 were taught to write, and 84 were 'cyphering'. In 1815 it was decided to appoint one or more monthly visitors from among the subscribers 'whose occasional presence may invigorate the attention, and thereby accelerate the improvement, of the children'.[83] A separate British School for girls was built behind the Crescent in 1834 and provided education for about 100 children in its first year.[84] In 1840 a new boys' school was built in Victoria Road. The old boys' school was thereupon given over to infants.[85]

The numbers at the boys' school kept up fairly well in comparison with those at the National School for boys in Church Terrace; there were said to be 300 in 1841, the first full year of the new British boys' school,

[67] Poll Books at Wisb. Mus.

[68] *V.C.H. Cambs.* ii, 417–18.

[69] Ibid. 338.

[70] Ibid. 342–3. A subscription list for 1717 is preserved in the Bedf. County Rec. Office, ref. FN 1232.

[71] [S.P.C.K.], *An Account of Charity Schools . . .*, 1709, 1710.

[72] *Rep. Com. Char.* (*Cambs.*), [103], H.C. (1837–8), xxiv, s.v. Wisb. St. Peter.

[73] Bp.'s Mun., Ely, B 6.

[74] Gardner, *Dir. Cambs.* (1851), 633–4.

[75] Gardner's 1851 Directory prints a list of endowments and legacies, p. 610.

[76] Three 'foundation bricks' were laid by Jobson, Edes, and Jeremiah Jackson on 5 May 1814. Some contemporary coins were deposited by Edes at the Fisher's Lane corner of the building (Jackson's Diary, 23 Mar. 1813, 5 May 1814 (Wisb. Mus.)).

[77] Report made by Nat. Soc. in 1846–7 (privately printed) (Archives of Nat. Soc.).

[78] Archives of Nat. Soc.

[79] M. of Ed. File no. 12/65.

[80] Ibid. no. 12/64.

[81] M. of Ed. File no. 12/74 (Queen's School).

[82] Gardner, *Dir. Cambs.* (1851), 611.

[83] *Rep. Brit. and Foreign Schools Soc. 1816*, 'Extracts from Local Reps. and Corresp.', 111.

[84] Ibid. *1835*, 'Extracts . . .', 97.

[85] Gardner, *Dir. Cambs.* (1851), 611.

and later figures were 261 (1858), 254 (1859), 310 (1869), and 292 (1870).[86] In 1858–9 about three-quarters of the boys were being taught to write and about half were learning arithmetic. Eighteen were learning 'mensuration and algebra'. All the boys learnt geography and 'object drawing', and nearly one-half were instructed in history and in grammar and composition.[87] The average school life seems to have been about three years. It is noteworthy that nearly half (48 out of 102) who left in 1858–9 took up 'field work'—even in a town of 10,000 population. Two became pupil-teachers.[88]

These British Schools were not so successful as the National Schools in raising subscriptions, and it was partly their failure to do so that led to the establishment of a Wisbech School Board in 1876.[89] The Board took over the British Schools, replacing those for girls and infants by a new school in Elm Road in 1878–9;[90] it also opened a small school 3 miles out of the town along the South Brink.[91] The old infant school building, 'a fair room but in an inconvenient part of the town', was converted into a school for art, science, and technology. The girls' school, which stood next a graveyard, was condemned and demolished in the late 1870's. The boys' school was enlarged in 1880 and 1895 to a capacity of 300. It was closed in 1928, but reopened in 1930 as an annexe to the Elm Road school and so remains.[92] The Elm Road school was twice enlarged in the 'nineties to accommodate about 350 children. In 1929, in spite of the opening of the Queen's School (q.v.), numbers had exceeded this figure.[93]

In 1913 it was revealed that the recognized accommodation in the five primary schools of the town exceeded the numbers on the books by only 2 per cent., a dangerously small margin in a growing town. The County Council proposed to build new schools for boys, girls, and infants, to close the Lower Hill Street and Victoria Road Schools, and convert the St. Augustine's School into one for girls only. In 1915 a site was obtained by compulsory purchase at the north end of Queen's Road but the war stopped the scheme. The project was revived in 1925 and after some local opposition a senior school was erected. The plans provided for 320 boys and 320 girls of 11 and over, and the new school was completed, at a cost of over £20,000, in 1928. It was named the Queen's School, after Queen Mary, and was the first of its kind in the Isle. New classrooms for handicrafts and domestic science were added in 1947 and the school now takes the senior children from Leverington and Gorefield as well as Wisbech itself.[94]

The county boundary of the Isle and Norfolk, based on the old course of the Well Stream, ran through the built-up area until 1934 (see above). Such streets as Norwich Road and Bowthorpe Road were divided haphazard between the two counties. In general the numbers of children resident in the area of one local education authority but attending the schools of the opposite authority roughly balanced each other, and no financial adjustment was ever made.[95] But at intervals considerable local feeling was aroused owing to a temporary disturbance of the balance. For example in *c.* 1907 some 40 more Wisbech children were attending Walsoken schools than there were Walsoken children in Wisbech schools, and this fact was brought forward against the proposed new school in Ramnoth Road (see below). Again, St. Augustine's School in Wisbech was for a long time specially attractive to Walsoken children, who contributed to its chronic overcrowding.

The following three Walsoken schools were brought into Wisbech Borough under the 1934 boundary changes.

Kirkgate Street School, Walsoken,[96] was opened in 1858 as a National School. The cost, including the teacher's house, amounted to £458, towards which the Society made a grant of £30. The site (part of the glebe) was given by the rector. The original accommodation was 116, increased by enlargements in 1891 and 1896 to 169, and reduced by the 1910 reassessment to 130. It was transferred to the Walsoken School Board in 1891, and became a junior and infants' school in 1927.

The Walsoken School Board was established in 1875,[97] and the following year built a school in Norwich Road.[98] By this time most of the population was concentrated in New Walsoken on the Wisbech border. This school was a large one. There were originally 345 places, to which 100 were added in 1906. Expansion, however, did not keep pace with the increase in the population. The Norwich Road site was too small for any further enlargements, and in 1909–11, after much controversy (see above) the Norfolk County Council built a new school in Ramnoth Road[99] for 220 boys. Thereupon the Norwich Road school was reserved for girls and infants, for 310 of whom there was accommodation in 1942. Between 1927 and 1948 Ramnoth Road also took the senior boys from Kirkgate Street, but from the latter year all senior children from Walsoken have been taken by the Queen's School.

At the end of the 18th century there were also about five dame schools for boys, and the same number for girls, in the town.[1] In the late 'forties there were 15 private academies in Wisbech, 7 of which took boarders.[2] In 1868–9 there were 5 dame schools in St. Augustine's parish.[3]

A Sunday school, to which the capital burgesses granted £20 a year, had been set up in 1786 for children unable to go to the charity schools.[4] It was rebuilt in Deadman's Lane (Alexandra Road) in 1809 and liberally endowed by Jobson.

In 1835 the ratio of children at school to the total population was 1 in 9·6 in Wisbech—slightly below the county as a whole (1 in 9·4) but well above the country at large (1 in 10·7).[5]

It is noteworthy that except for the Girls' High School (1904) on the North Brink, all the Wisbech

86 *Rep. Brit. and Foreign Schools Soc. 1859*, 'Extracts . . .', 94; ibid. *1860*, 'Extracts . . .', 101; ibid. *1872*, 'Extracts . . .', 86.

87 These proportions seem to have been roughly constant in the period 1860–70.

88 *Rep. Brit. and Foreign Schools Soc. 1859*, 'Extracts . . .', 94.

89 Corresp. in Archives of Nat. Soc. relating to school provision in Wisbech St. Augustine's parish. The actual date of establishment of the Board was 28 Nov. 1876 (*Lond. Gaz.* 1876, 6527; Kelly, *Dir. Cambs.* (1900)).

90 Kelly, *Dir. Cambs.* (1900).

91 This school was closed in 1929, and is now situated in Elm parish (q.v.).

92 M. of Ed. File no. 12/66.

93 Ibid. no. 12/61.

94 Ibid. no. 12/74.

95 Since about 1925 Norfolk C. C. has made itself responsible for the education of senior children, and the Isle for juniors, in parishes like Elm, Emneth, Outwell, and Upwell which lie on the county boundary

96 M. of Ed. File no. 12/81.

97 *Lond. Gaz.* 1875, 5809.

98 M. of Ed. File no. 12/82.

99 Ibid. no. 12/83.

1 Bp.'s Mun., Ely, B 6.

2 Gardner, *Dir. Cambs.* (1851), 651.

3 Archives of Nat. Soc.

4 *V.C.H. Cambs.* ii, 346.

5 Ibid. 352.

schools are on the right bank of the river. This was used as an argument against the Queen's Road site in the early 1920's.[6]

MISCELLANEOUS INSTITUTIONS

Wisbech was an assize town from at least 1515–16[7] until 1836 when the liberty of the Isle of Ely was abolished. Assizes were held half-yearly, alternately with Ely. By the late 14th century quarter sessions were being held at Wisbech.[8] It has remained a quarter-sessions town ever since; the January and July sessions for the Isle are held here, those in April and October at Ely.[9] The Shire Hall on the east side of the Market Place was pulled down in 1810; it was replaced by the existing Sessions House and Police Station on the South Brink.[10] Since the demolition of the Wisbech Jail (see below) the weekly Petty Sessions of the Wisbech district have also been held here; the daily magistrates' court for the borough was held in the room over the butter market until the demolition of that building (1854),[11] since then at the Town Hall.

From the end of the 16th century the parish registers record the burial of executed criminals. The gallows stood on the river bank a little below the Horseshoe corner.[12] A jail was built by the capital burgesses in 1616, when 10,000 bricks were ordered for the purpose at 15*s.* 8*d.* a thousand.[13] The keeper's salary was fixed in 1624 at £13 6*s.* 8*d.*[14] In 1681 Wisbech successfully sued the county for the custody of this jail, but relinquished its rights in 1757. In 1720 the wall of the jail was rebuilt, and at the same time it was resolved that the 'farm' of the prisoners should be for not less than £14 a year.[15] A new building, erected in 1807 next the site of the new shire hall on the South Brink, was used for French prisoners of war. This jail was closed in 1846 and replaced by a building in Gaol Lane (now Victoria Road) designed by Basevi. This was a small prison (forty-three cells) and was closed and demolished in 1878, under the Prisons Act of the preceding year.[16]

Wisbech was also in some degree the county town of the Isle until the building of the county hall at March in 1908. A proposal made at the second meeting of the County Council (21 February 1889)[17] to hold meetings at March, Wisbech, and Ely in rotation, was not carried into effect, but the clerk of the council, the treasurer, the education officer and the inspector of weights and measures all had their offices in Wisbech in the early days of the administrative county.[18]

The first recorded post office in Wisbech was in Upper Hill Street, in 1793.[19] In 1851 it was in Cornhill,[20] and the present building in Bridge Street dates from 1885–7. It has been enlarged in 1898, 1904, and 1930–8.[21] A private company operated a telegraph service until 1870, when it was taken over by the Post Office authorities.[22] Telephone service was provided from 1898 by the National Telephone Company.[23]

A weekly newspaper known as the *Lynn and Wisbech Packet* was started in 1800, but the first newspaper actually printed in Wisbech was the *Wisbech Observer* (1813). It had a brief existence and no copies are known to be extant.[24] Another weekly, the *Star in the East*, ran from 1836 to 1840. Its editor was James Hill, the well-known Unitarian,[25] and its publisher Neil Walker, one of the historians of Wisbech. 'The influence it exerted was of such a dubious character that but few deplored its decease.'[26] A rival *Wisbech Gazette* lasted six months only (October 1837 to April 1838). During 1877 a daily paper, the *Telegram*, ran for a period of five months.[27] The present local newspapers are the *Isle of Ely and Wisbech Advertiser*, *South Lincs., Cambs. and West Norfolk Journal*, issued on Wednesdays and Saturdays, and the *Wisbech Standard* (founded 1868), issued on Fridays. The former was founded in 1845 by John Gardiner, whose family controlled the paper for over a hundred years.[28] It assumed its present name in 1887.[29]

The first Wisbech bank was opened in 1782 by Jonathan Peckover at his general shop in High Street, and was moved twelve years later to Bank House, North Brink. It was a branch of the Norwich firm of Gurney, and is now represented by Barclays Bank.[30] A savings bank was established in 1818, and collected £435 4*s.* 6*d.* in deposits in its first year.[31] Friendly societies started in Wisbech about twenty years later, the Ancient Shepherds and Oddfellows in 1837–8, the Foresters in 1841.[32]

As far back as 1654 some of the burgesses deposited their books for common use in a room over the church porch; Secretary Thurloe contributed eighty-one volumes to the library thus inaugurated.[33] A Literary Society was founded in 1781. Its books were catalogued in 1804, with a supplement in 1813.[34] By 1824 the Society had 84 members and a library of 2,500 volumes. The Wisbech Museum was founded in 1835.[35] The Literary Society and Museum were amalgamated in 1877.[36] The museum is an exceptionally good one for

[6] M. of Ed. File no. 12/74.

[7] The earliest Plea Roll preserved in the Bp.'s Mun. of Ely is of 7 Hen. VIII (A. Gibbons, *Ely Episc. Rec.* 116).

[8] *V.C.H. Cambs.* iv.

[9] Kelly, *Dir. Cambs.* (1933). A significant commentary on the social structure of the Isle in the middle of the 19th century is given by the information in Gardner's 1851 Directory (p. 612) that the presiding magistrate at the Wisbech Quarter Sessions was generally the Revd. Henry Fardell, vicar of the town, and at Ely the Revd. W. G. Townley, rector of Upwell. In fact, though not in name, the secular power of the Church survived the reforms of 1836; until the rise of a landed aristocracy amongst the wealthy farmers, the incumbents of rich livings, like Fardell, Townley, and Peyton of Doddington wielded enormous local influence.

[10] Gardiner, *Hist. Wisb.* 307. The pillory stood on the roof of the old Shire Hall.

[11] Gardner, *Dir. Cambs.* (1851), 612.

[12] Anon. late 18th-century diary in Wisb. Mus. The gallows is said to have been rebuilt after 1758 with timber from the old Wisbech Bridge. Jackson records an execution at the Horseshoe in his Diary, 10 July 1819.

[13] Corp. Rec. iii, 139.

[14] Ibid. iv, 45.

[15] Corp. Rec. vi; Walker and Craddock, *Hist. Wisb.* 425–6.

[16] *2nd Rep. Commrs. Prisons* [C. 2442], 28, H.C. (1878–9) xxxiv. Its site is still visible in the blocks of houses of a different pattern from the rest between Victoria Road and Cannon Street (so named from a former drill ground).

[17] *Camb. Chron.* 22 Feb. 1889.

[18] Kelly, *Dir. Cambs.* (1900); M. of Ed. files.

[19] P.O. Estab. Bks. 1760.

[20] Gardner, *Dir. Cambs.* (1851), 649.

[21] G.P.O. H.Q. Min. 3486/1929.

[22] Ibid. E 3989/1869.

[23] N.T.C. Dir.

[24] Gardiner, *Hist. Wisb.* 66–67.

[25] Ibid. 67–68.

[26] Gardiner, *Hist. Wisb.* 66–67. It was apparently a Chartist paper.

[27] Ibid. 73.

[28] Infm. Mr. G. S. Gardiner, a member of the family.

[29] Gardiner, *Hist. Wisb.* 68–72.

[30] Ibid. 192–3; W. H. Bidwell, *Annals of an E. Anglian Bank*, 363–4.

[31] Min. Bk. in Wisb. Mus.

[32] Gardiner, *Hist. Wisb.* 228, 292, 294.

[33] Hist. MSS. Com., *9th Rep. App.* 293.

[34] Diary of Rev. J. Jackson, 15 Feb. 1813 (Wisb. Mus.).

[35] Gardiner, *Hist. Wisb.* 197–8. A long and amusing list of the miscellaneous contents of the Museum in its early days is given in Gardner, *Dir. Cambs.* (1851), 617–20.

[36] Gardiner, *Hist. Wisb.* 197–8.

a town the size of Wisbech. A branch of the county library was established in 1947 in Alexandra Road, in the premises formerly used by the Art School and earlier still by the infants of the British School.[37]

A movement for a working men's club started in 1863, the parent body being the Young Men's Total Abstinence Society. An inaugural meeting on 5 January 1864 produced £24 16*s.* in subscriptions. During the first year 90 members joined, and by 1866 the club had 267 members. Its first headquarters were in hired rooms in Upper Hill Street, but in 1867 the club moved to its existing premises at Alfred House, Lower Hill Street. The new building was purchased for £800, of which £600 was subscribed by the Peckover family, who throughout its existence have been generous patrons. A new hall was built in 1871–2 and a gymnasium in 1873. By 1875 the membership amounted to 800. In 1898 the Wisbech Working Men's Club and Institute was the most financially successful of all English working men's clubs, having 1,136 members, a revenue of £1,818 and investments of £5,901. In 1904 there were 1,304 members, 376 of them women. During 1912, 15,636 books were issued from the library.[38] The membership in 1937 was about 1,300.[39]

In 1869 a site of 19 acres on Lynn Road was purchased by the Corporation from the Ecclesiastical Commissioners for £2,400 and laid out as a public park.[40] The North Cambridgeshire Hospital, overlooking the park, was opened in 1873. The cost (£8,000) was defrayed by Miss M. E. Trafford-Southwell, a descendant of the family formerly resident at the castle. She also provided £6,000 for endowment, increased by subscriptions to £10,000. In 1937 there were seventy-five beds at the Hospital, including twenty in a children's wing built as a memorial of the jubilee of King George V.[41] The local District Nursing Association was founded in 1898.[42] An earlier effort in the same direction had been Mrs. Mayer's Asylum, founded under the will (1811) of Judith Mayer of Wisbech. Her intention was to provide an 'asylum' for poor persons afflicted with palsy, rheumatism, gout, blindness, or any other complaint which might render them objects of compassion, and for this purpose she bequeathed £500 for a building and £1,400 for its endowment. The courts decided against her plan in its original form, but an ordinary almshouse was built (1815) on the canal side and endowed under this charity, each of the 5 inmates receiving £10 a year and an allowance of coals.[43] William Watson, the historian of Wisbech, by his will dated 1832, gave £1,000 towards the establishment of a poor persons' dispensary.[44] No further references to this institution have been noticed.

The street lamps which excited admiration in 1796[45] and were so severely reduced in 1823[46] were replaced in 1832 by gas lighting, the quality of which was for a long time unsatisfactory and the prices high.[47] The Wisbech Gas Light and Coke Company, with works in Eastfield, was formed in 1859.[48] Its present large gas-holder is a landmark visible from afar in many directions.

Cleansing of the streets was first undertaken in 1721;[49] in 1788 it cost £16 a year.[50] The paving of the market place was carried out as far back as 1549,[51] but was not extended to cover the whole town until the Improvement Act of 1810, when special commissioners were appointed for this purpose.[52]

An extension to the churchyard was made in 1832, and the victims of the cholera epidemic buried in it. In 1840 it was consecrated and used to the exclusion of the churchyard proper. The chapel is from the designs of Willis and Basevi.[53]

A cemetery was opened on Leverington Road in 1835. It has an attractive chapel, designed by William Adams in 1848, in the form of a little Doric temple.[54] The churchyard and its extension were closed for interments in 1855, and the Leverington Road cemetery was supplemented by one at Mount Pleasant Bank in 1881. The ground for both cemeteries was left unconsecrated, so that no division into Church of England and Nonconformist portions was necessary.[55]

A theatre existed in North End in 1792.[56] A new one was built in 1793, as part of the redevelopment of the castle estate.[57] It was used by companies from Lincoln and Norwich for about fifty years, and also for public meetings.[58] Macready played at this theatre in 1836, when James Hill the banker and Unitarian was owner.[59] It was not successful, however, probably owing to its bad situation in a back lane, and it was used for the last time as a theatre in 1847. After its abandonment as a theatre the building was used for a time by the Wesleyan Methodist Reformers,[60] and later as an art school. The building still stands and is used by a spiritualist society.[61]

In the late 18th century race meetings were held on a course beside Elm Road, near the turning to Elm village. In 1775 there were four entries for a purse of £50, but two of the horses died in the second heat, held an hour after the first. The last meeting, held in 1780, showed a balance of £121 6*s.* 5*d.*[62]

A Wisbech preservation society called the Wisbech Society and Preservation Trust Ltd. was founded in 1938.[63]

WORTHIES The most famous native of Wisbech is Thomas Clarkson (1760–1846), the slavery abolitionist.[64] Others include Richard Huloet (*fl.* 1552), lexicographer,[65] John Crane (1572–1652), apothecary, a great benefactor to the town; William Godwin the elder (1756–1836),

[37] Infm. the late E. J. Rudsdale.
[38] Gardiner, *Jubilee of the Working Men's Inst. at Wisb. 1864–1914.*
[39] Kelly, *Dir. Cambs.* (1937).
[40] Gardiner, *Hist. Wisb.* 185.
[41] Ibid.; Kelly, *Dir. Cambs.* (1937).
[42] Gardiner, *Hist. Wisb.* 186.
[43] *Rep. Com. Char.* (*Cambs.*) [103], H.C. (1837–8), xxiv, s.v. Wisbech St. Peter.
[44] Gardner, *Dir. Cambs.* (1851), 636.
[45] iv, 802.
[46] See above, p. 257, note 64.
[47] Gardiner, *Hist. Wisb.* 182–4
[48] Ibid.
[49] Corp. Rec. v (15 Nov. 1721).
[50] Ibid. vii.
[51] £17 2*s.* 2*d.* from the sale of the goods of the Trinity Guild was used for this purpose. See above, p. 256, n. 33.
[52] 50 Geo. III c. ccvi.
[53] Gardner, *Dir. Cambs.* (1851), 603.
[54] Ibid. 624.
[55] Gardiner, *Hist. Wisb.* 187–8.
[56] It is shown on the map of this date in Wisb. Mus.
[57] The building is described in *The Theatre Notebk.* iv, 21–23.
[58] e.g. by Cobbett in 1830. Until the erection of the present Town Hall on Cornhill in 1811 it was the only public hall in the town.
[59] Gardiner, *Hist. Wisb.* 286; H.O. 129/7/193. Infm. the late E. J. Rudsdale.
[60] Gardner, *Dir. Cambs.* (1851), 615.
[61] Infm. the late E. J. Rudsdale.
[62] *Fenland N. & Q.* vi, 151–2. The discontinuance of Wisbech Races was probably due to the fact that about this time, when the draining of the Fens spoilt the game shooting, most of the local gentry left the neighbourhood, and their manor houses (e.g. White Hall on the North Brink, and Needham Hall and the former Vaux manor house in Elm) were converted into farm-houses.
[63] *Annual Rep.* (1951).
[64] A biography by E. L. Griggs (1936).
[65] *D.N.B.* All references in this section are to this source, unless otherwise stated.

author of *Political Justice*, and father of Mary Wollstonecraft, Shelley's wife; James Pegge (*fl.* 1832–46), the first Baptist missionary to India; and Octavia Hill (1838–1912), the housing reformer and philanthropist.[66] Other notable men and women connected with the town are Jane Stuart (d. 1745), said to have been a natural daughter of James II, who sought refuge at Wisbech after her father's downfall, earned a living as a worsted spinner and became a Quaker;[67] Henry Banyer (*fl.* 1739), medical writer, who practised in the town; and Richard Wright (1764–1836), Unitarian missionary, minister of the Particular Baptists from about 1790 to 1810. The vicars of Wisbech include John Warkworth (d. 1500), Master of Peterhouse and reputed author of a chronicle of the reign of Edward IV, John Saul Howson (1816–85), Dean of Chester, and John Scott, brother of Sir Gilbert Scott the architect.

CHARITIES[68] The charities of Wisbech are very considerable, as is the case with most boroughs of ancient standing. The following date back beyond the middle of the 19th century; those of Elizabeth Wright, Abraham Jobson, Judith Mayer, and William Watson have already been dealt with.[69]

John Crane, the apothecary, by his will dated 1651, gave the Black Bull Inn (later the 'Three Tuns' and now the New Inn) with its shops, oil mills, and other appurtenances to the capital burgesses, with proviso that half the rents should go towards augmenting the salary of the Master of the Grammar School, and half towards providing corn and firing for poor persons whom the capital burgesses should nominate. In 1837 the Grammar School master drew £20 from this source. The revenue from this charity was increased in 1666, when, on the inclosure of Wisbech High Fen, 8 acres were allotted in respect of the inn; this piece of land was let at £10 15*s.* in 1851. In 1802 the Corporation sold the inn for £1,615 to redeem the land tax on their estates in general. The Commissioners of 1837 questioned the validity of this sale, so that after that date the Corporation paid over to Crane's charity the amount of tax redeemed from the income of their general property.

Crane also directed his executors to buy an estate, the rent of which was to go annually in rotation to Cambridge University and to four corporate towns (of which Wisbech was to be one). The towns were to make loans of £20, to a maximum of £200, to young men to enable them to set up in business, and were to apply the remainder to an annual sermon and the relief of various classes of honest poor. The estate so purchased was at Fleet (Lincs.). In 1837 it produced £396, of which £180 was in that year (which was Wisbech's 'turn') lent to 9 persons, £16 given to 18 persons in reduced circumstances through no fault of their own, £50 given in small sums to 281 'poor men and women in want' and the remainder used to pay rates and taxes. The Corporation were said to be anxious to increase the amount paid in larger sums to the more deserving cases; the donor in his will had warned them to beware of 'dissembling hypocritical persons'.

William Holmes, by will dated 1656, gave £300 for investment by the capital burgesses, the interest to be lent to poor tradesmen in sums of £10 for periods of three years free of interest. In 1837 there were eighteen such loans outstanding.[70]

Thomas Parke in 1628 gave the Bell Inn for the use of the poor. In 1837 the revenue from the lease (£151 including an adjoining house) was used, in about equal amounts, to provide clothing for poor persons and for payment of rates and taxes. In 1842, with the permission of the Court of Chancery, the Corporation sold the property to carry out a street improvement and invested the proceeds in 3½ per cent. annuities, the interest on which was about £49 annually in 1851. Parke also gave 28 acres of land at Elm to augment the salary of the Grammar School master.

Dr. Henry Hawkins, a native of Wisbech who died in London in 1631, gave £300 for building almshouses. Six were built the following year, and were in use up to 1835. By this date they had fallen into decay and were demolished, the site and materials being sold by auction for £644 10s. Six other almshouses, known as King John's, were sold at the same time for £45 and demolished; they stood at the east end of the churchyard. To replace the two sets, the Corporation built six double almshouses for about £900 in King's Walk. These still exist.

A set of four almshouses, called Stermyn's Almshouses, was erected in 1614 with £100 bequeathed by Mrs. Jacomin Stermyn.[71] They were replaced by a set of six two-room almshouses in 1813. These still exist in Stermyn Street, between Church Terrace and the canal. Five one-story cottages in Love Lane, called Castle Almshouses, were built in 1813 by Joseph Medworth, the purchaser and developer of the castle estate and himself a Wisbech charity boy.

William 1st Viscount Saye and Sele gave £100 in 1656 for clothing for the poor. His family, more generally associated with Oxfordshire, held land in Waldersea in the 17th and 18th centuries. Another 17th-century bequest to provide clothing for poor persons was the gift (1695) by Bishop Patrick of the revenues derived by the see of Ely from the butchers' Shambles in Wisbech. The Shambles were bought outright by the Corporation in 1811, and the rents in 1837 amounted to £17 10*s.* Richard Royce gave 16 acres in Wisbech High Fen for the same purpose; the rent in 1837 was £12. Alice Throckmorton's charity, of £25 (1678) for the clothing of the poor, seems to have disappeared by the middle of the 19th century. Later clothing charities included the Dorcas charity, originally established in 1813 by voluntary subscription, and endowed by Dr. Jobson with £50 a year in 1827. In 1846 558 garments and 4,796 yards of various stuffs were sold under this scheme to 557 persons, the charity advancing half the price. George Swaine gave £400 3 per cent. Consols to the churchwardens in 1829 to provide clothing for the poor.

There were several other charities of a more miscellaneous character. The oldest were those of Bartholomew Edwards (£10 in 1583) and John Williamson (£40 in 1589) for the general benefit of the poor. William Skortred in 1603 gave 12 acres of

66 A biography by E. Moberly Bell (1942).

67 *Fenland N. & Q.* ii, 178–80.

68 The sources for this section are the Cambs. Report of the Charity Commissioners (103, H.C. (1837–8), xxiv), of which pp. 275–330 deal with Wisbech, and R. Gardner's *Directory of Cambs.* for 1851, which contains a useful summary and statements of receipts and payments for the principal charities during 1848–9.

69 See pp. 266, 269.

70 For the history of the other part of Holmes's charity, relating to the Grammar School, see *V.C.H. Cambs.* ii, 328.

71 She also left £100 for the rebuilding or repair of the market cross, which was accomplished in 1614 (see above, p. 263).

pasture in Sayer's Field, Wisbech St. Mary. In 1837 the rents amounted to £21 10*s*. and provided a hundredweight of coals and a quartern loaf each to about 300 poor. Secretary Thurloe gave £150 in 1656 for apprenticing and £50 for making a road from the Corn Market to the little sluice. £10 was spent on the former object in 1836. Richmond Girling (1668) gave the Wisbech poor a share in 40*s*. rent from land at Stradbroke (Suffolk), payable to Wisbech every tenth year. This payment lapsed in 1808 but was recovered, with arrears, in 1833 through the efforts of William Watson, Colonel of Volunteers, brewer and historian. The charity of John Baxter (1793), of all his property in the funds (£1,220 in Navy 5 per cent. stock), was for the benefit of disabled churchgoing poor, and was applied by the Corporation in the 1830's to those who had fallen from good circumstances into unforeseen distress. There were eight recipients in 1837. Loake's charity of £200 (1801) was also mainly for the benefit of the churchgoing poor, and was when the Commissioners reported given to five widows in 20*s*. clothing vouchers. The munificent Dr. Jobson, whose total gifts to charity amounted to upwards of £4,000, provided for the distribution of bibles, prayer and hymn books, and the 'promotion of psalmody in the chapel of ease', as well as his direct gifts for the benefit of the National and Sunday Schools. The flood of charitable donations was by no means stayed in the middle of the 19th century, for three benefactions were made between 1837 and 1851. Mrs. Elizabeth Stevens gave £500 in 1838 and John Johnson £200 in 1841 for poor widows. Steed Girdlestone gave £100 in 1841 for Christmas dinners for the poor; each family was to have 7 lb. of beef.

ANALYSIS OF HEARTH TAX ASSESSMENTS, ISLE OF ELY, 1662, 1666, AND 1674

Introductory Note

The Hearth Tax was granted to the Crown as a perpetual revenue in 1662, by Act 14 Car. II c. 10; it was abolished in 1689, by Act 1 Wm. & Mary c. 10. Broadly the Tax consisted of a half-yearly payment of 1*s*. for each hearth in the occupation of each person worth more than 20*s*. yearly who was a local (church and poor) ratepayer. It had a far wider incidence than any permanent direct tax previously imposed or to be imposed until the 19th century. The Hearth Tax Assessments consist of lists of persons with the number of hearths in their occupation. In all Assessments most entries represent separate houses or dwelling places and almost all entries represent separate households. Because of the wide incidence of the Tax, the lists, when analysed, provide useful information, not obtainable from other contemporary sources, about the size, character and importance of places. For that reason the following tables have been prepared. It must, however, be emphasized that the entries analysed are but entries in a Tax Assessment of the responsible heads of households or families, or of house owners, with hearths in their occupation. Statistics based on these entries cannot be translated automatically into population statistics and it is unsafe to try to do so.

The Tax was collected half yearly from Michaelmas 1662 to Lady Day 1689, but Assessments from the local areas of administration were only returnable to Quarter Sessions and the Exchequer, and accounts for such areas were only audited in the Exchequer, from Michaelmas 1662 to Lady Day 1666 and from Michaelmas 1669 to Lady Day 1674. It is usually for those two periods only that records of local administrations are preserved among the Exchequer records at the Public Record Office and that Assessments are preserved among the records of Quarter Sessions. In the other periods, when the Tax was managed by Farmers or Commissioners and local accounts were audited at their central Hearth Office, the records are commonly wanting. This is regrettable, since there are good grounds for believing that the records so produced contained more names than those which survive.

Within the two periods for which there are records eight Assessments were usually produced in each local area. The Tax was first administered by the sheriffs and under them there were two Assessments: for Michaelmas 1662, which served also for the succeeding year, and for Lady Day 1664. Under the Receivers, who next administered the Tax, there was one Assessment: for Michaelmas 1664, which was based on a revision of that of Michaelmas 1662. In most places this Assessment served also for the succeeding year. There was one Assessment under the mixed administration of Lady Day 1666. Under the second administration of the Receivers there were four Assessments, of which the earliest and latest usually served respectively for the year and a half ending at Michaelmas 1670 and for the year ending at Lady Day 1674. The Assessments of this last administration, within each local area, were usually made on the same principle. This principle was similar to that on which the Lady Day 1664 Assessment was prepared. It differed, however, from the principles on which the Michaelmas 1662, Michaelmas 1664 and Lady Day 1666 Assessments were prepared, and these in turn differed among themselves. Ready comparison between successive Assessments, where these survive, is therefore usually impossible. For this and other reasons it has been considered desirable to analyse all three of the Cambridgeshire Assessments which have been perfectly preserved. They are the Assessments for Michaelmas 1662, Lady Day 1666, and Lady Day 1674.

The roll for Michaelmas 1662 [E 179/84/436] was the first Assessment of the Tax. In Cambridgeshire as elsewhere, this Assessment was prepared very rapidly in the summer of 1662. The roll consists of lists made locally by parish constables and officials. These lists were enrolled by the Clerk of the Peace, certified by the Justices of the Peace at General Quarter Sessions on 17 July 1662 and delivered to the King's Remembrancer's Office of the Exchequer on 14 August 1662. Except for the Colleges at Cambridge and Ely each entry professes to represent a 'house'. For the Colleges the phrase 'severall Lodgeings and Chambers' replaces 'house'. Provision for enrolling, as distinct from granting, exemptions to those who were not, by poverty or non-payment of rates, *ipso facto* exempt was not made until 1663, by Act 15 Car. II c. 13. This Assessment, therefore, was intended to do no more than list all those legally bound to pay; though an unknown number of persons recorded in it was subsequently granted exemption certificates. In general it includes fewer names than either of the other Assessments but it provides the best means of comparing places in one county with those in another, since details of more of the Assessments for Michaelmas 1662 are preserved than of those of any one later date.

The books for Lady Day 1666 [E 179/244/22] are of a different character. By patent, 30 March 1666, the Tax had been farmed from Lady Day 1666, but, in order to effect repayment of outstanding loans by the City of London which had been secured on this revenue, the sub-farmers, for this one collection only, acted as Receivers and accounted in the Exchequer. But there were considerable difficulties over these accounts, and the Cambridgeshire books, like others which survive for this collection, were made out after the tax had been collected and were submitted directly to the Exchequer, without enrolment or certification by the Justices. The lists were made by the sub-farmer's officers and are drawn up under the headings of 'Paid' and 'Unpaid'; many of the persons included in the latter were no doubt legally entitled to exemption, but the lists make no attempt to indicate them. The entries give simply the names and numbers of hearths. The accuracy of the books was sworn to by George Wilmot, the sub-farmer, when his account was taken on 1 December 1671.

The roll for Lady Day 1674 [E 179/244/23] consists of lists made by the Receiver's sub-collectors, with the assistance of parish officials. It was enrolled by the Clerk of the Peace, certified by the Justices at General Quarter Sessions on 8 October 1674 and delivered to the Exchequer on 20 November 1674. The lists are divided into taxpayers and those legally exempt by certificate; for a few places which were parishes the lists of exempt are not given, while with some places which were chapelries they appear to be included in the exempt for the parent parish. No details are given of paupers, for they did not require a certificate to secure exemption, but an exhaustive study of overseers' records would be necessary before one could be sure that, as a rule, no paupers were included among the certified exempt. Among the taxpayers the owners of empty or recently destroyed houses and the occupiers of houses recently built or with recently added hearths in them are distinguished, while there are several entries where payment is said to be in arrears without any of these causes being added. Such hearths are entered to the left of the perpendicular line in the lists of taxpayers (normal hearths being entered to the right of that line), but are nevertheless included in the total of taxpaying hearths.

The three documents have been analysed on the same principle. Against each place is given, in separate columns, the total number of persons recorded as occupiers of from one to twelve hearths, while a further column gives the occupiers of thirteen or more hearths; final columns give the total number of entries and hearths in each place. For 1662 only one line is needed for each place. For 1666 three lines are generally needed, for those recorded as having paid, those unpaid and for the total of the two categories. For 1674 four lines are generally needed, for the taxpayers, for those among the taxpayers who owned empty, ruined, or newly built houses or whose hearths are entered to the left of the line, for the certified exempt and for the total of the three categories; except where, for a particular place, there is no list of exempt or no empty houses, &c. The 1662 and 1674 rolls list the University, the borough of Cambridge, the Isle, and Cambridgeshire in that sequence, the hundreds being in alphabetical order; the 1666 books list the University, Cambridge, Cambridgeshire, and the Isle in that sequence, the hundreds being in no particular order. Here the places, within hundreds, are given in alphabetical order of parishes but Thorney and Whittlesey, which in the 17th century seem still to have kept some of their character as extra-hundredal liberties, are shown separately. Attention is drawn in notes to places which in one document are represented by a separate list but in another are merged with the parent parish; to the few entries which plainly do not represent households or houses where persons normally lived; and to differences (commonest in 1674) between the total number of hearths in the recorded entries and the total given in the document.

Apart from the three documents already mentioned, there is a roll (slightly imperfect) for Michaelmas 1664 [E 179/84/437], which contains a very thorough revision of the 1662 Assessment; it was enrolled by the Clerk of the Peace and certified by Justices at General Quarter Sessions on 11 January 1666. It has not been analysed, partly because its revisionary method does not lend itself readily to compression into tables and partly because there is some inconsistency in its indication of the exempt. It is, however, almost as comprehensive as the 1674 Lady Day Assessment, for it gives a gross addition of 1,173 entries to the 3,793 recorded for the Isle in 1662 and a net total for the Isle (after deducting errors in the 1662 Assessment, houses ruined since then, &c.) of some 4,900 entries,[1] compared with the total 5,091 in 1674. Its lists should therefore not be neglected by those concerned with the population of a particular parish or district.

[1] The additions are: in Ely, 103; Ely hundred, 66; Thorney, 32; Whittlesey, 185; Wisbech hundred, 458 (of which 321 in Wisbech itself which includes Wisbech St. Mary, &c.); Witchford hundred, south part, 206; Witchford hundred, north part, 123. The deductions total 66: 43 entries of houses burnt or ruined whose inhabitants have left, 6 of houses being rebuilt after destruction and 17 entries said to be mistakes.

ANALYSIS OF ISLE OF ELY HEARTH TAX

1662 Michaelmas

	Totals of entries, with hearths													*Totals*	
	1	*2*	*3*	*4*	*5*	*6*	*7*	*8*	*9*	*10*	*11*	*12*	*over 12*	*Entries*	*Hearths*
ELY															
Trinity	166	111	58	22	17	6	6	4	1	2	2	..	1 of 13	396	909
St. Mary	90	43	24	16	4	3	4	1	1	..	..	..	1 of 13	187	408
College[1]	6	4	1	..	3	3	4	..	1	..	1	..	1 of 18	24	116
Chettisham[1]	13	10	2	1	1	..	..	1	..	..	..	..	..	28	46
Stuntney	6	13	3	2	..	..	..	..	..	..	..	..	1 of 15	25	64
Total	281	181	88	41	25	12	14	6	3	2	3	..	4	660	1,553
ELY HUNDRED															
Downham	69	42	11	..	3	..	..	..	..	1	..	..	..	126	211
Littleport	71	70	13	8	..	2	..	1	..	..	..	..	..	165	302
Total	140	112	24	8	3	2	..	1	..	1	..	..	..	291	513
THORNEY[2]	106	39	6	6	2	1	..	..	1	..	..	..	..	161	251
WHITTLESEY[2]	319	159	21	9	4	1	..	..	..	2	..	..	{1 of 16, 1 of 15}	517	813
WISBECH HUNDRED															
Elm	12	25	12	8	1	1	1	1	..	1	..	..	1 of 17	63	183
Leverington[3]	34	48	17	14	2	5	..	..	1	..	..	..	..	121	286
Newton	15	9	7	3	2	..	..	..	..	2	..	..	..	38	96
Outwell	20	21	4	3	1	1	..	1	..	..	..	..	1 of 13	52	118
Tydd St. Giles	24	19	8	3	1	..	..	1	..	..	..	..	..	56	111
Upwell	26	20	8	1	3	2	..	1	..	..	1	..	..	62	140
Welney	10	10	..	2	2	..	..	..	..	..	..	..	..	24	48
Wisbech[4]	100	113	58	46	19	19	9	4	3	4	1	3	{1 of 25, 1 of 14}	381	1,141
Total	241	265	114	80	31	28	10	8	4	7	2	3	4	797	2,123
WITCHFORD HUNDRED[5]															
Coveney[6]	23	17	1	3	..	..	..	..	..	..	..	..	1 of 13	45	85
Manea	21	15	2	..	..	..	..	..	..	..	..	..	..	38	57
Haddenham	72	82	17	8	6	3	1	2	..	1	..	..	1 of 17	193	417
Mepal	7	18	3	2	1	..	..	1	..	..	..	..	..	32	73
Stretham	42	30	15	4	1	2	..	..	..	..	..	..	..	94	180
Thetford	8	20	4	..	1	1	..	..	..	1	..	..	..	35	81
Sutton	45	80	10	6	..	1	..	1	..	..	..	..	..	143	273
Wentworth	3	9	3	1	..	1	..	..	1	..	..	..	..	18	49
Wilburton	23	24	6	5	..	2	1	..	..	1	..	..	1 of 18	63	156
Witcham	23	24	3	1	3	2	..	1	2	..	1	..	..	60	148
Witchford	26	18	4	3	2	..	..	..	..	..	..	..	..	53	96
Chatteris	79	91	10	6	1	..	1	..	..	..	..	1	..	189	339
Doddington	39	24	7	3	1	1	..	..	..	..	1	..	1 of 22	77	164
Benwick	18	13	1	..	..	..	..	..	..	..	..	..	..	32	47
March	125	63	23	6	6	1	2	..	..	..	..	..	..	226	394
Wimblington	40	18	5	5	..	..	1	..	..	..	..	..	..	69	118
Total	594	546	114	53	22	14	6	5	3	3	2	1	4	1,367	2,677

[1] *Ely.* The lists for the *College* and *Chettisham* are given separately after the list for *St. Mary* but they have no heading and the hearths are included in the total for *St. Mary.* Among the College entries are the *Porter's Lodge* and *Schoolhouse,* each with one hearth.

[2] *Whittlesey* and *Thorney* are listed at the end of Wisbech hundred and their hearths are included in its total. *Whittlesey* includes the inhabitants of *Coates, Eastrea* and *Pondsbridge,* who are listed separately in 1666.

[3] *Leverington* includes the inhabitants of *Parson Drove,* who are listed separately in 1666 and 1674.

[4] *Wisbech* includes the inhabitants of *High Fen, Murrow and Guyhirn, Tholomas Drove* and *Wisbech St. Mary,* who are separately listed in 1666.

[5] The places which later constituted the two divisions of this hundred are grouped together.

[6] *Coveney* includes the inhabitants of *Wardy Hill,* who are listed separately in 1666.

ANALYSIS OF HEARTH TAX ASSESSMENTS

1666 Lady Day

	Totals of entries, with hearths													Totals	
	1	2	3	4	5	6	7	8	9	10	11	12	over 12	Entries	Hearths
ELY															
Trinity, paid	221	122	50	28	25	8	6	5	4	2	1	2	{1 of 15, 1 of 13}	476	1,101
Stuntney, paid	4	11	2	2	..	..	..	..	..	..	..	..	1 of 15	20	55
both, unpaid	77	18	7	5	1	..	..	..	..	..	..	..	..	108	159
both, total[1]	302	151	59	35	26	8	6	5	4	2	1	2	3	604	1,315
St. Mary, paid	92	45	15	8	5	1	3	1	3	..	..	..	..	173	346
Chettisham, paid	7	9	2	1	..	..	..	1	..	..	..	..	..	20	43
both, unpaid	39	7	3	1	..	..	..	..	..	..	..	..	..	50	66
both, total	138	61	20	10	5	1	3	2	3	..	..	..	..	243	455
Total	440	212	79	45	31	9	9	7	7	2	1	3	3	847	1,770
ELY HUNDRED															
Downham, paid	56	51	3	..	2	..	..	..	..	1	..	..	..	113	187
unpaid	15	9	1	1	..	..	..	..	..	..	..	..	..	26	40
total	71	60	4	1	2	..	..	..	..	1	..	..	..	139	227
Littleport, paid	81	65	9	3	..	1	..	..	..	..	..	..	..	159	256
unpaid	38	7	2	..	..	..	..	..	..	..	..	..	..	47	58
total	119	72	11	3	..	1	..	..	..	..	..	..	..	206	314
Total	190	132	15	4	2	1	..	..	..	1	..	..	..	345	541
THORNEY[2]															
paid	116	29	18	1	3	1	..	..	..	1	..	..	..	169	263
unpaid	19	4	2	1	1	..	..	..	..	..	..	..	..	27	42
Total	135	33	20	2	4	1	..	..	..	1	..	..	..	196	305
WHITTLESEY[2]															
Whittlesey, paid	216	119	20	8	3	2	..	..	..	..	..	1	{1 of 16, 1 of 13}	371	614
unpaid	65	18	6	4	2	1	..	..	..	..	..	..	..	96	151
total	281	137	26	12	5	3	..	..	..	..	..	1	2	467	765
Coates, paid	40	7	1	..	1	..	..	..	..	..	..	..	..	49	62
unpaid	21	2	1	..	..	..	..	..	..	..	..	..	..	24	28
total	61	9	2	..	1	..	..	..	..	..	..	..	..	73	90
Eastrea, paid	28	22	2	1	..	1	..	..	..	..	..	..	..	54	88
unpaid	7	3	2	1	..	..	..	..	..	..	..	..	..	13	23
total	35	25	4	2	..	1	..	..	..	..	..	..	..	67	111
Pondsbridge, paid	55	8	..	..	..	..	..	..	..	..	..	..	..	63	71
unpaid	13	9	..	..	..	..	..	..	..	..	..	..	..	22	31
total	68	17	..	..	..	..	..	..	..	..	..	..	..	85	102
Total	445	188	32	14	6	4	..	..	..	..	..	1	2	692	1,068
WISBECH HUNDRED															
Elm, paid	21	35	4	5	2	4	..	1	1	1	..	..	1 of 13	75	197
unpaid	5	3	2	1	1	..	..	..	..	..	..	..	..	12	26
total	26	38	6	6	3	4	..	1	1	1	..	..	1	87	223
Leverington, paid	15	37	6	5	3	1	3	..	1	..	..	..	..	71	178
unpaid	9	6	3	..	..	..	..	..	..	..	..	..	..	18	30
total	24	43	9	5	3	1	3	..	1	..	..	..	..	89	208
Parson Drove, paid	20	21	3	3	..	..	..	..	..	..	..	..	..	47	83
unpaid	2	1	2	..	..	..	..	..	..	..	..	..	..	5	10
total	22	22	5	3	..	..	..	..	..	..	..	..	..	52	93
Newton, paid	8	12	6	..	3	1	..	..	..	..	1	..	..	31	82
unpaid	3	3	1	1	..	..	..	..	..	..	..	..	..	8	16
total	11	15	7	1	3	1	..	..	..	..	1	..	..	39	98
Outwell, paid	23	17	3	2	..	1	..	1	..	1	..	..	..	48	98
unpaid	2	1	1	1	..	1	..	..	..	..	..	..	..	6	17
total	25	18	4	3	..	2	..	1	..	1	..	..	..	54	115
Tydd St. Giles, paid	33	19	2	4	1	..	1	..	..	..	..	..	..	60	105
unpaid	7	4	1	2	..	..	..	..	..	..	..	..	..	14	26
total	40	23	3	6	1	..	1	..	..	..	..	..	..	74	131
Upwell, paid	44	17	3	..	5	2	..	..	..	..	1	..	..	72	135
unpaid	5	3	3	..	1	..	..	..	..	..	..	..	..	12	25
total	49	20	6	..	6	2	..	..	..	..	1	..	..	84	160

Continued

[1] *Ely.* The inhabitants of the college are included in the Trinity list but are not separated.

[2] *Whittlesey* and *Thorney* are listed at the beginning of Witchford hundred and their hearths are included in its total.

1666 Lady Day (cont.)

	Totals of entries, with hearths													Totals	
	1	2	3	4	5	6	7	8	9	10	11	12	over 12	Entries	Hearths
WISBECH HUNDRED, *Continued*															
Welney, paid	10	6	..	2	1	..	..	..	..	..	..	..	..	19	35
unpaid	7	4	1	..	..	..	..	..	..	..	..	..	..	12	18
total	17	10	1	2	1	..	..	..	..	..	..	..	..	31	53
Guyhirn, paid	26	4	1	..	..	..	..	..	..	..	..	..	..	31	37
unpaid	1	2	1	..	..	..	..	..	..	..	..	..	..	4	8
total	27	6	2	..	..	..	..	..	..	..	..	..	..	35	45
Murrow, paid	13	12	1	1	..	..	..	..	..	..	..	..	..	27	44
unpaid	2	2	1	..	..	..	..	..	..	..	..	..	..	5	9
total	15	14	2	1	..	..	..	..	..	..	..	..	..	32	53
Tholomas Drove, paid	4	6	..	1	..	..	..	..	..	..	..	..	..	11	20
unpaid	1	..	1	..	..	..	..	..	..	..	..	..	..	2	4
total	5	6	1	1	..	..	..	..	..	..	..	..	..	13	24
Wisbech High Fen, paid	12	2	..	..	..	..	..	..	..	..	..	..	..	14	16
Guyhirn, &c., total	59	28	5	2	..	..	..	..	..	..	..	..	..	94	138
Wisbech St. Mary, paid	28	29	11	8	1	2	1	..	1	1	..	..	..	82	194
unpaid	6	4	3	2	1	..	..	..	..	..	..	..	..	16	36
total	34	33	14	10	2	2	1	..	1	1	..	..	..	98	230
Wisbech St. Peter, paid	181	84	34	42	22	14	9	3	5	1	3	3	1 of 24	402	1,048
unpaid	22	10	6	4	2	1	..	..	..	..	..	..	..	45	92
total	203	94	40	46	24	15	9	3	5	1	3	3	1	447	1,140
Total	510	344	100	84	43	27	14	5	8	4	5	3	2	1,149	2,589[3]
WITCHFORD HUNDRED[4]															
Coveney, paid	15	4	1	2	..	1	..	..	..	..	..	..	..	23	40
unpaid	..	1	..	1	..	..	..	..	..	..	..	..	..	2	6
Wardy Hill, paid	14	7	..	..	..	..	..	..	..	..	..	..	..	21	28
unpaid	6	1	..	..	..	..	..	..	..	..	..	..	..	7	8
both, total	35	13	1	3	..	1	..	..	..	..	..	..	..	53	82
Manea, paid	38	10	2	1	..	..	..	..	..	..	..	..	..	51	68
unpaid	9	2	2	..	..	..	..	..	..	..	..	..	..	13	19
total	47	12	4	1	..	..	..	..	..	..	..	..	..	64	87
Haddenham, paid	89	82	10	6	..	1	4	..	..	1	..	1	..	194	363
unpaid	27	13	5	2	1	..	1	..	..	..	..	..	..	49	88
total	116	95	15	8	1	1	5	..	..	1	..	1	..	243	451
Mepal, paid	12	17	2	..	..	..	..	1	..	..	..	..	..	32	60
unpaid	2	1	1	..	..	..	..	..	..	..	..	..	..	4	7
total	14	18	3	..	..	..	..	1	..	..	..	..	..	36	67
Stretham, paid	47	27	7	2	..	..	..	..	..	1	..	..	..	84	140
unpaid	20	8	5	3	..	..	1	..	..	..	..	..	..	37	70
total	67	35	12	5	..	..	1	..	..	1	..	..	..	121	210
Thetford, paid	12	13	4	2	..	1	..	..	..	..	..	..	..	32	64
unpaid	..	2	1	2	..	..	..	..	..	..	..	..	..	5	15
total	12	15	5	4	..	1	..	..	..	..	..	..	..	37	79
Sutton, paid	67	55	7	3	..	..	1	1	..	..	..	..	..	134	225
unpaid	18	9	1	2	..	..	..	..	..	..	..	..	..	30	47
total	85	64	8	5	..	..	1	1	..	..	..	..	..	164	272
Wentworth, paid	12	6	3	1	..	1	..	..	..	..	..	..	..	23	43
unpaid	..	2	..	1	..	1	..	..	..	..	..	..	..	4	14
total	12	8	3	2	..	2	..	..	..	..	..	..	..	27	57
Wilburton, paid	20	16	3	4	..	2	2	..	..	1	..	..	1 of 18	49	131
unpaid	2	8	..	2	1	..	..	..	..	..	..	..	..	13	31
total	22	24	3	6	1	2	2	..	..	1	..	..	1	62	162
Witcham, paid	30	21	2	2	1	1	..	1	1	1	..	..	..	60	124
unpaid	3	3	1	1	..	..	..	..	..	..	..	..	..	8	16
total	33	24	3	3	1	1	..	..	..	..	..	..	..	68	140
Witchford, paid	24	15	4	1	1	..	..	..	..	..	..	..	..	45	75
unpaid	7	5	1	1	..	..	..	..	..	..	..	..	..	14	24
total	31	20	5	2	1	..	..	..	..	..	..	..	..	59	99
Chatteris, paid	132	64	4	1	..	..	..	1	..	1	..	..	1 of 18	204	312[5]
unpaid	6	3	1	2	1	..	..	..	..	..	..	..	..	13	28
total	138	67	5	3	1	..	..	1	..	1	..	..	1	217	340

Continued

[3] The total of hearths for the hundred is totted in the book as 2,254 paid+315 unpaid = 2,569; it is actually 2,272+317 = 2,589.
[4] Except for Chatteris, the places which later constituted the two divisions of this hundred are grouped together.
[5] Owing to a wrong addition on p. 46 of the book the total is given as 313.

1666 Lady Day (cont.)

	Totals of entries, with hearths													*Totals*	
	1	*2*	*3*	*4*	*5*	*6*	*7*	*8*	*9*	*10*	*11*	*12*	*over 12*	*Entries*	*Hearths*
WITCHFORD HUNDRED, *Continued*															
Doddington, paid	49	17	4	2	1	..	..	2	..	..	..	1	..	76	136[6]
unpaid	6	2	2	1	..	..	..	..	..	..	..	..	..	11	20
total	55	19	6	3	1	..	..	2	..	..	..	1	..	87	156
Benwick, paid	19	9	..	2	..	..	..	..	..	..	..	..	..	30	45
unpaid	6	2	..	..	..	..	..	..	..	..	..	..	..	8	10
total	25	4	..	2	..	..	..	..	..	..	..	..	..	38	55
March, paid	220	51	10	4	3	2	1	..	..	1	..	..	..	292	412
unpaid	12	5	3	3	..	..	..	..	..	..	..	..	..	23	43
total	232	56	13	7	3	2	1	..	..	1	..	..	..	315	455
Wimblington, paid	40	12	3	1	..	1	..	..	..	..	..	..	..	57	83
unpaid	7	5	1	1	..	..	..	..	..	..	..	..	..	14	24
total	47	17	4	2	..	1	..	..	..	..	..	..	..	71	107
Total	971	498	90	56	9	11	10	6	1	6	..	2	2	1,662	2,819

6 Owing to a wrong addition on p. 33 of the book the total is given as 135.

1674 Lady Day

	Totals of entries, with hearths													*Totals*	
	1	*2*	*3*	*4*	*5*	*6*	*7*	*8*	*9*	*10*	*11*	*12*	*over 12*	*Entries*	*Hearths*
ELY															
Trinity,[1] charged	162	121	64	36	23	12	9	6	7	..	2	4	{1 of 14, 1 of 13}	448	1,198[2]
empty, &c.	4[3]	9	..	..	1	..	..	..	..	..	..	..	..	13[3]	27[2,3]
exempt	47	..	..	..	..	..	..	..	..	..	..	..	..	47	47
Stuntney, charged	2	13	4	2	..	..	..	..	..	..	1	..	..	22	59
empty, &c.	1	..	..	..	..	..	..	..	..	..	..	..	..	1	1
total	216	143	68	38	24	12	9	6	7	..	3	4	2	531	1,332
St. Mary, charged	82	48	19	18	7	4	5	2	1	1	..	..	1 of 20	188	456
empty, &c.	1[4]	1	..	..	..	1[4]	..	..	..	..	..	..	..	1[4]	9[4]
exempt	37	..	..	..	..	..	..	..	..	..	..	..	..	37	37
Chettisham, charged	4	6	3	1	1	..	1	..	..	..	..	..	..	16	41
empty, &c.	..	3	..	..	..	..	..	..	..	..	..	..	..	3	6
total	124	58	22	19	8	5	6	2	1	1	..	..	1	245	449
Total	340	201	90	57	32	17	15	8	8	1	3	4	3	776	1,881
ELY HUNDRED															
Downham, charged	72	45	6	1[5]	2	2	1	..	..	..	..	..	..	129	213
empty, &c.	1	..	..	..	..	..	..	..	..	..	..	..	..	1	1
exempt	33	..	..	..	..	..	..	..	..	..	..	..	..	33	33
total	106	45	6	1	2	2	1	..	..	..	..	..	..	163	247
Littleport, charged	70	68[6]	15	7	1	2	..	..	..	..	..	..	..	163	296
empty, &c.	..	1	..	..	..	..	..	..	..	..	..	..	..	1	2
exempt	84	1	..	..	..	..	..	..	..	..	..	..	..	85	86
total	154	70	15	7	1	2	..	..	..	..	..	..	..	249	384
Total	260	115	21	8	3	4	1	..	..	..	..	..	..	412	631
THORNEY[7]															
charged	52	42	27	4	4	3	..	..	..	..	..	1	..	133	283[9]
empty, &c.	7[8]	..	1	..	1	..	..	..	..	..	..	..	..	9	15[9]
exempt	24	1	..	..	..	..	..	..	..	..	..	..	..	25	26
Total	83	43	28	4	5	3	..	..	..	..	..	1	..	167	324
WHITTLESEY[10]															
charged	227	208	55	14	7	2	1	..	3	..	1	2	2 of 14	522	1,008[11]
empty, &c.	11[12]	6[13]	2	..	..	..	..	..	..	..	..	..	..	19	29[11]
exempt	119	..	..	..	..	..	..	..	..	..	..	..	..	119	119
Total	357	214	57	14	7	2	1	..	3	..	1	2	2	660	1,156

1 *Ely, Trinity* includes the inhabitants of the College, but they are not separately listed; they include one hearth in *The Singing School* and two *In the Minster*.

2 The total is given as 1,227.

3 One newly built hearth was in an existing house.

4 One newly built hearth was in an existing house and six were in the Bishop's Palace.

5 These are said to be in two houses.

6 These include *The Quaker Meeting House*, 2 hearths.

7 *Thorney* is listed in the middle of *Witchford Hundred*.

8 These include one forge, in arrears.

9 The total is given as 299.

10 *Whittlesey* is listed in the middle of *Witchford Hundred*. It includes the inhabitants of *Coates*, *Eastrea* and *Pondsbridge*, separately listed in 1666.

11 The total is given as 1,044.

12 These include one hearth in a corn chamber, in arrears.

13 Two of these entries are said to be for two houses.

1674 Lady Day (cont.)

	Totals of entries, with hearths													*Totals*	
	1	*2*	*3*	*4*	*5*	*6*	*7*	*8*	*9*	*10*	*11*	*12*	*over 12*	*Entries*	*Hearths*
WISBECH HUNDRED															
Elm, charged	13	31	12	6	3	3	..	2	..	..	..	..	1 of 17 1 of 13	72	214
empty, &c.	..	2	..	..	..	..	..	..	..	..	..	..	..	2	4
exempt	17	..	..	..	..	..	..	..	..	..	..	..	..	17	17
total	30	33	12	6	3	3	..	2	..	..	..	..	2	91	235
Leverington, charged	29	37	5	11	1	3	1	1	1	..	..	..	..	89	209
empty, &c.	..	3	..	..	..	..	..	..	..	..	..	..	..	3	6
exempt	11	..	..	..	..	..	..	..	..	..	..	..	..	11	11
total	40	40	5	11	1	3	1	1	1	..	..	..	..	103	226
Parson Drove, charged	13	18	6	1	1	..	..	..	..	..	..	..	..	39	76
empty, &c.	3	2	..	..	..	..	..	..	..	..	..	..	..	5	7
exempt	14	..	..	..	..	..	..	..	..	..	..	..	..	14	14
total	30	20	6	1	1	..	..	..	..	..	..	..	..	58	97
Newton, charged	12	12	6	1	4	..	..	..	1	..	1	..	..	37	98
empty, &c.	..	2	..	..	..	..	..	..	..	..	..	..	..	2	4
exempt	6	..	..	..	..	..	..	..	..	..	..	..	..	6	6
total	18	14	6	1	4	..	..	..	1	..	1	..	..	45	108
Outwell, charged	13	16	5	1	1	1	..	1	..	..	..	..	1 of 13	39	96[14]
empty, &c.	..	1	..	..	..	..	..	..	..	..	..	..	..	1	2[14]
exempt	7	..	..	..	..	..	..	..	..	..	..	..	..	7	7
total	20	17	5	1	1	1	..	1	..	..	..	..	1	47	105
Tydd St. Giles, charged	19	21	7	4	2	..	1	..	..	..	..	..	..	54	115
empty, &c.	1	1	..	..	..	..	..	..	..	..	..	..	..	2	3
exempt	13	..	..	..	..	..	..	..	..	..	..	..	..	13	13
total	33	22	7	4	2	..	1	..	..	..	..	..	..	69	131
Upwell, charged	24	24	5	..	7	1	1	1	..	..	1	..	..	64	154
empty, &c.	3	3	..	..	..	..	..	..	..	..	..	..	..	6	9
exempt	26	..	..	..	..	..	..	..	..	..	..	..	..	26	26
total	53	27	5	..	7	1	1	1	..	..	1	..	..	96	189
Welney, charged	7	11	..	4	..	..	..	..	..	..	..	..	..	22	45
empty, &c.	1	2	..	..	..	..	..	..	..	..	..	..	..	3	5
total	8	13	..	4	..	..	..	..	..	..	..	..	..	25	50
Murrow, Guyhirn, &c.[15] charged	34	27	9	3	..	..	..	..	..	..	..	..	..	73	127
empty, &c.	1	1	1	..	..	..	..	..	..	..	..	..	..	3	6
exempt	31	..	..	..	..	..	..	..	..	..	..	..	..	31	31
total	66	28	10	3	..	..	..	..	..	..	..	..	..	107	164
Wisbech St. Mary, charged	28	35	9	9	1	4	1	1	1	..	..	..	1 of 14	90	228
empty, &c.	..	1	1	..	..	..	..	..	..	..	..	..	..	2	5
exempt	34	..	..	..	..	..	..	..	..	..	..	..	..	34	34
total	62	36	10	9	1	4	1	1	1	..	..	..	1	126	267
Wisbech St. Peter, charged	147[16]	94	53	44	22	22	11	4	5	3	2	1	1 of 22 1 of 13	410	1,165[17]
empty, &c.	1	5[18]	..	1	..	..	..	..	..	..	..	..	..	7	15[17]
exempt	105	..	..	..	..	..	..	..	..	..	..	..	..	105	105
total	253	99	53	45	22	22	11	4	5	3	2	1	2	522	1,285
Total	613	349	119	85	42	34	15	10	8	3	4	1	6	1,289	2,857
WITCHFORD HUNDRED															
Coveney,[19] charged	18	17	2	2	1	..	..	..	..	..	..	1	..	41	83
Manea, charged	22	13	5	3	..	..	..	..	..	..	..	..	..	43	75
empty, &c.	2	..	..	..	..	..	..	..	..	..	..	..	..	2	2
exempt	17	1	..	..	..	..	..	..	..	..	..	..	..	18	19
total	41	14	5	3	..	..	..	..	..	..	..	..	..	63	96
Haddenham, charged	72	84	23	9	6	4	5	..	1	1	..	..	1 of 17	206	470[20]
empty, &c.	1	3	1	..	..	..	..	..	..	..	..	..	..	5	10[20]
exempt	75	..	..	..	..	..	..	..	..	..	..	..	..	75	75
total	148	87	24	9	6	4	5	..	1	1	..	..	1	286	555
Mepal, charged	10	20	3	1	..	1	..	..	..	..	..	..	..	35	69

Continued

[14] The total is given as 97.
[15] *Murrow, Guyhirn, &c.* include the inhabitants of *Tholomas Drove,* and possibly those of *High Fen,* who are separately listed in 1666.
[16] These include *The Library, The madder mill* and *Old use,* each with one hearth.
[17] The total is given as 1,178.
[18] One of these is said to be for two tenements.
[19] *Coveney* includes the inhabitants of *Wardy Hill,* who are listed separately in 1666.
[20] The total is given as 474.

1674 Lady Day (cont.)

	Totals of entries, with hearths													Totals	
	1	*2*	*3*	*4*	*5*	*6*	*7*	*8*	*9*	*10*	*11*	*12*	*over 12*	*Entries*	*Hearths*
WITCHFORD HUNDRED, *Continued*															
Stretham, charged .	39	34	18	6	1	1	..	..	..	1	..	..	..	100	206[21]
empty, &c. . .	2	..	..	1	..	..	..	..	..	..	..	..	..	3	6[21]
exempt . . .	26	..	..	..	..	..	..	..	..	..	..	..	..	26	26
total . . .	67	34	18	7	1	1	..	..	..	1	..	..	..	129	238
Thetford, charged .	7	16	5	1	1	1	..	..	..	..	..	..	..	31	69
empty, &c. . .	1	1	..	..	1	..	..	..	..	..	..	..	..	3	8
exempt . . .	4	..	..	..	..	..	..	..	..	..	..	..	..	4	4
total . . .	12	17	5	1	2	1	..	..	..	..	..	..	..	38	81
Sutton, charged .	42	75	14	7	1	..	1	..	..	1	..	..	..	141	284
exempt . . .	50	1	..	..	..	..	..	..	..	..	..	..	..	51	52
total . . .	92	76	14	7	1	..	1	..	..	1	..	..	..	192	336
Wentworth, charged .	4	8	4	1	..	..	1	..	1	..	..	..	..	19	52
exempt . . .	6	..	..	..	..	..	..	..	..	..	..	..	..	6	6
total . . .	10	8	4	1	..	..	1	..	1	..	..	..	..	25	58
Wilburton, charged .	11	23	5	6	..	2	..	1	..	1	..	..	1 of 18	50	144[22]
empty, &c. . .	..	2	..	..	..	..	..	..	..	..	..	..	..	2	4[22]
exempt . . .	15	4	..	..	..	..	..	..	..	..	..	..	..	19	23
total . . .	26	29	5	6	..	2	..	1	..	1	..	..	1	71	171
Witcham, charged .	22	24	2	3	1	1	2	1	2	1	..	..	..	59	149
exempt . . .	9	1	..	..	..	..	..	..	..	..	..	..	..	10	11
total . . .	31	25	2	3	1	1	2	1	2	1	..	..	..	69	160
Witchford, charged .	11	20	6	2	2	..	..	..	..	..	..	..	..	41	87
empty, &c. . .	..	1	..	..	..	..	..	..	..	..	..	..	..	1	2
exempt . . .	22	..	..	..	..	..	..	..	..	..	..	..	..	22	22
total . . .	33	21	6	2	2	..	..	..	..	..	..	..	..	64	111
Chatteris, charged .	92	98	12	3	3	..	1	1	1	..	1	..	1 of 14	213	400[23]
exempt . . .	38	..	..	..	..	..	..	..	..	..	..	..	..	38	38[23]
total . . .	130	98	12	3	3	..	1	1	1	..	1	..	1	251	438
Doddington, charged .	37	21	9	2	1	1	..	..	..	..	1	..	1 of 18	73	154
exempt . . .	14	..	..	..	..	..	..	..	..	..	..	..	..	14	14
total . . .	51	21	9	2	1	1	..	..	..	..	1	..	1	87	168
Benwick, charged .	15	13	1	..	..	..	..	..	..	..	..	..	..	29	44
exempt . . .	8	..	..	..	..	..	..	..	..	..	..	..	..	8	8
total . . .	23	13	1	..	..	..	..	..	..	..	..	..	..	37	52
March, charged . .	139	72	30	6	4	3	1	..	1	..	..	..	..	256	451[24]
empty, &c. . .	7	5[25]	..	..	..	..	..	..	..	..	..	..	..	12	17[24]
exempt . . .	66	..	..	..	..	..	..	..	..	..	..	..	..	66	66
total . . .	212	77	30	6	4	3	1	..	1	..	..	..	..	334	534
Wimblington, charged	31	25	5	4	1	..	..	..	..	..	..	..	..	66	117[26]
empty, &c. . .	2	..	..	..	..	..	..	..	..	..	..	..	..	2	2[26]
total . . .	33	25	5	4	1	..	..	..	..	..	..	..	..	68	119
Total . . .	937	582	145	57	23	14	11	3	6	5	2	1	4	1,790	3,269

21 The total is given as 211.
23 The total is given as 398.
25 These include the *Guildhall and the School*, two hearths together, in arrears.
22 The total is given as 150.
24 The total is given as 467.
26 The total is given as 126.

Isle of Ely: Grand Totals

	Totals of entries, with hearths													*Totals*	
	1	2	3	4	5	6	7	8	9	10	11	12	over 12	*Entries*	*Hearths*
1662															
Total . . .	1,681	1,302	367	197	87	58	30	20	11	15	7	4	14	3,793	7,930
1666															
paid . . .	2,194	1,215	262	159	84	48	31	18	16	14	6	8	9	4,064	7,706
unpaid . . .	497	192	74	46	11	5	2	..	..	..	..	..	..	827	1,386
Total . . .	2,691	1,407	336	205	95	53	33	18	16	14	6	8	9	4,891	9,092
1674															
charged . .	1,582	1,440	454	223	109	73	43	21	25	9	10	9	15	4,010	8,935
empty . . .	32	40	5	2	2	..	..	..	..	..	..	..	..	81	145
ruined . . .	..	1	..	..	..	..	..	..	..	..	..	..	..	1	2
newly built . .	1	..	..	..	1	..	..	..	..	..	..	..	..	2	6
newly built (in other houses) . .	2	..	..	..	..	1	..	..	..	..	..	..	..	3[27]	8
arrears[28] . .	15	14	1	..	..	..	..	..	..	..	..	..	..	30	46
exempt . . .	958	9	..	..	..	..	..	..	..	..	..	..	..	967	976
Total . . .	2,590	1,504	460	225	112	74	43	21	25	9	10	9	15	5,091	10,118

27 These entries are not included in the grand total.

28 Some of these entries may represent empty houses; all appear to represent hearths on which the payment of duty was disputed.

THE VICTORIA HISTORY
OF THE
COUNTIES OF ENGLAND

A HISTORY OF
CAMBRIDGESHIRE
AND
THE ISLE OF ELY

INDEX TO
VOLUMES I–IV

EDITORIAL NOTE

THIS is a collective index to the *Victoria History of Cambridgeshire*, Volumes I–IV, apart from the Cambridgeshire section of Domesday Book which is separately indexed in Volume I. It presupposes that the corrections printed on page 77 have been made.

NOTE ON ABBREVIATIONS

The following abbreviations are used: abp., archbishop; adv., advowson; agric., agriculture; Alb., Albert; Alex., Alexander; Alf., Alfred; And., Andrew; A.S., Anglo-Saxon; Ant., Anthony; archdn., archdeacon; Arth., Arthur; Bart., Bartholomew; Beat., Beatrice; Ben., Benjamin; bp., bishop; bro., brother; cast., castle; Chanc., Chancellor; chant., chantry; chap., chapel; char., charity; Chas., Charles; Chris., Christopher; ch., church; chwdn., churchwarden; Clem., Clement; coll., college; ctss., countess; dau., daughter; dnry., deanery; dom. arch., domestic architecture; dchss., duchess; Edm., Edmund; Edw., Edward; Eliz., Elizabeth; Ern., Ernest; fam., family; Geoff., Geoffrey; Geo., George; geol., geology; Gil., Gilbert; grds., grandson; Hen., Henry; Herb., Herbert; hosp., hospital; Humph., Humphrey; hund., hundred; incl., inclosure; ind., industry; Jas., James; Jos., Joseph; Kath., Katherine; Laur., Laurence; Leon., Leonard; man., manor; Marg., Margaret; m., married; mkt., market; mquss., marquess; Mat., Matthew; Mich., Michael; Nat., Nathaniel; Nich. Nicholas; nonconf., nonconformity; par., parish; parl. rep., parliamentary representation; Perc., Percival; Phil., Philip; pop., population; prehist., prehistoric; pres., president; prof. professor; prot., protestant; rly., railway; rect., rectory; Reg., Reginald; rem., remains; Ric., Richard; riv., river; Rob., Robert; Rog., Roger; Rom., Roman; Rom. Cath., Roman Catholic; Sam., Samuel; sch., school; Sim., Simon; soc., society; s., son; stn., station; Steph., Stephen; Thos., Thomas; vic., vicarage; vct., viscount; vctss., viscountess; Wal., Walter; wid., widow; w. wife; Wm., William.

INDEX TO VOLUMES I–IV

INDEX TO VOLUMES I–IV

CORRIGENDA TO VOLUMES I–IV

Vol. I, page	32,	line 11 from end, *for* Flower *read* Fowler
Vol. II, page	x,	line 2 from end, *after* The Tudor Period and the Civil War *insert* by D. Cook, M.A., Ph.D.
,, ,,	7,	line 28, *for* 645 *read* 654
,, ,,	64,	line 13, *for* Hatley *read* Hatfield Broad Oak
,, ,,	111,	last line, *for* Walton *read* Walton (*recte* Malton)
,, ,,	112,	note 28, *for* 1818 *read* 1828, in error for Malton
,, ,,	168,	line 18, *for* Norfolk *read* Suffolk
,, ,,	211*b*,	line 17, *for* 1156 *read* 1056
,, ,,	224,	note 18, *for* Chesterfield *read* Chesterford
,, ,,	224,	note 18, *for* Elmington *read* Elmdon
,, ,,	261*a*,	line 7 from end, *for* Flaxflete *read* Faxfleet
,, ,,	288*b*,	line 24, *for* Wivelingham *read* Willingham
Vol. III, page	451*b*,	line 21, *for* Thomas *read* Henry
,, ,,	463*b*,	lines 1–2, *for* Winghole *read* Winghale
Vol. IV, page	29,	Key, No. 2, *delete* and U.D.C. Offices
,, ,,	31*b*,	line 16, *delete* the U.D.C. Offices besides
,, ,,	34,	note 85, *for* Eversley *read* Eversden
,, ,,	105*a*,	lines 25–26, *for* one of his own daughters *read* one of his step-daughters
,, ,,	105,	note 49, *should read* E 150/104/6; *Visit. Cambs.* (Harl. Soc. xli), 40.
,, ,,	117*b*,	line 18 from end, *for* 1930's *read* 1920's
,, ,,	136*b*,	line 12, *for* Theology *read* Geology
,, ,,	220,	note 28, *for* has been National Trust property since 1943 *read* is the property of Lord Fairhaven who in 1943 entered into a covenant with the National Trust for the preservation of the amenities of the park.
,, ,,	221*a*,	lines 9–10, *for* John Webb, the nephew and pupil of Inigo Jones *read* Peter Mills
,, ,,	221*a*,	line 27, *for* Cure *read* Cave
,, ,,	221*a*,	line 29, *for* in the north-west corner of the parish on the road to *read* just across the county and parish boundary to the north-west on the road to
,, ,,	221,	note 29, *insert at the beginning of note Country Life*, cxi, p. 1732.
,, ,,	233*a*,	lines 8–6 from end, *for* and a farmhouse in St. Mary's village about 120 yards west of the school are also noteworthy, though late in date (early 19th century). All these three *read* is dated 1723. Both
,, ,,	236*a*,	lines 18–16 from end, *for* Portions of the 17th century altar rails . . . which is *read* The rails of the main altar date from the 17th century. The altar in the south aisle is
,, ,,	236*a*,	line 12 from end, *delete* and a settee, all
,, ,,	237*a*,	line 13, *for* (1891, rebuilt 1926) *read* (1891, rebuilt on a new site 1926)
,, ,,	239	(Plan of Wisbech), *delete all road classification numbers*
,, ,,	240*a*,	lines 13–10 from end, *for* traffic, but the former . . . used very little *read* traffic, the Midland and Great Northern's Old Market line, and the Great Eastern's Wisbech Harbour line, both of which are still in use
,, ,,	240,	*delete note 19*
,, ,,	243*a*,	line 28, *for* (*c.* 1890) *read* (*c.* 1840)